COMPUTER LANGUAGES
FOR NUMERICAL CONTROL

COMPUTER LANGUAGES FOR NUMERICAL CONTROL

Proceedings of the Second IFIP/IFAC International Conference on Programming Languages for Machine Tools, PROLAMAT '73, Budapest, April 10-13, 1973

edited by

J. HATVANY

Division for Control of Discountinuous Processes,
Computer and Automation Institute,
Hungarian Academy of Sciences

1973

NORTH-HOLLAND PUBLISHING COMPANY – AMSTERDAM • LONDON
AMERICAN ELSEVIER PUBLISHING COMPANY, INC. – NEW YORK

Library of Congress Catalog Card Number: 73-83688
North-Holland ISBN: 0 7204 2088 1
American Elsevier ISBN: 0 444 10572 7

Published by:
NORTH-HOLLAND PUBLISHING COMPANY – AMSTERDAM
NORTH-HOLLAND PUBLISHING COMPANY, LTD. – LONDON

Distributed in the U.S.A. and Canada by:
AMERICAN ELSEVIER PUBLISHING COMPANY, INC.
52 VANDERBILT AVENUE, NEW YORK, N.Y. 10017

PRINTED IN THE NETHERLANDS

LIST OF CONTENTS

STANDARDIZATION

IMPLEMENTING NC PROGRAMS

SURFACE DESCRIPTION PROGRAMS

TECHNOLOGICAL PROGRAMMING AND PRODUCTION CONTROL

CNC, DNC AND POST-PROCESSING

CONCLUSION

1. INTRODUCTION

INTRODUCTION

From April 10th to 13th, 1973, the 280 participants from 25 countries who had come to PROLAMAT'73 in Budapest, Hungary, listened to 65 papers presented by authors from 14 nations. Those are the vital statistics of the Conference whose Proceedings this volume contains.

The name PROLAMAT /PRogramming LAnguages for MAchine Tools/ was coined to cover the theme of the 1st International Conference on Programming Languages for Numerically Controlled Machine Tools, held in Rome, Italy, 1969.
The success of this first international gathering led many participants to suggest that it should be followed at regular interoals by further PROLAMAT Conferences, and the Hungarian delegation in Rome offered to host the next one. The agreement of the sponsoring Federations, IFIP /International Federation for Information Processing/ and IFAC /International Federation of Automatic Control/ was duly obtained, and the date and venue agreed. The next PROLAMAT Conference is scheduled for Stirling, Scotland in 1976, so that the series now appears to be well established.

Full responsibility for the Conference was assumed by the two representative, international bodies listed below:

International Organizing Committee		International Papers Committee	
Williams, T.J.	/USA/, Chairman	Leslie, W.H.P.	/UK/, Chairman
Cohn, N.	/USA/	Barbance, J.C.	/France/
Coons, S.A.	/USA/	Budde, W.	/FRG/
Goransky, G.K.	/USSR/	Hatvany, J.	/Hungary/, Editor
Hatvany, J.	/Hungary/	Herzog, B.	/USA/
Holmdahl, G.	/Sweden/	Kochan, D.	/GDR/
Kádár, I.	/Hungary/	Remmelts, J.	/Netherlands/
Leslie, W.H.P.	/UK/	Simon, W.	/West-Berlin/
Nussey, I.D.	/UK/	Sohlenius, G.	/Sweden/
Opitz, H.	/FRG/	Tamm, B.G.	/USSR/
Semenkov, O.I.	/USSR/	Tarján, R.	/Hungary/
Togino, K.	/Japan/	Vámos, T.	/Hungary/
Vámos, T.	/Hungary/	Vymer, J.	/CSSR/
Weill, R.	/France/	Welbourn, D.B.	/UK/

/The local arrangements were made by a National Organizing Committee/ The International Committees first met in Ljubljana, Yugoslavia, on 19th August, 1971, then the International Papers Committee made a preliminary selection of 46 out of the 85 abstracts submitted, at a meeting in Paris, France, on 11th April, 1972. All the full papers were reviewed by two independent referees /not of the author's country/, and the final decisions were made by the Chairmen of the six Sessions. The Chairmen /nominated by the International Committees/, arranged their sessions as they wished and at the Conference each Session opened with a short survey of its field, presented by the Chairman. Beside the contributed and survey papers, there were also two invited lectures, one by Prof. T.J. Williams, the other by Mr. W.E. Mangold, and there were four informal, sign-on type round-table discussions.

In order that these Proceedings should be published as soon as possible, it was decided that they would contain only the original texts of all the papers /including the survey and invited papers/, without detailed transcripts of the discussions. Since, however, the session discussion periods and round-table exchanges were felt to have contained much interesting material, the Editor was charged with the task of outlining their contents. The comments following each Section, and those below on the Round-table discussions are therefore the Editor's own views on what took place - any attempt to check them with all contributors would have resulted in delaying the appearance of this volume till its content became obsolete. For similar reasons courtesy and professional titles and degrees have been omitted from the editorial content of these Proceedings, as this was felt to be a lesser evil than getting them wrong.

- * -

The round-table discussion on Adaptive Control, was led by Dr. M. Ács /Hungary/. Adaptive systems can be implemented in both hardware, and software - in the case of conventional machine tools without numerical control, only hardware systems can be considered, while adaptive controls with optimization /ACO/ are nowadays all based on software solutions. In between, there is a range of hybrids.

The three questions particularly discussed were sensors, ACO computer configurations and the economics of AC. The use of piezo-electric sensors in particular was commended and experiences described of their applications to milling machines and lathes.

On computer configurations, it was agreed that no very rigid line should be drawn between CNC and DNC systems. Computer-based adaptive control systems generally require core stores of 16-32 Kbyte, and the advantage of having double-precision and floating-point arithmetic facilities was stressed. A word of caution was put in against having more than one computer in the actual control loop /in the case of hierarchic systems/, as this would almost inevitably cause instability. With respect to AC economics, the conclusion was that ACO systems could only be offered economically as parts of DNC schemes.

The CAD/CAM Interface was the topic of a round-table discussion chaired by J. Vlietstra /Netherlands/, upon whose proposal it was agreed also to include the subject of Computer-Aided Testing /CAT/.

The participants concluded that at present Computer-Aided Manufacturing is a more advanced art in mechanical engineering than Computer-Aided Design. This is partly due to the fact that designers feel they are creative artists and are reluctant to cooperate, while the wide-spread usage of NC machines has made the introduction of CAM inevitable. The gap between CAD and CAM was estimated at about 10 years. At this time only the data base can form a suitable interface between CAD/CAM and CAT, but later other possibilities will arise. The interface problem is on the software level one of modularity and open-endedness.

The use of CAD is a necessity in some cases, e.g. the design of a mould for TV tubes. This can not be done by hand. To achieve further progress, it will be necessary to develop problem-oriented programs for many design applications. In IC layout design, for instance, a fully automatic, deterministic approach is not practically feasible and appropriate man-machine communication devices are rapidly gaining ground.

It was stated that there are two basic approaches to building Integrated Manufacturing Systems:

a./ to develop the different subsystems independently and then to interface them,

b./ to design the entire system as a single entity.

No agreement was reached on the relative merits of these approaches and a mixture of the two was also suggested. The first /so-called island/ method will create locally near-optimal modules but cause severe interface problems which have been known to result in failure. The time required for the development of a full IMS stepwise by the second method, was estimated at two years.

A discussion on the Economics of NC was held under the chairmanship of G. Hajós /Hungary/. The prime question was considered to be, whether the proportion of one-off, small-batch, medium series and mass-production was going to change significantly in the foreseeable future. It was contended that a real economic break-through for NC could only be expected if the proportion of small-batch and one-off production was to rise. This view was challenged by another approach, based on the assumption of a further rapid decline in the cost of electronics, leading to NC becoming a standard, automatically incorporated feature of all machine tools.

The manpower question was recognized to be the crucial one, where a cross-over point between rising wages and decreasing NC costs was said to be not far off. The next problem, however, would be to train and employ sufficient skilled personnel for the installation and maintenance of NC.

On the calculation of economic tool lives, it was pointed out that the present formulas yielded tool lives of a few minutes, which was palpably absurd. The urgent need for new materials was agreed. The question of permissible tool wear is intimately linked with the still unsolved problem of the in- process measurement of the workpiece. This would greatly accelerate the development of Numerical Control.

The Use of Preprocessing was discussed by a group headed by M.A. Sabin /United Kingdom/. It was stated that several papers at the Conference had dealt with the use of some sort or preprocessing

/e.g. Márkus, Pikler, French, Zaitsev/.

In a given environment it will often be possible to define for a particular family of parts, a simple data language which, processed by a FORTRAN-based pre-processor, will turn out an APT-like part program or even the CLDATA directly. When using a simple data language, the user may concentrate on semantics, while in using APT, the main effort is put into the syntax. One solution may well be the use of APT macros, as in the paper of Adams and French. The front-end data language referred to, may - among other possibilities - be a graphic one.

A question confronting the designer of a pre-processor is whether the object language of his system should be the APT part program format, or CLDATA. One argument in favour of a CLDATA output may be the need to program curves in three-dimensional space, of a type which can not be programmed in APT. There might also be cases where APT is highly inefficient.

- * -

In conclusion, the Editor wishes to thank all those who helped him in his work: Mrs. V. Simon, Secretary of the National Organizing Committee, the Session Chairmen and Secretaries, the Secretaries of the Round-table Discussions, the Staff of the Conference Bureau, and of course all the Authors of the Papers reproduced in this volume.

J. Hatvany

WELCOMING ADDRESS

G. Horgos
Cand. Tech. Sci.,
Minister of Metallurgy and Engineering of the
Hungarian People's Republic

On behalf of the Government of the Hungarian People's Republic and the Ministry of Metallurgy and Engineering I welcome you on the occasion of the opening of the PROLAMAT'73 Conference, held in our country and sponsored by the International Federation for Information Processing and the International Federation of Automatic Control.

First of all, may I express my gratitude to the organizing Societies and Institutes, to the Organising Committee and all persons involved for organizing this Conference, which I consider very important with respect to the development of our industry in its present stage. The managers of our industry will pay special attention to the papers presented here.

Our Government's program for industrial development and our long term policy is oriented towards increasing production intensively and steadily. Our possibilities for extensive industrial development are exhausted due to the lack of labour resources. It is therefore important for us to increase the profitability, effectiveness and productivity of the entire production process.

Due to the open character of our economy the further development of our international economic relations, international cooperation and the necessity to appear with marketable products on the international scene are questions of vital importance for us.

Hungary is a small country and we are unable to comply with all these requirements in all fields. We follow a selective industrial development policy and we have now to modernize the structure of our industry. We are emphasizing those branches of industry and groups of products, which harmonize well with our possibilities.

We consider the mechanical engineering industry to be particularly suitable for the adoption of the most advanced results of scientific and technical progress and for their utilization to the good of our economy.

Even within the mechanical engineering industry we can not reach the level of mass production for every product group, generally we manufacture in small and medium-size batches. It is evident that only automation and better organization can speed up the rate of production and its effectivness.

The importance of this Conference lies in the fact, that most of its papers are closely related with our present-day problems. The integrated systems of design, preparation, control and production are key elements in the improvement of our production processes.

Our Government and industrial management recognised in time that our industry, and especially our export-oriented machine tool industry, can comply with these requirements only by the use of computer and NC techniques.

Since 1967 we have taken appropriate measures and concentrated considerable resources on the introduction and propagation of NC techniques in Hungary.

At the Csepel Machine Tool Factory, with governmental assistance, we established an NC pilot plant, comprising different machine tools.In this way we established a facility for testing NC techniques in practice and amidst our own circumstances.

Based on this industrial experiment we commenced with a government-sponsored program to develop NC turning and milling machines, machining centers and other kinds of NC machine tools, approximately 20 different types in all.

Today our production and export in NC machine tools is quite considerable. Firms in the Federal Republic of Germany, Austria, Great Britain, France, Czechoslovakia and other countries among our buyers. Some of our machines were designed here in Hungary, some of them are based on licences.

We started the development of control units for NC machines in the Institute for Industrial Automation. This is now based on a foreign licence whose adoption is in progress and the full production of the control units can start soon.

Closely related with machine tool developments are the further improvement of the highly productive cutting tools, chucks and clamping devices. The production and development of these devices is also government-sponsored. We established an educational basis to instruct new, young technicians, who are able reliably to apply this new technique.

Our Hungarian participation in the integrated program of the Socialist countries for computer development provides the main basis for integrated production systems, because we manufacture the minicomputers of the range. Those computer aided design and production systems, which you can see at the time of the Conference are controlled by Hungarian mini-computers, produced by the VIDEOTON Works and by the Central Research Institute for Physics of the Hungarian Academy of Sciences.

This is the first occasion that we are demonstrating computer controlled machine tools for which the part programs were designed also by computer and the control tape was prepared by an interactive man-machine dialogue.

Parallel with establishing the bases for NC machine tool manufacture we also helped many of our enterprises to buy and to put into operation NC machines and thus to popularise NC techniques. These programs are under progress now and the first results come up to our expectations and hopes. Productivity and accuracy were remarkably improved, reliability became higher and the planned completion times for production became more secure using NC machine tools.

Recognising the benefits of NC techniques the interest of our enterprises for NC machines and their introduction increased. The enthusiasm of our young specialists contributed remarkably to the quick industrial introduction of these machines.

You can see from these facts that the scientific and material bases for manufacturing and applying computerized NC techniques in Hungary are well established. There are enthusiastic followers of this discipline in our country and due to their work this high level technical culture is expanding and produces more and more exportable products.

Our most important tasks are now: to modernise our NC techniques, to develop and apply integrated manufacturing systems and to create good conditions for programming and production preparation, for exploiting these high value machines.

To help this process the Computer and Automation Institue of the Hungarian Academy of Sciences became a member of the internationally known EXAPT-Verein, giving us the possibility to adapt the EXAPT part programming languages.

We established the Hungarian Association for Machine Tool Programming to coordinate all activities related to computerized programming. The members of this Association are our machine tool factories and our research institutes. Recognising the benefits of this joint effort many user enterprises, institutions and technical schools have joined the Association.

The Institute for Technology of the Machine Industry was commissioned to adapt part programming languages with governmental assistance. As a result of this the EXAPT and NELAPT systems are available in Hungary. This Institute also contributed to the part programming effort by elaborating its own FORTAP system, which contains wide-ranging technological services for turning machines.

These measures and planned efforts have established in Hungary the technical and computer oriented bases for the industrial application of NC part programming. There are post processors available for Hungarian-made NC machine tools and our guests will be able to see their application in practical work. We can offer wide-ranging software for our NC machine tools, which remarkably improves their effectivity.

Having established the possibilities for computer based NC programming partly by adaptation, partly by our own efforts we are now concentrating on developing a unified computer-aided part programming system for the Socialist countries. Our goal is to develop, in close cooperation with other countries, a processor family which consists of compatible members and also agrees with ISO standards. We consider it very important to design it in an easily adaptable form, exploiting the possibilities of technological standardization and well suited to the different computer configurations available to our users.

Concerning the development of computer controlled integrated manufacturing systems we are just at the beginning. International experience shows that besides the hardware and software problems, the requirements of this kind of system are high with respect to production planning and organisation and also for technological programming. We are expecting new ideas and help from this Conference in this field, too.

I hope, that my comments have convinced you that our Government has given top priority to the development of NC techniques and has made sacrifices to adapt them on a high level in Hungary. I have highlighted our main problems and this is the reason why our industrial management will pay special attention to the papers presented at this Conference.

This Conference helps us with our industrial development program at the very best time. Its mission is through the papers, discussions and international exchange of information to clarify some of the many problematic questions which we face.

I hope that the personal contacts of the participants will give an excellent possibility for further cooperation, which will advance the success of our program.

I am convinced that the Conference will be fruitful for every delegate and will assist in the further development of NC techniques.

May I open the Conference in the spirit of these ideas and thoughts and may I wish you successful and fruitful discussions.

CAM AND NC SOFTWARE SYSTEMS: NEEDS FOR AND BENEFITS FROM GENERALIZED AND MULTI-INDUSTRY STANDARDIZED LANGUAGES

Theodore J. Williams
Purdue Laboratory for Applied Industrial Control
Purdue University
West Lafayette, Indiana 47907 USA

ABSTRACT

Misjudgements by project personnel concerning project software requirements and capabilities have resulted in a high percentage of late and incomplete computer process control projects. By easing programming requirements through the promotion of the use of special high level languages and of specific program packages, programming language standardization activities promise to greatly ease the above mentioned difficulties. In addition, these same standardization activities have shown the applicability of the same programming techniques and many of the developed programs to a wide variety of manufacturing industries, including discrete manufacturing. Thus, on-line numerical control and computer aided manufacturing software become an important facet of a generalized process control software system.

INTRODUCTION

In a recent article /9/ reviewing the present status of direct digital control /ddc/ in the continuous fluid process industries such as petroleum refining and petrochemical manufacturing, S.J. Bailey, an editor of the technical journal, Control Engineering, makes the following observation:

> "To the hardy on-line experimenters who pioneered direct digital control /ddc/ back in 1961 and 1962, progress in this advanced computer control concept has been marginal. They point with pride to many ddc applications that have been successful - yet they concede that computers have not assumed direct command of on-line events in anything like the volume they had envisioned..."

COMPUTER LANGUAGES FOR NUMERICAL CONTROL, *J. Hatvany, editor*
North-Holland Publishing Company - Amsterdam–London

In another equally recent article /8/, Dr. R.H. Anderson of the University of Southern California makes the following predictions concerning the future applications of digital computers in discrete manufacturing:

> "Probably the fastest growing application of information sciences during the next decade will be computer based automation of the manufacturing process. All aspects of manufacture - design, prototyping, production engineering, part forming, assembly, inspection, material transfer and storage - will increasingly become <u>directly controlled by computer</u> /this author's emphasis/..."

This author was directly involved in the early work /16/ mentioned by Bailey and has written many articles concerning the automation of the continuous process industries /21,22/ which sound very much the same theme as does Anderson's paper. Let me then use this experience to call the reader's attention to the <u>errors</u> made by the continuous process industry enthusiasts and propose, first, how our colleagues in the discrete manufacturing industries can avoid the situation raised by Bailey or, second, how both can be rescued from it if it is already too late for a similar initial disenchantment to be avoided. These words are in no sense meant as a criticism or diminution of Anderson's thoughts or predictions. In fact, this author still agrees wholeheartedly with him as well as with the similar predictions made in his own earlier papers concerning the continuous process industries /21, 22/. Unfortunately, several, at that time totally unforeseen, developments have occurred which have delayed the materialization of the original pronouncements and can likewise seriously impede corresponding applications of digital control in the discrete manufacturing industries.

Let us begin by reviewing "what went wrong" and then propose some remedies. In keeping with the theme of this Conference, our discussion will be limited to the software aspects of the subject, hardware only being involved when it vitally affects the resulting programming.

WHAT WENT WRONG IN THE PROCESS INDUSTRIES AND A PROPOSED REMEDY TO THE CONDITION

The conditions described by Bailey /9/ in his article are not a sudden recent discovery by himself and others as he will quickly acknowledge. They have been building for the past several years. However, the situation has been strongly aggravated by the recent recession in United States industry. As a result of this latter economic factor, company managements have been strongly questioning the "pay-out" of computer control installations of all types. Their findings are summarized in Table I.

This disenchantment of management /particularly middle management/ with computer systems and its related causes as expressed in Table I is today the biggest impediment to a very widespread application of computer systems in the continuous process industries. It is,therefore, imperative that these conditions be corrected as soon as possible if digital control is ever to assume the dominant place in plant control schemes in the process industries for which its proponents hope. Likewise, great care must be exercised to assure that similar situations do not arise in the discrete manufacturing industries and others if indeed it is not already too late to prevent or correct them.

The lessons which we should then learn from the factors of Table I and which we should apply in all future projects, regardless of the industry involved are as presented in Table II. Most importantly here it is agreed that these lessons should be needed even at the decided expense of computer efficiency and increased hardware costs in order to avoid further project difficulties.

In view of the above there obviously has been a great deal of thought concerning the best remedies for these conditions. In these discussions it was readily agreed that a general computer control loop could be diagrammed as in Figure 1. The key here is that similar hardware in all situations promotes mass production with corresponding increases in system reliability and reduced costs. Similar hardware in different industries promotes the use of standardized programs for variable input and control output functions. The algorithm is specific to the application. It was also agreed that the general software system required for a computer control system could be pictured as in Figure 2. Finally, the general hierarchy computer control system needed for overall plant control could be illustrated as in Figure 3. It was further generally agreed that the so-called "remedies" of Table III would be necessary to correct the conditions previously listed.

In addition, and most important, it was felt by those involved that the generality noted in continuous process industry systems applied indeed to most, if not all, industries. If inter-industry cooperation could be developed in producing the "remedies" of Table III, then indeed the whole computer control field, not only that of the continuous process industries, would benefit.

TABLE I

A SUMMARY OF MANAGEMENT CONCLUSIONS CONCERNING COMPUTER CONTROL APPLICATIONS IN THE CONTINUOUS PROCESS INDUSTRIES

1. Computers are fascinating and intellectually inspiring subjects with which to work. Consequently, their proponents tend to be extremely, if not overly, enthusiastic concerning the computer's capabilities as well as their own individual skills. As a result there is a decided tendency /completely unpremeditated/ to over-estimate the rate of progress which is possible in completing projects, and in developing solutions to problems which may arise. Likewise, these individuals have tended to underestimate the difficulties, and the amount of equipment and time needed for these projects and problem solutions.
2. Early computer equipment was quite unreliable. In addition, the tendencies alluded to in Item 1 above resulted in projects being too large, too complex, and requiring too much machine speed for the available devices. While the reliability and speed problems have largely been corrected, memories of these earlier failures are hard to erase from the minds of management.
3. Computer enthusiasts have completely misjudged the acceptance of computer control by their uninitiated compatriots and the amount of training which is required to correct the resulting situations. Considering the usual small size of the "computer group" in most companies, complete acceptance by the on-site user group is absolutely necessary to assure the continuity of a project once the members of the computer installation group leave the project for their next assignment.
4. In many cases the extremely flexible and very sophisticated systems envisioned by early workers were /and are/ not necessary to accomplish the tasks required. Managements and the customer have not often demanded the qualities, amounts, variabilities, or intricacies of products which only computers could produce.

5. A corollary to Item 4 is the fact that there are generally other solutions which are available for most situations requiring any except the most sophisticated of computer control installations. Also, these other solutions, being extensions of presently available techniques, are more readily accepted by the non-computer trained individual and, in fact, are often sought out by him as counter factors to proposed computer applications.
6. The type of person who is most intrigued by computers tends to be the one who is most likely to be enthusiastic, innovative, and creative. These individuals see a ready opportunity to fulfill their creative needs through applications in the computer field. The result has been a tremendious proliferation of new machine designs, new codes, new programming languages, and - most serious of all - new reworkings of already solved problems for the sake on innovation alone or for only minor technical or economic gains.
7. Our machine-oriented ethics of efficiency and low cost have /and still do/ cause us to be much overconcerned as to whether our computer systems are planned and programmed so that their duty cycle efficiency is high, their memories are full, and as many plant operations as possible are crammed into them. While experience is tempering these factors, this condition plus that of Item 1 still prevails in far too many cases with the expected resulting effects on programming manpower requirements and on the difficulty of deciphering of programming errors and of making programming updates.
8. We have not yet solved the problem of man-machine communications to the satisfaction of all concerned. We have not yet been able to program the computer to allow it to answer all the problems which a wideranging human mind can pose for it concerning the operation of the plant in question.
9. The area where the results of the workings of Item 1, 6, and 7 combine to cause the most difficulties in project completion and user acceptance is in the area of programming management. Except for rare cases, projects still tend to be understaffed with resulting project delays and overruns. Likewise, innovation and the desire for efficiency and completeness combine to make project programming a much more difficult task that it need otherwise have been.
10. Contrary to the expectations and predictions of their propo -

nents, digital systems have not been as truly flexible as planned. Particularly, the changing and updating of their programs have proven much more difficult and error-prone than can be tolerated in most industrial situations.

11. It general the documentation of the programs of most systems has been woefully, sometimes totally, deficient, particularly when it comes to making it possible for a new engineer to pick up and modify a completed or partially completed project.

TABLE II

SOME LESSONS FOR ALL INDUSTRIES CONCERNING PROCESS COMPUTER CONTROL AS DERIVED FROM THE CONTINUOUS MANUFACTURING INDUSTRIES' EXPERIENCE

1. Despite our intellectual preoccupations with the study of "large scale systems," we are in general "mentally comfortable" only with relatively simple systems or with large systems which can be readily subdivided into distinct stand-alone entities which are themselves relatively simple. We should follow these same dictates in designing our industrial plant operational and control systems.
2. To ease the routine, repetitive, and monotonous tasks of implementing systems, including that of training, every possible action of this type should be relegated to a computer /although not necessarily to the one which is to be on-line/.
3. Man is the most error-prone component of a computer system, particularly during its development. Thus, his input should be kept to a minimum and it should be in a readily human interpretable form such as natural language as much as possible.
4. Every effort should be made to avoid "wastage" of the intellectual input to systems development by making use of general programming methods and systems and reusing previously developed systems unless grossly inadequate or inefficient.
5. Every effort should be made to make the documentation of a computer systems as "automatic" and computer-based as possible to assure its completion in the first place and its updating whenever changes are instituted in the system.

FIGURE 1

GENERAL ORGANIZATION OF A DIGITAL CONTROL SYSTEM

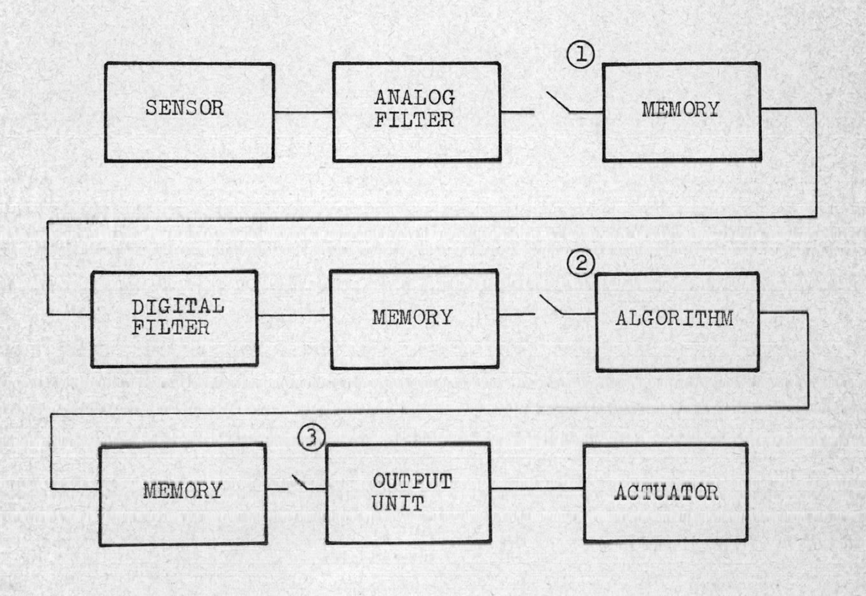

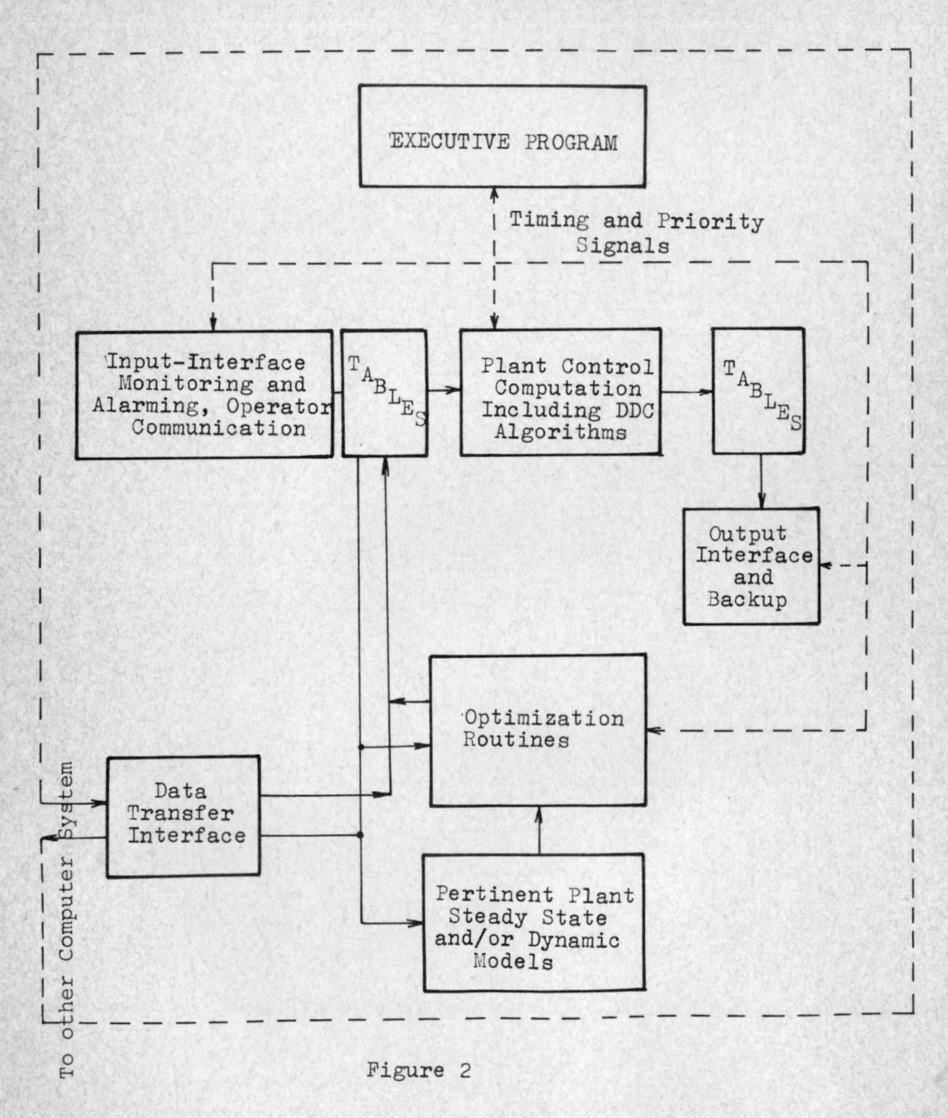

Figure 2

BLOCK DIAGRAM OF MODULAR INDUSTRIAL CONTROL PROGRAM SYSTEM

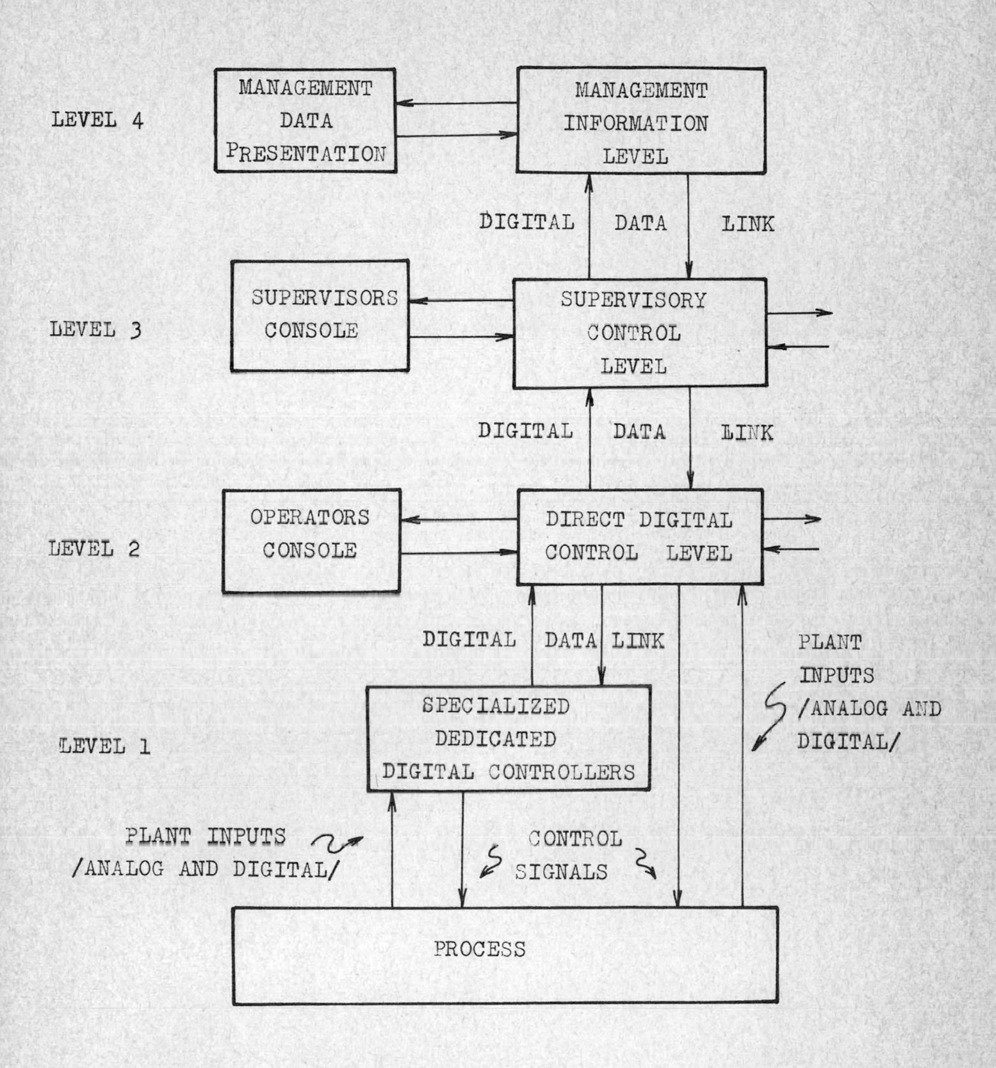

FIGURE 3

OUTLINE OF HIERARCHY ORGANIZATION FOR STANDARDIZED PROCESS COMPUTER CONTROL SYSTEM

TABLE III

SOME "REMEDIES" FOR THE COMPUTER CONTROL FIELD FOR THE CONTINUOUS PROCESS INDUSTRIES

1. Maximum use should be made of compiler or higher level languages, of host computer compilation of programs, and of preprogrammed packages for various applications to cut down personnel involvement.
2. Distributed computer systems should be used to the maximum extent economically practicable to keep individual application packages as simple as possible.
3. Acceptable cross industry standards for programming languages and for interface systems designs should be developed and adopted as rapidly as possible.

PRESENT STATUS OF THE DISCRETE MANUFACTURING INDUSTRY

A review is in order to develop the present status of computer control in the discrete manufacturing industries in relation to that of the continuous process control industries just mentioned. In this respect some of the comments of Mesniaeff /18/ and Anderson /8/ in their articles will be utilized here.

Computer Aided /CNC/ and Direct Numerical Control /DNC/

The first numerical control developments were directed toward elementary point-to-point controls and contouring. Then flexibility was considerably increased in the form of tool changers and machining centers. Concurrently there were related programming developments such as interpolation, multiple-axis control, improved feed rates, canned cycles, and related options.

Each improvement required additional electronic logic components, and size alone would have inhibited such growth had not integrated-circuit technology been incorporated. And even so, the cost of controllers, became the next stage in the development of automated manufacturing - the inclusion of memory allowed, in essence, time sharing of the electronic hardware with optional features being called up on a time-slice basis rather than requiring separate circuitry. And there were additional advantages of cnc, such as the possibility of field modification and the ability to perform algebraic computations.

Next in this natural evolution came dnc, or direct numerical control, where the emphasis in on computer-based communication for the automation of such nc support functions as program filing and retrieval, program editing, performance monitoring and tape elimination.

Beyond this level today are multistation manufacturing systems, which are based on a compromise between the discrete-part capabilities of nc and the high throughput advantages of the transfer line. A typical example is several nc machines designed and built as a task-oriented integrated package complete with a materials-handling and loading system. Raw materials are the input and a complete product mix the output. Examples of this concept are being installed today.

The last level of computer involvement, CAM /computer-aided manufacturing/, remains to be demonstrated, and potential applications are still speculation. The bottle-neck in manufacturing is the lack of suitable software, particularly with regard to the human interface problem. The next several years will require a major national-level goal-oriented effort in software development and sensing techniques. As the software picture becomes organized, the manufacturing plants should rapidly come around to true CAM systems.

CNC Systems

Two basic approaches have been used for cnc alternatives to conventional control:

1. A hybrid combination consisting of a computer and hardwired logic, where a general-purpose minicomputer replaces the functions of the conventional core unit and conventional

hardwired circuits are retained for the interpolation and position-loop logic.

2. A complete minicomputer system where all control functions, including interpolation and position-loop closure, are performed in the computer software.

At present, neither approach enjoys a significant advantage. The hybrid approach takes the position that hardware interpolation and loop-closure logic represent high volume items that are low in cost, have historically proven to be adequate and reliable, and provide data manipulation that would otherwise limit computer time for the core and option functions. Recent experience in the continuous process industries would indicate that these approaches may eventually win out.

Champions of complete software systems, on the other hand, claim that improved control methods result from "closing the loop" in software. One advantage is true acceleration and deceleration control. Elimination of interpolation and loop-closure hardware provides savings that can be used to justify a more powerful computer able to handle all necessary functions without being constipated.

In a sense, such a comparison is academic because both cnc approaches are more than adequate from the standpoint of today's machine-tool controls. The main advantage of cnc in general is flexibility. The control builder can adopt a single design to many machine tools, with customizing in easily changed software; the machine-tool builder can buy and service one control for an entire line of machine tools; and the user can end up with more features and options at a lower cost /above a certain minimum configuration/ and can add software improvements or modifications on-site rather than be exposed to expensive hardware retrofits if changes are required, as was the case with conventional nc.

Proliferation of these offerings will rise with the changing nature of construction practice. With conventional nc, a fairly substantial investment in specialized fabrication equipment and knowhow was required. Cnc, on the other hand, is an excellent example of borrowed technology where the hardware costs and control methods can be shared with other nonmachine-tool industries. It is not surprising that cnc systems are being offered by computer builders and small systems houses in addition to the established control builders. More than one machine-tool builder now has a subsidiary developing cnc controls which allows him to offer a complete true nc system, including machine tool and control, to the user.

DNC Systems

Returning to dnc, the reasons for development of dnc systems and the future of the same need to be further explored.

The nc machine program, as contained on the paper tape, is the visible downstream result of a complex upstream combination of programming, planning, scheduling, filing, and related activity. Once a proven tape is obtained, there is great reluctance to allow changes, particularly if the part is the least bit complex. Often rapid changes in th e machine program are either impracticable or else impossible. This is an interesting paradox, since nc is supposed to provide the flexibility required for highly engineered parts, where small lot sizes and proliferation of the design are the rule instead of the exception.

For example, a simple engineering change such as a larger fillet or a shift in the location of a hole cannot be obtained by simply informing the machinist to follow the new design while the drawings are being updated. More often than not, the engineering change requires a complete iteration through the entire part-program-generation process, including a rerun of the APT /or equivalent/ processor and postprocessor and a new debugging and proveout sequence. Add to this the problem of maintaining a current tape library when several programs are in a similar state of flux at the same time.

The point to be made is that such rigidity, if not controlled by meticulous planning, filing, programming, and scheduling, can easily destroy many of the advantages that are used to justify nc as a process.

Dnc was obviously foreseen as a way to avoid such rigidity by applying computer techniques directed at the systematic formalization of the information flow throughout the program generation, prove-out and documentation process. Needless to say, such an approach has advantages for the generation and proveout part programs as well as greatly aiding the changing and filing of existing programs.

In the process of improving information flow, it was inevitable that paper tape for information storage and transfer would develop into a bottleneck that had to be eliminated. Many early justifications for dnc stresses the mechanical problems associated with paper tape in an attempt to extoll the virtues of tapeless systems.

It now appears that the dnc effort is not unlike the fluidics effort of the early 60's. That is to say, the acceptance will certainly fall far short of the original forecasts. For example, only two or three of the systems presented in 1970 are still being actively marketed, and new developments in the two intervening years number less than three. Some of the more obvious reasons for the lack of commercial success of the present dnc systems are:

1. Premature. A quote from a vice president of a major machine-tool builder at the time of the 1970 show indicates that this word is valid: "We're all in the same business! What is worse, I'm afraid we are profoundly adept at displaying equipment for which the market is just beginning to develop."
   This concern was compounded by an economic decline which rendered it impossible to consider anything that remotely smelled of an experimental venture.
2. Lack of production definiton. The general are of on-line computer control of nc machine tools does not readily lent itself to product description. Manufacturers were caught attempting to "productize" a systems concept. As a result, each system emphasized certain features and excluded others. The result was a market left puzzled as to what dnc actually was.
3. Difficult justification. Commercial dnc systems are expensive. Several thousands of dollars of computer and peripheral equipment must be purchased above and beyond the cost of the nc machines. This expense can only be justified by significantly increasing the output of the machines. For some time to come, dnc will require difficult and uncertain justification, since it is neither fish nor fowl with regard to costing. To be sure, original costs and the maintenance and rentals are direct and

measurable. However, variables such as reduced scrappage, less downtime due to tape elimination, more efficient programming, less inventory, better operator communication, plus many others often lie at the interface between direct and overhead costs, and savings in such areas are still educated guesswork.

4. Management acceptance. Computer control in any form is new and requires reorganization at all levels, especially management, to be effective. Many companies have experienced severe problems with computer installations because they failed to recognize this fact. In the same line of thought, the lack of management understanding has made it impossible for many potential customers to explore any possible computer control advantages.
5. Lack of aerospace and defense buyers. A very real reason for the lack of dnc sales has been the inability of earospace and defense firms to purchase such equipment. In the past, particularly with nc, this segment of the industry has helped to pioneer new directions, and during the last period they have not been buying.
6. Insufficient backlogs. Something as expensive as a computer system is not going to be considered unless an extended period of backlog orders in an absolute certainty. In very few instances has this happened in the last two years.

In addition to these specific reasons for lack of acceptance of dnc, there is a general feeling that dnc is not an economic next step until it becomes part of a larger plant network. Manufacturing technology has progressed in the last two years to the pointwhere a significant number of potential dnc installations will in all probability be sidestepped and expansion capital used for integrated manufacturing or multistation manufacturing instead. Where then is the next step, computer-aided manufacturing, today?

In passing, the number of similar statements to those raised for the continuous process industries should be noted here.

Programmable Automation

The questions here are: /1/ On what time scale will computer-based automation of manufacturing proceed, and what form will it take:

/2/ what will be the impact of computer-based automation on product development and product sales and /3/ if there are significant benefits, such as reduced prices and shorter lead times, what technological developments are required to achieve computer-based automation so that those benefits can be realized?

There are, of course, no general answers they depend on many characteristics of the product and on the manufacturing method. There are three key attributes which distinguish programmable automation of manufacturing. These apply to all components of the manufacturing process - parts forming and machining, assembly, inspection, resource transfer and storage, and scheduling - which are:

1. Highly automated with human supervision, but requiring little human intervention in the routine manufacturing process /and are therefore not limited by human reaction times or by human variability/.

2. Flexible: all manufacturing components are programmable.

3. Highly integrated with computer-aided design and engineering facilities, accounting systems, and management information and manufacturing control systems, so that information acquisition from each of these areas contributes to the entire system.

Is complete programmable automation feasible today in a manufacturing facility? Probably not. But it is potentially closer than might be thought. To see how such a facility would operate and what developments are required for its realization, consider the following.

Developments Required

Programmable automation is a prediction of how a manufacturing facility will operate at some time in the future. The concept provides a useful framework within which to evaluate steps currently being taken toward computer-based automation. Three examples of such steps are: numerical control of machine tools industrial robots; and automated warehousing systems. Are current developments in these areas consistent with the overall concept of programmable automation described above?

Numerically controlled machine tools. There are three major areas of activity here. One involves dnc and cnc - how to distribute the computing power between a centralized computer and an on-site mini-computer. The argument does not functionally change the characteristics of computer control of machine tools. More important is the requirement that machine tools be capable of communicating with a centralized computer-based control system; both dnc and cnc permit this.

A second area of activity is the trend toward a cluster of different machine tools /e.g., boring, milling, grinding/ interconnected by a conveying system, with automatic loading and unloading of in-process materials from the individual machine tools. These systems permit a sequence of several different machining operations to be performed on a palletized part, under computer control, and without human intervention. These systems incorporate the programmable automation concept to a considerable extent for the metal-removal component of the manufacturing process.

The third area of activity involves adaptive feedback and in-process inspection - the use of sensors to monitor and control the machining process as it is being performed. Some increases in productivity will result from such adaptive control, but perhaps of greater importance to a highly automated facility is the ability of these sensors to detect such exceptional conditions such as tool breakage.

In summary, the current state of the art in the metal removal aspect of manufacturing is entirely consistent with programmable automation and, in fact, provides an example of such a system in microcosm.

Industrial robots. The current general of industrial robots are programmable "put and take" transfer mechanisms. They have no built-in feedback from tactile or visual sensors, and rely on absolute positional accuracy /about .050 inches in a working area having a diameter of about 20 feet/.

In a paper presented at the recent Second International Symposium on Industrial Robots, Mr. J. F. Engelberger, president of Unimation, Inc., /12/ said that second-generation Unimates will have a compatible computer interface, program memories of any desired size with random program selection, and the ability to synchronize with a moving target. He also predicts that a third-generation industrial robot, a "thinking robot with coordinated hand and eye," won't be needed at all in the industrial environment. Robots will retain their niche as a flexible interface between work stations and the conveying mechanism, but this does not require complex visual processing or "intelligence." However, some degree of sensing and feedback - especially tactile sensing - is clearly desirable to reduce the precision with which known parts must be presented to these robots. In current applications, the cost of special tooling to retain precise part orientation for a robot application is often greater than the cost of the robot itself; limited sensory feedback would dramatically reduce the requirement for such special tooling.

Automated warehousing systems. This technology is advanced to the point of being ready for integration into a programmable manufacturing facility. Most such warehousing systems rely on standardized pallets, and are capable of being interfaced to a computer-based inventory control system.

Given the above inroads of computers into manufacturing, it may seem that computer integrated manufacturing systems are well advanced. Actually, there are major blockages to be overcome, especially in the following areas:

1. Product design for automated assembly. Current design philosophies are too heavily biased toward the unique assembly capabilities of two-handed humans with their sophisticated visual, tactile, and force sensing systems.

2. Programming of such processes as assembly and inspection.

3. A work station specializing in visual processing, to perform subtle inspections, to provide initial orientation of parts, etc.

4. Programmable assembly and inspection machines.

5. An integrated software system capable of data collection, resource allocation, scheduling, control of production processes, and interface to accounting and management systems.

The last two development areas in the preceding list are of special importance.

The assembly process typically accounts for 50% of production costs. Especially in job shop operations, assembly is very labor-intensive. Consequently, it is subject to human variability and is not amenable to computerized coordination with other manufacturing operations. Therefore, programmable automated assembly machines are a vital component of an integrated manufacturing system. There are many trade-offs to be investigated in automated assembly:

1. A group of inexpensive, rather specialized assembly devices versus a sophisticated, more flexible machine.

2. Rigidity and mass for absolute positional accuracy /in the manner of current machine tools/ versus flexible manipulators using sensory feedback for incremental accuracy /in the manner of human assembly workers/.

3. The use of a general-purpose "wrist" capable of accepting special-purpose grippers and tools versus a general-purpose "hand" which obviates the need for a collection of specialized grippers.

4. The use of two or three "hands" working in concert versus one hand supplemented by work-holding fixtures and pallets.

Most likely there is no single set of right answers. Initially, more data is needed on the nature of the assembly task so that the spectrum of possibilities is more clearly understood. However, initial studies indicate that sophisticated visual systems and "artifical intelligence" are not needed for most assembly operations; the emphasis in development should be on simple vision devices and on low-bandwidth tactile and force sensing feedback.

The development of complete software systems for manufacturing facilities, is, of course, crucial to the concept of programmable automation. Perhaps the single most important development in programmable automation would be the implementation and distribution of the IBM COPICS /Communications Oriented Production Information and Control System/ /2/. The resulting <u>de facto</u> standardization of interfaces would bring to the area of manufacturing automation the same type of umbrella that has been in operation in the areas of computer peripherals and memory systems. Without such standardization, developments and products tend to become fragmented. /The only other alternative would seem to be an active role by such institutions as the National Bureau of Standards or the American National Standards Institute in setting interface standards for the modules of an automated production facility./

The software system governing an automated factory will be large and complex; however, the system appears /from the COPICS specifications/ to be highly modular, which will aid in its development. Such a system seems within the state-of-the-art in large software development efforts. It is important to realize, however, that the computer-software system must be used, and defined, directly by high-level manufacturing professionals, who may never have used computers before. Computer scientists will not be running the factory.

How Do We Get There From Here?

Naturally there will not be a sudden quantum jump in the computer-based automation of present manufacturing facilities. Neither will large numbers of shiny new factories be built soon incorporating all of these concepts. The capital investment in current production facilities requires an evolutionary approach to their modernization.

It is quite possible for programmable automation to evolve in existing production systems. Initially, the individual work stations may rely heavily on human intervention, with the worker gradually assuming more of a supervisory role as the capabilities of the machine improve. Also, the conveying system may gradually expand from an initial automated warehouse to provide material routing among increasing numbers of work stations. However, for there to be an orderly evolution toward programmable automation, it is vital that an entire manufacturing facility be studied as a system. Within that system, functional blocks must be identified and the interfaces between these blocks clearly delineated. Two obvious sources of interface specifications are the characteristics of the conveying mechanism - which interfaces with nearly all other mechanical functions - and of the information flow to and from the computer-based control system. During this evolution, it is important to work toward achieving the goal of total automation of the routine manufacturing operations. The maximum impact of computers in manufacturing will come from complete, real-time computer cognizance and control of all processes and resources, allowing the precise scheduling and allocation of those resources.

Unless extreme care is taken and unless the lessons learned in the continuous process industries are heeded, the same disenchantment of our managements and of our non-computer oriented associates can and probably will occur again. The same potential for large complex systems with intricate software and overloaded computers face us here as in the earlier systems. Surely, the preventatives or the remedies are the same here as for the continuous process and other industries.

PRESENT EFFORTS AT PROCESS CONTROL SYSTEM LANGUAGE STANDARDIZATION

As mentioned earlier, standardization has long been recognized as one means by which the planning, development, programming, installation, and operation of our plant control computer installations as well as the training of the personnel involved in all these phases can be organized and simplified. The development of APT and its variant languages by the machine tool industry is a very important example of this. The Instrument Society of America has been engaged in such activities for the past ten years, most recently in conjunction with the Purdue Laboratory for Applied Industrial Control of Purdue University, West Lafayette, Indiana /1,3/. While these efforts have involved both hardware and software topics the software project will be discussed here.

Through eight semiannual meetings the Purdue Workshop on Standardization of Industrial Computer Languages has proposed the following possible solutions to the programming problems raised above, and it has achieved the listed results.

1. The popularity of FORTRAN indicates its use as at least one of

the procedural languages to be used as the basis for a standardized set of process control languages. It has been the decision of the Workshop to extend the language to supply the missing functions necessary for process control use by a set of CALL statements. These proposed CALLS, after approval by the Workshop, will be formally standardized through the mechanisms of the Instrument Society of America. One Standard has already been issued by ISA /4,17/, another is being reviewed at this writing /5/, and a third and last one is under final development /6/.

2. A so-called Long Term Procedural Language is also being pursued. A set of Functional Requirements for this Language has been approved. Since the PL/1 language is in process of standardization by ANSI /the American National Standards Institute/ an extended subset of it /in the manner of the extended FORTRAN/ will be tested against these requirements /19/. Should it fail, another language will be tried.

3. The recognized need for a set of problem-oriented languages is being handled by the proposed development of a set of macro-compiler routines which will, when completed, allow the user to develop his own special language while still preserving the transportability capability which is so important for the ultimate success of the standardization effort.

4. To establish the tasks to be satisfied by the above languages, an overall set of Functional Requirements has been developed /11/.

5. In order that all Committees of the Workshop should have a common usage of the special terms of computer programming, the Glossary Committee of the Workshop has developed a Dictionary for Industrial Computer Programming which has been published by the Instrument Society of America /15/.

As mentioned several times before, it is the aim and desire of those involved in this effort that the Standards developed will have as universal an application as possible. Every possible precaution is being taken to assure this.

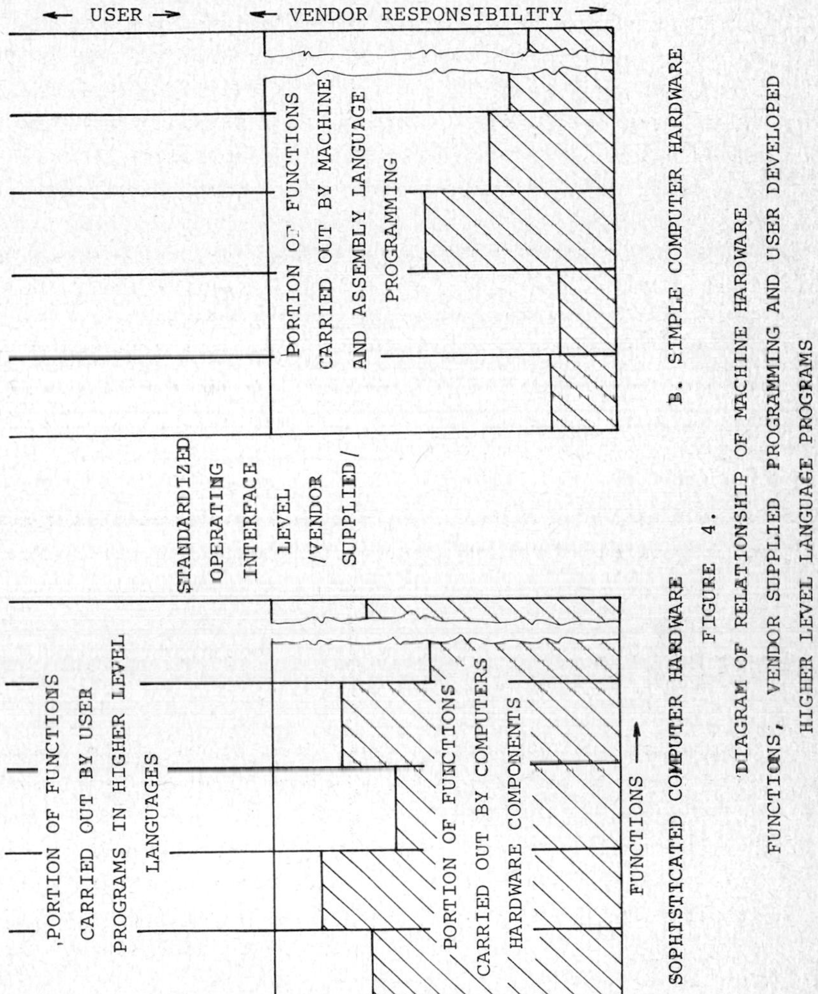

FIGURE 4.

DIAGRAM OF RELATIONSHIP OF MACHINE HARDWARE FUNCTIONS, VENDOR SUPPLIED PROGRAMMING AND USER DEVELOPED HIGHER LEVEL LANGUAGE PROGRAMS

The nearly total attention in these and similar efforts toward the use of higher level languages means that the vendor must be responsible for producing a combination of computer hardware and of operating system programs which will accept the user's programs written in the higher level languages in the most efficient manner. Figure 4 diagrams this situation, indicating that vendor hardware and software must supplement each other to achieve an equivalent result. A relatively simple computer requiring a much higher use of software accomplished functions would thus be equivalent, except for speed of operation, with a much more sophisticated and efficient computer with a correspondingly smaller operating system. Present trends in small computer architecture would indicate that the Type A machine of Figure 4 would probably be the more popular in the future. This is despite the fact that most present systems tend toward Type B.

The present desire on the part of both users and vendors for a simplification and clarification of the present morass of programming problems indicates that some standardization effort, the Purdue sponsored program, or another, must succeed in the relatively near future.

Future Possibilities and Associated Time Scales

The standardized FORTRAN extensions as described can be available in final form within the next one to two years. Some of those previously made have been implemented already be nearly a dozen computer vendors. The actual standardization process requires a longer period of time because of the formality involved. Commonly, three to four years total is required from proposal to final publication of the finished document. Thus, the 1974-75 period appears to be the key time for this effort.

The work of the other language committees of the Workshop are less formally developed than that of the FORTRAN Committee as mentioned just above. Successful completion of their plans could result in significant developments in the Long Term Procedural Language and in the Problem Oriented Languages are within the same time period as above.

In addition to its Instrument Society of America sponsorship, this effort recently received recognition from the International Federation for Information Processing /IFIP/ when the Workshop was designated as a Working Group of its Committee on Computer Applications in Technology. The Workshop is also being considered for similar recongition by the International Federation of Automatic Control /IFAC/.

As mentioned, this effort is achieving a very wide acceptance to date. Unfortunately, partly because of its Instrument Society of America origins and the personnel involved in its Committees, the effort is largely based on the needs of the continuous process industries. The input of interested discrete manufacturing industry personnel is very badly needed to assure its applicability across all industries. To provide the necessary input from other industries, it is hoped that one or more of the technical societies /United States or international/ active in the discrete manufacturing field will pick up cosponsorship of the standardization effort presently spearheaded by the Instrument Society of America and, in cooperation with it, make certain that a truly general set of languages is developed for the industrial control field. In addition, the committees of the Workshop are always in need of the help of willing workers. All who are interested are invited and urged to take part. The author may be contacted for details.

A PROPOSAL

As a set of conclusions to our discussion and as a counter to the developments listed in Table I, we can state that under most circumstances we should strive to achieve the following in our computer-based manufacturing systems regardless of the industry involved:

1. They should be kept as simple as possible, consistent with the result expected.

2. Functions should be distributed among a group of small computers, each carrying out a few relatively simple tasks, rather than being concentrated in one or a few large systems. This will become ever more important as costs of the small systems continue to decrease.

3. The computer language used by the engineer or programmer should be as near to a natural language as possible to permit self-documenting of the program to the maximum extent. This is incompatible with Item 2 above so host-system compiling will probably be necessary.

4. The computer system's source program should be transportable to other types of computer systems to avoid the necessity of reprogramming should the computer hardware be changed or a new but similar plant be built.

5. Dedicated, hardwired, component systems should be used whenever there is a question of break-even between hardwired and programmed systems.

6. The question of whether a particular computer is being efficiently used should not be a pertinent question in the systems analysis - only whether that particular application makes the proper economic payout considering both manpower and hardware costs.

As an additional desirable development for the future, we would list the following:

7. The discrete parts manufacturing industries should join the other classes of manufacturing industry to help achieve as much standardization of the programming art across all industries as is consistent with a continued technical development of computer systems.

REFERENCES

1. Anon., Minutes, Workshop on Standardization of Industrial Computer Languages, Purdue Laboratory for Applied Industrial Control, Purdue University, West Lafayette, Indiana: February 17-21; September 29-October 2, 1969; March 2-6; November 9-12, 1970; May 3-6; October 26-29, 1971; April 24-27; October 2-5, 1972.

2. Anon., Communications Oriented Production Information and Control System, 8 Vols. G320-1974 through G320-1981, International Business Machines Corp., White Plains, New York /March 1972/.

3. Anon., Minutes, ISA Computer Control Workshop, Purdue Laboratory for Applied Industrial Control, Purdue University, West Lafayette, Indiana May 22-24, November 13-16, 1972.

4. Anon., "Industrial Computer System Fortran Procedures for Executive Functions and Process Input-Output," Standard ISA - - S61.1, Instrument Society of America, Pittsburgh, Pennsylvania /1972/.

5. Anon., "Industrial Computer System Fortran Procedures for Handling Random Unformatted Files, Bit Manipulation, and Data and Time Information," Proposed Standard ISA - S61.2, Instrument Society of America, Pittsburgh, Pennsylvania /1972/.

6. Anon., "Wirking Paper, Industrial Computer Fortran Procedures for Task Management," Proposed Standard ISA - S61.3, Purdue Laboratory for Applied Industrial Control, Purdue University, West Lafayette, Indiana /1972/.

7. Alexander, Tom., "The Hard Road to Soft Automation," Fortune, 84, No. 1, 94-97, 147-150 /July 1971/.

8. Anderson, R. H., "Programmable Automation: The Bright Future of Computers in Manufacturing," Datamation, 18, No. 12, 46-52 /December 1972/.

9. Bailey, S. J., "Direct Digital Control - A Status Report," Control Engineering, 18, No. 11, 36-40 /November 1972/.

10. Biles, W. R., and Dragoo, R.E., Jr. "Pitfalls in Process Computer Control," Paper CC-72-73, NPRA Computer Conference, Philadelphia, Pennsylvania, November 26-29, 1972.

11. Curtis, R. L., "Functional Requirements for Industrial Computer Systems," Instrumentation Technology, 18, No. 11, 47-50 /November 1971/.

12. Engelberger, J. F., "Industrial Robots - Second Generation," Proceedings of the Seminar on 2nd Symposium on Industrial Robots, IIT Research Institute, Chicago, Illinois, May 16-18, 1972 /pp. 211-233/.

13. Evans, L., "Computerized Numerical Control Revisited," Control Engineering, 17, No. 8, 33-36 /August 1972/.

14. Ferrar, G. L., "The Attitude of Oil-Company Management About Computers," Paper CC-72-64, NPRA Computer Conference, Philadelphia, Pennsylvania, November 26-29, 1972.

15. Glossary Committee, Purdue Workshop on Standardization of Industrial Computer Languages, Dictionary of Industrial Digital Computer Terminology, Instrument Society of America, Pittsburgh, Pennsylvania /1972/.

16. Guisti, A. L., Otto, R. E., and Williams, T. J., "Direct Digital Computer Control," Control Engineering, 9, No. 6, 104-108 /June 1962/.

17. Kelly, E. A., "FORTRAN in Process Control Standardizing Extensions," Instrumentation Technology, 17, No. 5, 47-53 /May 1970/.

18. Mesniaeff, P. G., "The Technical Ins and Outs of Computerized Numerical Control - A Special Report," Control Engineering, 17, No. 3, 65-84 /March 1971/.

19. Pike, H. E., "Procedural Language Development at the Purdue Workshop on the Standardization of Industrial Computer Languages," Paper presented at the Fifth World Congress, International Federation of Automatic Control, Paris, France, June 1972.

20. Steering Committee, Workshop on Standardization of Industrial Computer Languages, "Industrial Control Computer Programming - The Coming Age of Language Standards," Automatica, 8, 499--500 /1972/.

21. Williams, T. J., "What to Expect From Direct Digital Control," Chemical Engineering, 71, No. 5, 97-104 /March 2, 1964/.

22. Williams, T. J., "Interface Requirements, Transducers and Computers for On-Line Systems," Automatica, 5, No. 4, 845-854 /1969/.

2. INTERACTIVE, CONVERSATIONAL AND GRAPHIC PROGRAMMING

INTERACTIVE, CONVERSATIONAL AND GRAPHIC PROGRAMMING

J. Hatvany
Computer and Automation Institute,
Hungarian Academy of Sciences, Budapest,
Hungary

The papers in this Section appropriately illustrate the three approaches to interactive part-programming which characterize the NC scene today. While the easiest way to categorize them is by reference to the hardware they use, this is only the most obvious manifestation of the differences in systems philosophy which underlie them.

According to this criterion they can be grouped as follows:

1. Minicomputer, stand-alone systems
2. Systems based on a large, central computer used in a time-sharing mode
3. Intelligent terminal systems, using a medium-sized host computer.

Each of these categories has its advantages and its drawbacks - the optimal trade-off for any particular environment will depend on a number of locally valid economic and technical factors.

The system presented by J.P. Crestin and J.P. Paillard belongs to the first category. In the first, experimental version a 32 K IBM/1130 was used with a Model 2250 graphic display, but the system is now being transferred to more modern, cheaper hardware, and the particular implementation is largely irrelevant to the philosophy of the approach.

The basic advantages of using a stand-alone minicomputer system are particularly attractive in a mechanical engineering environment. For a relatively small investment a factory is able to establish an in-house facility of its own, cheap and at the same time fully secure against unauthorized access to proprietary technological or product development information. The price to be paid for these benefits is in two factors, which may in fact be indifferent to many users: the

COMPUTER LANGUAGES FOR NUMERICAL CONTROL, *J. Hatvany, editor*
North-Holland Publishing Company - Amsterdam–London

limited service that the system will provide and the relatively long processing times required.

The main limitations of the system of Crestin and Paillard are that
- a/ it is confined to the simpler cases of $2\frac{1}{2}$ D machining,
- b/ the part-programming, text-editing, geometrical processing and display parts of the system are linearly sequential.

The latter limitation means, of course, that there can be no interaction with the part program from the graphic display, since none of the system elements offer reentry facilities. As for processing times, these are of the order of 1 minute in the experimental version, but will obviously decrease with transfer to faster hardware.

The systems of the second category are represented by three papers. D.G. Wilkinson uses a teletype terminal connected to a UNIVAC 1108, K.J. Davies has a storage tube terminal linked to an ATLAS, while J. Cremerius uses a refresh-type graphic terminal operating with a CDC 6000 /Cyber/ series computer. While these systems may at first sight appear to differ very greatly /ranging from the simplest terminal hardware to the most complicated, and hence from purely alphanumeric conversational part programming to full graphic interaction/, they have one feature in common: they all use large central computer systems in a time-sharing mode.

The main benefits from using a high-power computing system are that it affords facilities for a high degree of on-line interaction, almost instantaneous processing and any required degree of sophistication in the services offered /full, 3D, technology, etc./. It will also permit linking the part-programming phase with machining data base access, program library facilities, tool and material files, production control, and all the other aspects of an integrated information system.

The economic, organizational and security trade-off however, will depend very much on the size of the company, its existing computing facilities, geographical location and local tradition. In Europe, at any rate, very many of the thousands of small-to-medium firms which cannot hope to possess large computer systems of their own, are reluctant to take their part-programming outside and run it through a time-sharing bureau service. They are fearful of too much dependence on a remote computing centre and the communications link to it, and they are loath to entrust their confidential new product component descriptions to an outside organization's files.

The third category of systems - the use of an intelligent terminal with a medium-sized host computer - is represented by the paper of V.G. Zaitsev, who uses a local UM 1 computer linked to a MINSK 32 host. A similar philosophy underlies the system developed at the Computer and Automation Institute, Hungarian Academy of Sciences, which will be demonstrated to Conference participants, where a local VT 1010 B minicomputer with graphic display is hooked to a CDC 3300 medium host computer.

The advantage of the intelligent terminal approach is that it provides most of the facilities of the second category, while requiring only the medium-sized type of central computing facility for background work, which is now becoming generally available in medium-sized plants. Thus, while offering services considerably more sophisticated than those attainable with stand-alone minicomputers, the factory is able to keep its entire system in-house and link it to its general production control, management information and data base systems. The price to be paid for these advantages is the requirement for a relatively expensive terminal with reasonable processing power and local background storage, plus the multiplexing hardware and software for the central computer /one of these two components being additional to the requirements of the first two categories/. Furthermore, an intelligent terminal type system will have somewhat longer response times at certain points of interaction, than the large computer systems.

Even so brief a survey of the three main categories of interactive, conversational and graphic programming systems will, hopefully, have served to show that none can be declared intrinsically "superior" or more advanced than the others. In the particular environment of Hungary, for instance, where no very large central computing facilities are available, the choice is limited to categories 1 and 3. The first will be satisfactory where its implicit limitations are acceptable, but the dominant form for most of our factories will be the intelligent terminal approach. This alone, permits our part programming system to be linked to the other information systems of the production unit.

- x -

Two of the papers of Session I are concerned with completely new techniques in the field of NC part programming. C.A. Lang has built what he calls a model making machine - from the systems engineering point of view we might of course say it was a true 3D graphic display. The enermous utility of such a device is immediately apparent, and the examples in the paper serve well to illustrate this point. Yet the considerable model machining times and the lack of interaction through the model are limitations which could only be overcome if a true 3D optical display could at last be constructed. The obvious value of Lang's machine to the part programmer should help provide the rationale for the development of such a device.

Finally W. Nauck is presenting a novel means for the rapid, automatic input of complex geometrical shapes into a computer system, from a /scale/ model. This too, appears to be a valuable system component which could possibly be put to a highly interesting use as a geometrical feedback device in integrated manufacturing systems.

CONVERSATIONAL NELAPT, AN INTERACTIVE VERSION OF THE NELAPT PROCESSOR

D G WILKINSON
National Engineering Laboratory
East Kilbride, Glasgow, United Kingdom

Abstract: A processor is described which allows NELAPT part programs to be processed and edited in a conversational mode from a remote teletype terminal on-line to a time-sharing computer network. Part programs may be entered by typing, or from external files and a versatile editing facility is incorporated to allow the programmer to correct errors which are detected by the processor. Output files may be saved and examined by the programmer before a control tape is generated and if required complete listings can be obtained from a line printer.

1 INTRODUCTION

Conversational NELAPT is a computer program which can be used to assist in the production of NC control tapes for most types of NC machine tool. It gives the part programmer a great deal of control over the processing of his part program by performing all the computer tasks necessary when manipulating files of data in response to very simple commands. Part program statements can be checked immediately for errors in syntax or geometric inconsistency, providing him with written diagnostics and simple editing capabilities. The various stages of processing are under his control, he can examine any part of the CLDATA he wants and has the option of having it all printed on the line printer. Post-processing also comes under his direct control and he can have a machine control tape produced on his teletype within a very short time of commencing his conversation with the system. The use of a random access backup store for diagnostics and system messages makes it possible for users to choose their language (English, French or German) without the penalty of increased core size and the system is designed whenever possible to give the user the chance to correct any mistake or in extreme cases to terminate neatly. Conversational NELAPT although still capable of development is a very usable tool and should be ideally suited to the small NC installation. It uses the modules developed for the established batch processing version of NELAPT and reflects all the developments which have gone into that system.

The part programming language of NELAPT, like that of APT, conforms to the forthcoming ISO standard. It has a wide range of surface descriptions and includes a comprehensive area clearance language. It is fully described elsewhere (NEL, 1969) and will only be commented on briefly here.

A part program must commence with a PARTNO statement and end with FINI. It is completely free format and spaces are ignored with the exception of PARTNO, PPRINT and INSERT statements where a line is accepted exactly as typed. The only way in which the language of conversational NELAPT differs from the normal batch processing NELAPT is in the provision of extra vocabulary to allow the input listing to be suppressed. This is the FULIST/OUT statement, which is an addition to the existing FULIST/ON and FULIST/OFF statements.

2 THE BATCH PROCESSING NELAPT SYSTEM

To show the difference between the conversational and batch versions of NELAPT it is necessary to explain the batch processing method and the structure of NELAPT. As this has been done previously in detail (McWaters, 1970) only a brief description will be given here.

2.1 NELAPT Processor

The NELAPT processor is basically five separate modules each of which in the

COMPUTER LANGUAGES FOR NUMERICAL CONTROL, *J. Hatvany, editor*
North-Holland Publishing Company - Amsterdam–London

batch processing system processes the complete part program from PARTNO through to FINI noting errors as they are found and producing as output a serial intermediate file. This is shown in flow chart form in Fig. 1. The INPUT module reads the input one line (or card) at a time, performs rudimentary checks and uses the delimiters (commas, equals and slashes) to break down the input into a string of computer words. It writes a serial output file which is read by the DECODE module. The pass through DECODE completes the syntactic analysis of the part program, nests are broken down, MACROs and loops are expanded, arithmetic calculations are performed and the serial output file produced by DECODE is in the form of records where each record (composed of a string of couplets) is related to one logical part program statement. If errors are detected by DECODE, processing stops after the pass is completed.

2.2 Geometry

The GEOMETRY pass through the DECODE output file performs any geometric calculations which are required, generates the canonical form (mathematical standard form) of each geometric element which has been defined and, for each reference to a geometric element in a statement that requires further processing, inserts the appropriate canonical form in the serial output file. If any errors are detected by GEOMETRY, processing is stopped after the pass is completed.

2.3 Tool Selection

The TOOL SELECTION module reads the GEOMETRY output file and writes a new output file to which it adds any tool data required. (The tool data are extracted from the TOOL LIBRARY file external to NELAPT.)

2.4 Motion

The output file from TOOL SELECTION is read by MOTION which performs the cutter motion calculations and produces a CLDATA output file suitable for post-processing.

The post-processing is carried out by a separate control command as at NEL all processors produce standard ISO CLDATA and only one set of post-processors is required.

3 THE CONVERSATIONAL NELAPT SYSTEM

As mentioned in Section 1, the conversational NELAPT system uses the modules developed for the batch version of NELAPT, the principle difference being the manner in which the modules are organized and the addition of a control routine. Fig. 2 shows the communication paths in the conversational system and reveals the large organizational differences between it and the batch version of NELAPT. The only sequential intermediate files remaining are between GEOMETRY and MOTION and the CLDATA file. The CONTROL module contains buffer areas of storage which are used to transfer information between the various modules and external files. In addition to the new CONTROL module there is an EDIT module and two other modules, MACRO DEFINITION and MACRO EXPANSION which have been extracted from the DECODE module. Of the new files, the diagnostic and message files are random access, the addressable quantum being 28 computer words stored on drum on the Univac 1108. The other new files are all sequential files written using a normal FORTRAN write statement with a 12A6 format. This ensures compatibility with the DATA file structure standard on the Univac 1108 under the EXEC 8 operating system and allows these files to be conveniently handled independently of the NELAPT processor.

3.1 System Operation

The program flow from the user's point of view is explained in some detail in Section 5 and an associated flow chart is shown in Fig. 3.

The conversational system has three main functions: file handling, part program processing, and part program editing. The CONTROL module dynamically assigns files, reading source part program statements either from teletype or a previously established file and storing a completed part program on a new file, or replacing the source part program with an updated version.

3.2 Interactive Section

Part program statements from whatever source are processed individually through the INPUT, MACRO, DECODE and GEOMETRY modules. A mistake discovered by any of these modules is signalled to the CONTROL module which prints out an error message and ignores the incorrect statement. If the source statement was obtained from the teletype a corrected statement is solicited from the teletype, but if the source was a part program file, then control is transferred to the EDIT module. The signal to terminate the interactive phase of part program analysis is the FINI part program statement.

3.3 Motion

The sequential file output from GEOMETRY is now available to be processed by MOTION and assuming that the user decides to instigate this phase, a CLDATA file and print file are generated. The print file may be scanned by the EDIT module for selective listing by the user and in addition may be routed to the line printer for a complete listing if required. Post-processing is initiated by CONTROL after the user has indicated whether an output listing is required from the post-processor. (This feature is necessary because the output from the post-processor may not be formatted to suit teletype operation.)

After completion of post-processing, which is a separate operation, control passes from the post-processor back to the computer operating system. To produce a control tape the user must then type the control command; @TAPE.PUNCHT PUNTAP.

4 EDITING FACILITIES

Part programs being processed by Conversational NELAPT are temporarily stored as they are entered and may be edited whenever necessary. As will be described in Section 5 a permanent record of a part program may also be kept under a name designated by the user and this permanent record may also be edited by Conversational NELAPT if required. The editing mode may be entered in one of three ways: (a) by typing the command EDIT following the message *NOW* when entering a part program as described in Section 3; (b) by naming an existing part program file as described in Section 5.1; or (c) as a result of an error in a previously stored part program which is detected when it is being processed.

The response of NELAPT when the editing mode is entered is the same regardless of how it was entered. The message printed is **EDITING** *CHOOSE*.

4.1 Editing Commands

The user can now respond with one of the following commands;

a LIST/SOME LIST/ALL PUNCH
b INSERT REPLACE DELETE
c RUN RESUME QUIT

The commands (a) provide facilities for inspection of a part program or having it punched on tape at the terminal in use. Commands (b) allow the stored part program to be modified, and commands (c) provide exits from the editing mode.

After LIST/SOME has been typed the reply is *FROM WHICH STATEMENT? to which the user must respond with a number, then follows *HOW MANY STATEMENTS? and, after the user has typed another number, the required part program statements are listed. In the case of LIST/ALL the complete part program is listed and the response to PUNCH is similar but now the listing is preceded by the message **SWITCH ON TAPE PUNCH NOW**. After the part program has been punched the message **SWITCH OFF TAPE PUNCH NOW** reminds the user to do so. Following any of these three operations NELAPT will print the message *CHOOSE* after which another editing command must be typed.

The commands INSERT, REPLACE and DELETE invoke the identical responses *FROM WHICH STATEMENT? after which a number must be typed, then follows *HOW MANY STATEMENTS? After the number of statements is typed in the response depends on the command. In the case of DELETE the response is *CHOOSE*, indicating that the required number of statements have been deleted and another editing option may be exercised. In the case of INSERT or REPLACE the message *NOW* is printed, soliciting a part program statement, and is repeated after each new statement has been typed until the specified number of statements have been typed. Following this, the message *CHOO$E* will be printed. A new editing command should then be typed.

To exit from the editing mode the user must type one of the editing commands RUN, RESUME or QUIT. The command RUN will cause the stored part program to be processed as described in Section 3 from the PARTNO statement through to FINI or until an error is detected. The command RESUME is followed by *FROM WHICH STATEMENT? and, after a number has been typed, processing is resumed, listing the part program from the statement whose number was given through to FINI or until an error is detected. When in the editing mode, QUIT should be used if it is desired to make an exit from Conversational NELAPT without any further processing being performed. After this command the user will be asked if he wishes to save his part program (see Section 5) and following this Conversational NELAPT is terminated.

4.2 Editing Errors

If an incorrect editing command is typed following *CHOOSE* the message **INVALID EDITING COMMAND - TRY AGAIN** is printed followed by *CHOOSE*. When *FROM WHICH STATEMENT? has been followed by a number greater than the number of statements in the part program, the message **STATEMENT NUMBER OUT OF RANGE** is printed followed by *CHOOSE*. When the number given following the message *HOW MANY STATEMENTS? is greater than the number of statements remaining in the part program file, the part program is listed until the end of the file is reached when the following message is printed **END OF FILE ENCOUNTERED AT LINE 'n' where 'n' is the number of statements in the part program file. *CHOOSE* then allows another editing command to be given.

5 PROGRAM FLOW AND PART PROGRAM STORAGE

As with all computer programs which are processed in a multi-programming environment, the executive program of the computer must first be instructed to load and execute the required program. In the case of the Univac installation at NEL this is the single control instruction @XQT NELAPT, which is typed after logging into the computer. From this statement until the normal termination or the occurrence of the control character '@' as the first character of an input line (which indicates that a command follows for the computer executive system as described in Section 5.5c), Conversational NELAPT controls the program flow. A simplified flow chart of the system is shown in Fig. 3.

5.1 Initialization

To assist in following the flow of Conversational NELAPT, examples of 'conversations' are listed in the following tables. The user responses are underlined in each of the tables.

Table 1

```
@XQT NELAPT.
 **CONVERSATIONAL NELAPT - VERSION 006  NEL EAST KILBRIDE  GLASGOW**
 * ENGLISH?  FRANCAIS?  DEUTSCH?  SHORT?
ENGLISH
 *DO YOU WANT HELP?
NO
 *DO YOU WANT TO USE AN EXISTING FILE?
YES
 *PLEASE GIVE FILE NAME*
NOFILE
 **UNABLE TO ASSIGN FILE
            PROJECT*NOFILE.                ERROR  400010000000
 ** PLEASE TRY AGAIN **
 *PLEASE GIVE FILE NAME*
STILLNOFILE
 **UNABLE TO ASSIGN FILE
            PROJECT*STILLNOFILE.           ERROR  400010000000
 ** PLEASE TRY AGAIN **
 *PLEASE GIVE FILE NAME*
STILLNOFILE
 **UNABLE TO ASSIGN FILE
            PROJECT*STILLNOFILE.           ERROR  400010000000
 **CONVERSATIONAL NELAPT TERMINATED - SORRY**
```

If the reply to *DO YOU WANT HELP? had been YES a brief explanation of the system would have been printed. Where the user has no existing file (as in Table 2) a temporary file is established. Table 1 illustrates the result of the user giving an incorrect or non-existent file name. As can be seen, three chances are allowed to give an acceptable name before NELAPT terminates. The Appendix gives details of file-naming conventions and details of file-naming errors.

5.2 New Part Program

Table 2

```
 *DO YOU WANT TO USE AN EXISTING FILE?
NO
 *NOW*
PARTNO TEST PART PROGRAM
          1   PARTNO TEST PART PROGRAM
 *NOW*
FULIST/OUT
 *NOW*
P1=POINT/0,0,0
 *NOW*
P2=POINTE/0,0,0
      ONE OR MORE FORMAT ERRORS
 **ERROR IN STATEMENT    4
 *NOW*
P2=POINT/0,0,0
 *NOW*
L1=LINE/P1,P2
      SOLUTION IMPOSSIBLE
 **ERROR IN STATEMENT    5
 *NOW*
FINI
 *DO YOU WANT TO SAVE YOUR PART PROGRAM?
YES
 *PLEASE GIVE FILE NAME*
TESTPP
 *CUTTER MOTION CALCULATIONS?
NO
 **CONVERSATIONAL NELAPT NOW TERMINATED - CHEERIO**
```

In this case the user does not have an existing file so the part program statements (NEL, 1969) are processed as they are typed and if correct, stored on a temporary file. Note the use of FULIST/OUT to inhibit NELAPT from retyping the part program statements. This example illustrates how NELAPT detects syntax errors, ie incorrect spelling of POINT, and geometric errors, ie a line defined as passing through two points which are co-incident. The correct statements are finally stored on the permanent file named TESTPP.

5.3 Entry From the Editing Mode

Table 3

```
  *DO YOU WANT TO USE AN EXISTING FILE?
YES
  *PLEASE GIVE FILE NAME*
TESTPP
  **EDITING**
  *CHOOSE*
RUN
                1     PARTNO TEST PART PROGRAM
  *NOW*
CLPRNT
  *NOW*
MACHIN/NELCNC,1
  *NOW*
GOTO/P1
  *NOW*
GOTO/2,2,4
  *NOW*
FINI
  *DO YOU WANT TO SAVE YOUR PART PROGRAM?
NO
  *CUTTER MOTION CALCULATIONS?
YES
  **INITIATING CUTTER MOTION CALCULATIONS**
  **NO. OF ERRORS DETECTED =  0
  **NO. OF CUTTER LOCATION RECORDS = 10
  *DO YOU WANT TO EXAMINE CLPRNT?
NO
  *DO YOU WANT CLPRNT VIA LINE PRINTER?
NO
  *DO YOU WANT TO USE A STANDARD POST PROCESSOR?
YES
  *DO YOU WANT TO EXAMINE PPRINT?
YES
  **CONVERSATIONAL NELAPT NOW TERMINATED - CHEERIO**
  POSTPROCESSOR LISTING
  TEST PART PROGRAM
  N001G01X00000  Y00000                                    F000
  N002   X02000  Y02000  Z04000                            F000
```

In this example the part program file created in Table 2 is used as an existing file and further part program statements are added. Note that, after the user has given the name of the file, NELAPT automatically enters the editing mode as described in Section 4.

5.4 Error in Statement

If an error is detected in a part program statement, an error message is printed followed by the number of the statement in which the error occurred, with the erroneous statement being disregarded. The action taken next by NELAPT depends on whether the source part program statements are being input from a teletype or a computer file. If the source is a computer file, NELAPT immediately switches to the edit mode as described in Section 4 and this allows the part program to be corrected. If the source statement was input from a teletype the response is *NOW* and the correct statement may then be typed.

5.5 Completion of Part Program

The response of *NOW* after each part program statement will continue until terminated by one of the following user replies:

a the part program statement FINI (described in the following paragraph);
b the command EDIT (described in Section 4);
c a computer system control command.

In the unlikely event of a user typing as the first character of a line the character used by the computer executive program to identify control commands ('@' on the Univac 1108), NELAPT cannot accept any further instructions and has to relinquish control to the computer executive program. In this event the message printed will be **CONTROL CARD IN RUN STREAM** **CONVERSATIONAL NELAPT TERMINATED - SORRY**.

5.6 Response to FINI

After a FINI part program statement has been typed the response is *DO YOU WANT TO SAVE YOUR PART PROGRAM? If the answer is NO, processing continues as described in Section 5.7 and after NELAPT is terminated the temporary file containing the part program is discarded. If the answer is YES the response is *PLEASE GIVE FILE NAME*. The reply to this should be the name under which the user wishes to store his part program. (Although the installation at which NELAPT is being used may specify the file-naming conventions, NELAPT allows any name of up to twelve characters to be used.) If an existing file has been used as input the name of this file may be given again at this stage and the original part program will be replaced with the updated part program. If the user gives any other existing file name, NELAPT disregards it, to protect the contents of these other files, and replies with the warning number 5 described in the Appendix. If the new file name is accepted by NELAPT, processing continues as described in Section 5.7. The Appendix describes the cases in which the file name is unacceptable and following each such case after printing a warning message NELAPT allows the user two more attempts to give an acceptable file name. After the third unsuccessful attempt NELAPT terminates with the message **CONVERSATIONAL NELAPT TERMINATED - SORRY**.

5.7 Cutter Motion Calculations

At this stage the part program has been completely checked for syntax and geometry errors and may now be further processed to produce the cutter location data (CLDATA). NELAPT therefore asks *CUTTER MOTION CALCULATIONS? If the reply to this is NO, NELAPT terminates with the message **CONVERSATIONAL NELAPT NOW TERMINATED - CHEERIO**. If the reply was YES, the following messages are printed **INITIATING CUTTER MOTION CALCULATIONS** **NO. OF ERRORS DETECTED = m **NO. OF CUTTER LOCATION RECORDS = n *DO YOU WANT TO EXAMINE CLPRNT? Where 'm' is the number of errors found during the calculation of the cutter path and 'n' is the number of CLDATA records calculated. If the user answers NO to the above question, processing continues as described in Section 5.8; if the reply is YES he is then asked *FROM WHICH STATEMENT? to which the user must reply with a number. The next question *HOW MANY STATEMENTS? also requires a number in reply, following which the required number of CLPRNT statements are typed out and the user is again asked *DO YOU WANT TO EXAMINE CLPRNT?

Eventually, when the user answers NO, the next question is *DO YOU WANT CLPRNT VIA LINE PRINTER? A reply of YES to this will result in all the cutter path coordinates being listed on the line printer attached to the computer. This listing can then be sent to, or collected by the user.

5.8 Post-processing

At NEL a wide range of post-processors are available and can be accessed by a variety of processors including APT and Conversational NELAPT. To optimize the costs of NC processing, the NEL library of post-processors has been split into two files, one containing large post-processors and one containing small post-processors. The smaller post-processors are those most commonly used and following the reply to the question asked in the previous paragraph, Conversational NELAPT asks *DO YOU WANT TO USE A STANDARD POST-PROCESSOR? (where a STANDARD POST-PROCESSOR is one of those on the file of smaller post-processors). A reply of NO here would cause a normal termination of NELAPT as described in the following paragraph, after which a computer control statement executing the file of large post-processors may be typed in. Normally however, the reply would be YES which is followed by the question *DO YOU WANT TO EXAMINE PPRINT? (where PPRINT refers to the post-processor printout). The answer NO will suppress any post-processor listing, whilst YES causes this listing to be typed after the normal termination of Conversational NELAPT. The ability to suppress the output listing of a post-processor is necessary as in many cases this output is not formatted for teletype output.

The response of NELAPT to the previous answer is now **CONVERSATIONAL NELAPT NOW TERMINATED - CHEERIO** If post-processing has been requested, this will occur following the above message and may result in further printing. As post-processing is not at present performed in a conversational manner at NEL the additional computer control statement @TAPE.PUNCHT PUNTAP. must be typed to initiate the punching of a machine-tool control tape after post-processing is completed. One of the standard post-processors available at NEL is for an on-line plotter and to use this it is only necessary to answer YES to the question DO YOU WANT TO USE A STANDARD POST-PROCESSOR? The CLDATA will then be plotted as long as the correct MACHIN statement was included in the part program.

6 CONCLUSIONS

Table 4 is a listing of a run through Conversational NELAPT which uses a special diagnostic feature for printing all the computer control statements normally hidden and thereby revealing the extent to which NELAPT shields the user. Perhaps the most significant fact brought out by the development of Conversational NELAPT is the ease with which an existing modular program, which was designed for batch processing can be made interactive on a large modern computer. Only about 5 per cent of the new coding written for this system is in assembly language, 95 per cent being FORTRAN with very few modifications to existing code being necessary.

It is difficult to establish a cost for using an NC computer program which permits comparison with other NC programs due to the large number of variables that enter into such comparison, but prospective users rightfully demand some method of predicting their costs. One example which gives some guidance in the case of Conversational NELAPT is a lathe part program which includes some contouring and thread cutting and has been processed from a remote terminal using the NEL computer at the commercial charging rate. The part program comprised 107 statements and produced 123 blocks of control tape. From entering the part program to having the control tape punched took 30 minutes of terminal time during the peak load period of the computer and the computing cost was £7.5. Assuming the user was over 50 miles from the computer and the maximum telephone charges were applicable, the telephone cost would be £2.5 making a total cost to the user of £10. Higher data transmission speeds than the 10 characters per second used in this case would of course considerably reduce these costs, but the terminal equipment required is then more expensive.

Table 4

```
@XQT,C NELAPT.

@FREE            6
@USE       6,PRINT$
@FREE           17
@ASG,A      PROJECT*MESSAGES.
@USE       8,PROJECT*MESSAGES.
@ASG,A      PROJECT*DIAGNOSTICS.
@USE       7,PROJECT*DIAGNOSTICS.

 **CONVERSATIONAL NELAPT - VERSION 006  NEL EAST KILBRIDE  GLASGOW**
 * ENGLISH?  FRANCAIS?  DEUTSCH?  SHORT?
ENGLISH
 *DO YOU WANT HELP?
NO
 *DO YOU WANT TO USE AN EXISTING FILE?
YES
 *PLEASE GIVE FILE NAME*
TESTPP

@ASG,AX             PROJECT*TESTPP.
@USE       19,      PROJECT*TESTPP.
@ASG,T          16
@ASG,T          17
@ASG,T          21
@ASG,T          23
@ASG,T          25

 **EDITING**
 *CHOOSE*
QUIT
 *DO YOU WANT TO SAVE YOUR PART PROGRAM?
YES
 *PLEASE GIVE FILE NAME*
NEWPP

@FREE       PROJECT*NEWPP.
@ASG,CP                          PROJECT*NEWPP.,F2 .
@FREE       PROJECT*NEWPP.
@ASG,AX     PROJECT*NEWPP.
@USE      19,PROJECT*NEWPP.
@FREE           19
@FREE           21
@FREE           16
@FREE           23
@FREE           25

 **CONVERSATIONAL NELAPT NOW TERMINATED - CHEERIO**

@FREE       PROJECT*DIAGNOSTICS.
@FREE       PROJECT*MESSAGES.
```

Although a very usable tool as it stands, Conversational NELAPT is under active development on two fronts. The conversational aspect is being developed to make it even easier to use and less verbose. Selective plotting of the CLDATA is planned and should be a very useful feature, whilst interactive treatment of the cutter motion calculations is being considered. The other area of development from which Conversational NELAPT benefits are the new features incorporated in the batch version of NELAPT. These currently include facilities to increase the utilization of NELAPT at the design stage so that work can be handled from design through to manufacture. The development of file-handling capabilities for conversational use has been of benefit to the batch version of NELAPT in this instance.

Only the users can decide whether a conversational NC program is an attractive method of producing NC control tapes, but NELAPT even in its present form must be seriously considered by anyone who feels that they can justify the use of a computer in this field.

ACKNOWLEDGEMENT

This paper is published by permission of the Director, National Engineering Laboratory, Department of Trade and Industry. It is British Crown copyright.

REFERENCES

NEL. 2CL part-programming reference manual. First revision. NEL Report No 424. East Kilbride, Glasgow: National Engineering Laboratory, 1969.

McWATERS, J. F. and HENDERSON, W. T. K. The NEL 2C,L processor. In LESLIE, W. H. P. (Ed.). Numerical control programming languages, pp 153-176. Amsterdam: North Holland Pub. Co., 1970.

APPENDIX

FILE NAMING

When asked by Conversational NELAPT to provide a file name, the user should supply this name in accordance with Univac file-naming conventions.

The full name of a file is of the form PROJECT*NAME. or PROJECT*NAME PROJECT and NAME are identifiers consisting of letters and/or digits and have a maximum length of 12 characters each. The end of a file name is signified by a space or a period, any characters coming after a space or a period being ignored.

It is usually sufficient for the user to supply only the file name. That is, he need only specify NAME and Conversational NELAPT will append the project supplied on the RUN control statement used in logging on to the computer.

For example, if the project named on the user's RUN statement were PROJ1 and if the user gave the name of his file as FILE1, the conversational program would attempt to reference the file PROJ1*FILE1.

However, if the user wishes to name a file under a project other than that specified by the RUN statement he should supply the complete name, eg PROJ2*FILE1.

Where more than one programmer in one firm may be involved in a project it is useful for each to prefix his file name by a personal number allocated to ensure that he never creates the same file name as another member of the project. If, for example, he alone uses the number 7, then there will be no chance of anyone else creating a name like 7FILE1 or 7CAMPLATE3.

FILE DIAGNOSTICS

The file name supplied by the user is checked by Conversational NELAPT and if invalid an explanatory diagnostic is printed. The possible diagnostics are as follows:

1 FILE NAME INVALID - MUST NOT EXCEED 12 CHARACTERS

eg LARGEFILENAME
IMMENSEPROJECT*LARGEFILENAME
IMMENSEPROJECT*FILENAME
PROJECT*LARGEFILENAME.

2 FILE NAME INVALID - '.' OR '*' MUST NOT BE 1ST CHAR. OF NAME

eg *FILENAME
*.
.
.FILENAME
.PROJECT*FILENAME.

3 FILE NAME INVALID - INCOMPLETE NAME GIVEN

eg PROJECT*
no filename given
filename not terminated by space or period.

4 UNABLE TO ASSIGN FILE

PROJECT*NAME ERROR 400010000000

This diagnostic occurs when the user specifies as an existing file one which does not exist at that time.

5 UNABLE TO ASSIGN FILE

PROJECT*NAME ERROR 440000400000

This diagnostic occurs when the user specifies as a new file one which exists at that time.

In each of the above cases, the user is given the chance to try again and is asked once more to specify a filename. Three opportunities in all are given for the user to specify a valid filename. If his third attempt fails, Conversational NELAPT will terminate with the message CONVERSATIONAL NELAPT TERMINATED - SORRY.

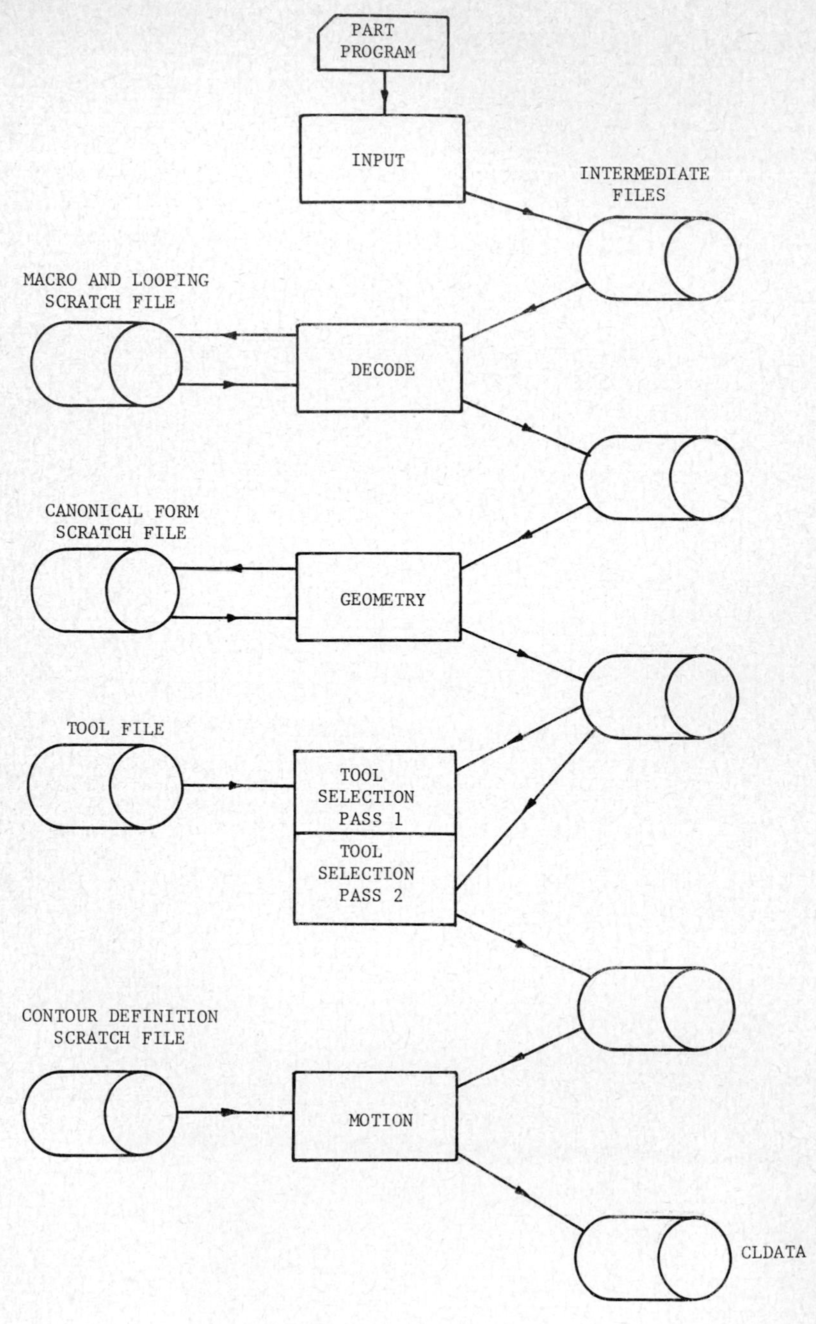

FIG. 1 NELAPT BATCH PROCESSOR

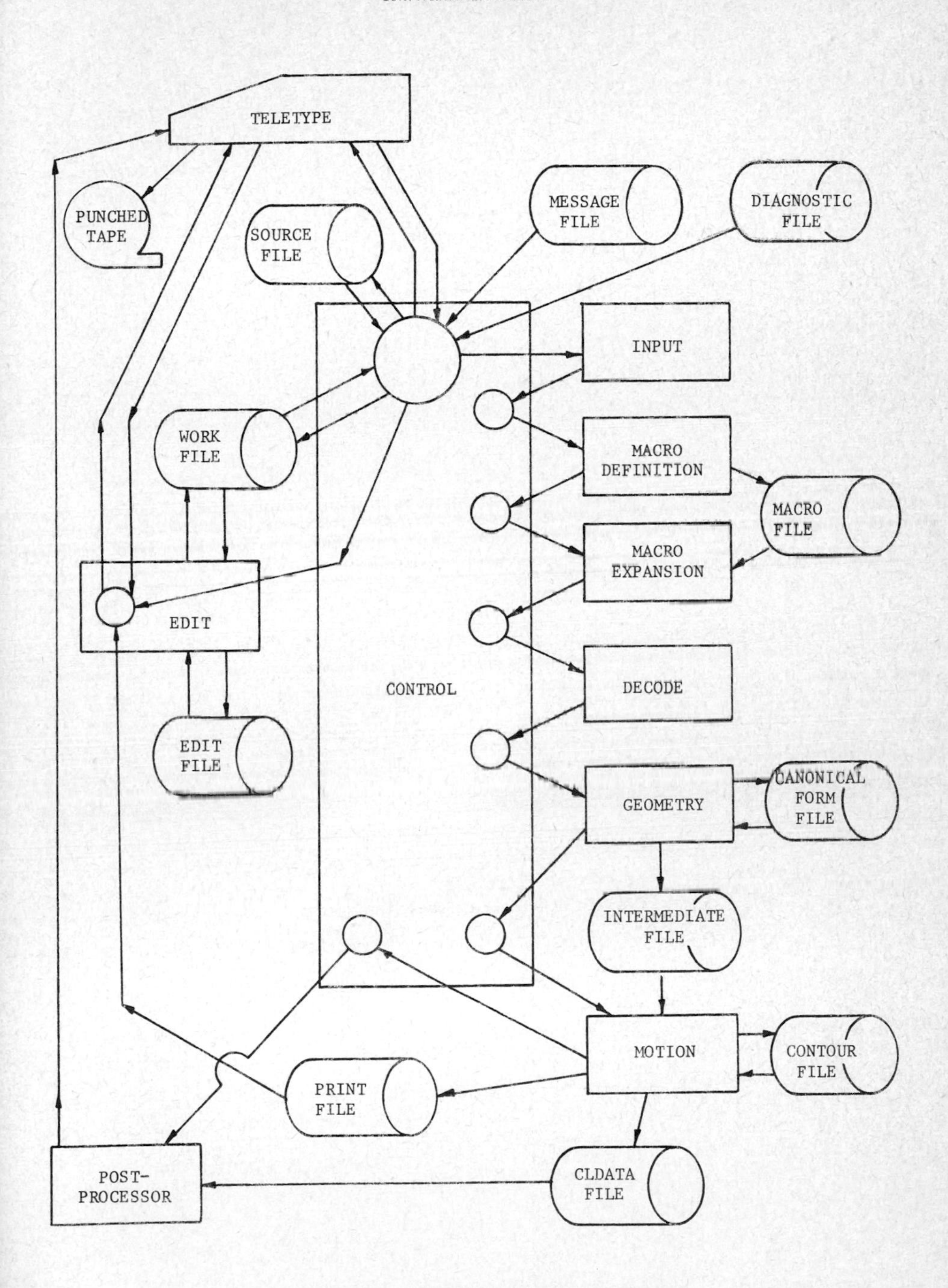

FIG. 2 CONVERSATIONAL NELAPT SYSTEM

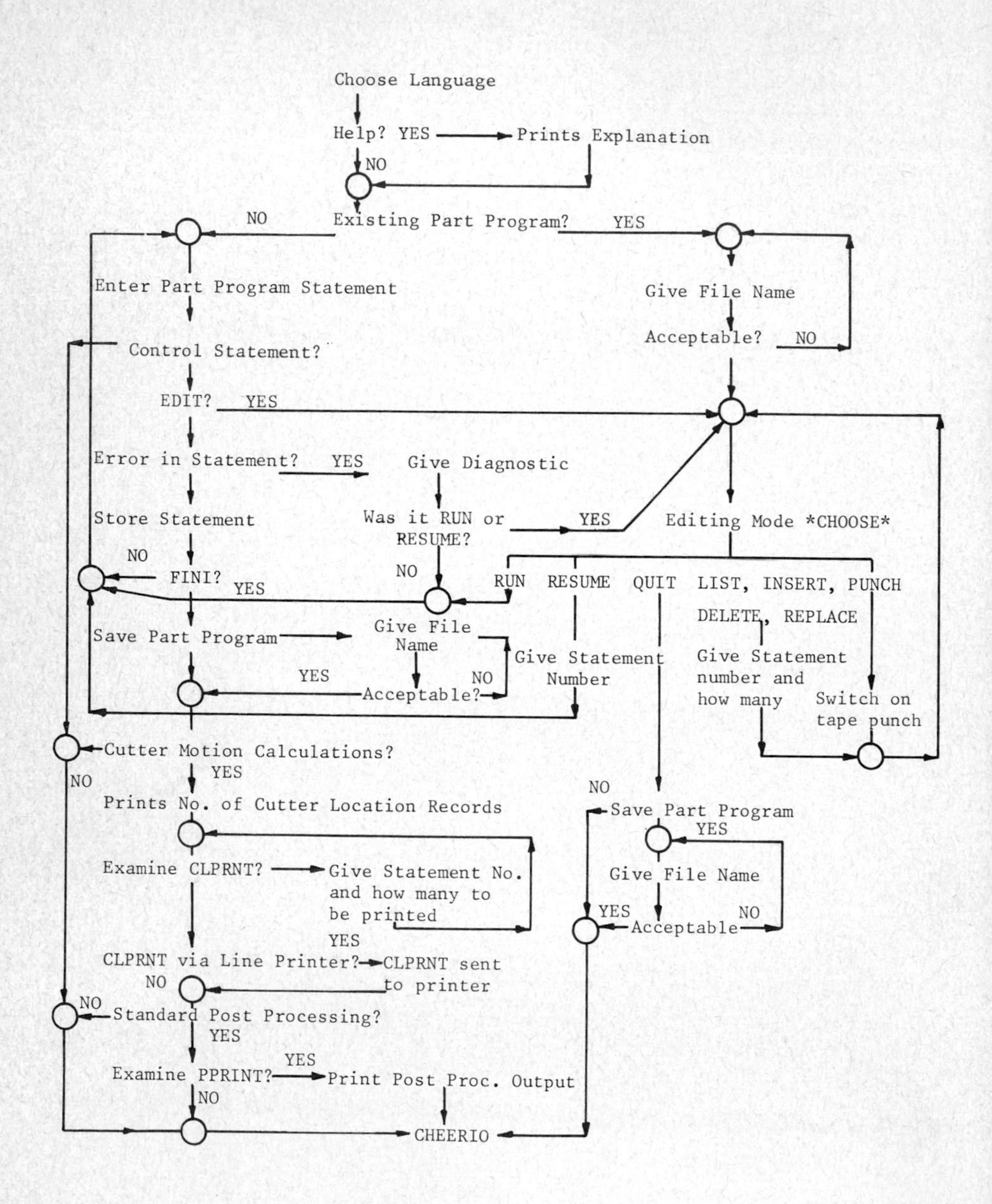

FIG. 3 USER'S FLOW CHART

GNC - A Graphical NC Processor

K. J. DAVIES

Computer-Aided Design Centre
Cambridge, England.

Summary

A description is given of the purpose, use and features of a graphical part-programming system for 2D continuous machining.

Major features of the system include the use of low cost computer graphics terminals at remote locations from a multi-access computer system and the individual treatment of the various stages of NC processing.

Introduction

In October 1971, a joint project was established to develop a prototype NC part programming system, utilising low cost computer graphics terminals connected to a multi-access computer system. The work has been centred at the Computer-Aided Design (CAD) Centre, Cambridge and the three industrial partners involved in the project, British Aircraft Corporation (Weybridge), Midcast Numerical Control (South Wales) Limited and Plessey Numerical Controls Limited have each made valuable contributions.

As with a number of NC graphics systems, the GNC system is only intended to deal with 2 and $2\frac{1}{2}$D continuously machined components, though some point to point facilities may be added later. Initial development has been based on the use of storage tube terminals such as that shown in Figure 1. Terminal costs are therefore well within the £5000 category and connection with the computer system is via the switched public telephone network.

FIGURE 1.

COMPUTER LANGUAGES FOR NUMERICAL CONTROL, *J. Hatvany, editor*
North-Holland Publishing Company - Amsterdam–London

The main objectives of this graphical NC system are:

i) To simplify part programming procedures.

ii) To give less opportunity for part programming errors.

iii) To make it easier and cheaper to correct errors.

iv) To drastically reduce lead time in producing control tapes.

The basic structure of this NC graphics system differs from others in that component geometry definition is divorced from cutter path definition. The 3 stages of NC processing are kept quite separate and are dealt with by at least 3 different programs called up by the user:

i) Bounded geometry definition.

ii) Machining sequence definition.

iii) Post-processing.

The successful operation of this system (or collection of computer programs) does rely on a number of general timesharing system features. These include the availability of a fairly sophisticated filing system and file editing program, and the ability to execute interactive or non-interactive programs from a remote terminal.

The definition of bounded geometry (program (i)) is carried out by a non-interactive program which describes 2D composite curves using all the usual points, lines and circle definitions and constructions. The processed geometry may be viewed graphically at the terminal and any shape errors noted. The user can then edit his geometry data file to correct illegal geometry definitions, syntax errors or make shape changes. The geometry program may then be quickly re-run and the cycle repeated till the user is satisfied. It is possible that several program re-runs can be carried out in one hour.

Only when the user is completely satisfied that the geometry is error free will he define any machining sequences. Program (ii) is an interactive program which accepts the geometric description of the curves output from the first program. This enables sequences of cutter movements to be specified in relation to this predefined geometry by way of a simple command language, and use of the graphical facilities of the terminal. The user need not refer directly to component co-ordinates or to geometry parts lesser than complete composite curves. All the supplementary information required on the final control tape can also be specified and is finally processed with cutter path information to produce an APT compatible CLFILE.

Post processing and tape production (program (iii)) is a routine processing task which can now be carried out using post-processors from the APT or 2CL systems. However, since the user has had the opportunity to check and correct geometry at an early stage, and has specified and animated the tool path graphically, few errors should now exist and a plot of the final output is hardly necessary.

If modifications are required at a later date to the geometry or machining sequences, then by retaining the simple intermediate data files (kept in readable character form) the programs (i) and (ii) may be re-entered readily and new control tapes produced.

In this way it is possible that error free production tapes can be produced in hours rather than days, and with the minimum of computer processing.

Bounded geometry definition program

The program used to define bounded geometry is an adaption of a program developed by BAC Weybridge. The bounded geometry is in the form of composite curves (known as K-curves) which are defined and stored as a series of straight line and circular arc spans. An extended version of the highly developed Profiledata geometry program is used to define basic points, lines and circles, which may then be used in K-curve definitions.

Geometry terms

The very terse Profiledata geometry symbols and qualifiers are listed below:

P	Point	S	Line
C	Circle	W	Symbolic distance
A	Anti-tangential	L	Left
R	Right	H	Horizontal
V	Vertical	D	Distance
T	Tangential	N	Near
F	Far	CC	Circle centre point
A	Angle	Q	Perpendicular
R	Repeat		

The usual geometry conventions concerning direction of infinite length lines and complete circles (the 'unbounded' geometry) apply. The terms left (L) and right (R) are used to distinguish between each side of a line (Figure 2). The terms

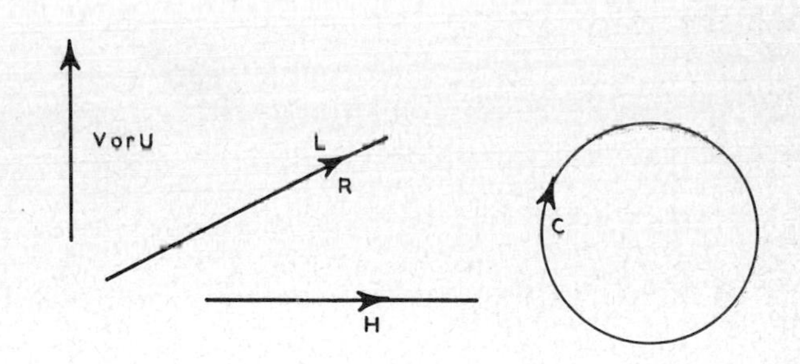

FIGURE 2.

near (N) and far (F) are used to distinguish points of intersection of a line and circle (when looking in the direction of the line) and also to distinguish between the two possible circles which may be defined as tangential to a line and passing through a point (Figure 3). A straight line and circle are tangential (T) if their directions are the same as the point of tangency and anti-tangential (A) if not (Figure 4).

This extended version of Profiledata geometry includes the facility for symbolic distances which may be used to replace any explicit value in a definition. The present version of the program allows 22 different ways of defining a point, 21 line definitions, 19 circle definitions and 14 distance definitions.

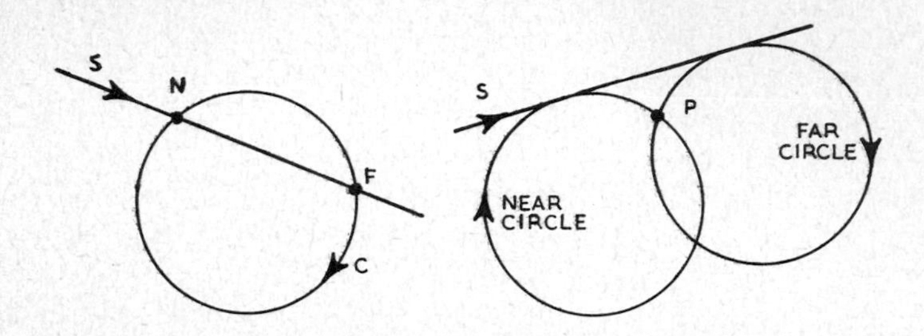

FIGURE 3.

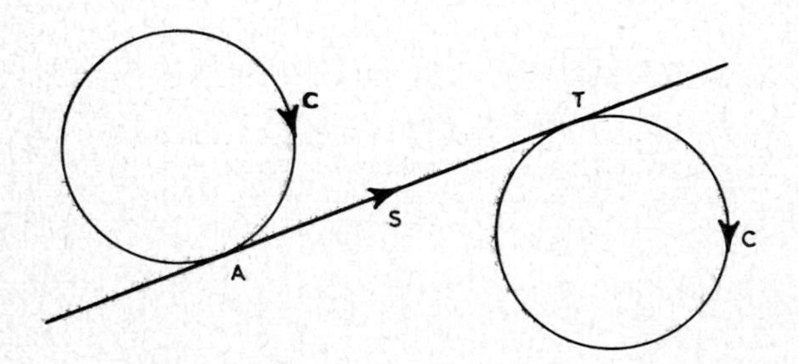

FIGURE 4.

K-curve terms

The K-curve terms are equally terse and as can be seen from the list of symbols below, several are common with geometry terms.

K	K-curve	P	Point
S	Line	C	Circle
A	Anti-tangential	T	Tangential
CP	Centre point	L	Left
R	Right	N	Near
F	Far	CF	Circle fillet
R	Modal fillet radius	EK	End of K-curve

Each K-curve is known by a symbolic name, consisting of letter K and an integer number. Each K-curve has a continuous direction implied by the order of definition. Points, lines and circles referred to within a K-curve definition must either have been previously defined, or a single depth of nesting of definition may be used. In this case the identifying number is replaced by the definition enclosed in brackets.

When defining a curve segment or span using a line or circle, the term A or T must be used to indicate which part of the line or circle is required (Figure 5), i.e. T when the K-curve direction corresponds with that of the line or circle and A for the opposing direction.

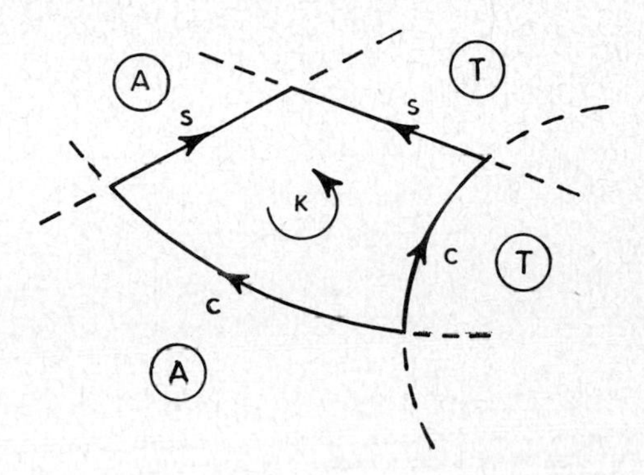

FIGURE 5.

Each K-curve definition must start and finish with a point (though the K-curve need not necessarily be a closed loop). Successive point definitions may also be used to indicate 'implicit' lines. A defined line may be terminated with a point or an intersection with another line or circle (Figure 6). As for geometry, intersections between lines and circles must be qualified by N or F terms.

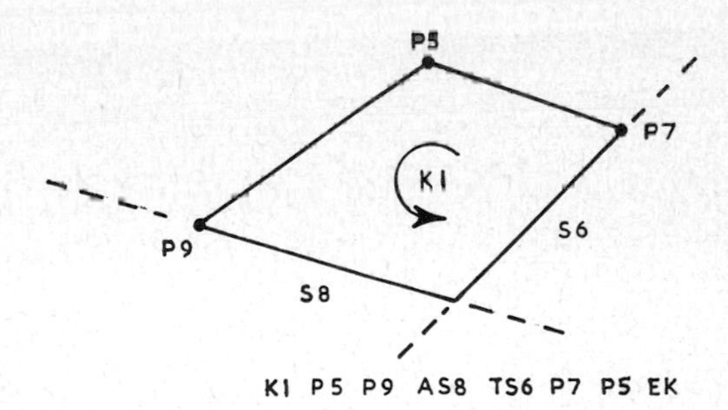

FIGURE 6.

Circular arc spans may be specified by quoting a start point, a centre and a finish point, or by using a defined circle and terminating at a point or intersection with a line or another circle. Circle to circle intersections must again be identified by L or R (Figure 7).

K-curve definitions allow circle fillets (CF) to be simply inserted between any combination of line and circle spans. When the same radius is to be used at several intersections, a modal value may be inserted using the code R at the start of the K-curve definition (Figure 8).

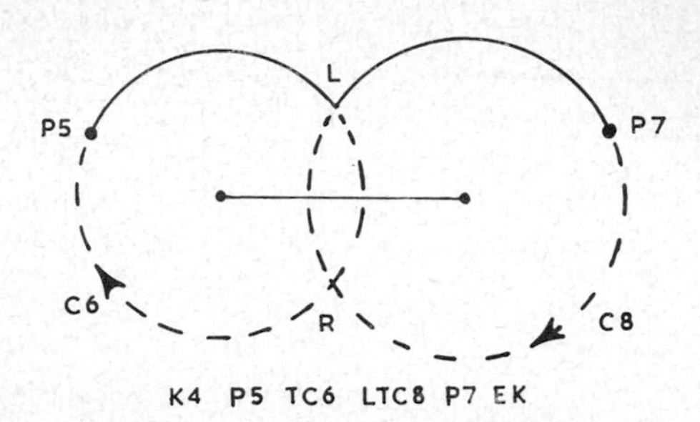

FIGURE 7.

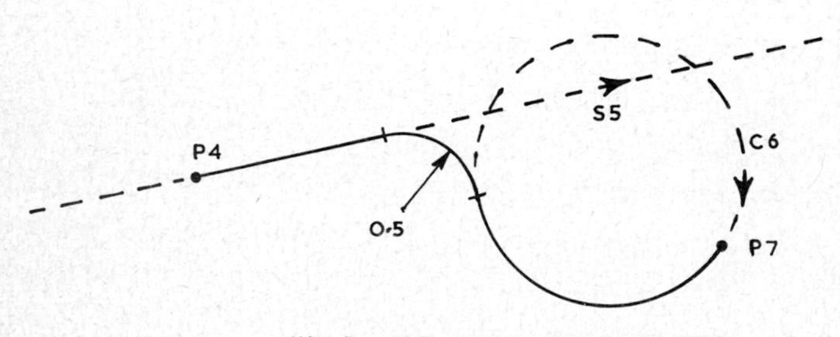

FIGURE 8.

Any modal value can be temporarily over-ridden by specifying a value after the CF term.

K-curve definitions may continue over several lines of data, and are terminated by the term EK. Figure 9 is an example of the use of K-curves to describe the bounded geometry of a pocket.

The output file of this program, which is passed on for further processing, is a character file consisting of three real numbers of each span of a K-curve. The three numbers represent the X and Y co-ordinates of an end point and a 'bulge' factor for the span (tan θ in Figure 10). This canonical form is used as a convenient basis for further processing.

Machining Sequence Definition Program

The facilities of the interactive machining sequence definition program may first be used to visually check the K-curves previously defined and then to carry out any re-orientation of K-curves to suit the positioning of the components in the co-ordinate axes of the machine tool.

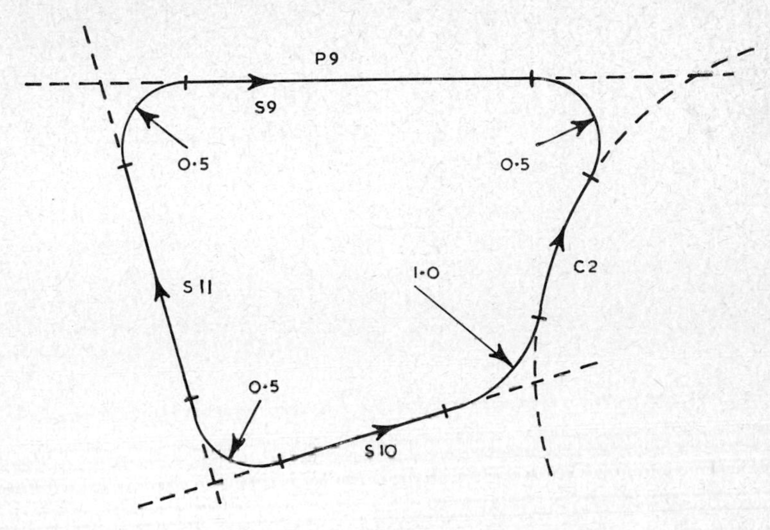

FIGURE 9.

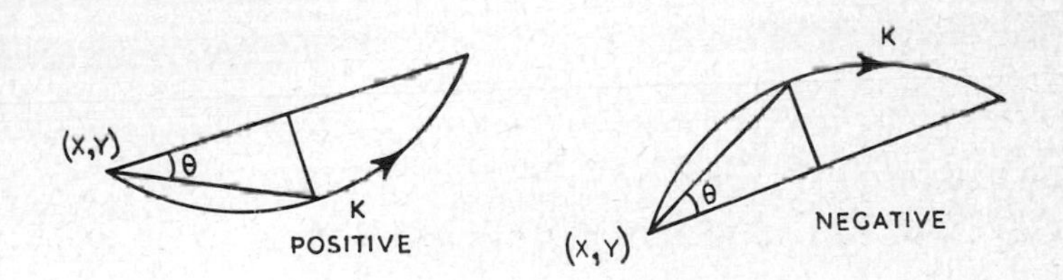

FIGURE 10.

The character file output from the first program is first read in. All subsequent program 'commands' and data are now input from the terminal, and the program responds to each command before the next is given. Visual checking of profiles can be carried out using the general picture transformation commands to alter picture scales and window position, together with an explicit DRAW command which causes all 'current' K-curves to be drawn on the screen (Figure 11). The span annotation is required by subsequent cutter movement commands. If shape errors are noted then an exit may be made from the program and the K-curve redefined with program (i).

Each K-curve has a transformation matrix associated with it, initially set to unity when the K-curve is read in. K-curves may then be re-orientated by using the commands TRANSLATE, ROTATE and MIRROR. K-curves may be repeated at different orientations and assigned new names (INSTANCE command) giving a limited macro facility to geometry definition.

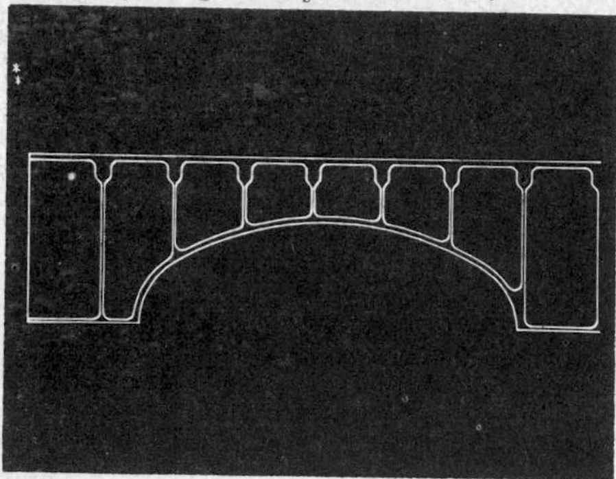

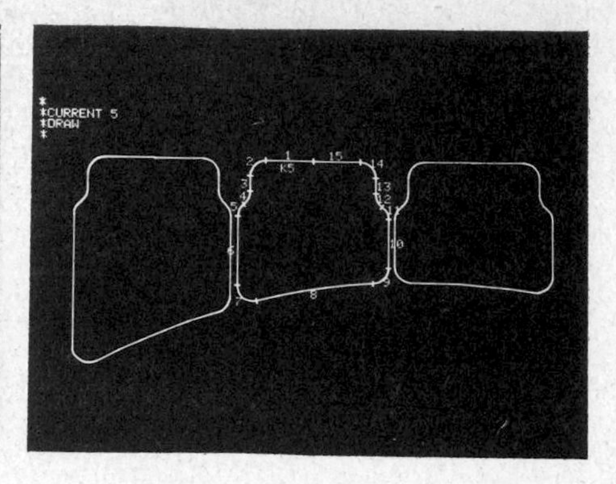

FIGURE 11.

Sequence definition

Having established the geometry of the profiles within the program, the user can commence the definition of a machining sequence. The command NEWSEQUENCE (only the first 3 letters of any command are significant) is the entry into the sequence definition mode. Subsequent machining sequence commands are checked, animated if necessary and then stored if successful. The CLFILE will later be generated from these. All other picture manipulation commands (DRAW, MAGNIFY, SHIFT, etc.) are still available but are not stored.

A small area at the top of the display screen is devoted to displaying feedrates, Z-plane heights and the cutter dimensions which will be assumed by the machining functions. K-curves have no particular Z-height associated with them and are always assumed to be in the XY plane. Movements of the cutter in Z are generally independent of XY movements (which may follow K-curves) except when using the RAMP facility. Usually the four feedrates (a rapid feedrate and cutting feedrate in each of the horizontal and vertical planes), cutter details (diameter and corner radius) and Z-heights of the clearance plane and working plane are defined at the beginning of the sequence . They may be altered individually at any time later in the sequence.

Cutter movements

The majority of cutter movements are defined by the commands LINEMILL (in association with OFFSET) and PROFILE. The initial cutter position is defined in machine tool co-ordinates by the MSP command (machine setting point). Later in the sequence the command HOME will cause the cutter to return to this point.

LINEMILL utilises the cursor facility of the terminal. On the storage tube terminal the cross hairs of the cursor may be moved about the screen by either a joystick device or a pair of thumb wheels. The command LINEMILL switches on the cursor, and the co-ordinates of the present cursor position are sent to the computer when a single character is depressed on the keyboard. Different characters are used to represent different functions with LINEMILL, for example, to indicate fresh air moves, roughing moves in pockets and zig-zag clearance moves. The cursor positions given represent the end points of straight lines moves, the centre of the cutter being indicated. The cutter path is indicated by a dashed line and is drawn before the cursor is re-enabled to give the next point.

The accuracy of the co-ordinates given by the cursor depends on the resolution of the device and the current scale of picture (the co-ordinates are transformed back with reference to the picture to give true machine tool co-ordinates) but the accuracy is usually sufficient for these types of moves.

When clearing pockets bounded by K-curves, the facility is given to compute and display composite curves which are offset from the original K-curves. The amount of offset is the cutter radius plus some specified value which is usually the allowance one would make for a finishing cut. Offset curves are not annotated and may have a smaller or greater number of spans than the original. The main use of the offset curve displayed in this manner is to indicate a boundary beyond which the cutter centre path indicated by the cursor should not stray if the final profile is not to be interfered with (Figure 12).

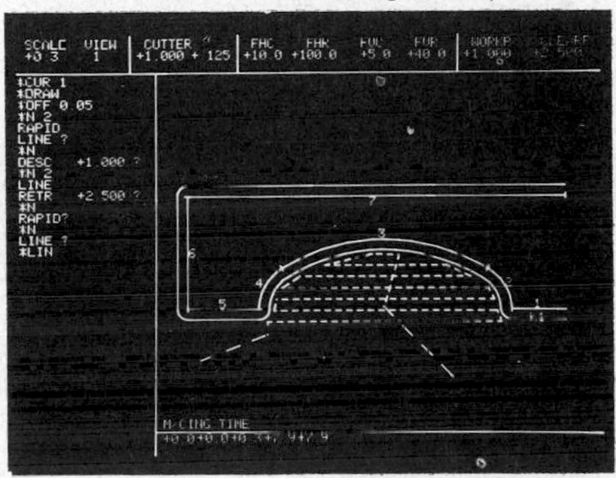

FIGURE 12.

Although indicating cutter movements with the cursor does give freedom to specify fresh air moves and area clearances quickly, unfettered by any rigorous algorithm, help can be given, if required, to ensure that any zig-zag movements cause correct overlap and are in fact parallel passes. The PROFILE command enables the user to drive a cutter accurately around a profile at the specified offset. At present the cutter can be driven from any span end point to any other span end point, and in either direction if the K-curve is a closed loop. Certain modification of the approach path to the first point on the K-curve are available such that the cutter makes a smooth entry onto the K-curve. Leaving the K-curve can also be treated in a similar manner. Immediately the command is processed, the cutter path is computed and displayed (Figure 13).

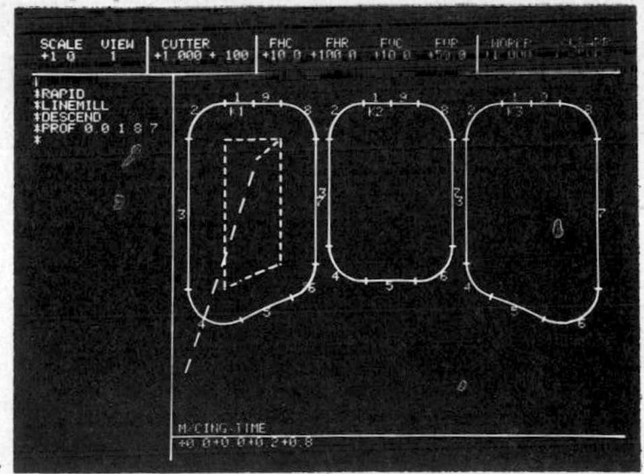

FIGURE 13.

The two qualifying commands RAMP and RAPID apply to the first cutter movement after the commands. RAMP will cause a gradual descent to the working plane height during the movement to new XY co-ordinates, whilst RAPID causes the appropriate rapid feedrate to be used. A range of supplementary commands is available to enable instructions for coolant, spindle speed selection and other auxiliary functions to be inserted in the appropriate place.

Editing sequences

A major feature of the program is that the sequences of cutter movements (a sequence being just a convenient self-contained string of cutter movements and associated data) may be edited by a very simple procedure.

The sequence to be edited must have been selected as 'current' by the use of the NEWSEQUENCE command as described above or by the SEQUENCE command (selects the existing sequence named). A new sequence can also be an edit of an existing sequence. Three edit commands are available:

i) N - Execute the next command(s) in the sequence (i.e. LINEMILL, PROFILE, etc.).
ii) D - Delete the next command(s).
iii) R - Restart the sequence.

New commands may be added to the sequence at the current position between any of these three commands - and since all intermediate commands must be executed when editing, then the screen animation will be up to date for any insertions. Sequences are finished by the commands FILE, ABANDON or DELETE which are self-explanatory.

The sequence may be checked in the XZ and YZ views. The command VIEW specifying the view required, automatically erases the screen and restarts the sequence. A complete sequence viewed in XY and XZ is shown in Figure 14 (below).

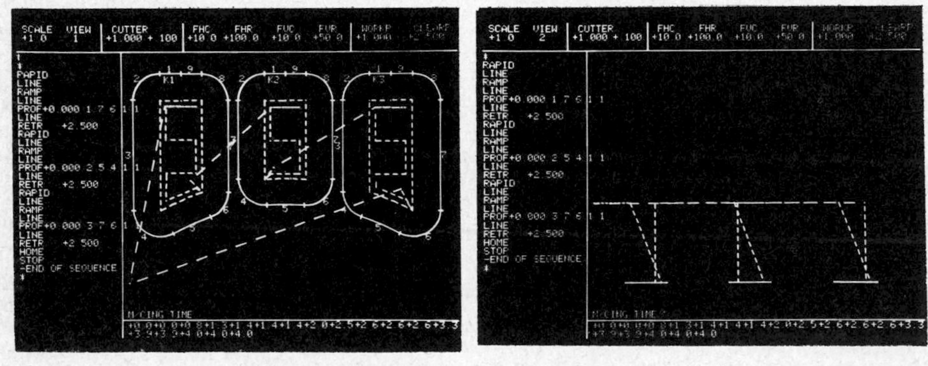

When an exit is made from the program the sequences, which are stored virtually as input, are re-processed in numerical order to produce an APT CLFILE.

Post-processing and tape production

From the outset of the project to establish this system the decision was made to avoid duplication of effort and utilize existing software - where suitable and where available. This was especially so in the area of post-processing. The first post-processor implemented at the CAD Centre was an APT post-processor for the Plessey control system of the machine which Midcast intended to use to carry out trials of the graphical NC system. This experience ensured that future use of existing post-processors can be organised in the most efficient manner.

The final control tape can be produced either at the Computer Centre or at the remote terminal. Without tape punching facilities graphics terminal costs can be reduced to approximately £2000, to operate reasonably at 300 baud over the switched telephone network. However, for minimum cost there will be a delay of 24 hours before tapes produced at the Computer Centre arrive by post.

At £5000 a storage tube terminal may be supplemented by a printer and tape punch - all operating at 300 baud on one line. There is an additional advantage to this terminal arrangement besides local tape production. The interactive machining sequence program is easier to operate when the conversation with the program is recorded on the printer, leaving the storage tube completely free for the graphics. Against this advantage one has to weight any difficulties with transmission errors and different tape codes required.

Conclusions

With a very limited amount of additional software effort, a prototype system has been developed which satisfies the four main objectives quoted in the Introduction. User costs are easily identified, as the terminal investment is limited and multi-access computing resources are utilised on a 'pay for what you use' basis. This should suit well the smaller NC machining specialists and jobbing shops. Ease of programming may well reduce the size of economical batches for NC machining.

The multi-access system required to support the interactive machining sequence program must have a reasonable response. No matter how this is provided, the potential users should be prepared to pay more for such a service than normal remote timesharing charges. Substantial gains are, however, still to be made when total costs are compared with conventional NC processing.

The approach of using separate programs for the various stages of processing, and relying on the timesharing system features to connect the simple interfaces, means that extending the capabilities of the 'system' should not be difficult. For example, opportunities may arise to develop an interactive bounded geometry program or to extend the geometry definition capability to the fitting of a sequence of circular arcs through two dimensional empirical data points. Both may sensibly be accomplished by quite separate programs, maintaining the simple interfaces.

The modular programming approach has also been followed within the interactive machining sequence program. Use is made of a well tried graphics system (GINO) and a small general purpose language processor - both of which have been used in many other applications. Even the datastructure and subroutine construction employed (the program is FORTRAN based) may well encourage use in other more specific CAD applications such as press tool design and shape layout problems.

Acknowledgement

My thanks to the individuals in the companies involved in the joint project for their contributions, and also to those at the CAD Centre whose ideas and efforts have helped this project.

A SMALL GRAPHICAL SYSTEM FOR PROGRAMMING NC MACHINES

J.P. CRESTIN and J.P. PAILLARD +
Ecole Nationale Supérieure de Techniques Avancées, Paris

Abstract : The graphical system presented here uses NC programming language Mini-ifapt. It permits interactive text editing of the source language, and makes it possible to display on a cathode ray tube a previously defined part with superimposed on the view the path of the cutter corresponding to some given NC program, to get by use of light-pen or keyboard some information concerning technological or geometrical data and to modify the image of the part as a consequence of the drilling operations.

1. INTRODUCTION

The great importance of using graphics to develop and debug numerical command (NC) programs is now beyond any doubt (see [1] for instance). To this effect, two types of systems are presently used : on the one hand, integrated design systems, such as NCG ([2]) or APT IGS, where the programmer defines directly from an interactive graphic terminal his part and the machining cycles ; on the other hand, post processors for plotters which display the path of the cutter. Former systems are very powerful but demand huge computer means ([3]) and only large companies can afford them ; latter systems are now frequently used but they only permit imperfect proofs of programs.

Recently, the cost of interactive graphic terminals has greatly decreased and the use of small computers for NC has been growing. These reasons made us develop a NC graphical system based on these means and which would take place between the two previously defined types.

Its main caracteristics are presently :

- the use of a simple NC programming language ($2\frac{1}{2}$ axis) and of its compiler without any change (this permits an easy implementation of other similar languages).

- the ability to display the path of the cutter, its envelope in case of drilling operations, and technological data of some interest to the programmer.

- the superposition on it of the image of a previously defined and machined part

- commands enabling the programmer to display any view of the part with or without hidden lines and the coordinates of any point of this view, and to interrupt the interpertation of the CLFILE at any moment to make changes in the source program.

Without going into the details of the flowchart, in the second part we give the basic concepts of the system with the methods we use ; in the third part, we describe how a programmer can use it and give its present performances ; the last part is devoted to possible future extensions of this system which must be considered, in its present state, as experimental.

* Present affiliation : O.B.M. PARIS

2. BASIC PRINCIPLES AND SYSTEM DESCRIPTION

The system was programmed at the Ecole Nationale Supérieure de Techniques Avancées on an IBM 1130 computer, connected with magnetic discs, a Benson plotter and a graphic display IBM 2250 Model 4 (i.e. without its own storage) with a light pen and function and alphanumeric keyboards.

It is a compartively old and expensive configuration, but is has a full graphic software available and is a good representative of the present possibilities of a small computer connected with an interactive graphic terminal.

All programs are written in Fortran IV and GSP (Graphic Subroutine Package) ([4]) for the graphic part, so that they will be easily implemented on other computers. This choice together with the fact that IBM 1130 has no hardware floating point arithmetic available explains the slowness of the execution of definite commands or calculations in the present version.

The NC programming language used is Minifapt ([5]). It is a subset of APT allowing point to point machining as well as plane and circular milling.

2.1. Composition of the system

The system includes

- the minifapt compiler

- a text-editor permitting correction of source programs

- a program for creation of three-dimensional geometries

- a program for interpretation of CLFILEs

- a display program

- a supervisory system enabling the programmer to activate the previous modules and containing a system for filing parts and Minifapt programs records.

Except the last one, all these modules stay usually on disc core and call upon frequent overlays during execution. Most of them being very common, we shall just insist upon the definition of parts and the display program.

2.2. Geometrical definitions

The core images of parts are three-dimensional. This permits easy projections along any coordinate axis and by means of a small extension should enable us to get skew projections and to apply geometrical transformations on parts.

For the sake of simplicity, the only geometrical elements we have considered are points, line segments, plane polygonal faces, and cylinders. These elements are sufficient to represent, at least approximately (in case of embossings for instance), most parts that are machined with Minifapt.

A point is defined by its coordinates a line segment by its end points, a face by its size and a cylinder by its axis (a line segment), its radius and the faces which intersect it. The bottom of a blind hole is implicitely defined by the corresponding end point of its axis. These data can be entered dynamically from the terminal. Presently, no checking is made at their entry, checking the planarity of a face for instance. If an error occurs here, then the results of following steps (determination of hidden lines for instances) is unpredictable.

After the interpretation of a point to point CLFILE, the programmer can request a modification of the stored image of the part to take into account the result of the machining. In this case, he must indicate the characteristics of the cutters (diameter ...) and the program determines, for each cylinder corresponding to the machining, its intersections with the faces and the cylinders of the part and modifies consequently its geometric definition. In fact, intersections between two cylinders are presently considered only if they are parallel, and intersections between a face and a cylinder only if they are parallel or if the face intersects the axis of the cylinder. In most cases, this simplifying hypothesis does not seem to be to disturbing here, as shown on figure 1.

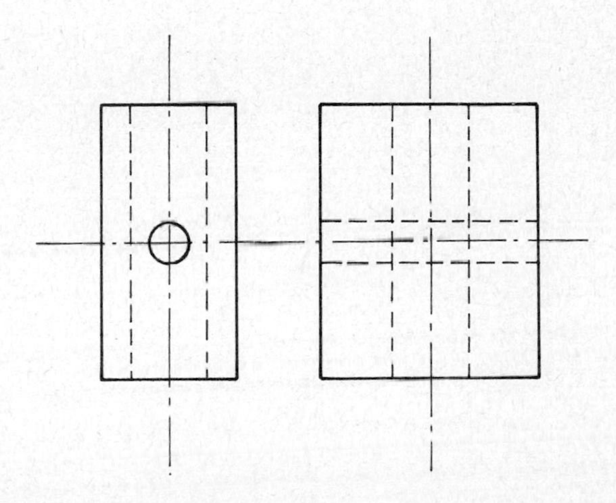

FIGURE 1

Front and top views of a cube with two holes intersecting at 90°.

2.3. Displaying views of a part

After any number of machining cycles, a projection along one of the three coordinate axis of a part can be displayed. Stock removal is taken into account only in the case of point to point machining (see § 2.2. above).

When the programmer does not request determination of hidden lines, in order for instance to shorten the calculation time, the displaying algorithm is quite simple. The display program creates a display list, using the core image of the part and omitting the coordinate corresponding to the projection axis. The scale and the center of the view are given by the programmer from the terminal, so that he can view enlargings of the part, which is easily done in GSP.

Upon request, it is used an algorithm for the <u>determination of hidden lines.</u> We give here its principle in the case of lines hidden by faces. The respective positions of each edge IJ and each face F are examined, except when one or the other is parallel to the projection axis. The intersection M of the plane of F with the line segment IJ and the intersections X_1, X_2, ... X_n of the projection i j of IJ with the projections of the sides of F are calculated (Figure 2).

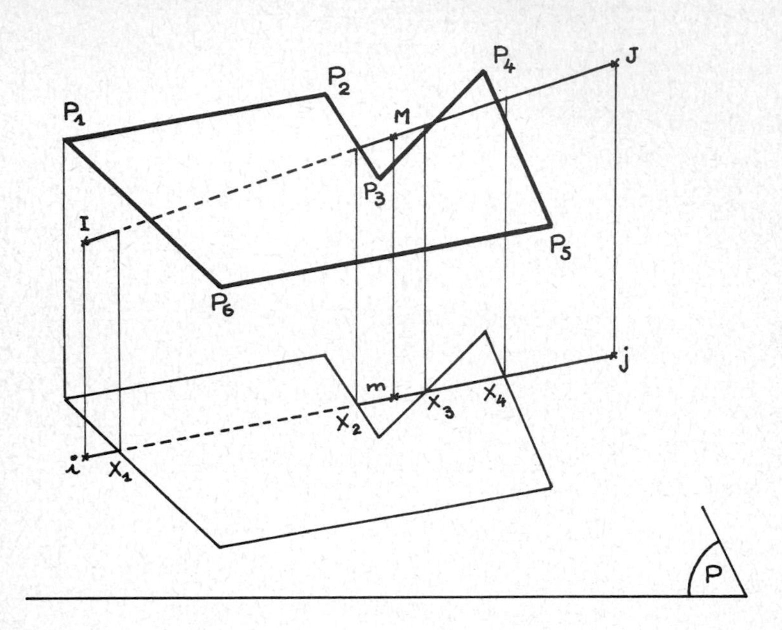

FIGURE 2

Then by comparison of the third coordinate of I, J and M, it is determined which side of M is below F. This side is hidden. Suppose now for instance that M lies between I and J and that m, X_p, X_{p-1}, .. i lie in consecutive order on the hidden part of the projection of IJ. The line segments (X_p, X_{p-1}), (X_{p-2}, X_{p-3}), ... (X_{p-2k}, X_{p-2k-1}) are hidden since m is always visible through face F.

Joining together all these elementary hidden line segments, we get the hidden lines of the view.

To determine the order of intersection points, we use the parametric representation of coordinates in the projection plane :

$$x = x_I + \lambda\ (x_J - x_I)$$

$$y = y_I + \lambda\ (y_J - y_I)$$

where (x_I, y_I) and (x_J, y_J) are the coordinates of I and J. The sequence of points on the line i j corresponds to the numerical sequence of their parameters. Points with values of λ lying outside [0,1] lie outside i j.

For cylinders, the generatrices are always considered hidden, and the determination of line segments visible through a hole is done only when the axis of the hole is parallel to the projection axis.

Even simplified like this, the algorithm is rather long, so that it is only executed upon request of the programmer.

3. USING THE SYSTEM

Upon request, a menu is displayed at the terminal and enables us to choose the next task to be performed. We exhibit here a typical sequence of tasks needed to develop a NC program.

First, the programmer may define the part to be machined and display a view of it with dotted hidden lines (figure 3).

FIGURE 3

Front view of a part with hidden lines

Then, using the text-editor, he can enter the Minifapt programm corresponding to the intended machining (figure 4).

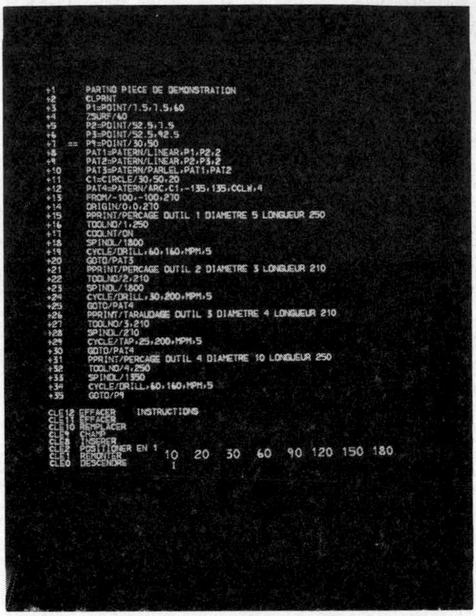

```
+1       PARTNO PIECE DE DEMONSTRATION
+2       CLPRNT
+3       P1=POINT/7.5,7.5,60
+4       ZSURF/60
+5       P2=POINT/52.5,7.5
+6       P3=POINT/52.5,42.5
+7   ==  P9=POINT/30,50
+8       PAT1=PATERN/LINEAR,P1,P2,2
+9       PAT2=PATERN/LINEAR,P2,P3,2
+10      PAT3=PATERN/PARLEL,PAT1,PAT2
+11      C1=CIRCLE/30,50,20
+12      PAT4=PATERN/ARC,C1,-135,135,CCLW,4
+13      FROM/-100,-100,270
+14      ORIGIN/0,0,270
+15      PPRINT/PERCAGE OUTIL 1 DIAMETRE 5 LONGUEUR 250
+16      TOOLNO/1,250
+17      COOLNT/ON
+18      SPINDL/1800
+19      CYCLE/DRILL,60,160,MPM,5
+20      GOTO/PAT3
+21      PPRINT/PERCAGE OUTIL 2 DIAMETRE 3 LONGUEUR 210
+22      TOOLNO/2,210
+23      SPINDL/1800
+24      CYCLE/DRILL,30,200,MPM,5
+25      GOTO/PAT4
+26      PPRINT/TARAUDAGE OUTIL 3 DIAMETRE 4 LONGUEUR 210
+27      TOOLNO/3,210
+28      SPINDL/210
+29      CYCLE/TAP,25,200,MPM,5
+30      GOTO/PAT4
+31      PPRINT/PERCAGE OUTIL 4 DIAMETRE 10 LONGUEUR 250
+32      TOOLNO/4,250
+33      SPINDL/1350
+34      CYCLE/DRILL,60,160,MPM,5
+35      GOTO/P9

CLE12 EFFACER    INSTRUCTIONS
CLE13 EFFACER
CLE10 REMPLACER
CLE9  CHAMP
CLE8  INSERER
CLE2  POSITIONER EN 1
CLE1  REMONTER          10  20  30  60  90 120 150 180
CLE0  DESCENDRE          I
```

FIGURE 4

Then he has the compiling of this program done and displays the previous image superimposed with the path of the cutter. The machining cycles are successively displayed with the following data : present cycle (drilling, milling, tapping ...), number of the tool, feed-speed, spindle-speed, auxiliary functions (sprinkling, ...) (figure 5).

Pointing at some point with light-pen, the programmer obtains its coordinates (Figure 2). After any machining cycle, he can stop the interpretation of the CLFILE and proceed with any other task (for instance text-editing to modify the source program).

Performances

The present system may be considered as rather slow, for the reasons explained in § 2.1.

For instance, creating from cards the core image of the part of figure 3 takes 70s, and displaying for the first time the view of figure 5 with calculation of hidden lines takes 1 mn.

In fact, as frames stay on the screen during the calculations, the processing times do not seem very long for the programmer. On the other hand the computer time is not very expensive for such a small system (50 $ per hour), and ought

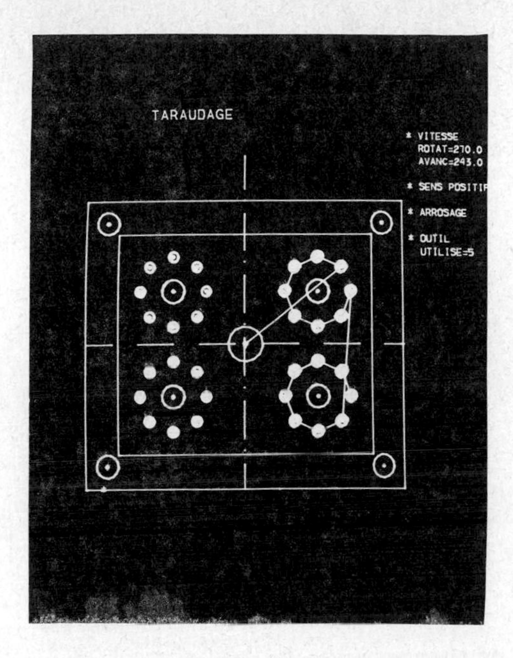

FIGURE 5

Top view of the part of figure 3, without hidden-lines during the interpretation of the CLFILE

to be still less expensive for more recent minicomputers, so that the time gained to develop and debug a program justifies largely the computer time spent.

Needed Storage

The programs use a common section of 8 K. The size of the temporary programs goes up to 4 K (Mini-ifapt compiler or program for the calculation of hidden lines) The permanent programs, comprising the graphical operating system of the computer occupy 17 K.

4. PROSPECTS

The present system has been designed to allow easy implementation on some other minicomputer connected to new graphic terminals, and using other NC programming languages of the same type. To gain computer time, we can consider to program in machine language a large part of the system.

Many improvements can be made : viewing skew projections, using better graphical symbolisms , getting a processing as elaborate for milling as now for drilling, though the latter point should demand much more sophisticated programs.

To conclude, the main purpose of the system presented here has been to investigate the capabilities for NC programming of small computers connected to graphic displays. This system is still under development Outside such a universal system as described here, we think that the principles and the methods we have used can be successfully applied to specialised NC programming systems which should include computer aided design programs and have so a still better efficiency.

REFERENCES

[1] D. PRINCE : "Interactive Graphics for Computer-aided design", Addison-Wesley, Reading 1971

[2] Numerical Control Graphics, IBM, Program Library 360 D.23.4.002

[3] J.J. LAVICK : "Computer-aided design at Mc Donnell DOUGLAS", in Advance Computer Graphics, ed. by R.D. PARSLOW and R.E. GREEN, Plenum Press, London 1971.

[4] Graphic Subroutine Package for basic Fortran IV, IBM Systems Reference Library 1130-25-C27-6934-1

[5] Miniifapt , Notice de Présentation, ADEPA

GRAPHIC COMMUNICATION SYSTEM FOR DESIGNING NC MACHINE TOOL CONTROL PROGRAMS

V. G. ZAITSEV
Institute of Automation, Academy of Sciences
of the Ukrainian SSR, Kiev, USSR

The modern automatic programming methods provide for a time-consuming process of manually preparing the primary program with subsequent introduction of changes in this program when debugging it on an electronic computer. The programmers i.e. program designers, as a rule, are at the same time experienced technologists. Therefore, the problems of extensively raising their efficiency are of great importance.

Let us designate the duration of the general cycle of preparing the program by $T_{ц}$, then

$$T_{ц} = T_{п} + T_{о} \qquad /I/$$

where $T_{п}$ = programming time,
$T_{о}$ = program debugging time.

One of the most effective methods of sharply reducing the duration $T_{ц}$ is the utilization of a direct dialogue with a cathode-ray tube display unit.

In doing this there is a practical possibility of organizing the dialogue in a number of ways which conventionally can be combined into two large groups.

The methods based on the existing automatic programming systems, which make it possible to reduce the time $T_{ц}$ at the expense of the reduction of the debugging time $T_{о}$, should be referred to the first group.

The methods of the second group imply the reduction of time $T_{ц}$ both by reducing the time $T_{о}$ and at the expense of the reduction of $T_{п}$ by utilizing fundamentally new programming methods.

COMPUTER LANGUAGES FOR NUMERICAL CONTROL, *J. Hatvany, editor*
North-Holland Publishing Company - Amsterdam–London

Let us examine several variants of organizing the process of program design applicable to each of the two mentioned groups.

The simplest method of the first group, which can be referred to the packet type of data processing, consists in the following: The work of the programmer is organized at a special console connected with a computer on which the program automation system has been realized. The console comprises: alphabet-numeric and graphic displays and the data typeout input device. During operation the programmer supplies the computer with the aid of a typewriter with the text of previously prepared programs using the accepted automatic programming language, which is simultaneously displayed on the alpha-numeric display screen. As the text is accumulated and the screen information capacity is exhausted, the operator clears the screen, retaining the information which was previously supplied to the memory, as if replacing the earlier written page by a new one. After completing the procedure of describing and supplying the data page-by-page into the computer, the latter supplies the operator with the list of pages on which errors were made with the aid of the alpha-numeric display screen correspondingly on each page, in addition to the text of the present program, the errors are corrected.

After making the corrections the program is again supplied to the computer, in case the errors are absent, a trajectory of the cutting tool progress is shown on the graphic display. As a result of the check-up of the geometrical information a necessity may arise to correct again the text of the program. A new trajectory will be obtained etc.

A certain modification of the above-described method consists in the fact that the geometry and the program itself can be partially checked, for example by describing each separate element of the tool trajectory.

To the second group we can refer similar methods[1] when the operator draws the programmed part on the screen of the graphic CRT display,giving the required dimensions and chooses the type of the machine tool and the machining technology out of those found in the "Catalogs" of the automatic programming system.

In the last case the operator should follow a definite working procedure the correct performance of which is monitored by the system reporting in due time of the errors that were made. Simultaneously with the drawing of the programmed part on the CRT the computer calculates the program and after the programming of the profile the programmer can record the subprogram on a punched tape, used later on to control the machine tool

The problem of organising the process of designing the control programs under conditions of a dialogue with the computer implies the solution of two comparatively isolated sets of questions: directly linked with the hardware and software of the dialogue and the solution of the main programming automation problems /in a number of cases the adaptation of the existing programming automation system to the new operating conditions/.

The combination of the hardware and the software which provide for the solution of the dialogue problem shall be referred to below as the graphic communication system.

A graphic communication system, intended to solve the above-mentioned problems has been developed by the Kiev Institute of Automation [2]. The operation of the system was displayed at the "Machine Tools - 72" Exhibition in Moscow in April 1972. It is organized as a multipanel system which makes it possible to design simultaneously up four programs under the time-sharing mode. The system is operated by a special electronic control computer УМІ-НХ which is linked with the central electronic computer "Minsk-32". The central computer solves the basic problem of the program design, the control computer provides the dialogue conditions, supply and display of the graphic data on the display screens. The programmer console comprises alpha-numeric and graphic displays, input-output device for typewriter /ПМ/ and light pen. The system is provided with a centralized memory /ОЗУ/ in which all the information of the displays for every programmer console is stored. From the point of view of the structural organization the system is characterized by the following features:

1. The data is supplied to the control computer from the typewriter or computer under the symbol-by-symbol mode.
2. The control computer has the simplest interruption system and

provides for exchange of information the length of which is equal to the machine word. The exchange of the phrases, consisting of several words, is accomplished in accordance with a sub-program.

3. The data from the light pen makes up a phrase of a variable length.
4. The exchange between the control computer and the memory "O3Y" is accomplished by means of phrases.
5. The representation on the display screens is coded by a combination of graphic instructions. Each of these instructions is realized in the form of a phrase consisting of a fixed number of words. The length of the phrase is equal to the word length of the memory "O3Y".
6. In order to maintain an unflickering representation the graphic instructions are delivered cyclically to the input of each display. They are delivered in the asynchronous mode at the presence of interrogation. In doing so one word of the memory "O3Y" is delivered to the display input at every cycle.
7. The graphic instructions, which are to be realized on the given display, are disposed in the "O3Y" in a dense array having some starting address α_H and final address α_K. Every new design instruction is added at the end of the array.
8. The realization of each graphic instruction into a sequence of the CRT electronic beam deflection signals is accomplished by the hardware inside every display.

In a general case the graphic representation in the system has the form of an orientated graph [3] G(V), where V = set of peaks. The topology of G(V) is described by list m. In addition each rib a, b where $a \in V$ and $b \in V$ and also peak Vi is given a corresponding value p(a,b) and p(Vi). By values p(a,b) and p(Vi) are understood random records which characterize in a certain manner the ribs or the peaks of the graph correspondingly.

A concept of unit H is introduced, by which a graph having a final number of peaks is understood, this graph can be a sub-graph of a larger graph i.e. $H \subset G$. The concept of the unit is quite useful in such a case when it is necessary to describe graphs with a larger number of peaks, containing a certain number of similar sub-graphs. Here it is possible to reduce the volume of the information necessary to set the graph. Unit Hi can likewise be accompanied by a certain text record p(Hi). Values p(a,b), p(Vi) and p(Hi) form a certain set p which in combination with list m characterizes the representation.

The presentation of graphic representation and symbolic information on the CRT screens is connected with the necessity to set a law for shifting the electronic beam on the screen surface. This law can be of some degree of complexity and can directly depend on the representation peculiarities. Proceeding from the specific character of the program control problems, the following laws of the beam motion have been determined:

a/ Plotting a point /set of connected points/;
b/ Section of straight or broken line;
c/ Circle;
d/ Arc of circle.

Plotting each of the named simplest representation elements in the system is linked with the concept of the graphic instruction which was mentioned above. Instructions for plotting points, straight lines, circles and circular arcs are correspondingly distinguished. The list of instructions is supplemented by the text instruction.

In order to conditionally represent the units a number of graphic instructions are combined into sub-programs called the graphic subprograms.

Every graphic instruction is subdivided into two parts; the instruction proper and the data array. The instruction proper consists of the instruction code /degree of brightness, whether the image blinks or not/, starting coordinates for plotting X and Y. The instructions for plotting the points, straight lines and text are realized as group instructions. The instruction proper has the following structure shape:

$$K \; ; \; \Pi \; ; \; X \; ; \; Y \; ; \; N \qquad /2/$$

where K = instruction code,
Π = brightness,
X, Y = starting coordinates,
N = number of separate elements.

The data array for every instruction consists of additional information, necessary to plot the representation. For example, in the case of the instruction for plotting a broken line, it is the coordinates of all the bending points, set in a direct sequence,for the point plotting instruction - coordinates increment codes which set the next points, for the text instructions - sequence of codes of the displayed symbols etc.

The representation set by the combination of the graphic instructions is directly displayed on CRT screens and is observed by the programmer. Information structures m and p are not manifested on the screen and serve to describe the intercommunication between separate representation elements which are set by the programmer.

The general data structure is organized in the system in such a manner that every graph rib has a corresponding graphic instruction for plotting a point or a straight line, to the peak - points or circles, to the set elements of p - text instruction. For example, we can say that peak Vi of graph G(V) is graphically represented by graphic instruction KΠi, certain address αi etc.

Therefore, the general data structure is represented on the one side by a combination of graphic instructions and on the other hand by a generalized list m^* consisting of list m, lists m' and m" unambigously connecting the graphic instruction addresses with the peaks and ribs of the graph and the elements of set p respectively.

The general data structure of the system is given in Fig . I.

Fig. 2. is shown as an example.

In this drawing two points TKI and TK2 are interconnected with the aid of an arc of radios RI = 50. As mentioned earlier TK1 and TK2 represent the peaks of the graph, $p(V_1)$ corresponds to notation "TK1", $p(V_2)$- "TK2" peaks V_1 and V_2 are conventionally preset by the circle instructions, rib V_1, V_2 is described by the arc instruction, value $p(V_1, V_2)$ is represented by the text "RI = 50".

It is not difficult to see that when organizing the dialogue according to the first set of methods it is sufficient for the programmer to use only the level of the graphic instructions and the information set by m^* can be omitted. Consequently, we can speak of say two levels of the dialogue implying under the dialog, according to the first set of methods, the lower level dialogue, and according to the second set of methods, the upper level dialogue.

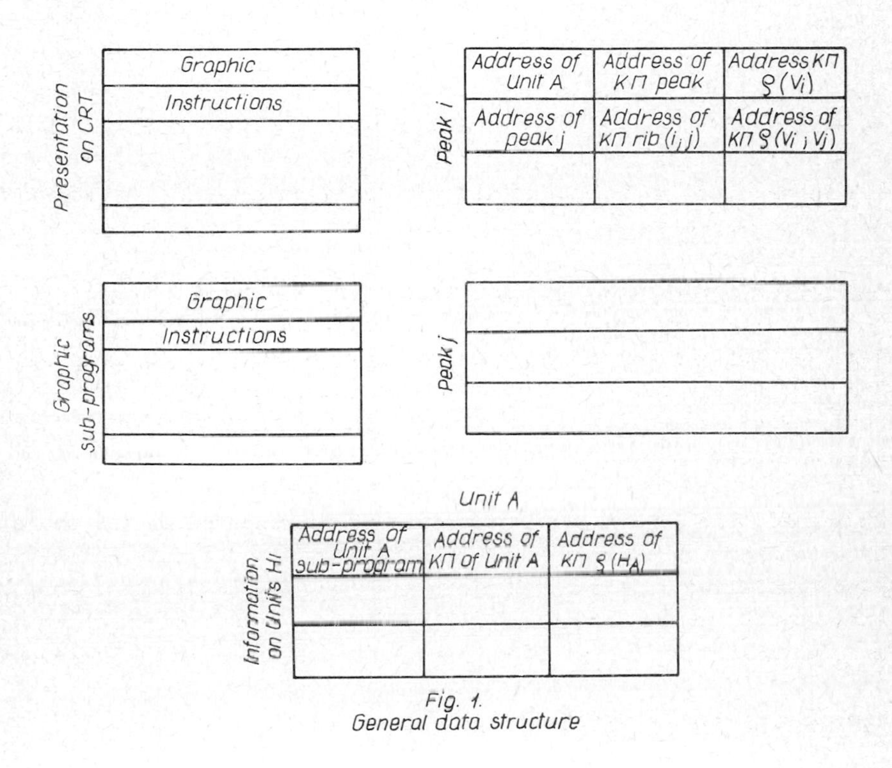

Fig. 1.
General data structure

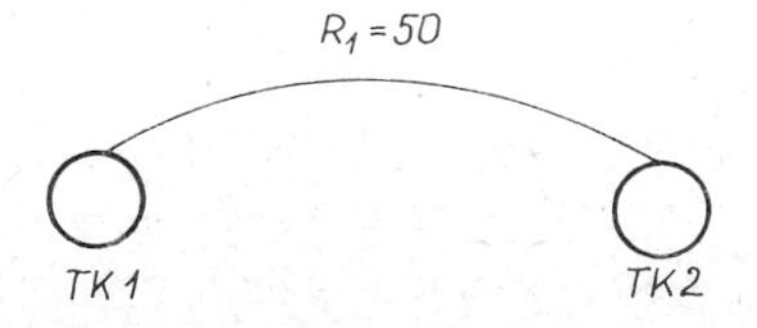

Fig. 2.
Example of representation

The contact of the operator with graphic communication system and the graphic communication system with the central computer is accomplished with the aid of a specialized algorithmic language of the graphic programming system СИПРОГ . The graphic communication language is only a means for setting, editing the graphic representations and exchange of the data with the central computer. To describe the program, just like before, the own problem-orientated language is used /for example APT, САП-2 when organizing the dialogue according to the first set of methods/. Therefore, the programmer is faced with a problem of the bilingual programming. Thus, it is important to set the relationship between these two languages during the process of conductiong the dialogue. Let us designate the graphic communication language by GL and the programming language by PL.

In the process of describing the program the programmer should use language PL but when graphically presenting such a description it is necessary to use only language GL. For example, applicable to the first set of methods of describing the program in the terms of language PL the text is "meaningless" in the terms of language GL. Since language PL is a means of programming and language GL is a means of dialogue, the peculiarity of the latter should be the requirement of the naturalism when designing and editing the graphic data. In the ideal case language GL should be easily adaptable to the requirements of a concrete investigation, should require a minimum capacity of the memory for the program - translator and be compatible, if possible, with any automatic programming systems. The above-mentioned language СИПРОГ meets well these requirements.

In accordance with the two levels of the dialogue, mentioned above, the language also has two levels. The lower language level operates with the design instructions. The upper language level makes it possible to work with a full graphic data system. A typewriter and a light pen are used as the technical means for making up the language phrases.

The following operations can be carried out in the lower language level:

1. Plot a point, draw a line, write a letter, line of letter, draw a circle, arc of a circle and write a text.

2. Remove any graphic instruction in accordance with the preset parameters.
3. Edit a text.
4. Unambigously determine a part of the representation by the light pen.
5. Mark /change brilliance/ separate representation elements.
6. Redistribute the memory for the representation elements.
7. Bring out on the screen and remove separate representations simultaneously retaining information about them in the memory.
8. Exchange data with the central computer.

The upper language level provides for additional possibilities such as:

1. To turn the representation element around a given point.
2. To shift the element.
3. To design and regulate the system and the lists when the representation composition changes.
4. To identify the representation elements utilizing the analysis of the lists.
5. To change the scale.
6. To combine the representation elements set in the unit and to give them the required designations.

The formal description of language СИПРОГ is sufficiently well given in [4] . To get familiar with the language we are giving a brief description of its lower level.

1.1. <u>Symbols</u>

⟨symbol⟩ :: = ⟨letter⟩ | ⟨digit⟩ | ⟨delimiter⟩ | ⟨designation⟩

⟨letter⟩ :: = А | Б | В | Г | Д | Е | Ж | З | И | Й | К | Л | М |
Н | О | П | Р | С | Т | У | Ф | Х | Ц | Ч | Ш | Щ |
Ы | Ь | Э | К | Я | Ъ | D | F | G | I | J | L | N |
Q | R | S | U | V | W | Z

⟨digit⟩ :: = 0 | 1 | 2 | 3 | 4 | 5 | 6 | 7 | 8 | 9

⟨designation⟩ :: = + | − | ж | — | ю | ↑ | X | > | < | [|] | _ | , |
% | ≤ | ≡ | ¬ | ⊃ | ∨ | ≠ | ÷ | ⊂ | ⊃ | ! | ∧ | ≥ |
○ | → | ← | ? | ↓ ∅ | + - ▽ | ◇ | ı | /

1.2. Identifiers

```
⟨identifier⟩    :: = ⟨letter⟩ | ⟨identifier⟩ |
                    ⟨letter⟩ | ⟨identifier⟩
                    ⟨digit ⟩
⟨name of graphic instruction⟩  :: = ⟨identifier⟩
⟨variable⟩    :: = ⟨identifier⟩
⟨designated record⟩ :: =  ⟨identifier⟩
⟨decoding⟩  :: = ⟨identifier⟩
```

Certain variables are of special meaning and they should not be used in other designations:
X - coordinate X, Y - coordinate Y, R - radius,
K - coordinates, ПК - coordinate increment, A1 and A2
 - arc parameters, p = composition of text.

The names of graphic instructions are chosen in the form of a two-letter designation:
TK - point plotting instruction; ПР - straight line plotting instruction; КР - circle drawing instruction; TT - text writing instruction; ЭЛ - element.

Decodings of ПЕР and ОТМ have a fixed meaning:
ПЕР - by pen; ОТМ - by mark.

1.3. Numbers

```
⟨number⟩   :: =  ⟨whole number without sign⟩
⟨value of variable⟩  :: = ⟨digit⟩ | ⟨whole number without sign⟩
                        ⟨digit⟩
```

1.4. Expressions

```
⟨expression ⟩ :: = ⟨simple expression⟩ ⟨compound expression⟩
⟨simple expression⟩   :: =  ⟨graphic instruction⟩
⟨graphic instruction⟩ :: =  ⟨name of graphic instruction⟩
                          ( ⟨list of parameters⟩ )
⟨list of parameters ⟩  :: = ⟨variable with numerical value
                           variable with decoding⟩ | ⟨list
                           of parameters⟩ ⟨variable with decoding⟩ |
                          ⟨list of parameters⟩  ⟨variable with
                           numerical value⟩
```

```
⟨variable with numerical value⟩  :: = ⟨variable⟩ = ⟨number⟩
⟨variable with decoding⟩ :: = ⟨variable⟩ = ⟨decoding⟩
⟨compound expression⟩    :: = ⟨simple expression⟩ ;
                              ⟨simple expression⟩
```

Some expressions and their meaning contents are given in Table 1. as an example.

1.5. Operators

The operator is the action unit. A total of six operators is found in the lower level of the language.

```
⟨operator⟩   :: = ⟨starting operator⟩   ⟨design operator⟩
                  ⟨exclusion operator⟩  ⟨mark operator⟩
                  ⟨correction operator⟩ ⟨input operator⟩
⟨operator⟩   :: = ⟨designed record⟩     ⟨representation
                                         desoription⟩
⟨representation description⟩  :: = ⟨simple expression⟩
                                    ⟨compoind expression⟩
```

1.5.1. Starting Operator

This operator makes it possible to display on the CRT screens the contents of two memory pages assigned to describe the representation on the CRT screens. When doing so two operating modes are possible. In the first mode the selected arrays are displayed with simultaneour retention of information in the arrays which were displayed earlier. In the second mode the latter are cleared:

```
⟨designated record⟩  :: = START

First mode:   START ␣ ⟨point plotting instruction⟩

Second mode:  START ␣ ⟨straight line plotting instruction⟩
```

1.5.2. Design Operator

Provides representation design on CRT in accordance with the preset representation description:

```
⟨designated record⟩  :: =   DESIGN
```

1.5.4. Exclusion Operator

The use of this operator is linked with the exlusion of graphic instructions /or separate instruction elements/ set by representation description. Earlier marked instructions can be excluded:

⟨designated record⟩ :: = EXCLUDE

1.5.5. Correction Operator

The action of this operator extends to the text instructions. Its use is linked with the necessity to edit the text in the form of inserts of some arlier prepared texts made in the edited text. In doing so provision is made for the shifting of all the symbols in the edited text, which follow the insert by a given value:

⟨designated record⟩ :: = CORRECT

The correction is made by consecutive indication by the light pen of the previous symbol before the insert and any symbol in the insert text.

1.5.6. Input Operator

All the graphic instructions, preset by the representation description, are searched for, an appropriate expression in language СИПРОГ is shaped and the data is supplied into the central computer in the symbol-by-symbol mode:

⟨designated record⟩ :: = DATA

1.6. Operator Sequence Setting

⟨operator sequence⟩ :: = ⟨operator⟩ · ⊔
⟨operator⟩ | ⟨operator sequence⟩ · ⊔
⟨operator⟩

When operating with the system, the programmer chooses definite operating modes depending on the language text, using for this purpose the alpha-numeric keys or the light pen.

The alpha-numeric information reaches the computer in the form of a sequence of bytes, the transmissions from the light pen differ by the information character depending on the chosen operating mode. It is subdivided into three main categories:

1. Numerical values of coordinates X and Y;
2. Address α i and number of the element of data design instruction the representation of which is indicated by the light pen;
3. Increment codes when presetting point TK (K = ПЕР)group instruction.

On receiving the information from the operator the program-translator shapes or edits the design instructions which describe the representation. The translator is designed on the principle of interpretation. It can be conventionally subdivided into two parts viz. the one which accumulates and analyses the input information and the executive one which presets and edits the representation and transmits the data to the central computer.

The executive part is realized in the form of separate sub-programs which are commutated in a definite sequence depending on the requirements of the input description. Such a design of the executive part is easy to interprete in the form of K - network [5], under whose units /non-commutated units/ we understand separate sub-programs and under the commutated units we understand certain binary variables q i Fig. 3. shows the random K - network which serves as an illustration of the above-said. The values of functions q i are determined at every stage of the input language analysis.

The analysis of the language text is done according to the tabular principle. The Table is given in Fig.3.

By this table it is easy to follow all the stages of translation:

1. Determination of the type of operator;
2. Varieties of graphic instructions;
3. Name of parameter and its numerical value;
4. Taking decision whether the operator's action spreads to the next graphic instruction or whether another operator follows.

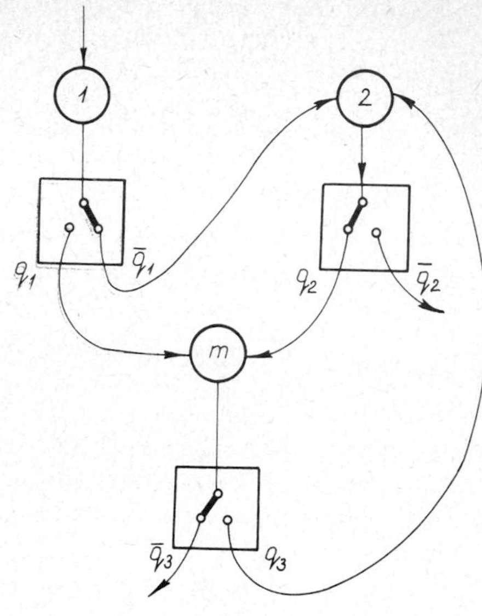

Fig. 3.
Example of K-network

The combination of the values of functions q i for every programmer console is determined by the, so called, translators's conditions field.

Under the multi-console operating conditions the time division is ensured by a special program-dispatcher. The dispatcher practically realizes a circular cyclic algorithm with a priority in attending to the input information from the light pen. The difference as compared with the circular cyclic algorithms consists in the fact that the interruption in attending each console cannot be accomplished in a random time moment t i , at the expiration of the attendance datum τ but it accomplished only in the case when the count ends by one of the sub-programs of the executive part. The latter makes it possible to substantially retain the volume of the intermediately stored data, practically reducing it to the storing of the field of the translator's conditions.

The central computer is similar to an additional operator's console having the highest absolute priority.

Table I.

LIST OF SOME PERMISSIBLE EXPRESSIONS AND THEIR MEANING CONTENTS

No.	Expression	Meaning contents
1.	2.	3.
1.	ТК (Х=0I00 У=02I0)	Point with coordinates X=0100, Y=0210.
2.	ТК (Х=0020 У=00I0 ПК=I327056)	Set of points with starting coordinates X=0020, Y=0010 and subsequent coordinate increment codes of 1,3,2,7,0,5,6.
3.	ТК (К=ПЕР)	Point or set of points set by the light pen.
4.	ПР(Х=00I0 У=0020 Х=0040 У=0060)	Section of straight line whose starting and terminal coordinates are preset.
5.	ПР (К=ПЕР)	Straight line /broken line/ whose starting, terminal and bending points are succesively indicated by the light pen.
6.	КР (Х=0I00 У=0050 R=0020)	Circle with radius R=0020 and centre coordinates X=0100, Y=0050.
7.	КР (R=0050 К=ПЕР)	Circle with radius R=0050 and oentre coordinates set by the light pen.
8.	ДО (Х=0I00 У=0020 =0I00 AI=0030 A2=0045)	Circular arc with radius R=0100 and coordinates X=0100, Y=0020 and arc parameters A1=0030 and A2=0045.
9.	ТТ (Х=00I0 У=0005 Р= СИСТЕМА	Text "СИСТЕМА" with starting coordinates X=0010, Y=0005.
10.	ТТ (К=ПЕР)	Text indicated by light pen.
11.	ДО (К=ПЕР)	Circular arc indicated by light pen.
12.	ЭЛ (К=ПЕР)	Separate graphic instruction element indicated by light pen.
13.	КП (К=ПЕР)	Graphic instruction indicated by light pen.
14.	ТК (К=ОТМ)	Marked point instruction.

1.	2.	3.
15. КР (К=ОТМ)		Marked circle instruction.
16. ТТ (К=ОТМ)		Marked text instruction.
17. КП (К=ОТМ)		Random marked graphic instruction.

Table 2.

REFERENCES AND DELIMITERS

Delimiters	References	Maximum number of symbols
␣	НАЧАЛО, ПОСТРОИТЬ, ИСКЛЮЧИТЬ, ОТМЕТИТЬ, КОРРЕКТИРОВАТЬ, ДАННЫЕ	14
(	ПР, ТК, КР, ДО, ТТ, КП, ЭЛ,	2
=	Х, У, К, АI, А2, Р, R, ПК,	2
)	ПЕР, ОТМ, Machine zero	3
␣	. , ;	1

LITERATURE

1. Kritzan Igor, Kurimsky Robert, CRT-programming cuts cost, Amer. Mach., 1970, 114, No.19, 71-74.

2. Б.Б. Т и м о ф е е в, В.Г. З а й ц е в и др., Многопультовая система графической связи для разработки управляющих программ к станкам с программным управлением, "Уникальные приборы", №IO, СЭВ, Москва, I972 г.

3. В.Г. З а й ц е в, Представление данных в системе графической связи, Сб. "Промышленная системология", "ТехнIка", Киев, I97I.

4. В.Г. З а й ц е в, Язык графической связи СИПРОГ, Сб. "Промышленная системология", "ТехнIка", Киев, I972.

5. В.Т. К у л и к, Алгоритмизация обьектов управления, "Наукова думка", Киев, I968.

APT/IGS: STATE OF THE ART IN NC GRAPHICS

J. CREMERIUS
Control Data Corporation, Graphics Applications
USA

INTRODUCTION

From the advent of the first CRT display capable of displaying vectors, the computer industry has been besieged with ideas and invisioned advances concerning the possible uses of a graphics console in a production environment. A multitude of attempts have been made to achieve this goal, but as of yet many have fallen short. Programs designed by software experts have, as a rule, been found to be lacking in either content or usefulness. Either they could not do the entire job or they could not be used by the users due to poor design. In an environment requiring cost and production efficiency, such as the one in which Numerical Control operates, any advancement of the magnitude foreseen is looked upon with real excitement and enthusiasm. It was in this atmosphere that the concept and design of Control Data's APT/IGS system came about. The system name stands for Automatically Programmed Tools/Interactive Graphics System. As released the system is a production tool used to interactively define geometry and generate APT cutter location output, at a graphics console, for use on Numerical control machines. APT/IGS is completely APT III compatible and provides all the capabilities of the APT system using an interactive graphics control.

BENEFITS OF THE SYSTEM

There are many benefits to be gained from using online graphics for programs not even connected with NC. Plotting applications previously requiring off-line, time consuming plotters, are well suited for conversion to online graphics. Large decreases in the amount of turnaround time required can also be achieved. NC, however, presents many other possibilities for cost reduction using graphics. As a part proceeds from initial design to the actual cutting, many people influence the amount of cost and time expended to complete it. Such variables as parts programmer productivity, scrapped or modified parts, span time and down or idle machines are among the areas which can be improved upon by graphics. A few of the more significant benefits to be derived from the use of the APT/IGS system are listed below:

- Elimination of programming manuscripts
- Optional elimination of APT source card keypunching
- Elimination of the need for offline plotting
- Greatly reduced span time for part generation
- Increased parts programmer efficiency
- Better utilization of the NC machine shop
- Fewer computer runs
- Simpler and quicker part program modifications
- Ability to optimize existing part program

REASON FOR DEVELOPMENT

It was the intent of the APT/IGS System design to ease the Parts Programmer's job of producing a completed part and to reduce both the total job span time as well as the number of trial runs in

COMPUTER LANGUAGES FOR NUMERICAL CONTROL, *J. Hatvany, editor*
North-Holland Publishing Company - Amsterdam–London

producing a good part.

In a batch environment, the typical parts programmer uses the following procedure {see Figure 1} to create the desired part:

1. With the use of engineering data and part drawings, he generates an APT source deck.
2. He then submits this deck for processing through the APT batch processor. It may take from five to ten submittals before an error-free run is made.
3. If no errors have been encountered, a CL {Cutter Location} tape is generated by the APT system.
4. The output is then taken to an offline plotter for further tape verification.
5. Assuming that no obvious errors are located the part is rerun through the APT system and a Post Processor to obtain an NC tape.
6. This tape is taken to the machine tool and trial cuts are performed.
7. Assuming no errors are observed, a part is cut.
8. If this part is not correct or a fixture is damaged, the parts programmer must resubmit a corrected source deck until the desired result is achieved.

As can be seen, depending on the complexity of the part, many computer runs can be required as well as the scrapping of several parts.

APT/IGS was designed to shorten and refine the procedure used to produce a part. Figure 2 outlines a typical part as it would be generated by APT/IGS;

1. The parts programmer brings his drawings and optional APT source deck to the 274 graphics console. The cards, if any, are input to the 6000/CYBER or the 1700, which in turn submits them to the 6000/CYBER.
2. The parts programmer now generates the desired part at the console. This includes part geometry as well as the necessary motion statements to correctly cut the part. He sees each cut as it would be performed and therefore can immediately eliminate errors that he may have generated. All the iterations of the batch mode are replaced by the interaction with the 274 console.
3. The intent is to create a correct part the first time and thereby generate a cost savings of computer time, parts programmer time and span time to the machine tool.

METHOD OF DEVELOPMENT

All of the aforementioned improvements would obviously be sought after by any production oriented NC machine shop. Yet in the planning stages of APT/IGS there remained the nagging remembrance that systems of this nature have been built before and many had failed. It was for this reason that a different design approach was undertaken utilizing parts programmers, those people who would be users of the system, to effectively design the system's overall needs. They were asked to describe their needs and the methods they most preferred to satisfy these needs. To accomplish this, users from among the APT customers of Control Data Corporation were solicited for the initial design stages. The results from this effort were then turned over to the software people with the instruction that designs were not be discarded just because they were found to be difficult or lacking in thoroughness.

HARDWARE CONFIGURATION

The selection of the graphics console is a very important ingredient in the system design. In order to provide a significant capability, the high performance 274 CRT was chosen over some of the more inexpensive lower performance tubes available. The 274 graphics console is supplied with a 4K or 8K display buffer and is connected to a remote 1700 computer system which acts as the controller. The 1700 is in turn connected to a CYBER computer system by high speed communication equipment. The final configuration selected is shown in Table 1 below. For completeness, the target computer and other related hardware are also included.

TABLE 1

CDC 6000/CYBER Computer	Host
CDC 1700 Computer System	Remote
CDC 274 Graphics Console	{20" diameter CRT with lightpen, function keyboard and alphanumeric keyboard}
CDC 1744 Graphics Controller	
CDC 1747 Data Set Controller	
CDC 1706 Data Channel	

Supplied with the 274 is a lightpen to be used for detecting lightbuttons from the CRT. In addition, there are optional function and alphanumeric keyboards which are not required by APT/IGS.

SOFTWARE REQUIREMENTS

APT/IGS is designed to be highly modular for better program organization. This design was also helpful in aiding the parceling out of the system implementation among the various companies participating in the effort. To provide easier understanding, coding, and modification, FORTRAN was chosen as the programming language to be used. In addition, the system utilizes a Plex data structure which provides quick random access to millions of words of mass storage. Large amounts of data accessed rapidly as well as fast response time were early design stipulations. To provide the response time needed, a task or overlay loader is used. The task loader is built to function best when loading small, compact modules of the variety used in APT/IGS.

SCREEN LAYOUT

To allow for an understandable display, the 20 inch diameter screen is divided up into working areas in the following manner:

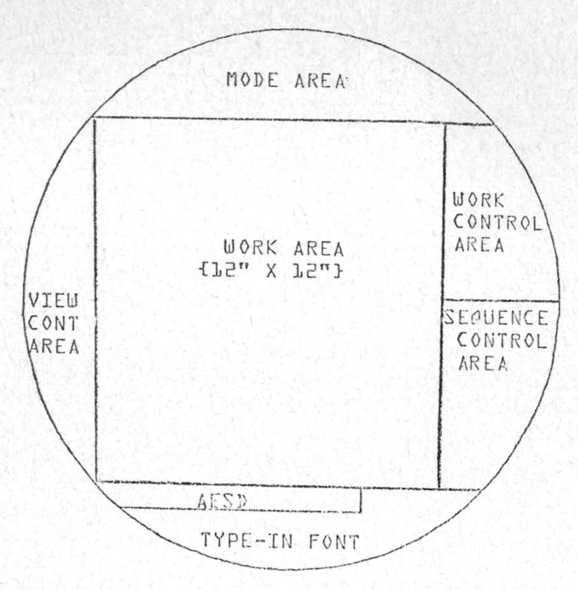

1. Screen layout of the CDC 274 graphics console as used by APT/IGS.

The 12" X 12" Work Area is used to display all geometry and cutter paths as well as certain alphanumeric data which the user deems usable.

The Sequence and Work Control Areas are used to display permanent and secondary lightbuttons which the parts programmer uses to converse interactively with the host computer. In response to his requests certain menus are displayed in the Work Control Area.

The Mode Area is used to display information which is useful during the generation of the program. Such information would include Cutter parameters, Feedrate, Coolant, Drive, Part and Check surfaces and the PARTNO.

The View Control Area is used to perform rotation, translation and zooming of the geometry and cutter paths within the Work Area.

The AESD area is an area used to display the last five APT statements executed by APT/IGS.

A type-in font is also provided for use if the alphanumeric keyboard is not available.

SYSTEM FEATURES

In the use of the aforementioned areas three very important considerations must be recognized. First, the system provides a tool which operates in a manner which the parts programmer is familiar with. Secondly, it provides him with all the capabilities provided by the APT III system already in production use. Thirdly, it allows him to work in a fairly random manner which is somewhat characteristic of the normal parts programming environment.

The first consideration was achieved by designing APT/IGS to produce as its primary output APT source cards. The user is able to read cards already in existence, create new ones interactively or merely type them in as he chooses. In this way, he can generate the geometry and cutter paths of the program in a manner very similar to his normal mode of work. At the same time, the large data base within the system provides instant access to the information which the programmer needs to effectively produce the part. Information such as canonical forms, card images, and view selection are just a few of the facilities available.

In supporting the capabilities of the current APT III system, APT/IGS gives full 3-D cut capability rather than the usual 2 or 2 1/2-D. To do this, the system was broken down into two distinct parts. First, that portion which, for clarity purposes, will henceforth be called the interactive portion of the system, and secondly, the portion of the system provided by the existent APT III processor. The interactive portion of APT/IGS will display and verify all cuts made in the 2-D mode. If a 3-D cut is to be performed and the resultant cutter path desired, the user need only select the APTIT lightbutton which will obtain the necessary cut vecoors from the APT processor. In this manner the system is able to provide full 3-D cut in a timely manner. The 2-D cutter paths generated by the interactive portion of the system are really pseudo cutter paths. That is, they are accurate estimates of the actual cutter path but the user must process the whole part through the APT processor to actually generate the cutter paths required to generate an NC tape. This method was chosen because the APT processor is a processor which has constantly been refined throughout its 13 years of existence and its power is not easily duplicated.

CLIPP PROCESSOR

The method used to go to and from the APT processor is called "CLIPP" {Convenient Length Incremental Part Program} processing. This concept allows processing source statements through APT in quantities of from one to the whole deck in any increment chosen by the user. He does not need to reprocess the statements preceding the statement to be processed unless an error has been generated. A file is maintained relative to each CLIPP which effectively saves pertinent information required for the execution of CLIPPs to be processed later.

All geometry capabilities which are provided by APT III are recognized and displayed by APT/IGS. Therefore when combined with the motion capability, the user describes the necessary geometry and then cuts the part one statement at a time. After each motion statement is processed the cutter path is shown providing the user with instantaneous knowledge of any invalid cutter paths. In addition to providing for the generation of all APT geometry and motion statements, APT/IGS also provides for the creation of Post Processor, Cutter, and other APT recognized statements. Thus the user has at his disposal all the tools and power of the APT system on an interactive console.

To be truly interactive the system allows the user random interaction, as opposed to a structured operating pattern. Experience has shown that any interactive process contains some degree of randomness. For instance, if a line was being generated interactively and it was discovered that one of the necessary points was missing, the user could interrupt the line creation, create the point and pick up where he had left off generating the line.

GENERATING A PART PROGRAM

In an effort to better describe the power of the APT/IGS system it seems useful at this point to briefly describe the sequence of a simple part generation. Beginning with only the engineering drawing the user initiates the APT/IGS program by signing onto the console and attaching the APT/IGS program. {See Figure 4.} He then would indicate the part identifier and input initial conditions such as tolerance and cutter parameters. Following this he selects the appropriate lightbutton to generate the display of the lightbutton menu showing all the various geometry types. {See Figure 5.} From this menu he selectes, for instance, the POINT lightbutton which results in the display of all of the accepted APT point formats.{See Figure 6.} If he wished to generate a point at the intersection of two lines, his concern would be directed to fulfilling the requirements of the respective format; i.e.,

$$POINT/INTOF,{\times}LI^{1}, {\times}LI^{2}$$

This format indicates that lines can be supplied either by selecting the desired lines from the geometry display in the Work Area, and/or by typing in the symbolic name{s}. Assuming the correct data was input, the line statement would be formed and then displayed if the line was found to intersect the 12 X 12 inch Work Area. In every case there is a prompting message indicating the next expected action but this is not a required action. The only time random action is not allowed is upon recognition of an error by the system. The error is displayed in a blinking status and must be "picked" by the user before he can continue further. With the use of the available formats and type-in line the user is able to create all the geometry required to represent the part. {See Figure 7.}

At this point if the part was fairly complex and, therefore, the screen fairly full of geometry, he would use the APT/IGS segmentation or bounding feature to create a more understandable display. By selecting the proper lightbutton the display from Figure 8 is brought up.

By selecting all the pieces of geometry he wishes segmented and picking INTERSECT he can direct the system to locate and mark all the intersections of the geometry selected. {Also shown in Figure 8.}

Now he can pick portions of geometry he wishes made invisible by lightpen detecting the desired geometry segment and the INVISIBLE lightbutton. The portions of the surfaces picked are redisplayed as dashed lines. Having selected all portions to be made invisible, the user indicates completion and the part is redisplayed with all selected portions of the geometry made invisible. {Figures 9 and 10}

The part is now in a more understandable and easier to visualize form, thus making the generation of the cut sequence a simpler task.

CUTTING THE PART

APT/IGS provides two different methods for interactively generating motion commands. The first is simply supplying the Drive, Part, and Check surfaces by lightpen detect or type-ins. A statement is generated and checked for validity prior to displaying the generated cutter display. {Figure 11}

The other method is a simpler one and presents even less chance for error. By placing the tracking cross next to the desired check surface or, in fact, on the check surface, the parts programmer can tell the system that he wishes to use the previous check surface as the new drive surface, and the surface closest to the tracking cross as his new check surface. Both of the methods described provide a very high degree of efficiency, eliminating the more common errors of going in an incorrect direction or using the wrong symbolic name for a controlling surface. Again, if it is determined that the motion statement is that of a 3-D cut, no pseudo cutter path is generated. To actually see the cutter path the user must use the CLIPP processor to process the statement. This is not, however, an absolute necessity for the user may decide not to go to the APT processor until the part program is completed.

At any point in the processing the user can change the viewing plane or degree of rotation with the View Control parameters or he can make use of an APT/IGS feature appropriately entitled the "ROADMAP". {See Figure 12.} This concept allows the user to supply a four letter "grid code" defining the four corners of the desired display. For example, suppose the complete part was described in a grid of squares whose columns went from A to J and whose rows went from A to J. The parts programmer could request that portion of the part, defined in the Work Area. The system would then blow up the contents of the rectangular area ACAD to fill the 12X12 inch Work Area. Thus the user is able to quickly change the view whenever he so desires without the necessity of inputing actual frame dimensions. The user can at any time call up the display of selected canonical forms, scan through the part program, retrace cutter paths, or make use of any of the thirteen different editing features provided by APT/IGS. Deletion, move, copy, forward and backward search, and list are some of the more frequently used editing capabilities. The system also supports the generation of nested definitions which can be inserted into geometry definitions just as easily as symbolic names. {Figs. 13 and 14}

Once the part has been completely processed using the described procedure the user selects the APTIT lightbutton to finish the processing through the APT processor and create an NC tape. If the length of time required to produce the part program exceeds the available daily console time, the user can invoke the Checkpoint/Restart feature of APT/IGS. This feature provides the capability to resume processing the following day with virtually no duplicated effort.

As the parts programmer continues to use the APT/IGS system he sooen becomes increasingly quick and efficient. The system gives the feeling of being able to stand on the machine tool and correctly direct the tool in a real time mode. The parts programmer constantly benefits from the instant feedback of possible error conditions. Instant access to the host computer serves to encourage a better programming environment. Turnaround time, many times considered the largest hurdle to an effective NC operation, is cut drastically. When modifications become necessary and a Mylar tape is required in a hurry, the on-line status of the graphics console becomes even more important. It is not necessary to resetup the machine tool with another part as a newly modified tape can be provided in a very short order.

In addition to its obvious usefulness in decreasing the cost of the total NC operation, APT/IGS can also be used effectively as a teaching aid for the training of new parts programmers. Armed with a tool such as this, the new programmer learns more quickly of the various capabilities provided by the APT language.

The future outlook for APT/IGS development is one of constant improvement. A second version is being considered which would, in contrast to Version 1, interface with the CDC APT IV processor soon to be released.

SUMMARY

To summarize, the development goal of APT/IGS was to increase the efficiency of the parts programmer and cut the costs required to generate a finished part. By accomplishing these goals, APT/IGS has also encouraged better, more controlled utilization of the machine shop. Fewer setups are required due to faulty or scrapped parts. Part modification is made easier and quicker as well as more reliable. These savings can be classified as direct reductions in cost to completion and there are further indirect savings to be gained, by eliminating the need for offline plotting and the many time consuming iterations between the computer and the programmer. The APT/IGS system allows the parts programmer to complete a job {from start to finish}. This eliminates the time lost in the thinking process of reorientating himself when working on several parts at a time.

Test sites for APT/IGS have indicated that the effort expended for development was well spent. Estimates of manpower savings have been significant and have exceeded the original design goals.

Control Data feels that the APT/IGS processor is a required step in the building of a total Computer Aided Manufacturing System.

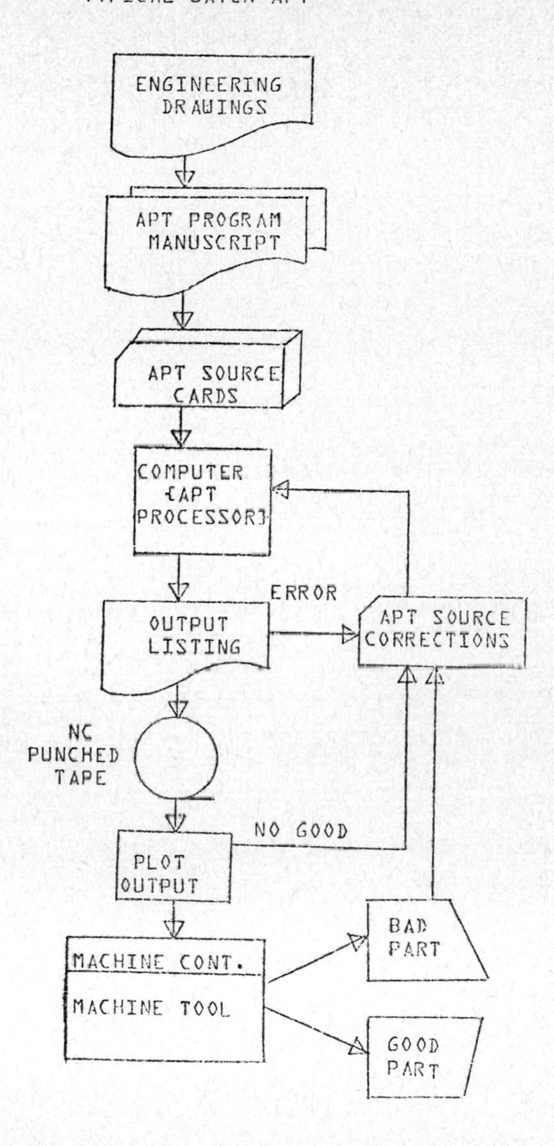

2. Flow diagram of the procedure used to generate and cut a part using Batch APT.

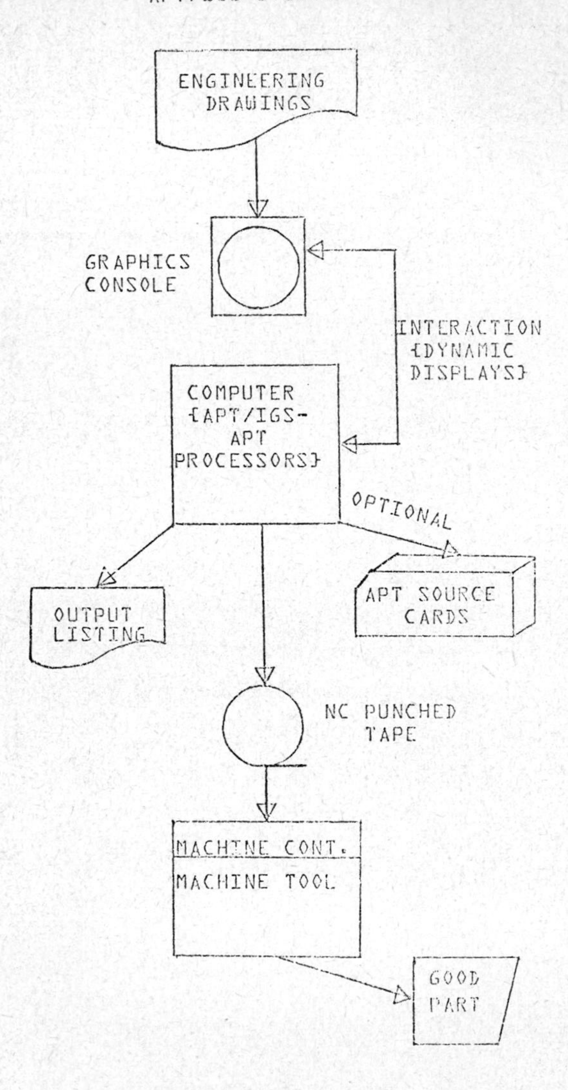

3. Flow diagram of the procedure used to generate and cut a part using the APT/IGS system.

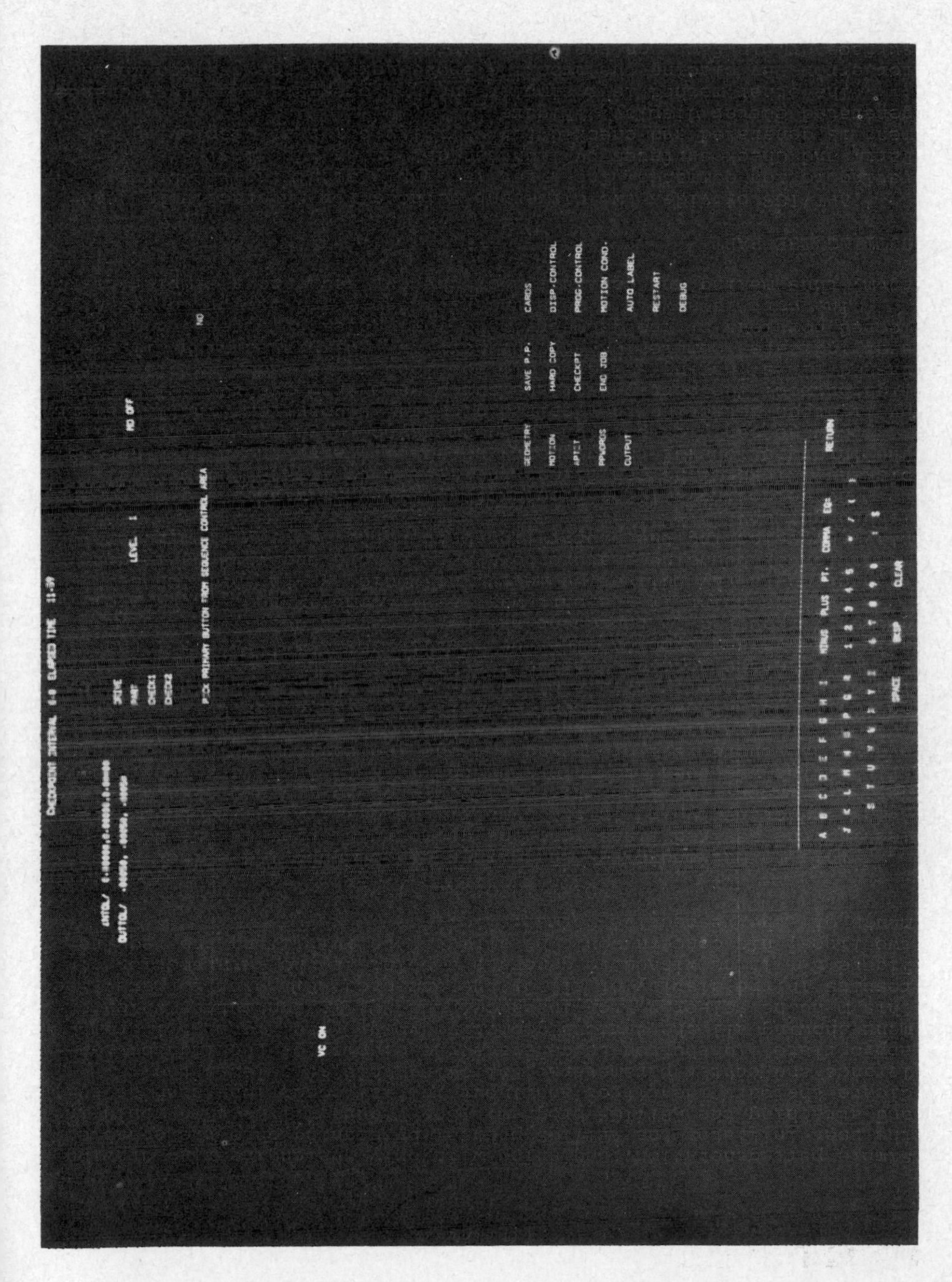

4. Actual view of 274 console upon initialization of the APT/IGS system.

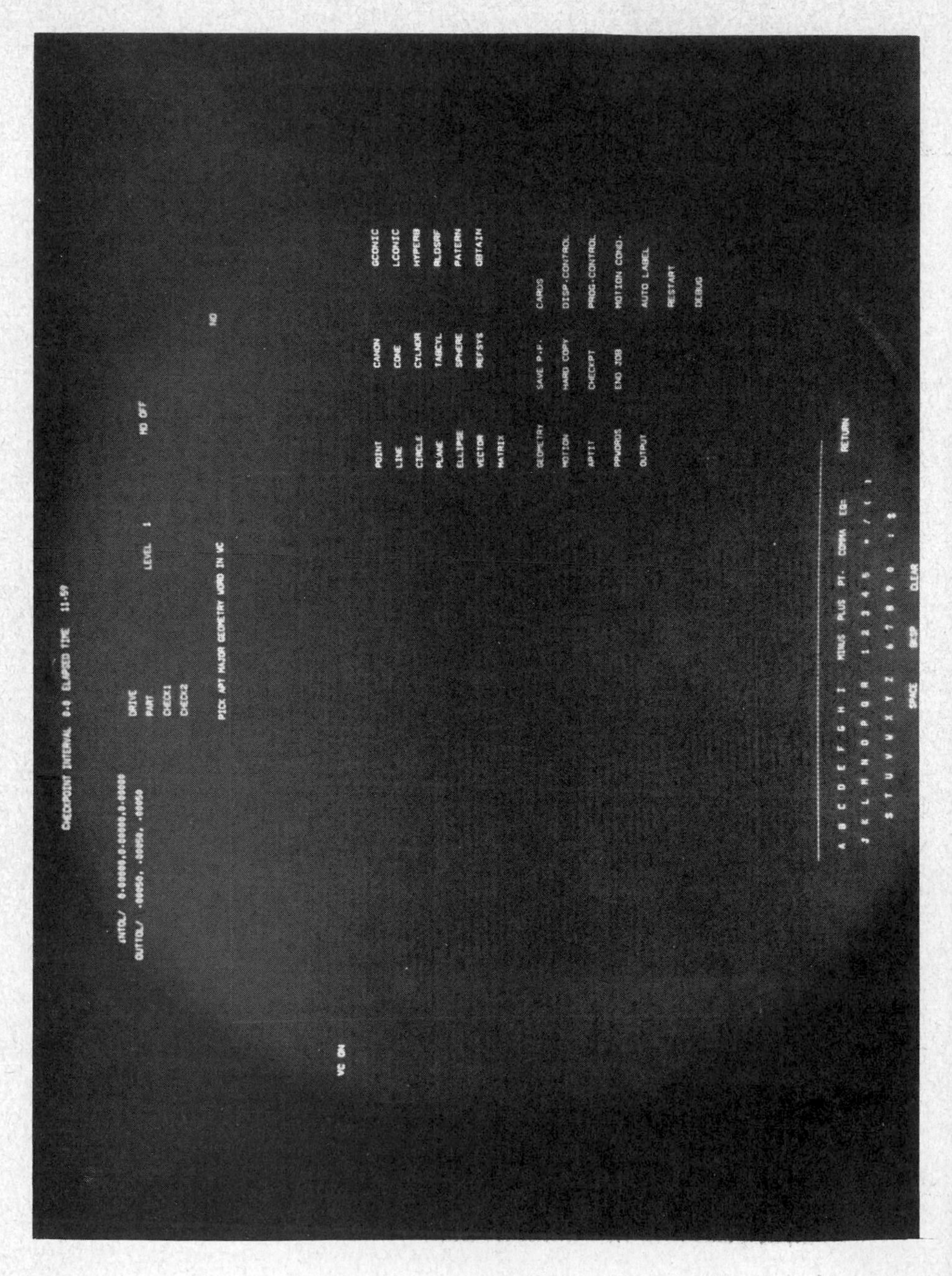

5. View showing menu used to initiate construction of a geometry definition.

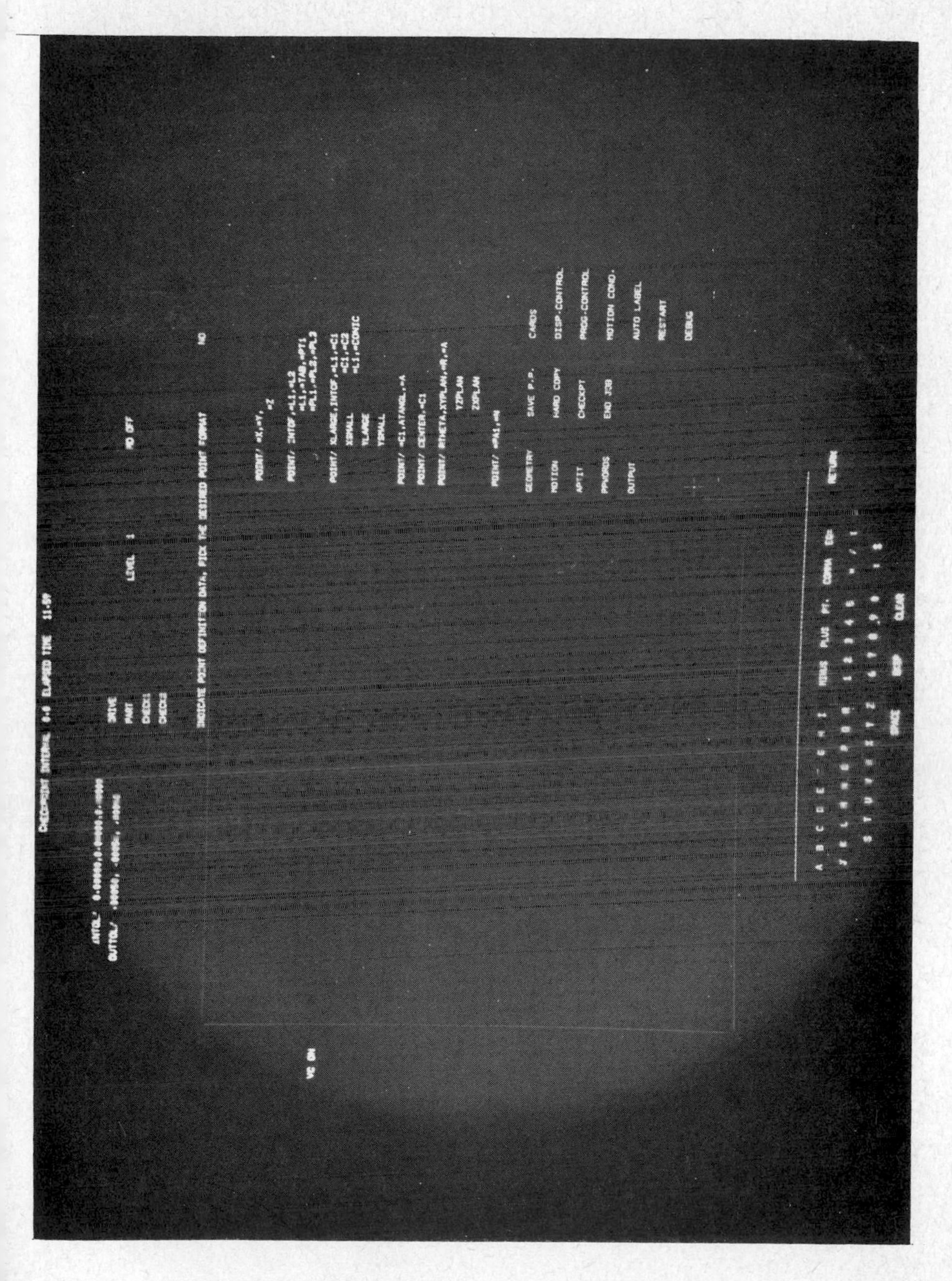

6. View showing POINT formats displayed after lightpen detection of the POINT button shown in Figure 5.

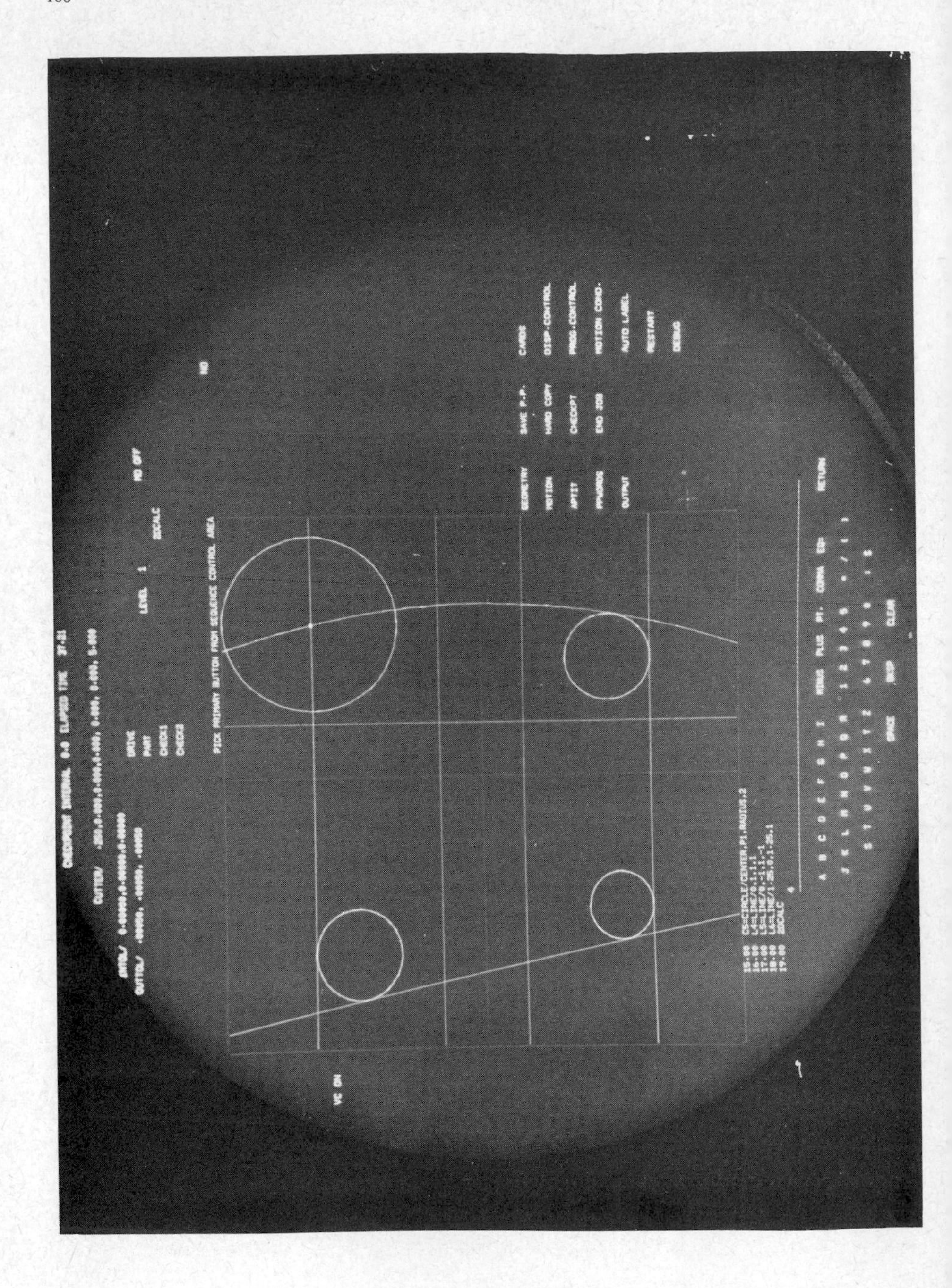

7. Display of geometry used to describe a simple part.

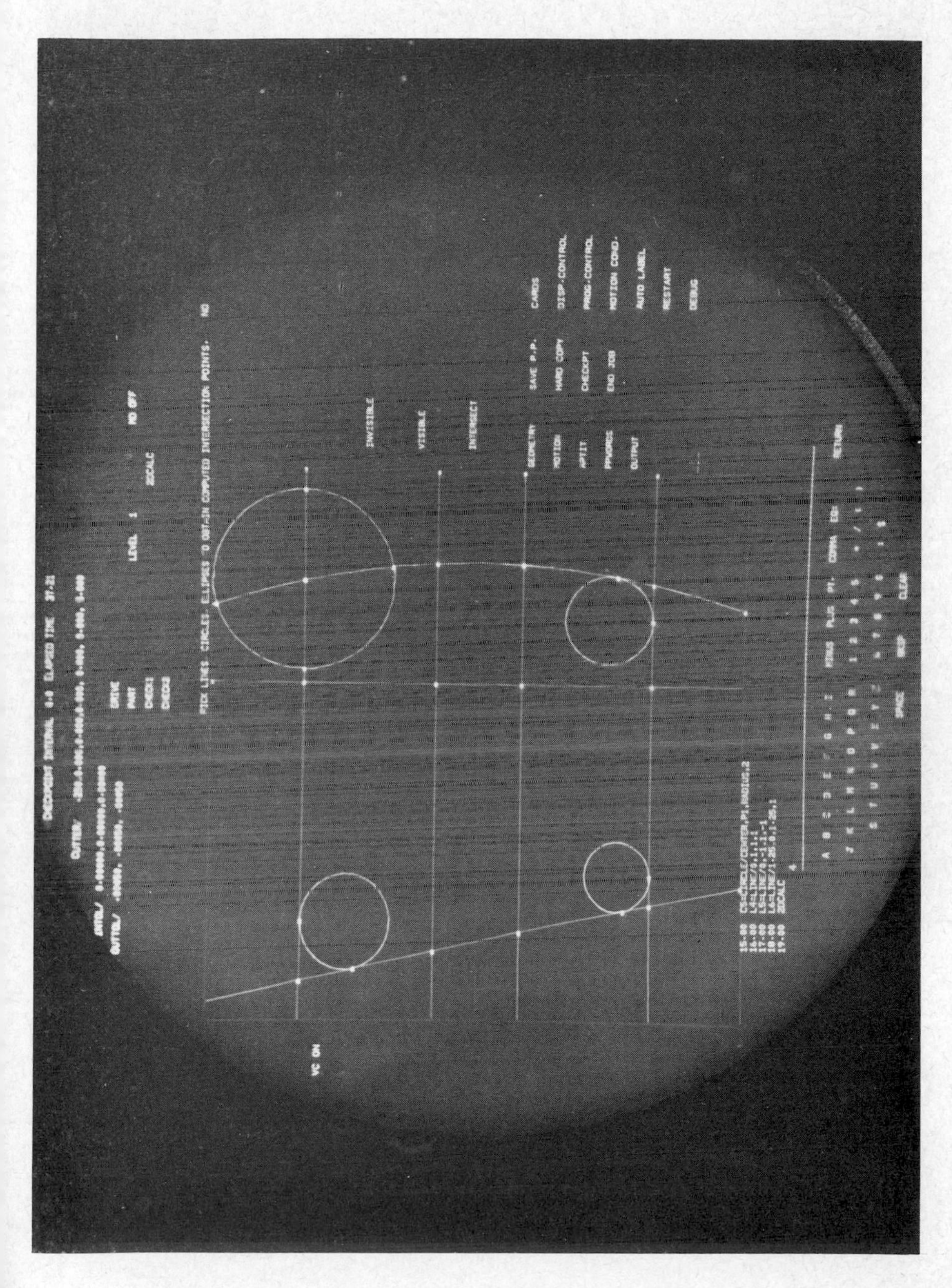

8. Display of segmentation menu and requested geometry intersections idicated by asterisks.

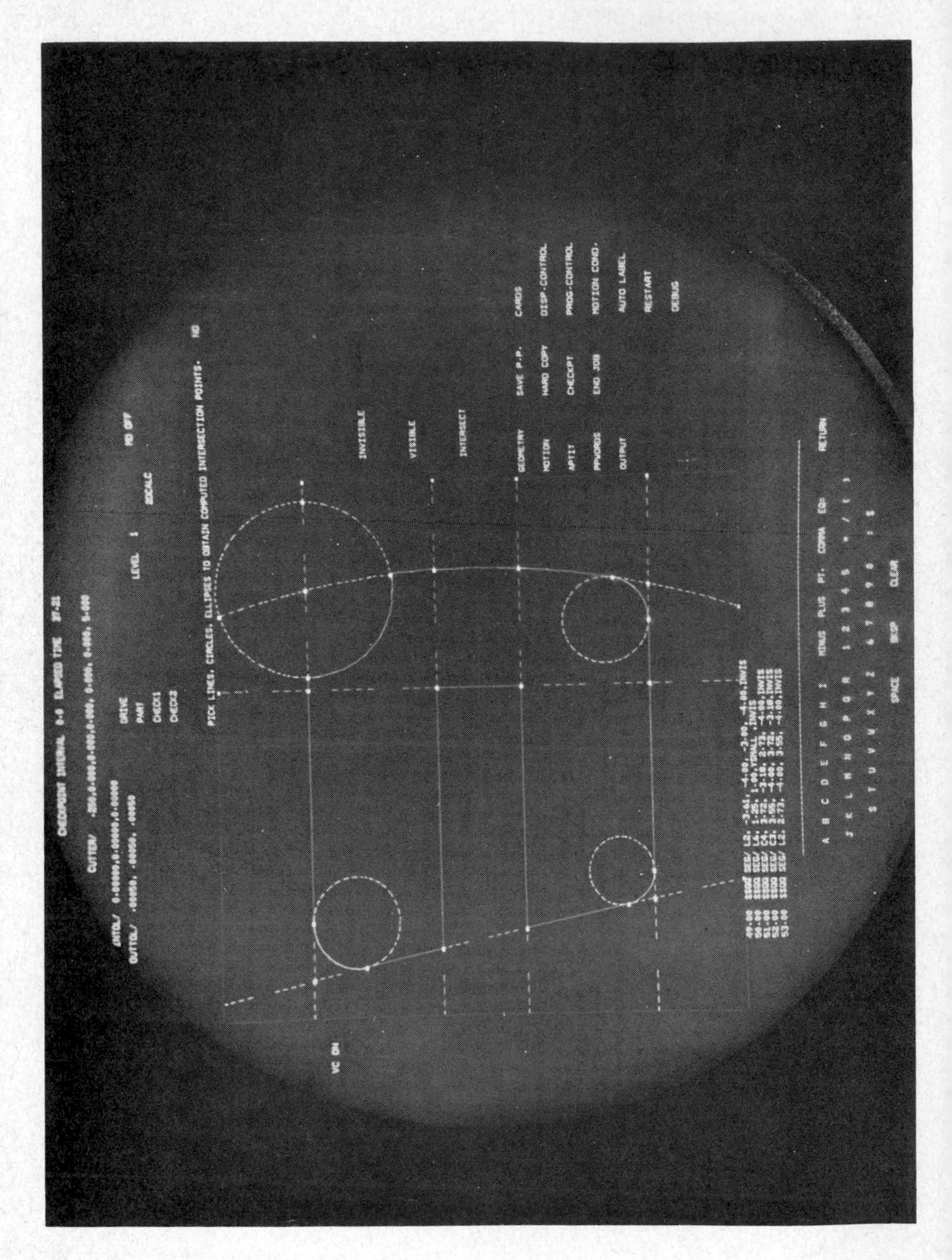

9. Partially segmented geometry as shown by "temporary" dashed lines.

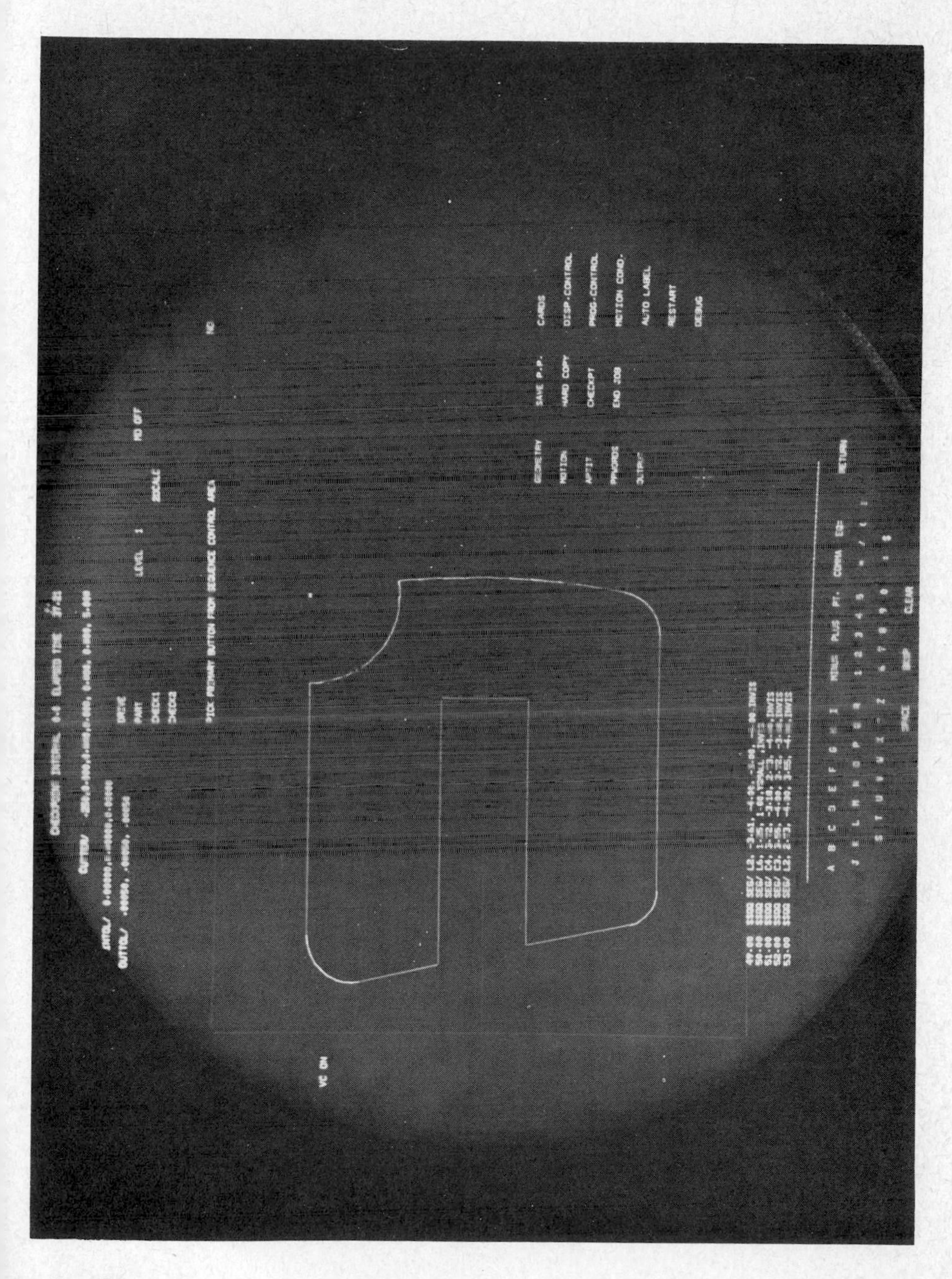

1o. Fully segmented part display.

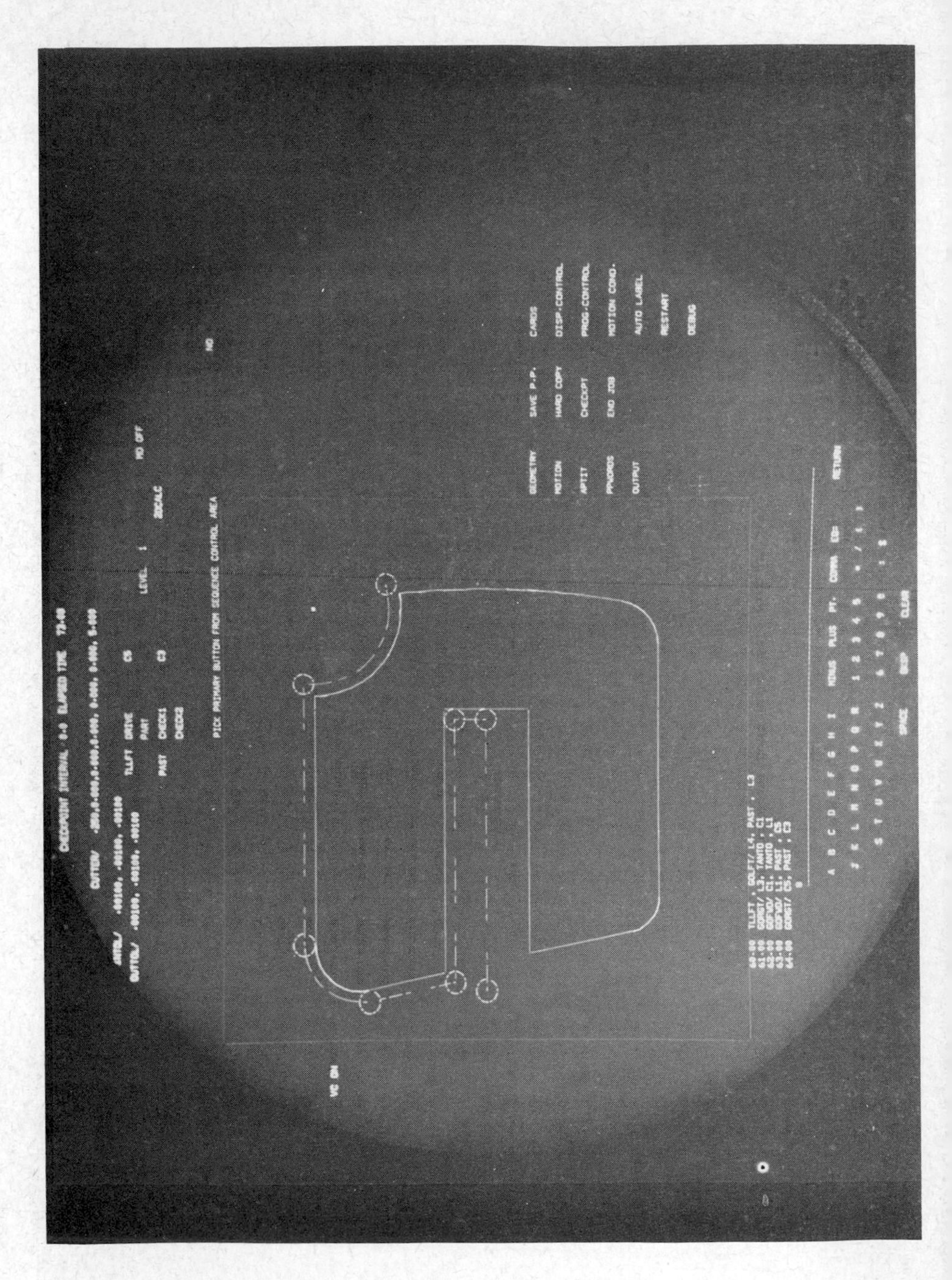

11. View showing the partial cut of a part.

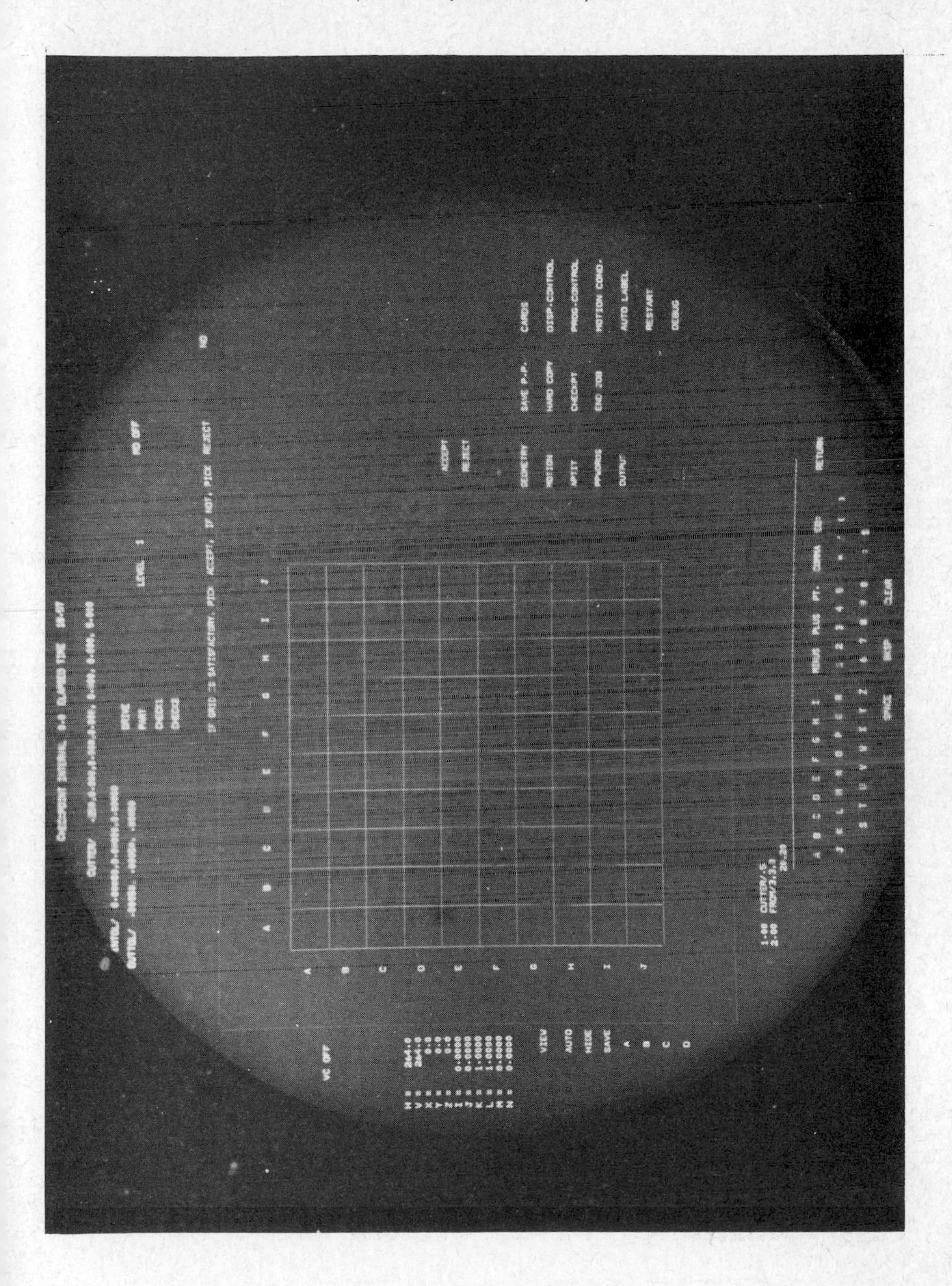

12. Initial display when defining a new ROADMAP.

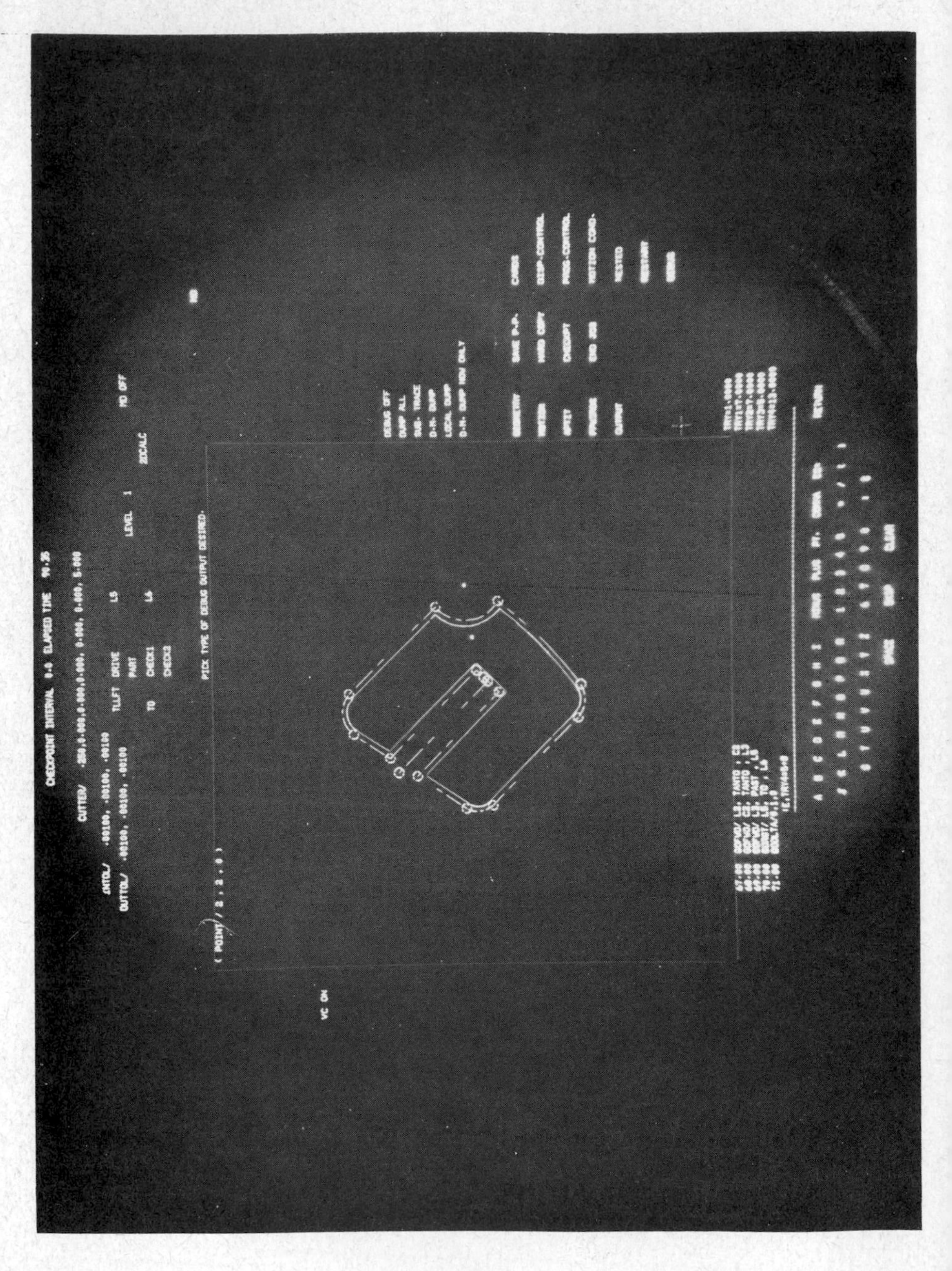

13. View showing a temporary nested definition at upper left hand portion of Work Area.

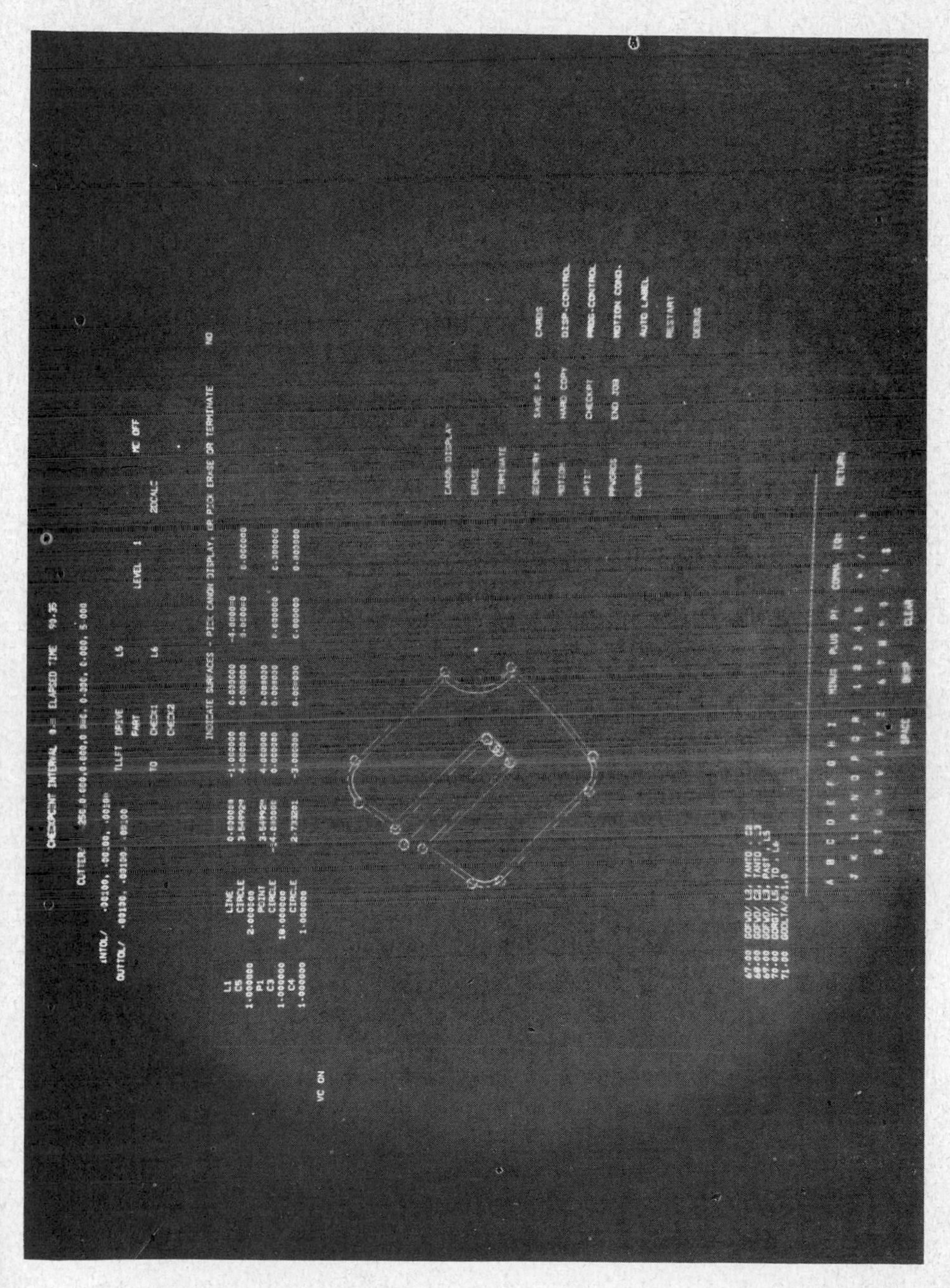

14. Canonical forms of surfaces indicated by user lightpen detect.

A THREE-DIMENSIONAL MODEL MAKING MACHINE

CHARLES A. LANG
Computer Laboratory
University of Cambridge
Cambridge, England
September 1972

Abstract: A computer controlled machine for making models in rigid plastic foam is described. It is intended to be used as a graphical output device to make models from computer stored geometries of shapes designed with the aid of a computer. Generation of tool paths makes use of standard graphics software.

1. INTRODUCTION

Computer displays and plotters have been used for many years to draw two-dimensional pictures of three-dimensional shapes. Unlike drawing by hand a variety of projections such as orthogonal, isometric or perspective may be computed and drawn with equal ease. Drawing the projected shape as a "wire frame" is the simplest computationally, but visualization may be enhanced by drawing the picture depth modulated, with hidden lines removed, or as a black and white or coloured half tone. Any of these visualizations may be drawn as a stereo pair. Any one picture however will show the shape from a single viewpoint, so difficulties may still arise in visualizing all the details of the shape. This problem of visualizing a computer stored shape motivated the construction of a computer peripheral device to make three-dimensional models. It was inspired by the use of foam models in the UNISURF (Bézier 1968) car body design system. Now that it is working we find that models give an excellent appreciation of shape, especially as they may be viewed from any direction and tilted to the light to reveal subtleties not apparent in computer drawn pictures of the same shape.

The machine was constructed primarily to make models for our Computer-Aided Design project. In our work on the design and production of mechanical engineering components we needed to be able to make models of computer stored shapes as a part of the design process, particularly for components with doubly curved surfaces. We forsee the day when industrial design offices will be equipped with a model making machine that will be used to give a better appreciation of shapes not only to their designers, but also to other designers working on associated components, and to management. On a more general tack the machine is being used to establish the usefulness of such a machine as a general purpose peripheral device. Other applications that have come our way are the cutting of surfaces

COMPUTER LANGUAGES FOR NUMERICAL CONTROL, *J. Hatvany, editor*
North-Holland Publishing Company - Amsterdam–London

represented by mathematical functions, three-dimensional histograms *[1], terrain models and busts of human heads.

2. HARDWARE

2.1 General Description

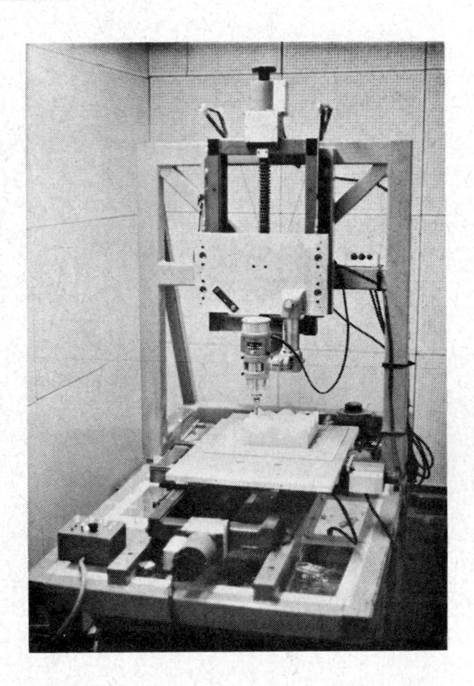

Fig 1. General view of the machine

Basically the machine is a lightweight 3-axis milling machine designed for cutting rigid plastic foam. Each axis is driven by a Slosyn HS50E*[2] stepping motor on open loop control. There are 200 steps/revolution and the tables move 1/1000th inch per step. The machine is controlled directly from a small, 8K core, PDP7 computer. Maximum table traverse is 12"×9"×9". The X and Y axes move

---

*[1] Bell Telephone Laboratories have cut histograms in plastic foam on a conventional numerically controlled metal cutting milling machine.

*[2] Manufactured by the Precise Division Rockwell Manufacturing Co.

horizontally, the Z axis vertically*. Attached to the table on the Z axis is a "power quill" which may be rotated about two axes manually. This is a high speed (max 45,000 rpm off load, 24,000 rpm on load) ½hp motor fitted with a ¼" collet. The frame is constructed from 3" square steel tubes, welded and normalized. The aluminium X-Y table has holes drilled on a 3" grid, with steel screw threads inserted. The workpiece is fixed by sticking it, or screwing it with nylon screws to a piece of standard ⅛" pegboard which has holes drilled on a ¾" grid. The pegboard is fixed to the X-Y table with four screws. Many of these features can be seen in Fig 1.

So far we have used cylindrical end and cylindrical radius end milling cutters up to ½" diameter though we plan to try other types of cutter, such as woodworking routing cutters. Cutters have been chosen that may be plunged directly into the foam, or be used to cut sideways, as this simplifies the computation of cutter paths. The machine is housed in a room with a filtered recirculating air system, and a dust sealed door with a large glass panel so that it may be observed in action.

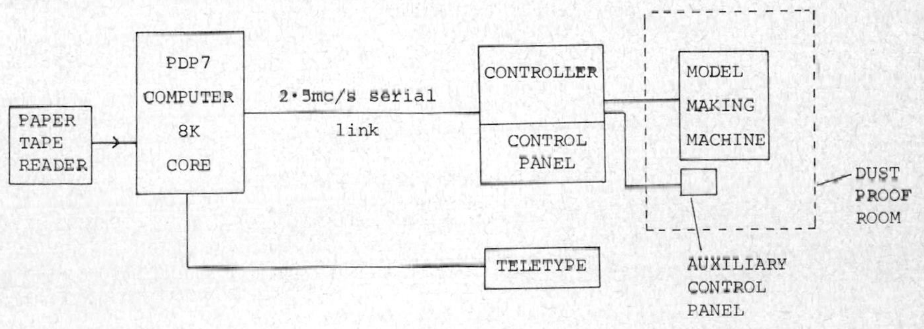

Fig 2. Block diagram of computer and model making machine

The machine is directly controlled from the computer via a simple hardwired controller.

2.2 The Hardwired Controller

The controller accepts two kinds of codewords from the computer, referred to as 'data' (which may cause table motion), and 'control'.

---

* Tables manufactured by Unimatic Engineers Ltd.

count (9 bits)	spare	1	X		Y		Z	
			step if=1	+ -	step if=1	+ -	step if=1	+ -

Data Codeword

spare (10 bits)	Disen- gage	0	Suction Motor		Cutter Motor		Enable	
			switch if=1	on off	switch if=1	on off	switch if=1	on off

Control Codeword

Fig 3. Format of data and control instructions

The data codeword specifies the axis, or axes, to be moved by one step, plus a count, which represents a delay before the next data codeword is executed. The motors are stepped before the delay. This allows interpolation, with or without acceleration of the motors, to be controlled by software in the computer. Fig 4 shows the codewords necessary to step the motors for a short increment $\Delta x=3$, $\Delta y=1$, $\Delta z=-2$.

		X	Y	Z	Codeword
	0	1			Step X, count=t_1
t	t_1			1	Step -Z, count=t_2-t_1
i	t_2	1	1		Step X,Y, count=t_3-t_2
m	t_3			1	Step -Z, count=t_4-t_3
e	t_4	1			Step X, count=0

Fig 4. Data instructions

The hardware buffers three data codewords to ensure that the stepping motors are never held up waiting for data. When a data codeword completes execution the data in the buffers is pushed down, the oldest one commences execution and the hardware signals the computer to send a new data codeword.

The control codeword permits the computer to enable or disable interrupts from the machine, turn the cutter and suction (for dust collection) motors on and off, and disengage the machine. Once disengaged, it must be engaged manually before it can be run.

Flags		Errors					Status				Spare (6 bits)
Data	Error	Logic	X | Y | Z Edges	Link Fail	Spare	Cutter Motor	Suction Motor	Enabled	Engaged		

Fig 5. Format of status information

Status information is sent back to the computer. The two flags can each cause an interrupt. One is a request for data as described above; the other signals an error. Additional status bits determine the type of error - a failure in the link to the controller, a traverse limit violation (detected by micro switches on the slides), or a logic error which is sent if a data codeword is transmitted when the buffers are full or if the buffers all become empty. The spare bits could be used to signal movement if ever the need is found to close the control loop.

The controller (less power supplies) fits into a box 4"×11"×17" the front of which is the control panel. Using this panel the tables, cutter and suction motors can be manually controlled. An auxiliary control panel is available next to the machine inside the dust-proof room.

3. SOFTWARE

3.1 Control program in the PDP7 computer

A user controls the machine by typing commands on the PDP7's teletype. These are read by a standard command interpreter (Cross 1972) used within our CAD group. After checking a command, control is passed to a subroutine to execute it. Tool paths punched onto paper tape (see section 3.2) are read into the PDP7 on command. The various facilities of the PDP7 program are most easily seen by examining the commands.

In the following description, n is a positive integer while x y z are possibly signed integers and s is a character string enclosed by two occurrences of the same delimiter (e.g. 'FRED').

Motion controlled from the TTY

MVX x	)
MVY y	) - Move given distance on one axis only
MVZ z	)
MV x y z	General 3-axis move (assumes z=0 if not given)
GOTO x y z	Move to position (x,y,z) in terms of the program's internal counters

Various machine controls

SPINDLE	ON	Start cutter motor (if under computer control)
	OFF	Stop cutter (ditto)
XYZ	SET x y z	Set internal counters
	PRINT	Print values of same ('TOOL COORDS')
FEED	SET n	Set feed rate to n steps/sec provided n is less

		than the allowed maximum (currently about 200)*[1]
	MAX	Set feed to allowed maximum
	PRINT	Print current feedrate

Execution of EIA tape produced by GINO (see section 3.2)

GINO	SCALE n	Set scale (X1 to X15).
	HEIGHT n	Set height of tool at start above GINO Z=0 (this controls the maximum depth of cut).
	CHECK	Read tape for syntax errors without motion.
	PLOT	Execute tape in 2D (X-Y). Z increments are ignored. 'Tool up' and 'Tool down' raise/lower by ·100" thus at the start the spring loaded pen (fixed in the collet) should be this amount above normal plotting height.
	CUT	Execute tape in 3D.
	RESUME s	To recover after tape error, taking s as the corrected line.
	CONTINUE	After loading a new tape, the tool path being on several tapes.

3.2 Producing Tool Path Tapes

Tool paths are presented to the PDP7 on paper tape in EIA variable-block non-interchangeable format, although with a small amount of programming the PDP7 could accept tapes in any code or format. To make the model making machine available in one of the common numerical control languages (e.g. APT), it would be necessary to write a post processor. For our purposes however, it has been convenient to regard the machine as yet another output device for our standard graphics subroutine package, GINO (Woodsford 1971). GINO, which is output device independent, has comprehensive facilities for specifying three-dimensional objects in terms of points, lines, arcs of circles and characters*[2], plus transformations (scaling, rotation, translation, projection) and windowing. The user specifies a tool path by providing a program which makes calls to GINO in accordance with data also supplied by him, thus generating a paper tape for the PDP7. If this data were in a standard form, then the combination of the program and GINO would be a

---

*[1] Acceleration of the motors is controlled from an acceleration table in the PDP7. Experiments show that the motors can be accelerated to about 400 steps/sec, but more powerful motors will be required to go faster.

*[2] Characters have not yet been implemented for the model making machine, but could be useful for engraving.

true post processor.

GINO generated EIA tapes use the following codes:

X,Y,Z	To specify incremental moves - fields are minimum width with leading zeros and trailing spaces suppressed e.g. X10Y-272Z30.
Hn	Each record is terminated by a check field Hn where n is the number of characters preceding the H e.g. X10Y-272Z30H11.
M00	Program stop (end of tool path).
M20 and M21	No cut - movements are at a preset height above the workpiece. (see GINO HEIGHT section 3.1) until the next M21. Invisible moves programmed with GINO thus cause the tool to rise above the workpiece and are then accumulated until the start point of the next visible move (indicated by M21). At that time the tool moves to the x,y of that point above the workpiece, then plunges vertically down to its z co-ordinate.
M30	End of tape (tool path continues on another tape).

The z co-ordinate must always by positive. If at any time during the generation of an EIA tape GINO detects it goes negative an error message is printed (HIT TABLE AT x,y) and moves are continued at z=0 until z becomes positive again.

3.3 Computing Tool Paths

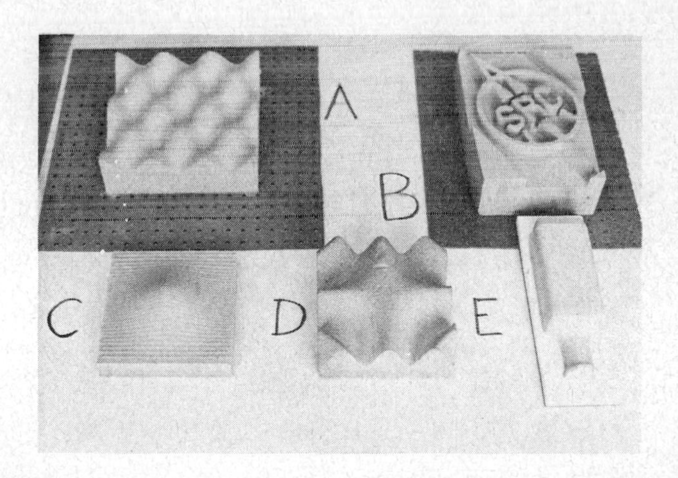

Fig 6. Models cut on the machine

GINO is used by all our programs which compute toolpaths. Its device independence is particularly useful, since before generating EIA tapes, the tool paths may be checked by viewing them on a plotter or display. The identical tool path generation program may be used, the only changes necessary being the specification of the output device, and the addition of calls to transformation subroutines to produce the desired view. Examples are shown in the discussion below.

3.3.1 B-Spline Surfaces

A new form of parametric surface, the B-spline surface (Riesenfeld 1972) is being investigated at the University of Syracuse, General Motors, this Laboratory and elsewhere. An existing B-spline surface program which generated a surface from a set of input data points, was first amended with GINO calls to generate pictures of the surfaces. Next an algorithm to compute surface normals was added. This is used to generate tool paths on an offset surface suitable for use with a cylindrical radius end cutter. Cuts are computed along lines of constant parameter. Options permit cutting to proceed in either parametric direction within any given range of parameters on the surface, and control of the parametric increment along the cutter path and the parametric spacing between successive cuts. Shapes C and E, Fig 6, are B-spline surfaces. Fig 7 shows normals along the cutter path for shape C. The complete program to generate the surface and compute the tool path is about 220 FORTRAN statements.

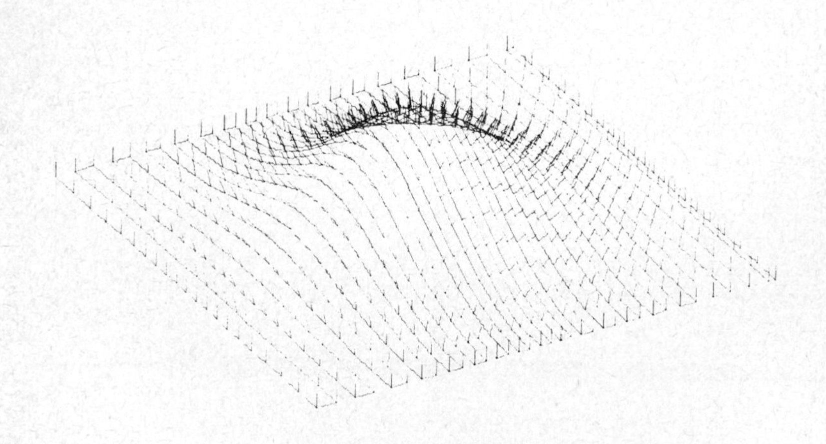

Fig 7. Normals along the tool path, shape C.

This program generates no roughing cuts. When the maximum excursion in z is greater than the greatest depth of cut permissible, then more than one pass is made with the same tool path tape, setting the z=0 plane at increasingly greater depths relative to the tool's initial position (see GINO HEIGHT section 3.1). Since the PDP7 program knows when the tool has risen to its initial height above the workpiece, it accumulates cut vectors above this height until the height is again reached, then moves directly to that point.

3.3.2 Surfaces Defined by Mathematical Functions

A FORTRAN program generates a tool path tape to cut a surface defined by an arbitrary FORTRAN function FUNC(X,Y). FORTRAN functions to give the derivatives dFUNC/dX and dFUNC/dY at a point X,Y may also be supplied, from which the surface normals and so the offset path is computed, to allow cutting with a cylindrical radius end cutter. If the latter functions are not given, the derivatives are computed numerically by taking a pair of points, one before, the other after, the point X,Y in the chosen direction. Depending upon the maximum excursion in Z and the diameter of the tool chosen the program computes a number of roughing cuts with a coarse feed between successive cuts, followed by a finishing cut with a fine feed. As in the B-spline surface program surface normals may be viewed as a check prior to cutting. Shape D in Fig 6 has a surface defined by FUNC=A+B*COS(C*X*Y).

3.3.3 Arbitrary Shapes

A very simple FORTRAN program reads in a tool path defined in terms of points, lines and arcs of circles. Moves to points, or along invisible lines are made above the workpiece, all other moves being made as specified. Circular interpolation is provided by GINO, which is used to generate the tool path tape. No offsets are computed. The tool path tape to cut the SRC badge, shape B Fig 6, was generated with this program, the data being prepared by hand.

4. NOTES ON THE MODELS

People for whom we have cut models have been enthusiastic about their quality. Models B and E, Fig 6, in particular, cut from urethane foams, have a very fine finish. Expanded PVC foam was used for A, C and D; it is less brittle than the urethane foams and so tends to have whiskers at the tops of the cusps if the spacing between successive cuts is large (e.g. equal to the tool radius). The bottle, E, has been painted with gloss polyurethane paint, giving it a very presentable appearance. None of our models have been hand finished; the lines left in the surface by the tool do not seem to detract from examination of the surface in any of the models we have cut so far.

In the future we plan to make models which cannot be cut from one direction only, but which must be remounted and cut from different directions. Further, we plan to build more complicated models from component parts cut on the model making machine.

5. CONCLUSIONS

This inexpensive machine has proved to be a worthwhile peripheral device. Although all the models could have been cut on conventional 3-axis numerically controlled machine tools it has been highly convenient to have the machine in the Computer Laboratory with similar access to it that we have to plotters. Direct control from the PDP7 has proved highly convenient as changes in control functions can be quickly made by altering the PDP7 program. Now the need is for a machine to cut larger models, at higher speeds.

6. ACKNOWLEDGEMENTS

Dr. D.J. Wheeler, P.J. Payne, P. Cross, N. Unwin, P.A. Woodsford and J. Pan-Sesar have all made major contributions to this project. L. Knapp, a summer visitor from the University of Syracuse, wrote the program described in 3.3.1, R.P. Parkins the one in 3.3.2 and Dr. A.R. Forrest the one in 3.3.3. Professor M.V. Wilkes has given constant encouragement.

7. REFERENCES

Bézier, P.E., (1968) How Renault uses numerical control for car body design and tooling, SAE 680010, Automotive Engineering Congress.

Cross, P., (1972) A family of command interpreters for Atlas 2, PDP7/9/15 and other computers. To be published as CAD Group Document No.64, University of Cambridge.

Riesenfeld, R.F., (1972) Bernstein-Bézier methods for the computer-aided design of free-form curves and surfaces. Ph.D. thesis, University of Syracuse, to appear.

Woodsford, P.A., (1971) The design and implementation of the GINO 3D graphics software package. Software - Practice and Experience, 1, No.4, p335.

COMPUTER PROGRAMMES FOR PHOTOGRAMMETRIC TECHNIQUES IN ENGINEERING

WILFRIED NAUCK
VEB Carl Zeiss JENA, 69 Jena, DDR

Abstract: An introduction into the basic principle of photogrammetric measurement is given and a system of computer programmes is described, by means of which the photogrammetric measuring data can be transformed into a form required by the engineer. Examples of application taken from propeller production and automobile body construction are discussed.

1. INTRODUCTION

Photogrammetric measuring techniques achieved great importance in topography. Numerous successful attempts were made to apply these in engineering (Lacmann 1950). It is shown in this report that by improved measuring instruments and by means of modern computer techniques photogrammetry can be used with advantage in certain computer-aided design and production processes.

2. PHOTOGRAMMETRIC MEASUREMENT ON TECHNICAL PARTS

Photogrammetric cameras which compared with normal cameras are distinguished by a low-distortion lens, by a known and stable geometry and a large format are used for photographing the object to be measured and producing object picture 1 and object picture 2 (Fig. 1), which differ from each other due to the different camera positions. Two image points (Fig. 1) with the image coordinates X_1, Z_1 as well as X_2, Z_2 correspond to each object point. From these image coordinates the space coordinates of the object points may be calculated, if the orientation of the photogrammetric cameras relative to each other is known or if it is determinated by special algorithms (Buchholtz 1960), (Finsterwalder-Hofman 1968).

Although, basically, the correlated image points can be measured separately from each other, special stereoplotting instruments are mostly used in practice with an optical system which provides a stereoscopic image of the measuring object, when the object images are viewed through an eyepiece. Additionally, there is a measuring mark visible, which allows of being displaced relative to the

COMPUTER LANGUAGES FOR NUMERICAL CONTROL, *J. Hatvany, editor*
North-Holland Publishing Company - Amsterdam–London

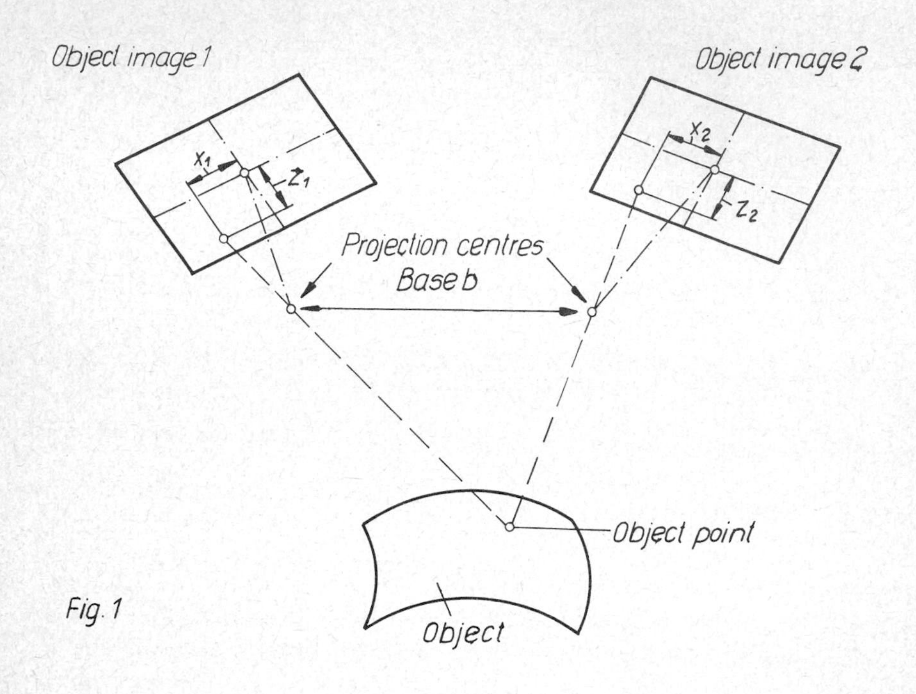

Fig. 1

stereoscopic image by operational elements. When the measuring mark has been brought into coincidence with the stereoscopic image of the object point to be measured, it is possible to record the correlated image coordinates (in the case of stereocomparators) or the space coordinates calculated by means of a mechanical on-line analogue computer (in the case of stereoscopic analogous plotting instruments).

Let us summarize the above: While in the conventional mechanical measuring technique measurements are made on the object itself, two photos are first taken of the measuring object in photogrammetry. These two pictures are put in a special stereoplotting instrument and a stereoscopic photograph is produced, on which measurements are made in a similar way as on the object itself when mechanically measured. Advantages and disadvantages of this technique are summarized in Section 5.3, Table 1.

3. DEMANDS OF THE ENGINEER ON PHOTOGRAMMETRY

Measurements of geometry of technical objects are necessary:

- For checking the geometry of parts (correction of the production process, if necessary).

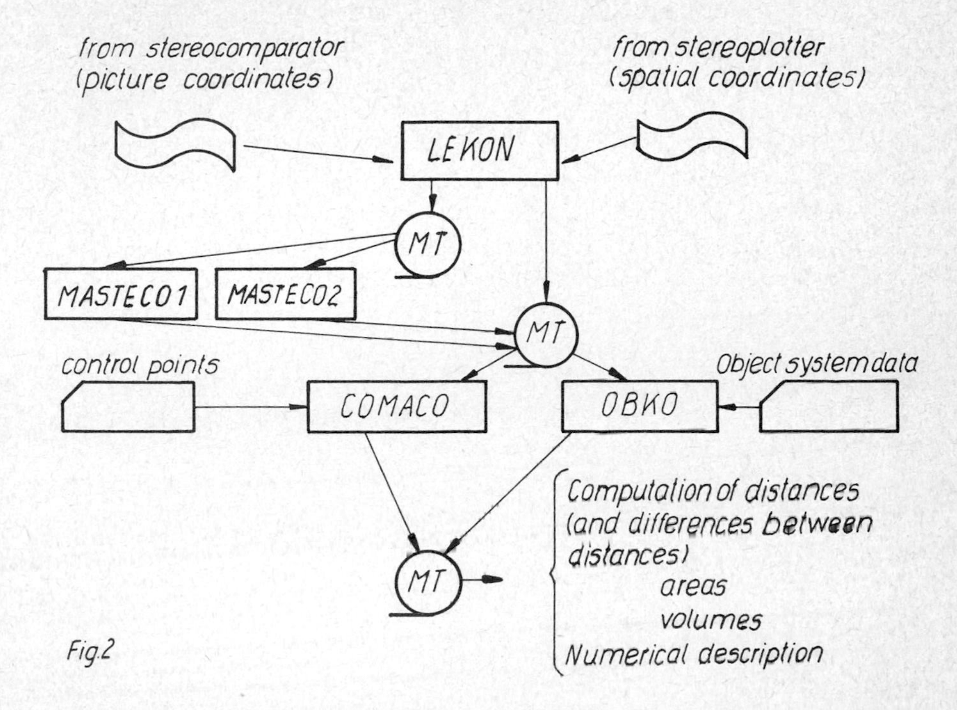

Fig.2

- For determining dimensions, when the object geometry is produced by means of empirical processes (trials, styling).
- For determining dimensions, when these have to be taken into consideration in subsequent production processes (and cannot be kept sufficiently constant).

The engineer will make the following demands on a measuring system:

- Representation of the measuring results in numerical form in an object-oriented coordinate system
- Elimination of systematic measuring errors
- Minimization of the measuring efforts.

These demands can only be met by supplementing the abovementioned instruments by a special programme system, which is to be described in the following section.

4. A PROGRAMME SYSTEM FOR THE APPLICATION OF PHOTOGRAMMETRY IN ENGINEERING

Figure 2 gives a survey of the programme system (designation

FOMAT IF). The punched tapes received from the stereocomparator (recording of the picture coordinates) or from the stereoscopic analogue plotting instrument (recording of the space coordinates) are read by the programme LEKON and their contents converted to a magnetic tape. Because of this separate programme (12 K bytes) the following programmes remain unaffected by code problems; therefore they could completely be written in FORTRAN (ICL 1900). Picture coordinates are processed by the MASTECO programmes into space coordinates; when using MASTECO 1 (32 K bytes) the photogrammetric cameras must have the following positions relative to each other during the exposure: The connecting line of the two projection centres (line b in Fig. 1) and the two camera axes (normals on the photographs at the image centres) lie in one plane; furthermore, the angle included by the camera axes must be known (frequently the cameras have devices which render possible the parallel orientation of the camera axes with high accuracy).

If the orientation of the photogrammetric cameras relative to each other is not known, it can be determined by means of the programme MASTECO 2 (44 K bytes). The basis of the algorithm used here is the fact that the rays coming from corresponding image points through the projection centres must intersect in one point (object point). By keeping the one camera constant (computationally) and moving the second the orientation that existed during the exposure may be reproduced. Subsequently the space coordinates of the object points can be calculated also in this case.

When the orientation is known, smaller errors may be expected, but it cannot always be realized.

The task to specify the space coordinates in an object-orientated system is solved by the following programmes:

1. COMACO (44 K bytes): If some (at least three) points of the measuring object in the object coordinate system are known, it is possible by comparing the known and the photogrammetrically measured coordinates of the points to calculate by means of the spatial Helmert transformation a transformation matrix as well as scale factors. Then all points are transformed into the object system.

2. OBKO (40 K bytes): In design processes, in which an empirically designed model is the basis of the construction, it is expedient to specify the object system corresponding to some regulation, e.g. in such a way that the symmetry plane is the coordinate plane or that

the coordinate system is established by three object points, whose coordinates are, however, not known. In these cases the programme OBKO allows to determine the object coordinate system according to the above given criteria in relation to the measuring coordinate system and to establish the measuring points in the object coordinate system.

5. APPLICATIONS OF PHOTOGRAMMETRIC MEASURING TECHNIQUES IN ENGINEERING

For the engineer accurate space coordinates of object points are an intermediate result, which is required for the calculation of length, areas, or volumes. Two interesting applications will be described.

5.1 PRODUCTION OF SCREW-PROPELLERS

The following aims are envisaged:

1. Improvement of the geometry of the casting mould and of the technology of the casting process to reduce the machining

2. Measurements on each propeller during the manufacturing process to assure the proper machining.

Figure 3 shows the process for the second case. The following explanations are necessary: A characteristic of the propeller production is that the casting process is not mastered so that a propeller can be milled from the casting blank according to a fixed and unvariable programme for the respective propeller type. For that reason numerically controlled milling machines for propellers (Kobayashi 1966) contain additional input possibilities, by which errors of the propeller geometry (offset in rotary and axial direction of the propeller blades) can be taken into consideration. Photogrammetric instruments, the programme system FOMAT IF and other programmes are used in the following way:

- Measurement of one side of the casting blank, calculation of the coordinates for drilling the hub from characteristic points of the propeller.
- After boring measurement of the second side of the propeller, production of a point model of this propeller
- Production of a correlated point model of the constructed pro-

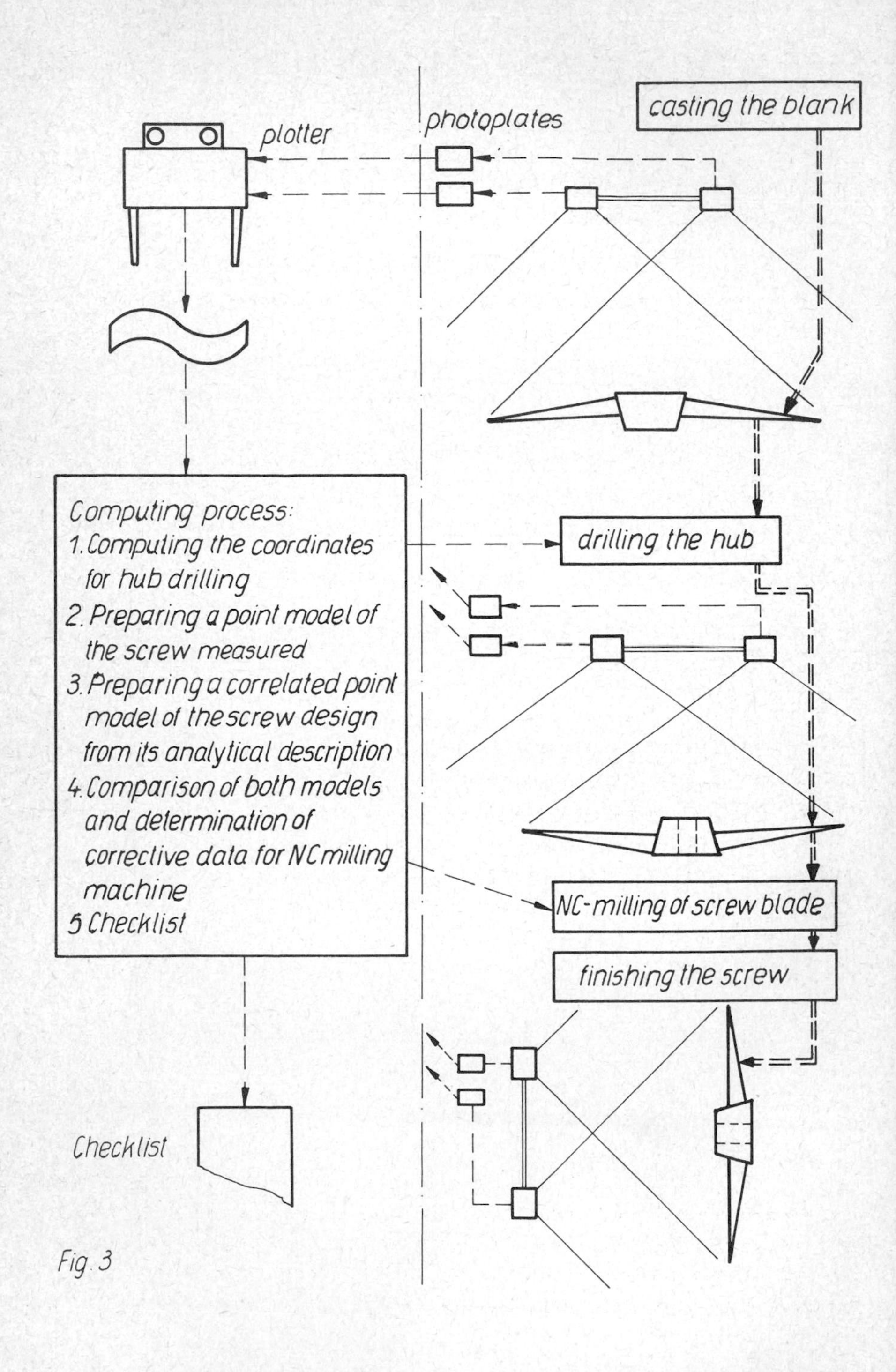
casting the blank
photoplates
plotter
Computing process:
1. Computing the coordinates for hub drilling
2. Preparing a point model of the screw measured
3. Preparing a correlated point model of the screw design from its analytical description
4. Comparison of both models and determination of corrective data for NC milling machine
5 Checklist
drilling the hub
NC-milling of screw blade
finishing the screw
Checklist

Fig. 3

peller from an analytical description of the constructed propeller
- Comparison of both models, determination of the instantaneous correction data for each blade of the propeller for the numerically controlled milling
- Final measurement as a certificate for authorities and buyers.

Summary of the advantages of the technique:

- To carry out measurements it is not necessary to transport the propeller to the measuring set-up, but the measuring equipment is brought to the place of the propeller
- A precise alignment between measuring equipment and measuring object is not necessary.
- A minimum time is required for measuring the propeller.

These advantages are counterbalanced by rather comprehensive calculating processes, which however become practicable by using computers.

5.2 APPLICATIONS IN THE DESIGN, PRODUCTION AND CONTROL OF AUTOMOBILE BODIES

The construction process in the manufacture of automobile bodies is characterized by a permanent interaction between concrete models and technical drawings. Depending upon the type of the car and the customs of the manufacturer the geometrical data are more or less determined by concrete models. For example, the manufacturer of motor-lorries has made a new driver's cabin at the scale 1 : 1. For the fabrication of splines (templates) he requires numerous coordinates of sections of different points of the cabin in selected planes. The sections are marked on the model. The coordinates of any given number of points of the cabin were determined by photogrammetric measurement and issued on punched tape in a coordinate system that was specified by the car manufacturer. The further processing was made by technologists of the automobile manufacturer.

In another case a car model was produced at a reduced scale (Fig. 5). On the engine bonnet numerous points were measured in such groups that an analytical representation of the motor bonnet by Coons' patches (Coons 1967) could be carried out. By the use of the balancing calculation measuring errors and model errors could be eliminated to a certain degree (Shu 1969). These experiments demonstrate ways and means how by using imperfect but indispensable

Fig. 4: Photogrammetric camera equipment (Type IMK 10/1318 of VEB Carl Zeiss JENA)

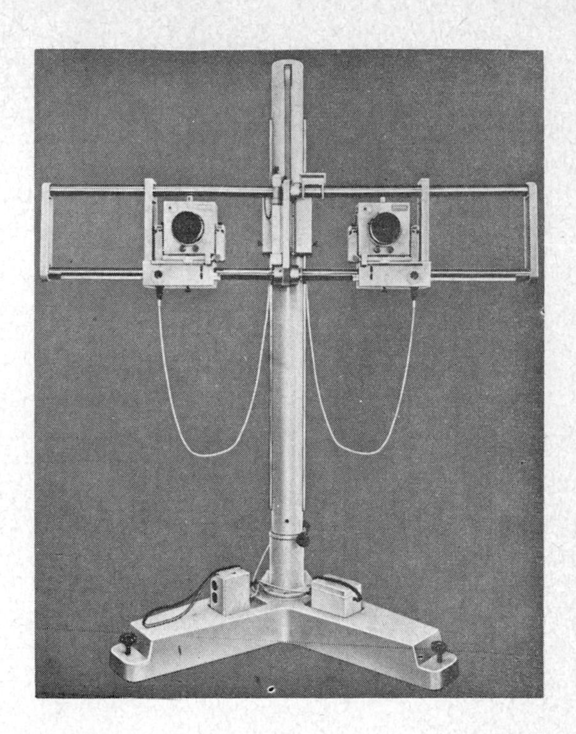

Fig. 5: Section of a copy of a photoplate with automobile model

Fig. 6: Photogrammetric plotting equipment (Type Stecometer of VEB Carl Zeiss JENA)

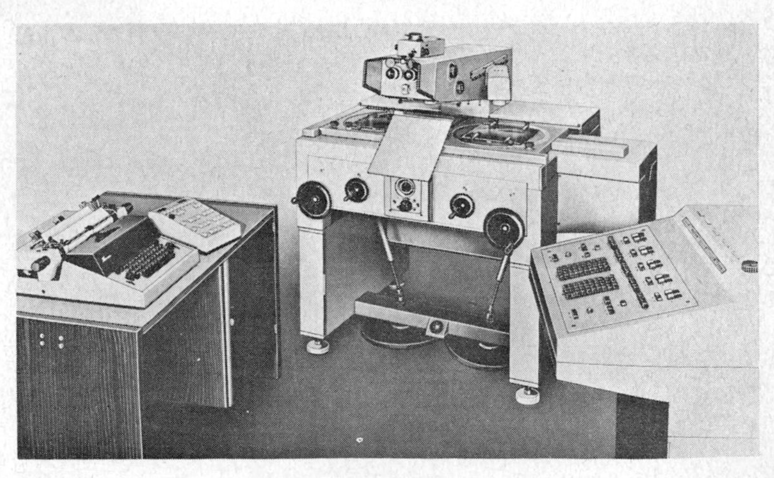

concrete models it is possible to achieve more perfect numerical models which then are the basis for producing punched control tapes (Schwegler 1971).

5.3 COMPARISON OF PHOTOGRAMMETRIC AND MECHANICAL-ELECTRONICAL MEASURING TECHNIQUES

In connection with the last mentioned applications it may be useful to deal with the relation of photogrammetric measuring techniques to mechanical-electronical measuring techniques for space coordinates. In Table 1 a side-by-side comparison is made, from which it may be seen that both techniques complement each other. On the other hand it has become clear in this contribution that numerous problems (orientation, numerical description) arise in the application of each method.

Instrument type / Parameter	Photogrammetric measuring instruments	Mechanical-electronical measuring instruments
Size of measuring object	Practically any size, for small objects poor relative accuracy	Depending upon the measuring instrument abt. 0.76 x 0.5 x 0.4 m to 5.5 x 2.5 x 2 m
Measuring accuracy	Error depending upon the camera distance, 0.2 mm minimum (for propeller measurement 1 mm is guaranteed)	Error depending upon the size of the measuring instrument 10 µm - 100 µm
Consistence of the measuring object	Optional	Solid
Transportability of the measuring object	Not necessary	Necessary
Dynamic measurement	Possible (measuring time min. 1/250 sec)	With standard equipment not possible
Degree of automation of the measurement	Low	High

Price of the instrument (referred to photogrammetric instruments)	1	2 - 4

6. CONCLUSIONS

In the technical preparation the amount of machines and instruments can be reduced, when processes of geometrical correlation are simulated on suitable numerical models in automatic computers. Photogrammetry is a means of producing such models. For their rational application a flexible software system is required. The manufacturer of modern measuring and machining equipments has to consider the possibilities of modern software in his development work in such a way that the tasks for hardware and software are optimally distributed. This requires a close and confidential cooperation with the user.

7. REFERENCES

Buchholtz, A. (1960), Photogrammetrie, VEB Verlag für Bauwesen Berlin

Coons, S.A. (1967), Surfaces for Computer-Aided Design of Space Forms, MAC-TR-41, Projekt MAC, MIT

Finsterwalder-Hofman (1968), Photogrammetrie, Walter de Gruyter und Co. Berlin

Kobayashi, K., Nozawa, R., Hanaoka, N. (1969), Numerically controlled machining of propellers, PROLAMAT-report Rome 1969, p. 152

Lacmann, O. (1950), Die Photogrammetrie in ihrer Anwendung auf nichttopographischen Gebieten, S. Hirzel Verlag Leipzig

Schwegler, H. (1971), NC-Fräsbearbeitung gekrümmter Flächen, Industrie-Anzeiger, 93, 5, p. 88

Shu, H., Hori, S., Mann, W.R., Little, R.N. (1969), The synthesis of sculptured surfaces, PROLAMAT-report Rome 1969, p. 162

DISCUSSION

The discussion centered around the papers of Davies, Cremerius and Lang.

Answering questions from Mr. U. Grupe, and Mr. Eitel, Mr. Davies pointed out that the accuracy of manipulations with the joy-stick depends on the scaling of the picture. It was sufficient for roughing cuts, while for smoothing the algorithm itself chooses the nearest point. The system was implemented on a UNIVAC 1108 computer operated in a normal time-sharing mode.

Asked to identify the pocketing routines used, Mr. Davies said there were really none - the job was done interactively, roughing by moving the cursor and finishing by giving a statement and the start- and end-point positions.

Mr. M.A. Sabin stated that the pocketing routines used by British Aircraft were compatible with GNC, and were available. He pointed out that the cutter movement representation used by Mr. Davies was an entirely novel one /offset profiles displayed and cutter shown as a point/.

Most of the questionsto Mr. J. Cremerius were concerned with the economic justification for using such a large system. In reply to Mr. Grupe, the author said that the larger the part program, the bigger the benefit of the interactive mode. Answering Mr. Guedj, he contended that if a smaller computer was used, there could have been no 3D capabilities. Response times now are less than half a second. While it is true that the screen appears to present a very large amount of information at once, the operator can choose the part of the screen on which he wants to operate and ignore the rest.

Replying to Mr. B. F. Hirsch, Mr. Cremerius said they had involved a number of practical part programmers in the development of the system, which represented a 40 man-year effort. In answer to Mr. J. Vlietstra, he defined an "intelligent tube" as a console that can carry out functions like rotation, zooming, etc. without the host computer.

Mr. C.A. Lang was engaged in lively discussion by J. Vlietstra, M. A. Sabin and B. Gott, who asked what distinguished the author's model-making machine from a production machine, and whether it could not be used for the latter purpose. Mr. Lang considered his machine to be a computer peripheral device. Manufacturing would require a far heavier, more expensive machine and raise many technological problems.

Mr. P.Bézier stated that his system could be used for both modelling and manufacturing, using the same program, but of course different hardware. In model-making, speed is more important than accuracy. In his system this was 150 mm/sec.

Asked by Mr. Sabin to comment on turn-around times, Mr. Lang said this depends not on technology, but on the complexity of the model. A model of medium complexity would take about 30 minutes to make.

3. USING NC PROGRAMS AND INTEGRATED CAD/CAM SYSTEMS

USING NC PROGRAMS AND INTEGRATED CAD/CAM SYSTEMS

B.G. TAMM
Institute of Cybernetics, Academy of Sciences,
Tallinn, Estonian SSR

Since the 1st International IFIP/IFAC PROLAMAT Conference held in Rome, numerical control has gone through a remarkable progress. It has grown out of single purpose programs for machine tools towards the integration of multifunctional systems for computer-aided design and computer-aided manufacturing. A brilliant proof to that are the contributions submitted to our session we are going to discuss.

Now we seem to have reached the point where few new programming systems are suggested for the classical metal cutting operations, because the well-known APT, EXAPT, NEL 2 C.L., SYMAP, etc. have proved sufficient for solving the principal problems of numerically controlled milling, drilling, turning and other basic cutting operations.

At present the generally accepted programming systems are being improved. On the one hand, additional subsystems are proposed, which together with the existing ones enable us to solve the problems more extensively and/or more thoroughly and, on the other hand, make it easier to adapt software for new metal cutting equipment as well as for new computer interfaces.

At the same time, however, only few attempts have been made to analyse the results obtained in order to reach some effective standardization level. The Purdue Workshop is a pleasant exception in this case. Though one must admit that for several reasons it is perhaps still too early to demand it. First, both the geography of Numerical Control /including CAD and CAM/ and the scope of the problems it deals with increase so rapidly that it is extremely difficult to view the situation from the international standpoint. Secondly, such a laborious job does not promise an immidate and direct profit for any particular organization or for any group of specialists.

If about ten years ago numerical control had access only to metal cutting - a typical process of computer-aided manufacturing, then during the last few years it has expanded in several directions. Computer-aided design is the most popular and promising branch of its development.

Several brilliant ideas have been expressed, written and imp-

COMPUTER LANGUAGES FOR NUMERICAL CONTROL, *J. Hatvany, editor*
North-Holland Publishing Company - Amsterdam–London

lemented concerning computer applications for designing different technical devices and engineering processes. Today, at our Session, we shall have a pleasant opportunity to hear about some of them.

Besides, computer-aided manufacturing is no longer limited to metal cutting, but can be used in cutting other materials as, for instance, glass and plastics, where also great accuracy is needed. This in its turn calls for inventing measuring machines and methods. And as we can see one of the possible solutions to automating the measuring process is again to use Numerical Control.

After these few introductory remarks let me now make some comments on the papers we are going to discuss later on and to put some questions in order to obtain more information from the authors, additional to that given in the contributions.

TIPS, the next smart name given to a programming system, though not Japanese, comes from Japan. The authors of an excellent paper we are going to hear about, Mr. Okino, Mr. Kakazu and Mr. Kubo of Hokkaido University, Sapporo, describe the TIPS-1 system as new software for manufacturing systems in general, including design and drawing.

The essence of the system is the serious attempt to link together the information processing of computer-aided design, drawing and manufacturing. TIPS-1 programs for design process, for calculation of area, volume, weight, centre of gravity, moment of inertia, stress and strain, for the drawing process are already available. For the manufacturing process there are NC tapes for positioning, turning and 3-dimensional milling.

The theoretical foundation for specifying cutter or pen locations, shape and surface descriptions, etc. is based on pattern description using set functions and the penalty method, both well described in this paper.

TIPS-1 programs on the system level are written in FORTRAN IV and its user-oriented language is APT-like, rather simple and mnemonic using its own set of canonical forms.

The working out of a system like TIPS-1 is no doubt a remarkable success. It is one of the latest achievements which put into practice the trend of integrating the CAD and CAM systems. The paper though otherwise very interesting, contains, however, little if any information about the size of the TIPS-1 programs, about the man-years needed to develop them and about the computer interface used for running TIPS programs. So the Chairman's request to the authors is that these items should be explained during the session report.

There are several local reasons why one should neglect the NC software available. For example, the program CADRIC presented by Mr. Jan Pejlare, which is one of the topics at our session discussion, was worked out because of a direct need for a specialized NC drilling system in one of the Swedish shipyards.

Another typical situation arises when the attempts to run an NC processor on the available hardware installation fail. This was the case with the Computer Centre of C.K.D. Prague when they discovered the NEL NC processors to be too large for handling at their ICL 1905 installation.

It is good to notice that if somebody in engineering is compelled to rework something, he always tries to do it in a somewhat creative manner. This is what specialists from the Prague Computer Centre did: in such a situation, they started to develop their own processor based on the ideas of the well-known 2C.L. During this work they divided the task of evaluation of the three-dimensional continuous path between the processor and post-processor by increasing the rolè the post-processor correspondigly. In addition several new macros were invented as well as new facilities for testing the part programs.

These original results reported by Mr. Macurek and Mr. Vencovsky make up for the somewhat didactic character of the first part of the report.

Question: In the future you certainly will find yourself in a position, where you have to run your NC software in several workshops by different computers, some of them probably larger than ICL 1905, some perceivably smaller. Can you imagine your system in those circumstances, considering the need to supply the technological parts?

In the paper entitled "Communication in Computer-Aided Design and Computer-Aided Manufacturing" Mr. Lacoste and Mr. Rothenberg from Technical University, Aachen, describe an integrated system for controlling planning, design and manufacturing in a workshop.

The project of the system consists of three main parts: workshop data acquisition, dialogue programming and input and handling of workpiece information. The authors have pointed out the great importance of supplying CAD and CAM systems with screen displays and light pen technique. This enables us to use interactive programming and provides the designer or operator with direct access to the process run.

In NC programming the software is based on EXAPT-2, which seems a natural and good solution.

The man-machine communication system of the workshop is even doubly interactive: first during the workpiece description and planning preparation, and for the second time during the run of the EXAPT programs. This makes the system rather flexible and increases its effectiveness a good deal. The experience of Aachen Technical University will surely be of considerable interest to everybody dealing with CAD and CAM systems.

"CAD of Mechanical Components with Volume Building Bricks" - that is the title of the contribution submitted by Mr. Braid and Mr. Lang from the University of Cambridge, U.K. It is a program for a designer for building the shape of a 2 1/2 dimensional or 3D mechanical component by adding or subtracting volumes. It can be operated interactively or in batch mode. It is based on a number of primitive volumes of unit size, so-called Building Bricks, and on three main operators by which these building bricks can be combined. The operators are negation, merging and intersection.

The designer equipped with a screen display is given a rather powerful and mnemonic programming apparatus for designing complicated 3D geometric parts using these bricks.

It seems that both the primitive volumes and the modifying operators are well chosen, which makes the designing process rather simple and flexible.

As a result of the work of the designer and programmer, two descriptions of the part are obtained - one is the stored data-structure and the other is a list of shape-giving commands. The authors promise to develop another program with additional information about machining for cutting the shape in hard plastic foam. A very good thing for the designer to check whether the shape is correct, and a natural way to put into practice the interactive principle in a man-machine system.

I am sure the authors can give us some new information about this program during our Session.

One of the recent applications concerning programming of NC measuring is discussed in two papers coming from the GDR. The report "Problems Concerning the Automatic Programming of NC Measuring Machines" is written by Mr. Frohberg and Mr. Hörnlein from the

"Fritz Heckert Industries", Karl-Marx-Stadt and specified as a contribution for discussion in our Session.

The division of the proposed programming language into two parts, scanning and evaluation, makes the system "off-line". Owing to this the measuring machines are not dependent on fixed movement cycles induced by the nature of a particular measuring problem,the surfaces and bores of the part can be treated separately.

It is significant that the programs for measuring on the machines with point-to-point control are designed as a part of AUTOTECH/SYMAP programming system, thus enlarging the technical possibilities of an already well known system.

To my mind a feature of vital importance is that the programs have been executed on several different series of computers, as, for instance, IBM 360, ICT, R and Minsk-series computers. This together with the more than 20 actually existing post-processors commendably underlines the trend to international cooperation, which is one of the aims of our conference.

Since the principles of the system are well enough described in the written report, I appeal to the authors to show the audience the main sources of lowering the cost of programs and the frequency of errors.

The other report devoted to NC measuring is Mr. Lotze's (from the Technical University of Dresden) "A Problem-Oriented Programming Language for the Length-Measuring Techniques" for converting point coordinates into the desired length and angular dimensions, for detecting deviations from the true shape and carrying out the instructions for the measuring machine.

The system is developed by the author in cooperation with the company of Carl Zeiss Jena. The language parmits us to write mnemonic source programs which have to be run twice. During the first run the measuring statements are generated and during the second run the measuring data are evaluated.

Since the computer-aided measuring problem is rather new and first treated at PROLAMAT, I should like to ask you for a brief comment to be able to compare your system with that of Mr. Frohberg and Mr. Hörnlein.

The paper presented by Mr. Markus and Mr. Cser of the Institute for Technology of Mechanical Engineering, Budapest, follows the trend to extend the NEL 2 C,L system, to make it handy for using in the particular conditions of Hungary, to adapt it for the corresponding equipment and ICL 1900 hardware. This is the aim of the TEVE system - one of the subjects of the paper. The other one is a program for evaluating the machining time-data. It is typical that, depending on the available machining equipment and computer hardware, different economic indices characterize the total economic effect of the process. Furthermore, some economic indices can be considered of different importance, depending on what company or country employs them.

These I consider the main reasons why our Hungarian colleagues designed a special set of programs for evaluating economic calculations. We shall hear about the usefulness of such kinds of NC software today.

"Calculation of Drilling Coordinates"-CADRIC - as was mentioned already - is the name of an NC drilling program coming from The Swedish Shipbuilders' Computing Centre, Göteborg, and presented by Mr. J. Pejlare.

The purpose for designing such a system was quite concrete, it was ordered by a shipyard for drilling tubeplates for heat exchangers. It is oriented for simultaneous drilling order with 1, 2 or 3 drills. Both a polar and an orthogonal coordinate network can be used for the description of node patterns. Inner and outer boundaries for the node pattern network can be defined. They may be of circular or polygon form.

To sum up: CADRIC is a very narrowly oriented system, I should call it target-oriented. It is plain and simple, easy to use, effective for solving the particular problem it was meant for. According to the report it reduced the time needed for preparation of the production of a typical tubeplate from 80 hours to three plus a few minutes of computer time. That seems to be an excellent result.

Mr. Morelj from "Siemens", Munich, introduces us to a new application of NC, a programming system for NC-machines used in production of electrical sub-assemblies named "PERFEKT".

A variety of NC installations is involved in the production of electrical sub-assemblies: plotters, drilling machines, "wire-wrap" machines, inserting machines, testing machines. It is important to handle the data acquisition in a right manner. Each printed circuit board has 500-1000 points and the corresponding connections. The same input data are used for artwork generation, drilling and continuity-testing. CAD programs simulate the network, define the physical details and generate the manufacturing data of the assembly, so that all the information about the product is stored in a readable form for the computer.

The program structure corresponds to the well-known scheme: input through a set of preprocessors - a processor - output through a set of post-processors. The "PERFEKT" processor performs five prinicpal tasks and takes care of the verification of the control information.

Experience proves "PERFEKT" to be fully feasible. Manpower reduction is 85% and cost reduction 50 to 70%, compared with manual programming. I hope Mr. Morelj will show us some more interesting examples he has solved with his system.

A fairly considerable number of processors for NC, CAD and CAM has been designed and implemented so far. A large amount of literature precisely describes what the systems are doing: what you have to put in, which button you have to press, and what result you get. However, in only very few publications can one find the answer to the question "How do the processors work?"

When designing a processor the system programmer has to consider certain different restrictions and conditions. They determine the whole set-up: the nature and size of the program building blocks, the intra- and interconnections, the ways of handling different data structures, etc. All this is clear to the authors of a particular system but not to bystanders. It appears to be not so simple to explain all those details to others, if only because of the deficiency of the language of the programming theory.

In the paper "On the Processor Design Problems for Specialized Programming Systems", the co-authors of several processors developed in the Soviet Union, Mr. and Mrs. Pruuden from the Institute of Cybernetics, Tallinn, present the results of their experience in determining some principles of designing problem-oriented processors.

Since up to now most workshops in the USSR were equipped with small computers, a serious restriction for the system architect was the limited size of the core memory. A good example how to overcome this difficulty is the APROKS processor for NC flame cutters. The processor is skilfully divided into sections with intermediate languages between them, in order not to overload the core memory and to minimize the number of accesses to the secondary storage.

In the second part of the contribution the structure of a self-extending integrated programming system is described. The language of each subsystem is treated by a common table-driven system processor. The process of generating new subsystems or altering the existing ones is essentially simplified by the use of a special problem-oriented language, processed by the same system processor.

Again I should like to ask the author to give some data about the size of the integrated system, the character of the subsystems and the computer hardware it is programmed for.

TIPS-1; TECHNICAL INFORMATION PROCESSING SYSTEM FOR COMPUTER-AIDED DESIGN, DRAWING AND MANUFACTURING

Norio OKINO, Yukinori KAKAZU, and Hiroshi KUBO
Faculty of Engineering, Hokkaido University,
N.13, W.8, Sapporo, Japan.

Abstract: A software system which can collectively execute CAD and CAM in a through process is reported in this paper. This system is based on the new methodologies "Pattern description by set function" and "Penalty method for pattern processing" and is intended for practical use.

1. INTRODUCTION

Information processing techniques in the field of CAD(Computer-Aided Design and Drawing) and CAM(Computer-Aided Manufacturing) have seen a great development in recent years and we have many practical programming languages for NC machining and NC drawing. These approaches, however, are only the prelude in challenging the advancing developments in this field and there are many problems which remain unsolved.

One of the challenge points is the problem to link together CAD and CAM. Since, properly speaking, desing, drawing and manufacturing have a close connection to each other, these should be treated in a through process.

The system, TIPS-1, described in this paper is designed for the purpose of lumping the information processings of design, drawing and manufacturing together.

TIPS is written in capitals from Technical Information Processing System. Version-1 of TIPS,i.e., TIPS-1 is applied to the design of the three(or two) dimensional shape of a mechanical part and its drawing and NC machining which includes layout and scheduling for production.

System planning and language design of TIPS-1 were carried out by the first author using the basic routines developed by the second and third authors. Many application routines of TIPS-1 are very actively being coded by the working group for TIPS and began steadily to yield results. The whole aspect of the technique on TIPS-1 is summarized in this paper.

2. FUNDAMENTAL FORMATION OF TIPS-1 SYSTEM

2.1, A Procedure of Total Information Processing for Computer-Aided Design, Drawing and Manufacturing.

TIPS-1 is deduced from a human designers procedure called "Cut and Try" or "Trial and Error". As shown in Fig.1, after a designer describes the imaginary shape originated in his brain into TIPS-1 language as the input data and this is fed into the computer, he calls some subroutines from the TIPS-1 library according to his demands and can confirm some characteristics of the shape.

For example, he can check whether his image is practical by selecting a subroutine for automatic drawing and can calculate the weight,moment of inertia or stress and strain destribution etc. manufacturing layout is carried out in the same process too and NC Tape , production drawing and machining specifications are produced.

This process is a kind of simulation and is repeated untill the specifications are satisfied by modifying the input data, after evaluating the results.

COMPUTER LANGUAGES FOR NUMERICAL CONTROL, *J. Hatvany, editor*
North-Holland Publishing Company - Amsterdam–London

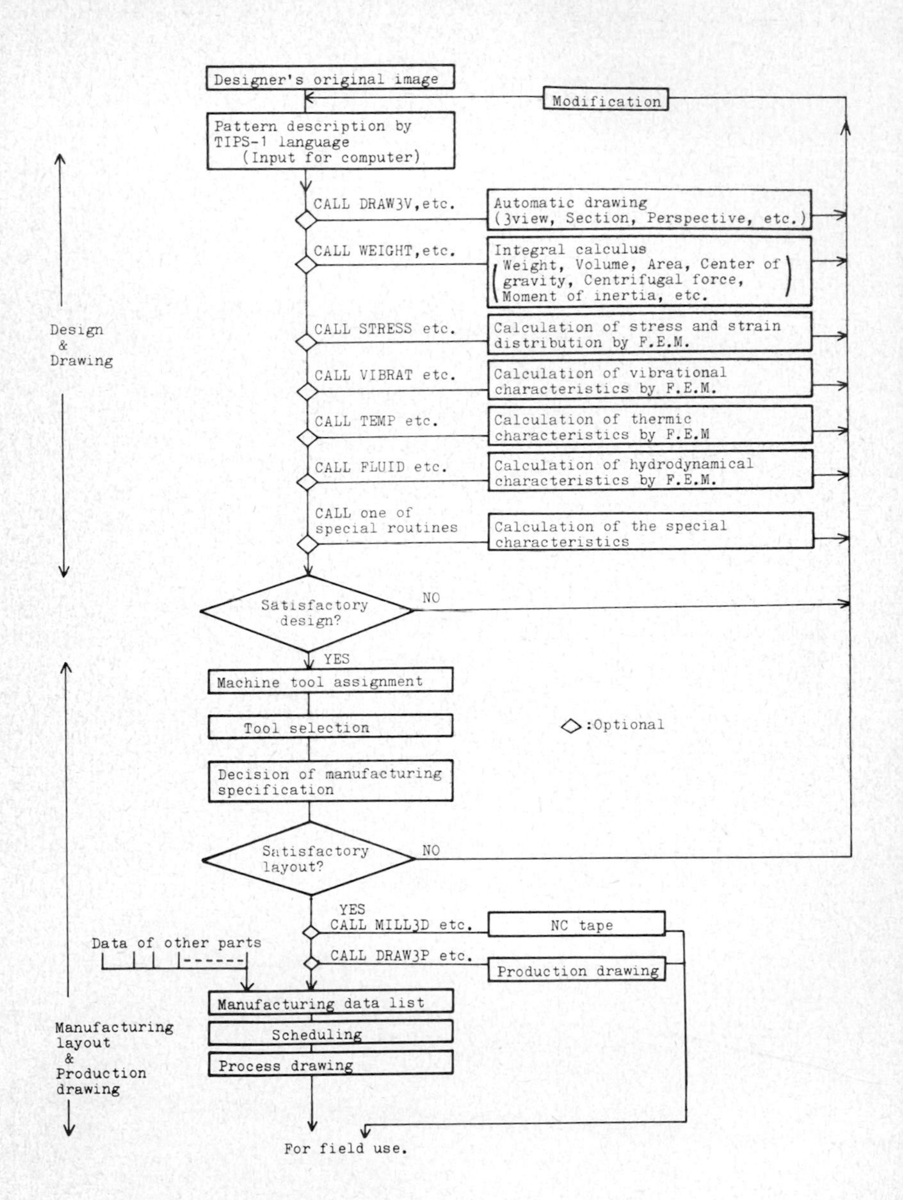

Fig. 1, Block diagram of TIPS-1

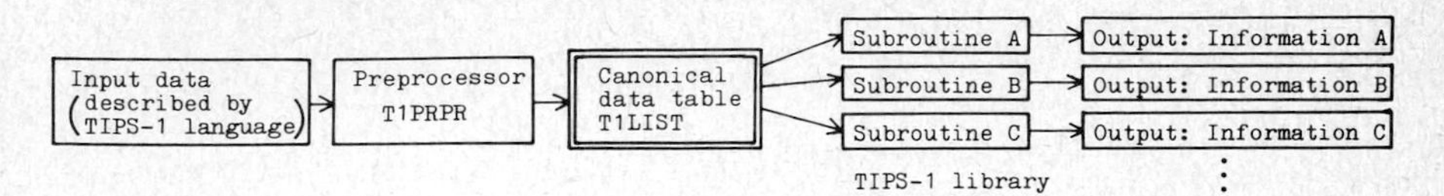

Fig.2, Structure of TIPS-1

2.2 System Structure

Computer programs of TIPS-1 are described by FORTRAN IV and can be run through usual FORTRAN O.S. without special compiling systems. As shown in Fig. 2, input data is stored in the adjustable array "T1LIST", after being put into canonical form by preprocessor T1PRPR. T1LIST is the common data table from which all subroutines shown in Fig. 1 select their necessary data.

There are three kinds of programs in TIPS-1 system, (1) main routine; consists of almost CALL statements except a few specification statements, (2) subroutines; preprocessor TIPRPR, utilities, special routines for output of necessary information and additional routines, (3) input data; described by TIPS-1 language. The movement during these routines is illustrated in Fig. 3.

3. THE DESCRIPTION METHOD OF A SHAPE

3.1 Using Set Function

In order to enable the system seen in Fig. 2 and 3 to come into existence, it is necessary to describe a shape into input data in which all information about the shape is included.

The method using the set function that is developed for TIPS-1, is one of the effective techniques which serve the purpose mentioned above. The principle is easily understandable by thinking in the following sequence.

(1) Partitioning a part into some segments.

A shape S can be expressed as the union of every set, S_1, S_2 ···, S_m as shown by an example in Fig. 4.

$$S=S_1 \cup S_2 \cup S_3 \cdots\cdots \cup S_m = \bigcup_{j=1}^{m} S_j \quad (1)$$

These sets S_j are called "Segment".

(2) A Segment is composed of several elements.

Let G_{ij} called

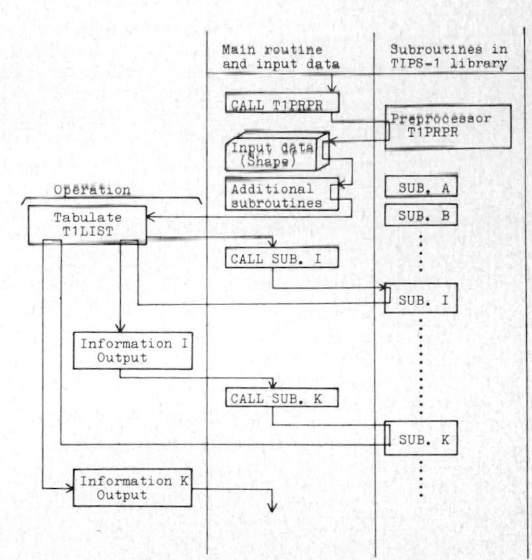

Fig.3, Process flow in TIPS-1

"Element" denotes a domain defined by an inequality equation $G_{ij}(\divideontimes)\geq 0$. A Segment S_j is expressed as the intersection of every set, G_{1j}, G_{2j}, $\cdots$, G_{nj} as seen in Fig. 5.

$$S_j = G_{1j} \cap G_{2j} \cap \cdots\cdots \cap G_{nj} = \bigcap_{i=1}^{n} G_{ij} \tag{2}$$

(3) Pattern Matrix

Substituting eq.(2) into eq.(1),

$$S = \bigcup_{j=1}^{m} \left\{ \bigcap_{i=1}^{n} G_{ij} \right\} \tag{3}$$

This is the general equation expressing a shape or a pattern by the set function. Let this form into a matrix as the following equation and name it "Pattern matrix".

$$S = \begin{pmatrix} G_{11} \\ G_{21} \\ \vdots \\ G_{n1} \end{pmatrix} \cup \begin{pmatrix} G_{12} \\ G_{22} \\ \vdots \\ G_{n2} \end{pmatrix} \cup \cdots\cdots \cup \begin{pmatrix} G_{1m} \\ G_{2m} \\ \vdots \\ G_{nm} \end{pmatrix} = \begin{pmatrix} G_{11} G_{12} \cdots\cdots G_{1m} \\ G_{21} G_{22} \cdots\cdots G_{2m} \\ \cdots\cdots\cdots\cdots\cdots \\ G_{n1} G_{n2} \cdots\cdots G_{nm} \end{pmatrix} \tag{4}$$

3.2 TIPS-1 Language

For field use, however, these formulas can not be easily used. TIPS-1 language is prepared for the convenience of practical use instead of direct expression by the formulas. Table 1 and 2 are the list of TIPS-1 language. The language adopts a fixed format according to the input format mentioned in the next section.

The canonical form shown in Table 1 consists of a few words which express simple patterns and it will be noted that the language include the word "CONTOR" that can give free patterns by accompanying individual FORTRAN programs.

The convenient form in Table 2 is APT like and is stored in T1LIST after it is converted into the canonical form by preprocessor T1PRPR. Therefore, an increase of words for special uses is possible by appending the conversion routines to the preprocessor.

Table 3 illustrates the language neatly so that the meaning of a word can be understood at a glance.

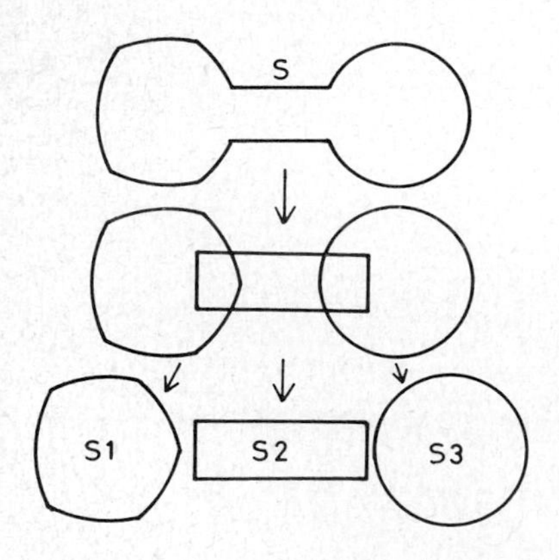

Fig.4, Partitioning S into three Segments.

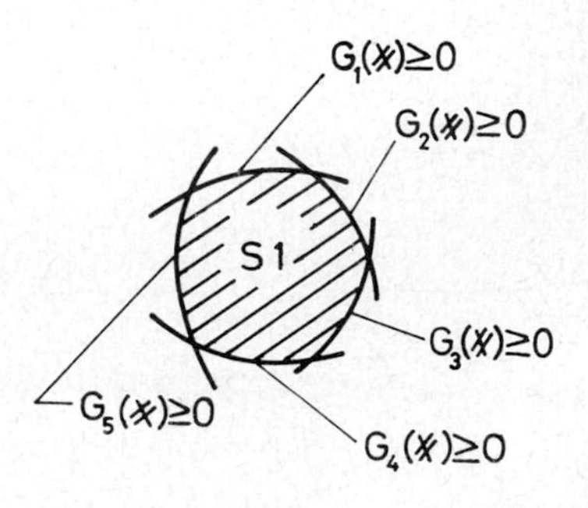

Fig.5, A Segment (Intersection of several Elements)

Table 1, Canonical form in TIPS-1 language

(The number which corresponds to that in Table 3)

i=2	3	4	5	6	7~10	11	12	13	14	15	16	17	18	19	N
POINT						x1	y1	z1							1-0
LINE						x1	y1	z1							2-0
CIRCLE		r				x1	y1	(z1)							3-0
TRIANGL						x1	y1	(z1)	x2	y2	(z2)	x3	y3	(z3)	4-0
RECTANGL		α	L1	L2		x1	y1	(z1)							5-0
ANGLE		α				x1	y1	(z1)							6-0
GANGLE						x1	y1	(z1)	x2	y2	(z2)	x3	y3	(z3)	6-2
PLANE						x1	y1	z1	x2	y2	z2	x3	y3	z3	7-0
CYLINDER		r				x1	y1	z1	x2	y2	z2				8-0
THREAD		r	p			x1	y1	(z1)							9-0
HOLE		r				x1	y1	(z1)							10-0
ELLIPS		α	a	b		x1	y1	(z1)							11-0
PARABOLA		α	a			x1	y1	(z1)							12-0
HYPERBOL		α	a	b		x1	y1	(z1)							13-0
CUBIC		α1				x1	y1	(z1)	x2	y2	(z2)	x3	y3	(z3)	14-0
SPHERE		r				x1	y1	z1							15-0
CONE		α				x1	y1	z1	x2	y2	z2				16-0
ELIPSOID		a	b	c		x1	y1	z1							17-0
ELIPARD		a	b			x1	y1	z1							18-0
HYPBOLD1		a	b	c		x1	y1	z1							19-0
HYPBOLD2		a	b	c		x1	y1	z1							20-0
ROUND		r				x1	y1	(z1)	x2	y2	(z2)	x3	y3	(z3)	21-0
CONTORNN															
CONTOR	ROTATE		S1	S2	S3	x1	y1	z1	x2	y2	z2				22-1
CONTOR	REVOLV	α	S1	S2	S3	x1	y1	(z1)							
CONTOR	MIRROR		S1	S2	S3	x1	y1	z1	x2	y2	z2				
ARROW		α	L			x1	y1	(z1)							24-1
ARROWARC		α	β	r		x1	y1	(z1)							24-2

note: The expression x1,y1,(z1)means x1,y1 or x1,z1 or y1,z1
S1,S2,---; Segment names

Table 2, Convenient form in TIPS-1 language

i=2	3	4	5	6	7	8	9	10	11~13	14~16	17~19	N
POINT	*	x1	y1	z1								1-0
POINT	APT	INTOF	L1	L2								1-1
POINT	APT	C1	ATANGL	α								1-2
POINT	APT	P1	RIGHT	TANTO	C1							1-3
POINT	MULTI	x1	y1	z1	x2	y2	z2		P3	P4	P5	
LINE	*	x1	y1	x2	y2							2-0
LINE	SEGMET	x1	y1	x2	y2							2-1
LINE	APT	P1	RIGHT	TANTO	C1							2-2
LINE	APT	RIGHT	TANTO	C1	RIGHT	TANTO	C2					2-3
LINE	APT	P1	ATANGL	α								2-4
CIRCLE	*	r	x1	y1								3-0
CIRCLE	APT	CENTER	P1	TANTO	L1							3-1
CIRCLE	APT	CENTER	P1	P2								3-2
CIRCLE	APT	XSMALL	L1	XSMALL	L2	RADIUS	r					3-3
CIRCLE	DAPT	r	YSMALL	IN	C1	IN	C2					3-4
CIRCLE	MULTI	r	x1	y1	x2	y2			P2	P3	P4	
CIRCLE	LOFTIG								P1	P2	P3	3-5
RECTANGL	*	α	L1	L2	x1	y1						5-0
ANGLE	*	α	x1	y1	(z1)							6-0
ANGLE	SEGMET	α	L1	L2	x1	y1	(z1)					6-1
PLANE	DAPT	INCLUDE	LINE	L1								7-1
PLANE	DAPT	TANTO	CYLINDER	CY1	XLARGE							7-2
PLANE	APT	P1	PARLEL	PL1								7-3
PLANE	APT	PERPTO	PL1	P1	P2							7-4
CYLINDER	DAPT	r	LINE	L1								8-1
CYLINDER	APT	P1	ax	ay	az	r						
HOLE	*	r	x1	y1	(z1)							10-0
HOLE	MULTI	r	x1	y1	x2	y2			P3	P4	P5	
ELIPS	LOFTIG	α2	α3						P1	P2	P3	11-1
PARABOLA	LOFTIG	α1							P1	P2	P3	12-1
CUBIC	LOFTIG	P1	P2	P3	P4							14-1
SPHERE	LOFTIG	P1	P2	P3	P4							15-1
ROUND	*	r	L1	L2								21-1
ARROW	*	α	L	x1	y1	(z1)						24-1

Columns 11~13, 14~16, 17~19: Numerical coordinate

Table 3, An illustration of TIPS-1 language (Correspond to N)

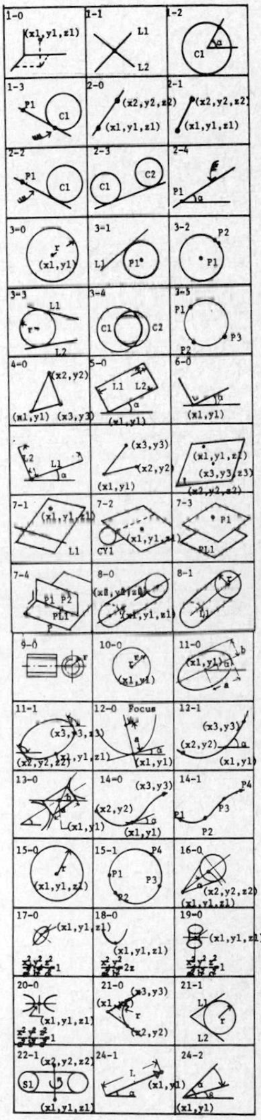

4. INPUT FORMAT IN TIPS-1 SYSTEM

There are three kinds of input cards in the TIPS-1 system, they are, Segment card, Part card and Supplement card. The chief input in TIPS-1 is the information about the shape of a part and the Segment card is used for this purpose.

The format to describe Segments of a part is based on that of T1LIST and it is moreover founded on the format of Pattern matrix. That is, a column of T1LIST corresponds with that of Pattern matrix (Fig. 6). Data for one column are given by three cards as shown in Fig. 7 and are stored alphabetically into T1LIST as seen in Fig. 6. A Segment defined by a column of T1LIST is given as the intersection of an Element of a pattern whose boundary is given by TIPS-1 language in i=1~20 and an Element of rectangular prism whose boundary is given by the lower and upper limit of x, y, z in i=21~26 (See Fig. 8).

T1LIST usually defines the union of every Segment, but if necessary, the intersection of several Segments can be defined too as shown in Fig. 9, by entering the same label into the last columns, i=27~29, of every C-card.

Part card is used for the description of part name, some comments and a blank dimension whose format is the same as that of Segment card. This is distinguished from other cards by the card key PA, PB and PC. Supplement card is utilized when a Segment needs more data about manufacturing or drawing. They are selected by the card key which is not A, B, C and PA, PB, PC, and are stored in one dimensional array, T1SUPL. The data format of this card depends on every subroutine of manufacturing or drawing.

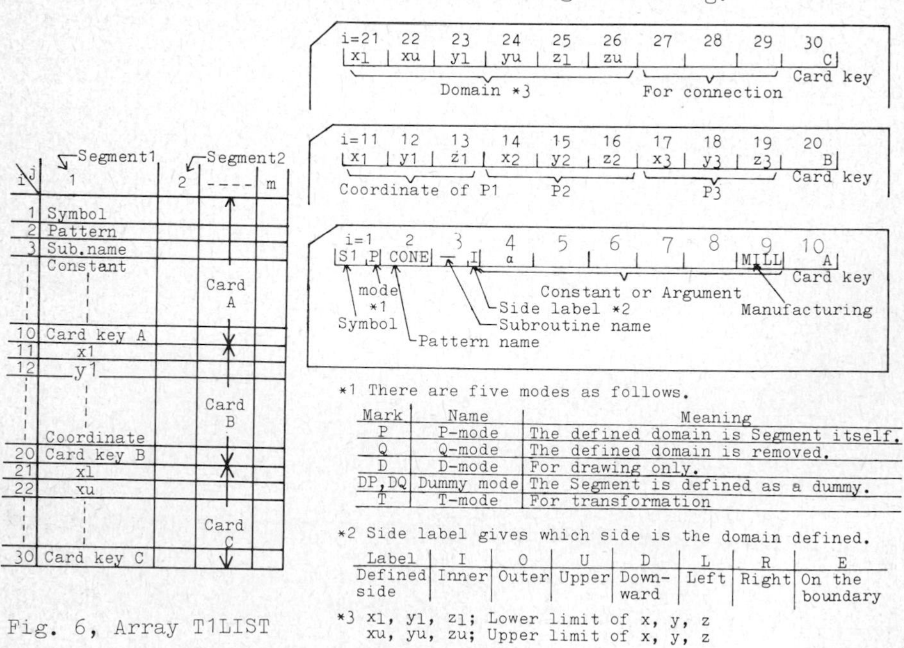

Fig. 6, Array T1LIST

*1 There are five modes as follows.

Mark	Name	Meaning
P	P-mode	The defined domain is Segment itself.
Q	Q-mode	The defined domain is removed.
D	D-mode	For drawing only.
DP, DQ	Dummy mode	The Segment is defined as a dummy.
T	T-mode	For transformation

*2 Side label gives which side is the domain defined.

Label	I	O	U	D	L	R	E
Defined side	Inner	Outer	Upper	Down-ward	Left	Right	On the boundary

*3 xl, yl, zl; Lower limit of x, y, z
xu, yu, zu; Upper limit of x, y, z

Fig. 7, Format of Segment card

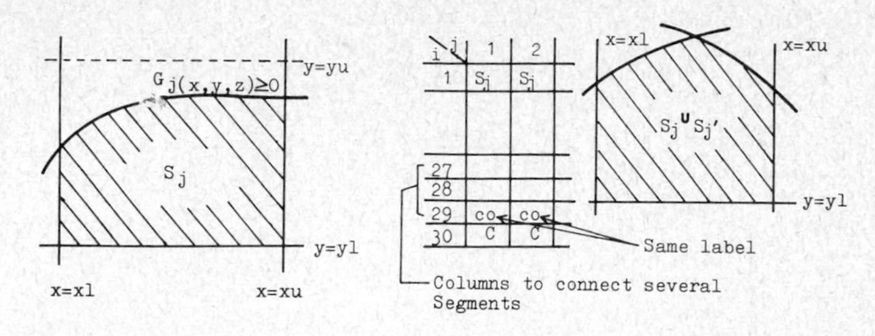

Fig.8, Typical pattern of Segment

Fig.9, Intersection of two Segments

Example; Input data examples for a test piece and a turbine blade are shown as follows; Example 1, a test piece is coded in Table 4 where two rectangles, RE1 and RE2 are defined as P mode and circle C1 and parabola PA1 are given as Q mode with MIRROR image respectively. The shape of a turbine blade in Fig. 10 is coded as Table 5 and it should be noted that this is an example of a three dimensional shape having a free surface.

5. UTILITY ROUTINES

The TIPS-1 system has several utility routines in which FPATEN and MESH3D are important. FPATEN(J, X, Y, Z, F) calculates $F=G_J(x, y, z)$ when coordinate (x, y, z) is given (see Fig. 8) and is used to find the relation between a point (x, y, z) and the domain defined by $G_J(x, y, z) \geq 0$. In general, FPATEN is used by the flowchart in Fig. 12 where job A or B is selected according to the calculation of F.

MESH3D is the routine for partitioning a shape into the three dimensional lattice mesh as seen in Fig. 11. In the computer, a block in the lattice mesh has numerical numbers, 0, -1 and Segment number J, which have one to one correspondence to inside, outside and on the boundary of the shape. We can see the result of applying CALL MESH3D to example 1 and 2 on a sheet of line printer as in Fig. 13.

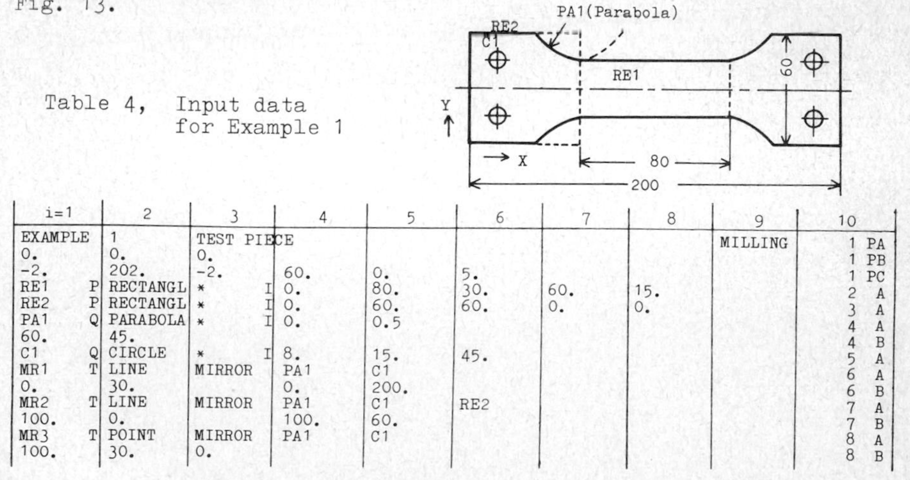

Table 4, Input data for Example 1

i=1	2	3	4	5	6	7	8	9	10
EXAMPLE	1	TEST PIECE						MILLING	1 PA
0.	0.	0.							1 PB
-2.	202.	-2.	60.	0.	5.				1 PC
RE1 P	RECTANGL	* I	0.	80.	30.	60.	15.		2 A
RE2 P	RECTANGL	* I	0.	60.	60.	0.	0.		3 A
PA1 Q	PARABOLA	* I	0.	0.5					4 A
60.	45.								4 B
C1 Q	CIRCLE	* I	8.	15.	45.				5 A
MR1 T	LINE	MIRROR	PA1	C1					6 A
0.	30.		0.	200.					6 B
MR2 T	LINE	MIRROR	PA1	C1	RE2				7 A
100.	0.		100.	60.					7 B
MR3 T	POINT	MIRROR	PA1	C1					8 A
100.	30.	0.							8 B

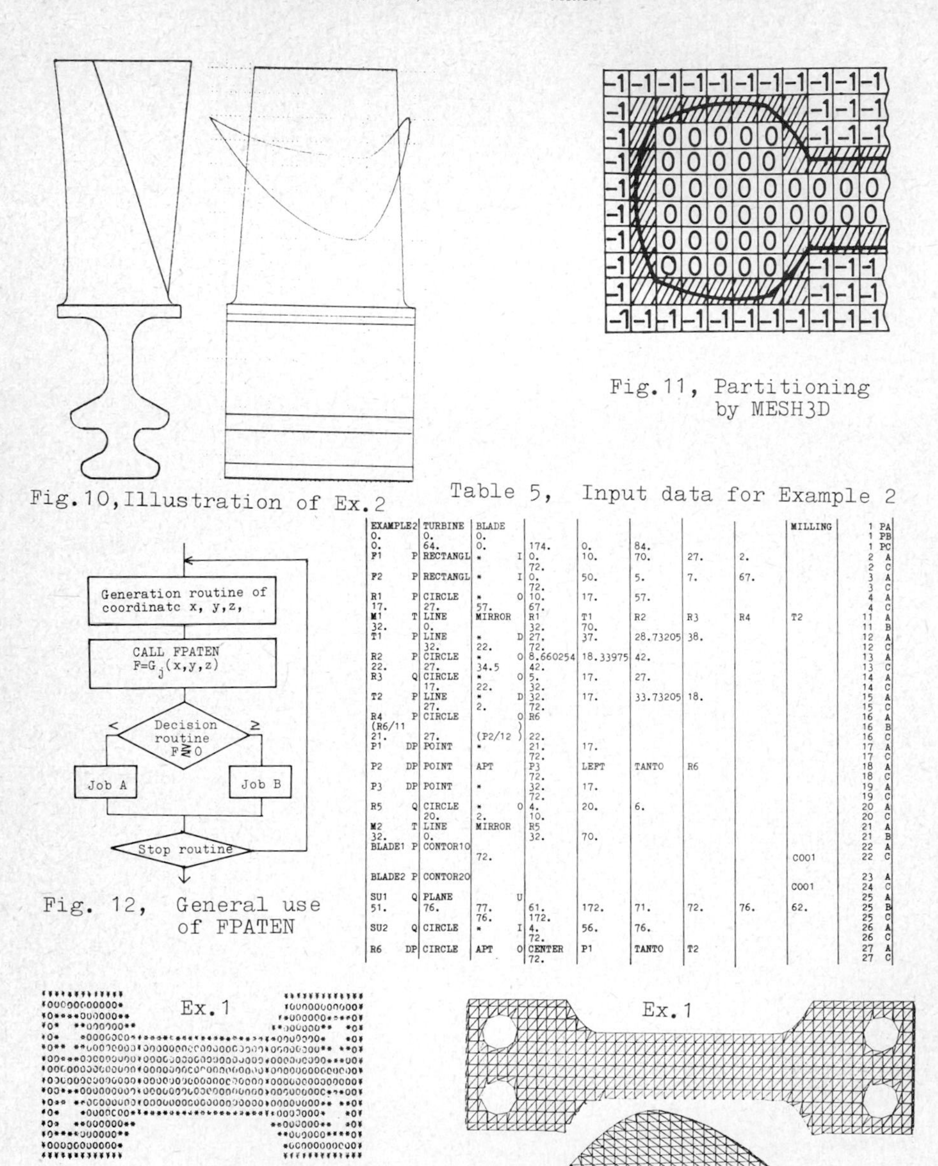

Fig.10, Illustration of Ex.2

Fig.11, Partitioning by MESH3D

Fig. 12, General use of FPATEN

Fig. 13, Applying MESH3D to Ex.1 & 2

Fig. 14, Automatic sectioning for F.E.M.

Table 5, Input data for Example 2

EXAMPLE2	TURBINE	BLADE						MILLING	1 PA
0.	0.	0.							1 PB
0.	64.	0.	174.	0.	84.				1 PC
F1 P	RECTANGL	* I	0.	10.	70.	27.	2.		2 A
			72.						2 C
F2 P	RECTANGL	* I	0.	50.	5.	7.	67.		3 A
			72.						3 C
R1 P	CIRCLE	* O	10.	17.	57.				4 A
17.	27.	57.	67.						4 C
M1 T	LINE	MIRROR	R1	T1	R2	R3	R4	T2	11 A
32.	0.		32.	70.					11 B
T1 P	LINE	* D	27.	37.	28.73205	38.			12 A
	32.	22.	72.						12 C
R2 P	CIRCLE	* O	8.660254	18.33975	42.				13 A
22.	27.	34.5	42.						13 C
R3 Q	CIRCLE	* O	5.	17.	27.				14 A
	17.	22.	32.						14 C
T2 P	LINE	* D	32.	17.	33.73205	18.			15 A
	27.	2.	72.						15 C
R4 P	CIRCLE	O	R6						16 A
(R6/11		}							16 B
21.	27.	(P2/12 }	22.						16 C
P1 DP	POINT	*	21.	17.					17 A
			72.						17 C
P2 DP	POINT	APT	P3	LEFT	TANTO	R6			18 A
			72.						18 C
P3 DP	POINT	*	32.	17.					19 A
			72.						19 C
R5 Q	CIRCLE	* O	4.	20.	6.				20 A
	20.	2.	10.						20 C
M2 T	LINE	MIRROR	R5						21 A
32.	0.		32.	70.					21 B
BLADE1 P	CONTOR10								22 A
		72.						C001	22 C
BLADE2 P	CONTOR20								23 A
								C001	24 C
SU1 Q	PLANE	U							25 A
51.	76.	77.	61.	172.	71.	72.	76.	62.	25 B
		76.	172.						25 C
SU2 Q	CIRCLE	* I	4.	56.	76.				26 A
			72.						26 C
R6 DP	CIRCLE	APT O	CENTER	P1	TANTO	T2			27 A
			72.						27 C

6. DESIGN ROUTINES

Working group for TIPS-1 develops every routine for design as seen in Fig. 1. We introduce a part of these programs completed.

6.1 Integral calculus routine for weight, volume, area, center of gravity, centrifugal force and moment of inertia, etc.

The partitioning quadrature and the Monte Carlo method are used for these calculations. As seen in Fig. 11, mesh blocks stored zero are gathered at first and then every shadowed block is apportioned by the Monte Carlo method. The routine FPATEN is combined to use the Monte Carlo method by putting the random generation routine of x, y, z into the first block in Fig. 12.

6.2 Calculation routine for stress and strain distribution.

The finite element method (F.E.M) can be applied to design problems for a part having an arbitrary shape. Since many practical programs for F.E.M have already been developed, they are adopted into TIPS-1 after linking them to the automatic sectioning routine FEM2D which makes a network of triangles. MESH3D and FPATEN are used in this routine too, that is, a mesh block stored zero is divided into two triangles and a shadowed block is approximated by one or two triangles whose vertex is on the intersection of mesh lattice and boundary of the shape. This intersection is founded by the seeking method using FPATEN. Fig. 14 is an example obtained by using FEM2D.

7. DRAWING AND MANUFACTURING ROUTINES

Many routines according to Fig. 1 for drawing and manufacturing are now being developed but in this paper we can not describe other routines except a drawing routine DRAW3V for 3 view and a manufacturing routine MILL3D for 3 dimensional NC milling because there is not enough space for describing them.

A new technique which we call the Penalty method is specially developed for DRAW3V and MILL3D. Fig. 15 is a geometrical view of the Penalty fixed around the pattern which is defined by input data. The Penalty P is given by the following equations.

$$P = \prod_{J} \left\{ \sum_{I} \left| R_{IJ}(x, y, z) \right| \right\} \tag{5}$$

where

$$R_{IJ}(x, y, z) = \begin{cases} 0 & \text{when } G_{IJ} \geq 0 \\ G_{IJ}(x, y, z) & \text{when } G_{IJ} < 0 \end{cases} \tag{6}$$

$G_{IJ}(x, y, z)$ is obtained by using FPATEN. Since eq. (5) is computed in every shadowed mesh block, the required repetition times for $\prod$ and $\sum$ are usually only a few.

The Penalty method is applied to DRAW3V and MILL3D as follows.

(1) For rapid approach; The NC commands for the first approach of cutter or pen from preset point to the boundary of the pattern are produced by the search vector of minimum seeking along the penalty surface (Fig. 16).

(2) For the final path; The path of cutter or pen along the boundary of the pattern is given after zigzag trials to find whether or not intersecting occurs between cutter (or pen) and pattern (Fig. 17). The utility FPATEN is also applied to this trial by substituting the zigzag generation routine into the first block of Fig. 12.

(3) For the collision check; Since a Penalty can be fixed around the intersection of work and cutter or other machine parts, our computer finds the existence of collision by checking the

intersection through the seeking process along the penalty surface. DRAW3V and MILL3D are produced by using the combined effects of MESH3D, FPATEN, T1LIST and the Penalty method. Fig. 10 is the result of CALL DRAW3V. Applied examples of MILL3D is shown in Fig. 17.

8. CONCLUSION

The outline of TIPS-1 has been described above. There is, of course, insufficient space for a complete description of the system but important points are as follows.

(1) A practical construction of software for a through process of design, drawing and manufacturing has been proposed.

(2) For the description of arbitrary shapes, the method using a set function has been proposed.

(3) For field use, TIPS-1 language can be easily used instead of mathematical expression.

(4) According to the system shown in Fig. 1, we can use many routines already developed which produce automatically corresponding information for design, drawing and manufacturing when input data describing arbitrary shape of a part are given. Basic routines whose name are written in this paper were successful in a field test.

The authors are also deeply indebted to the menbers of the working group of TIPS-1 for their considerable assistance with the computer programs.

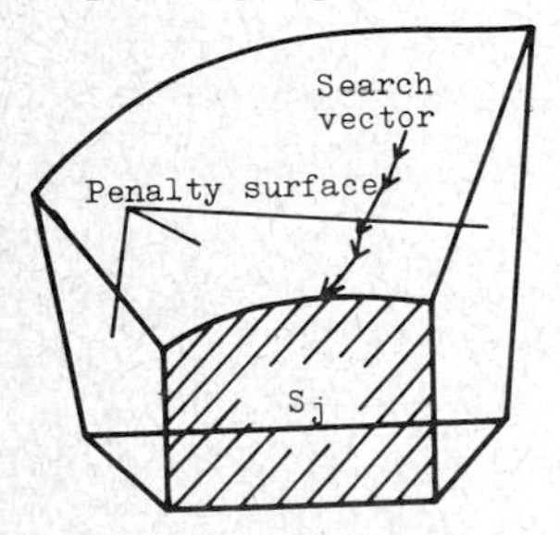

Fig.15, Geometrical view of Penalty surface.

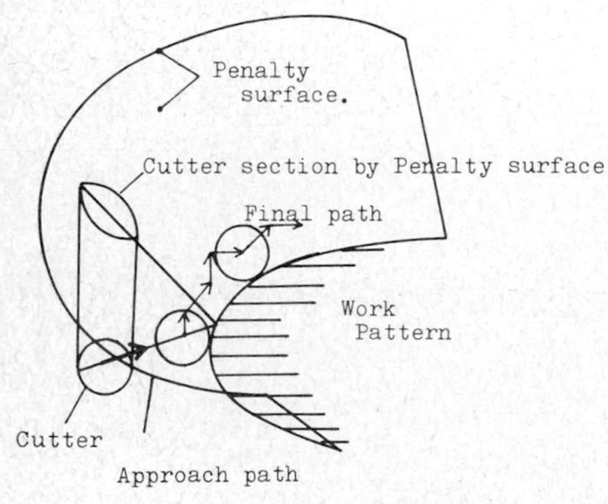

Fig16, Detection of cutter path along the Penalty surface

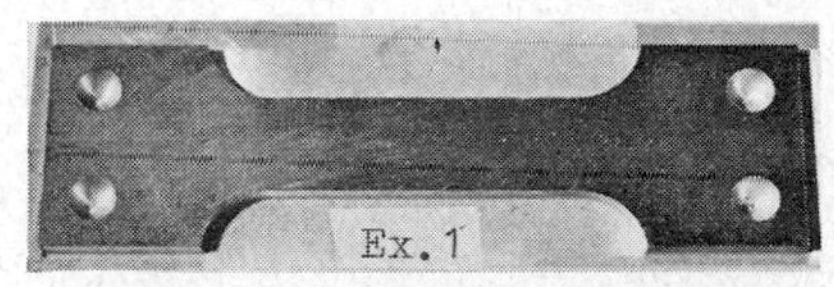

Fig. 17, NC milling of Example 1 & 2

NC PROGRAMMING IN C.K.D. PRAHA

I. MACUREK and J. VENCOVSKY
Computer Centre of C.K.D. Praha,
Czechoslovakia

Abstract: This paper deals with the application of NC in C.K.D. Praha which is one of the biggest engineering manufacturers in C.S.S.R. occupying significant place among NC users in the country. The authors have developed NC processor for ICL 1900 computers. Some of its features are presented for comparison with the NEL NC system.

1. INTRODUCTION

Engineering manufacture of C.K.D. Praha is in small-to-medium batch quantities and as such is one of the biggest in C.S.S.R. NC machine tools and the Computer Centre with ICL 1905 are now firmly established as essential part of engineering production and make the automation of many manufacturing operations economically feasible.

The majority of NC problems being solved in C.K.D. Praha belongs to the 3C or 2C,L categories on milling machines or to the 2C type of work on lathes.

The control tapes are produced mostly on the ICL 1905 configuration and rarely also on the ZUSE 25 small computer. Both the NC machine tools and their control units are of Czechoslovakian origin. An example can be seen in fig. 1. This NC lathe is manufactured by the firm SKODA Plzen and is used for production of electrical rotating machines.

Two different methodes of generating the control tapes are presented in the paper. The first is applied to cutting of complicated shapes, e.g. twisted contours of the impeller wheels or turbine blades. The second method is used for solving some of the 3C type problems often encountered in engineering production and all 2C,L and 2C problems. This method, developed in the Computer Centre of C.K.D. Praha, follows similar idea to that which is behind the

COMPUTER LANGUAGES FOR NUMERICAL CONTROL, *J. Hatvany, editor*
North-Holland Publishing Company - Amsterdam–London

Fig. 1. NC workshop for production of electrical rotating machines.

NEL NC processors.

While our processor is not so sophisticated when compared with the NEL ones, it does not require so much of core store. Moreover, it enables the user to solve great many problems of continuous path work in the three coordinate system with minimum effort. We show in the paper how a simple modification of the 2C,L NEL NC subsystem would give it the same 3C features as our processor has.

Photographs of several parts produced by application of NC based on the above mentioned methods are presented.

2. SINGLE-PURPOSE NC COMPUTER PROGRAMS

Our NC team in C.K.D. Praha started in 1967 with writing single-purpose NC programs. First we developed software for computation of the tool path and for putting the commands for controlling the NC machine onto the paper tape.

The user is supposed to write his own FORTRAN program for his particular case and by calling appropriate parts of the software he is able to solve complicated 3C problems in a short time.

The information fed into the program are the geometric dimensions of the part which is to be produced. The output is the control punched tape and the listing enabling to check the manufacturing process.

The software was written for ICL 1905 and for NC milling machines manufactured by the Czechoslovakian firm TOS KURIM.

We use this approach till today in cases where the method presented in the next paragraphs is not suitable. It gives good results but the user must have reasonable experience in computer programming. A medium experienced programmer needed about three days to write the FORTRAN master program and get the control tape for the impeller shown in fig. 2.

The single-purpose NC computer programs could not achieve widespread use by mechanical designers not skilled in digital programming. In order to overcome this snag we tried to utilize the NEL NC problem oriented languages.

Fig. 2. Machining of the mixed flow impeller.

Fig. 3. Some parts machined with the aid of the C.K.D. NC processor.

3. NC SYSTEM DEVELOPED IN C.K.D. PRAHA

After some trials we came to conclusion that our ICL 1905 instalation cannot handle the advanced versions of NEL NC processors. Because of this and the necessity to solve a number of 3C problems for which we were unable to succesfully implement the NEL NC processor we decided to develop NC system of our own.

Our system is based on the general ideas of 2C,L and 2C processors. It works with similar geometrical definitions and arithmetics and allows the utilization of arrays and index variables and looping. The motion statements include the BEVEL and ROUND feasibilities even if they are off the contour definition. The NC system itself is composed of processor and post processor as the NEL NC systems are. It produces the standard CLDATA file. Some exceptions will be discussed later.

Each of the four sections of the processor occupies less than 20K of the ICL 1900 words in the core store. The ICL 1900 version of FORTRAN has been used.

Some examples of the parts machined with the aid of the above described NC system are presented in fig. 3.

In the next paragraphs we shall deal with some features of our processor which make it different from the NEL NC.

3.1 Three dimensional continuous path

On many parts produced by milling machines we can find that the bottom of the shaped area is a cone, cylinder, sphere etc. or a portion of the basic surfaces.

Of course, the APT processor can handle these situations. But we think we have found quite simple way how to produce such parts.

We can choose from two conceptions. One of them is oriented to processor and the other to post processor. Let us take the latter, as we regard it more advantageous.

Consider post processor statements

```
                  | definition of the surface |
    SURFACE /     | SWITCH ON                 |
                  | SWITCH OFF                |

    APPROX  /  value of the approximation in the 3rd coord.
```

and define corresponding post processor records CLDATA.

To produce a three dimensional path we need to control the cutter along the two-dimensional projection of the path into the plane parallel to XY and using post processor statements we define the surface which should be shaped as the part surface.

Suppose so far, that the statement SURFACE defines one of the before mentioned basic mathematical surfaces.

The post processor then contains two general routines

(1) for calculation of the tool height Z for given values of X and Y so that the tool touches the defined part surface,
(2) for chopping the tool path, contained in the CLDATA record of the type 5000, into the sequence of elemental paths with regard to accuracy given by the statement APPROX.

Both routines can be easily programmed in FORTRAN. Routine (2) calls

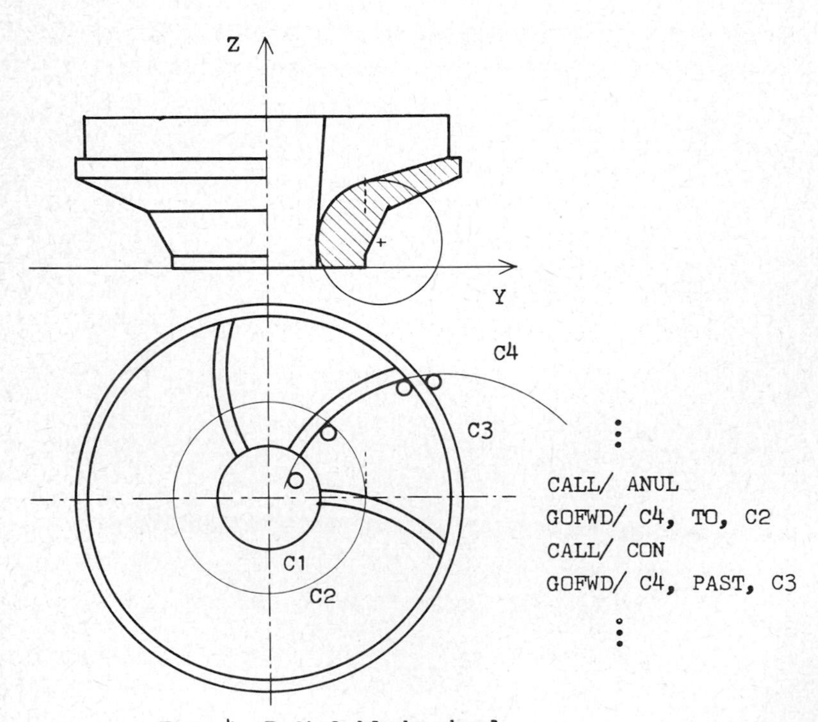

Fig. 4. Radial blade wheel.

routine (1).

Once defined, the surface can be activated in any section of the tool path using the SWITCH ON and SWITCH OFF modifiers.

The surfaces can be named if their definitions are put into the macro definitions. For example, assume the definition of the cone surface be in the macro CON and of the annular surface in the macro ANUL. In the process of milling of the radial impeller shown in fig. 4 the post processor adds the Z coord. no matter what the kind of control of the first two coordinates X and Y is.

This approach oriented to post processor has the following advantages

(a) modification of a processor is very easy and can even be avoided if we regard the definitions of surfaces and of approximation satisfactory e.g. by the statement PPFUN,

(b) the part program can be tested without the sometimes lengthy computations of the 3rd coordinate.

The more complex surfaces can be included into the system as well, but in a different way. The statement

SURFACE / EXTERNAL

transfers the control of the post processor into the subroutine EXTSURFDEF. This subroutine defines an external surface, the data, if required, could be read from the punched tape or cards placed after the FINISH statement of the part program.

The aforementioned routine for calculation of the tool height Z, when an external surface is defined, calls the function segment EXTSURFZ, which evaluates the Z coordinate of the tool so that the tool touches the external surface.

The segments EXTSURFDEF and EXTSURFZ are written by the NC system user himself for the surface which is to be produced. He inserts the segments into the post processor. In this way even machining of complicated parts, e.g. blades can be automatically controlled.

True, when an external surface is used, some FORTRAN programming is here still needed but the volume of the work required in FORTRAN has been vastly reduced.

3.2 Some features which facilitate testing of the part programs

Usually there are some logical mistakes in a newly written part program. We can discover them and test the part program more easily if the processor classifies the numerical values and canonical forms assigned to arithmetic and geometric variables and arrays as

- uncomputed
- dummy
- computed

At the very beginnig, each variable and array is uncomputed. After a successful computation of a geometric definition or of an arithmetic expression, the computed canonical form or value is obtained. When an error condition appears during the computation, then we get, as we call it, the "dummy" quantity as the result. Most often this is due to a senseless definition, e.g. when an attempt to compute the intersection point of the two previously defined lines which do not intersect is made, or a parameter in the definition is uncomputed or dummy, that is, no numerical values are assigned to it.

Let us have a look at the part program

```
        X = 20
         ...
    11) PO = POINT/ X, A
         ...
    12) P1 = POINT/ PO, DELTA, 1., 2.5
         ...
    13) GO TO/ P1
```

If parameter A has not been computed, then after processing definition 11 the variable PO is regarded as dummy. Processor prints out diagnostic of the error. This variable occuring then in the statement 12 makes it impossible to compute P1 as well as perform the motion in the statement 13. Since the dummy parameters PO and P1 are treated only as an obstacle for computing, no diagnostic is given here.

Moreover, the processor looks after every geometric variable and array to prevent overwriting their canonical forms.

When the part programmer wishes to overwrite a canonical form he has to inform the processor of his intention by using the statement

CANCEL / list

where the list is a sequence of geometric variables, arrays and array elements, successive items being separated by a comma.

The CANCEL statement is used in part program very rarely, because it is unusual to assign several canonical forms to the same variable or array.

3.3 Local names in macros

The names beginning with the letter Q we call local. They can be defined and used only inside the macro definitions. The same local names used in different macros are not related and represent different qualities.

The use of local names enables us to define complicated macros in an efficient way. These macros can often be devised in such a way that they can be used in various part programs. If this is the case, we put them into the library on the standard macro file. Fig. 5 shows an example of a simplified macro definition for drilling operation. Given quantity is the total depth D, the parameters MC and MT are the permissible depth of one cut and the maximum depth limited by the tool length, respectively.

```
DRILL = MACRO/ D
   ARIT/ QD, QI, QJ          ££ DECLARATION OF
                             ££ ARITHMETIC VARIABLES
   QD = D
   IF(D-MT)Q1,Q1,0
   QD = MT                   ££ MAX. POSSIBLE DEPTH
Q1)QI = ENTIER((QD+MC)/MC)  ££ HOW MANY TIMES ?
   QD = QD/QI                ££ HOW MUCH IN ONE CUT ?
   QJ = 1
Q2)GODLTA/(-QJ*QD)           ££ DRILLS WITH THE PREVIOUSLY
                             ££ DEFINED FEEDRATE
   RAPID
   GODLTA/(+QJ*QD)           ££ RETURNS THE TOOL
   QJ = QJ+1
   IF(QI-QJ)Q3,Q2,Q2
Q3)TERMAC
```

Fig. 5. A simplified macro definition using local variables QD, QI, QJ and labels Q1, Q2 and Q3.

There are no pattern definitions in our processor but we can achieve the same effect by using macro CALL statements. E.g. the macro CALL statement equivalent to the NEL definition

PAT1 = PATERN/ ARC, C1, 40, CCLW, 6

is

CALL/ PATARC, C1, 40, CCLW, 6, POIN

where POIN is the name of an array to which the computed points of the desired pattern are assigned.

Observe that our CALL statement differs in structure from that used in NEL NC languages.

3.4 Inlayed drive surface

Consider the tool movements in fig. 6 (a). It is not necessary for the cutter following the L1 surface to arrive up to point P and only then continue along L2. An auxiliary drive surface can be inlayed between L1 and L2 surfaces as can be seen in fig. 6 (b).

Using the inlayed drive surface results in shorter tool path

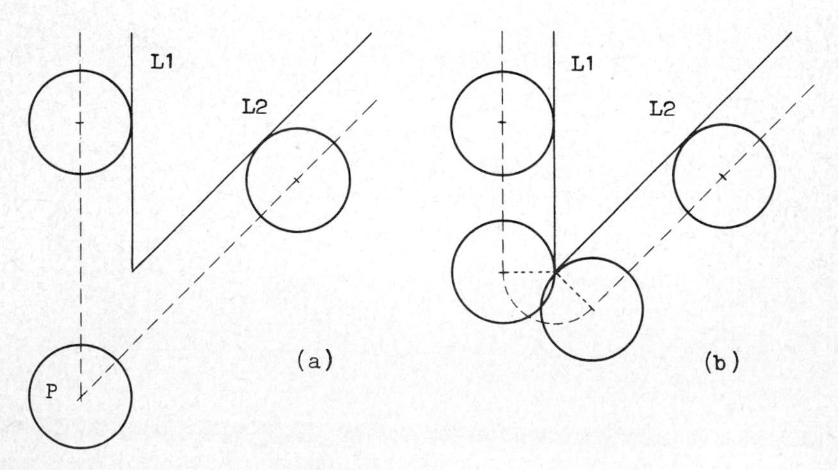

Fig. 6. Transfer of the tool from drive surface L1 to L2
(a) direct,
(b) using the inlayed auxiliary drive surface.

and a smaller manoeuvring space for the cutter is required.

Provided the drive surface used for the transfer of the tool exists, it is automatically inserted into the tool path by the processor when the tool is controlled by the statements GOFWD, GOBACK, GORGT and GOLFT.

The auxiliary drive surface is a section of a circle.

4. CONCLUSION

A new processor and post processors have been developed in the NC department of C.K.D. Praha. The NC system in conjunction with some single-purpose NC computer programs can be used for solving all the NC problems encountered in automatic production of mechanical parts in C.K.D. Praha.

The devised NC system has been successfully applied in NC production of shafts of electrical rotating machines and in a number of milling machine jobs.

In future we plan to supplement the processor with a technological section.

COMMUNICATION IN COMPUTER AIDED DESIGN AND COMPUTER AIDED MANUFACTURING

Dr.-Ing. J.-P. Lacoste and Dipl.-Ing. R. Rothenberg
Laboratorium für Werkzeugmaschinen u. Betriebslehre (WZL), TH Aachen, FRG

1. INTRODUCTION

The production process from the product planning over the design and workplanning down to manufacturing is represented by the elaboration, processing and transmission of workpiece information. While the workpiece data elaboration in the design process are unchanged over the whole production process, there come further steady instructions which are necessary for the manufacture of the workpiece. The extent of workpiece and manufacturing informations grows steadily in the progress of production-process. The workpiece information are produced essentially in the design process, the workplanning has to add the necessary manufacturing data as e.g. work sequence, cutting tools, clamping devices and cutting values. More and more programs are available for fullfilling some tasks automatically in the production process. By this the preparation of the input data in a suitable form grows up as the examination of the result-data and the correction in the given cases. The more complex the computer programs are, the more their use is full of exertions, time consuming and pertaining to mistakes.

The handling of the connected amount of information is only possible by a suitable system configuration (hardware) and by actual information systems (software). These are systems with communication facilities between man and computer.

In different fields of the production partial solutions exist already which allow the automation of some tasks, but which need for their use the manual preparation of input data. Efficient automation of production process can be achieved if the existing subsystems with their interdependence are integrated as a closed system. A complete automation of the production process is technically possible, but economically not suitable. Either the volume and the complexity of the required algorithms stand not in relation to the realisable profit or the developed system is of limited use. Hence it follows the requirements to design an integrated system in which the running process can be examined and controlled by a useful communication between man and computer. Besides information processing and handling the data preparation represents an extensive problem in an automatic planning system.

In this frame the WZL has initiated research projects which will be described briefly in this paper.
They are

- Workshop data acquisition
- Dialogue programming
- Input and handling of workpiece information.

2. WORKSHOP DATA ACQUISITION (WDA)

By collection of suitable data from the manufacture and their useful preparation, analysis, documentation and archivying will improve sensibly the existing organisation of the company. DNC systems are due to the direct-connecting of the com-

COMPUTER LANGUAGES FOR NUMERICAL CONTROL, *J. Hatvany, editor*
North-Holland Publishing Company - Amsterdam–London

puter with the manufacturing device for the workshop data collection particularly predestinated. A special advantage lies within the rapid information processing. The collected data permit the compression of the large amount of data to actual management-reports in short-periods by the computer. Thereby it results an extensive increase of the transparency of the production flow.

The principle of a workshop data collection system is represented in fig. 1.

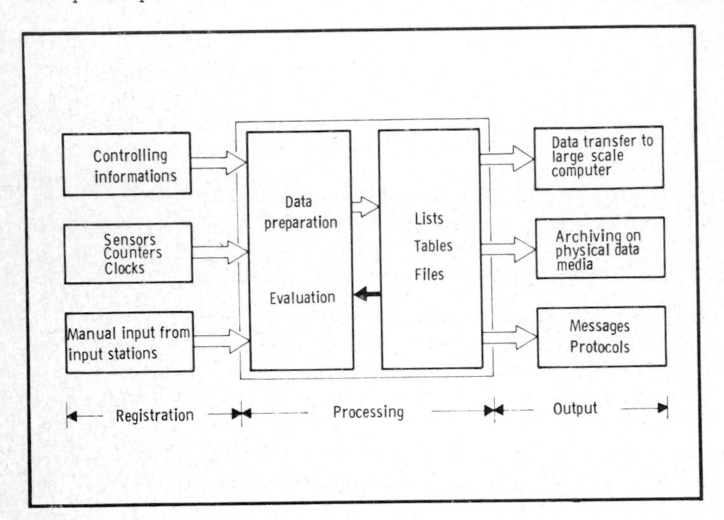

Fig. 1: Workshop data acquisition

The left side of the figure shows the source of collecting data. These data can be divided into three groups:
The first group comprises the data which are available in the programs and in the files of the DNC system or can be diverted out of these.
The second group comprises information which are entered in the computer by calling from sensors, clocks or counters.
The third group comprises such information which cannot be collected or only automatically collected after a gap of time. Therefore they must be given in by the operator by a suitable data terminal. These data terminals have to be easy in use and very reliable. Decade switches and keyboards fullfill these conditions.

The collected workshop data are prepared (i.e. they are brought in the suitable form for further evaluation, processing and then placed in suitable lists, tables and files).

The evalution of workshop data can be achieved by different aspects. The right side of fig. 1 indicates some examples. Therefore it is possible to transmit these data in compressed form for further processing on a higher production level. These data are not of interest immediately, but they are very important later on e.g. in the production-accounting. These data are to be recorded on suitable physical media e.g. magnetic tapes or punched cards.

At the WZL one WDA system is under development with a mini computer and a DNC computer. The mini computer is set as CNC-Computer and also as a data collecting center for several input terminals of workshop informations. The data will be transmitted from the controll-programs and far from the input terminals to the system. In the mini computer the data are sorted out, prepared and stored on a mass storage (disc) via data line between mini and DNC computer. The stored informations are available for later evaluations under different points of view. This system design with a preconnected satellite computer allows to discharge the DNC computer which will be used primarly by the direct numerical control. The WDA-system allows also status requests and reports over a display unit and to print out day or shift reports over line printer.

A complete WDA-System is a presupposition for an order assignement in the sense of production control. That means, that billeting of the machines and the order dependent data are transmitted from a production planning program to the DNC-computer. The DNC system has the task to transmit the information to the different machines. After message of completions a new task will be assigned to the machine, when the release conditions are checked. From a controllcenter it is possible to inquire the reason of any interruptions. When e.g. a machine breaks down the center can be asked for the reason and the duration of the interruption and also for the situation of the machine capacity. If necessary it is possible to change manually the priority of orders to be machined.

A better preservation of the delivery times, a shortening of the turn around times and better machine balancing will be obtained in this system through the dialogue facilities between man and computer, because any deviation of the predetermined workflow are immediately registered and a decision can be made with support of the most actual workshop information.

The given examples show, that the efficiency of DNC-systems can be increased considerably by the behaviour of an effective workshop data collection and evolutions.

3. DIALOGUE PROGRAMMING

The handling of data represents a steady growing problem in automatic planning systems. With the complexity of programs the amount of data and the interdependency between the data increases rapidly. Data handling will be therefore more and more tiresome, time-wasting and succeptible to mistakes. Real time data processing systems are excellently suitable for information exchange between computer and programer. Presuppositions are suitable active peripherical devices like display units, where the information can be given out and processed in graphical form. In this way the high visual perceptive facility of the human being is used during the work with electronic data processing systems.
Fig. 2 shows the task of a dialogue system in an integrated production process. The many tasks in the production process have different requests on type and amount of data. At present time nearly all subsystem data must be completed by manual input and be reentered for the next subsystem.

The manual completion and correction will be necessary in many cases. For that purpose a general dialogue system for the production process is to offer which permits a sure and rapid input, examination and change of planning data.

Input by a display unit allows to avoid all formal and - to a certain extent - logical mistakes. The more complex the planning system is the more important is the input-information without mistakes.

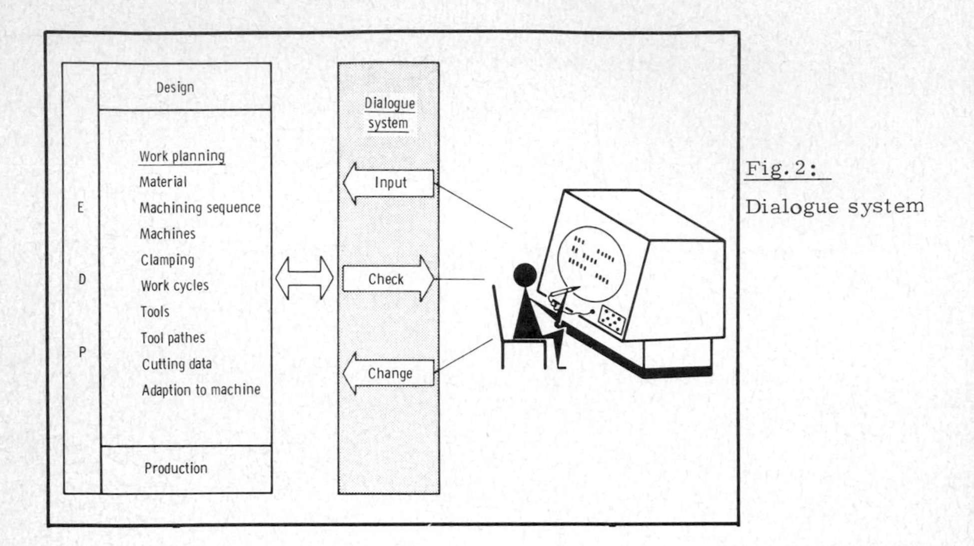

Fig. 2:
Dialogue system

With the menu technique the data input is fullfilled without any problem because the programmer is required only to choose different possibilities on the display unit.

At the WZL a menu technique processor is developed which can be used for any given planning tasks. As a first example the dialogue programming of numerically controlled machine tools with EXAPT 1 and EXAPT 2 could be realised. (Fig. 3). Working only with the workpiece drawing and the display unit, the programmer replies and completes the propositions offered by the computer over which the computer delivers the complete part program.

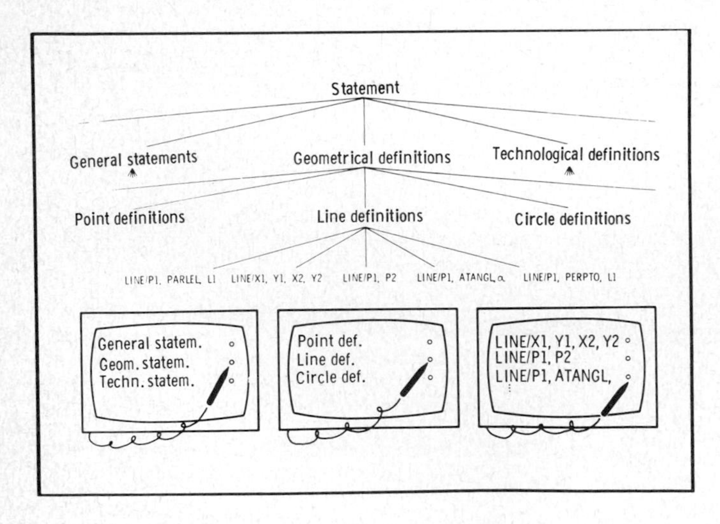

Fig. 3:
Data input by menu technique with a display

The part program represents the input-information for the NC-Processor, which is implemented on a large computer and generates the control informations for the NC-machine. Thus the connected computer configuration which is described in fig. 4 can produce on one side part programs which are transfered to the large computer. On the other side NC-processor interfaces can be transfered for examining and changing by the display unit. In this way the check of control information has become simple.

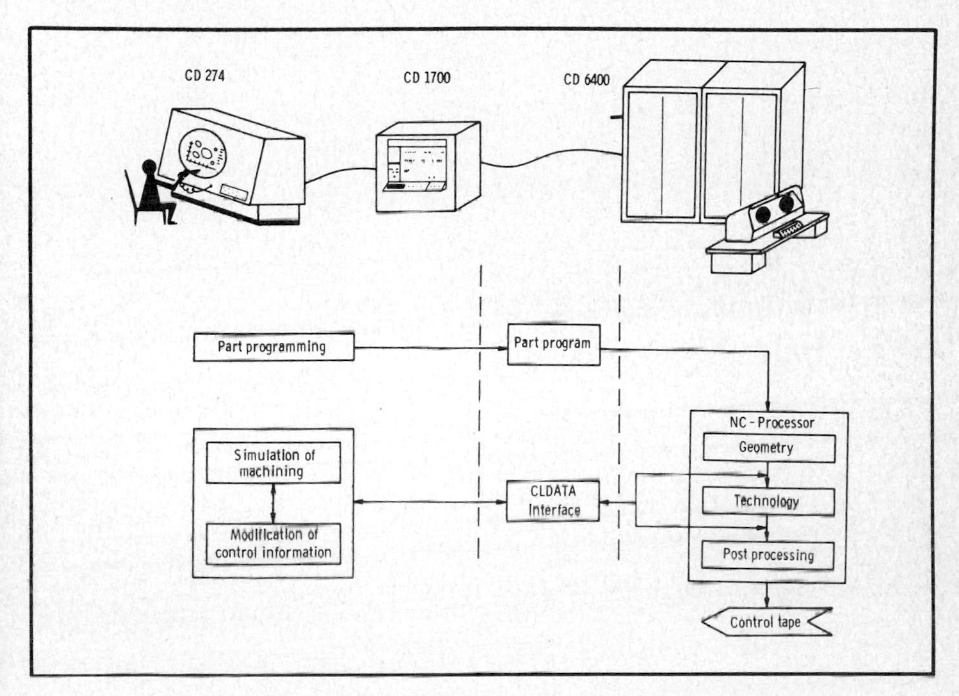

Fig. 4: Interactive processor

With the light pen the chosen display element can be identified and the corresponding workcycle can be directly influenced. Display units differ by the high output speed that can be delayed to any slower speed through a given simple command and suits particularly for simulation of tool motions. Fig. 5 shows the working cycle for drilling operation on a round plate. The lines with the arrows represent the tool motions. By choosing a drilling position or a hole with the light pen the corresponding measurements are represented in form of numerical values on the display range and changes can be made by input of new values.

The discussed dialogue system is not only designed for EXAPT system but for any input language on the basis of APT-Syntax and CLDATA interfaces relating to the ISO first draft proposal TC 97/SC5/WG1/139.

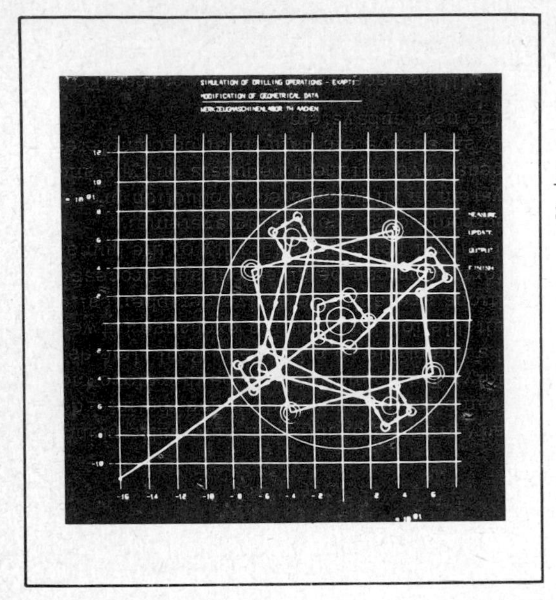

Fig. 5:

Simulation of tool motion

4. INPUT AND HANDLING OF WORKPIECE DATA

As another main point the problem of computer-oriented workpiece description shall be discussed. The meaning of this task for the realisation of an integrated planning system indicates fig. 6. Infact the amount of input information decreases by increasing automation, because the previous automatically elaborated informations are available in the computer for the next program. But at any level the problem of a computer oriented workpiece description and representation arise. An uniform and for every steps suitable description represents an extensive integration factor. Besides the uniform storage of workpiece information it allows to check and correct at any level of the automated production process.

In the WZL such a system is developed, which stores the geometrical, technological and organisational data of workpieces which are composed of planes and cylindrical surfaces. For testing its performance the system was implemented on an active display system. Investigations at the WZL have shown, that the described system - represented in fig. 7 with input-output over an active display unit - is particularly suitable for description and handling of workpiece data. Standardized form-elements are offered over the display unit, which are assembled by light pen, function- and alphanumerical keyboard. The accuracy of the position will be achieved by input of numerical values. In such a system it is necessary for practical use,that the standardized form-elements will be generated by the user in order to keep their amount small. By the introduction of the system a catalogue of standardized tool elements will be generated according to the given workpiece-spectrum and given to the designer for his workpiece description. Rationalisation efforts will be achieved on that way, because the designer is enabled to select standardized form elements only. Possibilities of the described system are shown in fig. 8.

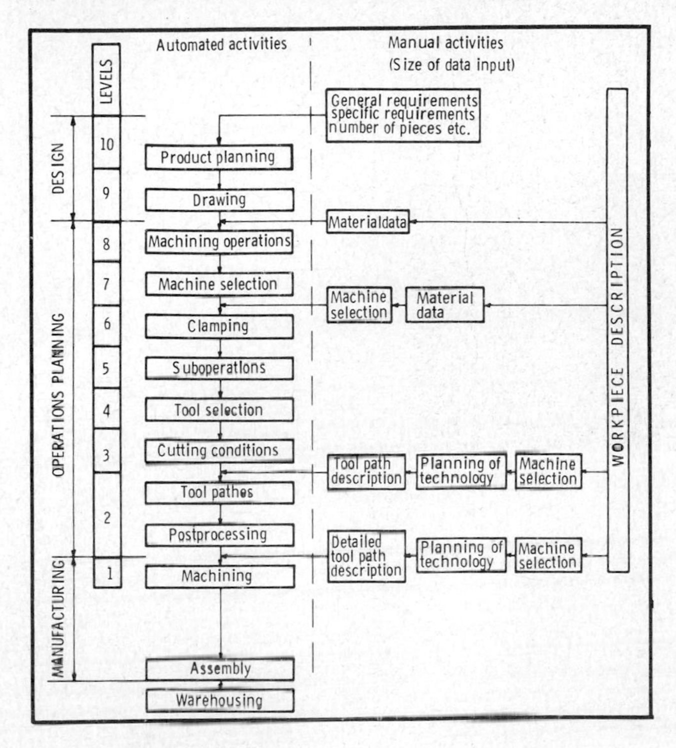

Fig. 6:

Importance of a workpiece description for the production process

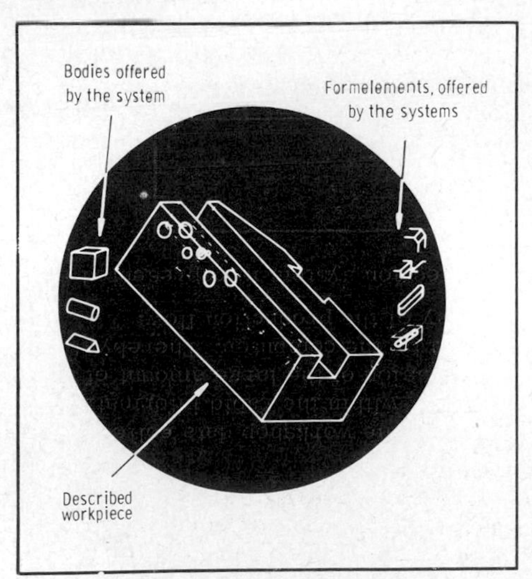

Fig. 7:

Workpiece description with standardized workpiece elements

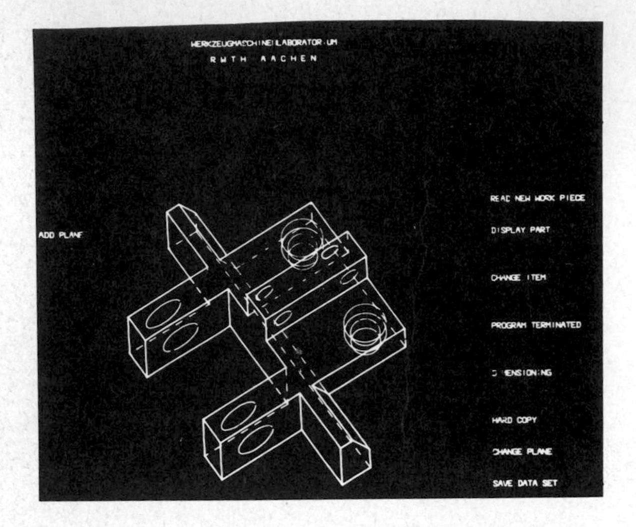

Fig. 8

Example of a Workpiece displayed on the CRT

The system for workpiece description was coupled to the available system EXAPT 2 (fig. 9). From the stored information the required lists of contour elements for EXAPT 2 will be prepared automatically and handed to EXAPT 2. The already described program (3.) for data input, check and change permits to put in every missing data for EXAPT 2-required instructions for technology and organisation - by the display unit, to examine intermediate results of NC-processor and change up to a error free control punch tape.

5. SUMMARY

A fully automated system is not within reality. Since available data are mostly incomplete, inexact and not sufficiently actual. Data omissions are to be closed by human manipulation by means of suitable communication media between man and computer. Interactive systems with graphical display are ideally appropriated for that purpose and allow the inclusion of man in the information process on all levels.

A communication system between man and computer enables data input, process control and necessary corrections of data.

Besides the efforts to improve communication in design and production planning process and considerable progress is made in the computer aided manufacturing on one hand and workshop data collection on the other by means of appropriated data links.

Suitable communication capabilities between man and machine provide the neccessary feed backs between workshop floor and planning for an efficient computer aided manufacturing, scheduling, balancing as well as management reporting and represent an important presupposition on the way to the automatic factory.

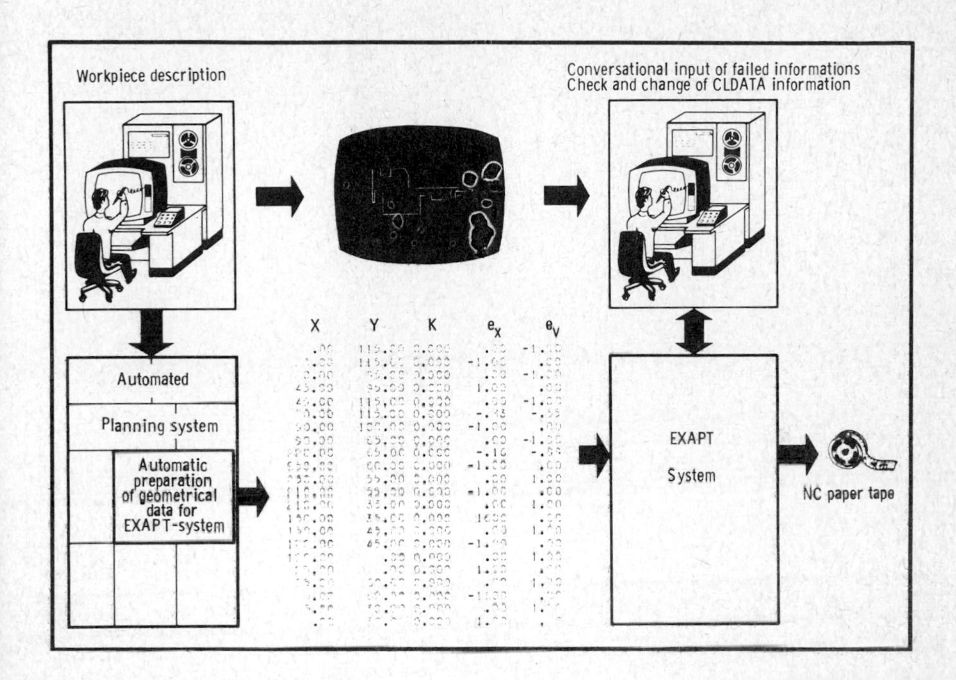

Fig. 9: Coupling of workpiece description and EXAPT 2

REFERENCES

Bäck, U., (1972), Not published reports, WZL, TH Aachen

Baum, M., (1972), Bildschirmunterstützte Informationshandhabung bei automatisierten Planungssystemen, Dr.-Ing.-thesis, TH Aachen

Lacoste, J.-P., (1972), Rechnerangepaßte Werkstückdarstellung für den automatisierten Produktionsprozeß, Dr.-Ing.-thesis, TH Aachen

Rothenberg, R., (1972), Not published reports, WZL, TH Aachen

COMPUTER-AIDED DESIGN OF MECHANICAL COMPONENTS WITH VOLUME BUILDING BRICKS

I.C. BRAID and C.A. LANG
Computer Laboratory
University of Cambridge
Cambridge, England
September 1972

Abstract: The paper describes a program which a designer can use to build up the shape of a 2½D or 3D mechanical component by adding or subtracting volumes. The program will plot or display line drawings of the component. It also generates a stored description of the component's shape, from which geometric information for manufacture of the component by NC can be derived.

1. INTRODUCTION

The customary representation of shape by means of drawings showing orthogonal views of engineering components has considerable disadvantages for NC. Drawings cannot be read direct by computer, and even when digitized, require human interpreters (part programmers) to deduce what shape is depicted. Additional machining information is of course necessary for the generation of NC tapes but the bulk of the input to a part program is geometric data. If this data is supplied direct from a stored, three dimensional description of shape, the part programmer's task is much reduced.

We have written a program which can be used to design the shape of a mechanical component described in terms of volumes. The program produces a stored, three dimensional description of the shape sufficient to define the geometry for the production of NC tapes.

To introduce the program an imaginary design sequence is first presented. Later, the mathematical basis of the shape representation, the algorithms employed in the program, and its linkage to NC are discussed.

2. DESIGNING A COMPONENT

The designer sits before a teletypewriter and a display, both connected to a computer. We assume for the moment that he is designing a new component rather than modifying an existing design. He has a repertoire of six built-in primitive volume building bricks (*primitives*) as shown in Fig 1. To begin he gives the coordinates of a point from which all components are to be viewed, by typing PROJECT 10 15 20 for the view point (10,15,20). He then calls up a primitive cube and gives it a name such as BLOCK. The cube is positioned at the origin with its

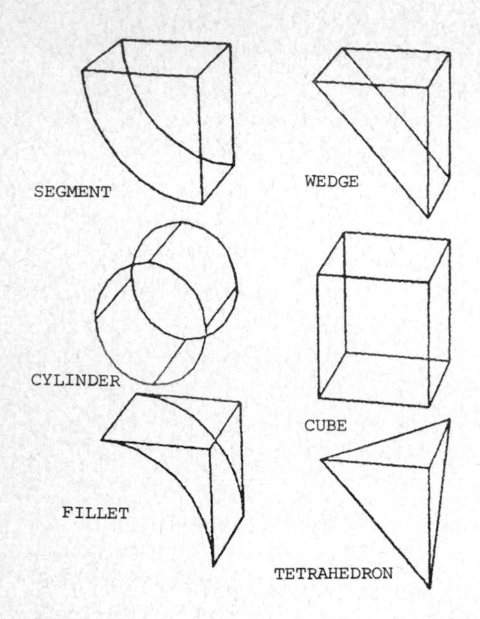

Fig 1. Six built-in primitives

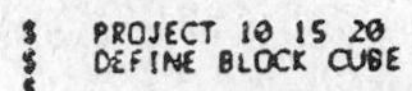

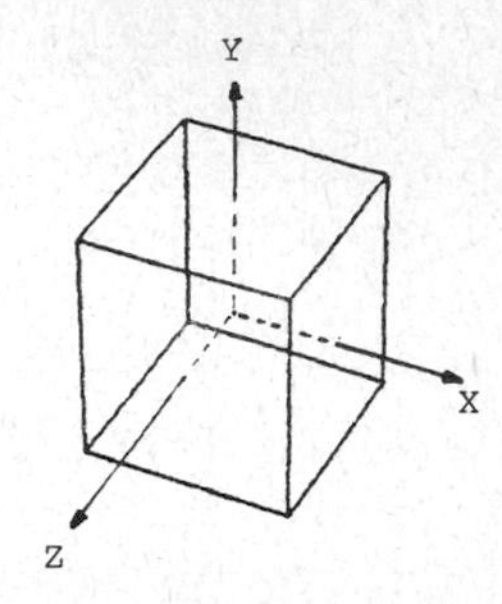

Fig 2. Primitive cube

vertices at (±1,±1,±1) and is displayed as shown in Fig 2 (where axes and coordinates have been drawn). Using the MODIFY command, he moves the cube by -1 along the y-axis and scales it by a factor of 2 in the x direction and 1·5 in the z direction. He then defines two more cubes and modifies them so that the three volumes are arranged as in Fig 3. (B2 is a *copy* of B1 - see below)

He now types the command COMBINE. The volumes are stuck together at the faces where they touch and all traces of the joins are removed (Fig 4).

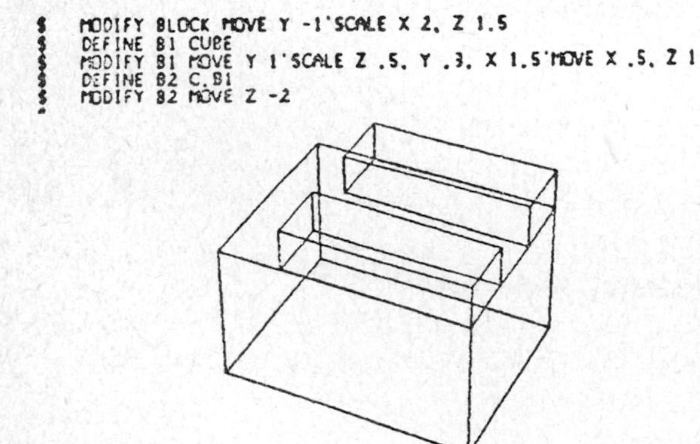

$ MODIFY BLOCK MOVE Y -1'SCALE X 2, Z 1.5
$ DEFINE B1 CUBE
$ MODIFY B1 MOVE Y 1'SCALE Z .5, Y .3, X 1.5'MOVE X .5, Z 1
$ DEFINE B2 C.B1
$ MODIFY B2 MOVE Z -2

Fig 3. Three modified cubes

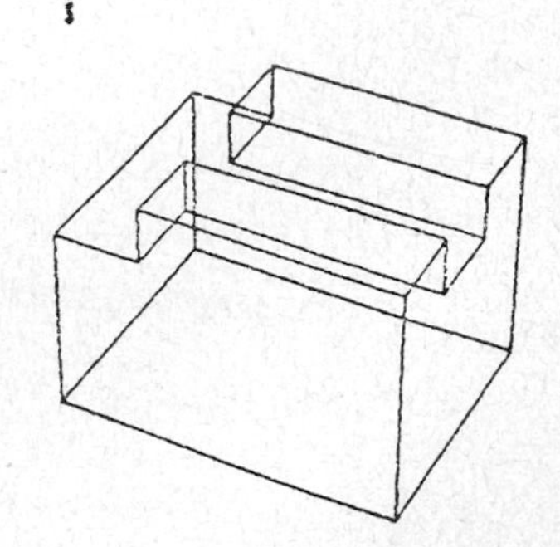

Fig 4. An object

The single, non-primitive volume which results is termed an *object* and is given the name of the oldest constituent volume, in this case BLOCK. B1 and B2 cease to exist.

Next a hole of radius ·25 is to be drilled vertically through the centre of the object. A primitive cylinder is defined and called C1. Initially its axis lies along the z-axis, its radius is 1 and its back and front faces are at $z=\pm1$. Using the MODIFY command it is first scaled to a radius of ·25, lengthened by a factor of 2·5 and then rotated by 90° about the x-axis (Fig 5). Note that although the cylinder passes through the object, the program does not at this stage detect the interference. Before it can be used to drill a hole, the cylinder must be made negative by the command NEGATE C1. This reverses the sense of the unit normals to each face, i.e. all the faces of the cylinder face *inwards*. The command INTERSECT C1 BLOCK makes the hole, and BLOCK appears as shown in Fig 6.

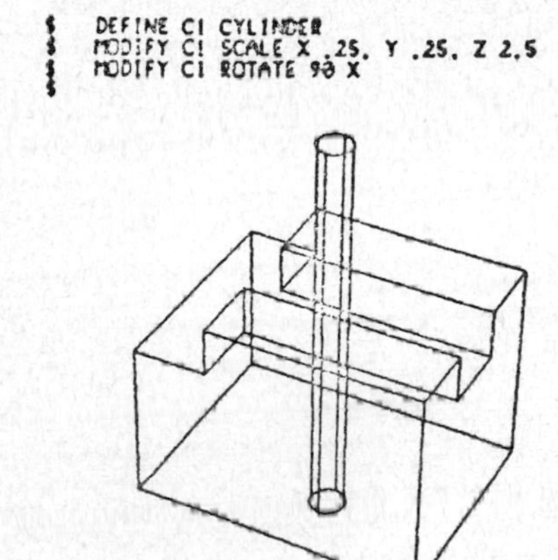

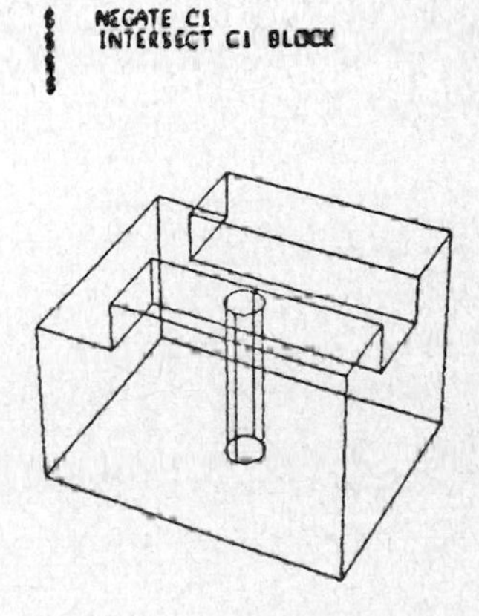

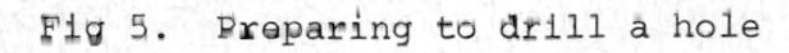
Fig 5. Preparing to drill a hole

Fig 6. After drilling

Intersections can be made with an arbitrary object and a negative cylinder or cube. Figs. 7 and 8 show BLOCK before and after an intersection with a modified negative cube. In Fig 7 the view point has been altered twice to show two orthogonal views. Fig 8 shows the object with only those faces which face towards the view point (now at (10,15,35)) displayed. Silhouette lines are added to curved faces which are partly visible. This operation is rapid and is termed local hidden line removal (hence the command LHL). Full hidden line removal and shaded pictures can be computed by further processing. The command SAVE places a copy of the complete object in a file on disc. To retrieve the object, the user would type GET CAD/ICB/FILE BLOCK.

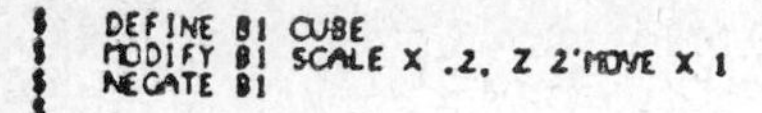

$ INTERSECT B1 BLOCK
$ SAVE CAD/ICB/FILE BLOCK
$ LHL BLOCK
$

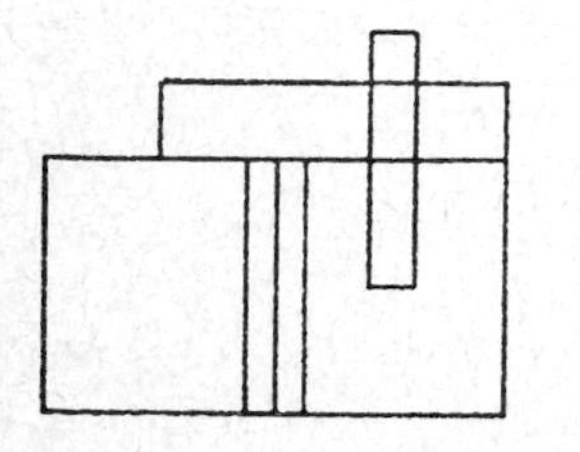

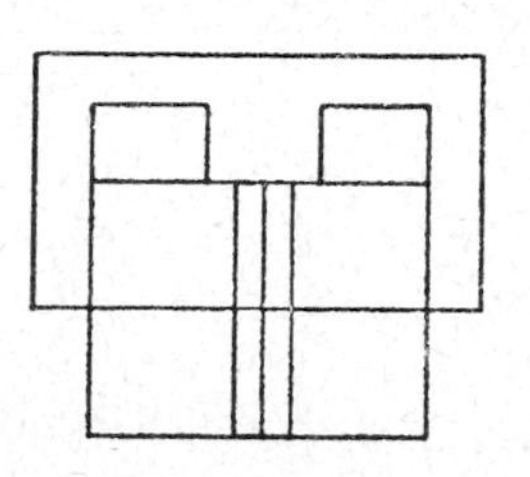

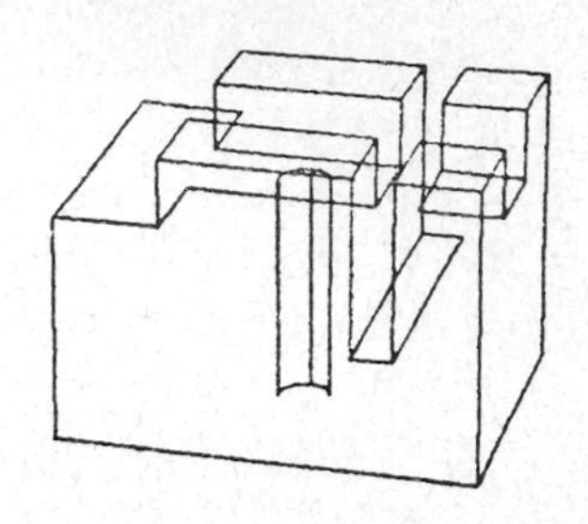

Fig 7. Preparing to make a slot.

Fig 8. Hidden lines (locally) removed

The designer now wishes to make a mould for BLOCK. First the object is made negative. Then it is enclosed in a positive volume (in this case a cube). The enclosing volume is positioned so that its lower face is coplanar with the lower face of BLOCK. As the object has been made negative, the normal to its lower face will point upwards i.e. along the positive y-axis, while the normal to the lower face of the enclosing cube has the same direction but opposite sense i.e. along the negative y-axis. Thus if COMBINE is typed, the two faces will be merged and the mould will be obtained (Fig 9). In Fig 10 the mould has been turned over, sectioned (with another negative cube) and all hidden lines have been removed.

```
$ DEFINE B1 CUBE
$ MODIFY B1 SCALE Y 2, X 2.5, Z 2
$ NEGATE BLOCK
$ COMBINE
$
$
```

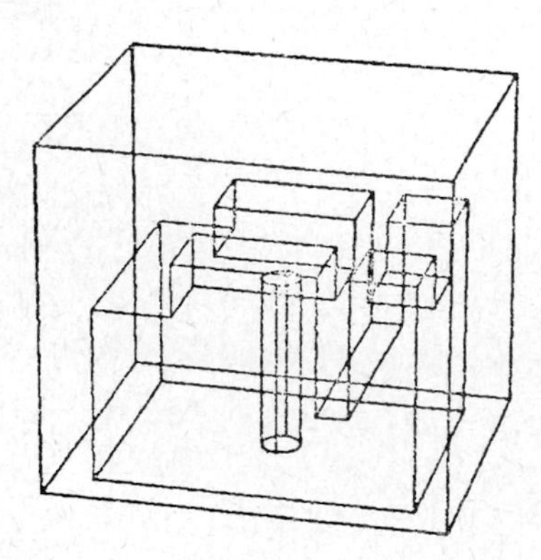

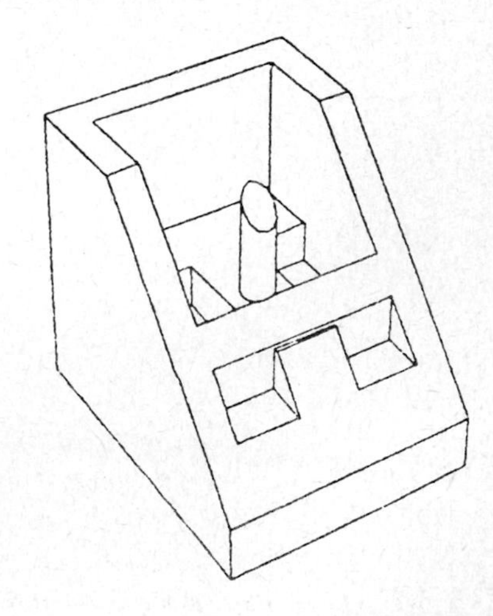

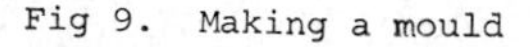

Fig 9. Making a mould

Fig 10. All hidden lines removed

To summarise the program as described so far; there are six primitive volumes, and three operations namely NEGATE which may be applied to any object, COMBINE which will stick together (merge) any number of objects at the flat faces where they touch, and INTERSECT which applies to an arbitrary object and a negative cylinder or cube. Any object can be moved, rotated or scaled. Quite short sequences of these commands are sufficient to construct complex shapes.

A more direct form of input is provided for 2½D objects. Termed *perimeter objects*, they are defined by giving a sequence of straight lines and arcs in the xy-plane. For each, the program constructs an object of constant width; its single outer perimeter is determined by the sequence of lines and arcs and its back and front faces are at z=-1 and z=1 respectively. Once defined, the perimeter object is treated as an ordinary object. It can be made negative, can be combined with other objects or can be intersected with negative cubes or cylinders. Perimeter objects make the insertion of pockets in other objects particularly straightforward (Figs 11 and 12).

Fig 11 also shows the sequence of lines and arcs given to specify the perimeter. The sequence starts and ends with an S. P specifies the coordinates of a point, L a line from the current position to the coordinate point given with the L. A defines an anti-clockwise arc, B rounds off a corner to a given radius, F is a straight line tangent to two whole circles given by W followed by the whole circle centre point and radius (made negative if the circle is clockwise in direction). The specification of perimeters is loosely based on the Ferranti simplified cutting sequence.

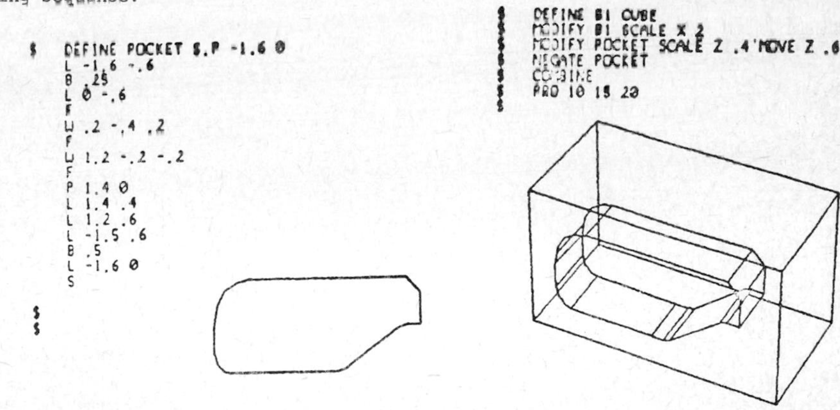

Fig 11. A perimeter object (view point at +∞ on z-axis)

Fig 12. Making a pocket

To conclude this section, several other commands are mentioned and an example of the design of a complex shape appears in Fig 13.

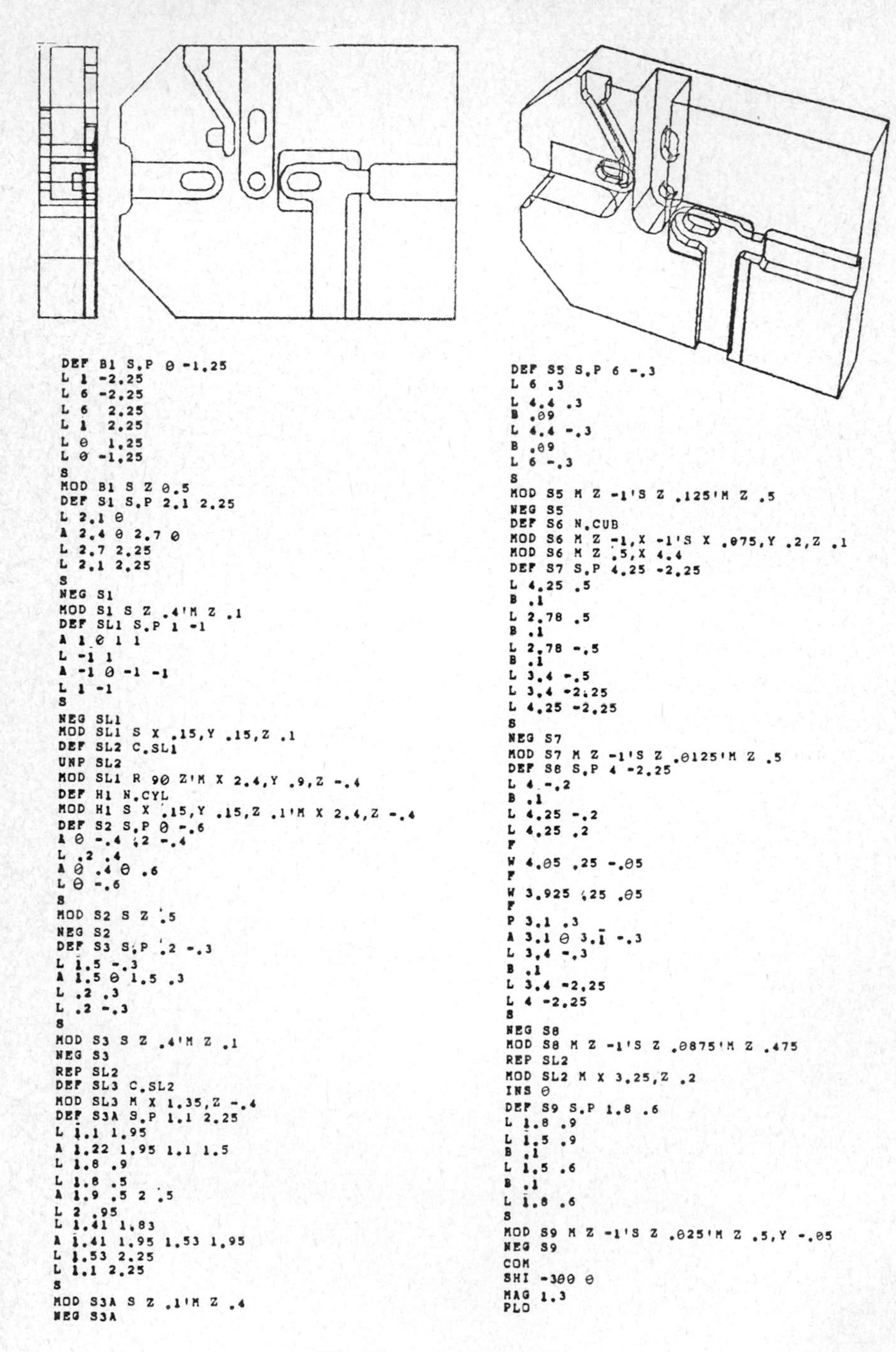

```
DEF B1 S.P 0 -1.25
L 1 -2.25
L 6 -2.25
L 6  2.25
L 1  2.25
L 0  1.25
L 0 -1.25
S
MOD B1 S Z 0.5
DEF S1 S.P 2.1 2.25
L 2.1 0
A 2.4 0 2.7 0
L 2.7 2.25
L 2.1 2.25
S
NEG S1
MOD S1 S Z .4'M Z .1
DEF SL1 S.P 1 -1
A 1 0 1 1
L -1 1
A -1 0 -1 -1
L 1 -1
S
NEG SL1
MOD SL1 S X .15,Y .15,Z .1
DEF SL2 C.SL1
UNP SL2
MOD SL1 R 90 Z'M X 2.4,Y .9,Z -.4
DEF H1 N.CYL
MOD H1 S X .15,Y .15,Z .1'M X 2.4,Z -.4
DEF S2 S.P 0 -.6
A 0 -.4 .2 -.4
L .2 .4
A 0 .4 0 .6
L 0 -.6
S
MOD S2 S Z .5
NEG S2
DEF S3 S.P .2 -.3
L 1.5 -.3
A 1.5 0 1.5 .3
L .2 .3
L .2 -.3
S
MOD S3 S Z .4'M Z .1
NEG S3
REP SL2
DEF SL3 C.SL2
MOD SL3 M X 1.35,Z -.4
DEF S3A S.P 1.1 2.25
L 1.1 1.95
A 1.22 1.95 1.1 1.5
L 1.8 .9
L 1.8 .5
A 1.9 .5 2 .5
L 2 .95
L 1.41 1.83
A 1.41 1.95 1.53 1.95
L 1.53 2.25
L 1.1 2.25
S
MOD S3A S Z .1'M Z .4
NEG S3A
```

```
DEF S5 S.P 6 -.3
L 6 .3
L 4.4 .3
B .09
L 4.4 -.3
B .09
L 6 -.3
S
MOD S5 M Z -1'S Z .125'M Z .5
NEG S5
DEF S6 N.CUB
MOD S6 M Z -1,X -1'S X .075,Y .2,Z .1
MOD S6 M Z .5,X 4.4
DEF S7 S.P 4.25 -2.25
L 4.25 .5
B .1
L 2.78 .5
B .1
L 2.78 -.5
B .1
L 3.4 -.5
L 3.4 -2.25
L 4.25 -2.25
S
NEG S7
MOD S7 M Z -1'S Z .0125'M Z .5
DEF S8 S.P 4 -2.25
L 4 -.2
B .1
L 4.25 -.2
L 4.25 .2
F
W 4.05 .25 -.05
F
W 3.925 .25 .05
F
P 3.1 .3
A 3.1 0 3.1 -.3
L 3.4 -.3
B .1
L 3.4 -2.25
L 4 -2.25
S
NEG S8
MOD S8 M Z -1'S Z .0875'M Z .475
REP SL2
MOD SL2 M X 3.25,Z .2
INS 0
DEF S9 S.P 1.8 .6
L 1.8 .9
L 1.5 .9
B .1
L 1.5 .6
B .1
L 1.8 .6
S
MOD S9 M Z -1'S Z .025'M Z .5,Y -.05
NEG S9
COM
SHI -300 0
MAG 1.3
PLO
```

Fig 13. A complex shape

An object can be copied e.g. DEFINE POCKET5 = C.POCKET4. If an object is copied, and the copy is scaled by a factor of -1 in one or more directions, a mirror image of the original will be obtained.

The viewed picture can be magnified by typing MAGNIFY followed by a scale factor. The point of the picture lying at the centre of the screen does not move during magnification. Equally a picture can be shifted up and down or to right and left using the SHIFT command. These two commands apply just to the 2D picture and allow the designer to zoom in on parts of the picture which is itself unchanged. PROJECT, on the other hand, changes the view point, alters the perspective transformation and so requires the picture to be computed afresh from the stored, 3D description of the shape.

The program can be directed to read commands from a file by typing INSERT. When all commands have been read, it returns to await further instructions from the user. Experience has shown that because of the irreversible nature of the COMBINE and INTERSECT operations, it is often easier to modify an object by editing the commands used to produce it, and executing them afresh, rather than by further addition or subtraction of volumes. Similarly, when using the program it is advisable to save an object (or objects) at several points during their design so that if an error occurs, it is only necessary to revert to the last saved version of the object, and to proceed again from there.

3. THE STORED SHAPE DESCRIPTION

Although the user has the impression that he is designing exclusively with volumes, shape is held internally in terms of finite bounding surfaces. The surface description based on the storage of object faces, edges and vertices, is also convenient for displaying or plotting pictures of the objects. The need for rapid response to the commands of an online user has strongly influenced the design of the data storage scheme.

Consider the primitive cube. It has 6 faces, 12 edges and 8 vertices. Each vertex is stored as three floating point numbers (x,y,z) which, since the cube has not as yet been *modified* (scaled, rotated or translated) will be one of ($\pm 1, \pm 1, \pm 1$). The vertices are held in a simple list and can thus be referred to by a number v ($1 \leq v \leq 8$). The edges likewise are held in a simple list; each edge is given by a pair of vertex numbers (v_1, v_2) and is referred to by an edge number e ($1 \leq e \leq 12$) which indicates its position in the list. Thus to draw a cube, the edge list is scanned once, and, for each edge, the vertex numbers are found, leading in turn to the vertex coordinates which are then transformed appropriately before being projected and combined into vectors for display.

Since any primitive (or object) can be linearly transformed, a single matrix T is stored with the cube. If the cube is modified, a modifying matrix M is calculated and the modification effected by multiplying the matrix T and also every vertex coordinate triple by M. The relationship of vertex numbers to edge numbers remains unchanged.

Another list of face entries (6 for a cube) is also kept. Each entry contains the face equation held as a reference to the *unmodified* primitive face and a reference to the matrix T which together are sufficient to determine the equation of the *modified* face of the primitive. The advantage of this arrangement (especially for curved faces) is that the unmodified face equations can be stored only once regardless of how many instances of a primitive actually exist. Each face entry also contains a pointer to a list of edge numbers giving the edges surrounding the face. The relation between faces and edge numbers is also invariant under modifications of the primitive or object. Fig 14 shows a diagram of the arrangement of data for a cube. The edge numbers in the edge lists are dashed when they refer to an edge whose direction in the face is opposite to its stored direction, i.e. its direction is v_2 to v_1. Fig 14 also gives six of the unmodified face equations which are stored only once.

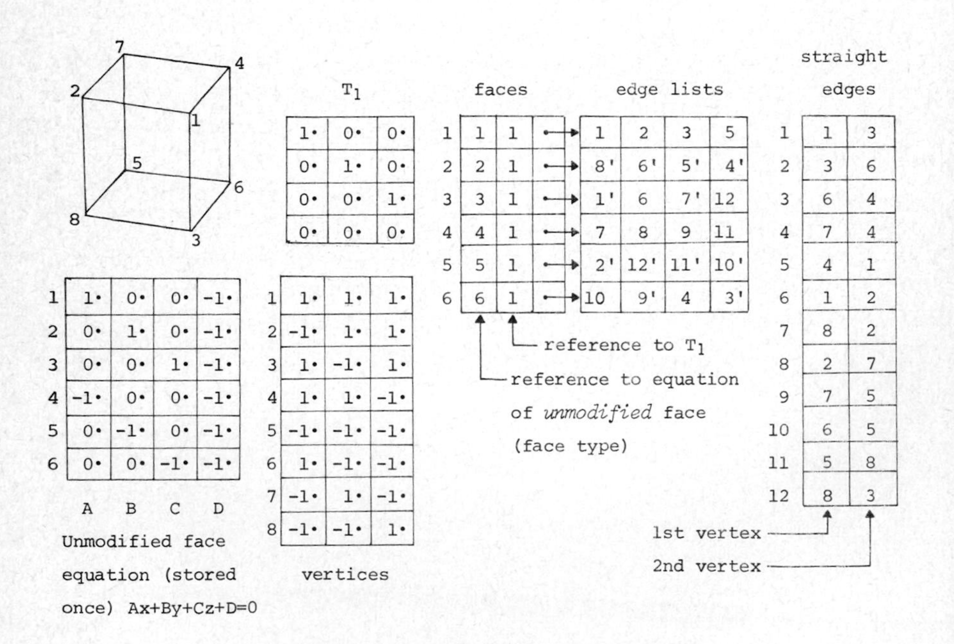

Fig 14. Data structure for a primitive cube (unmodified)

The representation of curved faces and edges is based on the analytic, parametric representation of the unit circle

$$x = \frac{(1-\sqrt{2})t^2 + (\sqrt{2}-2)t + 1}{(2-\sqrt{2})t^2 + (\sqrt{2}-2)t + 1} \qquad y = \frac{(1-\sqrt{2})t^2 + \sqrt{2}t}{(2-\sqrt{2})t^2 + (\sqrt{2}-2)t+1}$$

which is related to the better known form

$$x = \frac{1-t^2}{1+t^2} , \qquad y = \frac{2t}{1+t^2}$$

but gives a closer match between t and angle $a = \frac{2}{\pi}$ artan $\frac{y}{x}$ in the range $0 \le t \le 1$. From a computational point of view, the parametric representation is attractive (Forrest, 1969) because it avoids the calculation of square roots, cosines or sines, yet it is exact (disregarding rounding errors).

By making z depend on a second parameter s, z = 1-2s, the equations of the first quadrant curved face of the unit cylinder are obtained (Fig 15). After an arbitrary linear transformation, the equation of a curved face has the form

$$[x\ y\ z\ 1] = [\frac{t^2}{\Delta(t)}\ \frac{t}{\Delta(t)}\ \frac{1}{\Delta(t)}\ s\ 1] \begin{bmatrix} k_1 & k_2 & k_3 & 0 \\ k_4 & k_5 & k_6 & 0 \\ k_7 & k_8 & k_9 & 0 \\ k_{10} & k_{11} & k_{12} & 0 \\ k_{13} & k_{14} & k_{15} & 1 \end{bmatrix}$$

where $0 \le s \le 1$, $0 \le t \le 1$, and $\Delta(t) = (2-\sqrt{2})t^2 + (\sqrt{2}-2)t + 1$. A curved face equation could therefore be stored by saving the 15 quantities k_1 to k_{15}. In fact, groups of 15 coefficients are stored for only the few curved faces occurring in

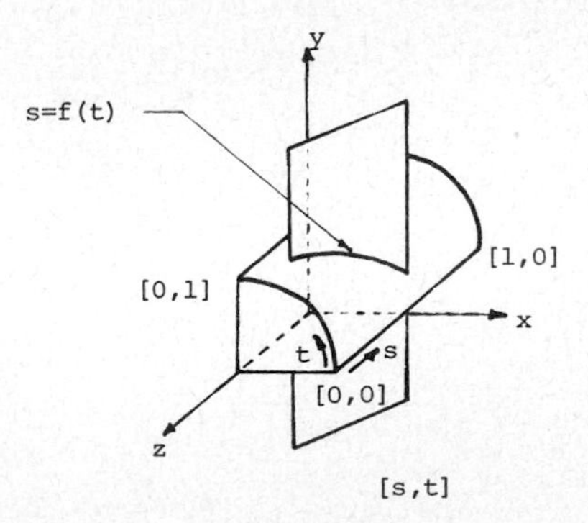

Fig 15. A curved face

unmodified primitives. Every instance of a curved face, like instances of flat faces, is held as a reference to an unmodified face and a reference to a modifying transformation. When the curved face is intersected by a plane Ax+By+Cz+D=0, the intersection curve can be expressed as

$$s = (L_1 t^2 + L_2 t + L_3)/\Delta(t)$$

together, for a finite curve, with the values of t (t_1, t_2) at the ends of the curve. Thus any curved edge can be held by storing a reference to the curved face on which it lies, (L_1, L_2, L_3) and (t_1, t_2). These quantities are invariant under modification. In practice, the vertex numbers at the ends of a curved edge are also stored.

Similar rational quadratic espressions would serve to describe conical surfaces but now the denominator will no longer be constant and at least five constants $(L_1, L_2, L_3, L_4, L_5)$ will have to be stored for each curve of intersection between a conical surface and a plane.

The space curves which result from the intersection of two cylindrical surfaces cannot be expressed in rational polynomial form and so, for the present, the program cannot deal with intersections of this type.

The data structure of a primitive cube has been described in detail, and the additions required for a primitive cylinder have been indicated. The same structure is sufficient to describe any object which has cylindrical or flat faces, and edges which are straight lines or circular arcs. Unlike a primitive, an object may have more than one transformation stored in its data structure. As will be seen in the next section, when two primitives are combined, some of the faces of the resulting object derive from one of the primitives, the other faces from the other primitive. Thus the transformations of both primitives must be included in the data for the final object.

4. ALGORITHMS FOR NEGATE, COMBINE AND INTERSECT

When an object is negated, each reference to an unmodified face equation is altered so that it refers to an unmodified face equation for the face of opposite sign. For a flat face this is equivalent to changing the equation from Ax+By+Cz+D=0 to -Ax-By-Cz-D=0, i.e. the same surface but the unit normal $A\underline{i}+B\underline{j}+C\underline{k}$ has opposite sign. For a curved face, s is replaced by 1-s, and a new face equation derived to represent the same surface with reversed unit normals. Edge lists are modified so that the *sequence* of edges of each outer boundary (and of each hole within) is reversed, thus preserving the convention that the outer boundary of a face, viewed from without, is described in anti-clockwise sequence, holes in clockwise sequence (see Fig 14). The entries (L_1, L_2, L_3) for curved edges must also be adjusted to take account of the change s→1-s. If an object is negated twice, it returns to its original form.

In COMBINE, all flat faces are compared, two at a time, and whenever two are found whose unit normals $\hat{n}_1$ and $\hat{n}_2$ are such that $\hat{n}_1 = -\hat{n}_2$, they are *merged*. The faces need not be convex or simply connected: either face can have a boundary with re-entrant angles and any number of holes within the boundary. First all the edges of one face are compared with all the edges of the other. Whenever they cross, the edges concerned are split at the point(s) of intersection. Secondly all the edges, and pieces of edges resulting from splits, are reconnected to form new faces to replace the merging faces. Fig 16(i) shows a cube and a cylinder which have been merged. Here the two original coplanar faces, each having four edges, become 8 small triangular faces, 4 with unit normal $\hat{n}_1$, 4 with unit normal $\hat{n}_2$. Fig 16(ii) shows a case where no faces result from the two merging faces. In

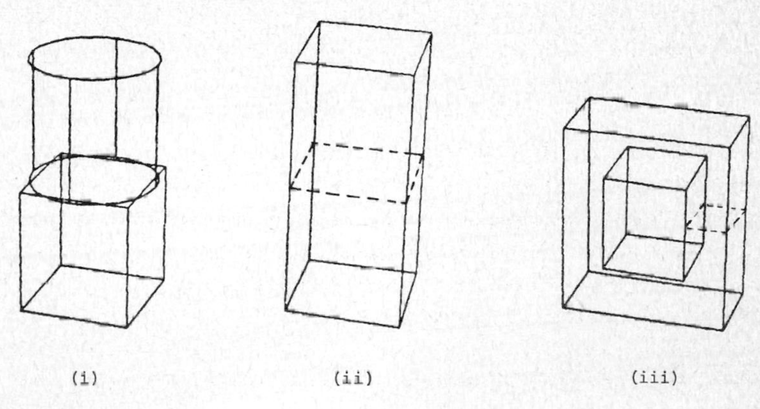

Fig 16. Merged faces

addition, the 8 vertical faces must be combined two at a time, into 4 faces (the dashed edges are deleted), and the vertical edges must be made into single edges running from the top face to the bottom. In Fig 16(iii), two faces belonging to the same object are merged. The merging faces disappear completely and the back and front faces each gain a hole.

Since INTERSECT requires one object to be a primitive cylinder or cube, it is always possible to transform both objects so that the primitive returns to its unmodified state at the origin. The primitive cube then has its faces orthogonal to the axes and lying within a sphere radius $\sqrt{3}$, centred at 0; the primitive cylinder has its back and front faces at $z=\pm 1$, its curved faces pass through the unit circle in the xy plane and it lies within a sphere of radius $\sqrt{2}$, centred at 0. Having transformed both objects in this way, INTERSECT proceeds to test each object face against the sphere enclosing the primitive, and if the face lies outside the sphere, the face and all its edges are marked as non-intersecting since they cannot possibly intersect the primitive. Next, all the unmarked faces and edges of the object are compared, face with edge, with the faces and edges of the primitive; intersections are recorded and edges are split. Finally new faces are formed from the old faces, edges and pieces of edges. Whilst making new faces, new edges must be made i.e. edges which lie along the line of intersection between a face of the object and a face of the primitive (see Fig 15).

The COMBINE and INTERSECT algorithms are complex. For further details, see Braid (1972).

5. LINKAGE TO NC

When a component has been designed, two equivalent descriptions of it are available. One is the stored data structure, the other which occupies approximately $^{1}{}_{10}$th of the space, is the list of commands used to create the shape. At present we are developing a program which will take the stored shape description and a minimum of additional information about machining, and will cut the shape in hard plastic foam on a computer-controlled, 3D model-making machine (Lang 1973) which we have built. The model will quickly show the designer whether the shape is correct. Later, similar techniques will be used to produce an NC tape for manufacture of the component.

6. CONCLUSION

The shape design program reported here makes the design of mechanical components and the production of drawings a quick and efficient procedure. Experience to date shows it to be easy to use. The example in Fig 13 took approximately three hours to design, starting from a dimensioned sketch, and needed about three minutes computation time. Newcomers to the program have learnt to use it after 30 minutes instruction. More important for NC, the program produces a systematic, stored 3D shape description which can be read direct by programs for generating NC tapes. Thus the part programming effort is greatly reduced by the elimination of the need to define the geometry of the shape.

REFERENCES

Braid, I.C., (1972) Ph.D. thesis (in preparation).

Forrest, A.R., (1969) Pertinent concepts in computer graphics, (University of Illinois Press, ed. M. Faiman and J. Nievegelt) pp.36-39.

Lang, C.A., (1973) A three-dimensional model-making machine, (Prolamat Conference Proceedings, 1973).

Problem Oriented Programming Language for Length Measurement

by Werner Lotze
Technical University of Dresden, GDR

Abstract

Universal and efficient utilization of coordinate measuring instruments (2 D and 3 D) requires computers and programme systems for geometric calculations that are easy to handle. This paper presents a programme system especially developed to convert point coordinates into wanted length and angular dimensions which detects any deviation with respect to true shape and position and automatically creates the required instructions for the measuring machine.

1. Introduction

It is the task of length measuring technique to quantitatively log the geometry of a workpiece (dimension, shape and position) and to compare it with the nominal values and tolerances specified in the drawing. Measuring equipment developed up to now generally aimed at the reading of the wanted quantity directly on the measuring instrument without additional computation. Plain manual measuring devices are common usage as far as elementary measuring tasks are concerned (e.g., cylinder diameter etc.). But if more difficult measurements have to be undertaken under otherwise similar conditions (cam disks, gear casings etc.), complex and expensive special equipment becomes a must.

As an alternative, multi-purpose two- or three-coordinate measuring instruments recommend themselves for this task. But they only permit direct measurements and the wanted quantities have to be calculated from the measured values.

Whenever complex measuring tasks have to be handled, the high instrumentation outlay may therefore be replaced with expanded computational effort which, however, would require a computer. Small fixed-level problems can be solved with programmable desk calculators and on-line operation between computer and coordinate measuring machines now is a known procedure. On the other hand, however, comprehensive evaluation of coordinate measuring values requires a larger computer which must also enable the creation of any desired evaluating programme. A programme system that permits both evaluation and preparation of information for performing such measurements shall be described in the following sections.

2. Geometric and Metrological Problems

Statements about the geometry of a workpiece have to be based on length and angular dimensions. As far as parts with practically ideal geometric boundary surfaces are concerned, all geometric elements of a body can be mutually related with the aid of analytical geometry. In compliance with the kinematics of most machine tools the following geometric elements (form elements) will be sufficient for describing more than 4/5 of all workpieces in question, i.e.:-

point
straight line, plane
circle, cylinder, cone, sphere
involute, helix
general line and surface

Because multi-purpose coordinate measuring equipment only measures point coordinates, all form elements of a higher order have to be derived therefrom. Actual workpieces will then contain a combination of these form elements. The parameters of the latter, their distance etc. then are the wanted measuring results.

Such a procedure shall be explained by way of the dovetail profile shown in Fig. 1.

The function determining length L 1 is not embodied in the workpiece but merely represents the distance between the two virtual points P 1 and P 2. These points must previously be determined as intersection points of straight lines. The lines in their turn have to be based on measuring points Q. But once these straight lines are already obtained it will be possible to determine the further wanted dimensions W 1, W 2 and L 2.

Hence, evaluation after point coordinate measurement can be made as outlined in Fig. 2. Apart from an analysis of geometric form elements, the concept of a complete problem oriented language further calls for due consideration of various metrological boundary conditions whereby the latter exert a pronounced influence on the formulation of the individual mathematical algorithms. This includes:

- Coordinate system of measuring machine, test specimen and computation

- Constant error compensation of measuring machine or method (e.g., contact difference of ball-shaped probe array with permanent magnet)
- Facility for probe replacement with measurements at surfaces of different accessibility
- Facility to change the workpiece position for measurement from various sides with secured dimensional continuity
- Elimination of workpiece alignment in the coordinate system of the measuring machine
- Protection of data sequence
- Manual or automatic control of the measuring machine
- On-line or off-line operation with the computer

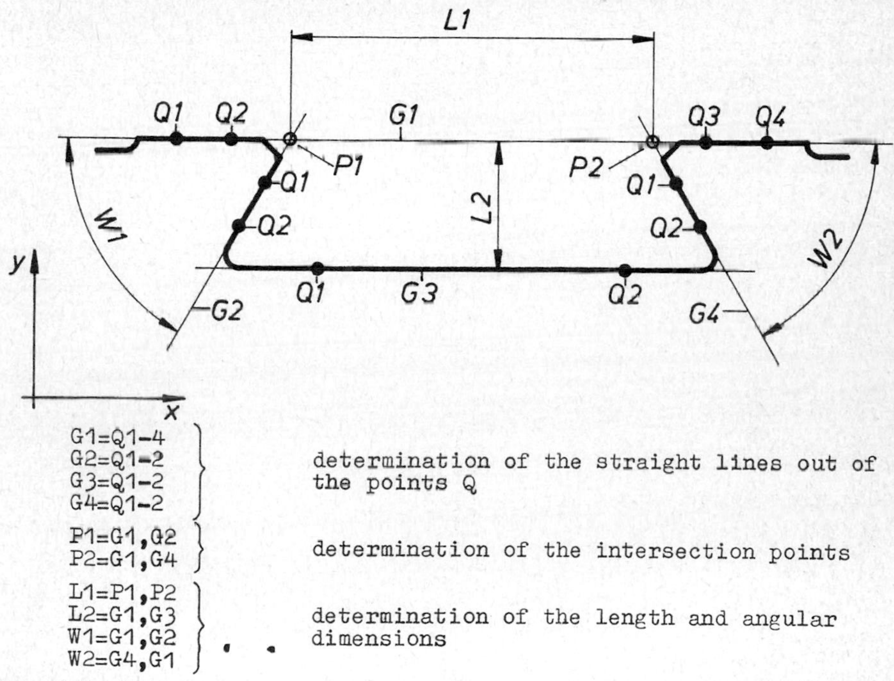

Fig.1,2. Drawing and schematic notation of a computer programme for a dovetail profile.

3. Basic Concept of the Problem Oriented Language

Based on the discussed analytical and metrological conditions, and in close cooperation with VEB Carl Zeiss Jena, we have developed a universal programme system, i.e., the problem oriented MAUS language (Meß-AUswertungs-Sprache = Measurement Evaluation Language).

The overall concept of this programme system can be taken from the schematic in Fig. 3.

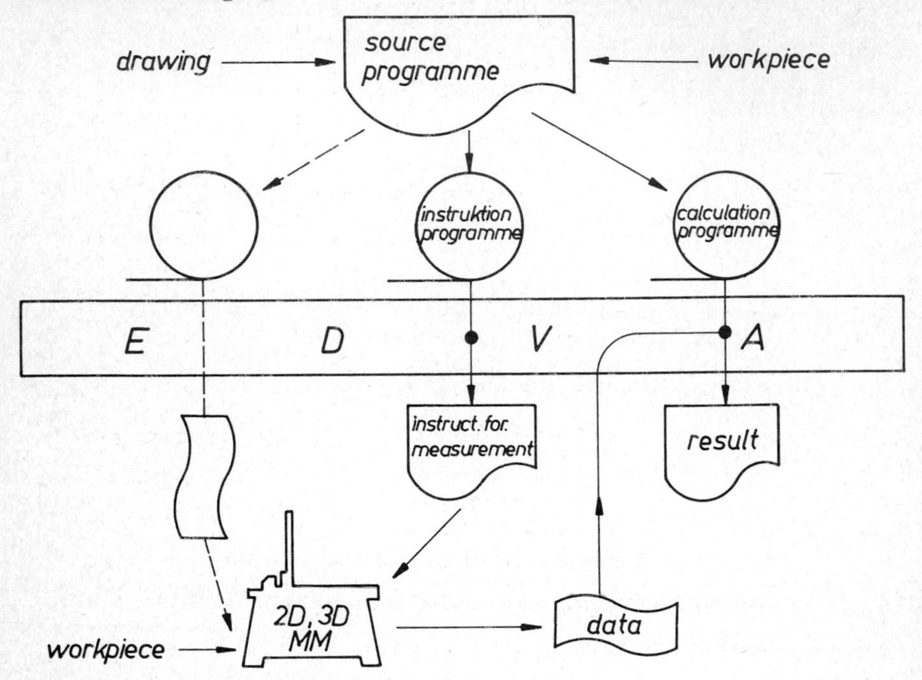

Fig. 3. Function and integration of the MAUS measurement evaluation language into the measuring process.

Starting from the drawing or workpiece, and based on a previously made programm-considered drawing, the source programme for measurement evaluation can now be written. This source programme requires two runs through the computer.

During the first run, all statements are generated for the logical creation of measuring data pertaining to each individual form element on the measuring machine. The output of these statements either is in plain writing for manually controlled measuring machines, or supplied as punched tape for both typewriter or data acquisition device of the latter.

In the second run of the source programm, measuring data evaluation takes place by calculation according to the flow chart in Fig. 4.

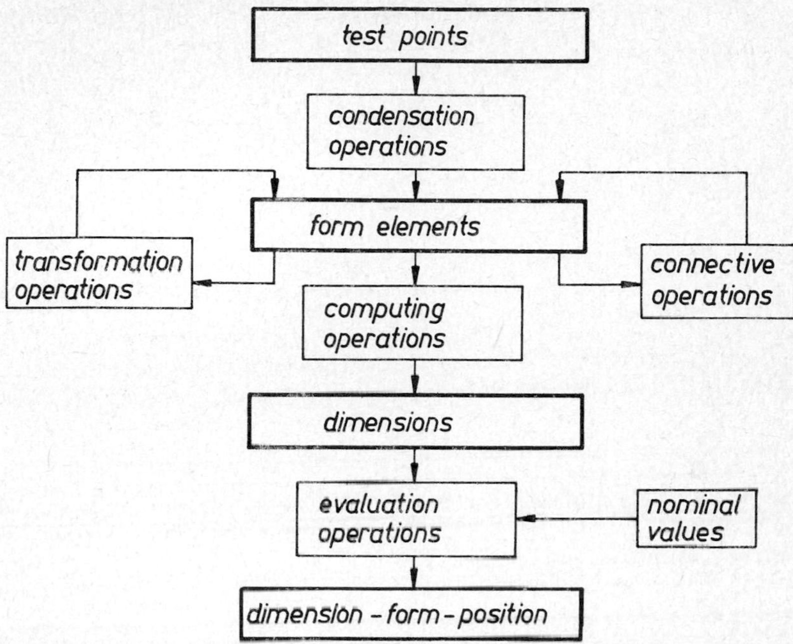

Fig. 4. Course of numerical measuring evaluation.

Condensing operations then generate geometric form elements (straight line, plane, circle etc.) from the test point coordinates which are subsequently stored. As a result, the memory will contain a quantified image of the measured body. With the aid of analytical geometry, new form elements can now be derived through suitable connective operations (e.g., straight lines as intersection of two planes). Transform operations then serve to take process dependent constant measuring errors into account (e.g., contact differences of the ball-shaped probe - Fig. 5) or to introduce coordinate systems that are inherent in a body.

In the next stage, computing operations will determine the wanted length and angular dimensions (for example, expressed as distance between two points). With given nominal values and tolerances, evaluating operations finally serve to assess limit overshooting.

To realize such a programme system, the following requirements have to be met:

- Availability of various directly addressable operations for frequently occurring condensing, transforming, computing, connective and evaluating operations combined with special case protection (e.g., intersection point of parallel straight lines).

- Introduction of a storage organization that warrants both economy and simple access to related geometric elements.
- Introduction of a module structure with automatic internal generation so that from the entirety of available programme modules only the actually required modules have to be activated in the working memory of the computer.

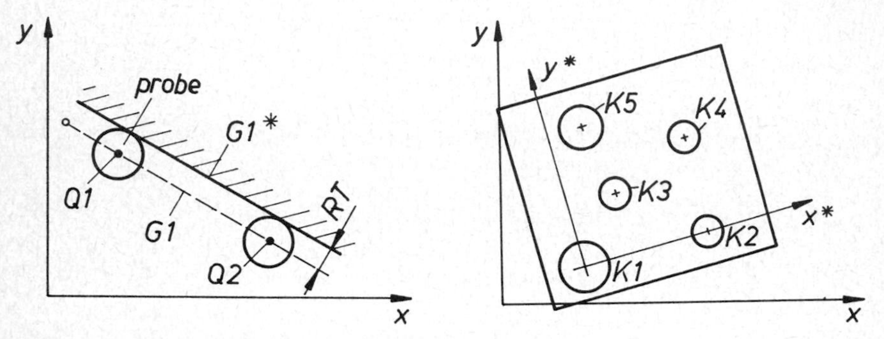

Fig. 5. Examples of transformation operations
(a) Correction of probe radius; derived from the measuring points Q 1 and Q 2 the straight line G 1 has to be shifted in the probe radius RT toward G 1*.
(b) Measuring points gained as machine coordinates (x, y) are transformed into the embodied coordinate system that has been defined by the bores K 1 and K 2.

The selection of mathematical facilities used for computation with form elements is closely linked with the mentioned three requirements. Some of the occurring problems shall now be outlined by way of the 'straight line' form element.

In analytical geometry a straight line is generally treated in its two-point form. In practice, however, this representation will fail as soon as angles close to 90° to the X-axis have to be dealt with. Still further complications will occur in the three-dimensional space.

Another point is the error propagation with all follow-up computations. From all straight line equations it is therefore only the parametric form - especially in its vectorial representation - which proves to be useful. But such a representation requires a somewhat greater storage capacity which becomes clearly evident from the following equations for the two-dimensional case:-

Hesse's standard form: $x \cos\alpha + y \sin\alpha - p = 0$

Vectorial representation: $\begin{pmatrix} x \\ y \end{pmatrix} = \begin{pmatrix} x_o \\ y_o \end{pmatrix} + \begin{pmatrix} a_x \\ a_y \end{pmatrix} t$

If the deviation from an ideal geometric straight line must be determined, more than two measuring points are required for compensation calculation (Fig. 6).

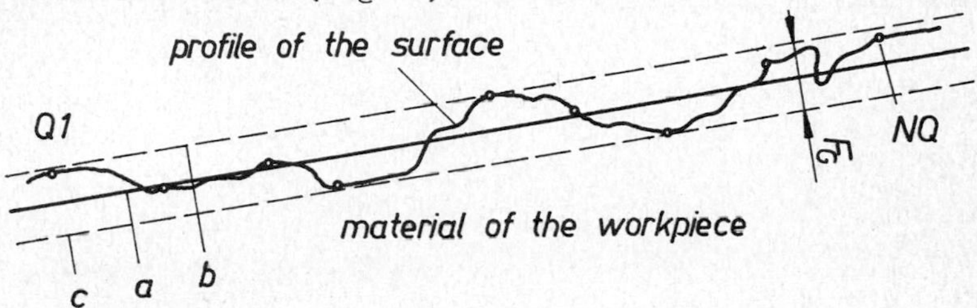

Fig. 6. Straight line from more than two measuring points
(a) Equalizing straight line
(b) Adjacent straight line
(c) Inscribed straight line
F_g Deviation from true shape

With various form elements, however, this calculation already imposes considerable difficulties although it still fails to meet Tailor's basic rules concerning the conjugation geometry which are of fundamental importance in length measurement.

Consideration of the above requirements and protection against faulty results caused by insufficiently spaced points eventually makes the procedure of straight line determination more comprehensive than a programm for an elementary straight line equation based on two points only. Still more complex are the algorithms for determining cylinders, cones or more general form elements.

With due consideration of the above, the presented programme system comprises about 30 K bytes. Thanks to its module structure and the internal generating facilities it can be implemented in computers having an 8 K byte working memory at 39 bit data length. Owing to prevailing numerical calculations, this programme system has been written in ALGOL 60 and organized in such a way that the source programme to be written immediately represents the actual main programme in procedure references. With known storage structure it is further possible to insert normal ALGOL text into the source programm to realize rather difficult or rare arithmetic operations. For elementary tasks the source programme may also be used with abstract notation.

4. Examples

Finally, an example shall be given on the application of the described programme system.

Here the dimensions of a workpiece as shown in Fig. 7 have to be determined.

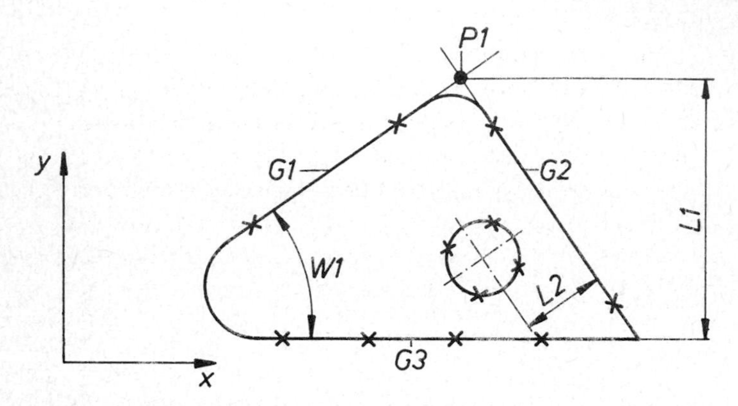

Fig. 7. Triangular slab with quantities to be measured.

The associated source programme (Fig. 8) has been written in free format mode. Additional explanations were added as textual statements after the slash.

```
     comment  source programm for triangular slab
     PR NR (1,2,ENDE);               programme number 1-2
     DIMENSION (2);                  two-dimensional problem
     KO DRUCK (2);                   control print-out
     G QQ (G,1,Q,1,Q,2,ENDE);        determination of straight line
     G QQ (G,2,Q,1,Q,2,ENDE);            "         "      "      "
     G NQ (G,3,Q,1,4,ENDE);              "         "      "      "
     K NQ (K,1,Q,1,4,ENDE);          determination of circle
     P GG (P,1,G,1,G,2,M1);          intersection point
     L PG (L,1,P,1,G,3);             determination of length
 M1: L PG (L,2,K,1,G,2);                 "         "    "
     W GG (W,1,G,1,G,2);             determination of angle
     DR NM(L,1,2);                   interstruction for print-out
     DR NM(W,1,1);                       "           "      "
ENDE:STOP;                           stop
```

Fig. 8. Complete source programme for a workpiece according to Fig. 7 (Text only for explanation).

In addition, the programme contains labels in order to either continue (M 1) or truncate (END) the computation at a suitable position after failure.

From the statement marked with M 1 - that serves to compute the distance D 2 - it can be seen that a point P can also be called-in as actual centre of a circle (K 1).

Should it be required to determine workpiece dimensions other than L 1, L 2 and W 1, statements for new connective and computing operations have to be written. However, further measurements need not be made in this case because the body to be measured has been covered already by G 1, G 2, G 3 and K 1.

Beyond the scope of arithmetic statements for the above mentioned elementary programme this system also permits a variety of organizational instructions. Hence it will be possible, for instance, to obtain very short source programmes with the almost fully automatic call-in of frequently occurring operations (i.e., error corrections, transformations etc.) and low programming expenditure simply by creating cycles and blocks for workpieces having many elements of the same kind.

Summary

Starting from the measuring points, the derived form elements, the dimensional elements to be created and desired/actual value comparison inclusive of the automatic address of measuring statements from a uniform source programme, this paper describes the set-up and application of a complete programme system for evaluating coordinate measurements.

Experience has shown that such a programme system essentially increases the economy and capacity of multi-purpose coordinate measuring equipment.

P E R F E K T
A PROGRAMMING SYSTEM FOR NC-MACHINES USED IN PRODUCTION OF ELECTRICAL SUB-ASSEMBLIES

A. MORELJ
SIEMENS A.G.
Munich, F R G

Abstract: A variety of special NC installations are used in the production of electrical sub-assemblies. This paper briefly describes a programming system designed for these types of NC installations, with the particular emphasis on the acquisition of input-data, the program structure and the verification of the control tapes.

1. INTRODUCTION

A variety of NC installations are involved in the production of electrical sub-assemblies:

- Plotters for generation of printed-circuit artwork for plug-ins and platters (backwiring)
- Multi-spindle drilling-machines for automatic hole drilling of plug-ins and platters
- "Wire-wrap"-machines for discrete wiring of platters
- Inserting-machines for insertion of components on plug-ins
- Testing-machines for continuity-testing of plug-ins and platters.

Control information needed for these applications is - except in few simple cases of drilling - usualy very large. For example a typical control tape for artwork generation contains some 80.000 to 400.000 characters. It is obvious that computer aids are inevitable, and for this purpose a programming system called PERFEKT was designed at our company. It is already in use for several years and it proved to be successful. The programs were written mainly in COBOL for the Siemens Series 4004.

COMPUTER LANGUAGES FOR NUMERICAL CONTROL, *J. Hatvany, editor*
North-Holland Publishing Company - Amsterdam-London

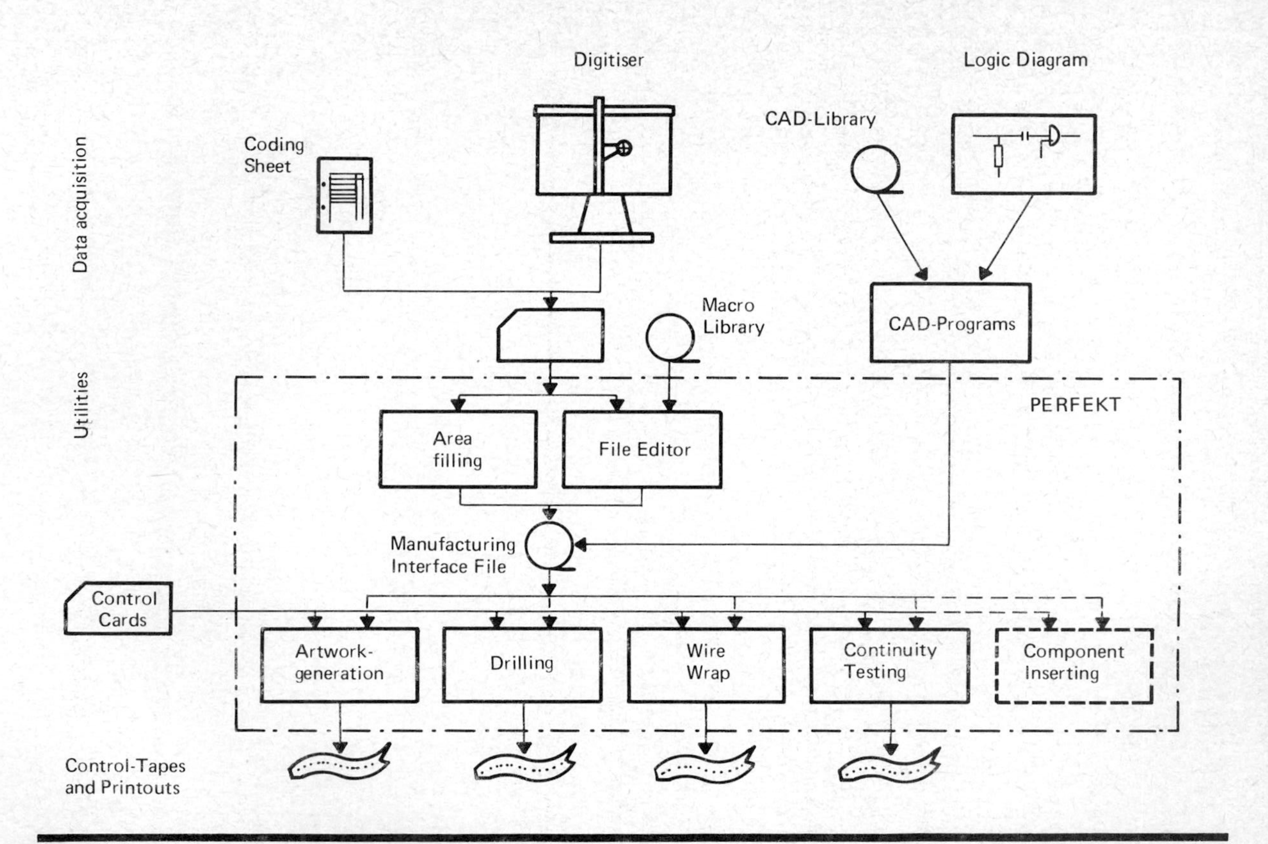

Fig. 1 Block Diagram of PERFEKT

As the block diagram (Fig. 1) indicates, this design was aimed at creating a system which:

- can equally satisfy the needs of a small or a large user
- can be easily adapted to environment changes
- has a clearly defined interface between development and manufacturing.

Some aspects of interest, for example acquisition of input-data, program structure and verification of control tapes, will be further discussed.

2. ACQUISITION OF INPUT-DATA

A typical signal layer of a printed circuit board (size 150 x 130 mm) contains about 300 pads and 150 connections. As 2 to 6 points are needed to describe one connection completely, the coordinates of 500 to 1000 points must be entered for each signal layer.

This is a large amount of input data and the data acquisition must therefore be thoroughly considered.

A first approach is to try to reduce the amount of data needed, and a second one to use some mechanical aids.

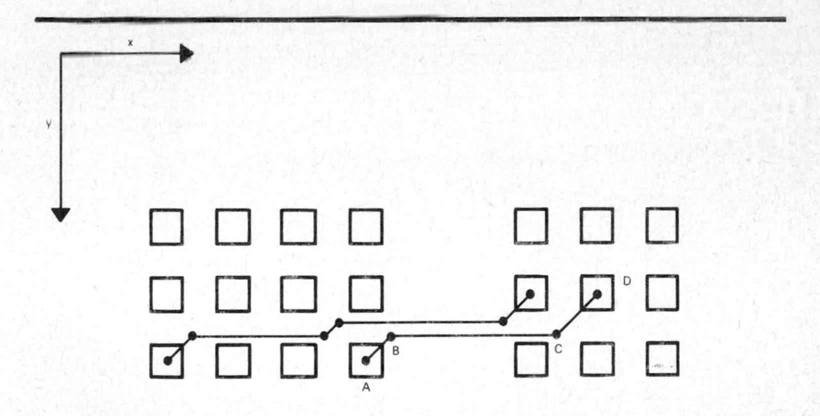

Fig. 2 Connections on PCB (sample)

The reduction of input-data could be achieved through standardisation. For example in the case of the connection of Fig. 2, the coordinates of 4 points must be entered in order to describe it completely. If the pad-entry-lines and the distance between the long line and the pads are constant, then points B and C can be derived from points A and D, thus requiring half of the input information. Through a further standardisation (e.g. possible component places, voltage pins and connections), the variable portion of a layer can be separated from the constant or standard one. The constant portion needs only be described once per type and stored as a macro on the library.

A simple macro call enters now a large amount of tested information. A further possibility is given by connections and voltage areas which can not be exposed all at once. Here, only the outer shape needs be entered, the motion of the exposure head inside this perimeter being generated automatically.

This and other measures help to reduce the input-data but do not eliminate the need of mechanical aids such as digizer.

It is clear that the same input-data are used for artwork generation, drilling and continuity-testing. This sharing of data makes the use of the system cheaper, and assures the compatibility - not the accuracy - of the control tapes.

Yet, in the case of CAD, we succeed in achieving a broader sharing of input-data. Here, the only inputs are the logical diagram and the construction type of assembly. CAD-programs simulates the network, defines the physical details and generate the manufacturing data of the assembly, so that all the informations about the product are stored in computer readable form.

Only a few control cards containing informations about the particularities of the manufacturing processes are still needed in order to generate the control tapes.

3. PROGRAM STRUCTURE AND PROCESSING

In order to accomodate the various users' requirements, the system must be able to process different forms of input-data (i. e. data generated by different systems, automatic, semi-manual or manual) and be able also of supplying control tapes suitable for different kinds (e. g. drilling, testing etc.) and types (e. g. single, multi-spindle-drills etc.) of NC-Machines. This requires a certain amount of modularity and clearly defined interfaces outside as well as inside a partial program such as shown in Fig. 3. Our experience has proven that this structure leads to an easier and cheaper maintenance.

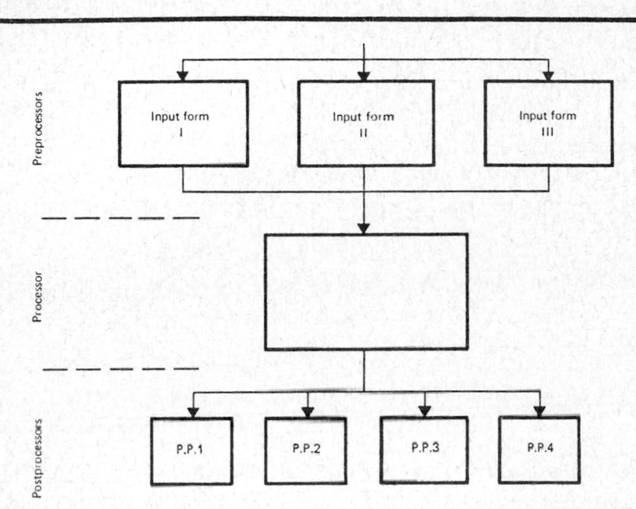

Fig. 3 Block structure of partial Programs

To illustrate the processor tasks we take a look at the partial program for drilling. The user must supply the hole coordinates of the first plug-in of the batch as well as the following control informations: target NC-machine, which hole-codes should be punched on each one of the control tapes, define the batch, the position of the batch on the machine table, the spindle positions, the number of drilled holes between toolchanges, and the number of control command blocks between two restart points on the tape.

The processor tasks are:

- to build up the batch
- to sort the holes according to the diameter code and hole coordinates
- to control if "multiple" holes exist (i. e. two or more holes having equal coordinates)
- to "minimize" the positioning time
- to print out the workshop sheets and other statistical data

The "minimize time" procedure is not an exact one. Oversimplified stated, hole to hole moves follow a meandershape. Investigation showed that a better "optimization" brings a very small reduction in machining time but increases considerably the computing time.

A further example, taken from the artworkgeneration should show the influence of the boundary conditions on the "minimize time" procedure. Supposing the typical configuration of a printed circuit board is to have a large quantity of printed wire connections with pads as terminating elements, there are two ways of producing this artwork:

a) Artwork elements i.e. pads and lines could be intermixed generated according to the "minimum time" principle (one pass)

b) Pads and lines could be generated in separate passes, after being sorted according to their apertures.

Method a) shows considerably more aperture changes than method b) but machining time is shorter. With the equipment in use in our manufacturing facilities, method a) leads to a prohibitive wear of the aperture wheel, so that is was decided to use method b).

4. VERIFICATION OF THE CONTROL INFORMATION

Since the input data, in spite of careful handling, are normaly not error free, the verification of the control information is important. Some aids are required to discover the logical errors; the formal errors are detected by the system PERFEKT.

In case of artworkgeneration it proved to be satisfactory to produce a control plot and do only a visual control. Control plots are not generated on the high precision plotter but on cheaper plotters like drumplotters. There are two possibilities:

- the control plotter accepts directly the same control information as the high precision plotter
- a computer program is needed to convert the control tapes for high precision plotter in a form acceptable to the control plotter.

The choice between these two possibilities depends on cost estimation. In case of a big volume job the first mentioned possibility is preferable because of its lower operating costs that normaly overweight the costs of the increased investment.

PERFEKT-generated control information for drilling shows such a small error rate that a separate verification would not be economical.

At the start of the production a first batch is drilled and checked visually against the artwork (positive and negative)*). The same method is also applied for the quality checks during production.

For technological reasons, some printed circuit boards have pads on all grid points, and only some of them are to be drilled. In such a case a drill control plot is generated and used instead of artwork.

Since the control information for testing is generated from the same input data as for the artworkgeneration, there is normaly no needs for a separate verification.

5. CONCLUSION

Experience in the use of PERFEKT shows that an integrated programming system for a product-group like PC-boards is fully feasible, provided that it is updated regularly to accommodate the technological changes. So it is inevitable to have standard interfaces, some program modularity and special features to catch up with various requirements (large versus small user).

At last some cost related aspects should be mentioned:
The development of PERFEKT has cost some 15 - 20 men years, and is now costing 2 - 3 men years for maintenance and updating.

Since our experience in generating control information has been conducted mainly with the aid of the system PERFEKT, a cost and time comparison with a hypothetical system generating this information manually is somewhat questionable. However, a rough estimation gives for the artworkgeneration and the drilling operation a manpower reduction of 85 % and a cost reduction of 50 % to 70 %, using the 100 % value for the manual programing.

*) Note all pads have a hole!

EXTENDING THE UTILIZATION OF AN NC PROCESSOR

by

T. Márkus and I. Cser

Institute for Technology of Mechanical
Engineering, Hungary, Budapest

Application of NC processors can be rendered more advantageous by means of various preprocessing programmes. Two programmes attached to the 2C,L system will be discussed in this paper; one of the programmes promotes preparation of production in a conventional production area while the other provides essential data for designing production economy based on NC technique.

The TEVE system

Planning and producing control cams required for manufacturing the workpieces form a considerable part of production preparation work for automatic turret lathes. To perform this work by more modern means a programme for NC machining of control cams /TEVE system/ was worked out in Institute for Technology of Mechanical Engineering, Budapest.

Developing TEVE system the general utilization was kept in view. This might be proved by the two stage construction of the system, the first stage produces conventional cam design and outputs cam geometry as a result; running both parts of the programme cam production preparation is also performed producing part programme required for manufacturing the workpiece on an NC mill. Different types of automatic turret lathes can be supplied. The programme reads in machine characteristics required for cam design from punched tape /adjustable cycle times, replacement of wheels, time and angle requirements regarding secondary motions, max. and min. cam radii, etc./.

Technological information on workpiece to be manufactured forms the main group of input data. Setting up the input data a dividing line in production engineer-computer relationship had to be found where human decision making /and choice of machinable workpieces/ will not yet be limited, however, rotine calculations are no longer performed manually. Here operation sequence, cutting parameters, slide travels and slide positions belonging to cutting operations should be written on FORTRAN data-sheet and then punched on card in accordance

COMPUTER LANGUAGES FOR NUMERICAL CONTROL, *J. Hatvany, editor*
North-Holland Publishing Company - Amsterdam–London

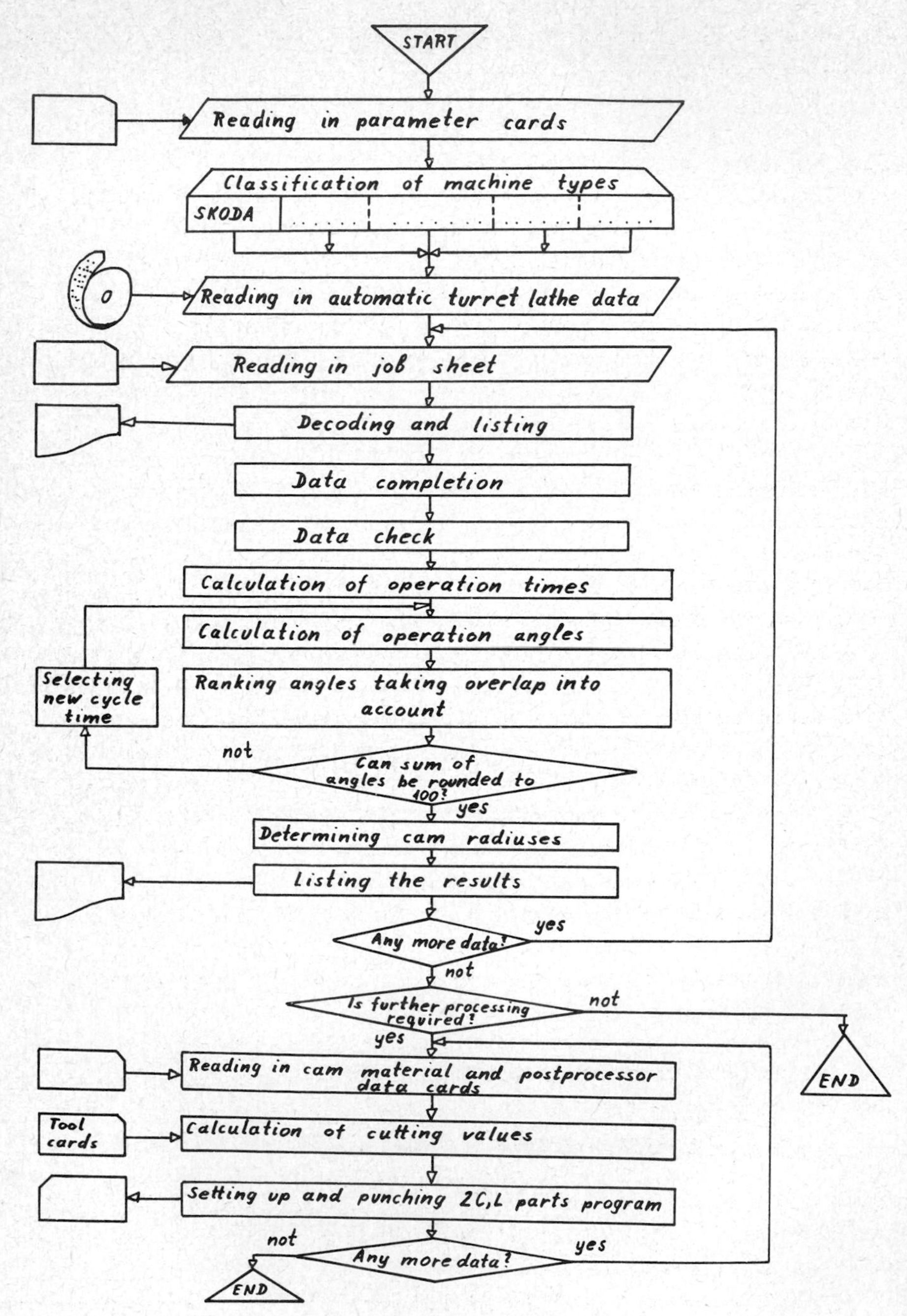

Fig.1: Schematic flow sheet of TEVE system

with syntactic and semantic rules of the input language. Printout of input data presented in Fig. 2. is illutrated in Fig. 3. Some detail of this will be discussed there.
To describe machining units a fixed word stock of five character is used and until the 32nd character position optional remarks may be written. Slide descriptions occur on 33-37 position. Machining units follow each other according to operation sequence independent of which slide performs machining. To indicate overlap among operations an overlap code is used on 4I-42 character position. This code corresponds to serial number of the operation that can overlap the operation in question, which is the last one in sequence. After overlap code cutting parameters and data concerning slide position should be given on fixed character field.

The programme of TEVE system is available in FORTRAN on ICL I900 series.
Its roughly outlined construction in Fig. I. After the cards containing data of machine tool type, number and stage of runnings, card numbers and workpiece identifiens were read in and the type of machine tool is identified processing begins with reading in machine data tape.
With decoding the programme prints input data /Fig. 3./.
The first part of program produces cam geometry /Fig. 4./.
Running the programme further on in order to set up commands written in the programming language, technological statements are also required. This work is carried out by a small technological processor, then geometrical and technological commands are specified, hence the 2C,L part programme of control cams is given by the system.

Control tapes for any machine type with 2C,L postprocessor can be obtained by running these part programmes on the processor and postprocessor so that production of control cams can have a background, which is not restricted.

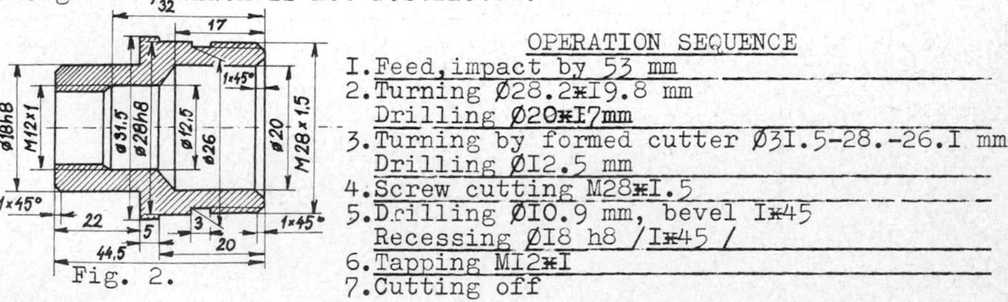

32
17
1×45°
ø18h8
M12×1
ø31.5
ø28h8
ø12.5
ø26
ø20
M28×1.5
1×45°
22
5
3
20
1×45°
44.5
OPERATION SEQUENCE
I.Feed,impact by 53 mm
2.Turning Ø28.2×I9.8 mm
Drilling Ø20×I7mm
3.Turning by formed cutter Ø3I.5-28.-26.I mm
Drilling ØI2.5 mm
4.Screw cutting M28×I.5
5.Drilling ØI0.9 mm, bevel I×45
Recessing ØI8 h8 /I×45 /
6.Tapping MI2×I
7.Cutting off

Fig. 2.

TEVE system

	NO.	Description of machining	Descr. of slide	Overlap code	Spindle speed	Feed	Screw cut. head extension	Travel data	Distance data
	NO	MŰVELET MEGNEVEZÉSE	SZÁN	FK	FORD	ELŐT	KIH	UT	TÁV
Impact	01	ÜTKÖZTETÉS	REVOL						135,5
Turret changeover	02	FVÁLTÁS							
Cutting:turn.dril.	03	VÁGÁS;ESZT.,FÚR ÁTM.20			0710	0.12		23,5	112.
Turret changeover	04	FVÁLTÁS							
Cutting:dril.dia.	05	VÁGÁS;FÚR ATM.12.5		06	0710	0.1		13,5	95.
Cutting:thread fin	06	VÁGÁS;MENETES RÉSZT SIMIT	HÁTSÓ	06	0710	0.02		33.	28.
Turret changeover	07	FVÁLTÁS,FORDVÁLTÁS	REVOL						
Screw cutting	08	MENETMETSZÉS			0090	1.5	6.5	15.	107.
Direction changeov	09	IRÁNYVÁLTÁS							
Screw cutter retur	10	MENETMETSZŐ VISSZAFUT		14	0710	1.5			
Turret changeover	11	FVÁLTÁS		14					
Cutting:dril.bevel	12	VÁGÁS;FÚR ATM.12.5,ÉLLET.		14	0710	0.08		17.	110.
Turret changeover	13	FVÁLTÁS		14				-42.	
Cutting:recessing	14	VÁGÁS;BESZÚRÁS	ELSŐ	14	0710	0.02		35.	18.
Speed changeover	15	FORDVÁLTÁS	REVOL						
Tapping	16	MENETFÚRÁS			0360	1.	5.	16.	116.
Direction change.	17	IRÁNYVÁLTÁS							
Screw cutter retur	18	MENETFÚRÓ VISSZAFUT		20	0710	1.			
Turret changeover	19	FVÁLTÁS		20				-8,5	
Cutting:cutting of	20	VÁGÁS;LESZÚRÁS	FELSŐ	20	0710	0.06		19.	10.
Moving away	21	TÁVOLODÁS							

REVOL - turret slide
HÁTSÓ - back slide
FELSŐ - upper slide
ELSŐ - front slide

Fig. 3.: Printing out input data

TEVE system

Cam data for turret slide:

MÜV. SORSZ	SZÁZADRÉSZ			TÁRCSASUGÁR			IV JELLEGE
1	0.0		1.5	72.0		72.0	KÖRPÁLYA
2	1.5		3.0	72.0		72.0	KÖRPÁLYA
3	3.0		20.5	72.0		95.5	SPIRÁLIS
4	20.5		22.0	95.5		99.0	EM.PARAB
5	22.0		34.5	99.0		112.5	SPIRÁLIS
7	34.5		36.5	112.5		85.5	SÜ.PARAB
8	36.5	38.5	44.0	85.5	90.0	94.0	SPI+SPI
9	44.0		44.5	94.0		94.0	KÖRPÁLYA
10	44.5	45.0	45.5	94.0	90.0	85.5	SPI+SPI
11	45.5		47.0	85.5		80.5	SÜ.PARAB
12	47.0		67.0	80.5		97.5	SPIRÁLIS
13	67.0		69.5	97.5		55.5	SÜ.PARAB
15	84.5		86.5	55.5		75.5	EM.PARAB
16	86.5	37.0	89.5	75.5	78.5	86.5	SPI+SPI
17	89.5		90.0	86.5		86.5	KÖRPÁLYA
18	90.0	91.0	91.5	86.5	78.5	75.5	SPI+SPI
19	91.5		93.0	75.5		67.0	SÜ.PARAB
BEF.IV	93.0		99.0	67.0		67.0	KÖRPÁLYA
BEF.IV	99.0		100.0	67.0		72.0	EM.PARAB

Cam data for front slide:

MÜV. SORSZ	SZÁZADRÉSZ		TÁRCSASUGÁR		IV JELLEGE
14	44.5	84.5	71.5	80.0	SPIRÁLIS

Cam data for back slide:

MÜV. SORSZ	SZÁZADRÉSZ		TÁRCSASUGÁR		IV JELLEGE
6	22.0	33.5	77.5	80.0	SPIRÁLIS

Cam data for upper slide:

MÜV. SORSZ	SZÁZADRÉSZ		TÁRCSASUGÁR		IV JELLEGE
20	90.0	97.0	75.5	80.0	SPIRÁLIS
21	97.0	100.0	80.0	44.0	SÜ.PARAB

Cycle time and tooth number of changeover wheels:

DARABIDŐ	VÁLTOKEREKEK FOGSZÁMA			
90.00	40	60	35	70

MÜV.SORSZ-operation No.
SZÁZADRÉSZ-angle
TÁRCSASUGÁR- cam radius
IV JELLEGE- curvature
KÖRPÁLYA- circular orbit
SPIRÁLIS- spiral orbit

EM.PARAB- ascending parabola
SÜ.PARAB- descending parabola

Fig. 4.:Printing out calculated cam data

Programme for machining time data calculation

In NC machining the selection of machine tool and the machining version that satisfies quality requirements and having best economic features has a great importance. Machining time data serve as data base for economic calculations. Time requirements of machining on a mill are determined technological time computing programme.

The programme can be run attached to 2C,L system or can be built into postprocessor. When running alone processing is expedient if a postprocessor is not yet available to a given type of machine /or it is not suitable for time data determination/ or if the target of running is to obtain preliminary information,therefore postprocessing is not necessary, and no control programme is required. E.g. if investment decisions will be carried out.
The programme that can be run alone is discussed in the following passage.

As input data two main groups of information are used:

- general validity technological and geometrical information concerning workpieces to be manufactured are provided by the 2C,L processing, the CLDATA;
- the data of NC mill that performs machining should be given in tabular form /see details in Fig. 6./ then from FORTRAN data-sheet they should be punched on cards.

Having these data the program adapts CLDATA information to given manufacturing conditions and on this basis actual time computing is done.

As a result the following information will be printed out:

- primary and secondary machining times, manual secondary time and total time requirement of machining;
- functions to be carried through manually /time requirements of control commands which can be omitted at will, STOP, OPSTOP, etc./;
- times of tools employed spent in cutting /this may be useful in determining tool sharpening cycles/.

The programme was made in FORTRAN for ICL 1900 series to 003 version of 2C,L processor. Its principal of operation is illustrated by the

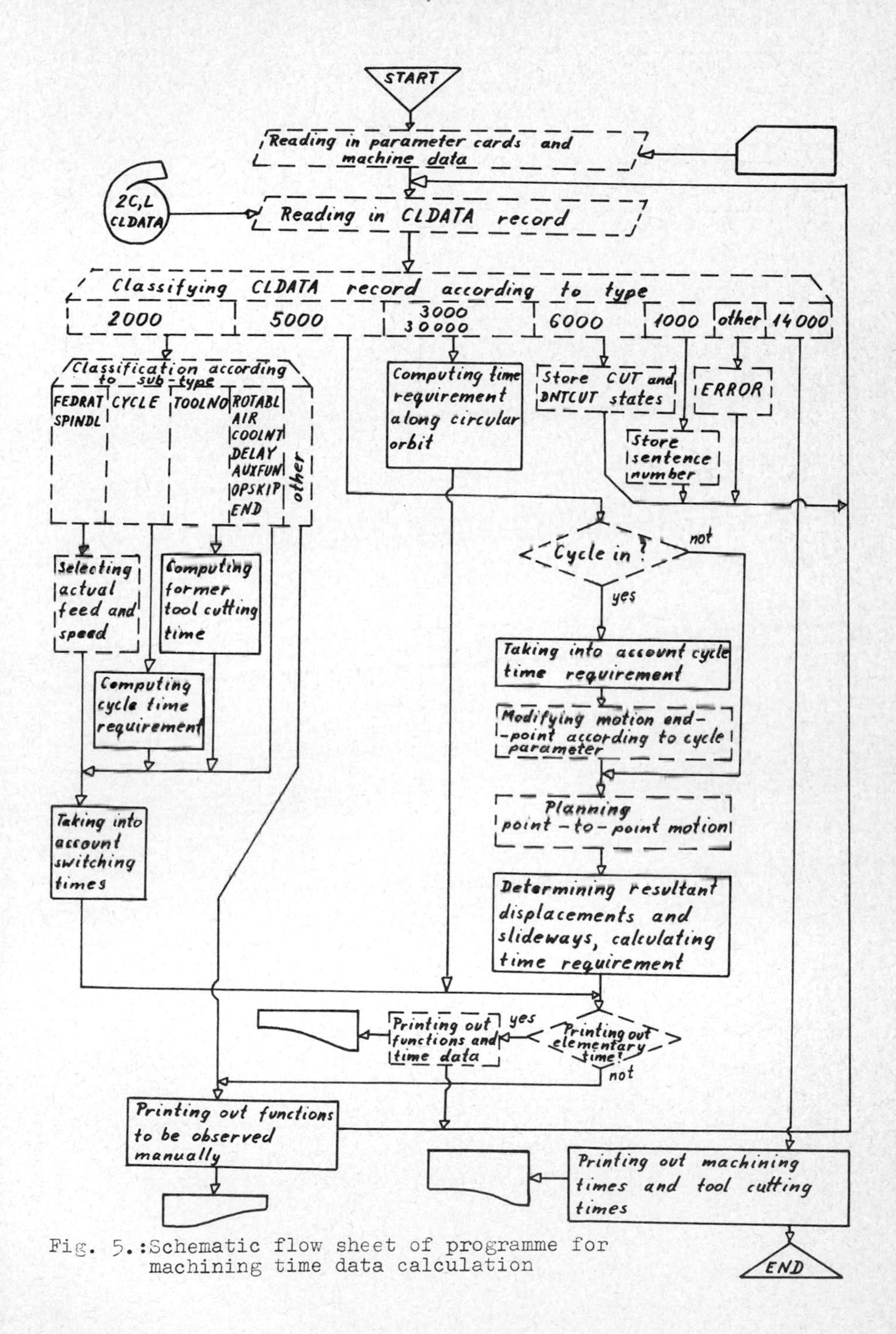

Fig. 5.:Schematic flow sheet of programme for machining time data calculation

INSTITUTE FOR TECHNOLOGY OF MECHANICAL ENGINEERING - BUDAPEST

DATA OF NC MACHINE-TOOL

No.	Parameter	Data values		
:	:	:	:	
I0.	Inching can be prohibilited	yes	no	
		I	2	
II.	Programming mode of feed	feed number	feed code	
		I	2	
I2.	Rapid travel positioning	continuous path cont-rol on more axes	line control on more axes	on one axis
		I	2	3
I3.	Rapid travel positioning by line control	simulta-neous control on X,Y and indipendent control on Z.	simulta-neous control on X,Y or X,Z or Y,Z axes	simulta-neous control on X,Y,Z axes
		I	2	3
I4.	Rapid travel positioning by continuous path control	only in XY plane	in XY or XZ or YZ plane	in XYZ space
		I	2	3
:	:	:	:	:
27.	Operating time /min./ of: feed			
28.	speed			
29.	tool			
:	:	:		

No.	spindle speed / rev/min /	No.	feed values / mm/min /	No.	feed of faceplate / mm/min /
:	:	:	:	:	:

INFORMATION ON INCHING

No.	feed values / mm/min /	length of inching / mm /	medium speed of inch. / mm/min /
:	:	:	:

Fig. 6.: Sample of data of NC machine-tool

flow-sheet in Fig. 5.

Processing starts with reading in machine tool data and parameter cards associated with information to be printed out, then continues with reading in and processing the individual CLDATA records. When processing motion statements the programme plans the way of how to reach individual motion end points and in case of a cycle the programme modifies motion end point according to specified safety distance, computes the resultant displacements and time requirements based on valid feeds. Time requirement of creeping motion depending on motion velocity is also taken into account here. Time requirements of switching commands are taken into account by the programme on the basis of tabulated machine tool data.

Ellaborating this programme modular principle was observed so that various functional sub-programmes could be built into other programmes without any difficulty.

In Fig. 5. through incorporating the blocks indicated by continuous lines any 2C,L postprocessors can be made suitable for determining time data.

REFERENCES

I. I. Cser and T. Márkus:
Control cam designing programme
/Research report, GTI I97I/
2. I.Cser and T. Márkus:
/Research report, GTI I97I/
Technological time data computing programme to 2C,L system

ON THE PROCESSOR DESIGN PROBLEMS FOR SPECIALIZED PROGRAMMING SYSTEMS

J.PRUUDEN and E.PRUUDEN
Institute of Cybernetics, Academy of Sciences
of the Estonian SSR, Tallinn, USSR

Abstract: The paper deals with some design problems of problem-oriented programming systems processors. Certain difficulties encountered in realizing of such processors are described, and some ways to surmount them are mentioned and illustrated by the examples of two systems elaborated at the Institute of Cybernetics of the Estonian Academy of Sciences.

1. INTRODUCTION

Most programming systems used in the process of preparing information for NC machine tools are typical specialized problem-oriented systems intended for solving engineering problems. Despite their numerous obvious and well-known positive characteristics considering the particular features of those using the computer as an subsidiary means in their everyday job, these systems still have some drawbacks often setting limits to their wide-spread application. The following ones should be mentioned as distinctive for the majority of specialized systems applied at present.

First, it is rather difficult to effectively realize the processors of such systems on small computers with limited capacity of the main storage. Secondly, most specialized programming systems and languages are "stand-alone" ones, not permitting their direct associate application in solving complex (e.g. design and manufacturing) problems. And - last but not least - it takes a great amount of effort and time to develop and test those systems.

This paper is concerned with some design problems of the specialized programming systems processors, especially with the elimination of the above-mentioned drawbacks. The paper is based on the experiences obtained in elaborating and applying such systems at the Institute of Cybernetics of the Academy of Sciences of the Estonian SSR.

2. SOME FEATURES OF SPECIALIZED PROCESSORS

Each engineering problem-oriented programming system consisting of a specialized language and processor is intended to solve a comparatively narrow problem class. The principal solving methods of these problems are fixed inside the processor. The problem to be solved is described in a problem-oriented language generally not on the level of calculation and logical procedures (such as by using a procedure oriented language like ALGOL, FORTRAN, etc.) but on the essentially larger problem units level. A fixed set of both purely mathematical and very special-purpose basic operations (procedures) is used to solve all the problems of same class.

These basic operations could be considered as the main operational modules of the given system. Hence, the modular structure of the considered systems processors follows as a conclusion.

The way to solve any particular problem is to select and execute the necessary basic operations in a strict order. The execution

COMPUTER LANGUAGES FOR NUMERICAL CONTROL, *J. Hatvany, editor*
North-Holland Publishing Company - Amsterdam–London

sequence could be prescribed generally either a) merely by the person solving the problem and describing these operations in a computer source program (just as it is done while using a procedure-oriented language), or b) entirely by the processor, or c) by both the person and the processor (the mixed mode). As a rule, the last one is used in the specialized engineering systems. At that, the higher the declarativity of the system input language, the greater the role of the processor in determining the necessary operations sequence.

With a few exceptions (e.g. the systems which are extensions to the existing procedure-oriented systems), the processors of specialized programming systems are generally the interpretive ones containing more or less separated interpretation block.

To a great extent, the system running efficiency is predestined by the choise of the processor structure. The way of effective organization of source program commands processing essentially depends on both the system input language complexity and the characteristics of the computer used.

In case of less declarative languages, the separate source program commands are the weakly bound ones. In the simplest case, each such command presents an individual autonomously executable unit, and the intercommand relations are realized only via executing results of these commands. The commands of those languages (e.g. described in (Fenves, 1965), (Roos, 1966), etc.) are generally processed by methods using for each command either a special analysing program, or an interpretation table treated by a common executive program. With these methods, the levels of command translation (and editing), command interpretation, and operational program modules execution could be discerned. The post-processing level is usually added to perform the final editing and display of the output results.

With more declarative languages, the semantic relations between separate commands are of great importance, which must be taken into account in the processor. The command sequence of such language often describes only the general aim of the problem and some intraproblem objects relations. Now, the particular command processing becomes more complicated depending also on some other source program commands (semantically related to the given command), and on the intrasystem intermediate results affecting the selection of the algorithms and the corresponding program modules. These factors can be considered using the method of internal parametric control of the processor as presented below. The method is based on the use of a finite set of logical variables to control the processor run during which the current values are assigned to the variables on the basis of the recognizes syntactical units of the source program, the analysis of input data, and some intermediate processing results. The method is illustrated by the examples given in the following sections of the paper.

The choice of processor structure also depends upon the characteristics of the computer. The methods of using the analysing programs or interpretation tables are well-suited for large computers vith sufficient main storage capacity. However, implementing the system with developed input language on small computers with limited main storage capacity, these methods often lead to a marked decrease of processor run speed and the whole system efficiency. This is conditioned by the great number of accesses to the tables and program modules residing in the secondary storage, the total size of which by far exceeds areas available for them in the main storage.

One of the possible ways to solve this problem is to use the above-mentioned internal parametric control method with simultaneous distribution of the source program processing into fixed number of autonomous stages, performed in consecutive order for the entire source program or for some larger constituents of it. At each stage, the total size of the program modules and subsidiary information must conform to that of the available main storage areas. This will be illustrated by the APROKS processor in Section 3.

As we know, the elaboration and possible extension of "stand-alone" specialized programming systems is a rather time-consuming and expensive process. This has led to attempts to develope "universal" processors destined to process several problem-oriented languages as well as "meta-systems" intended to generate specialized languages and systems. A number of interesting results have already been obtained in this field, such as AED (Ross, 1967), ICES (Roos, 1966), POLGEN (Burroughs, 1971), etc. Such an approach also permits to meet the problem of informational linking of individual specialized systems, desired in solving complex problems. In Section 4, one of such systems worked out at the Institute of Cybernetics of the Estonian Academy of Sciences is discussed in greater detail.

3. THE APROKS PROCESSOR

In (Pruuden, 1970), the general description of the APROKS language and system intended for preparing control tapes for numerically controlled flame cutters in shipbuilding is given. In this section, a more detailed discussion on the internal structure of that system processor is presented as an example of implementing a specialized language on a small computer without any high-speed auxiliary direct-access storage devices.

The input information for the processor is an APROKS language (Pruuden, 1971) source program describing the stock plate with nested parts and the cutting process. The structure of the APROKS source program is given in fig. 1. The complementary initial

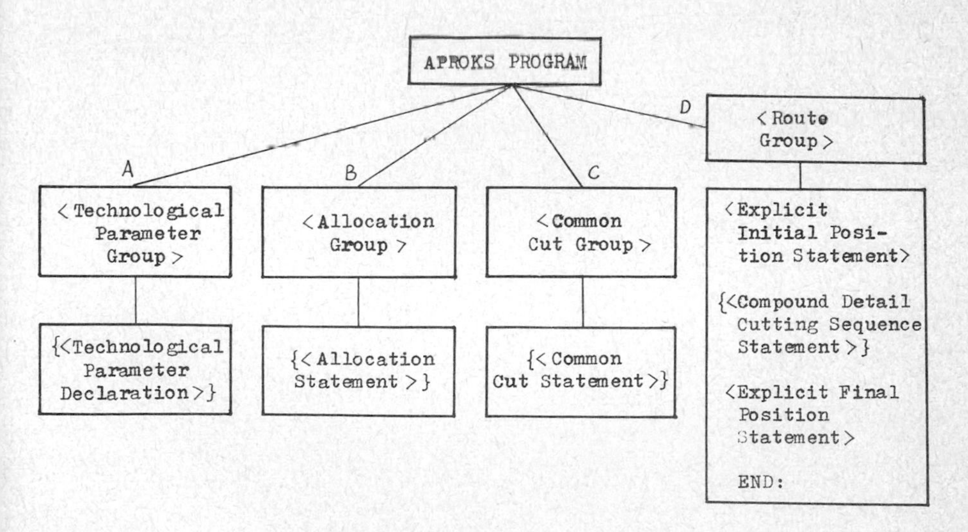

Fig. 1. APROKS program structure.

information for the processor is simplified geometrical data of the parts, calculated beforehand and stored in the secondary storage (at present, the APROKS system is linked to the GEOKODS system developted by G.G.Berezin (Leningrad, USSR) and used for calculating these data).

Taking into account the features of the computer used, the point of departure for processor development consisted in the requirements to minimize a) the number of accesses to the information in the secondary storage and b) the total size of the program modules and control information.

The logical structure of the APROKS processor is shown in fig. 2. In the processor, some larger syntactic units (groups of statements) are processed on several hierarchic levels. The APROKS language is a fairly declarative one, requiring semantic relations between separate commands to be considered. The internal parametric control method mentioned in Section 2 is used for this purpose in the system processor. On each processing level, the program modules are linked according to the source program statements and the intermediate data processing results. The complete functional schemata of the processor presents a directed graph with nodes corresponding to the operational program modules, and arcs corresponding to the possible linkages between these modules. To each arc, a logical condition (a Boolean function) is ascribed, depending on the values of some control variables. At each stage of the processor run, these variables take the values depending upon the recognized source program statements as well as on the intermediate data processing results. After executing any certain program module, the logical conditions on the arcs issuing from the node corresponding to this module are checked. If one of these conditions is satisfied, control will be transferred to the module conforming to the terminal node of the given arc. If none of the conditions is satisfied, the diagnostic check is performed.

Let us now consider in greater detail the source program processing levels in the APROKS system processor.

The processor starts with the syntax analysis of the APROKS source program. If there are statements of the first three groups (fig. 1) in the source program, these statements are treated first. As a result, the arrays of technological parameters, the location parameters of the parts on the stock plate, and the so-called part codewords will be formed in the main storage.

Now, the cyclic run of the processor on independent hierarchic levels follows. During each cycle, the complete output data forming process for one single part (i.e. the processing of one Compound Detail Cutting Sequence Statement) is accomplished. On the first level, the cutting order of the given part segments and their technological peculiarities are determined. The result is the sequence of cutter path location codewords and the technological data, stored in the main storage in the First Level Intermediate Language (I IL) form. On the next level, the parameters of the contour segments will be calculated on the basis of the results obtained on the first level, the simplified geometrical data of the part, and the allocation parameters. The results will be stored in the main storage in the form of the Second Level Intermediate Language (II IL). Each unit, of that information holds a certain number of machine words and consists of a contour segment codeword and corresponding numerical data. The arrangement of these units corresponds to the contour segments processing sequence. On the third level, the complementary contours are linked to the basic (rough) part contour and the final cutter path is formed. The results present the sequence of cutter path positions augmented by technological data, stored in the form of Third Level Intermediate

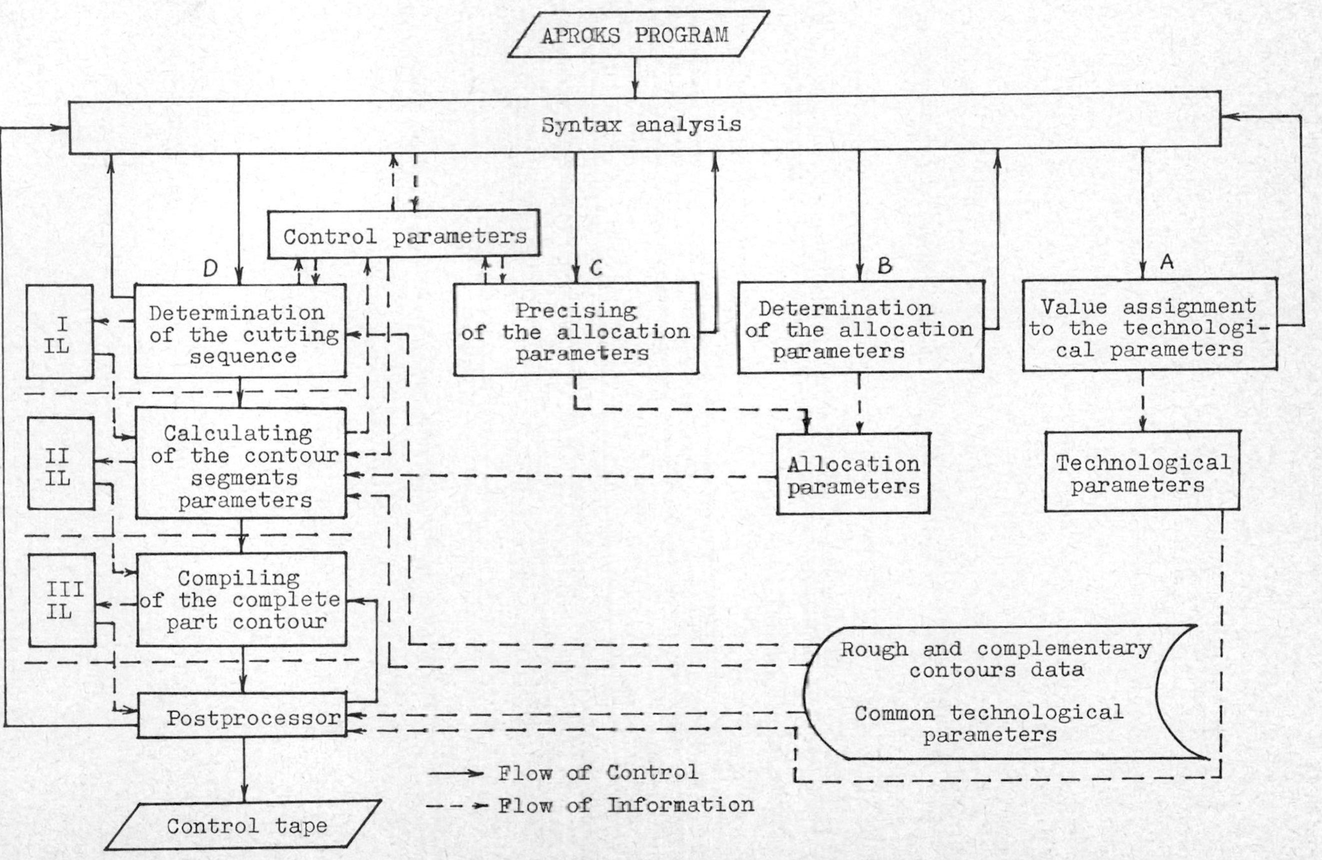

Fig. 2. APROKS processor structure.

Language (III IL) and containing the complete geometrical and technological data for cutting a single part. At the last level, the control tape data is formed and punched. After completing the forth-level processing, the cycle described is repeated for the next part specified in the source program, etc.

The APROKS processor is implemented on the Soviet Minsk 22 computer. The total size of processor programs is about 8000 words of 37 bits.

4. INTEGRATED PROGRAMMING SYSTEM

This section describes the general structure and functioning of the specialized integrated programming system (IPS), being under development at the Institute of Cybernetics of the Estonian Academy of Sciences and intended to run on Soviet third-generation computers.

The general structure of IPS resembles that of the ICES system. IPS could be described as a composite system including a number of information-connected special-purpose programming systems. Each (sub)system has its own specialized language and could be used both autonomously (for solving the corresponding isolated class of problems) and in common (for solving large complex problems). At the same time, IPS is also a meta-system with self-extension capabilities destined to generate new problem-oriented languages and systems, or to modify the existing ones.

The system consists of a) a set of specialized subsystems and b) a common system processor.

All the subsystems have the same internal structure, consisting of a subsystem language and a set of special processing means used by the system processor. These means include the command dictionary, the command interpretation tables, and a set of the program modules realizing the basic operations to solve the corresponding class of problems.

The system processor is intended to control the entire system run as well as that of the subsystems. The processor consists of the executive and a set of service programs (fig. 3).

The executive comprises three main blocks corresponding to the separate phases of the source program execution.

During the first phase, the input and preliminary analysis of the source program is performed by the first pass analysis block. As a result, the subsystem command dictionary is loaded into the main storage, the packed source program in the internal intermediate language is formed, and the label tabel is generated. This table is used for locating the labelled commands in the intermediate language program.

During the second phase (performed by the command interpretation block), the order of interpretation of the commands is determined and the interpretation of every separate command is accomplished. The interpretation process includes the command analysis, fixing of the command data, storing the necessary data into subsystem data area, and forming the initial sequence of the execution of the program modules. The intercommand relations are considered using the above-mentioned internal parametric control method and a set of control variables. Each command interpretation is controlled by the corresponding command table. Depending on the subsystem structure, either the entire source program, or a certain group of commands, or a separate command is interpreted at a time in this phase.

In the last phase, the execution of the necessary programs and the data output is realized by the execution block. The program execution sequence depends generally on the initial sequence fixed in the interpretation phase, and on the values of the control

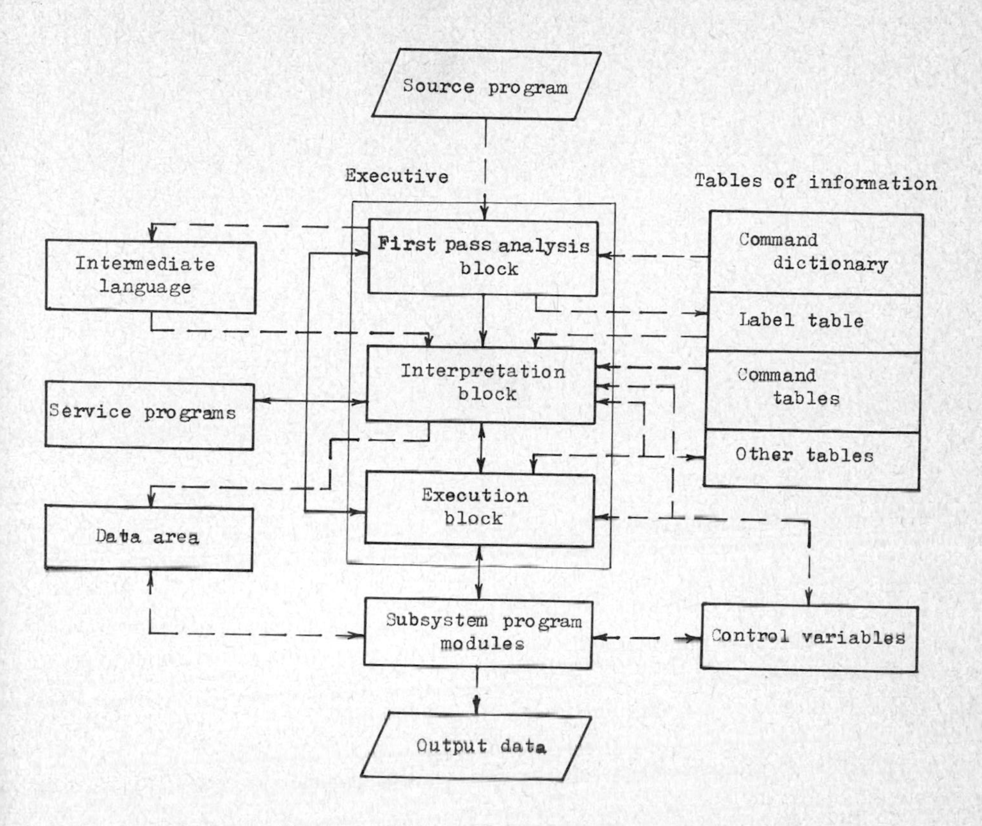

Fig. 3. IPS processor structure.

variables.

In all the processing phases, a set of service programs is used performing the tasks of system initiation, intrasystem data management, dynamic storage allocation, and some other subsidiary tasks.

The meta-system facilities of IPS are expressed by the special Language Generation Subsystem operating as all the other subsystems. This subsystem language is destined to describe the structure of the problem-oriented languages as well as their processing. As a result of running this subsystem, the special language processing means described above are obtained for a new (or modified) subsystem. To generate the subsystem program modules, a procedure-oriented language is used.

The above-mentioned capabilities permit to use IPS also to generate subsystems suitable for their integrated usage in computer-aided design and numerical control.

REFERENCES

Burroughs Corp., (1971), Problem-oriented language generator (POLGEN) reference manual, Burroughs B2500 and B3500 Systems.

Fenves, S.I., (1965), Problem-oriented computer languages, Appl. Mech. Rev. 18, 175.

Pruuden, J., (1970), APROKS - a programming system for NC flame cutters, in: Numerical Control Programming Languages (North-Holland Publishing Company, Amsterdam), 350.

Pruuden, J., (1971), APROKS - a specialized programming language (in Russian), Algorithms and Algorithmic Languages, Issue 5, (Computing Centre, Academy of Sciences of the USSR, Moscow), 79.

Roos, D., (1966), ICES System Design (The M.I.T. Press, Cambridge, Massachusetts).

Ross, D.T., (1967), The AED approach for generalized computer-aided design, Proc. ACM 22nd Natl. Conf., 367.

CADRIC - CALCULATION OF DRILLING COORDINATES

JAN PEJLARE
The Swedish Shipbuilders´ Computing Centre
Göteborg, Sweden

Abstracts: CADRIC (CAlculation of DRIlling Coordinates) is a computer program used for the description of node patterns. It will also enable checking of these patterns and generate paper tapes to be used in an N/C drilling machine.

1. GENERAL

CADRIC is a programsystem, which will produce the control data for an N/C three spindled drilling machine with a fixed distance between the spindles. The system was ordered by the shipyard of Eriksbergs Mekaniska Verkstads AB, where the drilling machine is used for the production of tubeplates to heat exchangers, in order to avoid the rather cumbersome exercise of manually programming the machine.

The Swedish Shipbuilders´ Computing Centre has been involved in N/C for a considerable time and gained experience from all levels in this area, like ADAPT compilerbuilding, the design of postprocessors and the wellknown VIKING system for the design and production of ship hulls.

It was decided that the CADRIC system should be a low-cost system, which in this case meant that it should not include any unnecessary facilities. It should also be very easy to use by untrained personnel and use a minimum amount of computertime. It was soon discovered that the development of a new control language or the use of an old one should not fullfill our requirements. Therefore CADRIC was designed as an ordinary computer program with the input data given in an exactly defined format, depending on what we want the program to do.

The tube plates contain a great number of holes to be drilled, a typical plate would have as many as 1700 holes. The holes are arranged in a regular pattern or network. The pattern is either formed symmetrically on circles in which case a polar coordinate system is suitable to describe the nodes of the network (fig. 1) or it can be formed regularly in columns and rows in which case an orthogonal coordinate system is applied (fig. 2). There can also be combinations of these two coordinate systems.

COMPUTER LANGUAGES FOR NUMERICAL CONTROL. *J. Hatvany, editor*
North-Holland Publishing Company - Amsterdam–London

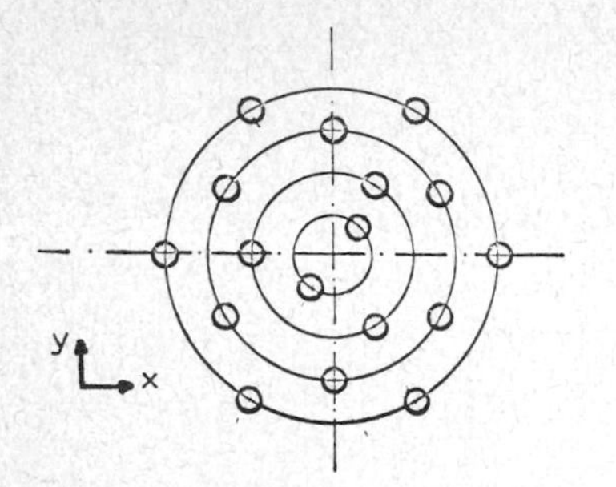

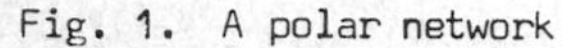

Fig. 1. A polar network

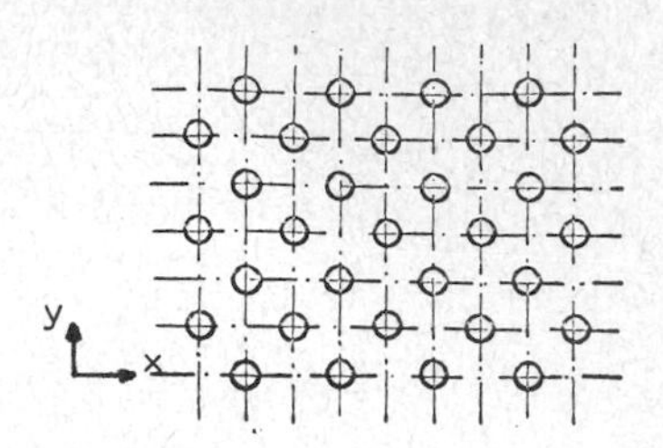

Fig. 2. An orthogonal network

The network can be distorted by deleting one or several nodes and further an outer and inner boundary can be defined, if necessary. The system will produce a printout of the coordinates of all drillholes, and a schematic picture of these in the form of a line printer listing, where an asterisk will show the approximate location of each hole. The drilling order of 3, 2 and 1 drill at the same time will be determined. The coordinates of the middle drill, which is the one to be controlled, will be printed together with a layout of the nodes to be drilled several at a time. A paper tape for the control of the drilling machine will be produced together with information about the drilling and positioning time and the total length of the control tape.

2. THE NETWORK

The nodes of a polar network are defined as points on concentric circles with a given center and radii.

One hole on a circle may be defined by giving the angle to the x-axis, or either the x- or y-coordinate of the hole with reference to the center. The rest of the holes are defined with an angle realtive to the first hole or by the number of holes on the circle.

The holes defined in the polar network can normally not be drilled several at a time.

The orthogonal network can consist of up to 154 columns. The number of nodes in each column may not exceed 78. The network is defined either by the distances between two nodes in the same column or in the same row respectively. The row distance should be chosen as a multiple of the spindle distance to enable the use of several drills at the same time.

The network can be modified by changing the distance between two particular columns (fig. 3). If it is required, all nodes to the right of a specified x-value can be moved vertically half the distance between two nodes in the same row. These modifications described here can be repeated several times if necessary.

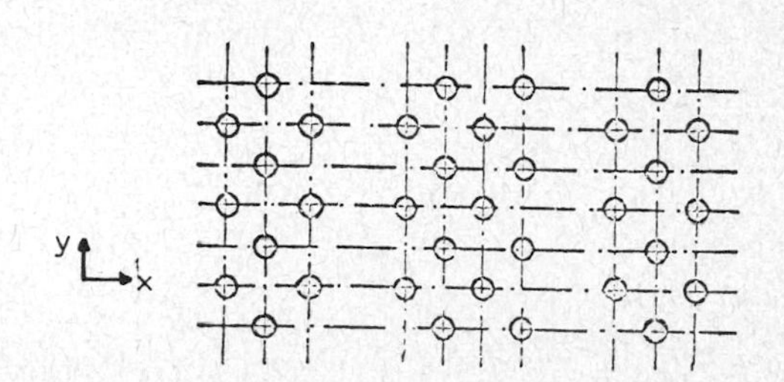

Fig. 3. The distances between columns can be changed in different ways.

The boundary of the network can consist of an outer and an inner boundary, which can be defined either by a circle or a polygon. The holes inside and outside the boundary will be deleted and consideration is taken to the diameter of the holes and the minimum allowed distance between the edge of a hole and the boundary (fig. 4).

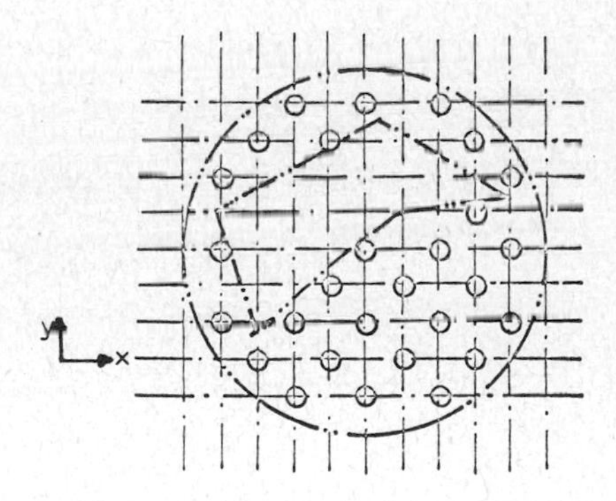

Fig. 4. Inner and outer boundaries can be defined.

If the boundary is given as a polygon, this must be convex in the x-direction and a maximum of 30 corners are permitted. The corner points has to be given in a counter-clockwise manner.

It is also possible to delete a complete row or column, which must be defined by a coordinate within a given tolerance.

If a certain hole should be replaced by a hole with another diameter, this can effect the neighbouring nodes within the new diameter plus a tolerance. This is performed by giving the coordinates of the hole.

3. PRINTOUTS

If wanted the coordinates of all valid nodes will be printed. A very useful tool for the constructing engineer for checking purposes is given by the schematic picture from the line printer of the valid nodes of the network (fig. 5). If the maximum number of nodes in one column is greater than 59, the picture is printed in two partitions. It must however be noted that due to the properties of the lineprinter the picture is slightly deformated.

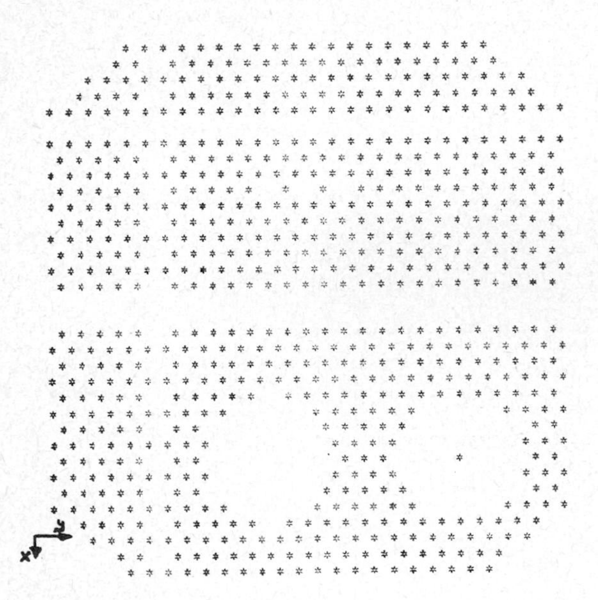

Fig. 5. A line printer picture of a network with most of the distortion facilities included. Note the coordinate system.

4. ECONOMY OF CADRIC

Before the advent of CADRIC, a total amount of about 80 hours were used for a typical tubeplate with about 1700 holes. This time included the calculations of coordinates, the construction of a special program-drawing, and the manual programming.

Now with CADRIC the time involved amounts 3 hours for the same piece of work. To this should be added approximately 3 minutes of computertime (IBM 370/145). It should also be noted that the possibilities of a programming error is now minimized. This is very important as the materical cost for a tubeplate is considerable ($2000).

REFERENCE

Weiss, P. and Zvolensky, L., (1970) CADRIC Version 1, Swedish Shipbuilders' Computing Centre, Report no. 48.

PROBLEMS CONCERNING THE AUTOMATIC PROGRAMMING OF NC MEASURING MACHINES

W. FROHBERG, H.-J. HÖRNLEIN

Forschungszentrum des Werkzeugmaschinenbaues
im Kombinat "Fritz Heckert" Karl-Marx-Stadt ,GDR

1. GENERAL INFORMATION ABOUT THE PROGRAMMING SYSTEM

In the GDR a system for the automatic programming of NC machines (NCM) by means of electronic data processing machines has been developed. This programming system, called AUTOTECH/SYMAP, consists of several independent parts. Depending on the different control methods for NCM and the respective technological tasks programs were broken down as follows:

- Programs for the continuous-path control machining on NCM (e.g. milling or grinding)
- Programs for the machining on drilling and grinding machines with point-to-point and straight-line control
- Programs for the machining on turning machines with straight-line control or continuous-path control
- Programs for the grinding on machines with point-to-point and straight-line control
- Programs for the measuring on machines with point-to-point control

In all these programs the preparation of the source programs and the internal processing in the computer is carried out according to a uniform basic principle which has been represented in the form of a flow chart in fig. 1.

Analogue to the programs listed above the programming language for the preparation of the source programs is subdivided into different language parts. They are all of a similar structure and have a unified vocabulary.

In developing the programming system fullest consideration was given to the requirements of the users. This expresses itself in the simple structure of the symbol language, the possibility to use a clearly arranged program form, the configuration of the vocabulary which has been adapted to the native language, the universal applicability made possible by the many ways of description and also in the technological orientation of the system with e.g. automatic computation of tool and cutting values.

The individual programmes have been carried out with one or several electronic data processing machines as for instance IBM 360/40, ICT 1904, R 300, Minsk 22. More than 20 post-processors for the system have so far been prepared, technologically tested and introduced. In the following, with the help of automatic measuring a program will be explained in detail.

2. COMPILER MEASURING AND ITS CONNECTIONS WITH DATA ACQUISITION AND DATA PROCESSING

The serial-model modification described below permits the measuring of prismatic works with axially parallel holes or bores, and faces as well as the provision of the data required for the evaluation of the measured values. The faces and holes of the body to be tested are regarded separately as individual form elements. The connection with regard to the position of the faces or holes and the determination of form deviations are evaluation operations and are done solely by computation.

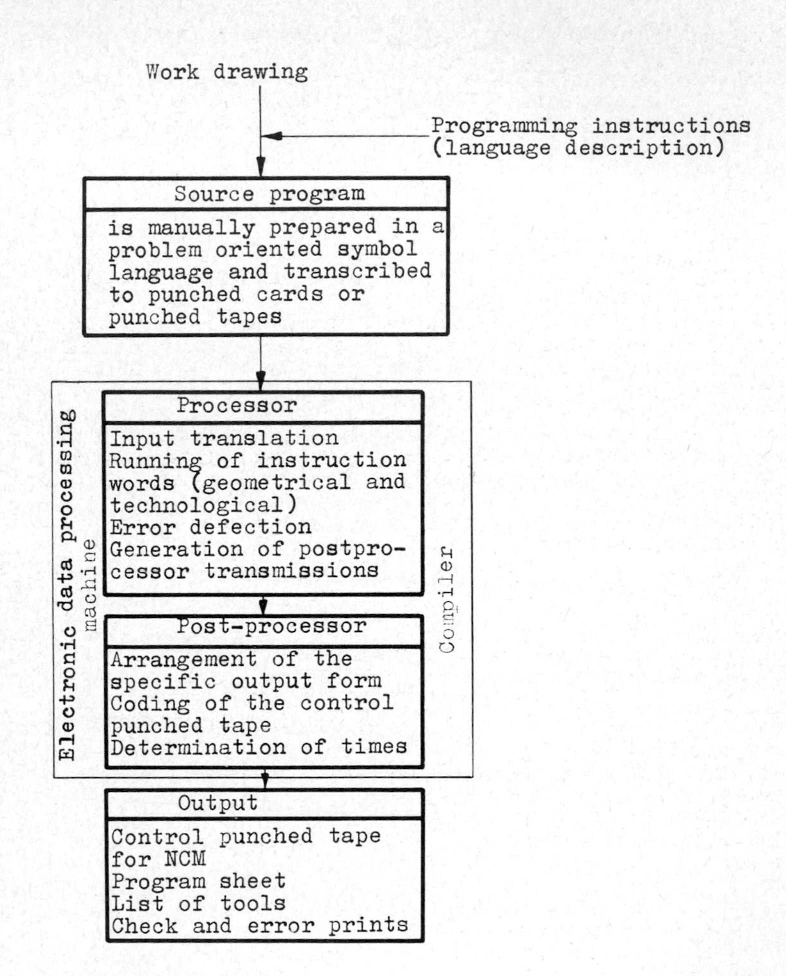

Fig. 1: Flow chart

This subdivision into individual form elements yields among others the following advantages:

- Measuring machines are not restricted to fixed motion cycles which result from the kind of the measuring task.
- The faces and holes can be measured in any order of succession.
- Repeated measuring of faces and holes due to a sequence of measuring tasks is no longer required as the subsequent evaluation is based on normal forms of the faces and holes that have to be formed.
- Necessary collision considerations are reduced to a minimum.

The provision of the automatic programming comprises the measuring and evaluation parts.

The technological operation data together with the geometrical instructions initiate the motion of the measuring machine.

Each measured face or hole is given a name to designate the measured values. The preliminary data for the evaluation of measured values are coded in the compiler-measuring system and constitute for the computer by which the measured values are processed an input information which addresses certain computing algorithms permitting thus an evaluation. Besides the preliminary data for the evaluation, the normal-form formation of form elements immediately after obtaining the measured values constitutes the basis for evaluation (fig. 2).

As a result of automatic programming we obtain a program sheet and a punched tape which contains the necessary machine instructions for measuring machine control, and the instructions for the evaluation of the measured values. The instructions for evaluation are supplied by the control of the measuring machine to the computer which evaluates the measured values obtained.

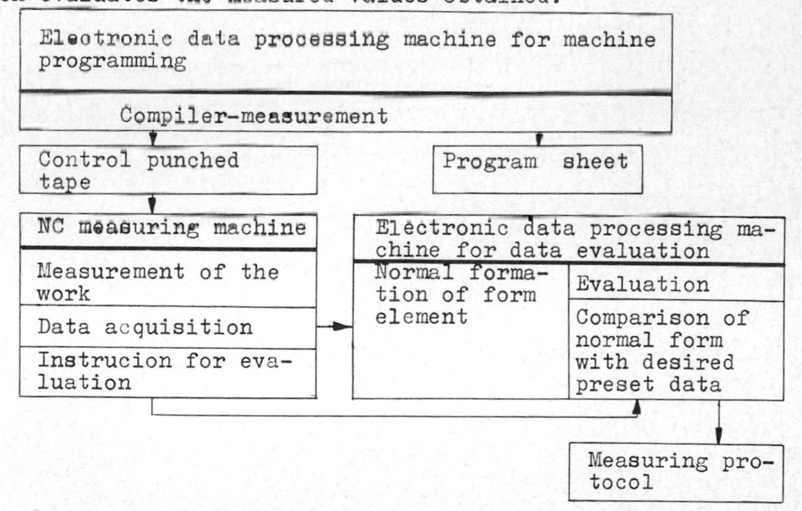

Fig. 2: Flow chart for measuring

3. SOURCE PROGRAM

The source program is the input program for the computer. It contains those data which are required for the preparation of the control information and the preliminary data for evaluation, and which have not already been directly stored in the processor.

The source program consists of the organisational part, the part for technological and geometrical preliminary data and the instructions for the evaluation of measured values.

In the organisational part there takes place the provision of technical instructions related to the program as

- Program number
- Input code
- Post-processor instructions listing the machine, kind of control and NCM code
- Block instruction
- Remarks
- Start instructions

In the part for technological and geometrical preliminary data the present measuring sequence is described. The technological data are provided in the individual operations VTEST, MED, MEFL and MEFLS. The positions of the holes and the measuring points on the surfaces are given by point definitions and point quantity definitions. This geometrical description has been taken from the language part which is valid for the machining on drilling and milling machines with point-to-point and straight-line control.

The technological part is followed by the part concerning the instructions for evaluation.

The source program can be prepared in a fixed and in a free form. For the first input method there exists a form with columns for instruction words, variables, definition words and parameters. In the case of the free form, there must always be a comma between the individual words and the parameters.

4. DESCRIPTION OF TECHNOLOGICAL OPERATIONS

Measurement of the work was realized with respect to language and program by four technological operations. To this end axial and radial calipers were used.

4.1. PRELIMINARY TEST

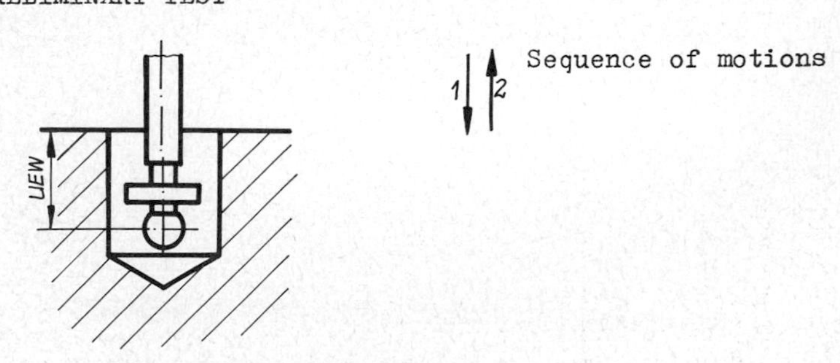

Fig. 3: Diagrammatic sketch VTEST

The axial caliper is lowered a predetermined path length into the hole. If the caliper does not respond up to the predetermined overtravel path (UEW), the hole can be measured without any danger by means of a radial caliper. In this technological operation the caliper position, diameter, clearance depth and the overtravel path are entered as parameters according to the given case. At this, the parameters of all technological operations are subdivided into compulsory and additional parameters.

The travel to the centre point of the hole in front of the surface is effected by giving the geometry following the technological provision of data.

4.2. MEASUREMENT OF THE DIAMETER (MED)

In this technological operation the motion sequence shown is realized by the program. It is possible to measure in one or several planes. The cycle also permits the measurement of holes open only on one side. If the depths of the individual working planes are not given by the source program, they are automatically determined depending on whether there is a through-hole or a bottom hole. As parameters for this technological operation are preset the caliper position, diameter, clearance depth, overtravel path, working depth of the bottom or through-hole, the number of working planes and, if necessary, the depths of the working planes. If the depths of the working planes are preset in the source program, a test is carried out in the program MED whether the radial caliper can travel to the predetermined depth without any danger.

Just as in the case of VTEST the travel to the centre point of the hole in front of the surface is effected by geometrical data.

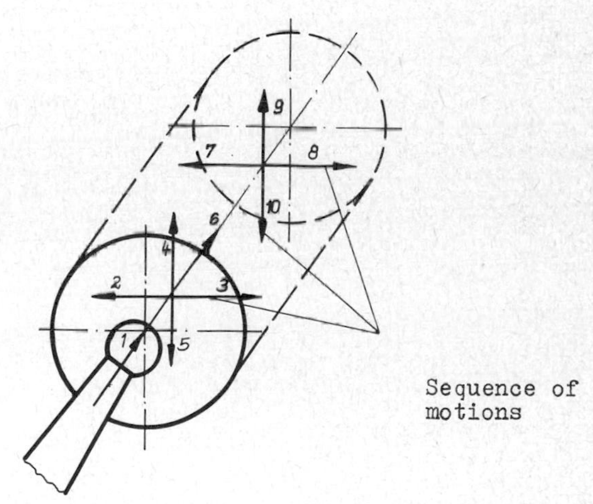

Fig. 4: Diagrammatic sketch MED

4.3. MEASUREMENT OF SURFACES (MEFL AND MEFLS)

In the surface measurement, the surface is measured point by point according to a scanning pattern provided in the geometrical part and having a certain distance between lines and columns. The program initiates the represented travel for each individually measured point.

As parameters are preset the caliper position, overtravel path (UEW), plane angle, plane angle of inclination (MEFLS) and a modifier which determines whether the travel sequence is realized by the lifting-off motion of the caliper or by moving the caliper axis.

The UEW magnitude confines the travel path of the measuring machine in the measuring direction. If, after travelling the overtravel path, the caliper point does not touch the work, the movement of the measuring axis is stopped. Thus the UEW acts as an additional safety device for the measuring machine.

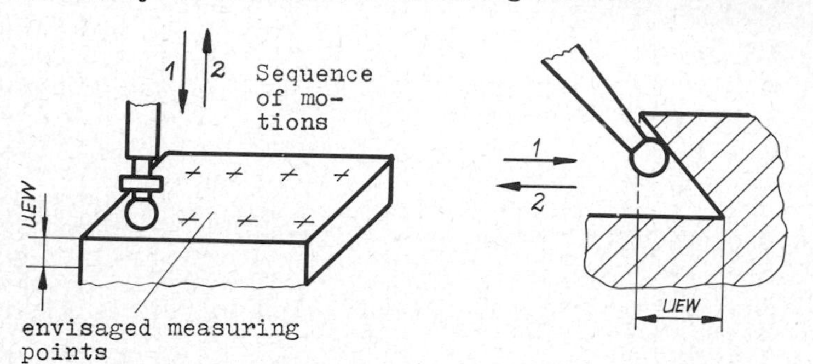

Fig. 5: Diagrammatic sketch MEFL

Fig. 6: Diagrammatic sketch MEFLS

5. REALIZATION OF 6-SIDE MEASUREMENT

To describe the centre points of the holes and the measuring points on the surfaces, clockwise Cartesian coordinate systems are used. The coordinate system of origin with its defined position with respect to the machine coordinate system constitutes the starting point. Starting from the coordinate system of origin, reference coordinate systems are provided by the instruction word DRLAGE. Each reference coordinate system has its radius of action for a certain work plane. with the positive Z-axis of this coordinate system always pointing outwards from that plane. Within the reference coordinate systems, the sequence of motions before the plane specified with the respective reference coordinate system, is determined by the provision of points and point quantities. For this positioning from plane to plane, i.e. from work side to work side, there is also used the instruction word DRLAGE in connection with the instruction word HIND. By HIND the computing program is provided with the highest work edge, i.e. obstructing edge with respect to the coordinate system of origin.

In the case of positioning from work side to work side, the following sequence of motions will result:

- Movement of the caliper point in the Z-direction of the coordinate system of origin (compare fig. 7, (1) on HIND + SIABST (SIABST being a safety distance provided by the program)
- Movement in the X-Y-direction of the coordinate system of origin by the displacement values given in the instruction DRLAGE taking into consideration SIABST (compare fig 7 (2).)

The further movement in front of the work side is then effected by provision of points or point quantities.

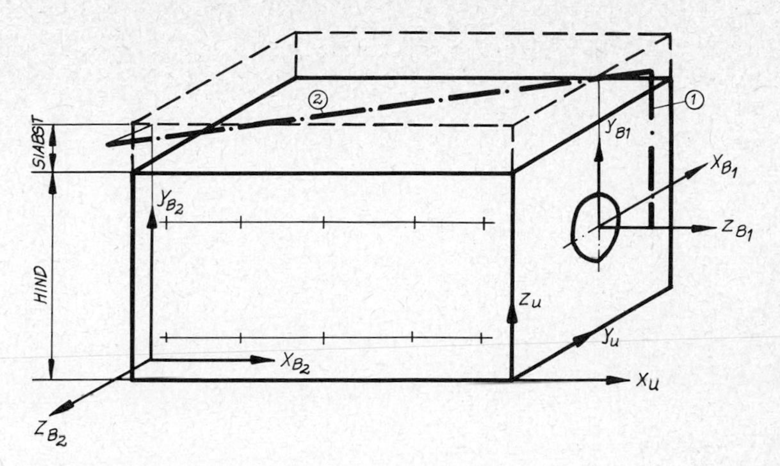

Fig. 7: Measurement of the work sides

6. INSTRUCTIONS FOR EVALUATION

As to the language and the program there are 16 evaluation operations being carried out, e.g. evaluation of diameter, coaxiality, shape deviations, distance, rectangularity.

For each evaluation operation there exists a certain definition word. For example, DBC means evaluation of the diameter and provides with the parameters hole number, desired diameter, upper and lower deviation the preliminary data for this measurement evaluation method. The representation of further evaluation operations can be taken from the source program example.

7. PROGRAM EXAMPLE

7.1. PROGRAM SKETCH

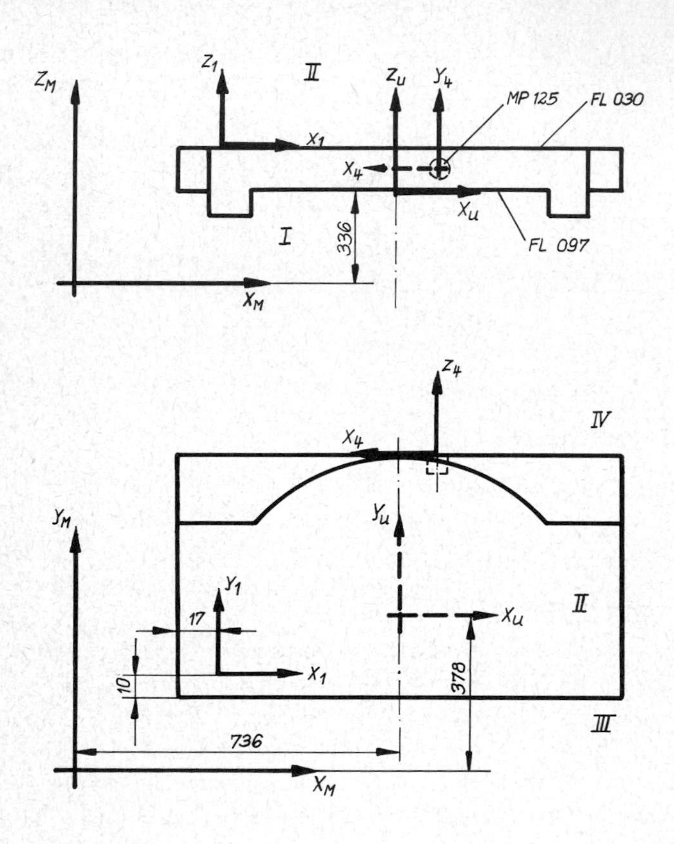

Displacement values

FL,MP	Variable	X	Y	Z	∡ x	∡ y	coord. index
FL 030	B1	-330	-153	85	0		1
MP 125	B2	50	302	50	90	180	4

Fig. 8: Program sketch

7.2. SOURCE PROGRAM

No.	Instruction	VAR	DEFW	1.parameter	2.parameter	3.parameter	4.parameter	5.parameter	6. parameter
1	PROGR	PRIS2	TITEL	518	ISO	VOIGT	16.7.71		
2			PPROC	PCMK09	PRS	ISO			
3	BLOCK		TMESS2						
	BEMERK	SWIVEL	SLIDE	203.17-0400-20.00.FINAL CHECK. PASSAGE 5					
5	EINRI	L 1	LAGE	736	378	336	0	0	
6	START	F 1	FTINF	1105	09300				
7	HIND	A 1	IST	185					
8	SIABST	A 2	IST	10					
9	DRLAGE	B 1	BXYZW	-330	-153	85	0		
10	BE	BL1	BLAGE	2					
11	TECH	T 1	MEFL	TST 71	DYNAM				
12	DEF	Q 1	QMATR	0	0	220	4	100	4
13		Q 2	QXYXYN	0	400	660	400	4	
14		Q 3	QQN	Q2	OHNE	1	4		
15	NAME	FL030							
16	OPENDE	Q 4	QSUM	Q1	Q3				
17	DRLAGE	B 2	BXYZW	50	302	50	90	180	
18	BE	BL2	BLAGE	1					
19	TECH	T 2	MED	TST 14	D 35 AD	10	ME 1	MEE 3	
20	NAME	MP125							
21	OPENDE	P 1	PXY	0	0				
22	AUSWER	AW1	FOABWFL	030	FZ 0.015				
23		AW2	PARALFL	030	FL 097 BZS 700		LZ 0.015		
24	....								

7.3. Post-processor print

```
PROGRAM NUMBER 518    PROGRAM CODE WORD   PRIS2   MEASURING MACHINE   C-MLPZ

PRT NUMBER: 110509300
--------------------------------------------------------------------------------------------------
No.   L  X/Y/Z/R/W  Y/Z/R/U  K/P/F  G   G   G   G   M   H H H H  A/V      B/U   C/O/    D/T          J
:001 L00                                            M12
N002 L05                                                H097     A0736000 B0378000
                                                                               C0336000
N003 L00         %Z1900000  K12    G22              M16
N004             %R1900000  K73                     M16
N005  %X0050000  Y1015000   K33                     M16
N006   W0950000
N007 L00 W0376000
:008                                    G25
N009 L00 %X0090000 %Y0310000 K22
N010               %Z0820000
....
....
....
....
....
N101 L01  X0982500 U10000   P2     G44       G30        H125                                  D+000000J+000000
N102 L00  X1010000
N103 L01  X1037500 U10000   P1                          H125                                  D+000000J+000000
N104 L00  X1010000
N105 L00                    K24    G43
N106 L01  Z0928500 U10000   P1                          H125                                  D+000000J+000000
N107 L00  Z0901000
N108 L01  Z0873500 U10000   P2          G23             H125                                  D+000000J+000000
N109 L00  70901000
N110 L00  Y1035000
:111 L19                                                H030                     0+0015
N112 L11                                                H030 H097                0+0015       T 0700000
N113 L28                                                H030 H125  V0035000 U+0000
                                                                                 0+0100
N114 L10                           G23                  H125     V0028000 U+0000
                                                                                 0+0021
N115 L00        %Z1326000          G22                 M16
....
```

8. CONCLUDING REMARKS

The compiler-measurement system was tested on NC measuring machines in a computer controlled machine system, and also applied in practice. By means of the compiler-measurement system the expenditure for the programming of NC measuring machines is reduced by about 70 %. One line of the source program results after the computer operation in about 5 control sentences on the NC punched tape. To compute one source program, about 3 min, computing time are required with an electronic data processing machine which has an available storage capacity of 100 K. Practice has shown that with automatic programming in addition to the time saved, the frequency of errors in the preparation of the NC program for the measuring machines is significantly reduced.

The use of automatic programming thus contributes considerably to the increase of labour productivity in work-preparing departments.

DISCUSSION

Mr. N. Okino, asked by Mr. C.A. Lang what additional information on the cutting sequence was required beyond the geometry description, replied that though the aim was to make this automatic, at present the manufacturing specifications are input manually. In the latest laboratory version a penalty method was used to decide the cutting sequence, but the cutter diameter was still input manually. Answering Mr. M.A. Sabin, Mr. Okino said that his system permitted shapes with pockets, free surfaces and other special patterns.

Mr. H. Eitel, who asked Mr. I. Macurek whether the parts shown on his slides had been made on 5-axis machine tools, was told that all the CKD machines were 3-axis ones only. In reply to Mr. B.G. Tamm, Mr. Macurek said that CKDAPT was written in FORTRAN and was transportable. Technological sections are being prepared.

A question to Mr. I.C. Braid from Mr. J. Vymer, related to the selection of the building bricks of his system. Mr. Braid said they had been selected to be a sufficient set, in fact they are more than that. The important choice was to provide faces which were flat or cylindrical and to provide operations for building objects composed of faces of this type. Asked by Mr. G. Hermann why the cone and sphere were not used, Mr. Braid replied that while cones could easily have been, spheres could not be included. Work is now in progress towards more general surfaces. The program now runs in 12 K words of 48 bites. For a complicated shape 7 minutes computing time was required.

Answering Mr. H.H. Winkler, Mr. Braid said the teletype was adequate as an input device for part description. The system is not a language processor and has no consecutive compiling and execution phases. Mr. M.A. Sabin asked whether, in cases where there was some duality in design, a part could be described in different ways at different stages. Mr. Braid replied that the data structure does not take notice of dualities. As for the input language, there should be ways of meeting this requirement, especially through the use of negative volumes.

Mr. A. Morelj was asked by Mr. S. Pilz, whether the PERFEKT system could be used for the design of integrated circuits or multi-layer printed circuit boards. Mr. Morelj said they had another, completely independent system for this.

In a question to Mr. T. Márkus, Mr. M.A. Sabin asked why generate a part program, instead of a CLTAPE. The author replied that the main reason was to keep the program small and not have to incorporate a motion section, which is anyway present in all the processors. Mr. D.G. Wilkinson pointed out that in his experience the hardest part of cam design on automatic lathes was the minimization of cycle times. Was this automatic here? Mr. Márkus said that at present all the technological data /distance of tool head, tool length, etc./ were input to the program.

Mr. H.H. Winkler asked Mr. W. Frohberg whether the measuring machine programming language was an extension of SYMAP, whether the machine was controlled conventionally or by CNC, and what happens if there are gross errors in the part. In the reply it was pointed out that the programming language described in the paper is for the evaluation of measurements and not for producing control tapes for NC measuring machines.

4. STANDARDIZATION

N/C LANGUAGE STANDARDIZATION IN I.S.O.

BY

WERNER E. MANGOLD *

SPERRY - UNIVAC CORPORATION

PHILADELPHIA, PENNSYLVANIA, U.S.A.

The need for standardization is clearly demonstrated by the continuing explosion of language developments. The criteria for standardization demands justification based on usage, compatibility, number of problem solutions and implementation feasibility.

When I last spoke to you at the 1969 Prolamat Conference in Rome we jokingly established that "A standard is a dull document! It is produced by a committee of dull people who argue forever. They consume reams of paper before they agree on a position which is already obsolote when it is adopted! A standard is an unfair way to limit competition! Often it is quickly pushed through a secret committee!" Etc., etc., etc.

Today I wish to give you a brief glance at our committee's history, organization and accomplishments since then.

1. STANDARDS

Standards developed under the procedures of ISO, The International Standards Organization, involve no secrecy and no exclusion of any interested parties. The procedure may often appear quite sluggish because of the requirement to canvass all interested parties and numerous provisions for review and appeal.

In the field of numerical control languages, we can show the opposite to standardization by simply listing some of the languages currently in existence. The following list comes to 33 N/C languages but there may well be twice that many now:

---

* Chairman, I.S.O. NC Language Committee, Manager, Manufacturing Systems, SPERRY-UNIVAC, P.O.Box 500, Phila., Pa., U.S.A.

ADAPT	AUTOPRESS	CLAM
APT	AUTOPROMPT	COCOMAT
UNIAPT	AUTOPIT	BSURF
IFAPT	AUTOSURF	MILMAP
MINIAPT	SPLIT	PAGET
EXAPT	SNAP	PMTZ
CINAP	CAMP	PROFILEDATA
APTLOFT	PRONTO	NUMERISCRIPT
AUTOSPOT	INCA	FMILL
AUTOMAP	SYMPAC	ACTION
AUTOPROPS	2CL	ZAP

2. THE INTERNATIONAL STANDARDS ORGANIZATION /I.S.O./

ISO was organized in 1946. It consists of the national standards bodies of 55 countries. The administration of ISO is in the hands of a council. The council appoints a general secretary who administers the general secretariete in Geneva, Switherland. The work of ISO is done by International committies. Approved standards are called ISO recommendations.

The objectives of ISO are:

1. To promote the development of standards throughout the world.
2. To further the coordination and unification of national standards.
3. To set up international standards providing in each case that no member body dissents.
4. To exchange information about the work of its member bodies and technical committies.
5. To cooperate with other international organizations which are interested in relating matters.

TC 97 is the ISO technical committee which is concerned with DATA PROCESSING standards for computers and information processing. It was established in 1961 at a conference on international standards for computers. TC 97 is responsible for standardizing the terminology, problem definition, programming languages and other data processing areas. The secretariat for TC 97 is held by the U.S.A.

TC 97 is made up uf 14 subcommitties, one of which, SC9, is for Numerical Control programming languages. The scope of the subcommitee is: "Standardization of Programming Languages for Numerical Control."

There are 10 participating member bodies:

France	Japan	Sweden
Germany	Romania	Switzerland
Italy	Hungary	United Kingdom
		U.S.A.

3. I.S.O. N.C. LANGUAGE COMMITTEE

The first meeting of the N.C. Language commitee took place on November 4, 1967 in Paris. It was then identified as Working group 1 within the programming language commitee S.C.5. This group held five more meetings since then: November, 1968 in Geneva, September, 1969 in Turin, September, 1970 in Berlin, and October, 1971 in London. The last meeting here in Budapest in April of 1973 was the first one under the helmet of S.C.9.

The following major accomplishments can be identified for each meeting:

PARIS '67

- Committee formation
- Definition of Scope and plan of work
- Identification and acceptance of candidate languages

GENEVE '68

- Establish working sessions for Sub-sets and Modular features, Input language and CLDATA
- Definition of the content of the CL Data Specs
- Establish liaison with SC-8 /N/C Hardware Committee/

TURIN '69

- Start the interim Ad Hoc Group meetings
- Rough draft release of CL Data major protion
- Definition of sub-sets and modular features
- Separation of input language and CL Data

BERLIN '70

- First attempt to create an input language document specifying all current APT III language which is also found in the other candidate languages
- Continuation of the effort to create a reference language
- Finalization of CL Data major portion standard and release for circulation at the sub-committee 5 level.
- Continuation of work on CL Data minor portion

LONDON '71

- First rough draft of input language document
- Survey of voting on CL Data major portion at the SC committee level

BUDAPEST '73

- Send CL Data major portion to T.C.97 into circulate for approval
- Complete minor portion of CL Data prior to next SC.9 meeting
- Complete basic input language document prior to next SC.9 meeting
- Study expanded scope to include languages for other industrial processes.

The task to come up with a recommendation for a standard international N/C language is monumentous. We have four proposals: APT /USA/, EXAPT /GERMANY/, 2CL /UNITED KINGDOM/, and IFAPT /FRANCE/. All languages are often called "APT LIKE".

One distinguished member of the committee once tried his hand at defining what is "APT LIKE" or "APT COMPATIBLE". He ran up against phrases like: compatible in language, compatible in structure, compatible on input language; similar to, like, etc., etc. As a result the following "simple" definition emerged:

> "An APT-like language is a program, built up of rules, which contain Fortran-like statements, English-like verbs and sequences of at least one English-like word and Arabic-like numbers. Where necessary they are separated by commas and/or a slash which does not have the meaning of an arithmetic operator and no numeric values may appear to the left of the just mentioned slash."

Now following this simple definition of an APT-like language let us look at some of the differences in the already mentioned languages:

1. VOCABULARY WORDS

APT:

DRILL REAM TAP BORE MILL	are minor post-processor keywords
TOOL	is a major keyword.

EXAPT:

DRILL REAM TAP BORE MILL	are major keywords
TOOL	is a minor keyword

2CL:

The listed keywords are used in the same way as in APT.

2. NAMES AND IDENTIFIERS

APT:

a. An identifier is a symbolic name, formed by a string of alphanumeric characters and it contains at least one letter.
b. A label is a symbolic name, formed by a string of alphanumeric characters, without any restriction.

EXAPT:

a. Same as in APT, but with the restriction that the first character is a letter.

b. Same as APT.

2CL:

Same as APT.

3. GEOMETRIC DEFINITIONS

APT:

```
a. Z S U R F / <plane>
b. P A T E R N/G R I D, <parameter list>
c. P A T E R N/R A N D O M, <parameter list>
d. T A B C Y L/N O Z,S P L I N E,<point> ,S L O P E, <scalar> ,
                -----------------,<point> ,S L O P E, <scalar>
```

EXAPT:

```
a. Z S U R F/ <scalar>  $$  where the scalar value defines a
                            plane parallel to the XY-plane.
b. P A T E R N/T R A F O, <parameter> list
c. P A T E R N/R A N D O M, <parameter> list $$ all the points
                            in the pattern must have same
                            z-value.
d. Not defined.
```

2CL:

```
a. Same as APT
b. P A T E R N/P A R L E L,  <parameter list>
c. Same as APT
d. T A B C Y L/N O Z,S P L I N E, <point> , T A N T O, <line> ,
               -----------------, <point> , T A N T O, <line>
```

4. MOTION STATEMENTS

APT:

a. F R O M/x,y,z
b. G O T O/x,y,z

EXAPT: Same as APT

2CL: Same as APT

---

5. SPECIAL STATEMENTS

APT:

a. R E F S Y S/ <matrix> $$ In a refsys-area only geometric definitions are allowed.
b. M A C R O - definition - S Y N and R E S E R V are not allowed in a M A C R O and cannot be used as actual parameters in the M A C R O-C A L L-statement.

EXAPT:

a. T R A S Y S/ <matrix> $$ The T R A S Y S also applies to motion-statements and moreover, every 3-coordinate defined by a Z S U R F-statement outside the T R A S Y S-area is set to zero.
b. M A C R O - No restriction in the use of S Y N and R E S E R V

2CL:

a. Not defined.
b. No restrictions.

6. INPUT-OUTPUT STATEMENTS

APT:

a. P A R T N O ' <string of characters> '
b. R E M A R K ' <string of characters> '
c. I N S E R T ' <string of characters> '
d. P R I N T ' <string of characters> '
e. When not using the quotation-marks the major words are fixed field words.

EXAPT:

a. P A R T N O/ <string of characters>
b. R E M A R K/ <string of characters>
c. I N S E R T/ <string of characters>
d. P P R I N T/ <string of characters>

2CL:

a. Same as APT.
b. Same as APT.
c. Not defined.
d. Same as APT.

As you can see standardization without a rigorous method of defining any language is very difficult.

5. ORIGINAL PLAN OF WORK OF I.S.O. N/C LANGUAGE COMMITTEE

The complete plan of work shapes up as follows:

Task 1: Determine if a candidate language falls within currently available criteria from ISO/TC 97/SC 5 to be applied in the standardization of a programming language.

Task 2: Study:
a/ Scope and organization
b/ Languages, current and future
c/ Subsets and Modular Features

Task 3: Define a reference language similar to APT from which proper subsets can be derived.

Task 4: The ISO N/C language committee will establish liaison with Subcommittee ISO/TC 97/SC 8 and other applicable ISO Technical Committees.

Task 1 mentions criteria for standardization. This pertains to the attributes of a candidate language such as its need, utility, and general acceptance. Following are just a few of the requirements:

a/ A substantial number of prospective users
b/ The language must accommodate a substantial number of the users problems
c/ The language should be compatible with other applicable practices
d/ A processor of the language must be implementable with hardware and software presently available.

6. MEETING RESULTS

The plenary meeting of the committee is usually broken up into four special sub - group meetings in order to pursue the program of work in a more efficient manner. These four groups, their tasks and accomplishments are as follows:

Group #1 - Subsets and modular features:

Task: /a/ To obtain a definition of subsets and modular features.
/b/ To define a nucleus and to approach the task not from the level of a reference language but from its elements.

The following definitions have been agreed on:

A reference language is an N/C language from which all further subsets or candidate languages are derived.
A subset is a complete N/C programming language capable of producing an output for a particular type of manufacturing requirement.

A modular feature is an additional capability or part of a subset which cannot be processed on its own.

An attempt was made to classify the language subsets into five main groups not including the complete Reference Language. The break down is as follows:

1. Point to Point
2. Two dimensional contouring
3. $2\frac{1}{2}$ dimensional contouring
4. 3 dimensional contouring
5. Multi axis contouring

A nucleus or minimum language required to process a simple part program for point to point machinery was defined by listing facilities such as geometric definitions and motion commands, administration and post processor commands.

Group # II - Numeric Meta Language Processing System /NUMEPS/

This group has been suspended. The original task, to explain the meta language as a tool and to make the work of the committee simplier was not fulfilled because it was not possible to find a single authority for the control of documentation, processing of the meta language and the allocation of code numbers. Two ground rules have been set up:

a/ Additions to the Reference Language shall be made at the request of any one of the sponsoring organizations

b/ Deletions from the Reference Language shall only be made with the agreement of all the sponsoring organizations.

The new Group # II task is the definition of the input language for technology features.

Group # III - CLDATA

Task: To recommend an I.S.O. standard CLDATA format.

Early agreement was reached that every candidate processor shall have the facility to produce an APT III CL tape, possibly <u>in addition</u> to its own CLDATA RECORD type. As a result, the many industry proven APT III postprocessors could be used on other APT-like processors, i.e., EXAPT, 2CL, IFAPT.

As ISO draft proposal called "N. C. Processor Output," Logical Structure and Major words, has been prepared by the Committe /SC5/1/ and upon first circulation to all the member bodies of SC-5 produced the following vote results: For the standards: /6/ Germany, Italy, Sweden, United Kingdom, Canada, /0/ Chechoslovakia /0/. Against the standard: /2/ France, U.S.A., /1/ Japan.

This proposal is now being passed on for a vote by the next higher committee, namely, TC-97. Approval is generally granted at that level if the lower level committee has previously passed it.

The standard for the major portions of an N.C. processor output will be used in conjunction with the minor elements proposal which is still being worked on.

The CL Data specification of the postprocessor standard is based on APT III working practice with consideration given to technology extensions. The ISO recommendation states that <u>each processor</u> using any N.C. programming language shall be capable of producing CL Data as defined in the standard. Each Postprocessor shall be capable of using the standardized CL Data as its input.

All major words are grouped into /14/ record types as follows:

RECORD TYPE	NAME
1000	Input sequence
2000	Post-processor instructions
3000	Surface data
5000	Tool position
6000	Tolerance or cutter information
9000	Axis mode; units
14000	"Fini" record
15000	Unsegmented tool path
16000	Workpiece contour description

RECORD TYPE	NAME
17000	Tool description
18000	Material description
19000	Machine description
28000 to 32000	Proprietary records.

There are a total of 60 major words as follows:

MAJOR WORD	INTEGER CODE	MAJOR WORD	INTEGER CODE
AIR	1011	CLEARP	1004
AUXFUN	1022	CLRSRF	1057
BREAK	16	COOLNT	1030
CHUCK	1073	COUPLE	1049
CLAMP	1074	CUTCOM	1007
CLDIST	1071	CYCLE	1054
DELAY	1010	GOHOME	17
DISPLY	1021	HEAD	1002
DRAFT	1059	INSERT	1046
DRESS	8	LEADER	1013
END	1	LETTER	1043
FEDRAT	1009	LINTOL	1067
LOADTL	1055	OPSTOP	3
MACHIN	1015	ORIGIN	1027
MCHTOL	1016	OVPLOT	1042
MODE	1003	PARTNO	1045
OFSTNO	1083	PENDWN	12
OPSKIP	1012	PENUP	11
PIERCE	1090	RAPID	5
PITCH	1050	RETRCT	7
PPFUN	1079	REWIND	1006
PPLOT	1014	ROTABL	1026
PPRINT	1044	ROTHED	1035
PREFUN	1048	SAFPOS	1094
SELCTL	1056	THREAD	1036
SEQNO	1019	TMARK	1005

MAJOR WORD	INTEGER CODE	MAJOR WORD	INTEGER CODE
SPINDL	1031	TOOLNO	1025
STAN	1080	TRANS	1037
STOP	2	TURRET	1033
SWITCH	6	UNLOAD	10

The proposed standard for the minor portion of the CL Data is nearing completion now. It was of course much more difficult to reach worldwide agreement on the large variety and number of combinations possible when the total postprocessor statement is considered.

The following example shows the major and minor portions of typical postprocessor input statements:

SPINDL/RPM, 5000, RANGE, 2

The document on minor elements defines, in the context of the major word, the minor elements which can be associated with each of these previously defined major words within a postprocessor statement.

It is also recognized that existing NC processors allow minor elements of postprocessor statements in any order without restriction. Postprocessors normally check the validity of particular element strings.

The integer code /IC/ numbers given in the standard are the codes which are used to represent the input language vocabulary key words on CLDATA.

GROUP IV - Input Language Comparison

Task: /a/ To arrive at similarities
/b/ To discuss and possibly resolve differences.

Short of an effective syntax comparison of all candidate languages this group has recorded identities. This area of agreement is now being enlarged as differences become resolved.

As a result of the April, '73 Budapest meeting we are now in the final stages of preparing a proposal document for submission to TC 97. The document is called "N.C. Processor Input, Basic Part Programming Language." In essence the document covers that part of the ultimate I.S.O. reference language which has received international acceptance to date.

7. GENERAL OBSERVATIONS

As long as computer manufactures supply an N/C language processor "free" a user is tempted to use it. However, because of the great variety of N/C tools the user himself must secure a post processor or other form of output tailoring.

In the United States, the momentum is for the APT language and APT post-processors. Users of the popular AUTOSPOT did bring about increased point to point accomodation in APT while they could not get new post processors for AUTOSPOT only.

The power of the European processors EXAMP, 2CL and IFAPT should become increasingly known in the U.S. as they are gaining in popularity in Europe. In summary it appears that the APT like languages are most popular today. The proponents of other language may have to demonstrate why one should not use what is sometimes referred to as the "defacto" standard.

8. THE FUTURE

There should be a continued effort in the area of workshop technology, area clearance and sculptured surfaces. One can foresee a steady accumulation of relative data as to the behavior of machines and tools instead of any sudden breakthrough in predictive algorythms.

Real innovations may show up as byproducts from the efforts in computer graphics and bounded geometry. Forinstance, the major conflict on a sculptured surface is on "SMOOTHNESS" both visually and in the form of stress distribution. The integrity of the cutter path against violation of boundary surfaces or other sectors of the same surface must also be guaranteed.

Most of the major computer manufacturers will offer only APT like languages and even if software is separately priced an independent organization can hardly afford to create and promote an alternative. Standardization in ISO should continue. After the acceptance of the first international output and basic input language recommendations a full superset reference language with defined subsets is the next target. Due to the time delays in adopting standards, provisions must be made to allow for the useful APT like languages of today and the future.

A truly standard language processor may or may never be implemented on any computer. But the existence of a standard will force a high level of compliance and strong justification for any variation.

9. CONCLUSION

The job to be done by a standards group is not an easy one. Often they are second guessed, criticized and even questioned about their true intentions. But the work is surely for the benefit of many concerns, countries and economies, so, in order to insure that the results are benefical to the greatest number we need the broadest possible representation.

We invite all member and observer nations, which do not yet participate in this activity to send their Numerical Control Language expert to the meetings of the I.S.O. N.C. language subcommittee.

The next meeting takes place in May '74 in Paris. It will be a 3 day meeting. To become an active member and voting participant one has to be assigned as an official delegate by a particular member nation's standards organization.

ISO 97/5/1 MEETINGS

1	Paris	Nov 67
2	Geneva	Nov 68
3	Turin	Sep 69
4	Berlin	Sep 70
5	London	Sep 71
6	Budapest	Apr 73

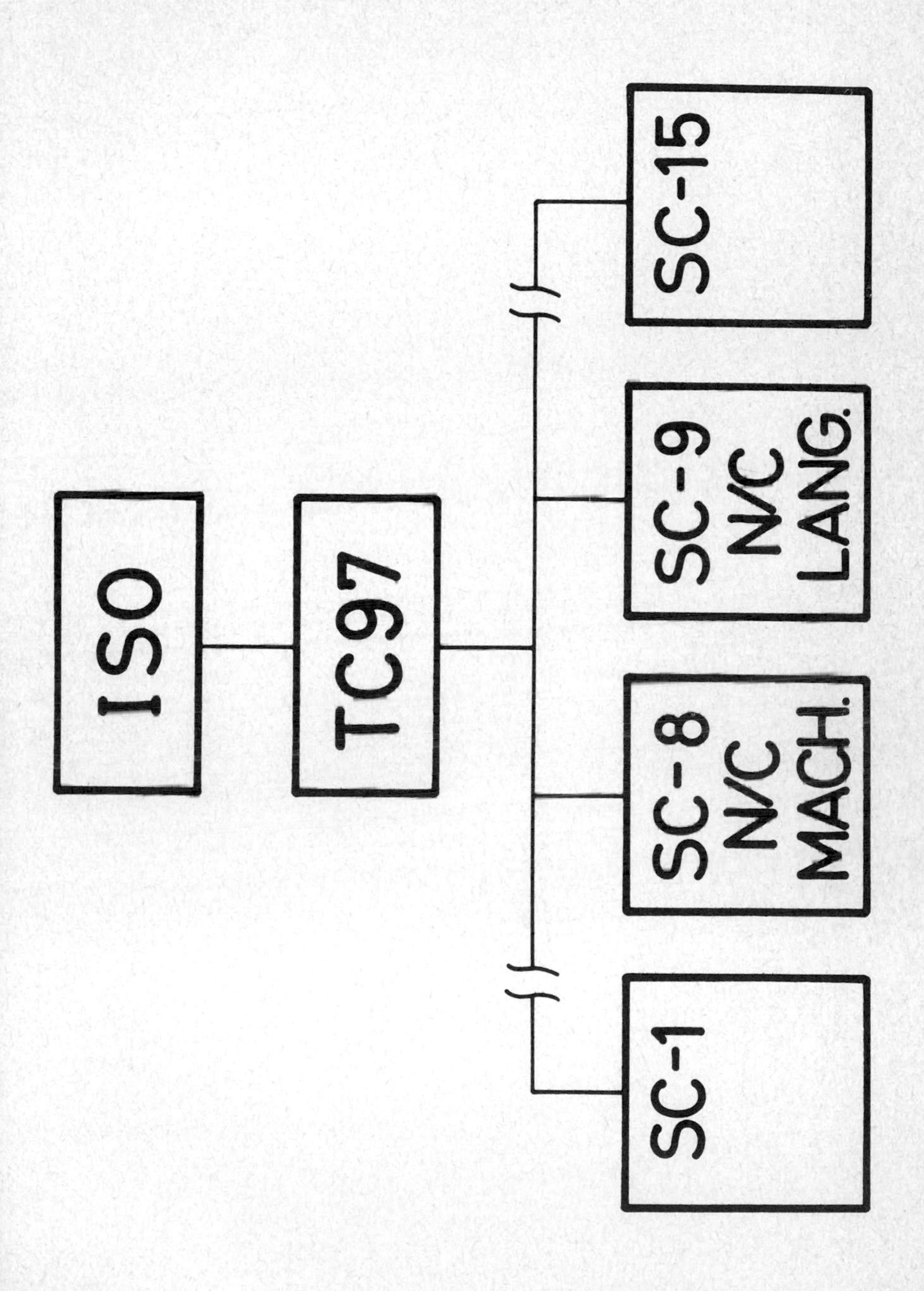
ISO
TC97
SC-1
SC-8
N/C
MACH.
SC-9
N/C
LANG.
SC-15

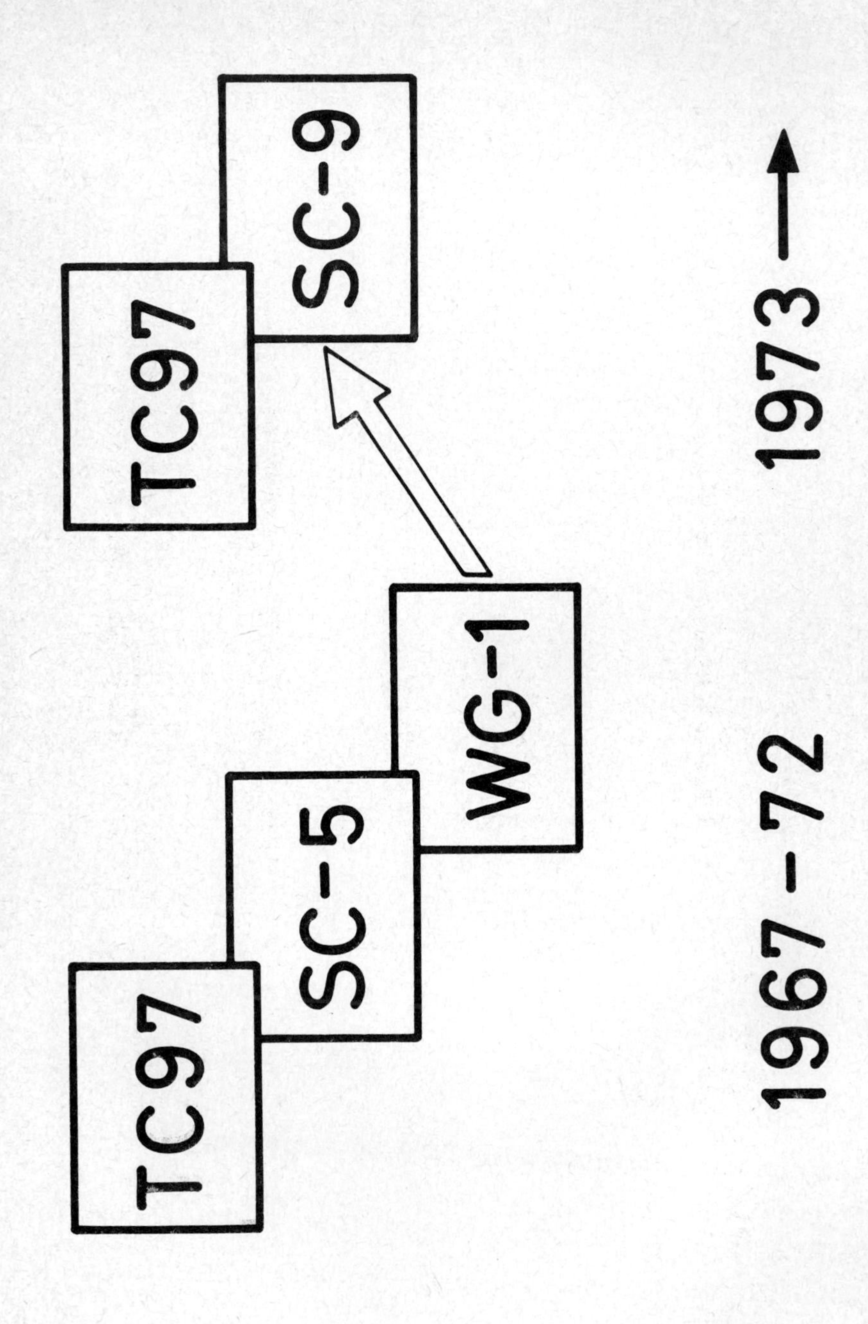
TC97
SC-5
WG-1
1967 - 72
TC97
SC-9
1973

ISO 97/5/1 RESOLUTIONS 1967–72

37 Proposed

37 Passed

218 "Yes" Votes(combined)

6 "No" Votes(combined)

165 DOCUMENTS (1967-72)

47	U.S.A. & secretariat
35	97/5/1
26	U.K.
18	Germany
17	France
6	Switzerland
5	Italy
4	ECMA
3	Japan
3	Sweden
1	Netherlands

VOCABULARY WORDS

APT and 2 CL :		EXAPT:
Minor P-P Keywords	DRILL REAM TAP BORE MILL	Major Keywords
Major Keyword	TOOL	Minor Keyword

IDENTIFIERS and LABELS

APT and 2 CL:

a. I D E N T I F I E R is a symbolic name
   . String of alphanumerics
   . At least one alphabetic

b. L A B E L is a symbolic name
   . String of alphanumerics
   . No restrictions

EXAPT:

a. First character must be alphabetic

b. Same

GEOMETRIC DEFINITIONS

<u>APT:</u>

a. ZSURF/ <plane>
b. PATERN/GRID, <parameters>
c. PATERN/RANDOM, <parameters>
d. TABCYL/NOZ, SPLINE, <point>, SLOPE, <scalar>,
. . . , <point>, SLOPE, <scalar>

<u>2 C L:</u>

a. Same as APT
b. PATERN/PARLEL, <parameters>
c. Same as APT
d. TABCYL/NOZ, SPLINE, <point>, TANTO, <line>,
. . . , <point>, TANTO, <line>

<u>EXAPT:</u>

a. ZSURF/ <scalar> /parallel to xy-plane/
b. PATERN/TRAFO, <parameters>
c. PATERN/RANDOM, <parameters>/same "Z" all points/
d. Not defined

SPECIAL STATEMENTS

APT:	a. R E F S Y S / <matrix>	/geometric definitions only/
	b. M A C R O definition	/no SYN and RESERV/
EXAPT:	a. T R A S Y S / <matrix>	/motion statements included/
	b. M A C R O definition	/no restriction/
2 CL:	a. Not defined	
	b. No restrictions	

ISO DRAFT PROPOSAL

NC PROCESSOR OUTPUT

Logical Structure
and
Major Words

CLDATA STANDARD

Vote Count, S.C.5

FOR	AGAINST	ABSTAINED
GERMANY	FRANCE	JAPAN
ITALY	USA	
SWEDEN		
UNITED KINGDOM		
CANADA /O/		
CZECHOSLOVAKIA /O/		

RECORD TYPE	NAME
1000	Input sequence
2000	Post-processor instructions
3000	Surface data
5000	Tool position
6000	Tolerance or cutter information
9000	Axis mode; units
14000	"Fini" record

RECORD TYPE	NAME
15000	Unsegmented tool path
16000	Workpiece contour description
17000	Tool description
18000	Material description
19000	Machine description
28000 to 32000	Proprietary records

MAJOR WORDS

MAJOR WORD	INTEGER CODE	MAJOR WORD	INTEGER CODE
AIR	1011	CLEARP	1004
AUXFUN	1022	CLRSRF	1057
BREAK	16	COOLNT	1030
CHUCK	1073	COUPLE	1049
CLAMP	1074	CUTCOM	1007
CLDIST	1071	CYCLE	1054

MAJOR WORDS /CONT'D/

MAJOR WORD	INTEGER CODE	MAJOR WORD	INTEGER CODE
DELAY	1010	GOHOME	17
DISPLY	1021	HEAD	1002
DRAFT	1059	INSERT	1046
DRESS	8	LEADER	1013
END	1	LETTER	1043
FEDRAT	1009	LINTOL	1067

MAJOR WORDS /CONT'D/

MAJOR WORD	INTEGER CODE	MAJOR WORD	INTEGER CODE
LOADTL	1055	OPSTOP	3
MACHIN	1015	ORIGIN	1027
MCHTOL	1016	OVPLOT	1042
MODE	1003	PARTNO	1045
OFSTNO	1083	PENDWN	12
OPSKIP	1012	PENUP	11

MAJOR WORDS /CONT'D/

MAJOR WORD	INTEGER CODE	MAJOR WORD	INTEGER CODE
PIERCE	1090	RAPID	5
PITCH	1050	RETRCT	7
PPFUN	1079	REWIND	1006
PPLOT	1014	ROTABL	1026
PPRINT	1044	ROTHED	1035
PREFUN	1048	SAFPOS	1094

MAJOR WORDS /CONT'D/

MAJOR WORD	INTEGER CODE	MAJOR WORD	INTEGER CODE
SELCTL	1056	THREAD	1036
SEQNO	1019	TMARK	1005
SPINDL	1031	TOOLNO	1025
STAN	1080	TRANS	1037
STOP	2	TURRET	1033
SWITCH	6	UNLOAD	10

5. IMPLEMENTING NC PROGRAMS

IMPLEMENTING NC PROGRAMS

D. McPHERSON
National Engineering Laboratory
East Kilbride Glasgow, GB

The papers collected together in this session cover several aspects of NC programs and languages, and it is difficult to find a single theme under which they may be discussed. However, the opening paper presented by Professor Williams provides such a theme; in one paragraph he said "The type of person who is most intrigued by computers tends to be the one who is most likely to be enthusiastic, innovative and creative... The result has been a tremendous proliferation of new machine designs, new codes, new programming languages and - most serious of all - new reworkings of already solved problems for the sake of innovation alone or for only minor technical or economic gains." The work described in these papers is considered against this background: do they describe worthwhile extensions to the wide range of facilities already existing; do they show how the problems of introducing new facilities might be eased; do they take into account the costs of running and maintaining the computer programs?

Mr Sabin in his paper discusses how to reduce part programming costs for aircraft components with a system centred on the general purpose APT language and processor. Pre-processors related to certain classes of components are proposed as a solution, and he shows how component geometry and method of machining is being handled in this way at the British Aircraft Corporation.

Dr Nussey and Mr Pinter present the philosophy behind the introduction of the IBM System/370 Numerical Control System. Based on experience gained with the System/360 NC systems, it contains extensions to the facilities covered by AUTOSPOT, ADAPT and APT/360. Three processors remain for positioning, simple contouring and complex contouring; these might be seen as "preferred subsets" of a large set of facilities, providing for three main categories of usage while limiting the high supporting costs which would be necessary for a large number of subsets.

COMPUTER LANGUAGES FOR NUMERICAL CONTROL, *J. Hatvany, editor*
North-Holland Publishing Company - Amsterdam–London

In contrast to this, the target for the LINK System described by Mr Galeotti is a single system which will run on small computers. The part programming language is APT compatible, but extended to include workshop technology language for drilling, boring, etc. The increasing availability of small computers and the demand for technological capability is claimed as justification for this new development.

Dr Budde and Mr Weissweiler describe the development towards modular systems. Users of EXAPT, for example, will be able to select the program modules they require and can enter their own "know-how" into the technology data files. The justification for these developments is seen as providing low-cost systems individually tailored to suit specific companies.

Modularity is also the main point of the paper by Dr Kochan and Dr Bock. The modules for a given application may be collected manually or alternatively loaded automatically. The aim is to reduce computing times and storage requirements.

Dr Schreiter describes extensions to SYMAP since the last PROLAMAT Conference. Commonly needed facilities are shown with examples of the use of SYMAP. The much-discussed topic of collision checking was covered by Dr Tempelhof, who showed that an extended multi-tool version of SYMAP gave considerable savings in part programming and production costs.

The paper by Mr Pikler described a conversational system implemented on small computers for checking spelling and syntax when programming in APT-like languages. The advantage to a user here is reduced processing costs through eliminating most errors, at an early stage in part program construction, on an inexpensive computer; overall turn-round is also reduced. This system /AIR/ was produced using the CHANGE language, as described by Mr Legendi, who also covered the development of a post-processor for a machine producing printed circuit cards.

Miss Paton gave a paper on how the NELAPT system is supported through system updates, documentation and user meetings. This paper shows how comprehensive such support should be to give a good service to users.

In conclusion, it appears that existing NC programs running on large computers are being supported and converted to run on new computers as well as being extended to meet user requirements. There is a trend towards making these processors highly modular to facilitate maintenance and the formation of sub-sets. NC software on small computers is being developed, and here the aim is to provide rapid and cheap processing for the limited needs of certain types of user. Probably both types of system will continue to be developed, but the increasing use of small computers is metalworking for on-line control may give rise to a demand for NC processors on such machines; this demand could evolve from tape and source language editing facilities towards full NC systems - even although limited in complexity they could suit many types of user who favour in-house processing.

THE USE OF THE CHANGE LANGUAGE FOR NC AND CAD LANGUAGES

T. LEGENDI

Institute for Computer Science and Automation
of the Hungarian Academy of Sciences /MTA-SZTAKI/

Abstract: The CHANGE language has been designed for defining and processing languages of medium complexity which may have non-linear mode of execution too. Two concrete applications are given in the paper suggesting general use in the NC and CAD areas.

1. ON THE MAIN FEATURES OF THE CHANGE LANGUAGE

The characterization of the language is restricted to the features important for the given applications.

1.1. MODE OF EXECUTION /THE MULTIPROCESSOR/

CHANGE programs may be executed by an unlimited number of processors. There are instructions in the language for defining new processors and stopping or suspending the existing ones.

Each processor has an instruction counter and an instruction counter modifier. Their actual values may be changed in programmed way too.

During an execution step each existing and not suspended processor is executing one instruction.

The value of instruction counter gives for its processor the number of the instruction to be executed. The new value of the instruction counter will be computed by adding the value of the instruction counter modifier /if only the executed instruction does not prescribe some other value/.

At the beginning of the processing there exists one processor with instruction counter set to +1, and with instruction counter modifier set to +1.

The multiprocessor has linear mode of execution if one processor is working and the value of its instruction counter modifier is +1 /or any other fixed positive value/. Any other mode of execution is non-linear.

There are two possibilities to have non-linear mode of execution /to start processors, to associate values differing from +1

COMPUTER LANGUAGES FOR NUMERICAL CONTROL, *J. Hatvany, editor*
North-Holland Publishing Company - Amsterdam–London

with the instruction counter modofier/. It is very effective to combine these facilities.

1.2. SELF-MODIFICATION

A CHANGE program can be modified during run-time by program without recompiling.

The language has two instructions for self-modification. The first allows to change the place of labeled instructions in the program. The other instruction allows to change a parameter to another one throughout the program. At the time of the change in both cases certain marked variables of the instruction /or parameter/ may be replaced by their actual value.

These instructions enable direct program-generation and running the generated program within the same running period /this run may be supervised by the generator program/.

1.3. EXTENSION

The language is extendible - new data types and new instructions may be defined and processed.

The syntax description of the extended instruction can be only simple. It may contain texts, parameters and repetition counters that describe a construction /group/ in a left-to-right order.

The semantic description is a CHANGE program /semantics/. It may contain any instruction of the language.

In semantics a special program part may describe modification of semantics as the formal parameters of semantics are changed to the parameters of a concrete extended instruction.

These modifications may depend not only on the actual parameters, but on the whole program inclunding the extended instruction /the language has instructions giving access to the standard internal program description/.

Semantics may have instructions for self-modification that are effective when the extended instruction is activated during the execution.

The extensions may be flexible due to possible modifications on different levels.

1.4. THE EXTENDED LANGUAGE MAY HAVE NON-LINEAR MODE OF EXECUTION

In the semantics of extensions the mode of execution may be

prescribed as non-linear.

The CAD and NC languages to be processed have in most cases a natural non-linear mode of execution. Interpretation of such languages by a linear language is cumbersome.

This paper suggests to approach these languages by non-linear extensions.

2. THE ADMAP POSTPROCESSOR

The NC equipment ADMAP developed by MTA SZTA serves for producing printed circuit cards. It has drawing and drilling heads moved by step-motors. The instruction set of ADMAP contains among others moving instructions, a drilling instruction, a pen-up/down instruction.

A control tape for ADMAP consists of the following parts:

the name of the printed card in legible punch

the name of the first technological operation in legible punch

the instructions to be executed during the first technological operation

the name of the second technological operations in legible punch...

The PP produces the control tape processing the card description which consists of two main parts.

The first is the list of technological operations to be performed.

The second part is the geometrical/technological description of cards. This part contains the name of the card, the sizes of the card and the descriptions of lines on the card.

A line description includes the list of points /a line is defined by points connected with straight lines/ and the number of the layer where the line is.

A point is defined by coordinates and a technological instruction defining whether the point must be touched and what sequence of ADMAP instructions must be performed there during different technological operations.

The PP partly optimizes the path length of heads during technological operations.

A simple card description:

```
DRAWING  1
DRAWING  2
DRILLING 3
```

```
NAME = CARD A
SIZE 150,100
T 10 10     The first line
K 10 20
T 15 10
N 20 10     The second line
K 20 25
FINISH
```

2.1. THE ADMAP/CH POSTPROCESSOR

A card description is equivalent to a program in ADMAP/CH.

The first instruction of a program is always PP/i/, i is an integer defining a "PP processor". The possible variants of PP processors will be discussed. Execution of instruction PP/i/ includes the syntax-checking of the following program. After this the control is automatically transferred to the NAME instruction resulting in the legible punch of the card name. Then, the following instruction, the SIZE instruction will be executed. /The coordinates of points are checked whether they are less or equal to the maximum size. In case of error the run will be stopped, if no error has place the control will be transferred automatically to the first technological operation./

The technological-operation instructions may be executed serially or parallel according to the PP/i/ instruction. The run will be automatically stopped after the execution of the last technological operation.

Execution of a technological-operation instruction begins with legible punching of the name of the operation. After that points /at drilling type operations/ or lines /at drawing type operations/ to be processed within the technological operation are listed.

This listing can be done by special execution of the geometrical/technological description part of the program. Special execution of a point-defining instruction means connecting it with the list if this instruction must be executed during the given technological operation /if not - the instruction works as a do - nothing instruction/.

After the listing a subroutine automatically changes the order within the list for optimizing the path length.

After the optimization the listed instructions /in optimized order/ will be executed. During this execution the path between points is incremented and expressed as ADMAP instructions and the technological instruction is also changed to ADMAP instructions.

The program is unchanged during the execution for the list contains the numbers of the instructions to be executed continuosly. In CHANGE it may be solved by direct programming of the instruction counter. Executing drawing /line/ type operations the optimization may require execution of a group of instructions in a reverse order. In CHANGE this can be solved assigning the value -1 to the instruction counter modifier.

If the modification of the program is allowed during individual technological operations /the program is held on secondary memory and at the beginning of each operation is read into the memory/ then sorting and optimizing may be done by overwriting the program. It is a typical CHANGE operation.

3. DIALOGUE MACHINES /DIALOGUE LANGUAGES/

The organization and flow of man-machine dialogues in interactive use of computers have common elements in different applications and in some degree are independent of the concrete tasks to be solved. These general features of dialogues may be described on dialogue languages.

A dialogue consists of dialogue steps. During a dialogue step information is given for computer /"question"/ then the computer sends information /"answer"/, that may depend on former dialogue steps.

The description of each possible flow of dialogue is the program of dialogue. A dialogue program is equivalent to a dialogue machine on which dialogues may run.

A dialogue machine /program/ may be programmed on a common programming language or on a dialogue language. Dialogue languages serve specially for writing dialogue programs.

Processors of dialogue languages may be programmed on a common programming language or on a general dialogue language /"dialogue-compiler compiler language"/. A general dialogue machine serves for running general dialogue programs. /It produces the processors of dialogue languages./

The system DISTAR-B /Forgács, 1973./ is a part of a general dialogue machine.

The dialogue language processor AIR /Pikler, 1973./ has been performed by the use of this system. Dialogue programs may be written in a so called direct code. There is an assembler input language under work.

A higher level input language may be defined by CHANGE extensions adding the capability for processing dynamic dialogue programs /that may be modified during dialogues/.

3.1. THE SYSTEM DISTAR-B

The system DISTAR-B contains common subroutines to assemble processors of dialogue languages.

Dialogue machines programmable in DISTAR-B languages can be thought of as an abstract computer that may take up dialogue states of finite number. In each state it executes some initial action, after that issues a question and later takes the answer and analyses it and then takes up a /new/ dialogue state.

3.2. THE DIALOGUE LANGUAGE AIR

The dialogue language AIR has been designed for interactive program writing in APT type languages using an alphanumeric display. It may be used fore more general purposes.

The dialogue machines programed in AIR will be referred to as AIR dialogue machines.

An AIR dialogue program must contain the description of each dialogue state. A description of a dialogue state contains the number of the dialogue state, a question to be issued /initial action/ and the possible answers with the corresponding /new/ dialogue states. A question is a string, an answer is a string or a symbolic notation of a string. In the latter case the symbolic notation must be assigned a value during the run of dialogues.

The questions may be of type menu. In this case the answer may be only a pointer /address of a line or cursor address/ to an element of the menu.

The answers during a dialogue to non-menu type questions may be strings fixed in the dialogue program or new strings to be assigned or having been assigned to symbolic answers.

3.3. THE REALIZATION OF DIALOGUE LANGUAGE AIR IN CHANGE

The definition of the AIR processor requires as a maximum a dozen of extended new instructions that are simple.

The description of a dialogue state in AIR/CH is a group of instructions.

The first is a labeled string /or string-type variable assigned a value before the dialogue/. The label is corresponding to the number of the dialogue state while the string to the question to be issued. This instruction will be followed by a group of instructions. Each corresponds to an answer and the related new dialogue state. In case of a non-menu type question, these instructions may consist of a string or string-type variable and a label. The string /variable/ defines the answer, the label defines the new dialogue state. In this case the string-type variable must be assigned a value before the dialogue. If we want to assign a value during the dialogue we must write a NEW group identifier /NEW POINT, NEW LINE,.../ before the string-type variable. The value will be assigned if only it is not existing already among the answere in the group /this causes an error call/ and in the future part of the dialogue will be used as a possible answer.

In case of a menu-type question the answer may be a line address /where the chosen answer is/ or a cursor address /where the chosen answer begins/.

Accordingly the basic word LINE or CURSOR and an integer type constant /or variable, assigned a value before the dialogue/ defines the answer and a label following it defines the new dialogue state.

A program part describing a dialogue state may be closed by an error instruction /ERROR, label/ which indicates a dialogue state if the answer during the dialogue in this state is wrong.

Without an error instruction the AIR/CH processor stops the dialogue with an error indication.

REFERENCES

Forgács, T. and Krammer, G.: /1971/ DISTAR-B általános dialógusrendszer, MTA AKI osztályközlemény.

Legendi, T.: /1972/ Az ADMAP nyomtatott áramköri lapokat készitő berendezés post-processor programja, Mérés és Automatika 2, 64.

Pikler, Gy.: /1973/ Minicomputer-based conversational program writing system, PROLAMAT 73.

A MINICOMPUTER-BASED, CONVERSATIONAL PROGRAM WRITING SYSTEM

G. PIKLER
Research Institute for Computer Science and Automation of the Hungarian Academy of Sciences,
Budapest, Hungary

1. INTRODUCTION

The purpose of creating an interactive part programming system was to overcome some of the difficulties of part programming in the APT-like NC programming systems. A study of the literature and of the experiences acquired in Hungary led to the conclusion that the main problems to be solved were as follows:

- The APT-like programming systems generally run on large computers. Running costs are increased considerably if errors committed in writing the part programs are discovered by the diagnostic programs of the processors themselves. International experience shows that an average of 3-4 trials and amendments are needed before an error-free run of the part-program takes place.
- The practice hitherto has been for the part programmer to be highly trained for the job. He must have a sound knowledge of the language in which he is going to write the program, be well versed in machining technology and be fully acquainted with the local workshop requirements.
- The part programmer has had to attend personally to a large number of tiresome details in writing his program.
- The use of interactive part programming systems based on big computers is often not economic, since it involves maintaining a terminal to a time-sharing system in an environment where this may not otherwise be justified.

The system which has been developed presents the following solutions to these problems:

- The system produces a syntactically completely faultless and semantically large error-free part program.

COMPUTER LANGUAGES FOR NUMERICAL CONTROL, *J. Hatvany, editor*
North-Holland Publishing Company - Amsterdam–London

- The part programmer does not have to have a detailed knowledge of the programming language he uses, he may write his part-program without knowing the language.
- The part programmer has to supply nothing beyond the data and parameters required for the given job. He is not concerned with the tiresome details of format, orthography, etc.
- Conversational part programming may take place on a minicomputer. This not only reduces the costs of the process but also makes it possible for the smaller firms to produce their part programs in--house. This can be contributory to deciding the perennial "In--house", versus "Out of House" debate.

The system designed to solve the problems discussed, requires a min. 20 kbyte minicomputer linked to an alphanumeric CRT display, a background disc store and a tape punch to output the resultant part programs.

2. THE MAIN REQUIREMENTS FOR THE PART-PROGRAMMING SYSTEM

Before developing an interactive system it is necessary to determine exactly for what type of user it is to be written. The permiss from which the present work departed was that the most economic part programming technique for a factory would be the one enabling the part program to be written as fast as possible, with a minimal knowledge of the language. The minimal language knowledge required for an advantageous use of the system was assumed to comprise the following:

- A general acquaintance with the structure of the language;
- The types of instruction used in the language /though not the exact form or syntactic rules of each instruction/;
- The use of special programming facilities in the language /e.g. MACRO, LOOP, etc./.

Any of the APT-like languages can be learnt to comply with these requirements in one or two days.

A knowledge of the type of user, the aims to be achieved ant the properties of the APT-like languages now make it possible precisely to specify the main requirements for the system:

- All part programs written in any APT-like language have to contain some indispensable instructions, such as PARTNO and MACHIN, moreover PART in EXAPT, CUTTER in 2C,L, etc. The system must make sure that the programmer always supplies these instructions.

- The part programmer must be helped by the system to find the most suitable instructions for his program, as quickly as possible. This task is solved by the use of a multi-level /three-level/ menu system in which the operator picks his choice by pointing. In this case the system must see to it that when one instruction has been written, the search for the next one should begin from the menu level which will lead to the next instruction by the shortest route.
- The part programmer must not be left to write the instruction he has selected by himself, since this would require perfect familiarity with the language and permit him to make many mistakes. The system must write each instruction itself, by putting questions to the part programmer to elicit the parameters figuring in the instruction /identifiers, numerical values, selection of appropriate modifiers/. The part programmer must answer each question and the instruction can then be faultlessly written.
- The type of answer that can be given in reply to the questions is determined by the APT-like language. The system should try to make the form of reply as simple as possible. Three types of answer have been adopted for the system:
  Pointing for selecting one answer from a number of alternatives presented to the par programmer;
  Alphanumeric entry for the requested identifier /only six character spaces are to be provided/, or for comments;
  Numerical entry for supplying numerical data.
- The interactive system must ensure that the parameters given by the part programmer should be systactically faultless and also that the most frequently committed semantic errors should be prevented from occurring in the program. Two methods are applied simultaneously to this end. One is that the mode and form of reply are chosen to make it impossible for the programmer to cause an error. The other is that the answers received by the system are submitted to diagnostoc testing. /These diagnostics are limited to those errors that cannot be detected by the first methods./ The following are the errors which are to be prevented by the two methods:

First method

1. The selection of the modifier from among several possible in a particular instruction is by pointing. /Questions appear as to the possible modifiers./
2. The identifiers in a part program may not be longer than six characters. The answer must therefore be limited to six spaces bounded by two marks.
3. The same method is to be used to limit the lengths of comments.

Second method

1. The part programmer must in every case reply to the question put to him.
2. The APT-like languages do not allow any identifier to begin with a number. This must be checked.
3. Identifiers used to give names to instructions may not occurr earlier in the program for a similar purpose.
4. Identifiers by means of which reference is made to an other instruction within an instruction, must have occurred previously in the program.

- Ergonomic considerations make it necessary for the programmer to learn immediately whether the system has accepted the information he has given it, or not. This can be done in a number of ways. In case of an erroneous reply it can present an error message which includes instructions for correcting the error, or if the dialogue so requires it can put a further question, or on conclusion of an instruction it can write out the completed instruction.
- A part from the above types of error, the replies may also include errors which the system is not able either to prevent or to check. The system must provide facilities for the programmer to correct these errors while program writing. The facilities offered must permit errors to be corrected as quickly as possible. Their nature will depend on where in the part program the error is noted. The hardware editing features permit errors to be corrected before the text is entered into the computer. If the error is detected within the cycle for writing the instruction, a facility must be provided for begining the cycle over again. /By pointing at a certain line on the screen , the cycle may be restarted, the information so far entered in that cycle is deleted./ If the error

is detected after the instruction is completed, the manipulating features of the system can be used to sorrect it.

- On completion of the part program, the system must perform certain checks:
  1. Elimination of any defining instructions in the program to which no reference is made in any other instruction.
  2. In the case of closed instruction chains which require not only an opening but also a closing instruction, a check to see that the latter is present.
  3. Whether the post-processor named in the MACHIN instruction is included in the list of post processors.
  4. Whether instructions which are forbidden or limited by the post-processor that has been called, occurr in the part-program.
- A part program which has passed through the final check must be printed out as a listing and also output on punched or magnetic tape.
- To make the system flexible and useful, it must include a number of manipulation facilities. These are;
  1. Any of the instructions previously written in the part-program must be callable on the screen.
  2. The instructions so far written must be sequentially displayable.
  3. There must be a facility for inserting one or more instructions between any two others.
  4. Any previously written instruction should be deletable.
  5. A list of all identifiers used in the part program must be callable on the screen.
  6. Certain characteristic instructions /e.g. FROM, ZSURF, etc./ in the part program should be displayable.
  7. Some APT-like systems permit the inputing of certain code numbers /e.g. the code number of the thread type in EXAPT 1/. These code numbers must be displayable on the screen.
  8. The flexibility of the system can be greatly increased if there is a facility allowing the part programmer also to write instructions directly, by-passing the interactive system.

The major part of the above requirements has been met by the system to be described - evidently to the extent permitted by the concrete language used.

3. THE STRUCTURE AND ORGANISATION OF THE SYSTEM

The satisfaction of the many requirements listed in the previous chapter may be represented by a rather complex graph /fig. 1./. The graph may be divided into four parts, according to their structure and functions. These four parts are:

1. Commencing part
2. Selection part
3. Instruction writing part
4. Feedback part

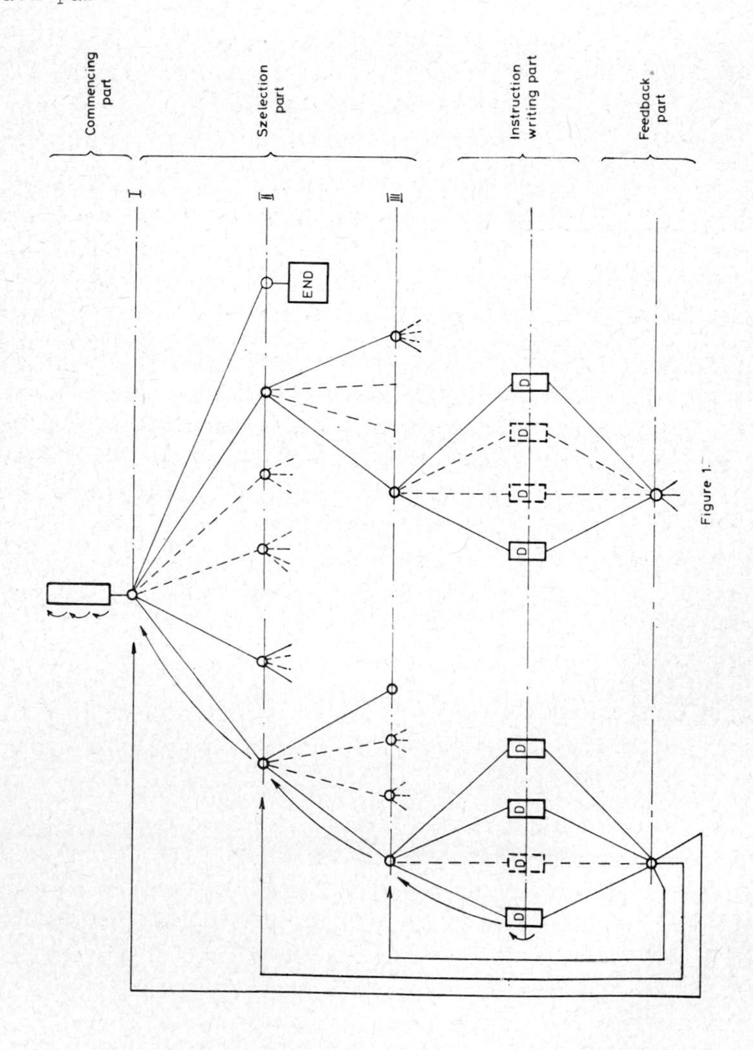

Figure 1.

The Commencing Part contains in sequential form dialogues for writing instructions. These are the dialogues of those instructions which it is obligatory to supply in every part program. They contain small menus and also questions to obtain data. Within the dialogues there are some small feedback sections to facilitate error correction.

The Selection Part resembles in its organisation a three-tiered tree structure. A maximally three -level tree structure is sufficient to enable the suitably grouped instructions in the APT-like languages to be easily found and selected. The nodes of the tree structure form the selection menus with which the required instructions may be chosen. The Selection Part is not an absolute tree structure, since it contains the feedback loops shown in the figure, from the lower level to a higher one. These are needed to provide correction facilities in case of erroneous selection.

The Instruction Writing Part is at the ends of the branches of the tree structure. This part contains sequentially the dialogues belonging to the various instructions. To facilitate error correction within the dialogues, they contain feedback loops.

The Feedback Part is situated in the last part of the graph. It may be regarded as a single-level tree structure whose branches point to various levels of the Selection Part. The organization of this part of the system was rendered necessary by the cyclic nature of part programming. It is not permissible for the selection of each instruction to take place from the first level of the Selection Part even when this is not necessary.

4. DISTRIBUTION OF THE INSTRUCTIONS OF APT-LIKE LANGUAGES WITHIN THE GRAPH

In the case of an interactive part programming system the distribution of the instructions is determined by the part programming requirements. These are:

- It is advantageous to maintain the sequence that has been found most suitable inpractical work. First those instructions are given which are obligatory for all programs, then the geometry, the technology, and after the instructions defining the technology or activising other functions, the tool motion statements. This order does not imply that it really has to be observed, but it does determine the criteria for grouping the instructions.

- Functionally similar types of instructions should, for ease of selection, be placed in the same group.
- Instructions which differ in function from the functions of a given group but which are often used in conjunction with those instructions, are best put in that group.
- Easy access must be provided to all instructions.

The above requirements were observed in forming the various groups of instructions and locating them in the graph.

The Commencing Part contains those instructions which absolutely must be provided for the part program to run through the processor at all.

The main groups of the instruction distribution are determined by the branches from the first level of the tree structure of the Selection Part. These are as follows:

- Geometrical instructiin group
- Technological instruction group
- Group of other technical instructions involved in programming
- Group of call and execute instructions
- Motion instruction group
- End of part program
- Group for selective manipulation facilities of interactive system. /This branch is not a member of the group of instructions./

The above groups provide the first level menu for the selection of instructions.

The various groups contain the following instruction types, which are independent of which member of the APT family the system is being prepared for:

- The geometrical group contains those instructions or groups of instructions which are intended to define the description of the geometry. It is useful also to include in this group the instructions determining transformation of the coordinate system.
- The group of technological instructions must include all those prescribing the technological data required for machining. These may include cycle working, single operations or instructions introducing further technological values.
- The group of other technical instructions involved in programming accommodates instructions intended to facilitate the work of the programmer /REMARK, MACRO, LOOP, etc./.

- The call and execute group consists of instructions which activise some defining instruction, or render some auxiliary machine tool function operative.
- Motion instructions include all those involved in making the tool move.
- The group for selective manipulation facilities contains the menu of the manipulation facilities described in the preceding chapter.

5. STRUCTURE OF THE DIALOGUE PROGRAM

The structure and operation of the interactive part programming system are shown on Fig. 2.

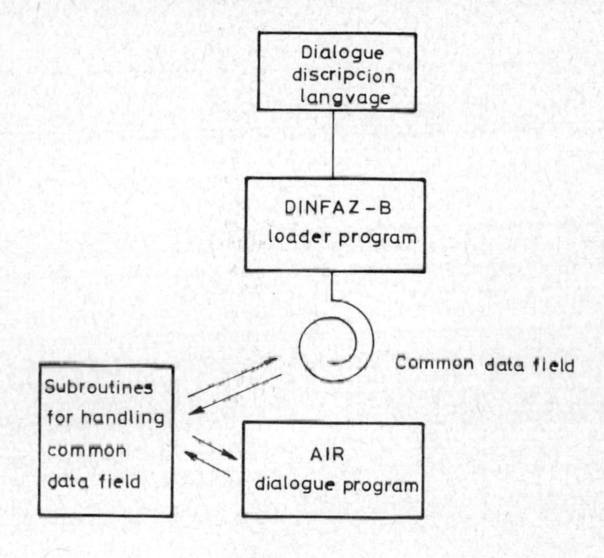

Figure 2.

The dialogue is described by means of a dialogue description language. The program written in the dialogue language is loaded by the DINFAZ-B loader program which locates it in suitably segmented form in the peripheral store. The dialogue is executed by the AIR program, on the basis of the information stored in the common data field. The AIR program communicates with the data located in the peripheral store by means of the subroutines for handling the common data field.

The dialogue desription language is a directily coded language which in its general form consists of two main parts. All the texts occurring in the dialogue have first to be put into the first part, in the form of records. The second part consists of status records. All status records contain two main parts, viz. the action part and the analysis part. The action part can store the data relevant to the question posed by the dialogue, while the analysis part provides the new status record numbers where the dialogue can be continued, depending on the reply given.

In the system being described, four types of status record were developed. The first type is used to describe those dialogues where a textual or numerical answer is expected of the programmer. The action part of the status record is as follows,

T	N	SP_1	SH_1	.....	SP_n	SH_n	VH	VT

Figure 3.,

where

T = 1 is the type number of the status record,
N is the number of texts written on the screen,
SP is a text pointer /the serial number of the text table/,
SH is the place of the text on the display screen,
VH is the place of the expected answer,
VT is the type of information expected /alphanumeric, numeric, or comment reply/.

The analytical part consists of intervals as follows:

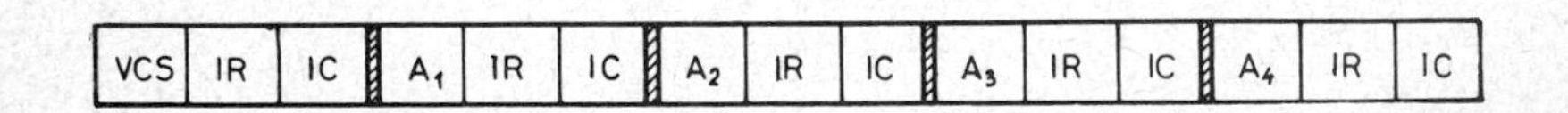

Figure 4.,

where

VCS is the cursor address upon pointing,
IR is the serial number of the next statusrecord of the condition is fulfilled,
IC is the index number of the connecting GOTO instruction in the AIR program,
A_1 is a code number for the case where the programmer does not provide an answer in the required location,
A_2 the identifier in the reply commences with a numeral /code number/,

A_3 the requested identifier is diagnostically erroneous /code number/,

A_4 code number of error-free answer.

In the case of a question requiring a number, the intervals begining A_2 and A_3 do not figure in the analysis part. The subroutine IANAL included in the AIR program analyses the answer received, and as a result gives it a code number A_1, A_2, A_3 or A_4. Upon comparing the code number thus received with the code numbers found in the intervals of the analytical part, the numbers in the second and third element of the interval indicate at which new status record the new dialogue is to be continued, and at which label, marked by which connecting GOTO index number found in the AIR program.

The second type of status record is used to describe the menu--type dialogues. The action part resembles the action part of the first type of status record, while the analytical part differs in that the places A_1 ... A_4 contain the screen line numbers of the answers to be given to the various elements of the menu, and that there is one more internal than the number of items on the menu.

The third type of status record srves to display text to which the system expects no answer. This type is used generally to write out error messages and to delete certain information from the screen. Its action part is the same as the action part of the first tape of status record, while the analytical part consists of only one interval comprising two elements. The first contains the serial number of the next status record, the second the index number of the connecting GOTO, figuring in the AIR program.

The fourth type of status record differs from the structure of the others. It is used to compose the AIR-type instructions. The first element of the action part denotes the type of the status record, the second element contains the number of words occurring in the instruction. The subsequent elements contain either the serial number of a text table, or a specific character upon whose presence the AIR program incorporates the information provided by the programmer in the instruction being composed. The status record may be built with two types of analytical parts, depending on whether the dialogue of the instruction to be composed has been completed, or further dialogues are necessary for its completion. In the first case the analytical part consist of three intervals as shown in Fig. 5.

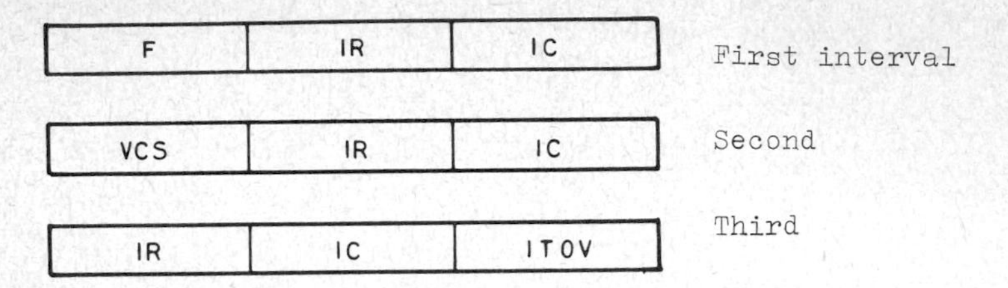

Figure 5.

Here in the first interval

F = 1 is a flag, showing that the composition of the instruction has been completed,

IR is the serial number of the statuc record,

IC is the index number of the connecting GOTO in the AIR program.

In the second interval

VCS is the cursor address upon positing,

IR is the status record serial number upon pointing,

IC is the index number of the connecting GOTO upon pointing.

In the third interval

IR is the serial number of the next status record, provided that the programmer accepts the instruction displayed on the screen,

IC is the index number of the connection GOTO,

ITOV= 1 if the part program has not ended
= 2 if the part program has ended.

The analytical part of the second type has only one interval which corresponds to the first interval shown in Fig. 5. The differences are that

F = 0, showing that the composition of the instruction is not yet completed,

IR is the serial number of the next status record,

IC is the index number of the connection GOTO.

The DINFAZ-B loader program permits the status records which describe the dialogue to be put into the background store in segmented form. This is important because only one segment at a time is read into the operative store and it thus becomes possible to create a system consisting of several segments. Each segment - like the dialogue description language - consists of two parts, the text table and the status records belonging to the segment. It is also possible to establish a common segment containing only a text table and residing permanently in the operative store. The common segment contains those text records which are used in all segment. The length of a segment is determined by the amount of space occupied in the operative store by the AIR program and the common segment.

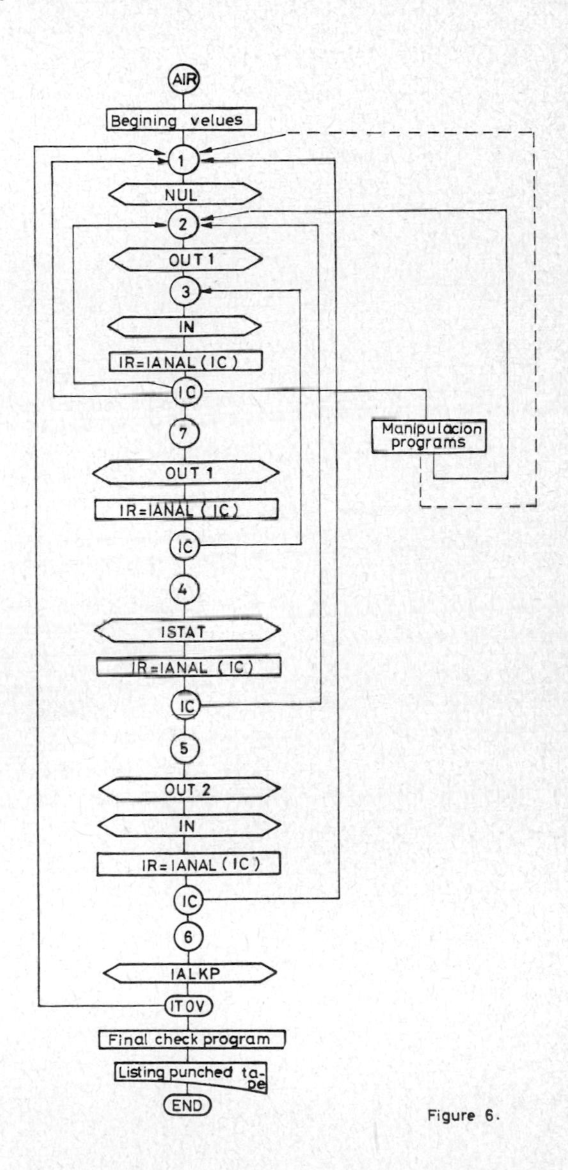

Figure 6.

The block diagram of the AIR program is shown in Fig. 6. This executes the dialogue between the system and the programmer, with the hepls of the status records. The main part of the program consists of seven program segments /subroutines and functions/, moreover the part which offer manipulation facilities and permit the programmer to carry out his final check. The program segments perform the following functions:

- Subroutine NUL. Deletes the screen contents and resets some common blocks to zero.
- Subroutine OUT1. Puts texts on the screen in accordance with the information found in the action part of the status record of setial number IR. /It operates on the action parts of the status records types 1, 2 and 3./
- Subroutine IN. Enter the information written by the part programmer.
- Function IANAL. Analyses and diagnoses the entered information, then according to the status records determines the new status record serial number IR and the new IC value.
- Subroutine ISTAT. Composes the instruction part of the part program belonging to the dialogue section concerned, according to the type 4 status record.
- Subroutine OUT2. Outputs the completely composed part program instruction on the display screen.
- Subroutine IALKP. Locates the instruction approved by the programmer in a common store file.

The operation of the AIR program commences with the subroutine NUL setting the initial values and deleting the contents of the screen. Using the status record with the next serial number IR, the OUT1 outputs the question on the screen. The subroutine IN reads the answer, which is evaluated by IANAL which determines the serial number IR of the new status record and the value IC belonging to it. If the question that is posed continues with a further question, the program jumps to label 2. If the operator has pointed to the return line, i.e. the feedback built into the dialogue, the program continues with label 1. In the case of an error message or an error-free answer, the subroutine OUT1, through label 7, carries out the error display or deletes the corresponding lines of the display according to the type 3 status record. If there is an error message, the program continues to run from label 3, while if the answer is error-free, it runs from label 4. The instruction belonging to the dialogue is composed by the subroutine ISTAT, according to the type 4 status record. If further dialogues also belong to the composition of the instruction, the program continues to run from label 2, through IANAL. When the complete instruction is composed, the program jumps to label 5. Subroutine OUT2 displays the composed instruction on the screen. The programmer either accepts the displayed instruction or not. If not, the program continues to run from label 1, deleting the composed instruction. If yes, the subroutine IALKP places the composed instruction in store and the writing of the part program continues either with label 1, or with the final check program, concluding with a listing.

6. SUMMARY

A part programming system developed for a concrete APT-like language consists of two parts. One part is the text-tables and status records describing the dialogues which permit the instructions to be composed, the other part consists of the loader program which reads the status records and the AIR program which carries out the dialogues. The second part is able to process status records belonging to any APT-like part programming language. The DINFAZ-B loader program and the AIR program are thus independent of the particular APT-like language used. The same can not be said of the status records. The dialogue must be designed separately for each APT-like language, creating the necessary status records and text tables.

The system has been implemented on two computers: the CII 10010 minicomputer with 20 Kbyte core and the CDC 3300 medium computer. The CDC 3300 version was written in FORTRAN IV and the display was simulated with a line printer and card reader. The CII 10010 version was written is the ASTROL assembly language. The status records written in the language described above, can be read by both version.

The AIR program, together with the disc and display control software, occupies about 8 Kbyte in the CII 10010 computer. Reserving 2 Kbyte of core for the common text field in the case of a concrete APT-like language, this leaves 10 Kbyte for a status record segment.

Part programming systems for two APT-like languages - EXAPT-1 and 2C,L - have been implemented according to the principles and conditions set out in this paper. The status records for the EXAPT-1 system occupy 9 segments, for the 2C, L system 11 segments. Each segment contains about 150 status records and 50 text records.

According to the experiences gained so far, a part program of medium size can be written in 15-20 minutes with the system. This program will be completely free of syntax errors and free of most kinds of semantic error.

7. REFERENCES

1./ Computer Programming Series
Digital Computer Graphics
American Data Processing, INC. Detroit, Michigan
2./ Butlin, G. A. and Hubbold, R. J. /1969/ A Scheme for Man-Machine Interactive Structural Analysis, Internationa Conference on Computer Aided Design 15-18 April.
3./ Green, R. E. /1970/ Computer Graphics, Computer Aided Design Spring
4./ Forgács, T., Gerhardt, G., Kocsis, Krammer, G. /1971/ Általános dialógus rendszer, MTA AKI
5./ Gutterman, M. M. /1964/ APT Numerical Description study, Illinois Istitute of Technology, Chicago, Illinois, 21 August, SCL-DC-64-130
6./ EXAPT 1 Part Programmer Reference Manual /1969/ Verein zur Förderung des EXAPT-Programmiersystems e.V Aachen, May
7./ Conversational 2C,L /1970/ NC news NEL East Kilbride, Glasgow
8./ 2C,L Part-programming Reference Manual, National Engineering Laboratory, NEL Report No. 424.

Collision-Free and Simultaneous Work Machining on Metal-Cutting Machine Tools for Large-Size Works and Resulting Requirements for Machine Programming of Same

KARL-HEINZ TEMPELHOF

Technische Hochschule Otto von Guericke Magdeburg

German Democratic Republic

1. Importance of subject investigated

Investigations by RÜMMLER (1972) into the coarse structure of the parts assortment in machine building revealed that the parts class BOXES/COLUMNS/BEDS (housing-type and prismatic parts) occupies a share of

8 to 12 % of drawings
5 to 10 % of annual output in numbers
40 to 45 % of manufacturing hours to be spent
45 to 50 % of production value obtained

out of all parts classes in machine building. Also LANGE and PLESCHAK (1971) put special stress on housing-type parts when assessing the question of rationalizing the technological preparation of manufacture with regard to workpieces, since the share of this type of parts within the whole assortment of parts will further increase in future as forecasts show.

Along with the improvement of conventional planes and bed-type millers as well as horizontal boring, drilling and milling machines, there have been recently developed manufacturing means with numerical point-to-point positioning and straight-line control systems for machining workpieces of this type. These means have been provided with several machining units allowing several processes in machining boreholes and/or plane surfaces as well as (partly simultaneous) machining of several faces of a workpiece to be implemented in one setup.

For the machine programming of point-to-point and straight-line controlled NC-machines for boring, drilling and milling operations there has been developed by GDR specialists the language section SYMAP (PS) at the extension level BOFR 1 (RICHTER, 1971). By means of this language section it is possible to clearly describe all boring, drilling and milling operations with regard to geometrical relations and conditions as well as to the machining processes required on NC milling machines, including even machining centres with a maximum of five machining units, simultaneous machining with two machining units each and two possibilities of positioning in z-direction (adjustment of spindle and cross-stay or table) as well as rotation of workpieces through three axes and tool change by hand, through turret head, or through tool magazine.

For the programming language SYMAP (PS) the first extension level has actually been implemented with regard to computing technique. Out of the number of problems not yet accounted for at this development level, it will be the aim of this paper to deal with collision-free machining of prismatic and housing-type workpieces and with some results of investigations made into this field.

COMPUTER LANGUAGES FOR NUMERICAL CONTROL, *J. Hatvany, editor*
North-Holland Publishing Company - Amsterdam–London

2. Considerations with regard to collision

Not mentioning here the intended action of a cutting tool into the workpiece to be machined, a collision between two or more members moving towards each other within the working space of a machine tool will occur when one point of the contours of each of the members moving simultaneosly towards each other will meet at the same coordinate values of the same coordinate system. In order to eliminate unintended collisions it is necessary to know for each machining application the duration of the machining operations, the geometrical dimensions of the sectional faces to be machined, their position relative to the reference coordinate system, the designs of the tools, of the workpiece, of the moving machine elements and that of the working space.

Influencing variables are

the workpiece (with obstacles such as recesses, protrusions, lugs etc.)
the tools (maximum diameter, unsupported length of tool, number of tools being simultaneously in action)
the machine tool (design of working space, dimensions and arrangement of moving machine elements)
the clamping elements (design, number, position, type of mounting on machine tool table).

When machining workpieces on metal-cutting machine tools for large-size workpieces with several machining units traversing in different axes there is, in the cases of uniaxial and simultaneous machining, the possibility of collision between the following partners:

I tool - tool
II tool - workpiece
III tool - elements of machine tool
IV tool - clamping elements
V elements of machine tool - workpiece
VI elements of machine tool - clamping elements
VII elements of machine tool - elements of machine tool

Especially in case of simultaneous machining, these possibilities of collision result from job-borne movements of the machining units relative to each other and from such conditions as they are created, for example, by the movements of the crossrail of openside and double-column bed-type milling machines.

The case I (tool - tool collision) is possible, if the diameters of the tools working on faces positioned orthogonally relative to each other during the machining process exceed a certain limit. The case workpiece - workpiece collision may be excluded, since there is no movement of the workpieces relative to each other with multiple clamping of the latter. Due to the function and conventional mounting of clamping elements, the case of clamping elements - clamping elements collision can be excluded as well. The collision case VII may occur, if the machine tool elements considered are arranged in a position offset by the angle α.

Early detection and avoidance of possible collisions between tool and workpiece (case II) are of utmost importance. Two cases will have to be distinguished: a) tool - workpiece collision during the machining process, i.e. during the active path of the tool, b)tool-workpiece collision during the passive path of the tool from one machining spot to another. Considerations as to collision case II should take into account workpiece-borne obstacles, which are referred to as geometrical obstacles. One speaks of a geometrical obstacle with extension in the direction of the positive or negative z-axis of the reference coordinate system (standing vertically on the machine-tool table), if there is a surface complex between the two machining spots, the geometric enveloping surfaces of which need not be machined, and if the machining unit arranged diametrically to the sectional face of the workpiece to be machined needs to be moved perpendicularly to the direction of feed of the machine-tool table.

The methods of mathematical description of obstacle contours, e. g. that of HAHN (1970), developed for other manufacturing processes and representing an essential prerequisite for extensive and correct collision considerations prior to executing the actual machining process, have been further developed by us for milling operations on bed-type milling machines. The principle of the Methods 1 through 3 described below consists in a crosswise comparison of the contours of the tools or machining units being positioned nearest to the corresponding collision partner in terms of both establishing the differences of the coordinate values concerned and performing a difference comparison. Method No. 4 has been newly developed by MARKMANN (1973).

Method No. 1: Determination of the ranges of possible collisions within the working space of the machine tool in case of simultaneous machining (Fig. 1):

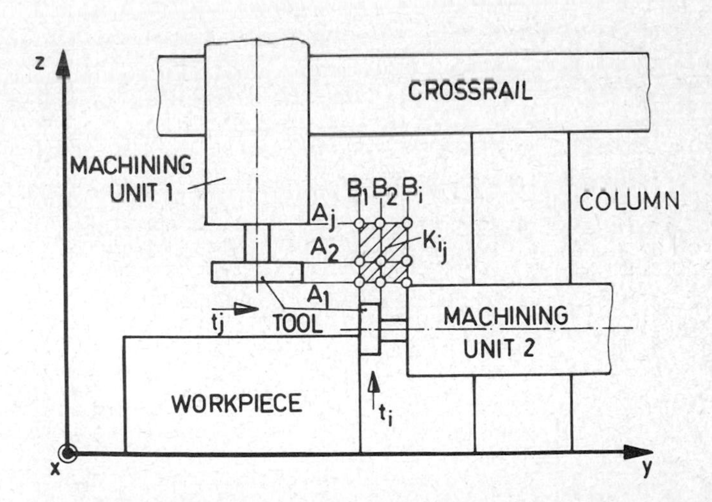

Fig. 1: Determination of the range of collision within the working space of the machine for simultaneous machining

Starting point of considerations is a projection-type representation of all partners being in final position of machining, of collision cases I, III and VII. Starting from the contours of each of these partners, observing the direction of the tool infeed (t_i, t_j), straight-line families A_j ($1 \leq j \leq m$) and B_i ($1 \leq i \leq n$) will be extended into the working space of the machine. The intersection points K_{ij} of the straight-line families A_j and B_i define the area of possible collisions of the cases I, III and VII.

Collision occurs, if

$$Z_{A_j} < Z_{K_{ij}} \quad \text{and} \quad Y_{B_i} < Y_{K_{ij}}$$

For this case the process-planning engineer or machine setter shall - with a view to preventing a collision - establish displacement rates by which the partner(s) entering the collision range shall be moved against directions t_i or (and) t_j.

Method No. 2: Comparing the coordinate values of the contours of the collision partners interacting within the working space of the machine tool (Fig. 2):

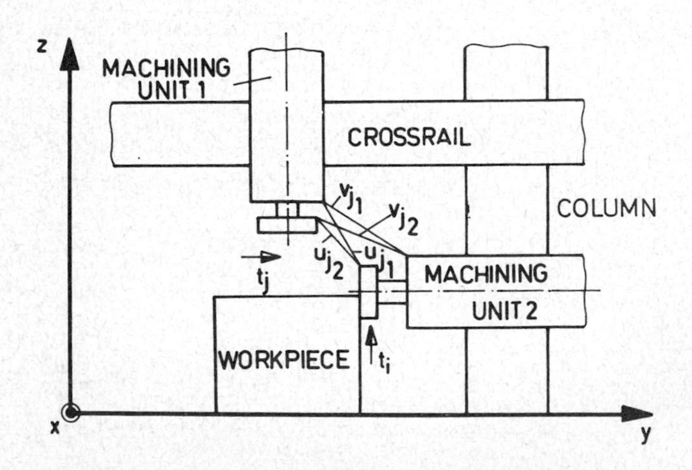

Fig. 2: Considerations of collision by comparing the coordinate values of the contours of collision partners interacting within the working space of the machine

Starting from the projection-type representation of the machining units and tools in their final position obtained on completion of machining operations, the intersection points S_{uj} and S_{vj} of the straight lines, which have been extended into the contours of the tools interacting within the working space of the machine tool, and subsequently the lengths of the distances u_j and v_j will be calculated.

Collision as to cases I, III and/or VII will occur, if one value of u_j or v_j becomes negative:

$$\sum_{j=1}^{n} u_j < \sum_{j=1}^{n} |u_j| \quad or \quad \sum_{j=1}^{n} v_j < \sum_{j=1}^{n} |v_j|$$

To obtain satisfactory results, this method requires considerations in several planes of projection. Its application for the purpose of detecting collision is disadvantageous.

Method No. 3: Iteration method (similar to HAHN (1970), Fig. 3):

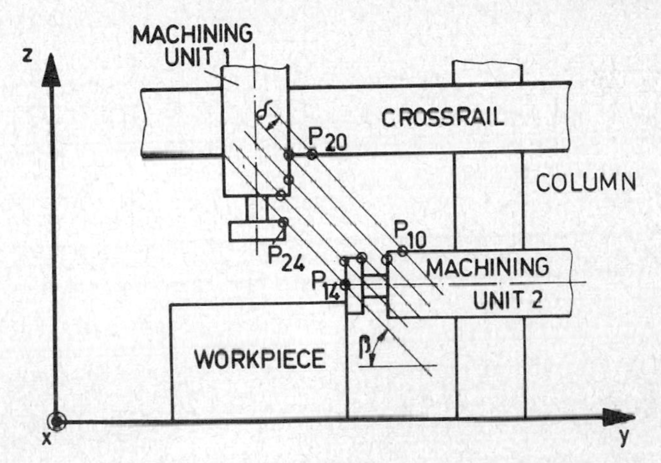

Fig. 3: Recognition of collision using the iteration method

With an angle β areas of parallel straight lines are imposed at the specified distance δ iteratively on the projection-type representation of the contours of tools and machining units, and the intersection points of these straight lines with the contour lines of the tools and machining units are calculated. Collision as to cases I, III and/or VII will occur, if coordinate values $Z_{2i} < Z_{1i}$ result for one of the intersection-point groups P_{1i} and P_{2i}. For this case, displacement rates to eliminate collision shall be established and further checking calculations be executed, as has already been described for Method No. 1.

This method has been developed with the aim to recognize possible collisions not only in the final position (as with

Methods No. 1 and 2) but during the movement of tools and machining units. Exactness of statements obtained by this method is dependent on raster distance σ (cf. Fig. 3), provided the geometry of the workpiece has been given with sufficient accuracy. Exact statements require great computational efforts.

Method No. 4: Introduction of planes for the purpose of determining obstacles (MARKMANN, 1973, Fig. 4):

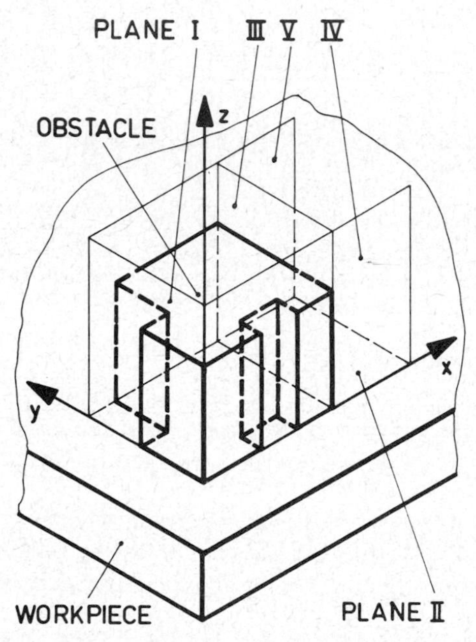

Fig. 4: Determination of obstacles by surrounding geometrical obstacles by means of adjacent planes

Planes are applied to geometrical obstacles occurring, in order to determine the location of these obstacles within the coordinate system (for the example in Fig. 4 the planes I, III, IV and V). The points of the geometrical obstacles to be marked will be projected into the xy-plane of this coordinate system, termed plane II. The traces of the planes I, III, IV and V applied to the obstacle form a rectangle (in special cases a square) in the projection plane II; this rectangle surrounds the obstacle, and its corner points are the limit points for collision-free machining of workpieces, too.

In accordance with this method developed for collision case II, geometric obstacles can be very advantageously described by the programming language SYMAP (PS). It can be applied under the following conditions:

a) The entire obstacle is surrounded by planes.

b) The projection plane II is always within or parallel to the working plane.
(In principle an inclination of the projection plane is possible; it is, however, not allowed for in the calculation model characterized in Section 3.)

c) Planes I, III, IV and V are perpendicular to projection plane II.

d) The traces of planes I and II run parallel to the y-axis, those of planes IV and V parallel to the x-axis.

3. Extension of possibilities for SYMAP

If the workpiece to be machined shows geometrical obstacles, the possibility of automatic selection of tools and technological operating parameters given among others for milling operations can only be utilized, if the part programming engineer has manually determined both the active and the passive paths of the tool in order to obtain collision-free machining (increased expenditure for obtaining the source programmes). To perform collision investigations for, say, milling tools being in action, the following procedure shall be adopted:

(1) Collecting the coordinates of the workpiece faces and obstacles to be machined on a coordinate acquisition sheet.

(2) Decision on primary milling direction depending on the surface dimensions and with a view to minimizing indexing times.

(3) Determination of tool depending on the machining case (coordinates of faces and obstacles).

(4) Calculation of the cutter paths required.

(5) Determination of tool overrunning perpendicular to cutter path.

(6) Decision as to whether for the workpiece faces adjacent to an obstacle a previous face milling along the boundary to the obstacle will be necessary ("free-milling").

(7) Determination of tool overrunning in the direction of cutter path.

(8) Final determination of the tool path which ensures collision-free machining of workpieces. Indication of the initial and final coordinates of the machining process for calculating the passive tool path.

When aiming at collision-free machining of workpieces there will - as MARKMANN (1973) stated - arise as a further sectional problem the necessity of including into the considerations of collision the feature of simultaneous machining, the calculation of the number and optimum distribution of clamping points and the optimization of traversing paths.

In order to relieve process planning engineers (part programming engineers) of such jobs, we developed an algorithm complex for ensuring collision-free machining of housing-type and prismatic workpieces on openside and double-column bed-type milling machines (MARKMANN (1973)).

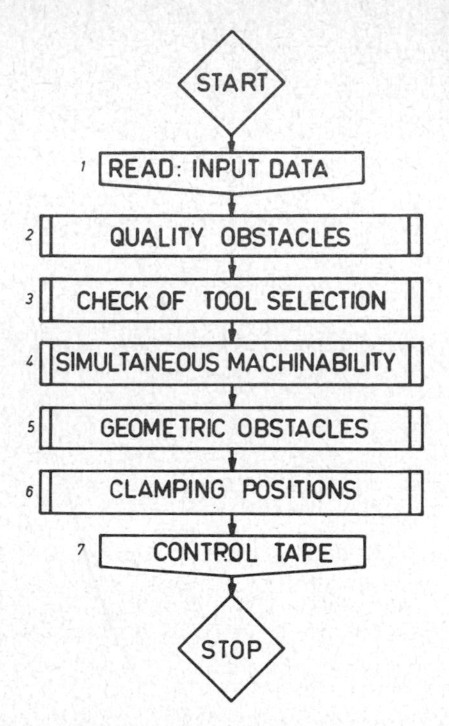

Fig. 5: Algorithm complex "Collision-free machining of housing-type and prismatic workpieces on openside and double-column bed-type milling machines"

We started from appropriate analyses of workpieces and tools as well as analyses of the machine tool types to be used for various machining operations. The algorithm complex is subdivided into blocks in accordance with the sectional problems covered by it (Fig. 5). It is based on Method No. 1 dealt with in Section 2 to cover collision cases I, III and VII, and on Method No. 4 to cover collision cases II and V. Collision cases IV and VI are allowed for by the block "clamping points".

In block "quality obstacles" starting from the coordinate values of both the obstacles and the workpiece faces to be machined as pre-set by the process planning engineer (part programming engineer) in the source programme the machining cycle of the operation 'Milling' is determined with a view to collision case II. A check is made to show in which coordinate direction the tool will hit an obstacle. By an analysis of the relative tool movements in four coordinate directions ($\pm$ x, $\pm$ y) out of 16 possible types of interfaces of the workpiece (sectional) face there will be determined the particular one and, following this, the primary direction of movement of the particular tool as well as the type (surface or line milling) and sequence of the machining operations.

The block "check of tool selection" effects the correction of tool diameters required in case of collision hazard, allowing for not only face-milling cutters and shell-end mills, but also numerous other types of milling cutters according to standards in the G D R (TGL).

In block "simultaneous machinability" the possibility of simultaneous machining is checked, regarding the collision cases I and III. At the same time the existing possibilities of simultaneous machining of faces, trains of lengths, and surfaces with polygonal geometrical contour are investigated. Because of the restricted traversing-path length of the sideheads of openside and double-column bed-type milling machines, the position of the workpiece on the machine table is of essential importance to simultaneous machining. Therefore the optimum position of the workpiece on the machine table is calculated. The corresponding dimensions are printed out by the computer and shall be adhered to by the operator to within ± 5 mm.

With the block "geometrical obstacles" the traversing path from one machining spot to the next is optimized and the shortest path for by-passing the obstacle is calculated. Three types of the passive tool path which can ensure collision-free movement to the next machining spot are distinguished:

HU : Passing over obstacle
HLU : By- passing the obstacle on the left
HRU : By-passing the obstacle on the right.

With a view to mastering the problem, the complex collision problem given with the particular manufacturing job is resolved into the collision cases mentioned in Section 2 for uniaxial machining as well as for machining of workpieces with several machining units (simultaneous machining) and solved by separate and step-by-step evaluation.

In block "clamping points" the number and dimensional distribution of the clamping points on the workpiece are determined through calculating the cutting power F_s, feed power F_v and passive power F_p with a view to collision-free machining and to minimizing down-time resulting from change of workpiece. In accordance with the size of the workpieces to be machined, it is largely the clamping methods using clamping bolts, holding straps, clamp dogs and pneumatic clamping elements that are allowed for.

4. Efficiency of utilization

By using the algorithm complex characterized in Section 3, the problem of by-passing geometrical obstacles during the cutting action can be solved by data-processing systems. This feature relieves the process-planning engineer (part programming engineer), in the course of establishing SYMAP source programmes, of jobs such as

- determination of collision-free active paths of tools
- settling the primary milling direction

- establishing the technological operations "line milling" and/or "surface milling" (since, with the algorithm, "surface milling" is generally prescribed, and in the case of the occurrence of obstacles during the machining process, "line milling " is automatically specified for "free-milling" (cf. Section 3 (6)) of the particular faces
- determination of minimum traversing paths for collision-free movement towards the individual machining spots, i.e. minimum and collision-free passive paths of tools.

In addition, for each manufacturing job there will be further simplification in establishing SYMAP source programmes by using a coordinate acquisition sheet which shall be fed into the computer together with the associated source programme. Thus the specification of tool path coordinates for the workpiece sectional faces to be machined is eliminated which, in the case of manual elaboration, would become the more expensive the more obstacles were adjacent to the surface to be machined. Likewise, the definition of a train of lengths as has been common so far and which required the indication of the coordinate values for collision-free paths in the parameters specification of the source programme is no longer necessary.

MARKMANN (1973) has proved for one application (milling a cross-rail) that the scope of the source programme and, consequently, efforts to be made by the process planning engineer (part programming engineer) for the elaboration of same will be reduced by approximately 25 % by the new extension level of SYMAP. A cost analysis made by the same author showed that there will be a reduction in prime cost by some 300 % when using the new extension level as compared with possibilities given so far. It is estimated that the reduction in prime cost will be even higher with a more sophisticated degree of the manufacturing jobs to be programmed.

REFERENCES

HAHN, J. (1970): Automatische Fertigungsplanung für kurvengesteuerte Einspindel-Drehautomaten. (Automatic Production Engineering for Cam-Controlled Sigle-Spindle Automatic Lathes.) TU Berlin: Dr.-Ing.-Diss.

LANGE, H. und PLESCHAK, F. (1971): Strategie für AUTEVO muß aus Prognose abgeleitet werden. (Strategy for AUTEVO Shall Be Derived from Forecast.) Fertigungstechnik und Betrieb 21, p. 11

MARKMANN, D. (1973): Kriterien für kollisionsfreie Großteilbearbeitung prismatischer und gehäuseförmiger Werkstücke auf Ein- und Zweiständer-Bettfräsmaschinen. (Criteria for the Collision-Free Machining of Large-Size Prismatic and Housing-Type Workpieces on Openside and Double-Column Bed-Type Milling Machines.) TH O. v. Guericke Magdeburg: Dr.-Ing.-Diss.

RICHTER, H., HELLMUTH, W., SCHOLZ, W., GENGMITH, S. (1971): AUTOTECH/SYMAP, SYMAP (PS) - BOFR 1 Sprachbe-

schreibung. (AUTOTECH/SYMAP, SYMAP (PS) -BOFR 1 Description of Language.)
Karl-Marx-Stadt: Großforschungszentrum des Werkzeugmaschinenbaues beim VEB Werkzeugmaschinenkombinat Fritz Heckert

RÜMMLER, G. (1972): Typung technologischer Prozesse unter dem Gesichtspunkt der Rationalisierung der technologischen Fertigungsvorbereitung (TV). (Standardization of Technological Processes with a View to Rationalizing the Technological Preparation of Production (TV).
Fertigungstechnik und Betrieb 22, p. 366

COMPUTER-AIDED PART PROGRAM GENERATION

M.A.SABIN
British Aircraft Corporation
Commercial Aircraft Division
Weybridge, Surrey, England.

Abstract: Special computer programs, written to simplify the part programming of specific classes of component can reduce part programming costs substantially. It has been found convenient and useful for such programs to generate part program text in a standard NC language. A comparison is made here with alternative approaches. The questions of which standard NC language, and whether the initial data needs to be structured into a language are considered.

1. INTRODUCTION

The idea that an NC processor has two parts, the main processor, and the post processors, is a very familiar one. The main processor is very general, performing functions and calculations common in the driving of all NC machines; the post processors, only one of which is used in the generation of each control tape, perform those functions which are special to a particular machine tool or control system. A completely general post processor would be very cumbersome, expensive to write and expensive to run.

Less familiar is the idea that similar arguments for specialization apply to the other end of the system, to the original part programmer's input.

A company typically makes a fairly small range of types of part. We make ribs, frames, skin panels, floor panels, access doors, cleats, and perhaps twenty others. The order of magnitude calculations made in the Appendix show that a lot less information is needed to specify a component than goes into the part program for that component. The factor is about ten to one on real information, with a further factor of about five to one giving some protection, through redundancy, against mispunchings and misreadings.

This ten to one reduction in part programming effort can be achieved by having special purpose computer programs which accept as their data information specifying the shape of a particular rib or frame, together with material or feedrate information. They then generate the part program which would otherwise have been written by hand.

We refer to such programs as preprocessors, and have come to accept that the case for having them in one form or another is as solid as that for having postprocessors.

Fig 1.

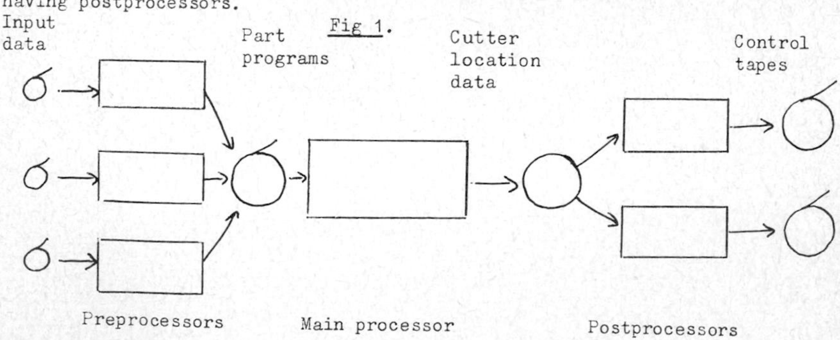

COMPUTER LANGUAGES FOR NUMERICAL CONTROL, *J. Hatvany, editor*
North-Holland Publishing Company - Amsterdam–London

2. ECONOMIC FACTORS

Our objective is simple - to save money by reducing part programming costs. The input data to a preprocessor can be considerably shorter, and requires considerably less thought to write, than the equivalent full part program. However, against this saving have to be set the costs both of writing and running the preprocessor. Careful choice of the component classes invested in, and of the generality of each preprocessor can maximize the return on investment.

Examination of the relative proportions of the various costs involved helps this choice. Consider the programming of fairly complex aircraft-type light alloy work. The ratio of part programming cost to computing is about ten to one if we allow for five or so runs per good tape. The cost of two large scale plots is about the same as the computer cost, as is the cost of a tape-proving run on the machine tool. The effect of a good preprocessor may well be to reduce the part programming to 10%, cut the number of main processor runs to three, and the number of plots to one. The cost of the preprocessor run itself will be of the same order as that of one shot through the main processor. The cost of the good tape is thus reduced from 13 to $3\frac{1}{2}$. However, the preprocessor, to be this good, may have taken a long while to write.

A simpler preprocessor, written in half the time, and at half the cost may leave the part programmer with twice as much to do, and may not make any savings at all in computer runs and plots. It will still reduce the cost of a good tape from 13 to $5\frac{1}{2}$, and so gain 80% of the return from 50% of the investment.

Moreover, it will have been earning its keep earlier, catching a larger fraction of the wave of work which probably triggered the writing in the first place.

This implies that very specific preprocessors, written as cheaply and quickly as possible, are most valuable. The more general program, however, will be potentially useful on many more components, and so may give a better return over a period of years.

3. EXAMPLES

BAC have written a number of preprocessors, with a wide range of generality. Examination of some of these illustrates the points discussed.

3.1 Scanning

Many aircraft parts have surfaces parallel to the outside skin, but are narrow enough for the heel line, and the variation of bevel angle along it, to form an adequate representation. Such surfaces could be cut using five-axis machine tools (or sometimes four-axis), but at the moment, for our sort of parts, it is more economic to scan the surface, using a barrel shaped cutter on a 3-axis machine tool.

Fig 2.

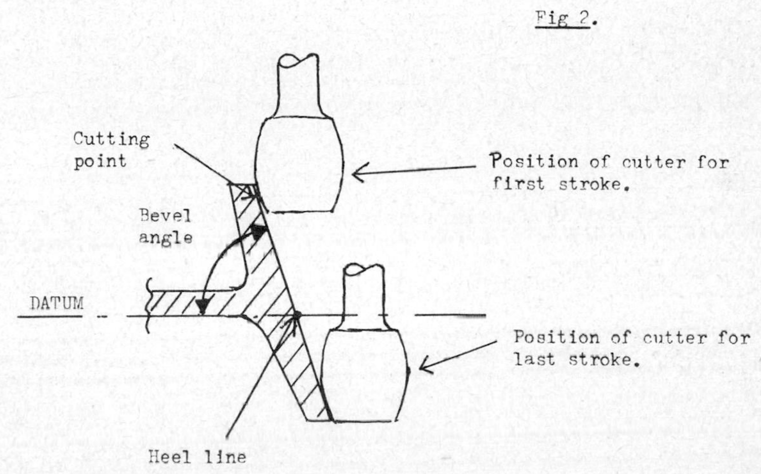

The separate cuts are all related to the heel line profile and to the bevel variation, but all are different. It should be noted that the most efficient cutting pattern is to have the cutting point, rather than the tool centre, move in parallel planes

Fig 3.

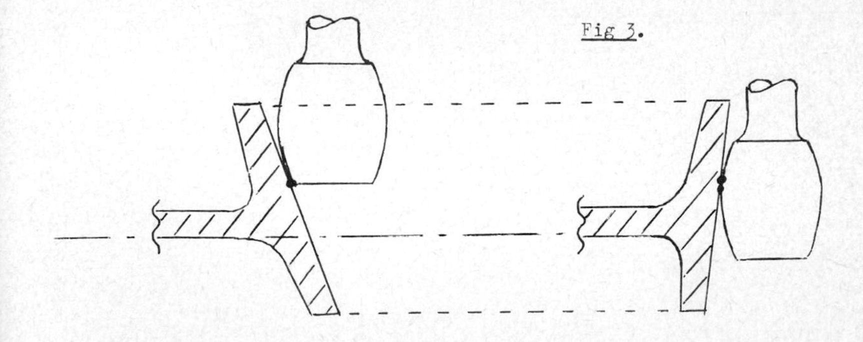

The input to this preprocessor is a simple table of ordinates and bevel angles, together with cutter information and details of feedrates, run-on radii, etc.. The expansion ratio depends on the number of strokes generated, but is typically between 10 and 20. However, the redundancy of the input is quite low, and so the programming time is not reduced by quite the same factor.

3.2 Macro Expansion

This program was not originally written as an NC preprocessor at all, but as a general purpose macro expansion facility. It has been used to generate Fortran program, data for technical calculation programs, and control tapes for drawing seat layout plans on a draughting machine, as well as for NC purposes.

An external macro expansion facility has certain advantages over a built-in one, such as exists in the APT-like languages. In particular, it can generate many part programs in one run as instances of a single macro, thus programming a whole family of parts at once. Minor advantages, occasionally useful in the hands of a skilled programmer, are the decoupling of the macro control syntax from the syntax of the generated text, recursion, and legal redefinition.

Such languages as GPM, TRAC, and ML1 are now becoming more widely available and can easily be applied in this way, so it should be less and less necessary in future for private macro expanders to be written.

3.3 Skeleton Expansion

A preprocessor written at BAC's Filton works with the same end in view was designed specifically to generate multiple copies of a part program varying at certain points by the insertion of different pieces of text. The input consisted of two parts - a skeleton part program, with the points of insertion named, and indicated by ≪ ≫ brackets

e.g. SPINDL/≪ SPIN ≫

and also a table of the replacement texts to be included in the various copies of the part program.

e.g.

VARIABLES \ TAPES	.	.	.	4	5
.					
.					
.					
.					
SPIN				38,CLW	40,CLW

The resulting part program for tape 4 would contain the statement

SPINDL/38,CLW

and for tape 5

SPINDL/40,CLW

The expansion ratio for this, as for typical uses of 3.2 above, varies between about three and ten to one. Higher ratios are quite conceivable, but do not seem to occur very often in practice.

3.4 Skin Mill

BAC find some advantage in machining wing skins from solid billets, leaving integral stringers and local thickenings or "lands" where loads from the ribs or internal equipment are coupled into the skin. Machines for gang-milling these panels have been in use since 1955, the earliest being controlled by cam bars, the latest by NC. The first application of NC at Weybridge, in 1959, was in the manufacture of such cams for the VC10 wing panels. Two preprocessors have been used in this area, one for part programming cams, the other for driving the NC mill directly.

The cam bar program automatically split the complete cutter run into standard length cams, and produced, as well as the NC part program, standard cam drawings. This reduced the cam drawing cost from 35 hours to 5 hours. Perhaps more important, it replaced a process in which numbers were copied from one sheet to another manually four times by one in which a single copying was sufficient.

The direct program had one difficult problem to solve. Imagine a 13 inch (330 mm) diameter cutter, spun at 3000 r.p.m. by a 120 horsepower (90kW) motor. The total width of cut in a gang may be 3 - 4 inches (80 - 100 mm) or even more. When cutting the top of a land the depth of cut may be $\frac{1}{4}$inch (6 mm) or less, but as the cutter rolls over the edge of the land the instantaneous depth of cut may rise to $2\frac{1}{2}$ inches (60 mm) for a short period. If the feedrate is not substantially reduced serious damage may result. On the other hand, reducing the feedrate to a standard safe value during the whole of the roll-over, which is what a part programmer would probably do, extends the floor to floor time noticeably. It is therefore desirable to compute a safe feedrate for each cut vector.

Fig 4.

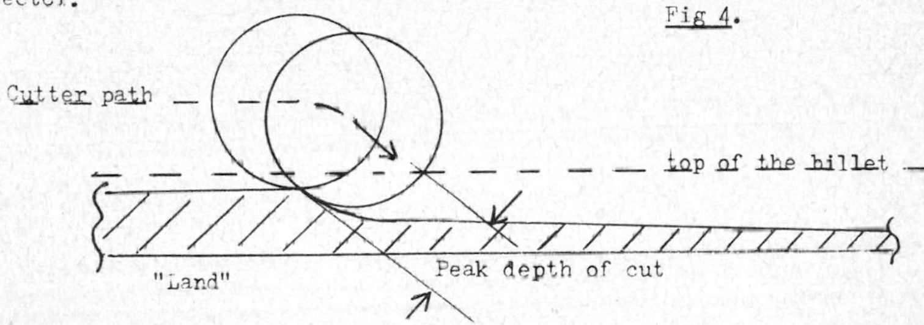

One serious mistake was made when we designed this program. Instead of generating a part program in a readily comprehensible form we tried to save computing time by interfacing into the main processor just before the tool offsets stage. This nearly doubled the cost of writing the program, and delayed its use substantially, because even with the usual listing and editing service programs it was difficult to read the intermediate interface. Certain faults proved difficult to trace which would have been easy if the output had been a part program in a familiar language. It also meant that the program had to be complete and 100% correct before it could be used at all.

If the interface had been in a familiar language the generated part programs could have been vetted and corrected by hand, and thus used, while there were still errors and omissions in the generator. This would not have saved as much as the correct generation, but any early return on investment is valuable in a cash flow analysis.

The equivalent expansion ratio of this program is about 20 to one, of which about half is due to the lower redundancy of the input data.

3.5 NMG link

The Numerical Master Geometry is the BAC program suite for designing and interrogating sculptured surfaces. It has two distinct links with NC, implemented in separate ways.

A) The sculpture milling of wind tunnel models. Here there is a great deal of computation done within the NMG which cannot be left to the main processor; the tool offsets when cutting a surface intersection, for example, or the automatic spacing of cuts to give the required cusp height. There is very little which can be left for the main processor to do, in fact, and so it is natural to link this type of machining at the cutter location level. The very high expansion ratio, together with the large bulk and low readability of a part program consisting almost entirely of short straight lines, made this the most effective coupling method.

B) The generation of profiles for such components as frames and ribs, which have a lot of other detail. Three facilities were implemented here:-

1) The output of definitions of points around the profile. This allowed further constructions to use profile points.

2) The output of motion statements driving the cutter round the profile.

and 3) The output of a profile in the correct format for input to the scanning preprocessor.

In each case the generated output formed only a part of the final part program. In order to avoid the error-prone editing of tapes or card decks we used a device which is well worth considering for inclusion in almost any preprocessor. Simply, the link program was transparent to all lines of input other than commands to itself. Any such line was copied directly to the output, thus allowing "manual" coding to be conveniently included with the generated text. From the part programmer's point of view the effect is the same as having an enhanced main processor.

For the actual generated text the expansion ratio is between five and fifty to one, depending on the length of the profile being processed, but the overall ratio is much lower because the actual profile machining is only a small part of the complete part program. Getting rid of the copying of numbers from computation sheets to coding forms is a more important effect.

3.6 K-curves

The most general preprocessor we have written so far is one for $2\frac{1}{2}$D machining of slab components with pockets. This has two programs interposed between the geometry and motion programs. The first allows the programmer to define composite profiles, made up of arcs of circles, straight line segments, and empirical curves (implemented by a circle spline technique which uses one circular arc only per data point.) The second expands commands such as

CUT K55,22&-23; CLR K19; CNR K19,27,54,23;

into the equivalent motion statements.

A powerful feature is the offsetting technique, which allows a curve with small fillet radii (or even sharp concave corners) to be machined correctly by, or cleared using a large diameter cutter. No multiple check surfaces or other detailed part programming is necessary.

The area clearance facility uses an algorithm which deals with islands economically, cutting one side first, then the other. This avoids the plunging associated with algorithms which complete one stroke at a time by lifting over islands.

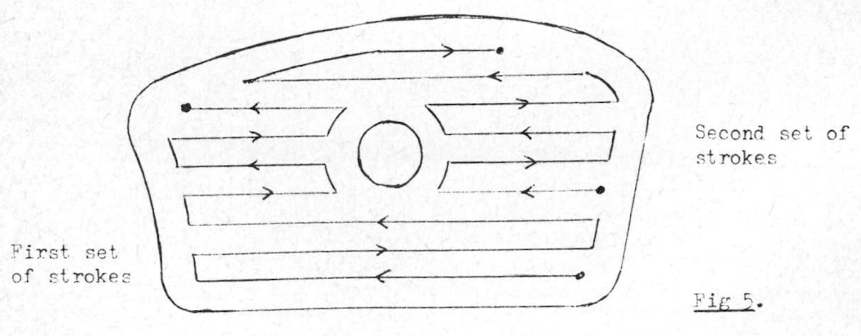

Fig 5.

Appropriate sink moves are generated automatically to enter pockets by ramping down at an angle.

When a large cutter has roughed a profile with tight corners, metal is left in those corners into which it would be unsafe to drive a small diameter finishing cutter. The CNR facility generates motion statements which drill out this material before the profile is finished.

A third program draws out the profiles as defined, fully annotated. This is a particularly useful debugging aid, because the drawing so produced can be superimposed on the component drawing, and all the lines should correspond. No programming whatever is necessary to produce this drawing; the profiles and annotation are all positioned and drawn automatically.

The initial version of this system has given expansion ratios of between two and four to one, and we hope to be able to improve this. This figure sounds low compared with others quoted in this paper; it is; but this preprocessor can be applied to at least half our work, and a figure of only two implies the reduction of our total part programming costs by 25% !

4. COMPARISON WITH OTHER APPROACHES

The preprocessor has a fundamentally distinct function from the main and post processors. It may be convenient, however, in some cases, to combine this function in the same program as the main, or even the post processing. We can identify five arrangements:-

A) The stand-alone program which itself generates a control tape.

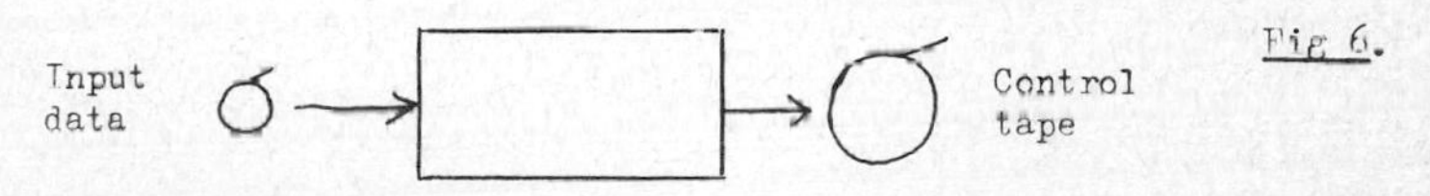

Fig 6.

B) The auxiliary processor which generates cutter location data for subsequent postprocessing.

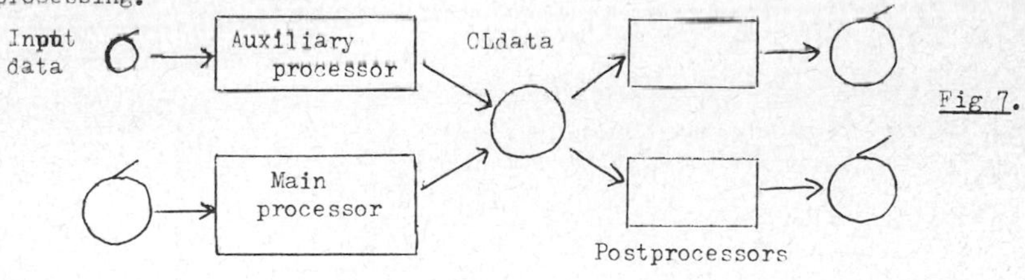

Fig 7.

C) Additions to the main processor.

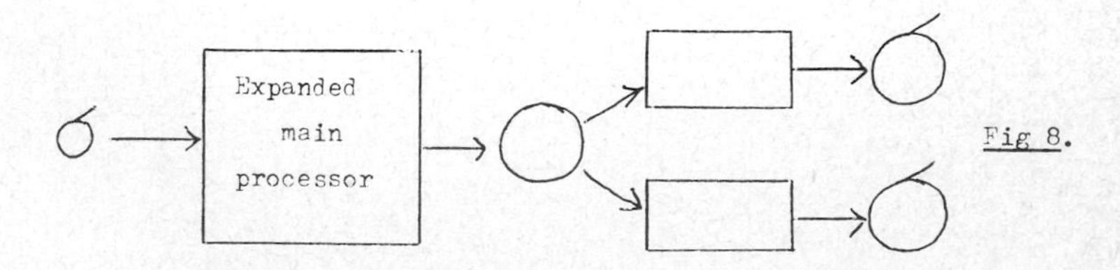

Fig 8.

D) System macros for the main pocessor.

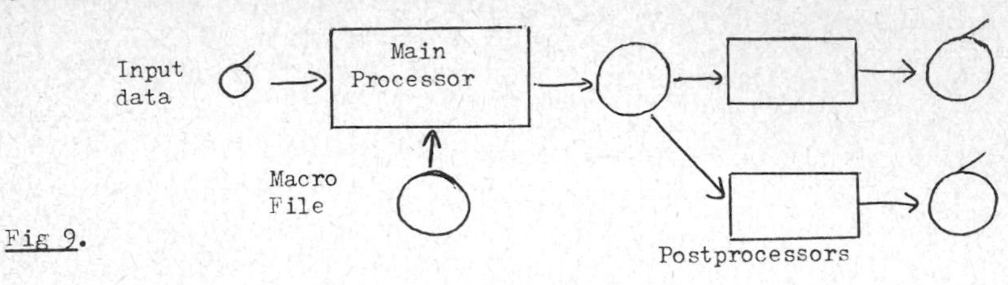

Fig 9.

E) Separate preprocessor.

see fig 1.

These approaches can be compared using various criteria,- capability, cost of writing, cost of running, and convenience in use.

The capability criterion is fundamental. Some facilities can only be implemented by A,B, or C, not D or E. These are the low-level facilities which actually increase the power and applicability of a language. Adding sculptured surfaces to APT, for example, has to be done by addition to the main processor itself, not just by a few system macros.

Most of the facilities we have considered above have been different; they have been adding to the high-level features, which allow otherwise lengthy coding to be replaced by much shorter and more convenient forms. These can be implemented by any of the five.

4.1 Stand-alone program

This is fully capable. Any control tape can be generated by such a program. The cost of writing is probably higher than that of a preprocessor, but it might not be by very much. The total cost of running will be the lowest of the five approaches. The principal disadvantages are;- that it is linked to one particular machine tool or control system; that it has to be complete and correct to be at all useful; that the control tape format itself has to be read to verify test runs; and that portions of standard part programming cannot be included in the control tape except by (error prone) splicing.

4.2 Auxiliary Processor

This is very similar to 4.1 except that the limitation to one machine tool or control system is removed, at the price of increased running costs. There are situations, such as surface sculpturing, where the other disadvantages are not particularly irksome.

4.3 Addition of facilities to the main processor

This is also fully capable. It is an option open only to those with a competent systems team who know their way round the main processor, and with control over the system being run. This effectively means those with in-house NC computing.

The cost of writing may well be higher than for preprocessors, because interactions with all the other parts of the processor have to be considered. The costs of running will be higher than 4.2 because the expanded main processor will be a much larger program than a special purpose auxiliary. It is possible that the costs of running standard part programs may be increased slightly by the addition. The advantage over 4.2 is that standard part programming may be easily included in the same control tape. This implies that the addition does not have to be complete to be useful, merely correct.

4.4 System macros

Provided that this approach is capable, it should have much lower writing costs than 4.3. The running costs will be much higher. It should be fairly easy to check a macro out, because full listing allows the expansion to be followed in source language terms. There are restrictions in some systems on the total amount of macro definition text. This may limit the power actually available in this option.

4.5 Preprocessors

From many points of view these are similar to system macros. They may be much more complex, because the facilities for logic, calculation, and file access in, for example, Fortran are much better than those in, say, APT. For the same reason the cost of writing should be lower. The running costs should be lower too, because the preprocessor need only usually be run once, whereas the macro expansion has to take place on every pass through the main processor. Also the calculations are performed more efficiently if compiled rather than interpreted.

The final advantages are that preprocessors can be written and run by even the smallest firm, and that the programs can be useful before they are 100% correct, because the text generated can be edited before the main processor is run. Unforeseen requirements and combinations of circumstances can also be taken care of in this way.

5. LANGUAGE QUESTIONS

5.1 The effect of the main processor language

Apart from the text expansion preprocessors 3.2 and 3.3 above, all the programs described have been oriented towards generating Profiledata, a language in which the cutter path is specified in terms of change points and circle centres rather than in terms of part, drive, and check surfaces. This feature has proved quite appropriate for automatic generation.

Shortly before this paper was written BAC decided to standardize on the use of an APT dialect for all part programming, and so we have been examining the problems of generating APT.

The scanning cuts problem would have been very difficult, because APT has no low-level facility for controlling the path of the cutting point, but BAC's Military Aircraft Division had added scanning facilities to the version of APT on which we are standardising, and so we do not need to solve this problem.

On the other hand, the APT generated by the K-curve program will be even easier to follow and to edit than the Profiledata we have been generating so far.

We have selected the Skin Mill program as being suitable for implementation by system macros.

Hopefully a lot more experience will have been gained before the date of the Prolamat conference itself.

5.2 Input data formats for preprocessors

The input data format for a preprocessor can, of course, be regarded as a programming language, and there are two points of view which can reasonably be held. The first, dogmatic view is that since APT is now the company's standard language all preprocessors should use the APT language processor. The other, pragmatic view is that preprocessor data seldom needs the power of nested definitions etc. which that language would provide, and that all the reasons for standardizing are satisfied by generating an APT program.

The author's personal preference is for the latter argument, but he is quite prepared to be convinced to the contrary by the discussion at this conference.

Acknowledgements:

The author wishes to thank all his colleagues who have written and used the programs described; also the British Aircraft Corporation for permission to present this paper. All the opinions herein are his own, rather than those of the Corporation.

APPENDIX: INFORMATION CALCULATIONS

In a part program the programmer needs to supply sufficient information to distinguish the cutter path he wants from all other paths. He actually supplies sufficient to distinguish his part program from all other text strings. If we assume 500 statements, 30 characters per statement, and 6 bits per character, the latter figure is about 100000 (10^5) bits. A typical redundancy is about five, and so the necessary information must be about 2 x 10^4 bits.

The information to distinguish the component to be machined from all others is much less; if there are 32 possible families of component, 100 logical decisions (is a particular feature present in this particular component?), and 100 key dimensions, each specified to an accuracy of one part in 10^6, it is just over 2 x 10^3 bits.

A reduction by a factor of 10 in the necessary information is thus theoretically possible. This is in fact a reasonably good measure of the amount of decision making the part programmer has to do. If the redundancy is kept at the same level as in normal part programming, there will be a similar reduction in the amount of text to be written out, punched, and read into the computer.

Now cutting part programming down by an order of magnitude looks so attractive that there must be a catch somewhere. This catch is, of course, that the control tape does contain sufficient information to distinguish a unique cutter path, and the full information has to be supplied to it somehow. It has to be supplied in the form of a computer program, which itself has to be written in its own source language, with some redundancy factor.

Although there are no theoretical estimates for the amount of necessary information in such a computer program, experience has indicated that the writing of such programs is less than ten times as expensive as a complex part program, and so one can expect a break-even point for the investment of about ten usages.

Reference:

The Mathematical Theory of Communication
C.E.Shannon and W.Weaver
University of Illinois Press
(paperback edition 1972)

SOME SPECIFIC TOPICS OF THE SYMAP CONCEPTION

HERBERT SCHREITER
Sektion Rechentechnik und Datenverarbeitung
Technische Hochschule Karl-Marx-Stadt, DDR

Abstract: The partial languages of the SYMAP language system distinguish themselves as opposed to other programming languages for NCM by a number of specific topics for machining description. The following facilities belong to them among other things: instruction compounds, definition cycles, the TECH-OP- and TECHKO-OP-relations, the PPROC-conception.

Always, if numeric controlled machinetools (NCM) are used in production, they must be programmed. And almost everywhere people have searched for methods to rationalize programming work, because it proved that manuel programming is too lengthy or unpracticable. According to the pattern of development of computer programming problemoriented languages were generated, by means of which the production tasks may be described much easier than in machine code of NCM. An electronic data processing machine is charged with manufacturing of the control program and different further informations important for the production. The procedure is wellknown in general. But the special programming systems differ of each other. The most frequent distinctive features are: structure of the data processing, source language, error treatment, technological content. They are not only connected with the size of the programming system and thus with the size of the electronic data processing machine necessary for implementation. For many users by reasons of computing costs and availability the fact plays an essential role, whether a small or middle computer is sufficient for automatic programming. In the following some specific forms of the language SYMAP (Kochan,1970) will be treated that exceed the normal demands on such a language. The partial languages of the SYMAP language system have been stated under that aspect that implementation on small or middle computers is possible. The biggest used computer until now is the IBM computer of the type 360-40 with 256 K byte for the implementation of SYMAP(PS), where 128 K byte are necessary for translation in using the operation system OS.

1. Instruction compounds in SYMAP(B)

Instructions are the sentences of the SYMAP language. Their nature may be descriptive or operative. That means, a sentence has an operative effect, if its contents describes immediately a detail of machining on the NCM, for example the application of a cutting process, the change of movement direction of a tool, the downfeed, the put in operation of the coolant and so on. Descriptive sentences have a indirect influence on the machining; they serve for definition of geometric elements, of tools, of cutting operations, of formed elements and others, or state global or local conditions, for example specification of the NCM, of control, of punched tape input-code, of reference coordinate system, of corner treatment in contour processing and so on.
In using the Backus-notation is valid:
<instruction> ::=<pure definition>|<declaration>|<statement>
The following form describes the general structure of an instruction, But it is not an extract of the formal definition of the syntax of SYMAP, that must be essentially more concret.
<instruction>::=<instruction word 1> | <instruction with reference to variables>
<instruction with reference to variables>::= <instruction word 2>,<argument>
<instruction with reference to variables>;<argument>

COMPUTER LANGUAGES FOR NUMERICAL CONTROL, *J. Hatvany, editor*
North-Holland Publishing Company - Amsterdam–London

⟨argument⟩::=⟨variable⟩|⟨definition of variable⟩
⟨definition of variable⟩::=⟨variable⟩,⟨definition word⟩,⟨parameter list⟩
The pure definitions belong to the instructions with reference to variables. They begin, exept definition cycles, with the instruction word DEF; DEF ∈ ⟨instruction word 2⟩. The argument is a definition of variable. In processing a pure definition the processor assigns the canonical form of the specialtype of variables concret values, but does not aktivate the elements defined in this way. The reference to defined elements is managed by statements or declarations of the class instruction with reference to variables, in which the arguments are only variables. If an argument represents a definition of variable, the variable in question is assigned before its call.
Instruction lists are basic components of each SYMAP program.
⟨instruction list⟩::=⟨instruction⟩|⟨instruction list⟩⟨instruction separator⟩
⟨instruction⟩
The structure of the program (Schreiter, 1969/1972) is not further dealt with in detail.
In SYMAP(B) may be programmed production tasks of contour processing on NCM with 2 1/2 d - continous path control. The description of the machining sequence is made in a similar way as it is known from ADAPT: lets imagine the contour of the part or the intended cut respectively as a route to drive over by a car within a road system. As the assistant driver you have to describe the driver, who is not familiar with the place, the way:
"...;
drive left into the street No. 1;
drive right into the street No.2;
drive right into a circle No. 3;
drive along the circle No. 4;
..."
The denotation of streets and circles are to be replaced by identifiers of curve elements of SYMAP(B) and the instructions sentences are to be replaced by instructions of this partial language.
It happens very frequently in part producing that contour sections repeat themselves, either because cutting is necessary in different z-planes or because the same geometric form is to be produced once again, because it occurs at the part in face of the original form shifted, rotadet, reflected or extended (see fig.1). For this cases SYMAP(B) provides the introduction of geometric formed elements in form of instruction compounds.

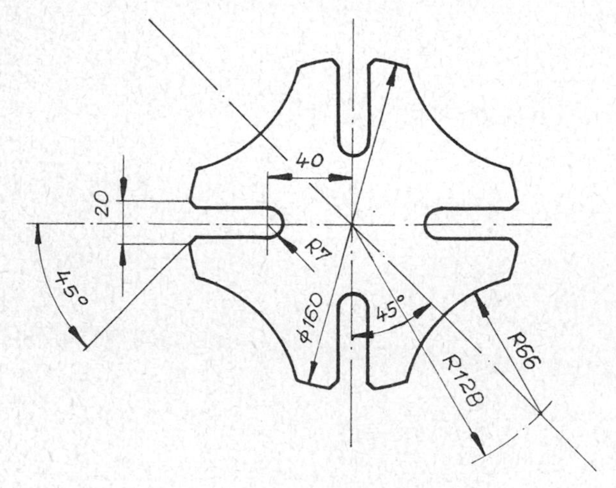

Fig. 1a. Part 1.

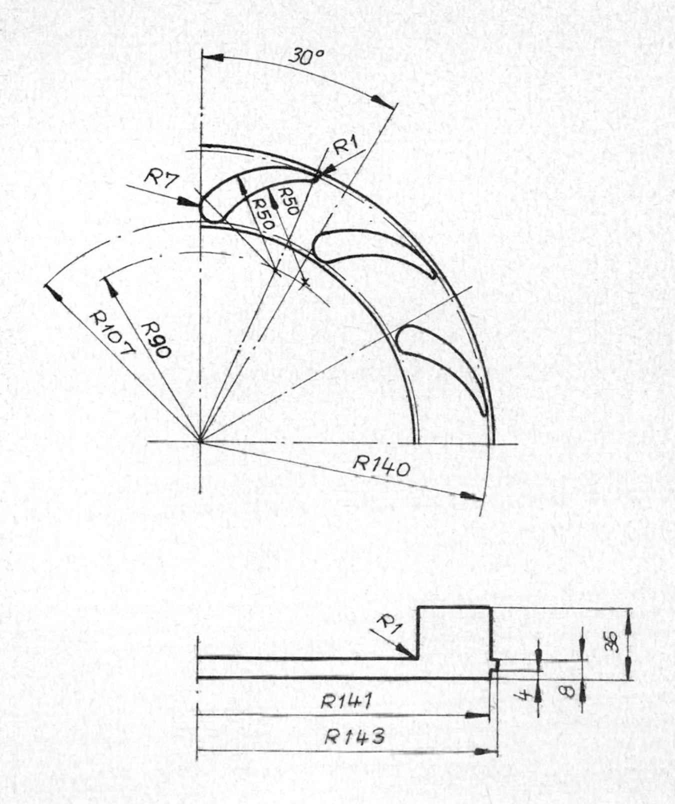

Fig. 1b. Part 2.

An instruction compound represents an oriented combined curve of the central path of the tool that consists of arc- and linesections, but to which initial and end element belong as circle and line respectively. In the case that the diameter of the tool is zero, the combined curve is laying on the contour (see fig.2).
Two kinds of instruction compounds are differed by their definition. The original form of an instruction compound is called primary. Bracing of instructions by the SYMAP words IVANF and IVENDE is typically for it. The primary instruction compound is defined in the following way:
<primary instruction compound>::= IVANF,<instruction compound variable>
<instruction separator><instruction list of the instruction compound><instruction separator> IVENDE.
Out of a few exeptions all kinds of instructions may occur in the instruction list of the instruction compound. In particular further primary instruction compounds may be contained, that provides an interleaving of arbitrary depth. The structure of bracing is like at <u>begin</u> and <u>end</u> in ALGOL 60.

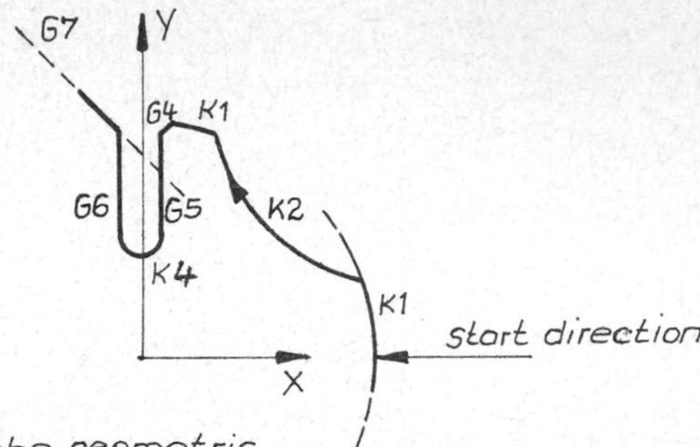

Fig. 2. Example for the geometric strukture of an instruktion compound.

The primary instruction compound of represented in figure 2 might be formulated in a part program in the following way:

```
    IVANF, I1; 1, BRTS1, K1;
               2, BLKS1, K2; K1; G4; G5;
               3, BVWS,  K4; G6;
               4, BLKS1, G7;
    IVENDE
```

During the program run the instructions of the list are processed like simple instructions. The combined curve of the central path of the tool generated by it - more exactly: its projection into the x-y-plane - is assigned including its boundary elements to the instruction compound variable in canonical form. Thus it only contains informations resulting of cutting statements. Other tool movements are not permissable within the instruction compound.
The preservation of the instruction compound as a part of the central path of the tool in a prepared form is a basic idea of the implementation. It allows to organize relatively simple according the program the further use of this geometric formed elements.
Such a use consists of the definition of a secondary instruction compound by a primary by means of geometric transformation or distance calculating. The syntactic definition is:

```
<definition of a secondary instruction compound>::=
                        <instruction compound variable>, IIU, <IIU-list>|
                        <instruction compound variable>, IIA, <IIA-list>
<IIU-list> ::=<instruction compound variable>,<transformation variable>
<IIA-list> ::=<instruction compound variable>,<arithmetic expression>,
              <direction modifier>
```

It is the question of a special case of the above definition of a variable. Whereas the IIU-list explains itself, there must be said to the IIA-list that the value of the arithmetic expression state the aquidistant distance of the derived combined curve to the defining combined curve. On the other hand the direction modifier means the relative position of the derived combined curve according to the orientation of the defining combined curve (RTS-right, LKS-left). The orientation of an instruction compound is transferred to the instruction compound derived from it. Thus the defining instruction compound may be secondary itself. This possibility is included in the syntactic definition. But each secondary instruction compound lastly goes back to a primary instruction compound, the orientation of which provides from the sequence of instructions within its instruction list.
The following program section contains the definitions of the secondary instruction compounds with the idendifiers I3 and I5 of figure 3:

```
...;
DEF, U2, DRE, P2, 90;
     I3, IIU, I1, U2;
```

```
   I5, IIA, I1, 4, LKS;
...
```

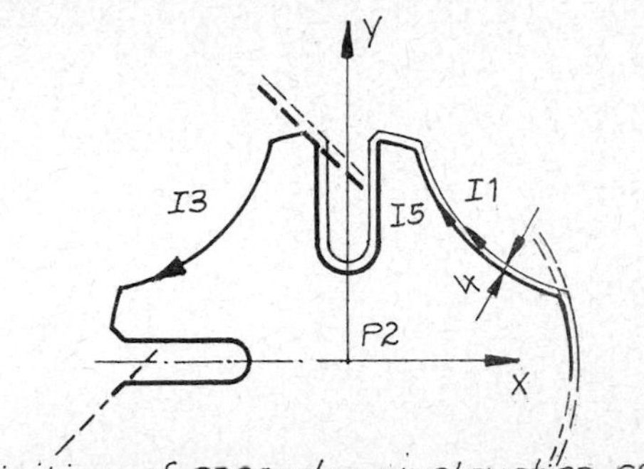

Fig. 3. Definition of secondary instruction compound

The combined curves of the central path of the tool represented by instruction compound variables may be added now by the part programmer at an arbitrary place determined by him as a continuation element to a path curve. For that he uses the call statements for instruction compounds.

Syntax:

<call statement for instruction compound> ::=
<instruction word for call-in the instruction compound>, <instruction compound argument> | <call statement for instruction compound> <argument separator> <instruction compound argument>

The instruction words for call-in the instruction compound regulate the addition of the instruction compound to a neighbour element. This addition may only be carry out with a boundary element of this instruction compound and is also defined as transition between the concerning boundary element and the neighbour element, for example circle - circle, circle - line, circle - point sequence and so on. But that is not treated here.

The instruction words mentioned decompose in two classes. Because the instruction compound variables represent combined curves oriented such a instruction word contains an information over it, whether the combined curve is added to the tool path according its orientation or in an opposite sense.

With it means:

DVWS, DRTS1, DLKS1, ... passing through, that is addition of the instruction compound according its orientation;

RVWS, RRTS1, RLKS1, ... return movement, that is addition of the instruction compound in the opposite sense.

VWS (along a tangential transition), RTS1 (branch off to the right side in the first cut-point), ... express moving-directions.

In the following the model explained is used to machine the part in figure 1a. It may be pretreated with an admeasure of 4mm.

The program part describes the geometry of the finishing cut.

```
...;
 6, START;
 7, PVON, P1, PXY, 110, 0;
 8, INRIP, P2, PXY, 0, 0;
 9, GVOR1, K1, KPR, P2, 80;
10, WERTS;
11, DEF, P3, PRW, 128, 45;
       G2, GABC, 1, 0, 10;
       P4, PGK, G2, K1, YGR;
       U1, SPGL, G3;
```

```
      U2, DRE, P2, 90;
      U3, DRE, P2, 180;
12, IVANF, I2;
13, IVANF, I1;
14, BRTS1, K1;
15, BLKS1, K2, KPR, P3, 66;
           K1;
           G4, GPW, P4, 45;
           G5, GABC, 1, 0, 7;
16, BVWS, K4, KXYR, 0, 40, 7;
          G6, GGA, G5, 14, XKL;
17, BLKS1, G7, GUG, U1, G4;
18, IVENDE;
19, DLKS1, I3, IIU, I1, U2;
20, IVENDE;
21, DLKS1, I4, IIU, I2, U3;
22, GHTR1, K1;
...;
```

2. Definition cycles

An other form of bracing than is used in defining primary instruction compounds is found in SYMAP(B) and (PS) in definition cycles. Thereby it is the question of special cases of pure definitions.
They are ordered syntactically in the following way:
<pure definition> ::=<straight on definition>|<definition cycle>
<definition cycle>::= ZYKLUS,<arithmetic argument><instruction separator> <definition list><instruction separator> ZYKEND
<arithmetic argument>::=<arithmetic variable>|<arithmetic definition>| <arithmetic expression>
<definition list> ::=<definition>|<definition list><argument separator> <definition>
A definition has a structure like the definition variables generally described above.
In processing a definition cycle in the case that the arithmetic argument is an arithmetic definition, the processor first execute the arithmetic definition, that means the value of the arithmetic variable is determinated. Always the sequence of definitions contained in the definition list is processed so much as the value of this variable or of this arithmetic expression respectively indicates. If this value is a fractional number, it is rounded to the next integer number.
The meaning of definition cycles lays in the possibility of iterative definition of elements of the language, but especially in the discretation by point sequences of curves given by formulas. By this the lengthy calculation of such point sequences is spared the programmer by means of the processor. Only by using these definition cycles the programming of machining cut and defining of point patterns along such curves can be done in SYMAP(B) or SYMAP(PS) respectively.
Example
Lets calculate and collect them to a point sequence 11 points with the x-coordinates x = -5(1)5 on the contour of a halfellipsis given after transformation of the coordinate system by the formula

$$y = \frac{2}{5}\sqrt{25 - x^2}$$

The according program section is:

```
DEF, A1, IST, -5; A2, IST, 0; A3, IST, 1;
     Q1, QXY, A1, A2;
ZYKLUS, 10; A1, IST, A1+A3;
            A2, IST, 2/5*SQRT (25-A1*A1);
            P1, PXY, A1, A2;
            Q1, QSUM, Q1, P1;
```

ZYKEND
In the contrary to primary instruction compounds there is not provided the interleaving of definition cycles.
There is estimated that the idea of the instuction compound and the definition cycle realized in SYMAP may also play a significant role in the extension of the language to problems of graphic data processing. The whole totality of the concept is not exhausted in NC-programming.

3. Application of machining operations on sites

A further point to be considered is the application of machining operations on different sites, for example, in producing of boring patterns, in producing of threads on different points of the part, in surface milling and so on. It is solved in SYMAP in a specific way. The call of machining operations is managed by the instruction words TECH and TECHKO respectively. In SYMAP(B) only TECH is permissable; the attached syntax is:

<call statement for machining operation>::= TECH,<machining argument> | <call statement for machining operation><argument separator> <machining argument>

An operation may include an available cutting process or for example only the establishing of a cutting variable. It is assiqued the technologic variable beeing in the argument in canonical form and it is valid in the following until it is abolished completely or partially by aktivating of other operations.

In SYMAP(PS) (Richter,1971; Schreiter,1972) there is an other case. This part of the language serves for programming of boring and milling machines and of machining centres with point-to-point and straight-line control. The following syntax is simplified insignificantly in face of the syntax-definition of SYMAP(PS):

<call statement for machining operations>::= TECH,<machining argument>| TECHKO, <machining argument>|<call statement for machining operations> <argument separator> <machining argument>

<operating range>::= OPENDE,<geometric argument> |<range list> <instructions separator> OPENDE,<geometric argument>

<range list> ::= OP, <geometric argument> | <range list><argument separator> <geometric argument>

<processing segment> ::= <call statement for machining operation><instructions separator><operating range> | <processing segment> <instructions separator> <operating range>

The different procedures of centralizing, drilling, counterboring, milling, reaming and thread-cutting belong to the machining single operations, which may be assigned to a according variable. They are called by the part programmer in that choice and order necessary for a special machining and thus they form a configuration (a cycle) prepared for application. Their application to an operating range, that means a set of machining sites, is executed dependent on the calling instruction word.
TECH implies the application of the first operation of the configuration to the complete operating range, then the second and so on until the configuration is exhausted.
On the other hand TECHKO implies the application of all operations of the configuration one after another an the first machining site of the operating range, then on the second and so on until the operating range is exhausted (see fig. 4).

The machining operations must be compatible with the sites, i.e. the application of point-to-point machining an line segments is false in the same way as the application of straight line machining on points or sequences of points. The syntactic definition of processing segment says that a configuration may also be applicaded to several operating ranges. The simplified syntax does not indicate the fact that statements, which does not injure the represented matter,may occur between call statement for machining operations and operating

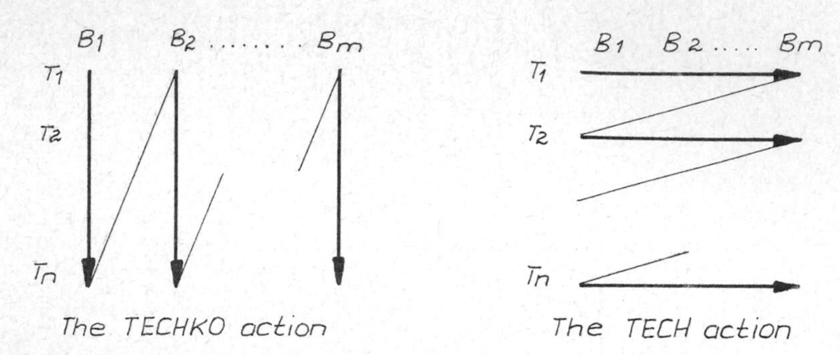

Fig. 4. Application of technological operations (T_i) to working positions (B_j)

range, between the single operating ranges and within an operating range. The instruction list of a SYMAP program is arranged in general in processing segments. A call statement for machining operations is not valid outside of the processing segment in which it is defined.
Befor treating an example let us mention that on a higher level of finishing the processor may be also worked with formed elements instead of machining operations. The syntax is analog; as instruction words FORM and FORMKO are used.
Example:
In the following is shown the TECH-OP-relation in a section of a SYMAP(PS) program for processing the part shown in fig. 5. The used NCM has a rotary table, which makes possible the multilateral machining.

```
...;
35, TECH, T3, BOSPIR, DV, 17.25, AD, 20;
          T4, GEWI, PG, 11, STEIG, 1.41, AD, 20, DB, 17.25, W, W37;
36, OPENDE, P3, PXY, 230, 20;
37, TECH, T6, BOSPIR, DV, 6.75, AG, 25;
38, OP, Q6, QSUM, Q1, Q2;
39, DRLAGE, B1;
40, OPENDE, Q7, QPRW, P2, 32, 135, 3, 270, AUS;
41, ZEBENE, Z1, Z, -20;
42, OP, Q8, QXY, 100, 200, -100, 200;
43, DRLAGE, B2;
44, OPENDE, Q9, QXY, 320, 320, 40, 320;
...
```

4. Postprocessor call
Lets deal with a last point.
The SYMAP compilers are structured in processors and postprocessors. According their contents both processing parts differ from which of APT- or EXAPT-translators. For this reason CLDATA has not been used as transfer language. There is a considerably detailed paper that confronts CLDATA and the transfer language developed for SYMAP (GFZ,1971). In detail there can not be reported about that on this place.
In the language postprocessor call is executed in the following way. The postprocessor instruction serves for specification of postprocessor.
Syntax:
<postprocessor instruction> ::= PPROC,<PPROC - list>
<PPROC-list> ::=<NCM>,<control>,<NCM-code>|<PPROC-list>, <text>
The PPROC-list contains that informations that save the adaption of control

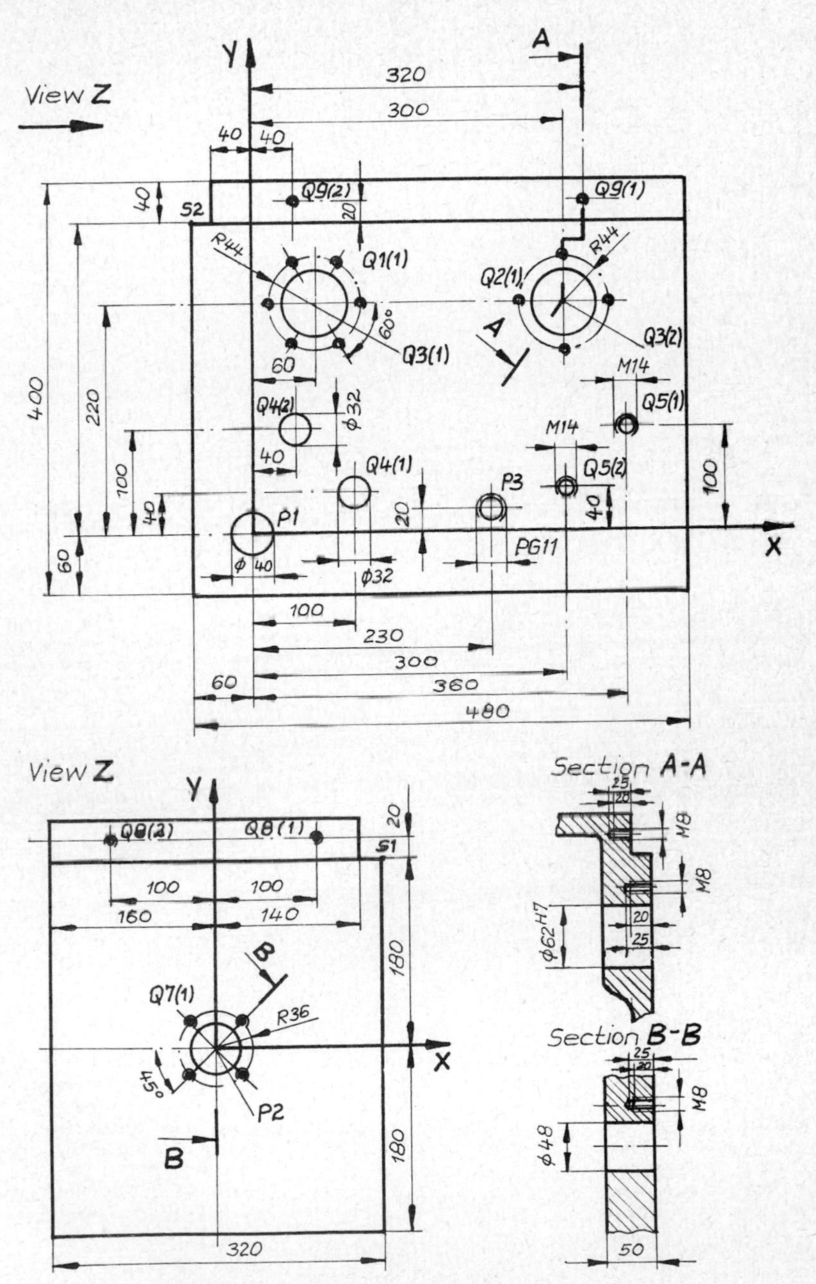

Fig. 5. Section of a part drawing

program to the importance of a special NCM. Each postprocessor instruction states the validity scope of a certain postprocessor in the source program. This is divided in blocks, within which partial tasks of the total processing problem may be programmed. The postprocessor instruction may occur at 3 different places:

a) immediately after the first instruction of a program, the program head, i.e. before the first block;

b) between two blocks, i.e. the blockend of the preceding and the blockhead of the next block;

c) within a block immediately after the blockhead. Where as possessing a local effect for the concerning block in case c), i.e. a control program for the given tool machine is generated, that contains only informations about the appropriate partial task, it has a global effect in the cases a) and b) : the control program contains informations about all partial tasks of the following blocks.

This fact is called multipostprocessing. Its practical meaning is emphasized thereby that control programs for test machines, for example drawing table may be also generated parallel to such for machine tools.
By that the enumeration of some specialities of SYMAP is finished. All language forms mentioned are implemented, not all in each compiler. The representation of such solvations, as are contained in the most problemoriented NCM-languages, was neglected conciously.

REFERENCES:

Kochan, D.(1970): The SYMAP programming system, In: Numerical Control Programming Languages, ed. by W.H.P. Leslie Amsterdam, London: North-Holland Publ.Co.

Schreiter, H. (1969) Sprachbeschreibung SYMAP(B). In: SYMAP, ed.by "Forschungsgemeinschaft" Automat. und Rat. d. technol.Fertigungsvorber.; Jena: VEB Carl Zeiss, Hauptabt. TV

Schreiter, H., Riedel, K., Schubert, G., Spielberg, D., Wetzel, J.(1972); SYMAP- eine Sprache für numerisch gesteuerte Werkzeugmaschinen/Einführung und SYMAP(B). Reihe Automatisierungstechnik Vol. 134 Berlin: VEB Verlag Technik

Richter, H., Hellmuth, W., Scholz, W., Hengmith, S. (1971): SYMAP(PS) - BORF1 - Sprachbeschreibung. GFZ-Information Nr.6, Karl-Marx-Stadt: GFZ Werkzeugmaschinen

Schreiter, H., Riedel, K., Schubert, G., Spielberg, D., Wetzel, J.(1972): SYMAP - eine Sprache für numerisch gesteuerte Werkzeugmaschinen/SYMAP(PS) und SYMAP(DB). Reihe Automatisierungstechnik Vol.135. Berlin: VEB Verlag Technik

Entwurf der DDR zur einheitlichen Übergabesprache Processor - Postprocessor. Karl-Marx-Stadt: GFZ Werkzeugmaschinen, Abt. 422, 1971.

The IBM System/370 Numerical Control System

I.D. Nussey (IBM United Kingdom Limited) and
H. R. Pinter (International Business Machines Corporation)

Abstract

This paper outlines the position of the IBM System/360 Numerical Control System, with regard to standards and numerical engineering, establishes further requirements and describes the features of the IBM System/370 Numerical Control System.

1. Numerical Control System Requirements

The IBM System/360 Numerical Control System (S/360 NCS) was announced soon after the advent of System/360. Four components were delivered between January 1967 and October 1968. These were APT/360, two versions of the bilingual processor AD-APT/AUTOSPOT for the major System/360 operating systems and a version of AUTOSPOT, capable of running under the Disk Operating System on System/360s with only 32K bytes of main storage. These programs are well known and soundly established; and their documentation widely available. They are not however the final word in NC support. Much as one might sometimes wish to stop the world, progress in NC part programming, in respect of both ends and means, has been rapid and incessant. The implementation of APT itself in the 1950s can be seen as one milestone, IBM's System/360 NCS in the middle 1960s as another and the German, French and British national developments of the later 1960s as yet one more. The wide gamut of systems presented at Prolamat 1969 (ref 1) has been extended, most significantly perhaps by the interpretive computer independent processors which offer generality and flexibility, though sometimes at some performance cost. At the same time, the use of computer aided NC has developed to an extent that was unforeseen in the 1950s, or even in 1968, when the papers for the first Prolamat were being written. The number of companies implementing some kind of system already runs well into four figures and by 1980 could easily exceed 10,000. The variety of computers, operating systems and peripherals on which these programs will run is considerable.

What must be done to ensure that the cost of these installations is kept to a minimum? There are three initial requirements. First, processors must attain some degree of standardisation in language. Doug Ross said at the Rome Prolamat meeting that the APT language was defined over a weekend, with the implication perhaps that it reflects it. Nevertheless it is a de facto standard. Thus for success in all but the simplest problem areas (where costs exceed benefits if an APT solution is used) or in major numerical engineering schemes where syntax directed language is fundamentally inappropriate it must be admitted that the use of an APT-like language is a prerequisite. It is important to stress that APT-like, rather than APT-compatible is the condition. The numerical control business is itself still developing rapidly and newer applications like grinding, turning and drafting require languages that reflect their particular needs, which include the accommodation of users with relatively limited skills and desire for problem oriented brevity. An APT-like base may be in some instances acceptable but there are nevertheless significant areas that cannot be so handled, reflecting the irrelevance of the current APT language to such problems as data base management, non-analytical shapes, interactive computing, graphic input systems and to the definition and solution of inherently unstructured problems, that represent at this time an area of almost entirely untapped opportunity to the computer user. Here one must pause before forcing APT into the solution, especially where the potential sophistication of the offering is not yet matched by the corresponding skill level in the user areas and where the nature, origins and volume of the input data are obviously inappropriate.

The second requirement is for an expanded CLDATA file, using the presently accepted standard but providing scope for extension, for example the addition of arc and technological data within the main line record types. It is to the credit of the organisations who so painfully push standards along that in the past few years they have made progress in this respect.

Standard CLDATA formats do ease post processor conversion as a user moves from one processor implementation to another.

A third requirement is the means of building and attaching problem oriented languages round the NC processors. It is becoming apparent that special hardware, especially engineering terminals, are important in this respect. Operating systems and primary application support products that permit easy access to the system during development and in production are also required. The NC processors themselves must offer easy to use hooks for attaching user designed systems.

System/360 NCS was basically satisfactory in these respects. During the period 1964 - 65, when the functional specifications were being fixed, there was a widespread feeling that the point to point market wanted a part programming language that was more obviously directed towards its problems than APT, but wished nevertheless to have a growth path upwards to the continuous path facilities of the APT-like languages. Although the resulting ADAPT/AUTOSPOT processor has gained good acceptance by customers, three factors led to a reconsideration of the strategy. One was the reluctance of all but a few machine tool companies to embark on what they saw to be two quite different post processor support operations. Another was an appreciation of certain technical advances in processor design. The APT/360 processor has performed well, with an initial concentration on reliability being followed by efforts to improve price performance and part programmer documentation. Only minor adjustments to capability were made, primarily in the user interface area. APT/ 360 user groups in the USA, France, Britain and Germany submitted suggestions for extensions. Practical and long term considerations meant that if these were to be adopted constructively, the system design would need to be restructured so that current and projected computer technology could be more fully exploited, and so that there would be in future the possibility of introducing further increments of geometric and technological capability more easily that the APT III philosophy admitted. The third reason for considering a new system was a wish to offer modular and compatible subsets and related supersets.

Considering now the second requirement, of minimising cost to the users, standardisation across the System/360 NC processors was not quite complete. Although in practice this lapse was no trouble to competent implementors, it is obviously something that should be eliminated.

The third requirement has been fulfilled quite successfully. Developments like the British Aircraft Corporation (Military Aircraft Division) Numerical Master Geometry Scheme, the Messerschmitt NC Graphics System (an improvement on the original Numerical Control Graphics program, NCG), numerous versions of the Boeing Seattle FMILL/APTLFT system for handling Master Dimension Data comprising the shapes of surfaces to be machined (FMILL fits bicubics to pseudo-rectangular grids and APTLFT computes cutter paths), the Boeing Wichita BSURF/BPOKET APT extension that can process surfaces defined by sections, boundaries and islands, are just a few of the APT/360 extensions to have been created sometimes simply using unit record data input, but often forming part of the larger design systems for aerofoil, automotive and blade design and manufacture, for example.

One example of a major integrated system is the McDonnell Douglas (Long Beach) method of using the surface definition data from engineering drawings for special machining, especially 5-axis machining of complex aircraft parts. The surface definition data is known as Master Dimension Information, or MDI. The following brief description of the system may serve to show the nature of a link between design and APT.

The primary function of aircraft lofting is to design the airframe surface shapes, taken for example from a relatively small skeleton frame clay mockup with relatively few definitive basic cross sections. The model is subject to many changes before a final shape is achieved. Coordinate measuring machines are used to digitise the shape. Fairing is then performed interactively with the aid of an IBM 2250 Graphic Display Unit, and the smoothed data is stored in a data base that the APT system accesses for machining. Integrity of data is a major feature of this lofting and loft data management system. The same information is accessed by Design, Manufacturing, Tooling and subcontractors. Another feature is the automatic drafting or NC machining of part surfaces to mould lines any constant distance inside or outside the lofted 'mesh' surfaces of the fuselage or nacelle and the 'ruled' surfaces of the wings and stabilisers. This is done through the special Douglas Section 2 DAC Arelem, which has other important capabilities. These include the derivation of data like surface normal vectors of each point derived by APT section 2 and geometric auxiliary view capabilities that minimise manual work by designers or draughtsmen. This is just one of many systems for surface machining associated with APT/360.

It has however not been easy to make system changes to the four processors without intimate knowledge of the programs, so that it may have seemed that only major companies could contemplate such action. New definitions were quite hard to introduce into any of the processors and implementation of a tool library in AUTOSPOT, for example, required a close study of the system logic specifications.

There have nevertheless been some major successes, two of which will be mentioned here. British Aircraft Corporation has shown that APT/360 can be tuned to a considerable degree by an individual customer if he knows exactly what he wants. BAC has obtained more or less the same performance from running APT in a 140K byte partition of an IBM System/360 that is achieved with the standard 224K version, simply by retaining those areas of capability that they had identified as relevant to their needs. The second example is an IBM Field Developed Program, CP-CMS/APT. CP-CMS is an abbreviation for Conversational Programming - Cambridge Monitor System, which supports the virtual machine approach on the System/360 Model 67. CP-CMS/APT was produced by the IBM Los Angeles Science Centre (ref 2). It offers full APT/360 language capability in an interactive time sharing environment, allowing convenient use of APT from low cost, low speed terminals such as the IBM 2741, and the full use of CMS facilities for part program edit and correction. Among the features of the system are the ability to review output selectively at the terminal (the high volume of APT output makes this an important feature for terminal users), while error messages are printed at the terminal as they occur. CMS features like the Context Editor, SPLIT, ALTER, COMBINE, etc., may be used to review, edit and manipulate the APT input and output files. The current status of the part program being executed can be obtained on request from the terminal. Post processors can be invoked during APT execution, or stand alone.

Various other APT-based time sharing systems have been developed by IBM for System/ 360. Among them are the German program TELEAPT which uses PL/1 as a base for an AD-APT implementation including point to point facilities, and a modularised post processor program to cut the post processor writing cost and time. This system runs under the IBM Data Centre CALL/360 Time Sharing Service, using Teletypes or IBM 2741s as input/output terminals. An experimental system which will have wider application as the APL program becomes more widely installed is TORA - Terminal Oriented APT - which uses the IBM 2741 communication terminal in association with an IBM 1055 paper tape punch and a storage tube switched electronically through a modified TSP/12. This program runs in an APL partition of an IBM System/360 or 370. This permits the user to run AD-APT part programs with optional animated display of the part geometry and cutter path; and provides punch tape output.

Both TORA and TELEAPT permit editing of files of part program source and output and within their respective design constraints are constructive approaches to terminal NC systems requiring continuous path capability.

To summarise the situation up to this point, since 1964 there has been significant investment by IBM and its customers in a comprehensive range of geometry-based NC processors, based initially on the batch processing capability of System/360 and subsequently in some attempts to explore the growing time sharing and terminal based data processing environment. Certain more advanced functions, like file handling, workshop technology and post processors, have been investigated with the caution that reflects experience and observation of inherently difficult problems.

So far, three requirements for an effective long term plan for minimising cost have been identified. To these should now be added the proper handling of the data processing environment, which has undergone significant changes in the past few years. These changes will be discussed with particular reference to System/370. Two of the more important of these are hardware related system control programs that provide virtual storage extensions to main memory, freeing programmers from traditional constraints and tasks, and direct access devices that offer speed, volumes and price performance orders of magnitude better than a few years ago.

One example is the cost of one million bytes of direct access storage, which is now only a few dollars a month. Terminal access has been enhanced through the time sharing products like IBM's Time Sharing Option (TSO) and System/370 CMS (Conversational Monitor System) which permits multiple users to run many different operating systems within an optimised time shared regime.

2. System/370 Numerical Control System

These four requirements together explain the need and justification for a further advance in numerical control processing. Hence the announcement of the IBM System/370 Numerical Control System (S/370 NCS), three components of which were announced in September 1972. The component names of the programs are APT-BP (Basic Positioning), APT-IC (Intermediate Contouring) and APT-AC (Advanced Contouring), whose functional capabilities may be thought of as supersets of AUTOSPOT, AD-APT and APT respectively. Availability to customers is planned from late 1973 onwards. This suite of programs offers a further contribution to the state of the art, almost entirely within the standards of today.

The next part of this paper summarises the major characteristics of the programs. The common features of the components will be reviewed first, followed by a description of the individual capabilities of each program. For the sake of brevity, it is assumed that the reader has access to or knowledge of the S/360 NCS part programming manuals.

Library Management and File Editing

A library feature is provided that allows for the storage and retrieval of user defined part programs, part program segments, procedures; data related to machine tools, cutters and material; and the canonical forms of geometric and non geometric entities from external files. Associated file editing functions include add, delete, replace and change. Members of files can be safeguarded by passwords if so wished. The user can include data from files and libraries in his part programs, as required. Changes to files may be permanent or held just for the duration of the current job. Source statements can be resequenced. An audit of modifications is provided.

Procedures

There will always be a requirement for users to extend NC processors. One example is the need for cycle capabilities that reflect the way they want to do point to point machining. Another is workshop technology, where users may wish to calculate feedrates and spindle speeds based on their own data on materials, cutters and machine tools and their own criteria for acceptability.

The procedure (PROC) permits a user to introduce these functions without knowledge of the processor system internals. A user can define and catalogue for future invocation any series of logical steps, known as a Procedure (PROC). This may be stored in the source library or in the canonical form file and may be updated through the Edit feature. The Procedure language includes GOTO, GODLTA, arithmetic IF, conditional JUMPTO, post processor statements, compute statements and special subscripts. The procedure is initialised through an EXEC statement and invoked as often as required. It is executed in the calculation phase of the processor and makes use of the data values available at that time.

As one example of the use of procedures, consider the range of drill sequences that could be required for single or multiple plate machining with material dependent break-ship criteria. A part programmer may use the PROCs supplied with the system, which correspond to the AUTOSPOT verbs except that unlike AUTOSPOT they will generate fully APT compatible CLFILE records. These may be used as they are or modified in part. Others may be created by the user. Procedures can call other procedures, for example to ream a series of holes that has previously been drilled.

Procedures have a complementary role to the traditional MACROs and have been designed to have much the same programming rules.

Machineability

This feature can enable a part programmer to obtain automatic calculation of feed rates and spindle speeds, based on tool material and on the characteristics and condition of the machine tool used for particular machining operation under particular machining conditions. It has been implemented through the procedure approach so that users may introduce standards for the type of machining operation they require, reflecting individual practices in 'optimising' feeds and speeds and in the classification and use of attributes of materials, machines, cutting criteria and tools.

A starter set of procedures has been included in the system, based on the work of J. J. Montaudouin (ref 3), whose Feed and Speed Technology System (FAST) has for some years been used by IBM Manufacturing. These include machining operation procedures (for spot drilling, counterboring, tapping, various milling types, reaming, etc), machine procedures for some of the NC machine tools for which FAST is used and other functions.

The boring operation may serve as an example of the power of this approach. The optimum feed and speed are calculated. This is done by estimating the rpm based on the surface speed acceptable for the material, adjusting it to the nearest speed available on the machine tool using the machine procedure and then calculating the feed based on the number of teeth and the acceptable feed per tooth for the tool. If the cut is for a finishing operation, appropriate adjustments are made to the surface cutting speed and feed per tooth prior to the relevant calculations. Part programmer override of computed data is readily performed. Error conditions are reported when the flute length of the tool is insufficient for the depth of cut or the rpm calculated is found to be below the lowest available on the machine tool.

In order to use the machineability system, the user is required to establish three permanent files. One is for materials and specifies the surface - cutter feed rates of the various types of machining operation when different tool materials are used. Another is a cutter file containing tool data and the third is a machine tool file which holds characteristics of the machines. The user is also required to create the machine tool procedures where these are not already provided, but he should find the examples easy to work from.

General Features

The system structure is flexible and provides greater file capacity than was the case in earlier systems. Full compatibility is provided with existing S/360 APT part programs, CLDATA and post processors. Improved performance is obtained over S/360 NCS programs, for equivalent functions, especially with respect to large part programs and there is provision for expansion and enhancement. An expanded PRINT statement provides an extensive input/output control, so that users may suppress or invoke such things as user definitions, cross reference lists and part programmer listings. Metric units are supported. One language is used throughout and there is total upward compatibility through the three components. A synonym print capability is available, at user option, which expands user-defined synonyms to the equivalent vocabulary entities. A new TRACUT (transformation of cutter location data) allows an additional level over S/360 NCS with or without the COPY feature.

Design Aid for Post Processors (DAPP)

Post processors are expensive and hard to obtain. DAPP offers opportunities to reduce cost and effort and for standardisation. As its name implies, DAPP is an aid rather than a complete solution, designed for point to point machine tools and contouring machine tools with up to three axes of continuous motion, using paper tape control media. DAPP comprises FORTRAN IV and assembler code and documentation. Through an easy to use questionnaire, the user is able to generate from 50% to 85% of the statements necessary for a post processor. The additional instructions are written by the DAPP implementor, as a Machine Tool Module (MTM). The complete post processor is an integration of the MTM, compiled questionnaire data and DAPP code. DAPP handles acceleration/deceleration and linear and circular interpolation. All the standard System/360 APT utilities are supplied with DAPP.

Diagnostic Aids

The part programmer can ask for one of three levels of diagnostic printout. The error messages are designed with the user in mind. The first level gives the error number. The second provides the number and an explanation. The third includes a detailed trace and, where possible, instructions regarding correction of the error. The user can alter the diagnostic level at any point in his part program. Diagnostic messages are written to a separate file, as well as within the part program lists, so that a user on a terminal can get enough information to correct his part program without having to print out the entire listing. A further diagnostic aid is one for cross referencing (XREF), which when invoked will print out each symbol used by the part programmer with a list of the lines in which it appeared.

Generic Definitions - APT-AC and APT-IC only

The generic capability is a new way of geometric definition. It offers convenient, efficient and consistent geometric data definition and removes some fundamental APT anomalies. It is based on the premise that each geometric entity can be defined by an ordered combination of points and scalars and/or vectors. In addition, points and vectors have several allowable forms through which they can be defined, for inclusion in generic definitions. Similarly, there are specific substitution rules and assumptions for the scalars used.

Advantages of using generic definitions include simplification of the part programmer's task - it reduces learning time and the number of definitions used for jobs, eliminates the need to memorize a long list of inconsistent definitions and enables diagnostics to be more precise and meaningful. Compute time will be shortened. A significant number of standard S/360 NCS methods of definition are common to those that can be created through the generic facility. All those that are not will still be processed by APT-IC and APT-AC, so that an experienced part programmer does not have to sacrifice what he knows.

Component Capability

Turning now to the characteristics of the individual processors, this is most easily done by comparing APT-BP and AUTOSPOT, APT-IC and AD-APT; and APT-AC and APT/360. Apart from the features reviewed above, all three new products have additional methods of definition of geometric entities over their predecessors.

APT-BP features additional to those in System/360 AUTOSPOT include new point, line, circle, plane, vector and pattern definitions, extended line/circle milling, synonym, APT polygonal pocketing, basic and expanded compute capability, auxiliary feedrate, contour motion commands up to three surfaces, point to point motion commands, expanded table capacity, system options for CLPRINT, NOTOST, etc., OBTAIN and acceptance of filleted end milling cutters.

APT-IC functions additional to AD-APT include additional geometric definitions, a generic definition capability that allows the user to vary the syntax of his geometric definitions by substitution, APT pocketing, expanded compute, expanded OBTAIN and TRANTO multiple check surface logic. Macros may be part of the canonical library.

APT-AC has additional geometric capabilities to those in APT/360. An extended synonym capability permits the substitution of any string of characters. The text definition permits the user to define a string of alphanumeric characters delimited by quotes as a canonical entity and an associated print and punch capability that allows the user to specify the form of his printed or punched output, using a TEXT statement. There is a canonical extract facility that permits a user to extract and use canonical parameters of a previously defined geometric definition in a compute statement. Bell shaped cutters and the torus definition are available. Documentated linkages for the more popular sculptured surface programs are provided.

In summary, APT-BP provides for point to point and simple line-circle contour machining, comparable to AUTOSPOT except that it accepts the input written in the APT syntax. APT-IC includes all the capabilities of APT-BP and extends the machining capability to the more complex 2D surfaces, including the use of a canted plane as a part surface. APT-AC includes all the features in APT-IC (and hence APT-BP) and extends the machining capability to 3D surface operations. It is comparable with APT/360, with additional functions.

3. Future Possibilities

The previous section summarised the features of the products IBM has designed for the NC needs of the 1970s, reflecting System/370 characteristics and seeking to satisfy the needs of NC users with respect to capability, performance and integration with design systems. It must be admitted that these programs do not answer all requirements. With reflection on the development of the System/360 processors it is probable that the majority of requirements will eventually be satisfied.

One example is that of handling sculptured surfaces. The virtual storage and virtual machine capabilities of System/370 eliminate the problem of data handling and user access. However, with the restrictions of current systems to specific industries and with the extreme difficulty of simplifying data input and manipulation, the development of a general system may be some way away.

A second need is for an NC turning processor, where it may be considered today that price performance from the point of view of part programming and computing cost has been subordinated to APT compatibility. A turning processor must have automatic cutter path generation. This problem is quite complicated, especially where multiple finishing of sets and check surfaces are involved. For user acceptance, it must be possible for roughing, finishing, grooving, drilling and threading to be handled automatically. It is also necessary for automatic updating of the blank profile to take place during area clearance, so that speeds and feeds are optimised. A further requirement is for tool, material, machine and other files along the lines of those provided by S/370 NCS. If all these problems can be solved (and to date, some progress has been made) there may be a better place in the market for a generalised turning processor than perhaps is the case today.

Acknowledgements

The authors thank the McDonnell Douglas (Long Beach, California) Loft Department for allowing them to describe their system. They acknowledge those other companies mentioned, British Aircraft Corporation, Boeing and Messerschmitt, whose numerical engineering programs have so significantly enhanced the IBM NC programs.

References

1. W H P Leslie (Ed), Numerical Control Programming Languages (1969), North Holland Publishing Company, London and Amsterdam.

2. R M Burkley, CMS APT, IBM Field Developed Program GB21-0432 (1972), Wheaton, Maryland.

3. N R Parsons (Ed), Numerical Control Machineability Systems (1971), Society of Manufacturing Engineers, Dearborn, Michigan.

THE LINK PROGRAMMING SYSTEM FOR N/C APPLICATIONS

MARCO GALEOTTI
Numerical Control Service-Systems Design Dept.
Cavaglià - ITALY

Abstract: A system is described which allows automatic preparation of tapes for N/C machine tools. Main features of the system are:

a) language similar to APT
b) point-to-point and basic contouring capabilities
c) built-in technology
d) minimum computing hardware requirement

The LINK system can be implemented on most available minicomputers with 16 K - 16 bits words core memory and 500 K words random access storage. Because of the minimum hardware required and the advanced design, the system greatly reduces the preparation costs of N/C tapes if compared with other languages, the system is built by: I) the LINK language which is fully described in these papers, and II) the LINK translator, whose structure is also analyzed here.

1. INTRODUCTION

Certainly there already are many N/C programming systems and very efficient too. Somebody could ask if there is a real need of building another system such as the LINK one. We don't claim this system is the best one, we only claim it is a powerful tool and with a cheaper cost as well.

During the first Prolamat conference held in Rome, 1969, two clear trends came out: first, an increasing demand for technological capabilities, and second, a growing number of N/C processors on small size computers.

Insertion of built-in technology within N/C processors has been more appreciative in Europe than in USA has. Probably the reason is the orientation of european market toward point-to-point applications whose technology is more suitable for standardization.

The LINK system has been mainly designed in order to satisfy both the need of N/C sophisticated programming systems with built-in technology, and, in the meantime, its easy implementation on small size computers.

2. SYSTEM MAIN FEATURES

When designing the LINK capabilities, many important characteristics belonging to the already existing systems have been conside-

COMPUTER LANGUAGES FOR NUMERICAL CONTROL, *J. Hatvany,editor*
North-Holland Publishing Company - Amsterdam–London

red. For instance, the language syntax is similar to the APT one, while allowing greater freedom to the part-programmers. Technological capabilities were partly deduced from EXAPT 1 and partly from ROMANCE and OPTAL sistems.
Main characteristics are the following:

The system has strong built-in technological capabilities. This means that the LINK system processor is able to select autonomously the correct sequence of tools required by a complex operation, and for each tool the properly feed as well as speed are computed

The system is suitable for both point-to-point and contouring operations.
A rate of 90% of all pieces usually worked on modern N/C machining centers requires point-to-point programming with minor contouring problems. The LINK system has been designed in order to be used easily in all-days pieces.

The system can run on small computers
There is an increasing trend toward the use of small low-cost computers for N/C application. This is mainly due to the full availability offered by a minicomputer to part-programmers. It follows a great time and money saving in the preparation of programs.
The LINK system processor has been written in FORTRAN IV basic language in order to allow its easy implementation on all computers with a minimum configuration of 32-K bytes + disk unit.

The LINK system language can be casily learned and used by part-programmers who already use APT-like languages (as APT, EXAPT, ADAPT, MINIAPT.... and so on). LINK language has a grammar and a sintax similar to APT, but the system is however much up-dated in concernig some others features.

LINK is a modular system
This means that the processor is segmented into a certain number of independent modules, which can be inserted or left-out from the system according to the users needs.
For instance, insertion or deletion of the so-called "presetting module" allows the system to be used with or without the need of establishing a file of preset tool lenghts.

The language grammar and syntax differ from APT in a few points. These points are, neverthless, relevant:

a) All definable elements such as points, lines, matrices, technologies, etc, may be named in a more restrictive way than APT does. All symbol names are made by one or more alphameric letters followed by a number. The part-programmer choice regards only the number combinations, while the letter must be strictly defined in a fixed format. For instance, all points definitions should start with the letter P. Valid symbols for points are therefore P0, P1, P514. This restriction has been accepted in order to get an efficient diagnostic at a very early stage of the translation process (that is during the sintactic scan). Note that a rate of 90% of part-programmers already uses this kind of symbology, even if using APT language.

b) Variables are used both for arithmetic computation and as a means for transferring real values to called MACRO. This allows arithmetic variables and MACRO parameters to form an unic items class. Moreover, there will be no need at all for partprogrammer to care about the difference between formal substituted parameters and variables. Note that the MACRO use restriction is irrelevant.

c) The geometric item indexing is allowed in a very simple way: where an integer is used, it can be replaced by a variable name. For instance, instead of P5 or L51 a part-programmer can use PV1 and LV12 respectively.

d) An items class called "qualified parameters" is here introduced in order to avoid any confusion when handling long numbers lists. A "qualified parameter" is made by a key-word followed by an equal sign (=) and by either a number or an arithmetic variable. Equal sign is optional. With this items class use, all numbers in a list are completely identified, they can be therefore written in any order. The traditional use of APT-like modifiers in front of all numbers would be not feasible.

e) The choice freedom allowed to the part-programmer when using APT-like geometric definitions has been introduced with the purpose of avoiding part-programs reruns due to non-sense reasons such as the writing of:

P1 = POINT/INTOF,XLARGE,L1,C2

instead of: P1 = POINT/XLARGE,INTOF,L1,L2

This freedom is permitted by the syntax translation original scheme done by the processor. Such a scheme is based on syntax tables called "meta-structures" described in the processor details paragraph.

f) Technological capabilities of the system are mostly based on the use of six tables. They are used by the processor for many purposes, namely the complex cycles splitting into single operations; the calculation of consecutive dependancies among diameters, depths, etc.; the feeds and speeds assignment, and so on.

3. THE LINK LANGUAGE

CHARACTERS SET the following characters are accepted by LINK:

26 alphameric letters (from A to Z)
10 digits (from 0 to 9)
9 special characters

"."	decimal point
","	comma
"+"	plus sign
"-"	minus sign
"o"	moltiplication sign
"/"	division sign
"="	equal sign
"("	left parenthesis
")"	right parenthesis

ELEMENTS

Charecters are combined in such way to form elements of the language.
Elements are divided into the following classes:

a) Operators/Separators. Defined by the special characters
[, + - * / = ( )]

b) Numbers. Defined by sequences of digits.
Decimal point (.) as well as plus or minus sign (+-) are optional.
Here is an example of valid numbers:
12, -12,3, +0,05, -.07, ...

c) Arithmetic variables. Defined by letter "V" followed by an integer number (0 ÷ 9999). Arithmetic variables are used to symbolize arithmetic quantities or values of aritmetic expressions. Arithmetic variables can be used everywhere numbers are usually required.

d) Symbols. Defined by a single alphameric letter or by a word followed by an integer number (0÷9999) or by an arithmetic variable. Symbols are used to assign names to relevant items of the LINK language.
The allowed symbols are:

Pn	or	PVn	for points
Ln	or	LVn	for lines
Cn	or	CVn	for circles
PATn	or	PATVn	for patterns
MATn	or	MATVn	for matrices
MACn			for MACRO's
Tn	or	TVn	for technological operation

hint! n must be an integer number

examples: P3,LV2,CN2,PATV55,MAT9999....

e) Qualified parameters (named here as identifiers). Defined by a combination of anyone word followed by both an equal sign and a number.

examples: Z=-50, D=30.8

Identifiers are used to assign arithmetic values to parameters defining technological operations. Numbers can be replaced by arithmetic variables.

examples: Z=V5, D=V3

Equal sign is optional, so that expressions like Z-35 or DV3 are allowed.
Here follows a list of allowed identifiers:

Zn	or	Z=n	for	Z-values
Dn	or	D=n	for	diameters
Fn	or	F=n	for	feed rates
Sn	or	S=n	for	speeds
LENn	or	LEN=n	for	lengths
DIMn	or	DIM=n	for	dimension
An	or	A=n	for	angles
Xn	or	X=n	for	X-axis values
Yn	or	Y=n	for	Y-axis values
Zn	or	Z=n	for	Z-axis values
Rn	or	R=n	for	rapid motions
Wn	or	W=n	for	W-axis values
Mn	or	M=n	for	M-functions

f) Modifiers. They are words which are included in the LINK system vocabulary and mainly used in geometric definition. In this case they specify which solution is required or are used to improve the readibility of the statement

Examples of modifiers are:
XSMALL, LEFT, TANTO,.....

g) Major words. They are part of the LINK system vocabulary and are used in geometrical and technological definitions to specify the kind of definition.

Examples of major words are:
POINT, MACRO, DRILL, ...

STATEMENTS

An organized sequence of elements is here called a statement. Each statement leads the system to a particular action. Statements are divided into the following classes:

a) Geometric definitions statements: their function is the description of geometric items.

b) Technological definitions statements : their function is the description of technological operations.

c) Operating instructions : their function is the combination of geometric and technologicaldefinitions in order to produce the real control orders.

d) Arithmetic computations : their function is the computation of arithmetic values and their assignment to arithmetic variables.

e) Logical and auxiliary statements : they allow MACRO's definition as well as their use, jumping and looping,copying, coordinates transformation, sorting criteria and so on.

STATEMENT SYNTAX

When using LINK language the part-programmer work is not limited to write statements in a fixed format. A particular characteristic of the system is a free format choice in writing statements. According to the programmer habits, both the order of the elements and the presence of some element are subject to variations.
In the description examples which follow, characters [ ] < > (symbolis not allowed by the language) are used as delimiters: elements written between <and> are optional; elements written between <and> can be in any order.

For example the description:
P3 = [POINT/] <[INTOF,] L1,L2
means that the statement can be written in many different ways:

P3 = POINT/INTOF,L1,L2 (APT compatible format)
P3 = INTOF,L1,L2
P3 = L1,INTOF,L2
P3 = L1,L2 (simplest possible format)

This other description example:

L1 = [LINE/] <LEFT,[TANTO,] C1>, <RIGHT, [TANTO,] C2>

indicates that following alternatives (among many others equally valid) are accepted:

L1 = LINE/LEFT, C1,RIGHT, C2

L1 = C1,LEFT,C2, RIGHT

Because of the above possibilities the LINK language can be casily used by part-programmers with some experience in APT-like languages. Part-programmers with a few hours of LINK language experience can fully exploit the possibility of writing simpler and shorter statements.

GEOMETRIC DEFINITIONS STATEMENTS

Their function is the description of geometric elements like:

points (9 available definitions)
lines (10 available definitions)
circles (12 available definitions)
patterns (12 available definitions)
matrices (7 available definitions)

All geometric definitions statements begin with a symbol followed by an equal sign (=) and by a sequence of elements. The description of the geometric item is stored by the system and thereafter

it can be recalled by means of the preassigned symbol.
Elements to the right of the equal sign can be symbols and numbers, operators, variables, identifiers, modifiers, major words according with the allowed formats.
Anywhere a real number is used, it can be replaced by a variable.
If the same symbol is used more than once at the left side of the equal sign (that is:if a symbol is redefined), only its last given definition is kept and available later on.

ARITHMETIC COMPUTATIONS

In many cases it is necessary to compute the value of geometrical or tecnological parameters by means of arithmetic expression.
The structure of arithmetic statements is:

variable = expression

Expressions can contain numbers, variables, arithmetic and geometric functions, arithmetic operators and other expressions included between parentheses ( )

Allowed operators are : + - * /

Arithmetic functions are:
SIN (SINE)
COS (COSINE)
TAN (TANGENT)
ATAN (ARCTANGENT)
SQRT (SQUARE ROOT)

The argument of arithmetic function can be either a number, or a variable, or any expression.

Geometric functions are: EXT

for obtaining one value from lines, points and circles canonical forms. The argument is either a symbol assigned to a line or to a point, or to a circle.

DIST (DISTANCE)

for the computation of minimum distance between points, lines and circles.
The arguments of DIST function are two preassigned symbols representing lines, points and circles.

examples:

```
V8= 100-EXT1 (P2)
V9= DIST (P1,C5)
V198= (((SQRT(A) + DIST(P1,L8))-EXT2(C9))
```

TECHNOLOGICAL DEFINITION STATEMENTS

Technological definitions assign a symbolic name to technological operations, like drilling, boring, tapping and so on.
The general structure of technological definition is:

Tn = CODE/parameter list

Where: "Tn" is an assigned name to the required operation;"CODE" indicates one of many possible words corresponding to a technological operation.
"Parameter list" is a list defined by a sequence of identifiers specifying diameters, depths, feeds, speeds, and so on.

For most operations the only needed identifiers are diameters and depths; all other cutting conditions are automatically kept by the LINK system processor from technological files.
In order to provide the part-programmer of a maximum possible flexibility, it's also allowed the overriding of the automatic assignements. This is done just by writing the identifiers relative to feed, speed, tool length and so on.
Technological definitions are stored by the system in form of CLFILE records and they can be recalled by WORK statement later on.

Technological operations are generally divided into four classes:

a) Single operations — They consist of single operations like drilling, boring, and so on.

b) Multiple operations — They consist of a sequence of single operations automatically generated by the processor. This means that the parameters of intermediate operations are computed by the processor, according to a few information concerning the required result. For example, the multiple operation "BOREH" (bored hole) corresponds to a sequence made by spot-drilling, drilling, boring; only the final diameter as well as depth must be given, while parameters relative to spot-drilling and drilling are automatically generated.
Note that all multiple operations refer to centered operations, that is all single operation apply to the same x,y coordinates.

c) Complex operations — They are similar to multiple operations with the difference that they are made by milling operation too. For example consider the milling operation of a "pocket". Required identifiers for this kind of operation are x-y dimension, angles, diameters, depth, and so on.

d) Special operations — Should for a part-programmer be impossible a successful use of the system automatic capabilities, he always could define special operations by means of a sequence of elementary orders. These orders could be X-Y-Z movements, spindle, activations, G and M-functions, recalling and so on. In this way it's possible to program with LINK language any problem which are usually not covered by standard built-in procedures.

LOGICAL AND AUXILIARY STATEMENTS

Logical instructions offer to part-programmers advanced technics, useful for writing programs.

It is often necessary to have some operations be done, depending on particular conditions : unconditioned and conditioned logical jumps tell the system either to execute or not a set of a given statements sequence. If such a set is to be repeated more than once, and, moreover, everytime with different values of parameters, then MACRO's can be defined and called into execution later on.

General structure of above statements is:

```
JUMPTO/ LAB1
IF/ Vn, LAB1,LAB2,LAB3
```

where LAB1,LAB2,LAB3 indicate labels, assigned by the part-programmer to some statements belonging to the program.

The statement structure used for handling MACRO's is:

a) MACN = MACRO

for example MAC5=MACRO or MAC12

this statement regins the MACRO body and names the particular MACRO.

b) TERMAC

This statement terminates the MACRO body

c) CALL/MACn

This statement calls for the execution of a MACRO

The MACRO body can contain all but MACRO and FINI statements. Parameters redefined by each call can be substituted by arithmetic variables. Before each call arithmetic statements can be used for assigning present values.

Auxiliary statements: this kind of statements helps the part-programmers in writing repetitive sequences of statements involving different system of coordinates.

TRACUT statement causes all motions computed under TRACUT action be rotated or mirrored according to the specified trasformation matrix.

COPY instruction allows motions statements sequence be automatically repeated by the system in different XYZ positions and angles. This is done by a repetition of the specified transformation matrix.

OPERATING INSTRUCTION STATEMENT

Once the part-programmer has written both geometric and tecnological definition, ha can then start to describe the relation between geometry and technology, that is the way in which they are associated. This is done by writing operating instruction statements. Their structure is:

WORK/(TECHNOLOGICAL LIST)/(GEOMETRIC LIST)

Technological list means a preassigned symbols sequence referring (T1,T2,...,TN) to predefined technological operations.
Geometric list consists of a preassigned symbols set referring to points and patterns. Instead of a point name, an explicit set of coordinates may be used. When some kinds of technologies are called for (milling,pocheting,...), additional modifiers may be added to define other technological cycle peculiarities.

for example:

WORK/ T1 / P1,P2,PAT130
WORK/ T1,T2/P1,X10,Y15,P9
WORK/ T6 / P1, CLW, 5.,P2,P3

Other operating instructions allow the part-programmer the assigning of priority and the definition of sorting kind to be done. All operations are split into single operations and then a sort is performed according to standard rules (for instance:
first spot-drilling and then drilling, boring and so on).
In addition, single operations may be ordered by the system either in a global way or by separate faces. The relative statement is:

SORT/OPER
and SORT/FACE

An other operating statement is:

VOLUME/ (GEOMETRIC PARAMETERS)

With this statement the part-programmer can define either cubic or cylindrical volumes to be saved during the piece machining.
After sorting by single operations, link system processor checks all motions and possibly produces by itself additional steps in order to by-pass the obstacle.

5. GENERAL STRUCTURE

The LINK processor has bean designed with the purpose of reducing the necessary core storage to a minimum of 32 K bytes, while maintaining processing time good peculiarities, as well as high quality diagnostic, and possibility of implementation on many existing mini-computers.
First version has been completely written in FORTRAN IV for a HEWLETT-PACKARD 2116 computer with 16 k of 16 bits words and 500 k words random access disk. A second version is being implemented

on a DIGITAL PDP11 computer with the same characteristics.
The processor has been logically segmented into five main sections

-.translation
-.geometry and computation
-.technology definition
-.operations task and sorting
-.volumes avoidance

each section is made by one or more phases, normally executed only once (with an exception in the second section).
The translation phase is made by two sub-phases: grammar analysis and sintax scan. Grammar analysis recognizes input characters strings and builds elements like numbers, variables, symbols, separators, etc. Sintactic scan is fully sintax directed and based on input single element (taken from the grammar sub-phase) with a "line" of sets called "meta-structures". Once the matching is achieved a successive actions set is executed (including indices flagging, processed items storing in one of 12 possible stacks, etc.), and then next input item is considered. If matching is not achieved, there will be no action at all. In both cases the present "line" addresses two different "next line".
A meta-structure "line" generally makes use of the following logic contents:
-.kind of grammar item
-.next line to be executed in case of mathing failure
-.next line to be executed in case of successful matching
-.available stack necessary for storing the item in case of successful matching
-.index to be set up in case of successful matching

In case of successful matching "next line" could indicate scan process is exhausted (that is a statement has been completely matched). The sintax sub-phase then quits the statements building work by mergig in a fixed order all stacks contents and by creating a "statement code" as function of all set up indices.
In case of matching failure "next line" could also indicate statement is not correct (that is, path within the meta-structure led to an impossibility), and that tested statement does not match with the present meta-structure but it could match with another one. In the latter case, "next line" indicates then which meta-structure is to be used for next attempt.
Thanks to meta-structure flexibility the system can accept input statements in many alternative ways. Parameters reordering as well as unimportant modifiers ignoring is simply done by addressing the stack storing the same parameters, regardless their different input formats. Diagnostic is greatly simplify by meta-structure task too. indeed a rate of 95% of all error messages regarding the translation phase is produced by meta-structure. When arithmetic statements are processed, stacks are used as push-down lists in order to get expressions codified in the reverse-polish notation.

Geometry and computation phase is in charge of all geometric cal-

culations and of logical processing for arithmetic computations, looping, MACRO's as well. When geometry and arithmetic are evaluated within the same logical phase, two possibilities arise. One concerns the use of arithmetic results to define geometric items; the other one allows to fetch from predefined geometric items some canonical parameters and use them as variables. This phase is actually made by two sub-phases (one for geometry and one for computation, looping, MACRO's). In order to allow the interaction between geometry and arithmetic without having a sub-phases continuous swapping, a scheme has been developed which allows the swapping only if needed.
Geometric definitions are temporarily stored until either a source program end or an "extraction" function is met. At that point the first sub-phase is exhausted and the second one comes in, all stored definitions are processed and then control is transferred again to the first sub-phase. Working in this way, swapping is notably reduced even if arithmetic and geometric statements are highly merged.

Technology phase processes technological definitions and stores the results (in the form of "canonical technologies") that will be used later on, together with the previously stored geometry.
Technology definitions processing involves the use of six tables:

a) Multiple operations table.
This table shows how to split multiple operations concerning a given material into single operations. Diameters, depths and other data needed by single operations are conputed as functions of available data referred to multiple operations. This is done by examining a conversion table.

b) Conversion table.
For a given material this table shows how to compute intermediate single operations parameters, starting from multiple operations available data.

c) Tool table.
For each tool identified by its geometrical dimensions, optimum feed and speed are supplied by this table

d) Material table.
For both a given material and a given tool range, this table lists the rate of modifications to be applied to feeds and speeds fetched from the tool list.

e) Tool lenght table.
For both a given material and a given tool range, this table indicates either one or more available setting lenghts. With the use of some modifiers a part-programmer can affect the choice; in absence of modifiers a default option lenght is used.
Whole action regarding the assignment of setting lengths can be removed if wanted. For convenience LINK system can be used in a simpler way, that is by setting the distance during the machining after each tool loading.

f) Single operations table.
This table is interrogated at definition process end in order to get for each single operation (at this stage already complete of all parameters) the corresponding CLFILE records sequence. Such a sequence is stored in the canonical technologies file.

The operation building and sorting phase processes the WORK statement. Each WORK order can cause the pairing of one or more geometric items with one or more technologies. Its meaning is: Apply the listed technologies to the listed geometric items.
The geometry list can contain point and pattern symbols for point-to-point operations; when milling technology is required it can contain line and circle symbols too. Pairing process is simply done by examining previously stored geometry and technology files. Immediately afterwards, TRACUT'and COPY's statements are executed. Operations sorting is then executed according to two available criteria: sorting by "face" or sorting by "operation".
In the first case the sorting key is made by:

-.face number
-.kind of operation
-.diameters range
while in the second case the key is made by:
-.kind of operation
-.face number
-.diameters range

The FACE statement could be used to inform the processor about rough pieces real position on the machin tool table. This statement could also be used for both machining pieces on index-table and for setting many pieces on the same table.

"Volume avoidance" phase processes the predefined CLFILE so that to perform some tool path corrections if needed.
As the operations sequence is determined by the processor during the sorting process, a part-programmer is not able to foresee all motions implied by that sequence. But he alternatively can specify one or more volumes (parallelograms or cylinders sections) using VOLUME statement.

DEVELOPMENT TRENDS OF COMPUTER-AIDED PROGRAMMING SYSTEMS

DR. -ING. W. BUDDE, DIPL. -ING. H. WEISSWEILER
EXAPT-Verein, Aachen, FRG

Abstract: Programming languages containing technology solve the cost- and personnel problems in connexion with the programming of NC-machines. Transfer of further tasks to the computer enhances the advantages. With the example of EXAPT ways and means are shown up for retaining the flexibility with a steadily growing degree of automation by using modular structure and user-oriented program algorithms.

1. INTRODUCTION

The development of manufacturing techniques is characterized by the trend to a higher productivity along with an increase of flexibility. This development does not serve an end in itself, as the trend of turnover - e.g. in mechanical engineering in the Federal Republic of Germany and based on the year 1960 - shows an increase of almost 700 % by the year 1985 (fig. 1). It is known that the expected production output has to be realized by an almost constant number of employees. In other industrial nations the situation is similar.

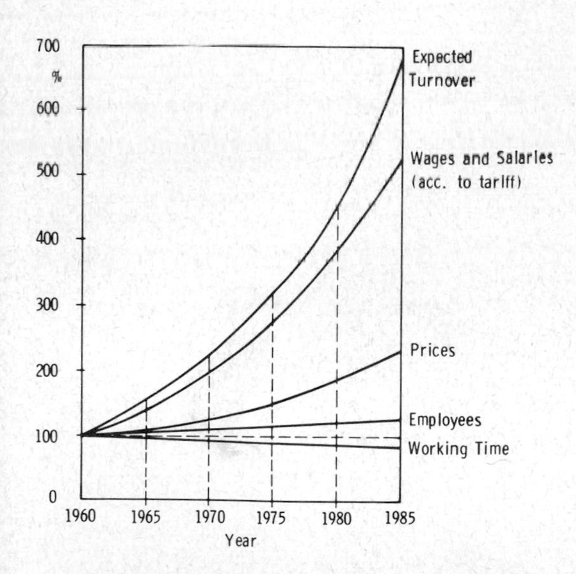

Fig. 1. Development of economic characteristics of mechanical engineering (shown by example of Western Germany)

For automating the manufacturing process as well as for solving the many different manufacturing tasks there is a steadily growing supply of manufacturing equipment, such as NC-machines, manufacturing systems, etc. For utilization of all advantages as well as for a full exploitation of the productivity of the dif-

COMPUTER LANGUAGES FOR NUMERICAL CONTROL, *J. Hatvany, editor*
North-Holland Publishing Company - Amsterdam–London

ferent units, carefully determined, exact planning data must be supplied by the work planning department, which is the section that precedes the manufacturing process. Today, limits of automation are not so much set by technical problems in the manufacturing field, but rather by unsolved organizational problems.

A typical example for this is the development in the field of NC-techniques.

2. TECHNOLOGY IN PROGRAMMING LANGUAGES

The use of numerically controlled machine tools shifts the work expenditure and responsibility for workpiece and manufacturing equipment from the workshop to the work planning department, as the manufacturing process must be predetermined in great detail by the part programmer and be transferred onto punched tapes.

The use of automatic programming systems makes the computer an effective aid also in the technical range of an industrial enterprise. Automatic programming systems open the possibility to transfer the manufacturing specialist, who was responsible for the control and operation on the shop floor to the production planning departments of the companies and to entrust him with the detailed preliminary planning of the manufacturing- and cutting process. Automatic determination of the machining technology by the computer leads to a reduction of the demands on the programmers to be used. This is a great contribution towards solving the shortage of personnel, as some few specialists with particular knowledge of the manufacturing process are in the position to support and employ effectively a greater number of part programmers. In addition, the following factors speak also in favour of transferring further technological tasks and decisions in the scope of NC-programming to the computer:

1. Relieving the part programmer of all routine work,
2. Storing empirical values of manufacturing specialists in the computer,
3. Increasing the accuracy of technological values,
4. Constant quality of technological values,
5. Successive improvement of technological data through a feedback from the shop floor,
6. Direct economic utilization of the latest technological developments by way of a central maintenance of the data stock maintained in the computer.

Programming systems that have the aforementioned properties provide the possibility to adapt planning methods and planning data centrally, e. g. to the latest scientific results, management decisions, or requirements imposed by the actual economic situation. Examples for this are the adaptation to market fluctuations or to changes of the type or costs of machine tools, tools, materials, etc. Figure 2 shows with the aid of a practical example for turning operations the economic advantages when the technological values are determined by the computer. By the use of a geometry-oriented programming language the programming time was reduced from 70 to 35 hours, i. e. by 50 % when compared with manual programming. The utilization of the technological efficiency of EXAPT 2, i. e. automatic cut distribution and determination of cutting values results in a reduction to 10 hours, i. e. to abt. 14 %. This shows the increase of productivity of individual programmers when using highly developed programming systems. Owing to automatically determined cutting data of a higher quality the manufacturing time was also reduced by 14 %.

However, when examining the development trends of programming languages one should also take into consideration the cost developments mentioned

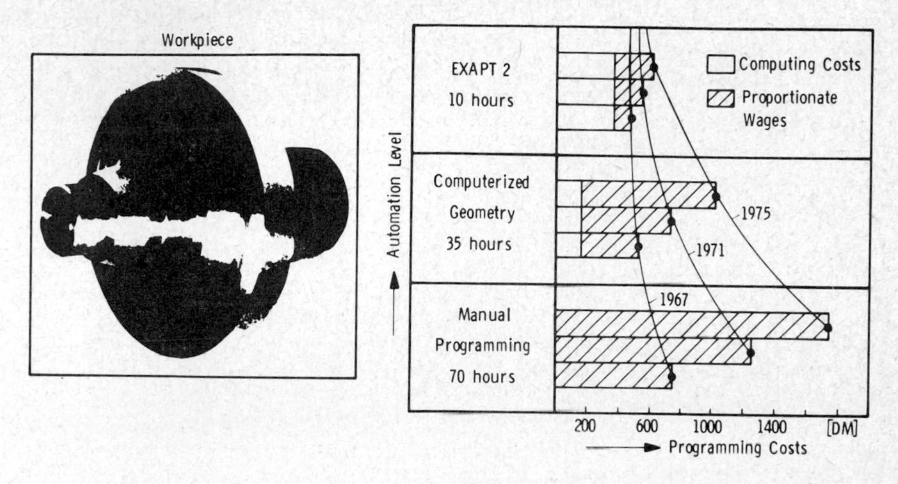

Fig. 2. Cost increase for manual and computer-assisted work

at the beginning. Let us have a look at the programming costs given in our example. Whereas the computing costs remain almost constant, or have a slightly decreasing tendency, the personnel costs continue to increase. In the year 1967 the cost saving when using EXAPT 2 at a rate of DM 10.80 per hour, personnel costs amounted to DM 288.-- as against manual programming. In the year 1971 the cost saving amounted already to DM 720.-- at a rate of DM 18.-- per hour, and in the year 1975 the saving is expected to amount to DM 1,140.-- at an average increase rate of personnel costs of abt. 8 % per year.

This proves the efficiency and economic advantages of highly automated programming systems containing technology. Owing to the prevailing cost situation, their significance in enhancing the productivity is going to increase further. The more tasks are transferred to the computer, the greater are the savings.

3. FURTHER DEVELOPMENT OF PROGRAMMING SYSTEMS

Figure 3 shows the current development state of different APT-like languages. As can be seen, highly automated programming methods containing technology are already available for the frequently occurring machining methods such as hole making, turning, and milling.

After all, abt. 90 % of the NC-machines installed in Germany can already be programmed with these languages. For multi-axial machining operations as well as for new ranges of application of numerical control such as nibbling, electro discharge machining, etc. there are, in addition, languages without specific technology such as APT, MINIAPT, and BASIC-EXAPT.

From the experience gathered over years of practical use, new requirements must be taken into consideration in the scope of the further development.

The standard systems are applied by many users. This fact justifies the high development costs involved. However, standardized systems and the steadily growing degree of automation in programming must not lead to the loss of flexibility. The system must be adapted to the requirements of the user and not vice versa. Therefore, the system must not have any rigid algorithms, but must be able to consider user-dependent experience. The system must be clearly sub-

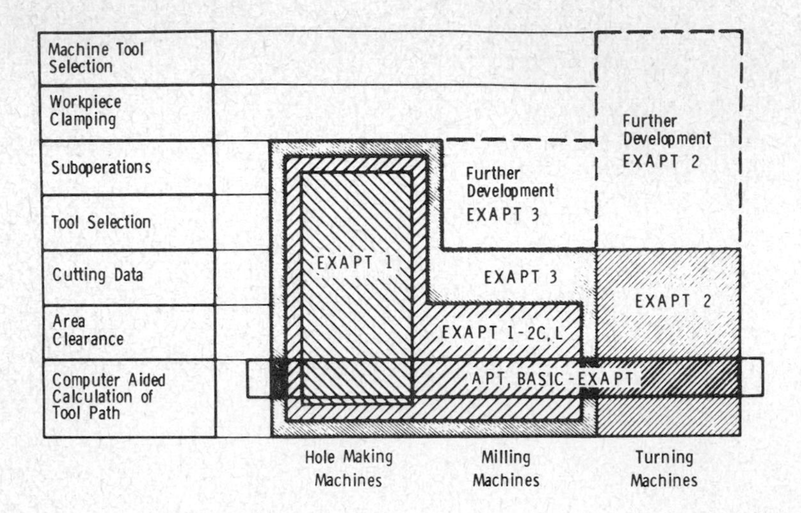

Fig. 3. Computer-assisted programming systems for NC-machine tools

divided into individually solvable functions and provide the facility to be linked to other systems, e.g. from the design section. It must be independent of specific computer types in order to give the user the possibility to choose the most adequate type of computer access and to make him independent of the operational readiness of a specific computer and/or the costing of the different computing centres. The advantages of the system must be accessible also to the conventional range of the manufacturing process. New developments in data processing, such as real-time-processing, dialog-programming, and screen units must be usable in connexion with the system.

In the following, an approach to solving the aforementioned requirements is shown in the scope of the development of EXAPT. Figure 4 shows the structure of the EXAPT-System and the influencing facilities on different sections based on the part spectrum, the accumulated experience in manufacturing techniques as well as on the existing workshop equipment.

3.1 CONSIDERATION OF COMPANY-SPECIFIC DATA

In programming systems that are machine tool- and user-independent, the company-specific data must not be fixed in the processing logic. The data must be supplied to the processing programs by means of files. Although these data, that are required also for the conventional manufacturing process, are available, they were never compiled as far as we know. They are a component part of the empirical stock of individual work planners and work preparation specialists. Although the necessity for the preparation of such files entails initial innovation problems, it provides on the other hand the advantages of a central compilation of data that can be used in many ways and can be evaluated also for other fields of application.

Owing to the feedback from the workshop, this data is automatically step by step brought to an optimum. Thus, the production process depends no longer on

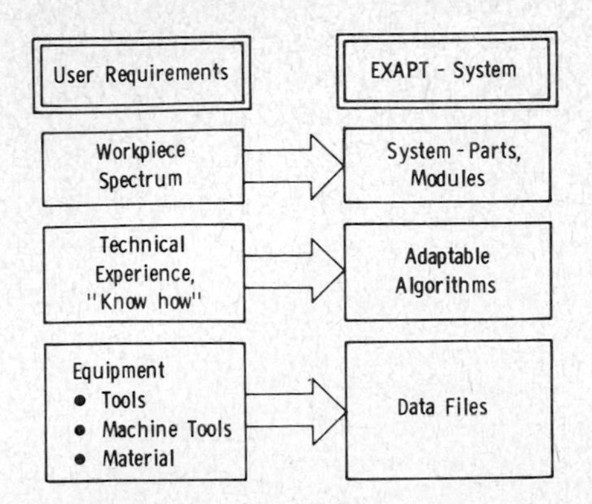

Fig. 4. User-oriented planning system
-Influencing parameters-

the judgement of individual work planners. The innovation problems connected with the preparation of such files can largely be solved by standard files.

3.2 GENERATION OF COMPANY-SPECIFIC WORK CYCLES

Just as there are company-specific data, there are also company-specific manufacturing methods that are not of a general type. They must not become a fixed part of the programming logic either. Whereas the company-specific data represent only boundary conditions of the methods, the algorithms, as shown in figure 4, constitute a higher level. They have an impact on the programming logic.

In the conception of programming systems it is necessary to make a strict difference between general and company-specific algorithms. The manufacturing methods of a general type become part of the programming logic. In the case of company-specific manufacturing methods it is however necessary to make use of flexible processing techniques, such as the technique of decision tables.

The decision tables are handled in the same way as the files. Also in this range, the innovation problems can be reduced by standard files. Figure 5 shows the determination of the standard cycle as well as of the specific cycle for the planned machining task.

3.3 USER-RELATED SYSTEMS

Owing to specific product groups and manufacturing procedures only parts of large, integrated data processing systems are of interest to the individual user. He is faced with the alternative of either using programming systems that are too extensive for him or of investing in company-specific and product-dependent own developments.

A way out of this situation is provided by the modular structure of the overall system. From this overall system that is tessellated out of many precisely defined function blocks, the user is in the position to compile an appropriate system to suit his specific needs. As the functions of the blocks are clearly delimited and can be easily overlooked, it is possible to complete the system by further, self-developed blocks or to replace already existing blocks by more appropriate ones. The extra expenditure caused by the conception of such a

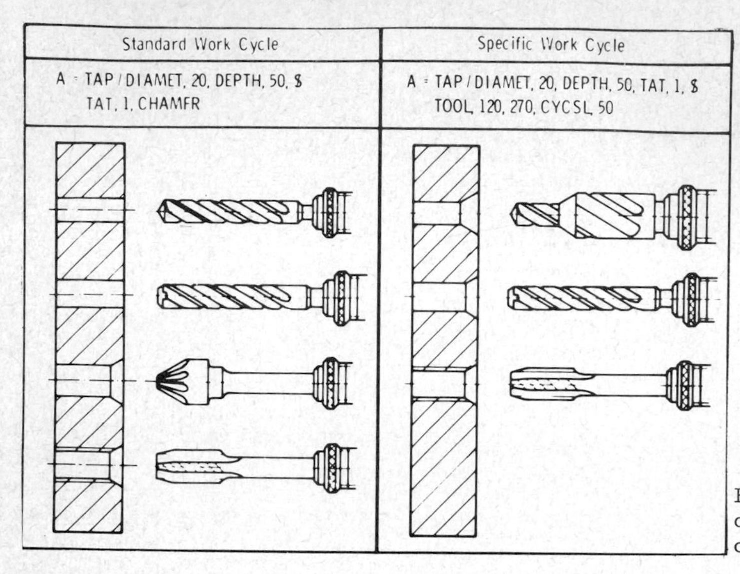
Standard Work Cycle
Specific Work Cycle
A = TAP / DIAMET, 20, DEPTH, 50, $
TAT, 1, CHAMFR
A = TAP / DIAMET, 20, DEPTH, 50, TAT, 1, $
TOOL, 120, 270, CYCSL, 50

Fig. 5. User-dependent work cycle generation

system is largely compensated by the following advantages:

1. Clear structurization of the overall system
2. Unified data structure of all modules
3. Reapplicability of modules
4. Great number of combination facilities
5. Small maintenance expenditure.

Figure 6 shows some general types of application. For the input different media are available (terminal, punched card, screen, CRT, etc.)

Different input languages (APT-like, fixed format, etc.) can be used, if appropriate input-translators are available.

The input program can call for standard programs, the so-called system-macros, that are stored in a pre-translated form in external libraries. When the translation is terminated, the technological data, such as Material-, Tool-, Machining-, Work Cycle-, and Clamping Device-Files are made available from external data stores.

The processing of part programs and the information flow for NC- and AC-machines are shown by arrows. When using AC-machine-tools, the determination of cutting values in the NC-Processor can be dropped. This applies also to the determination of cutter paths provided that the AC-System calculates automatically the tool paths. Graphical display on a plotter or screen can be omitted, but may be useful for verification of the part programming, as a replacement of test runs on the NC-machine tool or for the preparation of working papers for conventional manufacturing.

The processing of the part program is to take place pass by pass. In figure 6 each pass is represented by a modular block. The advantages of this structure are that each pass can be easily overlooked, as all input- and output-data are stored on external storage media, easy exchangability of individual modules, and the possibility to generate processors for the various tasks involving little expenditure. With a specific Master Program it is for instance possible to use the modules for building up a dialog-processor, that will permit a data flow with modification and interruption facilities on the different interfaces. For

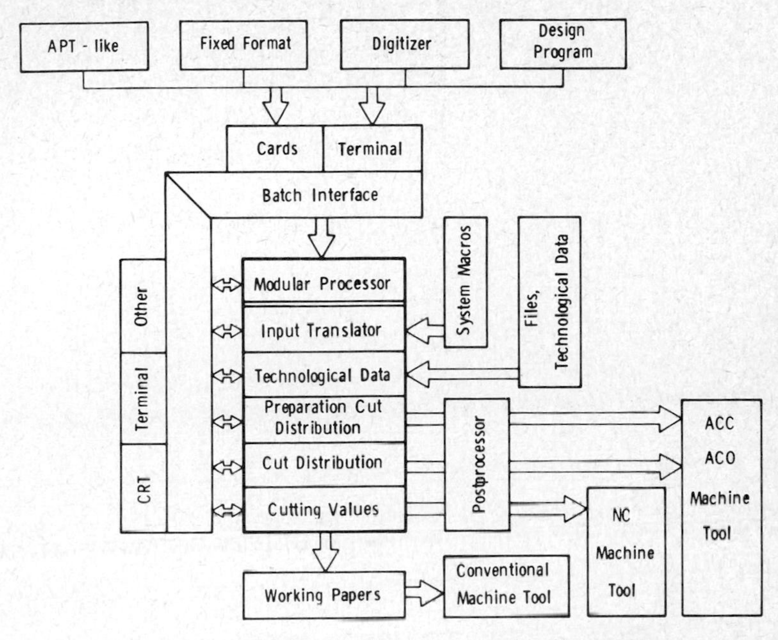

Fig. 6. Modular Processor System

the practical use of NC-processors it is important that the system be conceived in such a way that it can be handled easily by the user. Error messages must give the programmer so much information that an error analysis and correction do not require a processor specialist.

The input language used must be so clear and easy to be handled that subsequent modifications to part programs can be done with a minimum of expenditure. Special thought is given to modifications to the geometry of a part, to influencing the work sequence, and to machining details.

3.4 COMPUTER-RELATED SYSTEMS

To provide the industrial utilization of an NC-System it must be ensured that implementation on different computer types can be carried out. Besides the use of the standard computer language it is necessary to restrict oneself to generally valid I/O-methods of acces and to segmentation techniques. Greatest attention is to be paid to a minimum core storage requirement, as the applicability of the system depends on this parameter in the scope of multi-programming operating systems as are used today. In the case of a modular NC-Processor the core storage space lies close to the theoretical minimum, as only the data required at one time are resident in core and all other data are stored in external files.

4. FACILITIES OF UTILIZATION

Figure 7 shows the multilateral utilization facilities of the aforementioned system structure.

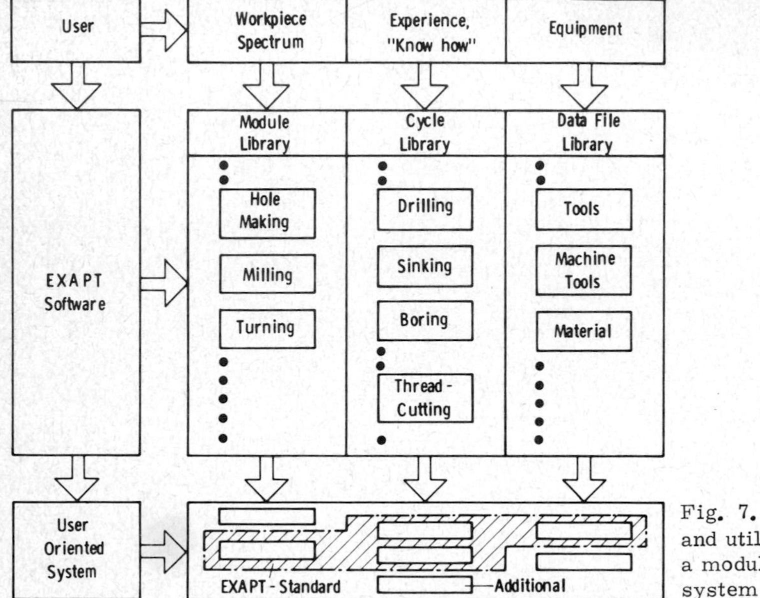

Fig. 7. Structure and utilization of a modular NC-system

The user has to produce a limited spectrum of workpieces. In the course of time a certain "know-how" has accumulated for this. The shop floor is tailored to the manufacturing profile and requirements.

The user selects the program blocks for his different manufacturing tasks from the program library. The universally applicable modules take the planning logic from the Cycle-Library and have, in addition, access to the Data-File-Library containing the characteristics for tool, machine-tool, and material. For the Module-, Cycle-, and Data-File-Library there are standard software and dataware available, for instance in EXAPT. A detailed documentation of the programs and standardization of interfaces and files guarantee the exchangability and extendibility as well as the multilateral utilization of the data stock. The maintenance of the libraries is done by the computer in batch or dialog access.

Cost-optimal, user-oriented systems are obtained by the combination of company-specific soft- and dataware with the standard system. Thus it becomes possible to utilize the advantages of data processing to a greater extent also by a greatest possible group of companies.

FLEXIBLE TECHNOLOGICAL PROGRAMMING SYSTEMS BY A GENERATING MODULAR STRUCTURE, SHOWN BY AN EXAMPLE OF COPY-TURNING

D. KOCHAN and H.-J. BOCK
Technical University Dresden, GRD

0. INTRODUCTION

Analysing the development of automatical programming in recent years, we can observe two characteristic main trends of the 62 programming languages which have become known so far.

Part of the first trend is the development of comprehensive programming systems being as universal as possible.

In addition, a large number of small special systems has been developed which are applicable to a scope being very limited in most cases.

Both trends are justified, with due regard to their specific advantages and disadvantages, their respective main drawbacks - insufficient flexibility and adaptability - having turned out to be a fundamental problem.

Therefore, further development is aimed at preparing flexible and adaptable processing programmes in spite of the complexity of technological decision processes. For that purpose, modular subdivision is desired.

1. THEORETICAL FUNDAMENTALS AND DEFINITIONS

A modular programming system is characterized by its high degree of subdivision into closed subsystems - the modules - and by the universal combinability of modules in specialized programmes for solving given problems.

In technical literature often the term "module" is defined and used very differently. In general, independence is regarded as the most important feature of a module. This criterion is also in the centre of the definition of modules in technological programming systems.

> "A module is an independent system of rules and data for solving given problems of different size whose content is defined so that redundancy and irrelevancy will be largely avoided and that it could be easily combined with other modules." [1]

COMPUTER LANGUAGES FOR NUMERICAL CONTROL, *J. Hatvany, editor*
North-Holland Publishing Company - Amsterdam–London

This definition holds likewise for algorithm and programme modules.

With regard to structure, the independence of a module is the basic requirement for the universal combinability of modules.

Not affected by the definitions of independence is the combined action with subordinate systems in the form of the interlocking of modules known from subprogramme technique.

Different technological tasks being characterized by different manufacturing conditions (e. g. means of manufacture, special operational conditions) are realized in each given case by a specific processing programme. Depending on the respective technological automation level, which is characterized by the relationship between the data to be determined automatically and manually, each processing programme consists of at least one or more modular complexes.

A modular complex contains all modules which are necessary for the solving of a technological problem within a technological subset. The technological subsets of the programming system are characterized by their forming of relatively independent parts of all rules for technological decision finding. The hierarchical structure of the modular system is shown in fig. 1.

Module size is determined by technological as well as computational-organizational influence criteria, generally both of them occurring in a complex way.

The technological scope constitutes the most important technological criterion for dimensioning a module. The larger the scope is, the more comprehensive is the multivalent applicability of the module concerned (see fig. 1, the technological subset "cut subdivision").

The range of validity and, connected therewith, the size of modules decrease within the processing programme towards its end due to the large number of specific machine details.

The most important criteria of module size and their influences on the trends are summarily shown in table 1.

The criteria and trends stated clearly show that the optimum size of modules can be found only as a compromise solution and that it will always depend on the concrete content.

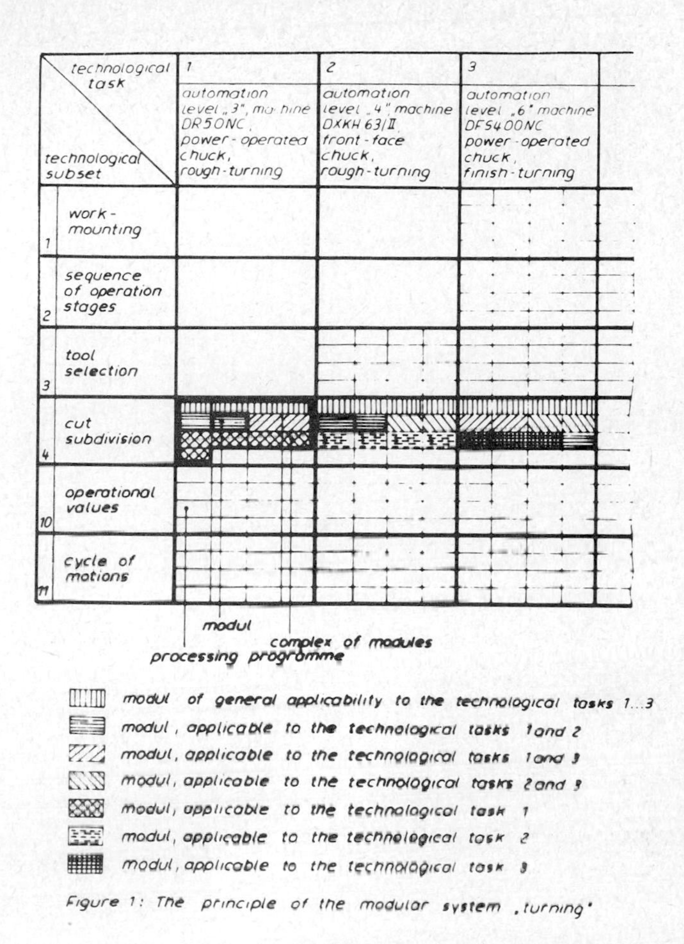

Figure 1: The principle of the modular system „turning"

Table 1
The most important criteria influencing module size

large modules	small modules
- simple programme organization - low expenditure of generating - a limited number of sub-programmes processible by computer-systems	- multivalent utilizability of modules - an efficient rate of utilization of working storages of computer-systems - variations of the scope of the technological rules - simple programming and testing - clearly arranged

Module formation is determined essentially by the criteria of module size and takes place in several phases (fig. 2). We distinguish between

- the completely new development,
- the formation of a new modular subdivision and
- the expansion

of a modular programming system.

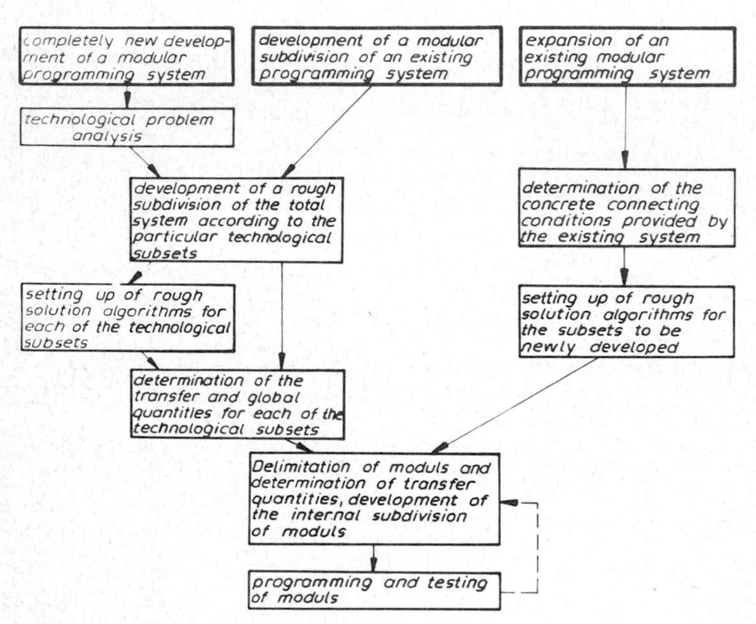

Figure 2: Stages of modul formation

During the designing and realizing of the modular programming system "turning", the geometrical and technological work data are stored in the so-called normal form in fields with random access, as it is the case with different programming systems, too.

This holds not only for workpiece data, but also the most important normalized data of manufacturing means are stored in global data fields to which access is possible from all modules. Only the data of the actual manufacturing means, i. e. of the manufacturing means necessary for the technological problem concerned, are stored into those fields.

Direct data transfer from module to module is justified only if similar parameters occur in only few modules.

2. PROGRAMME ORGANIZATION

The special feature of generating programming systems is that there are developed specific processing programmes being adapted to the respective machining task. Prerequisite to such a generating is the consistent modular structure of the programming system.

At least on organizational module - in the follo wing called "modular master" due to its function - is required for the problem-oriented and temporally consistent connection of modules for each processing programme. If for each technological subset and problem one modular master is present, there should be realized a higher degree of flexibility and reusability.

Fig. 3 shows an example of a simple modular master.

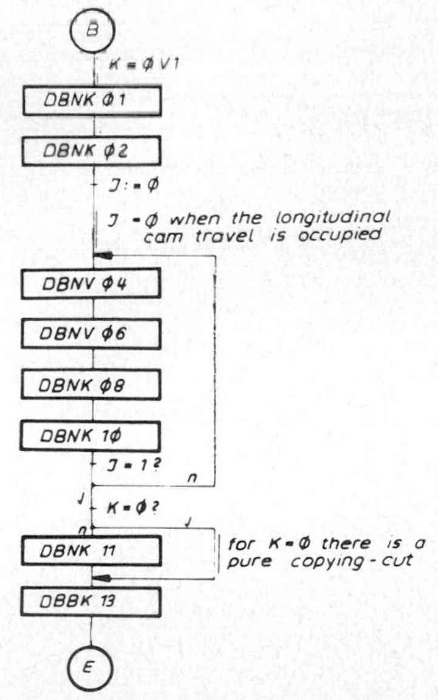

Figure 3: The master modular DLBK 19

Depending on whether the generating of the specific processing programmes is carried out inside or outside the computer-systems, a division is made into external and internal generating (table 2).

Table 2
Possible variants of the generating of technological processing programmes

relation to computer-systems	external	internal			
information-technical level	generator algorithm	generator algorithm		generator language	
form of use or availability	constant	constant	variable	constant	variable

2.1. EXTERNAL GENERATING

For external generating we proceed from the requirement that all modules belonging to the programming system should be available in an external storage medium, punched cards being most suitable.

Processing programmes for specific applications are establiched manually with or without a generator algorithm, depending on whether it is required by the size of the total programming system and the operator's knowledge.

Fig. 4 shows a part of a simplified generator algorithm.

Generally, compiled constant processing programmes ar used unchangedly for a longer period.

Therefore, the main field of application of external generating is the provision of processing programmes for users with relatively stable and delimitated manufacturing conditions and of compatible computing systems of small or medium capacity.

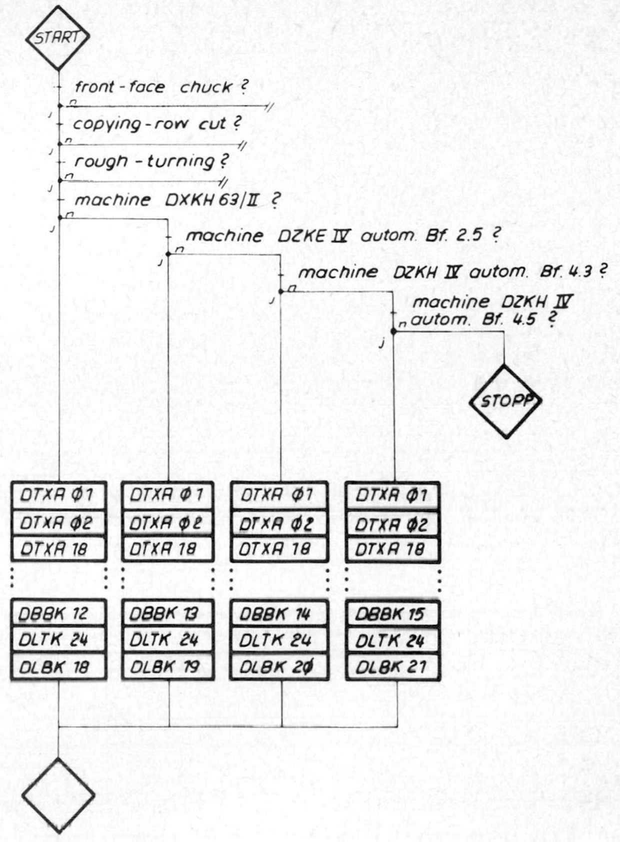

Figure 4: The generating algorithm „copy-turning" (part of it)

2.2. INTERNAL GENERATING

Internal generating is characterized by the feature that the establishing of specific processing programmes is carried out inside the computing system, computersystem-internal software being included.

Requisite to it is that all programme modules of the programming system are stored in an external storage of the system in the form of a module storage and that only the respective actual modules are loaded into the working storages.

The internal expenditure of organization during generating highly depends on the convenience of computer software. Standard software con be increasingly used for the programme transfer between the working and external storages.

Internal generating can be initiated in two ways:

- by means of an external generator algorithm,
- by means of a generator language and an internal generator programme.

For the form mentioned first, a generator algorithm - similar to figure 4 - is used, not each of the required modules having to be stated, but only the names or other coded data of the modular masters.

The short data for the designating and generating of the respective processing programme in the computer, having been determined manually by performing all operations of the generator algorithm, are fed into the computer in an appropriate and stipulated form. Then, the generating proper of the processing programme is carried out in the computer.

The way of generating described is especially appropriate to computer-systems of small or medium capacity, since for them the expenditure of storage and computation in the computing system for selecting the specific modules does not occur, because it is replaced by manual operations outside the computer-systems.

For internal generating with the generator language the technological task concerned is described definitely and completely by means of a special generator language, and the mechanical selection of the modules required is carried out in the computing system using the generator programme.

Depending on the total range of validity of the programming system, the generator programme may occupy much space. Because thus valuable working storage capacity would be occupied and also computing time would be extended, this form of generating is suitable especially for large-capacity computing systems.

The most developed version of mechanical internal generating will be given if the specific modules required are selected in the computer directly from the workpiece specification being in common use at present. It is assessed, however, that this variant - especially with the many possibilities which may occur for turning - could not be efficiently realized in the near future because of the high expenditure of generating required therefor in computers.

3. PRACTICAL REALIZATION

Initially, the checking and completing of the theoretical results was carried out with a part of the generating programming system "AUTOTECH turning", namely copy-turning. At present research is being performed on the total system "turning".

For the time being, mechanical determination is applicable only to the technological subsets of "performance characteristics" and "cycle of motion", but it is easily possible and also planned to bring this subsystem on higher levels of automation on the basis of its modular structure, the algorithm modules for the complex of "cut subdivision" already being available.

At present, the range of validity comprises four copying lathes. The great majority of programme modules however, possesses a higher range of validity. Thus, the subsystem can be expanded at relatively low expenditure.

The implementation was performed on a computer having a working storage capacity of 16 k words and being equipped with an alternate disk storage.

3.1. CONTENT AND STRUCTURE OF MODULAR COMPLEXES

A programming system is developed expendiently from the modular complex situated next to output, from the "cycle of motion" complex in the case considered, since a fast conversion into practice is guaranteed on the basis of a stepwise construction of the system.

Fig. 5 shows summarily the content and subdivision of the modular complex "cycle of motion", from which also the scope of the individual modules is evident.

For the technological subset "performance characteristics" at present about 25 modules are available, most of them being generally applicable to the machining by turning and only few being specifically designed for copying.

As an operational trial has shown, the determined technological performance characteristics resulted in reductions of basic time of 8 to 10 per cent as compared with the values determined manually so far. With the expansion of the range of validity of the modular complex by new mechines or groups of machines there became apparent the simple expansibility on the basis of the modular structure, because e. g. only few modules were to be newly made and, thus, within a short space of time the specific operational demands could be fully met.

technological subset \ machine		DXKH 63/II	DZKE IV autom Bf 25	DZKH IV autom Bf 43	DZKH IV autom Bf 45
1	computation of copying-lengths, rapid advance, rapid overrun	DBNK Ø1, DBXR 16, DBXR 17			
2	reduction of switch points	DBNK Ø2, DBXR 17			
3	transfer of the length data from the „technological field" to the „cam field"	DBNVØ3	DBNV Ø4		
4	assigning of the values to cam - row ranges	DBNVØ5	DBNV Ø6		
5	correction of the cam - row occupation	DBNK Ø7 DBXR 17	DBNK Ø8, DBXR		
6	printing of the cam-setting pattern	DBNVØ9	DBNV 1Ø		
7	computation and printing of the cutting-depth drum setting	DBNK 11			
8	completing of the cycle of motions and printing of the programme card	DBBK 12	DBBK 13	DBBK 14	DBBK 15
9	connecting of the moduls	DLBK 18	DLBK 19	DLBK 2Ø	DLBK 21

Figure 5: Modular complex „cycle of motions"

3.2. PROGRAMME GENERATION

Of the generating techniques stated, there have been tested in practice so far:

- external generating of constant processing programmes and
- internal generating of variable processing programmes with generator algorithm.

As the testing of the generating form mentioned first has shown, punched cards are more appropriate as information carriers than punched tapes tue to their simpler selectivity. The practical testing of the generating form mentioned second has also furnished its first positive results.

The names of the actual modular masters determined by means of the generator algorithm are read in and stored at the start of an expanded source programme. This is followed by the generating of the actual modules of the modular complexes of "performance characteristics" and "cycle of motion" and by the performing of all their operations.

Very short computing times have been obtained by the effective utilization of working storages (less than 1 min on an average). As fig. 5 shows, e. g. only 13 of the 24 modules of the "cycle of motion" complex will load the working storage at the same time. Results from quantitative comparisons, related to the efficiency of conventional programming systems and of the generating programming system "AUTOTECH copyturning", are not possible because a base is lacking.

In practical testing there became apparent a frequent occurrence of certain technological tasks. Thus, after a short initial period the parts programmer was able to dispense with the application of the generator algorithm, because he kept in mind the names of the modules concerned.

It can be summarily stated that the results and experience obtained from the development and performance of the generating modular programming system "AUTOTECH copy-turning" lead to the conclusion that a modular structure will be very suitable expecially for programming systems to be designed flexibly and expansibly, whereas a rigid structure is advantageous to programming systems being constant as to content and time.

[1] Bock, H.-J.
Investigations on the Structure of Flexible Programming Systems for the Computerized Manufacturing Preparation of Machine Tools, Shown by the Example of the Generating Modular Programming System "AUTOTECH Turning".
Thesis, Dresden Technical University, 1971

QUALITY ASSURANCE AND THE NELAPT PROCESSOR

E A S PATON
National Engineering Laboratory
East Kilbride, Glasgow, United Kingdom

Abstract: This paper describes the system by which NELAPT is systematically improved in a controlled manner.

1 INTRODUCTION

The range of numerically controlled machine tools for which the NELAPT processor is used extends from the simpler drilling machine tools to the more complex milling and turning machine tools. Consequently an NC processor which has such extensive capabilities is a very large computer program. No matter how well designed and how thoroughly checked out such a program is, errors are still likely to be discovered within it, particularly those areas exercised by new data (new types of part program) applied by users in a different application industry from those already using NELAPT. The task of managing such a large NC processor involves the constant debugging and updating of the code, the identification of alterations made and the execution of confidence checks on the reliability of the new releases.

2 THE NELAPT PROCESSOR

The NELAPT processor consists of a number of sections as detailed below.

SECTION	FORTRAN ROUTINES	ASSEMBLER ROUTINES	NO. OF STATEMENTS
INPUT	22	10	1413 + 424
DECODE	40	-	3182
GEOMETRY	284	-	9828
TOOL-LIBRARY	46	2	3731 + 73
CAWR1	33	-	1742
MOTION	216	2	22 523 + 59
(APT UTILITIES)	-	12	1154
UPDATE VOCAB	13	2	608 + 114
CAWR1 DATA PROG	4	-	372

Total number of FORTRAN statements = 43399

Total number of ASSEMBLER statements = 1824

In addition to the above code there are a number of pre-processed data files required for the successful execution of the processor.

On the 1108, which is a 36-bit word computer, the NELAPT processor is overlayed to enable it to occupy less than 52 000 words of core.

2.1 The Evolution of NELAPT on the NEL Univac 1108

In the present demand processing operating environment on the NEL Univac 1108 any one of the below listed versions of the NELAPT processors can be executed if so desired.

COMPUTER LANGUAGES FOR NUMERICAL CONTROL, *J. Hatvany, editor*
North-Holland Publishing Company - Amsterdam–London

VERSION	DATE	NC PROCESSOR	APPROX. SIZE (FORTRAN)
003	MAR70	2C,L	19 000
004	AUG70	2C,L	20 500
005	JAN71	2C,L	20 600
006	JUL71	2C,L	20 600
006	JUL71	NELAPT	38 900
007	JAN72	2C,L	20 700
007	JAN72	NELAPT	38 900
008	JUL72	NELAPT	43 400

There is also an experimental conversational NELAPT processor in use at NEL which currently corresponds in facilities to Version 007 of NELAPT.

Version 003 2C,L caters for a 2C,L subset of the APT language, the full language being aimed at 3C to 5C NC machines (see Appendix). 2C,L, which utilizes a selection of the APT input language, includes substantially all of the point, line, circle and tabcyl definitions as in APT, together with some extensions to provide wider 2C,L type facilities than are normally catered for in the larger APT processors. These are mainly to simplify the use of the tabcyl fitted curve facility and an additional area clearing facility which automatically provides many machine moves from one input statement.

Version 004 has the APT facilities of looping and subscripting, and introduces the ability to use special simplified calculations which avoid the production of cut vectors for circles and produce shorter control tapes where an NC controller is known to have circular interpolation facilities. The logic in area clearance is also improved in this release.

Versions 005 and 006 2C,L have code added to them to resolve many user problem areas, especially in tabcyl motion and area clearance.

Version 006 NELAPT is evolved from the integration of Version 005 2C,L, the milling program, the pattern features from a newly developed 2P,L drilling program and extra lathe part programming facilities from a newly developed 2C (see Appendix) turning program.

Versions 007 of 2C,L and NELAPT have an additional matrix multiplication facility to facilitate placing 2C,L shapes in 3C positions and a new line definition to permit the use of the X axis or Y axis as a valid definition. The method of handling post-processor vocabulary has been altered to simplify the introduction of new post-processor words which arise from time to time for new types of NC machine tools.

Version 007 2C,L is the final independent release of this processor on the NEL Univac 1108, future releases being extracted from the corresponding NELAPT processor.

Version 008 introduces a powerful new CONTUR/BIARC facility which is aimed at integrating NELAPT with the design process by enabling CAD (computer aided design) programs to output CONTURS (strings of straight lines, circles and/or fitted curves) ready for the production engineer to use with NELAPT, or for the design office to use the INPUT, DECODE and GEOMETRY phase of NELAPT for design calculations, leaving CONTURS for later NC machining. An optional workshop technology pass is available (that of CAWR1, standing for Computer Aided Workshop Rationalization) which provides for both stand-alone collection and retrieval of cutting information and also integrated retrieval of this information automatically from a NELAPT run.

2.2 NELAPT on Other Computers

Since the NELAPT processor is written mainly in ASA FORTRAN 4, it is virtually computer independent. The 2C,L processor is currently operational on the following computers.

Univac	1108
ICL	1900 series
CDC	6600
IBM	360 series
ICL	KDF9

Implementations are being carried out on the

Digital	PDP10
ICL	System 4 series
ICL	4130 (a restricted subset).

2C,L has been integrated into the British Shipbuilding Research Association ship design and production computer program (BRITSHIP).

2.3 The Modular NELAPT Processor

The NELAPT processor is designed to permit the extraction of a basic processor from the overall processor, and to construct from that basic processor, processors of differing complexity. The basic processor, or base module as it is frequently termed, includes the definitions for basic point, line and circle geometry and basic point-to-point motion commands. Processors aimed at specific requirements can be built up from this basic processor, for example:

a By adding to the base module the module with extended point, line and circle definitions allows the user extended point-to-point facilities with simple line milling capability.

b By adding the start-up and simple continuous path modules to (a), the user has a 2C,L type capability.

c The facility in (b) can be further extended by adding more complex geometry capabilities, for example empirical curves, either tabcyl and/or contur(biarc). These require the associated tabcyl and/or contur(biarc) motion modules.

d By retaining the simple motion module only, but expanding the geometry capability to include patterns, allows the user a comprehensive point-to-point process to be assembled, as in 2P,L.

e Similarly by adding the area clearance module to a processor having all continuous path capabilities gives a comprehensive type 2C,L processor.

f For the user interested in turning, no complex geometry and pattern modules are required, but special turning area clearance would be added, as in 2C.

It can be seen that this flexibility allows the user to construct a processor which meets his specific requirements.

The processors from which NELAPT was developed - 2C,L; 2P,L; 2C - represent three of the possible processors which can be generated from the overall NELAPT processor.

3. THE MANAGEMENT SYSTEM

A processor Management System at NEL ensures reliable releases of the NELAPT processor. At any time the NELAPT processor may be altered to implement new facilities, to improve existing implementation procedures or to correct failure areas. It is immaterial how frequently the processor is altered but, from both management and user point of view, it is desirable to have pre-scheduled release dates of the altered processor, say, twice yearly. Such releases are clearly identifiable in the Management System.

3.1 Programming Services

The nucleus for such a Management System is a central unit which is responsible primarily for issuing new releases of the NELAPT processor. This unit, called Programming Services at NEL, relies on information recorded on two standard forms to aid in their task, namely the Failure Report form, on which problems encountered by the user in the NELAPT system are indicated, and the Revision Form, on which alterations and/or proposed alterations to the NC system are recorded.

3.2 Failure Reports

When a part program (data for the processor, describing in APT language how a component, or part, is to be machined) fails to process correctly in the NELAPT processor the user completes a Failure Report form and sends it to Programming Services through the agreed arrangements at his site. On the form he gives a brief description of the problem, indicates which version of the processor he is using, on which computer, and whether or not the version has site modifications in it. When such a Failure Report is received by Programming Services, irrespective of its source, it is given a number so that it is uniquely identifiable in the Management System. The computer listing and card deck of the failed part program, together with the sketch or drawing of the part that failed, are filed by Programming Services so that they are available for reference for those investigating the Failure Report.

Failure Reports on the NELAPT processor are accepted by NEL irrespective of which type of computer the failure occurs on or how old the version of the NC processor is, or whether there are site modifications in their processor.

The initial analysis of the Failure Report determines in which of the following four categories the failure occurs:

a The facility has been used incorrectly in the part program.

b The part program processes correctly, at least to beyond the point of recorded failure, on Programming Services' Official Release of the processor on the NEL Univac 1108.

c The part program fails on the Official Release but processes correctly on Programming Services latest Field Test Release.

d The part program fails on the Field Test Release.

This analysis is discussed at the weekly Project Meeting at NEL between the computer programmers who maintain and enhance the NELAPT processor and the personnel from Programming Services who issue the new releases of the processor. If it is a part programming error this frequently means that the documentation on that specific facility is inadequate and requires improving. This should result in a Revision Form which indicates that the part programmer's manual should be altered, or even necessitates the publication of a Programming Note[1] on that facility. If the part program fails on the latest Field Test Release the computer programmer responsible for the specific area in which the failure occurs endeavours to resolve the problem and produce the necessary modifications for implementation in the next release of the processor. Programming Services records what action has been decided on the Failure Report at the Weekly Project Meeting. The submitter of a Failure Report is then duly notified of the decision made with respect to his failure.

3.3 Revision Forms

If a part programmer or computer programmer wishes to request, for example, a new facility in the NELAPT processor, or improved part programming documentation, such requests are recorded on a Revision Form which he sends to Programming Services where it is given a number so that it is uniquely identifiable in the Management System. If the proposal is supported sufficiently the necessary work is scheduled for programming.

The Revision Form was introduced primarily to identify alterations made to the NELAPT processor. The computer programmers who analyse problem areas, implement new features etc into the NC processor identify all such proposed or actual alterations on Revision Forms. Such Revision Forms received by Programming Services from the NEL computer programmers are accompanied by the following items with respect to the Univac 1108.

a A compiled listing of the altered/new routines.

b A detailed allocation of the collection of the elements where necessary (ie MAPs on the 1108 - relocatable and absolute).

c A listing and card deck of the data file of alterations.

d A listing of the vocabulary deck and tool-file data if alterations have been made to them.

e A part program listing as processed under the computer programmer's modified system.

f A card deck of the part program if the Revision Form is not associated with a specific Failure Report.

g Documentation alterations to the computer programmer's and part programmer's manuals (if necessary).

The listing of the routines, MAPs, data files, code alterations and part programs are filed by Programming Services, together with the supporting card decks. These items form the essential reference materials when a new release of the processor is being generated and tested.

Revision Forms received by Programming Services are generally in the 'proposed' state. At the Weekly Project Meetings they may be altered to the 'final' state, that is, for example, the alteration is to be implemented by Programming Services into the next release of the processor. 'Final' state Revision Forms are those which have been authorized by the officer in charge of the specific area of the processor with which the alteration is associated.

Alterations to specific lines of code in the NELAPT processor are identified in card column positions 74-80 by the Version Number and Release date (month and year) of the specific Official Release, eg JUL72** in columns 74-80 identifies changes made to Version 007 NELAPT JAN72 to produce Version 008 NELAPT dated July 1972. The first asterisk indicates that the NC processor is operational under the Univac 1108 EXEC 8 operating system, the second asterisk indicates that the changes are implemented in the NELAPT processor but not in the corresponding independent 2C,L processor.

The Revision Form is a multi-purpose form. Its varied uses are summarized in the categorization scheme currently in operation by Programming Services with regard to their written Status Report publications. Each category is identified by a single letter and more than one category may be applicable to one Revision Form. Twelve categories exist at present and are as follows.

C,P - Revision Forms which indicate alterations to the NELAPT documentation, ie the Computer Programmer's Reference Manual (CPRM) and the Part Programmer's

Reference Manual (PPRM), are identified by the letters 'C' and 'P' respectively.

F - If the Revision Form is associated with a Failure Report it is given a category 'F'.

I - 'I' indicates that the alterations on the Revision Form are related either to the implementation of a new facility or improving the existing implementation procedure within the processor or just correcting incorrect code for which no Failure Report has actually been submitted.

D - It is important that the user is informed of any new part programming diagnostics in the NELAPT processor. A Revision Form which is associated with the implementation of a new part programming diagnostic is categorized by the letter 'D'.

S - The Standard Test Deck and drawings may be altered, either to modify part programming statements or replace a complete part program etc; such changes are recorded on a Revision Form which in the categorization scheme is given the letter 'S'.

V,N - Official Releases in which Revision Form alterations are operational for the first time are identified in the Status Report by the letter 'V' followed by the version number, followed either by the letter 'N' or nothing, eg V6N identifies Version 006 NELAPT (July 1971), V6 identifies Version 006 of the independent 2C,L processor (July 1971). The significance of the letter N is that during the period when the two NC processors coexisted independently, NELAPT and 2C,L, if N is present then the alteration is made only to the NELAPT processor.

U - If the alteration refers for example to an ASSEMBLER routine or to a specific 1108 system processor, eg the Memory Allocation Processor (MAP), the Revision Form is categorized by the letter 'U', indicating that the change applies only to the Univac 1108 implementation of the NC processor.

W - A 'W' classification on a Revision Form in the Status Report indicates that the proposed alterations have either been given a low priority or they missed the cut-off date for that Official Release (being described in the Status Report) and will not be implemented until the subsequent Official Release.

X - As a result of implementing a specific Revision Form the Field Test Release may process correctly fewer part programs than does the current Official Release, in which case the alteration is not implemented in the next Field Test Release. Such a Revision Form is given the category 'X'. A revision Form may also be categorized 'X' if the proposed alterations are impossible to implement, for example, because of the structure of the NELAPT processor.

E - Revision Forms may be submitted by a user or implementer outside NEL, in which case it is given an 'E' classification. The 'E' category also applies to a Revision Form which resolves a Failure Report submitted by a user external to NEL.

3.4 New Releases

Programming Services issue two types of releases of the NELAPT processor, Field Test and Official Releases.

3.4.1 Field test releases

When programming Services receive a number of Revision Forms which specify alterations to the existing code, or the addition of new code, a new system is generated for in-house use. This release is termed a Field Test Release. Such releases serve two purposes:

a On-site part programmers have access to the most up-to-date version of the NC processor.

b Computer programmers can check their alterations to the system before an Official Release is generated.

Field Test Releases consist of the recompilation of altered/new routines only on the Univac 1108. A number of Field Test Releases may be issued between Official Releases. All Field Test Releases are generated from the current Official Release of the NC processor. The final Field Test Release is generated immediately prior to the next Official Release.

3.4.2 Official releases

The next Official Release of the NELAPT processor is produced by recompiling all the routines in the final Field Test Release and reallocating all the segments. Programming Services issue Official Releases of the NELAPT processor twice yearly, in January and July. The cut-off dates for alterations to the NC system are the 20th of the month preceding the Official Release. This means that alterations received by Programming Services after the cut-off dates are not included in the next immediate Official Release of the NC processor. The only Revision Forms submitted after the cut-off dates that are implemented are those necessary to resolve failures which arise in the final Field Test Release due to problems encountered in the new code being added to the processor. Only when all the alterations are merged into one system by Programming Services are such conflicts revealed. Programming Services acts as the coordinating unit with the computer programmers to resolve such failures before generating the Official Release. Each Official Release which is designated by a specific version number and date is uniquely identifiable in the Management System.

3.5 Confidence Checks

Whether it is a Field Test Release or an Official Release, Programming Services carry out the same checking-out procedure in an endeavour to assure the part programmer that the quality of the new release of the NELAPT processor is actually better than the previous release.

The first confidence check made on the new NC system is to run the twenty part programs which form what is termed the Standard Test Deck. Should any fail to process as expected Programming Services consults the computer programmers regarding the problem. A Failure Report is generated and processed as before, but with a top priority rating. The Standard Test Deck is under constant review so that it provides a comprehensive check on the NC processor as described in the part programming documentation and also checks any new facilities which have been added to it. Changes and proposed changes to the Test Deck are submitted on Revision Forms to Programming Services. The Standard Test Deck has proved effective in locating problem areas in the computer system's software.

To provide a more comprehensive checking-out procedure part programs referring to Failure Reports supposedly cleared, and Revision Forms of new facilities supposedly operational, in this new release are also processed. Should any fail to produce output listings the same as those obtained by the computer programmer under his modified system, the computer programmer is immediately consulted and the necessary action taken to identify and resolve the problem.

3.6 Project Meetings

A meeting takes place each week at NEL between the computer programmers who maintain and enhance the NELAPT processor and the personnel from Programming Services who issue the new releases of the processor. Each Failure Report and each Revision Form is discussed and a decision made on the proposed action necessary for each one. The content of the next Official Release is discussed and any major enhancements scheduled for inclusion are reviewed to ensure that the new code is operational by the cut-off date.

4. QUALITY ASSURANCE

Quality Assurance is more than just instituting confidence checks to new releases of the NELAPT processor. Anticipation of possible/probable sources of difficulties is an essential factor and techniques to minimise or eliminate them are very essential. Such techniques include the standardization procedure for the computer programming personnel engaged on the NELAPT processor and the retention of comprehensive records on Official Releases by Programming Services.

4.1 The Computer System's Software

The NC processor which is basically a large FORTRAN program is run under a powerful, versatile operating system. Changes to the operating system and/or FORTRAN compiler and library can affect the performance of the NELAPT processor. To minimize such adverse effects that such changes to the computer system's software may have, the computer programmers all use the same level of FORTRAN compiler, the same level of FORTRAN library and the same segmentation procedure. When changes to the computer system's software either necessitate alterations to the actual code in the NC processor, or just recompilation of the routines, this is performed on the current Official Release of the processor. This results in a new release which then replaces the current Official Release; it is identified by the same version number but with a different date. No other alterations are made to the NC processor at this time. The new release is checked out as before to ensure the quality of it.

4.2 Modified NELAPT Systems

As there are a number of personnel engaged in modifying different parts of the NELAPT processor it is essential that the computer programmers all use the same source/object code (symbolic/relocatable elements in Univac 1108 terminology) of the current Official Release issued by Programming Services as the base elements for generating their modified systems. They run the Standard Test Deck under their modified systems to ensure that the new code does not affect adversely the existing code in the processor. Similarly Programming Services generate Field Test Releases primarily to eliminate this problem before Official Releases are generated.

For Official Releases of the NELAPT processor, Programming Services recompile all routines and re-collect all segments, whether or not they have been altered. This is a precautionary measure to minimize problems which might arise if object code (relocatable elements on the Univac 1108) and segments were generated under different levels of the computer system's software.

An important by-product of this procedure is that the listing of the Official Release is the actual source code (symbolic elements on the Univac 1108) of that specific release.

4.3 NELAPT Records

Another aspect of Quality Assurance essential to the successful operation of the Management System is the maintaining of relevant records on the NELAPT processor. Programming Services retain magnetic tapes of all Official Releases, listings of all Official Releases, card decks and listings of the alterations necessary to produce the next level of Official Release from the present level, card decks and drawings of the Standard Test Deck together with their processed listings under the Official Releases, card decks and listings of Failure Report and Revision Form part programs as processed under the specific Official Release in which the code associated with them was implemented.

The importance of these records is reflected in a number of different ways. If a user is running under an earlier version of the NELAPT processor which may or may not have site modifications in it, and which may or may not be an 1108 implementation, NEL can help resolve problems for such a user because the failed part program can be run under the corresponding NEL Univac 1108 implementation. Therefore NEL can accept and process Failure Reports from such a user.

Similarly a non-1108 implementer may find differences, for example, between the processed Standard Test Deck part program listings as produced under his implementation and those produced by NEL under their corresponding version of the processor. This has proved a most useful source for checking the accuracy of the implementations of the NELAPT processor on different computers.

Outside implementers normally receive listings of the Official Release, the alterations from one release to the next, and the processed Standard Test Deck part programs if different from the previous release.

To identify quickly and collectively alterations made to one Official Release to produce the next Official Release, Programming Services generate a Documentary Routine entitled

'ABRIDGED ACCOUNT OF ALTERATIONS TO NELAPT'.

A listing of this FORTRAN comments routine is sent to outside implementers who receive the other listings mentioned above. The NELAPT system tape also has a copy of this routine on it. In the routine are details of the level of computer systems software under which the Official Release is generated on the NEL Univac 1108. The names of the routines which are altered to form this release of the NC processor are listed together with references to the Revision Form(s) and/or Failure Report(s) associated with the changes. Status Reports are included on Failure Reports, whether resolved or unresolved, on Revision Forms, stating which have been implemented, which have not, and on the Standard Test Deck, stating specifically whether they processed correctly under the new release, and which changes (if any) have been made to tho test deck part programs.

This Documentary Routine has proved invaluable to NEL staff working on the NELAPT project, especially when alterations made to earlier releases cause problems when new features are being implemented in a later release.

4.4 Users' Confidence in NELAPT

The current (August 1972) Official Release of the NELAPT processor on the NEL Univac 1108 is entitled

'NELAPT VERSION 008 JUL72** OFFICIAL NEL 1108 RELEASE'.

This title appears at the top of every page of printout of the processed part program and indicates to the part programmer the specific system release under which his part program is executed. This system identification technique applies to all Official Releases issued by Programming Services.

Should a user for example want to re-process a part program under a particular Official Release he must never doubt that the content of that release has altered in any way since his previous execution of it. If he suspects that it has been changed although the heading identification is the same this would defeat the purpose of any Quality Assurance on the NELAPT processor as his confidence in the system would be destroyed.

5. USERS' MEETINGS

No matter how effective the Management System is for a large computer independent program like the NELAPT processor and no matter how reliable the releases are, the Quality Assurance of it all stands or falls by the success of the NC processor in the Workshop. The most important people are therefore the users of the system.

This is why Technical Users' Meetings are scheduled at regular intervals. This is where the users have the opportunity to tell of their successes, but more important, of the failures they have experienced when using the NELAPT processor. If the processor is to be improved there must be a ceaseless feedback of information, for only by this cooperation between part programmer and computer programming personnel engaged on the NC project can there be a viable NELAPT processor. For such users' meetings Programming Services issue a written Status Report on the NELAPT processor, a report which endeavours to assure both user and implementer of the quality of the NC processor.

6. CONCLUSIONS

The Management System which was designed, developed, and implemented at NEL for the 2C,L milling program has proved successful during its short period of operation (approximately three years). Not only has the NELAPT processor evolved but so also the modularity feature within it, permitting the extraction of a number of NC processors with restricted capabilities. The number of implementations of the processor on different computers has also increased with the number of users accessing the system. Technical Users' Meetings have become an established event in the calendar.

Programming Services' experiences of operating the Management System have culminated in issuing regularly new reliable releases of the NELAPT processor, assuring both user and implementer of the quality of such releases.

ACKNOWLEDGEMENT

This paper is presented by permission of the Director, National Engineering Laboratory.

REFERENCES

1. Programming Notes are published by the NC Division at NEL at irregular intervals. They generally deal with the use of specific features in the NELAPT processor.

APPENDIX

ISO ABBREVIATIONS FOR NC SYSTEMS

1. The letters C and L are ISO abbreviations for NC systems capable of Contour cutting and Line cutting respectively. Thus 2C,L implies a capability of programming 2 axes of a machine tool to cut a Contour tool axis, whilst a third movement or axis can independently Line mill. This is by far the most common requirement for NC milling machines.

2. 3C - 3 axes controlled to move simultaneously to Contour.
   5C - 5 axes controlled to move simultaneously to Contour.

3. The letters C and L are as described in (1). P stands for Positioning (for punching and drilling). Thus 2P,L implies a capability to program for 2 axes Positioning of a machine tool whilst a third axis is suitable for straight Line drilling moves.

   2C - 2 axes controlled to move simultaneously to cut a Contour.

DISCUSSION

Several questions were addressed to Mr. T. Legendi. Mr.J.Vlietstra asked why the authors of CHANGE had not studied the results of the ALGOL 68 Group. The reply was that the work had aimed to produce a single, basic language in a limited time. Mr. H. Strempel wanted to know in what sense CHANGE was a postprocessor. Mr. Legendi said that it produces a control tape for the NC machine ADMAP, and in reply to Mr. D. Kochan said it had been implemented on a CDC 3300 computer in FORTRAN IV. The language had 200 instructions. Mr. M.A. Sabin asked whether the non-linearity described was similar to the parallel processing in ALGOL 68. On self-modification he remarked that the earliest computers had in fact used this technique, which was later found unsatisfactory because of the difficulty of finding out what goes wrong. Nowadays hardware facilities are used which allow the instruction text to remain in its original form throughout - this is particular advantageous in parallel processing. What real advantages were there in self-modification? Mr. Legendi said that CHANGE was very similar to SIMULA in giving a direct programming facility. Self-modification was a crucial point of the language which he believed would be much used in the future.

Mr. G. Pikler was asked by Mr. R. Sim, whether his program constructed postprocessor commands and replied that there was no special check on each postprocessor command.

Mr. K.H. Tempelhof, in reply to Mr. J.P. Crestin, who asked whether collision prevention could not be better solved by DNC, said that any of the usual information processing methods could be applied with his system.

Mr. H. Eitel asked Mr. M.A. Sabin whether it was not cheaper to have at least one 5-axis controlled machine tool instead of writing a number of preprocessors with a wide range of generality.Mr.Sabin said that scanning is cheaper but takes a little longer on a lower-cost machine tool. Mr. H. Schreiter remarked that while the use of a preprocessor was an attractive idea, he saw problems in the specification of the main processor. Were not some facilities /pockets, islands/ basic? The author replied that since sculptured surface data have all to be in core at one time, area clearance and pocketing do not fit very well into the main processor.

Mr. H. Zölzer asked Mr. I.D. Nussey how much work the user had to do if he wished to insert a procedure in the 370 NC System. Mr. Nussey said the procedure first had to be written, then an editing language was available and insertion was very easy. In reply to a question by Mr. J.P. Crestin on graphic aids, he said that because of cost problems, acceptance had been slower than expected. Now it is not so far away and storage tubes as editing terminals already had wide usage. Mr. V. Grupe asked what was planned with regard to turning and why the IBM extensions to APT violate the basic rules /e.g. character set and language structure/. Mr. Nussey said the processor described was not specified for turning. The rules had been broken because the users required it.

Mr. J. Vymer in a contribution on Mr. H. Weissweiler's paper pointed out that the more technology was put in the processor, the more customer- or user-oriented the system became. The APT input was not optimal, yet he was surprised that the EXAPT system was now adopting a fixed form input facility. Mr. Weissweiler replied that an optimum was never available. It could well be a combination of APT-like, fixed format and other inputs, such as the one now being developed. Mr. H. Zölzer added that this was not a clear fixed format like that in MITURN. EXAPT 2 will include group technology and this requires a different input stream, including a fixed sequence of parameters.

In a question to Mr. H.J. Bock, Mr. H.H. Brechtel asked how data handling and organization from one module to another was arranged. In the modular system are machine tool data considered in the processor, or postprocessor? The author replied that there was a "normal" form in which all modules can be accessed. Machine tool data can be considered in both places.

Mr. B. Gott asked Miss E.A.S. Paton how many people were employed full time in the support of NELAPT. How was feedback from users organized, and was this machinery adequate. Miss Paton said there were 4-6 people in NELAPT maintenance and service, feedback was through "NELAPT News," Users' Meetings and failure report forms. Mr. D. Kochan asked how easily different user-oriented programs could be locally generated. The author said this was possible, but the chief outlet was for fixed forms. Otherwise support would grow very difficult. Answering Mr. M.A. Sabin, she said very few of the failures reported turn out to have been part programmers' errors.

The discussion on Mr. W.E. Mangold's paper took place after he had had to depart, but the Session Chairman undertook to try and answer. Mr. K.J. Davies asked why the USA, the country where APT originated, voted against the CLDATA major word proposals based largely on APT III.

Mr. D. McPherson pointed out that the ANSI method of presentation differed from ISO in the documentation of the standard. ANSI used a numerical meta language on a computer. ISO has no responsible body to provide such a computer facility. Mr. J. Vlietstra added that the main problem had been the use of integer numbers for post-processor words and the alphanumeric representation of the words themselves. ANSI accepts both, while ISO CLDATA has only the integer version.

6. SURFACE DESCRIPTION PROGRAMS

SURFACE DESCRIPTION PROGRAMS

J. REMMELTS
Metaalinstituut TNO Apeldoorn,
The Netherlands

The subject covered by Prolamat is made up of a variety of disciplines, closely interconnected into a network, a fact which cannot be obscured by neatly categorizing the papers to be presented, in separate sessions, each with its own subjecttitle, suggesting a well-defined and limited area. Consequently, to select the session-themes and to class the papers in these categories has been a fairly arbitrary procedure. This is reflected in the choice of topics to be discussed, where pure surface description programs appear, side by side with systems in which surface description is an integrated part of the manufacturing process. Moreover, some papers which have been presented in other sessions contain, either implicity or formulated as such, a method of surface description.

It is notable that at the first Prolamat, three years ago in Rome, a session on surface description was not required: only one paper dealing with the subject proper was presented. Does this point to a considerable growth in interest over the last three years? If we restrict ourselves for the moment to the theoretical, the analytic side of the subject, the answer should be no. Truly fundamental publications, which were and still are, the basis of varied development- and application activity, date back to quite a few years before 1969, the year of Prolamat I. Since then, pioneering publications seem to have been rather scarce. But as to practical application of the systems developed, one can perceive a considerable expansion. This may have been stimulated, if not caused, by hardware developments which, for instance, produced new graphical I/O devices, both passive and interactive, providing increased capabilities at lower prices. So it is not astonishing to find that the greater part of the papers making up this session, are concerned with these graphic tools as a link in the man-machine dialogue.

For both visualization and interaction are badly needed provisions when working with the so called "sculptured surfaces".

It takes a profound knowledge and experience to be able to do surface fitting by handling sets of equations and by manipulating parameters only. All too soon the many parameters involved defeat any attempt at an interactive approach, convergeing to the predefined object function. But the criteria that dictate the acceptability of the mathematically defined sculptured surface are, by nature, inexact: one cannot express easthetical values in terms of mathematical logic.

Which leads one to the conclusion that the surface description obtained can only be judged upon by contemplating its graphical representation. These viewpoints are ably emphasized in more than one of the papers, notably by Mr. Nicolo and Mr. Ghezzi.

The use of the computer, whether interactive or not, in the design phase, is of course not limited to the problem of surface fitting. Stress- and strength-analysis are among the subjects where calculations are performed on the analytical representation of the object under consideration. In a way the final shape of the product is at that stage extant in numerical form, or becomes available in the course of this activity. The same is true in the final stage of the process from conception to finished product, the production proper. For instance, when the punched tape of an NC milling machine, containing the final information on the product under consideration, is produced. It is surprising then, that in the stages intermediate between the two representations of the object in computer memory just mentioned, one generally extracts the information from the computer, in the form of tables, drawings, etc., to perform a variety of manipulations, including calculations, ending up with a part program, containing the final information on the product, which is there upon reinserted into the computer for ultimate processing. Maybe one should not call this situation surprising.

After all CAD and NC-tape production pose problems of their own which vary widely and their development originated from completely unrelated viewpoints. But it would be surprising if the situation just described, would perpetuate. Which it will probably not, with the concept of data base gaining ground.
Mr. Flutter points this out in his paper stating:
'it hinges on having available an object definition which is

sufficiently powerful and flexible at every stage".
The same viewpoint is taken in both papers dealing with AUTOKON, where both Mr. Mehlun and Mr. Landmark emphasize the database-structure of their system.

On the theoretical side, the analytical formulation of "shapes"like carbodies and windtunnelmodels, poses mathematical problems of intricate nature. This is especially so when one realizes that the input data may contain uncertainties, or that part of the surface may be predefined in an analytical way.
Various approaches have been outlined in the past and are in use today. The bicubic model of GEMESH, the parametric bi-cubic FMILL, Coons' more general blending patches and Beziers polygonal representations of curves and surfaces, to name only a few. It is fortunate to have Prof. Bezier contributing his UNISURF concept. And equally interesting is the contribution of Mr. Nicolo, who in his paper takes up this approach and points to possible improvements for specific problem areas.

On the other hand, a number of papers are concerned with the application of Coons' patches. For instance Mr. Ghezzi introduces an extension to this principle, which opens the possibility of local adaption in specific areas, without undue increase in the number of patches. Here also, the importance of a suitable datastructure is emphasized.

Another paper mentioning Coons' approach, is the only by Mr. Hyodo, who describes the application to HAPT-3D. Here the APT-language is extended, a combination of a powerful system, refined through practical application, with newer techniques. Concurrently, the original APT-system itself, through its development organization CAM-I, is being extended with sculptured surfaces capabilities. Although no paper from that side was presented, the discussion offered an opportunity to hear from the CAM-I activity to get an indication of their line of attack.

Having mentioned APT, this progenitor of so many other languages, there is still another paper which deals with its possibilities. It

is of interest to note that in this case Prof. French does not so much deal with sculptured surfaces, as with the description of what might be called conventional ones, made up by combining the available basic geometry of APT. In these days, when modularity is much discussed, this way of using APT-macro's as building blocks for creating rather complicated shapes, seems very timely. Incidentally, this same principle of building blocks is exploited by Mr. Hyodo in a different manner.
His concept of CURVE and SURF combines elementary APT geometry entities into otherwise arbitrary curves and surfaces.

In the introduction a number of papers were indicated which do not deal with surface descriptions as such, but in which the mathematical synthesis forms a part of a larger manufacturing system. AUTOKON, mentioned earlier in other context, is a case in point, as is FORMELA, another Scandinavian contribution. Mr. Sandin describes a system /which seems to have its roots in the pre-computerera/ in which much practical experience has been accumulated. He also underlines the indispensability of a carefully designed data base.

Quite often, NC is identified with milling, turning and other metal removing processes. It should be appreciated that this limitation by no means reflects reality and that the application of NC outside the field of metal cutting plays an important role in industry. Therefore, the contributions from Mr. Hirano and from Mr. Ono are very appropriate, as they describe applications for flame cutting and pipe bending, respectively.

In summary it can be said that, although this session merely covers the subject of Surface Description, the menu is a varied one, yet with many ingredients common to several of the courses. Undoubtedly there is something to everybody's taste.

THE POLYSURF SYSTEM

A.G. FLUTTER
The University and the Computer Aided Design Centre, Cambridge, England.

Abstract: POLYSURF is a system for the computer aided design and manufacture of engineering objects which are composed largely of 3D curved surfaces. The POLYSURF routines enable a design engineer to assemble a mathematical model of such objects in terms of bounded surface geometry. The model can then be appraised, analysed and interactively adapted, and finally used as a basis for n/c manufacture.

1. INTRODUCTION

POLYSURF was written with a two-fold purpose. First to investigate the increasingly popular 'production process' of combining the disciplines of design, analysis and manufacture as closely as possible; secondly to implement various computer packages which would hopefully be suited to and give added power to this process.

The production process which is described in section 2 depends on having a central data bank which all the packages can access. With the computer hardware available (Shown in Fig. 1) this data bank resides on ATLAS disc. The principle things stored in the data bank are object descriptions as described in section 3. All the packages, several of which are described in section 4 to 8 accept input instructions from either an on-line ATLAS console or in batch mode from an ATLAS instruction file. The instructions cause data to be accessed from the data bank, processed in some way and any output information returned to the data bank. Some of the instructions available in the various packages are given in the relevent sections.

Several packages require either interactive graphics or drawing facilities. These are provided by a PDP7 satellite with 340 display and a Calcomp plotter

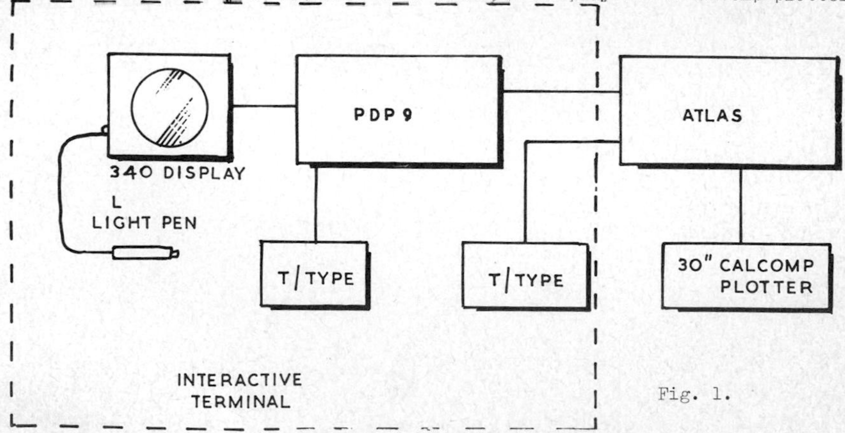

Fig. 1.

COMPUTER LANGUAGES FOR NUMERICAL CONTROL, *J. Hatvany, editor*
North-Holland Publishing Company - Amsterdam–London

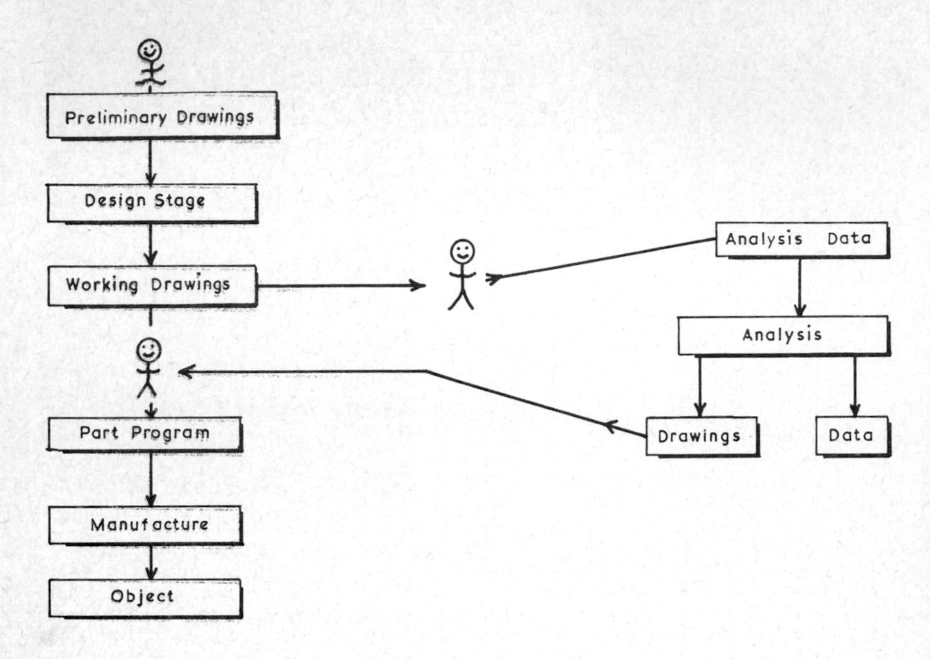

'A' INFORMATION FLOW IN TYPICAL PRODUCTION PROCESS

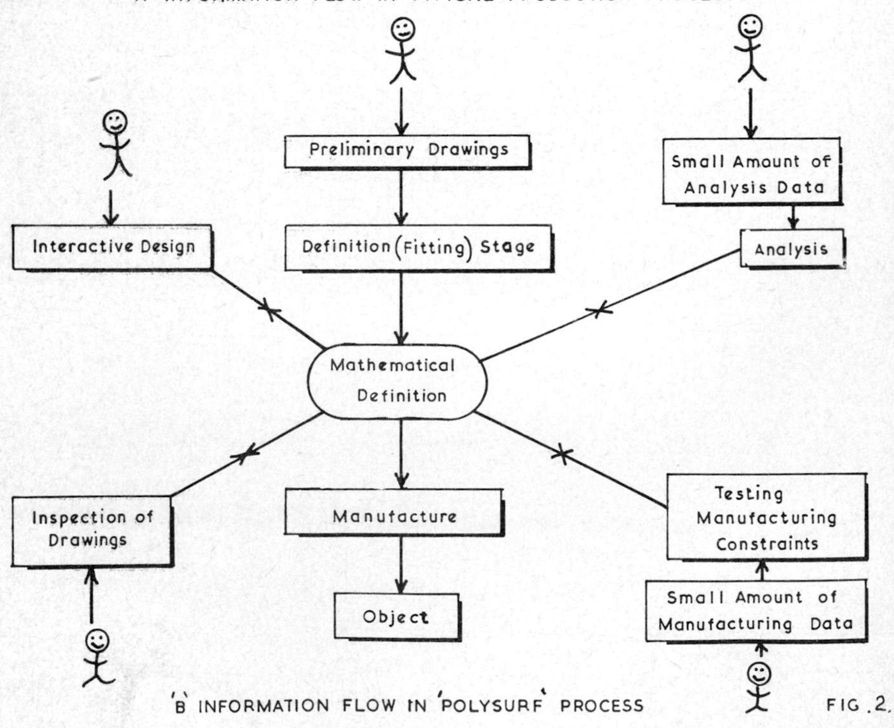

'B' INFORMATION FLOW IN 'POLYSURF' PROCESS

FIG. 2.

2. PRODUCTION PROCESSES

The processes used for the design analysis and manufacture of engineering objects vary so widely that stating general principles is perhaps dangerous. Whilst acknowleging this difficulty, an attempt is made in Fig. 1 to show a difference in principle between two production processes that are suitable for handling fairly complicated objects where design and analysis are relatively important. In process 'A' (the more common of the two shown), the object is redefined at each new stage in a form most suitable for the new stage. In process 'B', which has been adopted in POLYSURF, a single subjective object definition is used by all stages. This distinction becomes sharper and more significant when computer methods are used. In a typical process of type 'A', the object definition at the design stage is in the form of drawings. Even where the drawings are produced by computer methods, the object definition subsequently used is the drawing rather than the computer data it was produced from. At the analysis stage, input data is extracted from the designer's drawings in a form most suitable for the analysis undertaken. A variety of data and drawings are then used at the manufacturing stage to redefine the object in terms of geometrical part programming statements. This duplicity of definition is in contrast to process 'B' where all the stages use a single object definition, so that working drawings, analysis input data and geometrical part programs are avoided as far as possible.

Making comparisons between these two types of process is even more difficult than defining them. Suffice to say that from the limited experience gained with POLYSURF, process 'B' seems well suited for objects where the design and analysis stages are at least as important as manufacture. One other more concrete conclusion is that the success of process 'B' hinges on having available an object definition which is sufficiently powerful and flexible for use at every stage. In the case of objects composed of 3D curved surfaces, a mathematical definition in terms of bounded surface geometry fulfills this requirement.

3. BOUNDED SURFACE GEOMETRY

Bounded surface geometry for 3D curved surfaces may be conveniently based on surface equations of the general form:

$$\overline{P}\,(x, y, z) = \overline{F}\,(u, v) \qquad (1)$$

F is a vector function of u and v such that if u is varied whilst v has a fixed value, or v is varied whilst u has a fixed value, $\overline{P}$ traces a constant parameter line that is generally a 3D curve. If both u and v are varied simultaneously, P traces a 3D surface.

Apart from constant parameter lines, any functions of u and v having the form:-

$$f\,(u, v) = K \qquad (2)$$

may be used as boundaries of the surfaces.

If the function $\overline{F}$ is a bicubic then the surface generated is a bicubic surface and the constant parameter lines are cubic lines. If this surface is bounded by four constant parameter lines:

$$u = u_1, u = u_2, v = v_1, v = v_2 \qquad (3)$$

The resulting bounded surface is often called a bicubic patch.

In POLYSURF the function F is a piecewise bicubic so that the surface generated in a piecewise bicubic or patched surface and the constant parameter lines (patch boundaries) are piecewise cubics.

If there are m+1 discontinuities in the piecewise bicubic in u at u_1, u_1, u_2..... u_m, and n+1 discontinuities in v at v_o, v_1.........v_n then the surface generated will be a rectangular grid of m by n bicubic patches, where the internal patch boundaries are given by u_1, v_1 etc., and the boundaries of the entire surface by u_o, u_m, v_o, v_n.

The surfaces may be bounded by linear functions from (2) to produce trim lines:-

$$a\ u + b\ v = K \qquad (4)$$

The trim lines may be used to define complicated surface boundaries or for defining cut outs.

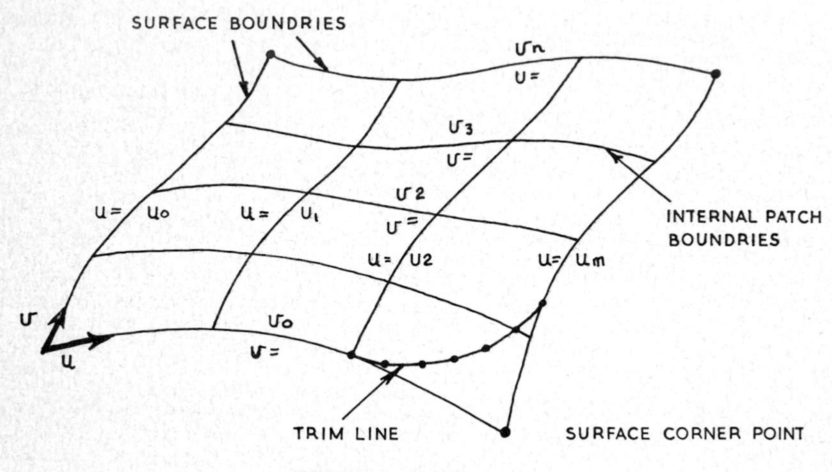

POLYSURF SURFACE

Fig. 3

The bounded patched surfaces defined above have several properties that make them particularly well suited for defining physical objects with 3D curved faces:-

(a) The accuracy to which a physical surface can be defined by means of patched surfaces depends upon the patch density (number of patches used). Consequently as the design is progressively refined, a more accurate definition can be produced, by increasing the patch density. What is more important is that for many engineering objects, sufficient accuracy can be obtained with acceptably low patch densities.

(b) A patched surface may either be continuous in slope everywhere, or have discontinuities of slope at any patch boundary. If all the internal patch boundaries are constrained to be continuous in curvature, the surfaces can also be continuous in curvature everywhere.

(c) A surface may be altered in such a way that the change is localised to a few patches.

(d) Since the surface definition is explicit, surface derivatives are easily calculated and a good basis for interrogation algorithms exists.

(e) With the bounding functions available (2) there is no restriction on the shapes of the surface boundaries. These bounding functions may be intersection lines between two patched surfaces or between a patched surface and any other surface.

Patched surfaces have proved to be a powerful form of mathematical surface definition for a production process of type 'B'. The POLYSURF routines (all of which rely on this surface definition) are described in the remaining sections.

4. THE FITTING PACKAGE - FITTING PATCHED SURFACES TO OBJECTS

Many surface design problems can be tackled by using a fitting routine to fit preliminary mathematical surfaces to input data. This can be followed by various interactive routines to allow the designer to gradually refine these surfaces.

In POLYSURF, the preliminary fitting routine is a least square fit to data points located whever the designer chooses on the object to be designed. To use this facility, the designer must first conceptually divide the object up into a number of surfaces. These surfaces may be of any size or complexity, but at this stage they must be treated as being continuous in slope. The surfaces should preferably have four continuous boundaries (although one or two may be degenerate) but if there are more they can be defined by using trim lines. The user must provide suitable input data for each such surface. The minimum data necessary are the co-ordinates of four surface corners. (In which case straight line boundaries and a linear surface would be fitted between them). Additional data points which may be provided are edge points, trim line points and interior points. The edge points must lie on the four-surface boundary lines, the trim line points must lie on trim lines and be given in the correct order, whilst the interior points may be located anywhere on the surface. At any of the specified data points a unit surface normal or a unit first derivative may be given. (Theoretically any order derivative is acceptable).

All of this data is placed in the data bank of ATLAS disc where it may be accessed by any POLYSURF package.

Typical data might be:

```
PTS 5                                    (name of data)
KA x, y, z, KB x, y, z                   (co-ordinates of 4
KC x, y, y, KD x, y, y                    corner points)

A x, y, y, A x, y, y etc.                (points on edge A)
D x, y, y, D x, y, y etc.                (points on edge D)

Tl x, y, y, Tl x, y, y etc.              (points on trim line No. 1)

P x, y, y, P x, y, y etc.                (any interior points)
P x, y, y N x, y, y                      (interior points with
P x, y, y, N x, y, y                      unit normal)
```

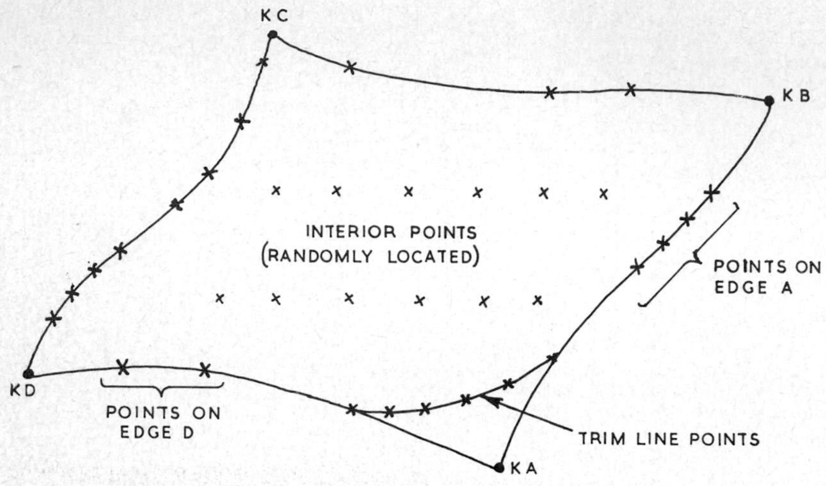

Fig. 4.

All the data supplied may then be fitted as closely as possible, to produce a preliminary surface design. The number of patches in the patched surface may be controlled by the user and may be progressively increased so as to produce the accuracy required. Suitable instruction for doing this might be:

```
(Load the fitting package)
Select PTS 5                 (suppose data is in points file No. 5)
Fit as 10 4 4 PTS 5          (this produces surface No. 10 of size
                              4 by 4 patches)
File SUR 10                  (puts the surface back into the data
                              bank
Set big plotter              (selects 30" plotter)
Plot SUR 10                  (plots the surface)
Plot PTS 5                   (plots the data points)
End
```

Mathematically, any of the constraints mentioned above is of the form:-

$$f_i \text{ (x and derivatives, y and derivatives) } \begin{array}{l} u = u_i \\ v = v_i \end{array} = K_i \quad (5)$$

which may be used as a constraint on equation (1). It is important to note that in equation (5) the parameter values u_i, v_i must be fixed. This is done in a limited way when the user specifies that some of the data points lie on boundary lines, since one parametric value will then be known. The remaining parametric values are iteratively calculated by the fitting program.

The fitting is done by a least square error fitting program, which fits surfaces of form (1) as accurately as possible to the constraints (5). The error function generally employed for fitting surfaces or curves to data points is:

the total error $E_i = F(u_i, v_i) - K_i$ (6)

in POLYSURF this error function has been modified to become:

the perpendicular error $E_i \cdot N_i = (F(u_i, v_i) - K_i) \cdot N_i$ (7)

where N_i is the unit surface normal at the ith point.

The choice of which of these two error functions to use depends partly on whether or not the designer wishes to input unit surface derivatives to the fitting algorithm. If he does, then error function (7) is much quicker at iterating good parametric values(u_i v_i for the ith data point). All these unknown parametric values are initially assigned by a calculation based on the distance the data points are from the four surface corners. A least square fit using error function (7) will usually produce a good parametric distribution within two or three iterations. This is substantially less than is often required with error function (6). Also the surface finally produced by (7) is usually a more accurate fit to the data than can be obtained by (6). However, the advantage of speed is lost if no unit derivatives are specified and the choice between the two error functions then becomes marginal.

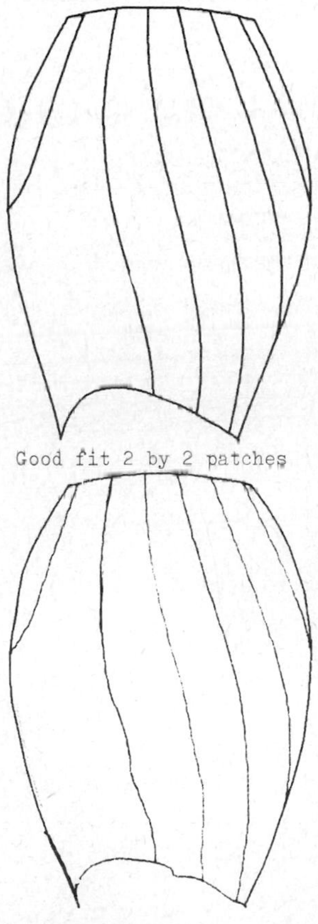

Good fit 2 by 2 patches

Instability 4 by 4 patches

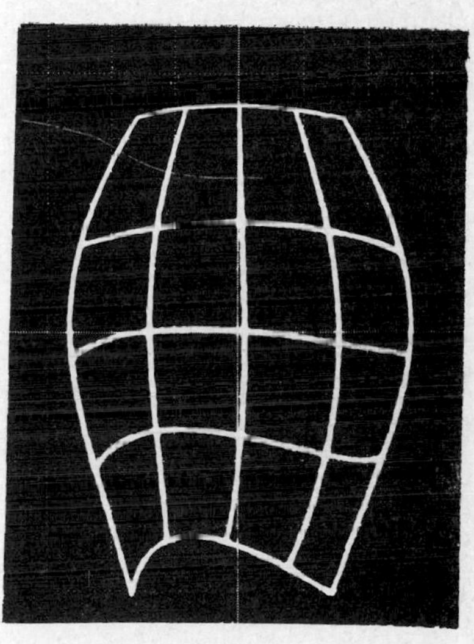

4 by 4 patches after removing errors by surface adaption

Fig. 5.

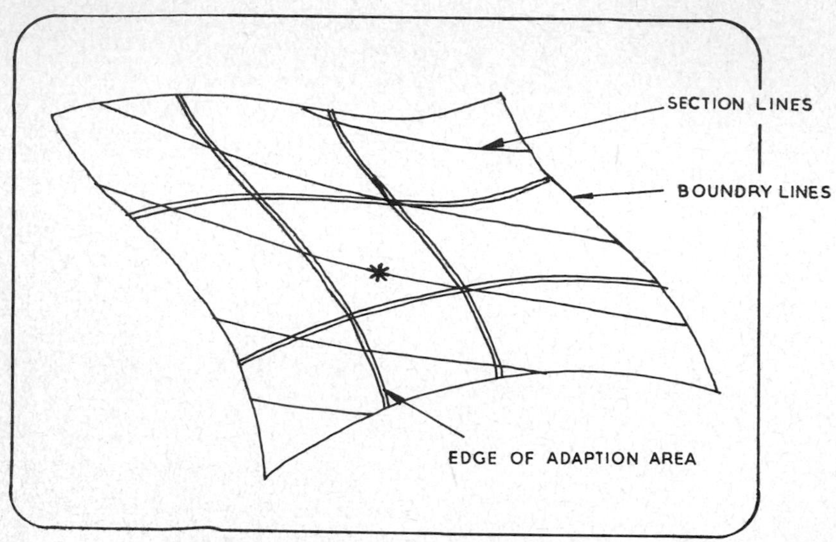

ADAPTION POINT (✱) AND ADAPTION AREA (SHEWN ═)

THE 340 DISPLAY WITH TELETYPE AND LIGHT PEN

Figs. 6.

The user can specify the adaption point anywhere on the surface by either a light pen 'hit' or by a teletype instruction. The adaption area is fixed by the program so that once an adaption point is chosen the adaption area is automatically displayed. For a given adaption point the user can alter the adaption area by increasing the patch density. Once the adaption point and area are fixed, movements of the adaption point may be specified by teletype instruction. This has the effect of dragging the surface within the adaption area so that the adaption point moves by the specified amount. Each such adaption takes about 1 second.

After several adaptions a new Calcomp drawing may be produced to show with greater resolution the changes made. If any relevent analysis programs were available they could also verify the new design. Suitable instructions for adapting a surface might be:

(Load the adaption package)	
Sel SUR 10	(the one produced by fitting in section 3 picture of surface appears on display)
Sey	(10 sections displayed parallel to XZ plane)
Scale 50	(scale 100 is full size)
Obx 200	(shift origin 2 inches)
(Make a pen 'hit' on surface)	(adaption point and area displayed)
Inx 100	(adaption point moved dragging surface 1 inch with it)
File as 11 SUR 10	(file adapted surface as surface 11)
Plot SUR 11	(plot adapted surface)
End	

The mathematics underlying this method is fairly simple. Equation (1) can be rewritten as

$$\overline{P} = \sum_{i=1}^{i=n} A_i \overline{F}_i (u, v) \qquad (8)$$

where A_i are n independent surface variables. Many surface design systems (including NMG (1) and MULTIPATCH (2)) allow the user to alter the variables A_i, which might typically be patch corner points or slopes. In POLYSURF, a suitable combination of all these surface variables is altered so as to produce the dragging effect described above. It was discovered after experimentation that this dragging effect did not always produce the required changes and facilities to alter patch corners, slopes and cross derivatives, would be a useful complement to the dragging facility.

Although sometimes necessary, to include a facility that allows the user to alter surface variables (such as patch corners) directly, ran against one of the principle aims of the system, which was to avoid the user needing to understand patch geometry, or even in fact realising that the surfaces he was fitting and adapting were composed of patches. This aim was embodied in other routines, by the use of randomly located data points for fitting instead of data points tied to patch boundaries, and the use of section lines rather than constant parameter lines in pictures. Sufficient experience has not been gained to discover if a designer without knowledge of surface patches could use the system as intended. However, the idea of trying to present an easy to use interface to the designer still seems sensible.

The propeller blade shown above is a simple example of a surface fitted by this process. About 200 data point constraints were fitted by a grid of 4 patches to an accuracy of better than .1%. When 16 patches were used the number of surface variables was greater than the number of data points and as a result an instability occurred, the result of which can be seen.

The inherent instability problem is one reason that fitting is only used as a preliminary design tool. However, the use of such a tool can often reduce the amount of subsequent interactive design time and it provides the valuable function of producing an initial patch layout. The experience gained with POLYSURF was that no matter how much effort was put into the batch mode fitting programs they could not always produce successful designs. User intervention of some sort is necessary. An attempt is made to provide this in the adaption routine.

5. THE ADAPTION PACKAGE - ADAPTION OF PATCHED SURFACES

For successful surface adaptions, two facilities must be made available to the designer. First a good method of visualising the surfaces and second a method of adapting the surfaces that is easy to use and understand.

Good visualisation is helped by the following features of the program:

(a) Large accurate drawings are available from the 30" Calcomp plotter to complement the pictures available on the PDP display (see Fig. 1).

(b) The pictures on the display can be scaled, rotated and shifted so that particular areas of the surfaces can be examined in detail.

(c) Both the plotter drawings and the display pictures have section lines drawn in addition to patch boundaries. It has been found in practice that it is easier to appreciate the shapes of the surfaces from section lines than from constant parameter lines. The set of sections parallel to the display screen are accurately maintained by PDP7 software during the adaption process described below.

The method of adapting the surfaces has been made as simple as possible so that a designer can use the system without needing to understand the mathematics involved.

The area of the surface to be adapted is defined in two ways:

(a) the adaption area. This area is shown on the display, bounded by double lines, and is the area affected by the adaption. Outside this area no changes occur at all.

(b) the adaption point. This is the point inside the adaption area which undergoes the greatest movement. Between the adaption point and the boundaries of the adaption area the amount of movement varies smoothly from a maximum to zero.

6. OFFSETTING PACKAGE - INPUT OF MANUFACTURING CONSTRAINTS

Since the mathematical surface definitions that are produced by the fitting and subsequent refinement procedures are used directly by the manufacturing package, they must also meet manufacturing constraints. These constraints will depend on the method of manufacture used, e.g. one important manufacturing constraint particularly relevent for n/c surface milling is the surface curvature constraint. The POLYSURF offsetting routine enables curvature to be checked and if necessary altered before the cutting routines are entered. The curvature condition which must not be violated is that when a cutter of radius r is used for milling, then the radius of curvature must never be less than this in concave regions. If this condition is violated then undercutting or gouging will occur. Although routines have been developed elsewhere for testing this condition mathematically they are usually expensive to use. The alternative adopted here is to display offset surfaces (toolpath surface) on the display. The offset surface, parallel to the design surface and distance r from it, is the locus of the centre of a ball ended cutter of radius r. If this offset surface has loops in it then gouging will occur.

Offset surfaces can be shown on the PDP display and examined for loops or lack of smoothness. Any trouble with moving the cutter centre along lines on this surface may then be anticipated, by either choosing a smaller cutter (and hence smaller offset), or by adapting the surface as previously described until the regions of high curvature are removed

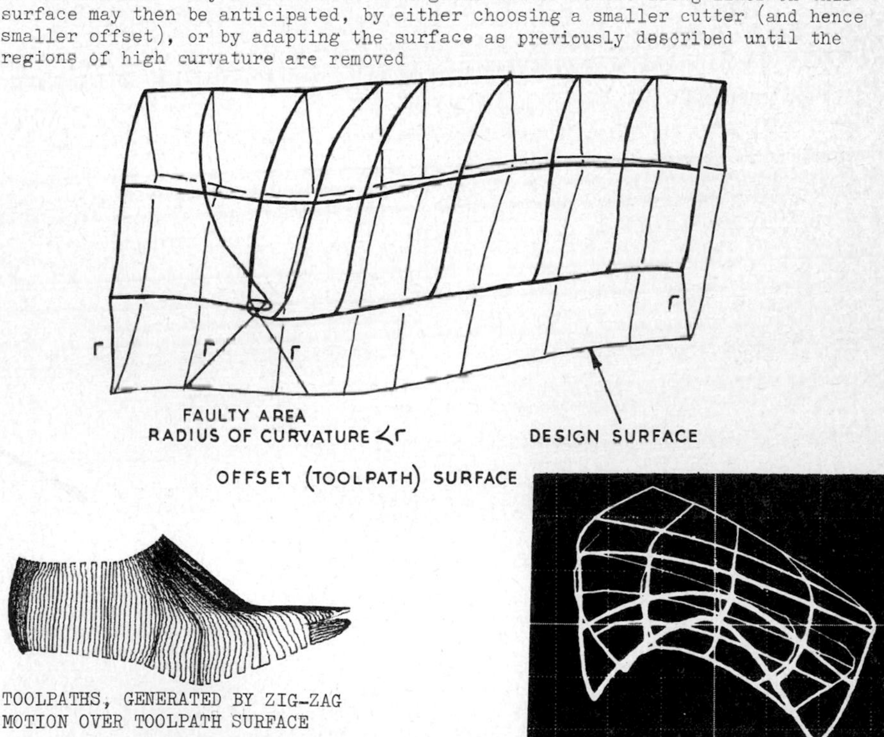

Figs. 7.

Suitable instructions for producing this offset surface might be:

```
(Lead offsetting package)
Sel SUR 11                          (Picture appears on display)
                                    (offset 1.5 inches to produce
Offset as 111 SUR 11 1.5            (surface 111. Both surfaces now
File SUR 111                         displayed.  Either may then be
Set big plotter                      adapted by adaption package)
Plot SUR 11 111
End
```

There are other manufacturing constraints which could be treated graphically in the same way and corrected at the design stage; such as the checking of draw angles on moulds and the checking of cutter clearances on awkward shaped objects.

7. ANALYSIS ROUTINES

Analysis routines could conveniently be used to analyse objects designed by the procedure described. Stress analysis of patched surfaces using triangular finite elements has been developed by Gill (3) and routines for calculating the volumes of objects enclosed by patched surfaces and planes exist in NMG (1). As stated earlier the more analysis routines which can be made available to interface directly to the object description (in this case in terms of bounded surface geometry) The more attractive it becomes to use such an object description.

8. THE MANUFACTURING PACKAGE

Several different methods of manufacture can be based on the mathematical definition described. For example full size section drawings could be produced to act as templates, or n/c tapes could be produced for surface milling.

A POLYSURF routine is available for producing n/c tapes for surface milling. The geometrical information is obtained from the object description and a few additional pieces of information are supplied in a part program. The program contains information as to which surfaces are to be cut and in what order, the size of the cutter, and the mounting position of the billet of material.

The toolpatchs are produced by offsetting the design surfaces by the radius of the cutter to produce toolpath surfaces. The cutting algorithm then produces a zig-zag toolpath by tracing to and fro over each surface specified. The processor and post-processor are integrated so that the output from this in an n/c tape. The metal shoe-last shown below was milled using the POLYSURF manufacturing routines.

The manufacturing routines could easily be extended to allow for a variety of cutters, but at present a ball-ended cutter is assumed. In cutting the shce last two roughing cuts were taken with a square ended cutter before the final cut was taken with a $\frac{1}{2}$" radius ball end. To ensure that a good surface finish is produced and that the cutter is not overloaded during roughing the user can preset the maximum feed rates in traversing and plunging.

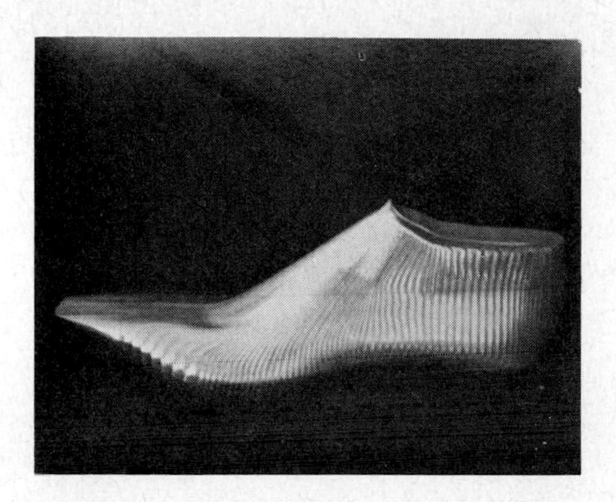

FIG. 8 SHOE LAST CUT FROM ALUMINIUM BILLET

Suitable instructions for cutting one surface on the shoe last (there are 3 altogether) might be:

```
(Load offsetting package)
Sel SUR 4                    (suppose this is one side of last)
Offset as 104 SUR4 0.5       (offset 0.5 inches)
File SUR 104

(Load cutting package)
Set Ferranti                 (post processing done for Ferranti
Sel SUR 104                   control system)
Mount 4                      (6 mounting positions are allowed)
Start 2                      (cutter starts cutting at corner 2)
Parallel 20                  (20 parallel cuts in zig-zag toolpath)
End
```

A suitable n/c tape will then be produced.

Note. This is a somewhat simplified description. Also the version of the n/c package currently available only accepts instructions in fixed format.

ACKNOWLEDGMENT: My thanks to the many people at the University and the CAD Centre who have helped to bring this work to fruition.

REFERENCES

1. SABIN, M. 'An Existing System in the Aircraft Industry. The British Aircraft Corporation Numerical Master Geometry System'. Proc. Roy. Soc. Lond. A, Vol. 321 (1971) pp 197-205.

2. ARMIT, A.P. 'A Multipatch Design System for Coons' Patches'. IEE Conference Prbbiation Number 51 (April 1969) pp 152-161.

3. GILL, J. Computer Aided Design of Shell Structures using the Finite Element Method, Cambridge University Ph.D. Thesis (1972).

UNISURF SYSTEM

Principles, Programme, Language

Pierre E. BEZIER

Professeur au Conservatoire National des Arts et Métiers, Paris

Directeur à la Régie Nationale des Usines Renault
Billancourt, France

ABSTRACT : UNISURF is aimed at computer-assisted conception of sculptured surfaces. It is also used for numerical definition of shapes previously defined, whether they are the result of experiments based on physical measurements, or they materialize purely subjective aesthetic conceptions.

1. INTRODUCTION

1.1. Existing languages

During the Prolamat meeting held in Rome in 1969, our colleague, Mr Mangold, pointed that 105 languages, at least, were in actual existence, and we may safely assume that zero-growth has not been reached yet in this matter.

The high quantity of solutions met is proof of a redundancy of effort. It is to be regretted indeed, but the Science and Technics History gives so many examples of such facts that we cannot help thinking it is inevitable in spite of the progress made in the field of knowledge transmission and exchange.

The variety of solutions met is probably due to the fact that the problem concerned is actually manifold. No overall solution is likely to appear, since its cost would be unacceptable. This is why so many methods are proposed.

1.2. UNISURF basic features

While elaborating the hundred and sixth system, UNISURF to name it, we had in mind a device to conceive directly sculptured shapes and, secondly, to translate existing objects.

Such system has to comply with some requirements :

a) be easily handled by operators (designers or stylists) whose knowledge of mathematics would be limited to basic geometry

b) materialize rapidly any shape defined with numbers, delays being seconds for curves, minutes for surfaces

   In fact, the complete definition of a shape can only be reached by trial and error, and an interactive method is a must.

c) adapt itself to basic descriptive geometry operations such as scaling, translations, rotations, plane sections, intersections, and so on ...

d) have a reasonable cost, for purchase as well as for maintenance.

Our system is based upon the existence of :

- a simple geometrical definition of curves and surfaces
- high performance drawing and milling machines
- direct computer control
- a programme and its relevant language

COMPUTER LANGUAGES FOR NUMERICAL CONTROL, *J. Hatvany, editor*
North-Holland Publishing Company - Amsterdam–London

2. MATHEMATICAL SOLUTION

Curves and surfaces are expressed by means of parameters. The advantage of this method is widely known and, therefore, our choice is self-explanatory.

2.1. Curves

2.1.1. Classical expression

The expression

$$\bar{P}(u) = \sum_{0}^{m} \bar{b}_i . u^i \qquad (u \in [0,1]) \qquad (1)$$

is used to compute the coordinates of odd points located on a curve, and deduct from them the trajectory of the tracer (pen or tool).

2.1.2. Characteristic polygon

Expression (1) is equivalent to

$$\bar{P}(u) = \sum_{0}^{m} \bar{a}_i . \frac{(-u)^i}{(i-1)!} . \frac{d^{i-1}\left[\frac{(1-u)^m - 1}{u}\right]}{du^{i-1}} \qquad (2)$$

In this case, vectors a_i, placed end to end in their index order, build up a polygon the shape of which can easily be related to the curve (Fig. 1). This is a great asset.

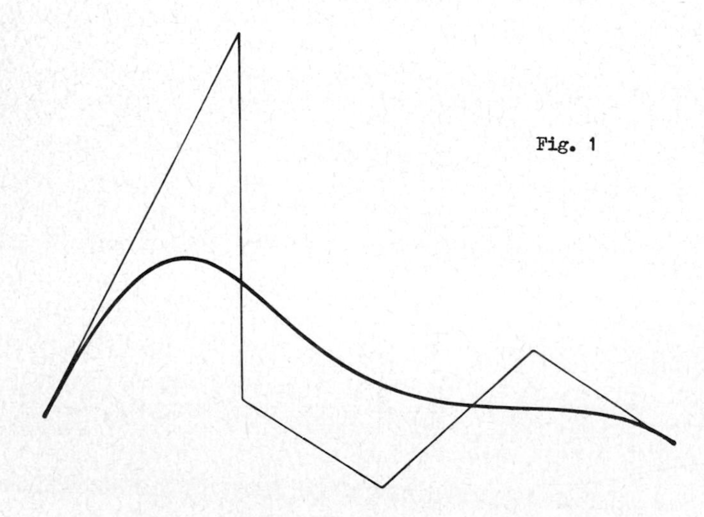

Fig. 1

Moreover, the boundary conditions, corresponding to the values 0 and 1 of the parameter, can be easily deducted : tangents are colinear with vectors a_1 and a_n and curvatures only depend respectively on a_1 and a_2, and on a_{n-1} and a_n.

The matrix used to deduct vectors b_i from vectors a_i is expressed by :

$$\left\| M_{ab_m} \right\| = \left\| (-1)^{l+c} \cdot C_{l-2}^{l-c} \cdot C_m^{l-1} \right\| \tag{3}$$

l and c being respectively the line and the column to which the general term belongs.

2.1.3. Alteration

To alter the shape of a curve, the operator shifts one or several apexes of the characteristic polygon. He rapidly knows by intuition the relation between the displacement of one apex and the correlative alteration of the curve.

We can also use a more scientific approach : supposing s_i represents the vector having for origin that of the set of reference and for extremity Apex A_i of the characteristic polygon, any point is expressed by

$$\bar{P}(u) = \sum_0^m \bar{s}_i \left[\frac{n!}{i!(n-i)!} \times u^i \, (1-u)^{n-i} \right] \tag{4}$$

The scalars are obviously Bernstein's functions (Fig. 2).

Fig. 2

2.1.4. Order increase

If the designer cannot obtain exactly the shape he wants by alteration of the polygon apexes, he must increase the number of legs of the polygon to get a better approximation. If this number is raised from "m" to "m+p", the new apexes are determined as follows :

$$\left\| M_{a\,b_{m+p}} \right\|^{-1} \times \left. \begin{Bmatrix} \bar{b}_i \\ 0 \end{Bmatrix} \begin{matrix} m \\ p \end{matrix} \right. = \left\{ \bar{a}'_i \right\}_{m+p} \tag{5}$$

2.1.5. Changing limits

The programme is set with a parameter varying from 0 to 1. We can use an other parameter w linked to u by such a special relation that when w varies from 0 to 1, parameter u will vary between two odd values u_0 and u_1.

Theoretically, the function connecting u and w has no other condition to fulfill, but practically the relation between u and w must be a bijection in

$$u \in [u_0, u_1] \ , \quad w \in [0,1]$$

If (u_0 and u_1) $\in [0,1]$, the initial curve is cut down

If (u_0 or u_1) $\notin [0,1]$, the initial curve is lengthened

If function $u = u_0 + (u_1 - u_0)\, w$ is selected, the parameter w has the same order as u. Although it is not compulsory, it is much better to comply with this condition.

A matrix translates the b_{ij} related with u into β_{ij} related with w.

2.2. Surfaces

2.2.1. Classical expression

The coordinates of an odd point is given, in relation with the two parameters u and v, by the expression

$$\vec{P}(u,v) = \sum_{\substack{i=0 \\ j=0}}^{\substack{i=m \\ j=n}} \vec{b}_{ij} \cdot u^i v^j \qquad (u,v) \in [0,1] \tag{6}$$

2.2.2. Characteristic mesh

Considering vectors b_{ij} do not help to imagine the shape of the surface defined this way, we therefore have to use, as for curves, an expression which would give the operator an intuitive knowledge of shapes.

By similarity with the curve characteristic polygons, a characteristic mesh defines a biparametric patch. Such patch is generated by a curve progressing in space and changing shape at the same time. The apexes of the boundary polygon move along the curves defined by the transverse polygons of the mesh.

The mathematic expression

$$\vec{P}(u,v) = \Phi\,(\vec{a}_{ij})$$

related with vectors a_{ij} and functions of formula (2) does not raise any interest because it is too complex.

2.2.3. Alteration

If $\mathcal{B}$ represents Bernstein's functions used in formula (4), we can express the position of an odd point of a biparametric patch as follows :

$$\bar{P}(u,v) = \sum_{\substack{i=0\\j=0}}^{\substack{m=0\\n=0}} \bar{s}_{ij} \cdot \mathcal{B}_{\substack{i\\m}}(u) \cdot \mathcal{B}_{\substack{j\\n}}(v) \tag{7}$$

The knowledge of functions $\mathcal{B}$ properties helps to alter rapidly the shape of a patch, but a little experience easily makes up for it.

2.2.4. Order increase

To change a (m×n) order network into an equivalent, but higher (m+p)(n+q) order, network the following operation is performed :

$$\left\| M^{a\,b}_{(m+p),(n+q)} \right\|^{-1} \times \left\| \begin{matrix} \|\bar{b}_{ij}\| & \|0\| \\ \|0\| & \|0\| \end{matrix} \right\|_{\substack{m\\p}}^{\;} = \left\| \bar{a}'_{ij} \right\|_{m+p} \quad (n,\ q,\ n+q) \tag{8}$$

2.2.5. Transposant

Any relation $\psi(u,v) = 0$ represents a curve on a patch.

If u and v are expressed by means of a parameter w

$$u = \sum_{0}^{p} c_i w^i \qquad\qquad v = \sum_{0}^{p} d_i w^i \tag{9}$$

the odd point of the curve defined this way is given by :

$$\bar{P}(w) = \sum_{\substack{i=0\\j=0}}^{\substack{n=m\\j=n}} \bar{b}_{ij} \left(\sum_{0}^{p} c_k w^k\right)^i \left(\sum_{0}^{p} d_l w^l\right)^j \tag{10}$$

which becomes, after simplification

$$\bar{P}(w) = \sum_{0}^{(m+n)p} \bar{e}_i w^i \tag{11}$$

It is a (m+n).p order curve. This expression, even if (m+n).p is large, is simpler to use than the intersection of the surface P(u,v) with a projection cylinder.

2.2.6. Sectioning

A patch being cut down by a transposant line (Fig. 3), we can work out the equation of the fraction included between this line and certain boundaries of the patch.

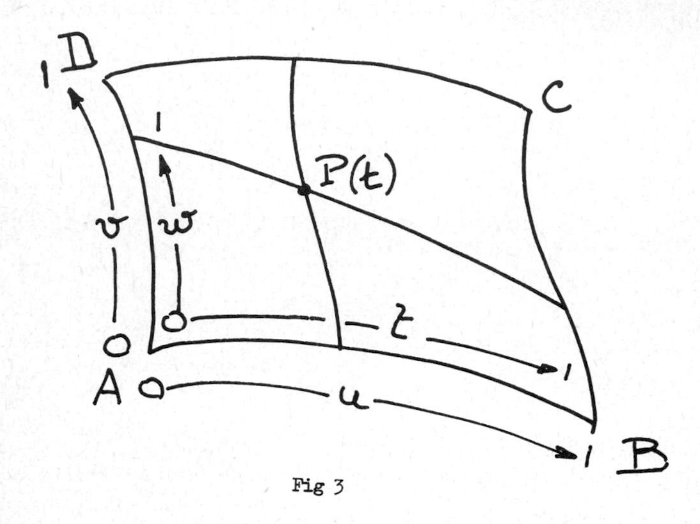

Fig 3

Supposing it is defined by relation (9), and it meets once all generatrices $P(u, c^{te})$ (Fig. 3), the patch is expressed by

$$\bar{P}(v,w) = \sum_{\substack{i=0 \\ j=0}}^{\substack{i=m \\ j=n}} \bar{b}_{ij} \left(\sum_0^p c_k w^k\right)^i \cdot \left[\left(\sum_0^p d_l w^l\right) v\right]^j \qquad (12)$$

functions u(w) and v(w) being of p order, the patch (12) belongs to the n × (m+n)p order.

If the transposant is isoparametric, it is expressed by

$$u = w \qquad\qquad v = v_0$$

and relation (12) finds itself more simple.

2.3. Blending

2.3.1. Curves

Two vectors having one common end are tangent if the last vector of the characteristic polygon of one curve is colinear with the first vector of the characteristic polygon of the other curve (Fig. 4).

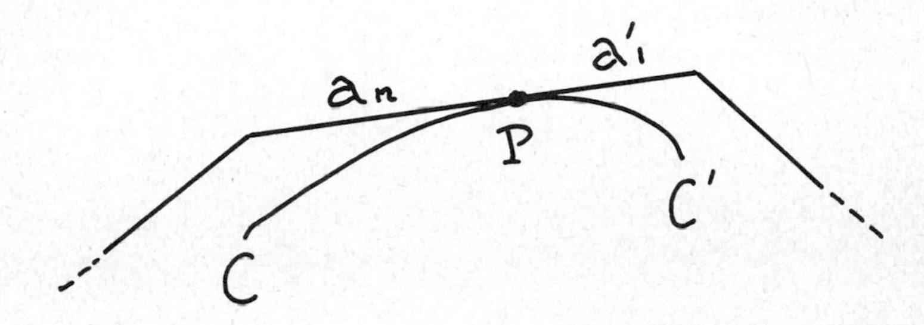

Fig. 4

If the two curves are to be osculatory, that means to have the same curvature center at the contact point, the condition to fulfill is expressed by the relations :

$$\bar{a}'_1 = g.\bar{a}_n \qquad (g > 0)$$
$$\bar{a}'_2 = h\bar{a}_n + k\bar{a}_{n-1} \qquad (k < 0) \qquad (13)$$
$$(g)^2 = k\,\frac{n}{n'} \cdot \frac{n'-1}{n-1}$$

in which

a'_1, a'_2 are respectively the first and the second vector of the polygon of one curve

a_n and a_{n-1} the last and last but one vectors of the polygon of the other curve

n and n' the number of legs of the polygons

g, h and k arbitrary scalars (Fig. 5)

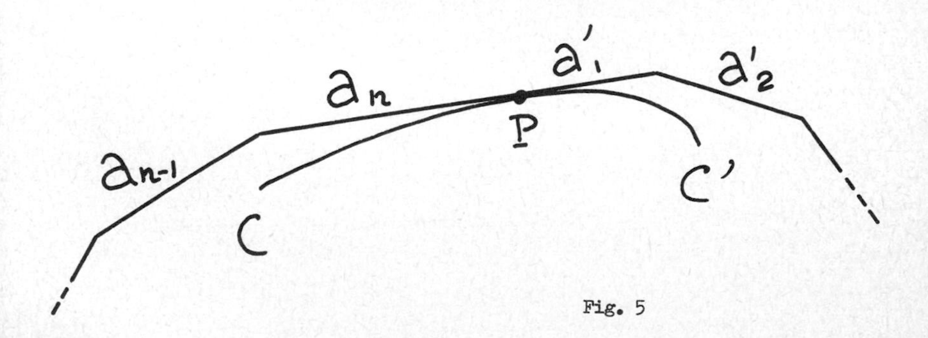

Fig. 5

2.3.2. Patches

2.3.2.1. Patches tangency

When two patches have a common boundary, P(u,0) their tangency is expressed by the general condition :

$$\left[\frac{\partial \bar{P}}{\partial v}(u,0) \wedge \frac{\partial \bar{P}}{\partial u}(u,0)\right] \times \frac{\partial \bar{P}'}{\partial v}(u,0) \equiv 0 \qquad (14)$$

This formula is not easy to handle or compute. It is best replaced by

$$\frac{\partial \bar{P}'}{\partial v}(u,0) = h(u)\frac{\partial \bar{P}}{\partial v}(u,0) + k(u)\frac{\partial \bar{P}}{\partial u}(u,0) \qquad (15)$$

in which h and k are arbitrary functions of u.

If h and k are respectively p and q order polynomials, and if S is of mth order respective to u, patch S' is generally, respective to u, of an equal order to the greater of the two values (m+p) or (m+q-1).

2.3.2.2. Tangency of several patches

When several patches are tangent along lines meeting in one common point, the conditions to fulfill involve, overmore the tangents coplanarity, the curvature of the limit boundaries and the twist vector of the patches on their common point.

The solution is rather tedious to find except if the patches number is limited to 3 or 4.

2.3.2.3. Blending by means of ancillary patches

While using the general tangency conditions of parametric surfaces, it is easy to define automatically the patches able to connect several surfaces previously defined and having no common boundaries (Fig. 6).

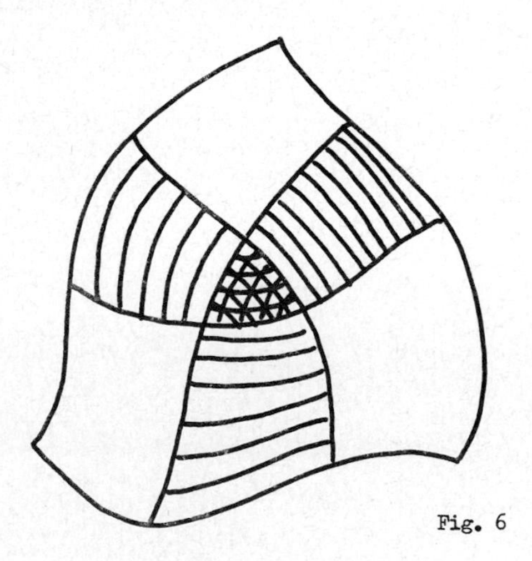

Fig. 6

If they intersect one another, they can be cut down "according to taste" and we are brought back to the preceeding case.

3. DESCRIPTIVE TRANSFORMATIONS

Matrices handle easily any parametric expression of curves and surfaces, to perform descriptive geometry operations such as translations, rotations and scaling.

There is no difficulty in dealing with mirror image and intersections.

4. PROGRAMME

4.1. Possibilities

Our programme has been created in order to

a) allow an operator, stylist or designer, to express his ideas by means of commands giving geometric elements, descriptive operations or motion order.

b) command, on-line, the workings of drawing machines, CRT's, and milling machines.

Moreover, the programme includes several processess to obtain automatically the curves and surfaces definition adequate to such constraints as points (accurate or approximative), slopes, tangencies, and so on ...

The whole programme is compatible with APT and it uses the same words whenever possible.

4.2. Control of the machines

This part of the programme deals with the servo control of the machines motion. It includes, particularly, cutter compensation for ball-end or toroid cutters.

In the case of milling machines, data carrying media are punched tapes. It is the same for pre-determined drafting such as scaling, rotations, perpectives, and so on..

4.3. Interactive language

It is used during creative work, to control drawing machines.

It first defines a certain number of geometric elements (points, curves, patches,...) by means of standard orders directly usable by the drawing programme.

Moreover, such language can express boundary constraints, blendings, etc.. These data are processed by particular modules that will turn them into standard orders. A comment will recall the constraints type.

During interactive work, standard orders are automatically stored into a direct access file. Once the definition is complete, orders are put in a record file, and handled automatically for sorting, storing and retrieving the required orders to start on a new job. All these operations are performed in the backgound.

Of course, descriptive geometry orders are included in the interactive language.

Some complementary orders help to choose the number of generatrices of the patch we want to draw, the line type, the colour ...

The programme also produces the punched tape that will control automatically drawing and milling machines.

4.4. Programme set up

The programme is split into several parts :

a) a monitor for decoding orders and switching them towards the corresponding sub-programmes. It ensures orders dispatching among different machines.

It is expressed in assembler language and processed on-line.

b) a programme to compute coordinates of points located on the trajectory, and carry them to the control programme of the machines.

Sub-programmes about descriptive operations are included in it.

c) a collection of special programmes to solve particular problems such as complex definitions, approximation algorithms, descriptive geometry special operations. Their development will complete the basic programme according to the users' needs.

Some programmes are expressed in FORTRAN and processed in the background. Others are expressed in assembler language and processed on-line.

d) a symbolic language for the preparation of the tape controlling the UNISURF surfaces milling.

Orders are the same as those used by drawing machines operators, and the relevant files are directly used.

This language will be, sooner or later, implemented on small computers, and we forecast the edition of post processor programmes.

5. CONCLUSION

The proper and actual UNISURF language has been developped after several years use of a simplified version. Many advantages appeared as we progressed towards its actual form :

a) Characteristic polygons and meshes are easily deducted from the shapes to obtain.

b) Curves and surfaces of higher orders than 3 can be used, availing a larger diversity of shapes and reducing the number of patches required bo build a complex surface.

c) The size of computers is relatively small

i.e. 16 K (16 bits) to control a milling machine
16 K (16 bits) to control a drawing machine
40 K (16 bits) to control 3 or 4 drawing machines
Archives and auxiliary programmes storage require a 3 M octets disk

d) The length of the tapes is short, avoiding reels changes and consequently relieving tape readers from a heavy burden.

i.e. a complex wing of a car will require about 30 m tape.

e) It is used either for conception or translation work.

INTERACTIVE CURVE FITTING

V. Nicolò and M. Piccini

FIAT - Direzione Sistemi e Informatica - Dipartimento
Calcolo e Tecniche Avanzate - Servizio Ingegneria
del Software

ABSTRACT: Intercative curve fitting is one of the main problems met when trying to fulfill the task of converting the description of the shape of a car body model into a computer compatible form. The solution proposed by Bézier has gained wide attention and acceptance mainly because of the idea of relating the shape of the curve to a polygon that mimics the curve itself. It is noted that the degree to which the polygon mimics the associated curve decreases as the number of points of inflexion, sharp bends or other odd behaviours require higher order polygons.

Some solution proposed by other authors to overcome this problem is briefly discussed and a contribution to the problem is given.

1. INTRODUCTION

The work reported has been developed within a project aiming to the computer aided design and manufacturing of car body dies.

In such a context, in fact, the ability to describe three dimensional shapes, either curves or surfaces, in a computer compatible form is of the greatest relevance. In other words the ability is required to summarize the shapes by few parameters, from which the coordinates and the local differential characteristics of the shape can be easily computed at any point, so to make efficient use of storage and computing time.

The shape to be represented in the computer may either be still in the stylist's mind or exist as a phisical model, - for instance a clay model - . In the first case the interaction between the stylist and the computer is the only choice, but also in the other case the use of interaction may prove profitable.

In the last instance, in fact, the input data to the computer are usually a great number of measured points which may be affected with measuring errors, sequence errors, errors due to the transcription from a memory support to another and - last but not least - imperfections of the model. Moreover: 1) due to the esthetical nature of the car body shapes, the interest is usually placed on the achievement of the cverall shape, - i.e., the one that human eyes put

COMPUTER LANGUAGES FOR NUMERICAL CONTROL, *J. Hatvany, editor*
North-Holland Publishing Company - Amsterdam–London

through the given points - , more than on the actual closeness of the fit to each measured point;

2) the nature of some of the possible errors is not really random.

In this situation, to obtain a satisfactory non-interactive fitting, an analytical criterion should be available through which: the errors should be eliminated, the "unwanted" features of the shape disregarded and the "meaningful" features retained by a minimal number of parameters.

The task of finding such an analytical criterion appears difficult expecially when some degree of generality is required. On the contrary the idea of taking advantage of the human ability to judge the quality of the shape achieved at each iteration of an interactive procedure is very attractive.

In the last case the operator would not only express his satisfaction upon the achieved shape but would go further by instructing the computer towards a better shape.

It is obvious that a good starting approximation and the means given to the operator to lead the computer towards the final shape are of the greatest importance for the efficiency of the procedure.

As far as curves are concerned the procedure based on the algorithm proposed by Bézier has gained wide attention and acceptance. In fact, the Bézier algorithm relates the shape of the curve to a polygon. The curve shares with it the end points and the end slopes, while the polygon itself recalls, exaggeratelу, the shape of the curve. So the operator can modify the shape of the curve manipulating the easy to understand parameters that define the shape of the polygon and is rid of any concern about the analytical form of the curve.

If the results of the modifications are displayed on line, the operator rapidally acquires the feeling of the dependance of the curve on the polygon and becomes able to model the polygon so that the wanted curve is produced after few iterations.

Since the information on the shape of the curve are completely given by the associated polygon, it is trivial that the number of the sides of the polygon is linearly related to the number of parameters that have to be stored to keep the information on the shape of the curve.

It is useful to give to the operator the possibility of increasing or decreasing the number of degrees of freedom at his disposal, when he feels the need of doing that, either because of the complexity of the shape or because he feels embarassed by the variety of choises he can make.

With Bézier curve this can be done either by increasing the number of the polygon sides, without varying the latest shape achieved, or decreasing the number of the sides, with minimal change to the latest shape achieved

Through the use of these possibilities in normal cases the operator is able to obtain a final curve that will also be described by a minimal number of parameters.

The use of this features of Bézier algorithm has led to some successful applications (1968). It has to be noted, anyway, that, - through Bézier choice of the functions that relate the curve shape to the polygon, - the degree, to which the polygon mimics the associated curve, decreases as the number of points of inflexion, sharp bends or other kinds of odd behaviours require higher order polygons.

A number of suggestions have been made by other authors to overcome this problem and are briefly reviewed in the next section.

The authors' approach to the problem is presented in section 3. Section 4. is dedicated to the description of an interactive implementation of the method and to the presentation of some results. Section 5. concludes the paper.

2. SOME EXTENSIONS OF BEZIER ALGORITHM

Bézier (1970) deals with his curves specifying their shapes in terms of the sides of the polygon, taken as vectors. Forrest (1970) has worked out a different analytic form of Bézier's curves specifying their shape in terms of the corner points of the polygon. Since the two forms are equivalent, we will refer to the last one.

Let P_o, P_1,....., P_n be the ordered sequence of the polygon corner points, the associated Bézier curve is of the n-th order and is given by the vector parametric equation:

$$(2.1) \qquad Q_n(t) = \sum_{i=o}^{n} P_i J_{n,i}(t) \qquad t \in [0,1]$$

where $J_{n,i}(t)$ denotes the weight function of the i-th point in a sequence of n.

For the set of J-s, the symmetric behaviour in the range of the parameter, the invariance to transforms and the ability to reduce exactely to the end points, for t=o and t=1, are required.

For Bézier curves $J_{n,i}$ is defined in the following way

$$(2.2) \qquad J_{n,i} = \binom{n}{i} t^i (1-t)^{n-i} \qquad t \in [0,1]$$

it reaches its maximum for t=i/n and is always positive within the range of the parameter, but decreases slowly as the parameter departs from the value i/n. (Fig. 1a)

Some of the consequences of such a behaviour are shown in Fig. 2, where the distance of the polygon from the curve and the presence of loops in the polygon denote poor local control of the shape.

An increase of local control may be achieved through subdivision of the curve in shorter spans, taking care that tangents continuity conditions (required for surface work) are met. The increase of control is paid, in this case, by an increase of complexity in the use of curves made from several spans.

A different approach to the problem leads to the definition of a set of J-s different from that defined by eq. (2.2).

Some authors (Forrest, 1970 ; Gordon, 1972) have observed that Bézier curves and Bernestein approximation are closely related. As matter of fact, Bézier curves may be regarded as Bernestein n-th order approximation to the vector valued polygonal function F(t):

$$(2.3) \qquad F(t) = n\left[\left(\frac{i+1}{n} - t\right) P_i + \left(t - \frac{i}{n}\right) P_{i+1}\right]$$

for $t \in \left[\frac{i}{n}, \frac{i+1}{n}\right]$, where i ranges from 0 to n-1.

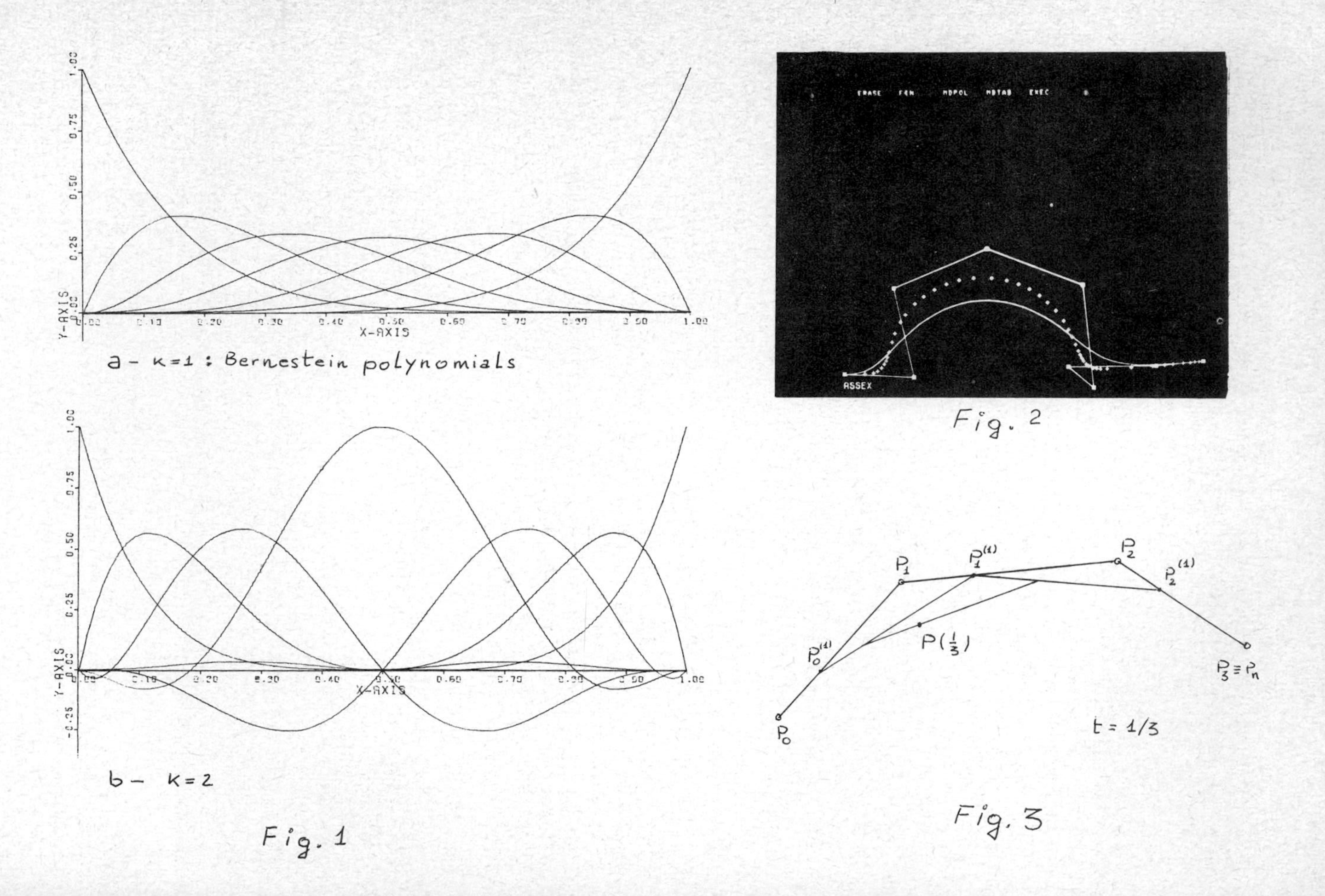
Y-AXIS
X-AXIS
a - k=1 : Bernestein polynomials
b - k=2
Fig. 1
ERASE FEN MDPOL MDTAB EXEC
ASSEX
Fig. 2
P_0
$P_0^{(1)}$
P_1
$P_1^{(1)}$
$P(\frac{1}{3})$
P_2
$P_2^{(1)}$
$P_3 \equiv P_n$
t = 1/3
Fig. 3

Hence, for the properties of Bernestein approximations (Davis, 1963; Gordon 1972) , to obtain curves which better mimic the behaviour of the polygon, one may simply increase the degree of the Bernestein approximation without changing the primitive function, i.e. the polygon. The order of the approximation $m \gg n$ would be set by the operator in order to control the closeness of the curve to the polygon. The drawback to the method is, obviously, the increase of the order of the polynomals to be computed.

Another extension of Bézier algorithm (Gordon, 1972) is based on the observation that functions (2.2) may be regarded as binomial probability density functions. In fact, let t be the probability of occurrence of a given event in each of n Bernoulli trials of an experiment; then, the weight of the i-th corner point in a sequence of n+1 is equal to the probability that the given event occurs i times in n trials.

Gordon and al. (Gordon, 1972) describe a technique that employs conditional probability functions as weights. Losely, it means that the number of trials is supposed to be $m \gg n$, but only n+1 occurrences, suitably chosen among the m+1 possible ones, are considered and denoted by the integers $\nu_o = o, \nu_1, \ldots, \nu_n = m$

The weight of the i-th point in a sequence of n is then given by the conditional probability of the occurrence ν_i.

The conditional probability functions result to be rational of order m, nevertheless the amount of the necessary computation is less than in the previous case. Again m can be left at the operator choise: in the limit as $m \to \infty$ the curve coincides with the polygon.

While the last two techniques are based on the increase of the order of the functions, Gordon (1972) and Forrest (1972) announce very encouraging results with a new technique based on B-splines (spline generalization of Bernestein polynomials). This technique is not yet known to the authors being described in announced papers (Riesenfeld, 1972) and has been here mentioned to underline that the research is still in progress on this subject.

3. THE ALGORITHM

Bézier has provided for his curves an interesting geometric construction which has been sketched in Fig. 3. From the n+1 corner points, n points are derived on the sides of the initial polygon by a proportioning procedure in accordance with a given value of the parameter t. Joining the derived points a new polygon is obtain and the procedure is repeated. After n stages of reduction the point on the curve, corresponding to the given parameter value, is obtained.

The same derived points may be also obtained by linear interpolation between each couple of consecutive corner points. Based on this observation, a generalization of this procedure may be obtained by the use of higher order interpolation schemes, as described hereafter.

Let $P_o, P_1, \ldots, P_n$ be the ordered sequence of the polygon corner points and let $g_{ih}(t)$ denote the i-th function of such a set of functions that:

$$(3.1) \qquad \sum_{i=o}^{h} g_{ih}(t) = 1 \qquad t \in [0,1]$$

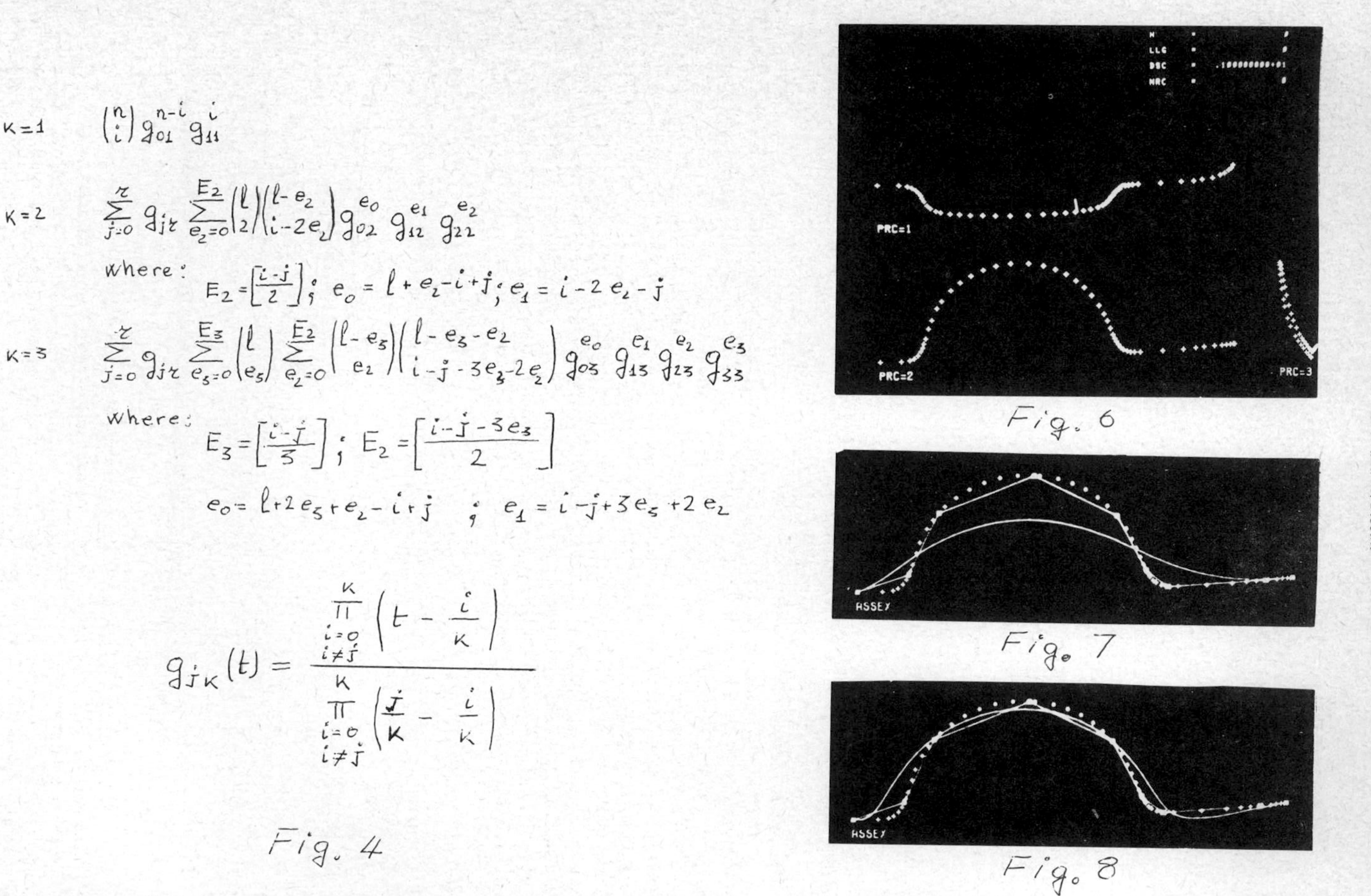

Fig. 4

Fig. 6

Fig. 7

Fig. 8

Moreover, suppose that sets of g-s have been defined for any value of h.

Once a value has been chosen for h, say k, the first stage of reduction is performed by the following basic interpolation scheme:

$$P_i^{(1)} = \sum_{j=o}^{k} g_{jk}(t)\ P_{j+i} \qquad t \in [0,1]$$

where i=o,1,....,n-k.

Suppose that:

(3.3) $n = lk + r$

where l and r are positive intergers and r is smaller than k. After l stages of reduction, one is left with r+1 points and is compelled to change the set of the g-s in order to reduce to one point. That is, the last reduction is performed as follows:

$$(3.4) \qquad Q(t) = \sum_{j=o}^{r} g_{jr}(t)\ P_j^{(1)} \qquad t \in [0,1]$$

Rewriting eq. (3.4) in terms of the n+1 initial points leads to:

$$(3.5) \qquad Q(t) = \sum_{i=o}^{n} P_i\ J_{n,i}^{(k)}(t) \qquad t \in [0,1]$$

Where the index k reminds of J-s dependence on the choise of the basic interpolation scheme.

The expansion of the J-s is shown in the Table of Fig.4, for k=1,2 and 3. In the same table is given the g-s form in the special case they are assumed to be Lagrange functions.

The general case and the detailed discussion of the mathematics are dealt with in the referenced bibliography (Nicolò, 1972). Because of the lack of space, only the most meaningful features of this technique will be here reported.

Condition (3.1) assures invariance to transforms. With the choise of Lagrange functions, the polygon and the curve share the end points, but the end tangents depend on the extreme k+1 vertices.

For k=1 the special case of Bézier curves is obtained. For $k > 1$ (see Fig.1b) $J_{n,i}^{(k)}$ decrease more rapidly than Bernestein polynomials as the parameter departs from the value at which the maximum is reached. A drawback is that negative values are, in such case, possible.

When $k > 1$ and n is a multiple of k, besides the end points also other k-1 internal vertices belong to the curve. For k=2 the vertex that belongs to the curve is $P_{n/2}$.

The $J_{n,i}^{(k)}$ are polynomials of degree non-exceeding n, no matter how the value of k is taken. Nevertheless, with evenly spaced corner points, higher values of k lead to curves closer to the polygons.

Unwanted behaviours due to uneven spacing of the points may occur for higher values of k.

4. THE INTERACTIVE PROGRAM

An interactive curve fitting program, that uses the algorithm of previous section, has been implemented in the evironment of the UNIVAC Time-sharing EXEC operative system. The program may be started from a keyboard terminal (an Olivetti Te 300 in our case) specifying the file where the measured points are to be found. These points are

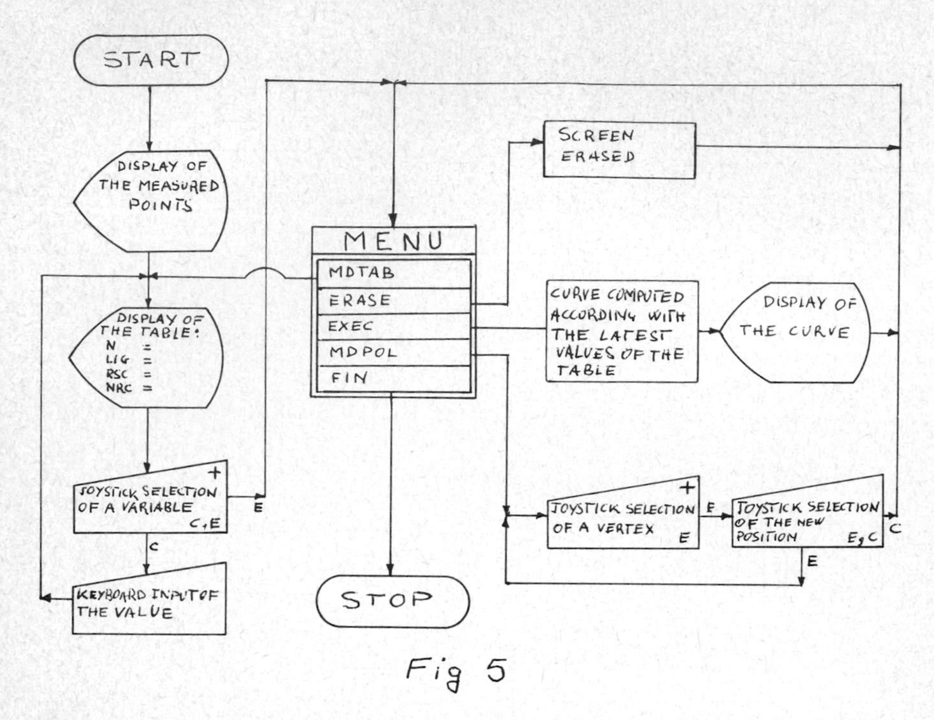

Fig 5

displayed in the three orthogonal views on a storage video display (a TEKTRONIX T 4002 in our case). The interactivity is obtained through a joystick and the TEKTRONIX keyboard.

The flow of the program is given in Fig. 5. After the measured points have been displayed, a table is presented to the operator who is given the possibility to set the values of the listed parameters.

They are: N, which is the number of sides of the polygon, LLG, which has the same meaning as k in the previous section, DSC, which is a scale factor (usually set to 1), and NRC, that is a code for the choise of the projection.

After the assignment phase, the operator may chose the wanted action through the menu. The selection of the menu action is made by the joystick.

The command EXEC causes the computation and the display of the curve, the polygon and the measured points, according to the actual values of the table and the latest choise of the polygon. The first time the position of the vertices is automatically taken with an even spacing among the measured points.

The command MDPOL enters the routine through which the operator is given the possibility of selecting a vertex with the joystick and move it to the next selected position of the joystick cross.

The command MDTAB enters the routine through which the operator is given the possibility of modifying the table. Note that, if the modification of the table concerns the order of the basic interpolating scheme, i.e., LLG, and/or an increase of the number of sides

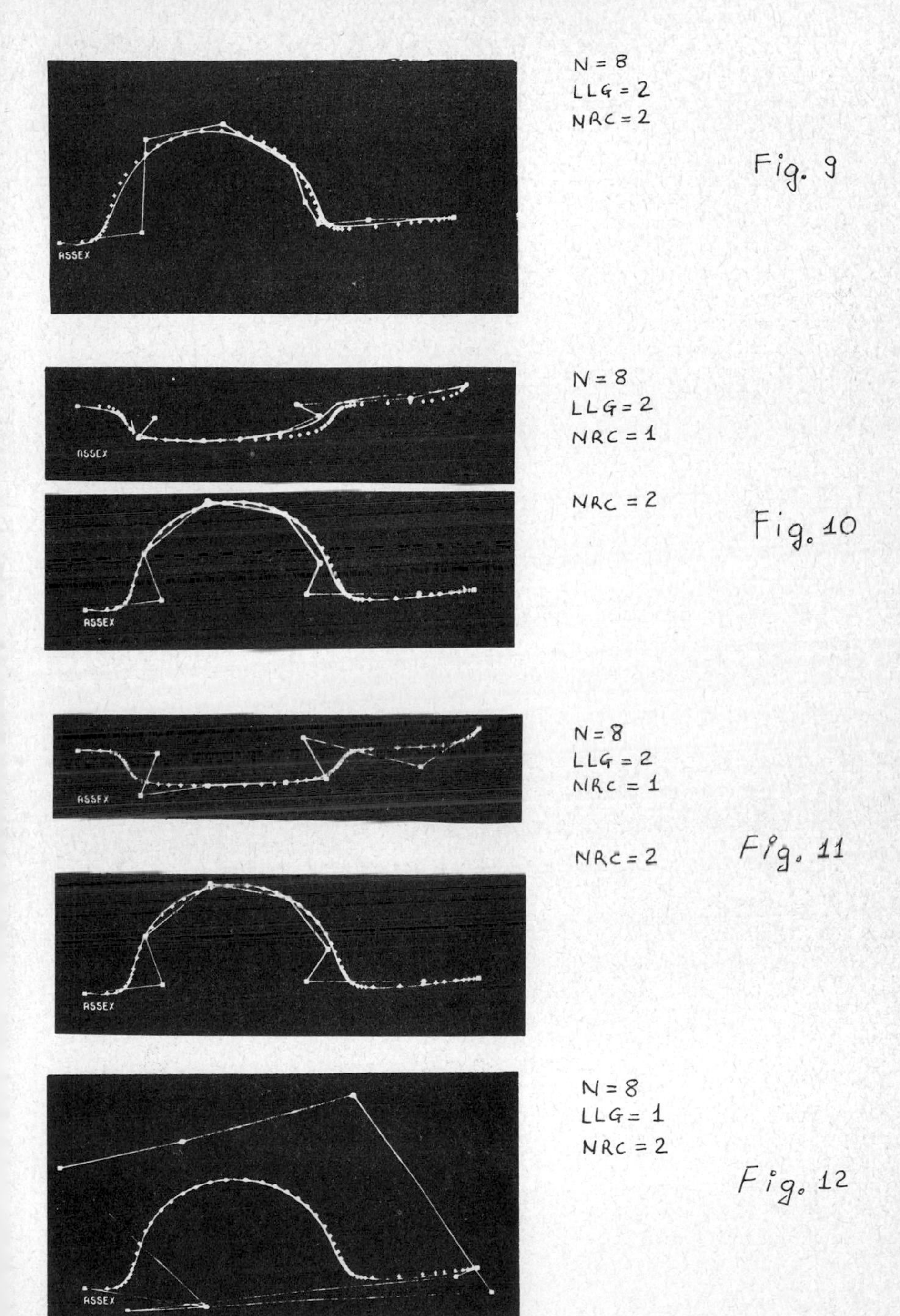

Fig. 9

Fig. 10

Fig. 11

Fig. 12

of the polygon, the shape of the curve remains unchanged after a new EXEC is performed.

The other menu commands are self-explanatory.

This simple structure of the program seems to be quite powerful.

After display of the input points (Fig. 6) the operator is able to decide which projection he will consider first. At this moment he can visually get an idea of the complexity of the curve and can chose the order consequently.

Since a good starting approximation speeds up the procedure he will probably start with a value of LLG equal to 2 or 3. In fact, as shown by comparison between Fig. 7 and 8, for the same initial choise of the polygon vertices among the measured points, with higher LLG the curve resembles much more the wanted shape.

The operator will then continue moving the vertices, increasing (or decreasing) the order of the polygon, displaying different views untill he is satisfied with the results.

Fig. 8,9,10,11 show the sequence of the shapes produced during a trial session on a real car body curve, a wheel base.

As the final step of the interactive fitting the operator will probably display the latest achieved shape with k=1 and obtain the Bézier polygon for the curve in order to check the end tangents and, if it is the case, correct them. This has been done in Fig. 12 where the complexity of Bézier polygon, as compared with the polygon of Fig. 11, is drammatically evident.

At the end of the session the operator may ask for a plotter drawing of the achieved curve.

5. CONCLUDING REMARKS

Following the definition of the problem and of the area where it fits, the nice features of the technique proposed by Bézier have been considered. They are the invariance to transforms, the intuitive results of variations of the polygon on the curve shape, the ease of computation of points and local analytical properties, the possibility of increasing the order of the polygon without changes in the latest achieved shape, etc.

Nevertheless it has been reamrked that situations occur where the need for an even more intuitive behaviour of the curve is felt. After a brief review of the status of the work in this field, a technique has been described that seems able to give the required degree of intuition to the behaviour of the curve, preserving the other desired features of the Bézier technique.

The final curves obtained are one span polynomial curves of limited degree. This situation seems favourable to their use as boundaries of surfaces in the context of a language for N.C. three dimensional countouring. In fact, for tool path generation and interference checking, a great deal of iteractive computation is needed, which gives a greatest value to semplicity in the use.

REFERENCES

Bézier, P., (1968), Procédé de definition numerique des curbes et surfaces non mathématiques; Systéme UNISURF. Automatisme 13, May

Bézier, P., (1968), How Renault uses numerical control for car body design and tooling. Society of Automobile Engineers, Paper SA SAE 680010.

Bézier, P., (1970), Emploi des machines a commande numerique. Massan & Cie. Eyrolles, Paris 1970.

Davis, P.J., (1963), Interpolation and approximation, Ginn-Blaisdell, New York.

Forrest, A.R., (1970), Interpolation and approximation by Bézier polynomials. University of Cambridge. C.A.D. Group Document No. 45 Oct. 1970.

Forrest, A.R., (1972), A new curve form for computer-aided design. University of Cambridge. C.A.D. Group Document No.66, June 1972.

Gordon, W.J. and R.F. Riesenfeld, (1972), Bernestein - Bézier methods for computer-aided design of free-form curves and surfaces. General Motors Research Laboratories, GMR 1176, March 1972.

Nicolò, V. (1972) Una estensione dell'algoritmo di Bézier per il disegno e l'approsimazione iterattiva delle curve. Rapporto PACS, FIAT Servizio Ingegneria del Software, to appear.

Riesenfeld, R.F., (1972), Bernestein-Bézier Methods for the computer-aided design of free-form curves and surfaces. Ph.D. Thesis, Syracuse University, to appear 1972.

HAPT-3D: A PROGRAMMING SYSTEM FOR NUMERICAL CONTROL

YOSHIHIRO HYODO
Production Engineering Research Laboratory of Hitachi Ltd.
Yokohama, JAPAN

Abstract: HAPT-3D is a computer programming system for numerically controlled milling machines. Its main object is to process a surface defined by its contour curves and sectional curves. The language (for part programming) of HAPT-3D is similar to that of APT. The HAPT-3D system is composed of three parts: a translator, an executive processor and a post processor. The processing method of curves and surfaces features this system. All curves and surfaces are expressed as a function of a parameter (curves) or two parameters (surfaces). In spite of the variety of their mathematical expression according to their types, common processing methods are used in the computer program. As this system has automatic cutter path generation routines, a part programmer need not describe detailed cutter motion statements.

1. INTRODUCTION

Among many computer programs for numerical control that have been already developed by many companies, APT is a well-known large scale general purpose program. It is powerful but it is not almighty. For example, even by APT, it is impossible or difficult to define and process such non-mathematical surfaces that are often observed at automobile bodies and certain parts of electrical appliances. Of course, APT can define and process some kinds of non-mathematical surfaces such as tabulated cylinders, and sculptured surfaces (recently), but it is more suitable and powerful when it is used for processing of mathematical surfaces.

HAPT-3D is a computer programming system for numerically controlled milling, whose main object is to process non-mathematical surfaces that are defined by their contour and sectional curves.

2. LANGUAGE FOR PART PROGRAMMING

In order to make clear the object and features of HAPT-3D, the language for part programming will be explained.

There are two types of part programming method: fill-up-the-form type and language type. The former is suitable for field use because it greatly decreases the work of part programmers, but it is not versatile because of the fixed form. On the contrary, the latter makes the part programming cost much labor, but its free format insures the versatility. The language type was adopted for HAPT-3D attaching importance to the versatility. The APT-like language was chosen because APT was greatly used in the world and it was based by the international standard language for part programming. For the common functions to both systems, the same expressions as APT's were adopted, and new expressions were created originally for peculiar functions of HAPT-3D.

The major keywords for the definition of geometric elements of both systems are compared in table 1. The POINT, LINE, CIRCLE and PLANE are common to both. The CURVE and SURF (surface) of HAPT-3D contrast with the ELLIPS (ellipse), HYPERB (hyperbola), GCONIC (general conic), TABCYL (tabulated cylinder), etc. of APT. The unique but vague expressions CURVE and SURF, irrespective of mathematical nature, are used for all kinds of curves and surfaces.

Some peculiar examples are given in the following.

An example of a curve that is composed of several lines and arcs of circle is

COMPUTER LANGUAGES FOR NUMERICAL CONTROL, *J. Hatvany, editor*
North-Holland Publishing Company - Amsterdam–London

Table 1. The comparison of the keywords for geometry

Common	HAPT-3D	APT
POINT LINE PLANE CIRCLE VECTOR MATRIX	CURVE SURF	ELLIPS HYPERB GCONIC LCONIC SPHERE CYLNDR CONE QADRIC TABCYL POLCON RLDSRF PATERN

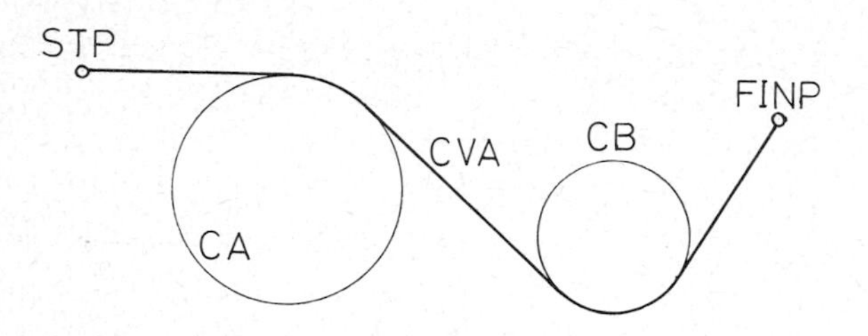

CVA = CURVE/STP, CLW, CA, CCLW, CB, FINP

Fig. 1. An example of a curve that is composed of several lines and arcs of circle

shown in fig. 1. Curves of this kinds are often observed in part geometry. Curve CVA is defined as a curve which starts from point STP, goes to circle CA tangentially and turns along CA clockwise, then goes to circle CB on the line tangential to both CA and CB, turns along CB counterclockwise, and finally goes straight to point FINP.

An example of a space curve defined by its two curves on different projection planes is shown in fig. 2. As a space curve is expressed by two views in a plan, this definition is often used when the input information is got from a plan. In fig. 3, curve CC is defined as a curve expressed by two curves CA and CB on xy-plane and yz-plane respectively.

Examples of the definition of surfaces are shown in fig. 3. Both surfaces SA and SB are defined by their four contour curves. Surface SAB is composed of two surfaces SA and SB.

Figure 4 shows the example of a simple part that HAPT-3D can process.

3. STRUCTURE OF THE PROCESSOR

The computer program of HAPT-3D is composed of three sections. The first section is a translator. It translates into intermediate language a part program written with APT-like language and given with the form of cards deck. At the same

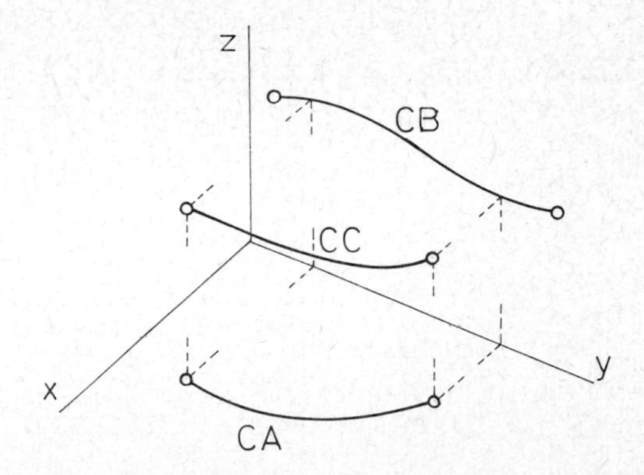

CC = CURVE/XYPLAN, CA, YZPLAN, CB

Fig. 2. An example of a space curve defined by two projection curves

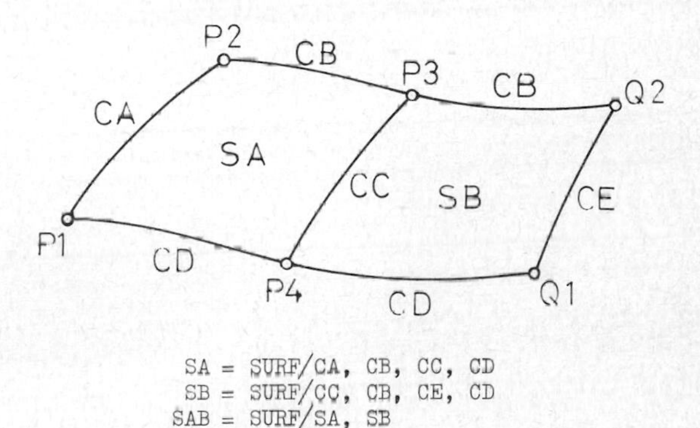

SA = SURF/CA, CB, CC, CD
SB = SURF/CC, CB, CE, CD
SAB = SURF/SA, SB

Fig. 3. Examples of surfaces

time, it performs grammatical diagnosis and the compilation of instructions such as the automatic insertion of instructions for the transformation of geometric elements. After all the statements of a part program were translated, the intermediate language text is written into ILFILE which is on a magnetic tape or disc. If any error was detected, the writing into ILFILE and the further execution of program are suppressed.

The second section is an executor. It reads a part program in ILFILE and executes it. It calculates the canonical form of all geometrical elements that are defined with various means. Also it calculates the motion of cutter and writes cutter path data into CLFILE which is also on a magnetic tape or disc. For certain cutting modes of a surface, as automatic cutter path generation routines are implemented, a part programmer need not describe detailed cutter motion statements.

The third section is a post processor. It reads CLFILE and generates a control tape whose format fits the machine designated by a part programmer.

The processor is written as far as possible in FORTRAN IV language in order to insure the independency from computers. Because of the complexity of the

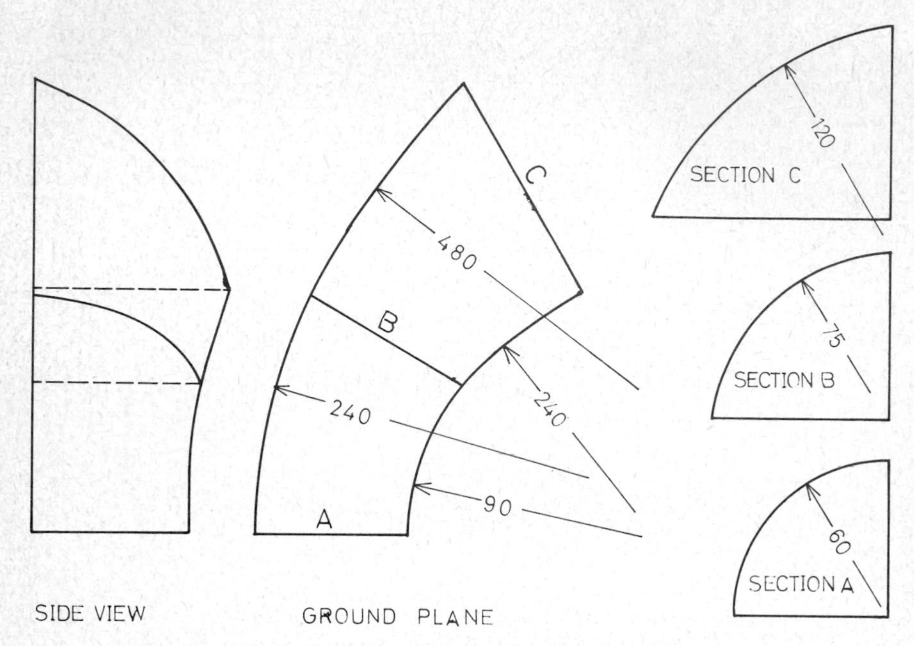

Fig. 4. The example of a simple part that HAPT-3D can process

processing of surfaces, the processor needs usually an internal memory of 256 kilo bytes or more.

The detailed description of the processing method of curves and surfaces will be described in the following sections because of its characteristic features, while that of the other parts of the processor will be omitted.

4. EXPRESSIONS OF CURVE AND SURFACE

4.1 Generalities

In HAPT-3D, a curve is generally considered as a space curve. The mathematical expression of a curve is a vector function of a parameter. Let it be $\mathbf{P}(t)$ where t is a parameter which continuously varies from its lower limit t_L to its upper limit t_H. If t is fixed, $\mathbf{P}(t)$ indicates a point, and if t changes from t_L to t_H continuously, the point describes a curve. That is, a curve is expressed by

$$\mathbf{P}(t) \qquad (t_L \leqq t \leqq t_H) \tag{1}$$

As t is limited between t_L and t_H, $\mathbf{P}(t)$ is a finite curve. In this paper, the notation $\mathbf{P}(t)$ is used to express both a curve and the point on a curve where a parameter is fixed to t.

The point $\mathbf{P}(t_L)$ is called the initial point of curve $\mathbf{P}(t)$, and $\mathbf{P}(t_H)$, the terminal point. A curve is oriented in the direction where t augments. The derivative of $\mathbf{P}(t)$

$$\mathbf{T}(t) = \frac{d}{dt}\mathbf{P}(t) \tag{2}$$

is called a tangent vector to the curve at point $\mathbf{P}(t)$. Because the differentiation is made with respect to arbitrarily chosen parameter t, the length of $\mathbf{T}(t)$

is not unity, but the direction of $\mathbf{T}(t)$ coincides with the direction of the tangent line at the same point.

A curve is divided into several segments as shown in fig. 5. The division is

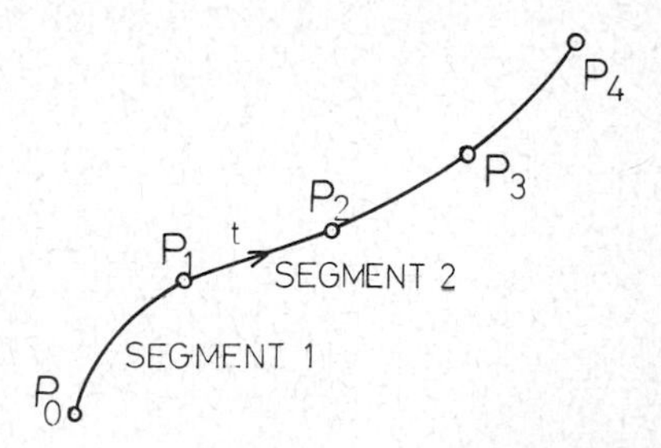

Fig. 5. A curve and segments

so performed that, in the same segment, the variation of the shape is rather simple and the mathematical expression remains the same. As regards the whole curve, complex variations may occour and the expression does not always remains the same. At a node (boundary point of two segments), the direction of the tangent is continuous, but the continuity of the curvature is not insured. As described above, the expression for a whole curve is not a single mathematical expression, but it is considered as if it was a single expression.

In the same manner, a surface is expressed by a vector function of two parameters, that is,

$$\mathbf{P}(u, v) \qquad (u_L \leqq u \leqq u_H, \quad v_L \leqq v \leqq v_H) \tag{3}$$

If parameters u and v are fixed, $\mathbf{P}(u,v)$ becomes a point on the surface, and if u and v vary continuously from u_L to u_H and from v_L to v_H respectively, the loci of $\mathbf{P}(u,v)$ cover the surface. The surface of HAPT-3D is a rectangle in the u-v plane.

At any point $\mathbf{P}(u,v)$ on the surface,

$$\mathbf{T}_u(u,v)=\frac{\partial}{\partial u}\mathbf{P}(u,v) \tag{4}$$

$$\mathbf{T}_v(u,v)=\frac{\partial}{\partial v}\mathbf{P}(u,v) \tag{5}$$

are called a u-tangent vector and a v-tangent vector, respectively. $\mathbf{T}_u$ and $\mathbf{T}_v$ have the same property as the tangent of a curve described above. And

$$\mathbf{W}(u,v)=\frac{\partial^2}{\partial u \partial v}\mathbf{P}(u,v) \tag{6}$$

is called a twist vector.

A surface is also devided into several segments called patches (fig.6). A patch is a unit of mathematical expression. A boundary curve of patches satisfies the condition of curve described above. A side line of patch corresponds to a segment of curve.

The actual shape of a curve (or a surface) depends on the type of function. Though there are many types of function, in the computer program, a curve (or a surface) is called by a common procedure independent from the types of function. Therefore, an application routine, a program for computing the intersection point of a curve and a surface for example, is written without any care of the types of curve and surface.

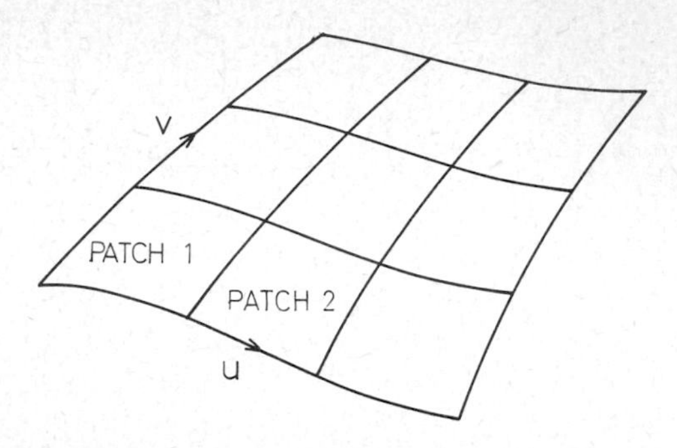

Fig. 6. A surface and patches

4.2 Canonical Forms

The canonical form of a curve and surface is composed of a head and some bodies. The form of the head is common to all kinds of curves or surfaces, but the form of the bodies varies according to the kinds.

The head of the canonical form of a curve is

$$(k, \ell, m, n, t_L, t_H, \Delta t)$$

where k is the code of the kind of the curve or function; ℓ , m , n are the indices for the bodies; t_L , t_H are the lower and upper limits of parameter t ; Δt is the increment of t for the span of a segment.

The head of the canonical form of a surface is

$$(k, \ell, m, n, u_L, u_H, \Delta u, v_L, v_H, \Delta v)$$

where k is the code of the kind of the surface or function; ℓ , m , n are the indices for the bodies; u_L and u_H are the lower and upper limits of parameter u ; Δu is the increment of u for the span of a patch in the direction of u ; v_L and v_H are the lower and upper limits of parameter v ; Δv is the increment of v for the span of a patch in the direction of v.

4.3 Mathematical Expressions

Some types of functions of curve and surface will be described in concrete forms.

4.3.1 Mathematical Curve and Surface

If a curve or a surface is directly defined by a function of a parameter or two parameters, the function is used for the mathematical expression without any modification.

4.3.2 Non-Mathematical Surface

As described above, the main object of HAPT-3D is to process a non-mathematical surface. At such a surface, the mathematical expression is decided at every patch by the Coons' blending functions.

Namely, if the case of a curve is described first, the mathematical expressions of the segment shown in fig. 7 are

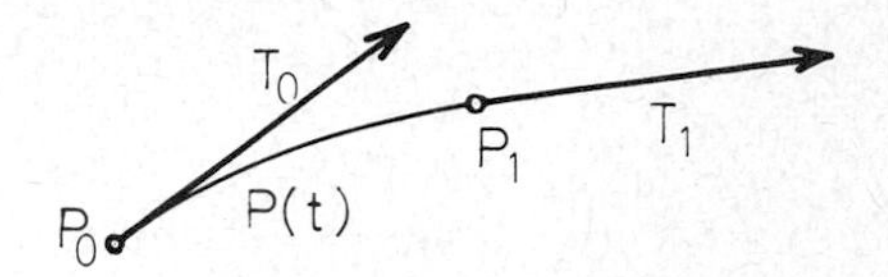

Fig. 7. A segment of curve

$$\mathbf{P}(t)=f(t)\mathbf{P}_0+f(1-t)\mathbf{P}_1+g(t)\mathbf{T}_0-g(1-t)\mathbf{T}_1 \tag{7}$$

$$f(t)=2t^3-3t^2+1 \tag{8}$$

$$g(t)=t^3-2t^2+t \tag{9}$$

where $f(t)$ and $g(t)$ are blending functions; $\mathbf{P}_0$ and $\mathbf{P}_1$ are the initial and terminal points of the segment, respectively; $\mathbf{T}_0$ and $\mathbf{T}_1$ are the tangent vectors at $\mathbf{P}_0$ and $\mathbf{P}_1$ respectively; t is a local parameter so chosen as $t=0$ at $\mathbf{P}_0$ and $t=1$ at $\mathbf{P}_1$.

Then in the case of a surface, the mathematical expressions of the patch shown in fig. 8 are

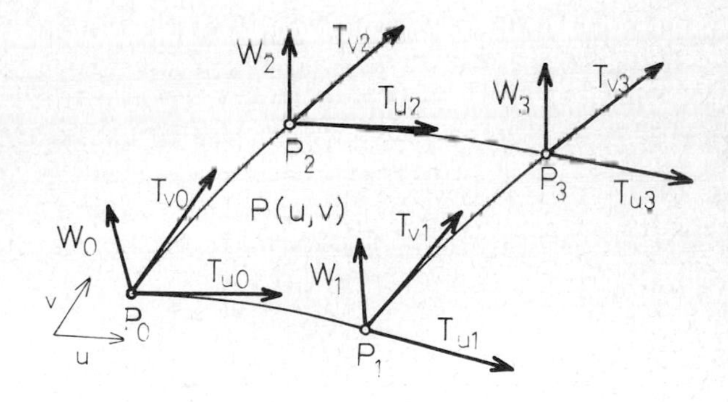

Fig. 8. A patch

$$\begin{aligned}\mathbf{P}(u,v)&\\
&=F(u,v)\mathbf{P}_0+F(1-u,v)\mathbf{P}_1+F(u,1-v)\mathbf{P}_2+F(1-u,1-v)\mathbf{P}_3\\
&+G(u,v)\mathbf{T}_{u0}-G(1-u,v)\mathbf{T}_{u1}+G(u,1-v)\mathbf{T}_{u2}-G(1-u,1-v)\mathbf{T}_{u3}\\
&+G(v,u)\mathbf{T}_{v0}+G(v,1-u)\mathbf{T}_{v1}-G(1-v,u)\mathbf{T}_{v2}-G(1-v,1-u)\mathbf{T}_{v3}\\
&+H(u,v)\mathbf{W}_0-H(1-u,v)\mathbf{W}_1-H(u,1-v)\mathbf{W}_2+H(1-u,1-v)\mathbf{W}_3\end{aligned} \tag{10}$$

$$F(u,v)=f(u)\,f(v) \tag{11}$$

$$G(u,v)=g(u)\,f(v) \tag{12}$$

$$H(u,v)=g(u)\,g(v) \tag{13}$$

where $F(u,v)$, $G(u,v)$, $H(u,v)$ are composite blending functions; f and g are the same functions to eq. (8) and (9); $\mathbf{P}_0$, $\mathbf{P}_1$, $\mathbf{P}_2$, $\mathbf{P}_3$ are the corner points of the patch; $\mathbf{T}_{ui}$, $\mathbf{T}_{vi}$, $\mathbf{W}_i$ are a u-tangent vector, a v-tangent vector, a twist vector at point $\mathbf{P}_i$ ($i=0\sim3$), respectively; and local parameter u and v are so chosen that $\mathbf{P}_0=\mathbf{P}(0,0)$, $\mathbf{P}_1=\mathbf{P}(1,0)$, $\mathbf{P}_2=\mathbf{P}(0,1)$ and $\mathbf{P}_3=\mathbf{P}(1,1)$. Sometimes $\mathbf{W}_i$'s are neglected because of their difficult calculation.

4.3.3 Equidistant Surface

Sometimes, there is a need to define a surface equidistant to a given surface. In this case, if the surface is approximated by a function, eq. (10) for example, the problem of the error of approximation will arise. So in HAPT-3D, the surface is calculated exactly as an equidistant surface to the given surface. Namely, its canonical form is equivalent to the following mathematical expression

$$\mathbf{P}(u,v)=\mathbf{Q}(u,v)+d\,\mathbf{N}(u,v) \qquad (u_L \leqq u \leqq u_H,\ v_L \leqq v \leqq v_H) \tag{14}$$

where $\mathbf{P}(u,v)$ is the function for the equidistant surface; $\mathbf{Q}(u,v)$ is the function of the given surface; $\mathbf{N}(u,v)$ is the normal to the given surface; d is the distance between two surfaces. $\mathbf{N}(u,v)$ is computed by

$$\mathbf{N}(u,v)=\mathbf{T}_u(u,v)\times\mathbf{T}_v(u,v) \tag{15}$$

5. INTERSECTIONS OF CURVES AND SURFACES

In order to show examples of the algorithm used in HAPT-3D, two examples of the calculation of the intersection of curves or surfaces will be given.

5.1 Intersection Points of Two Curves

As a curve is always regarded as a space curve, the intersection point of two curves is realized as a point where the distance between the two curves becomes minimum and tolerable (fig. 9). Let $\mathbf{P}_1(t_1)$ and $\mathbf{P}_2(t_2)$ be the expressions for both curves, where t_1 and t_2 are respective parameters, and put

$$S=\frac{1}{2}\{\mathbf{P}_1(t_1)-\mathbf{P}_2(t_2)\}^2 \tag{16}$$

Then S being a half square of the distance of two points $\mathbf{P}_1(t_1)$ and $\mathbf{P}_2(t_2)$, a necessary condition for S to be a minimum is

$$\frac{\partial S}{\partial t_1}=\{\mathbf{P}_1(t_1)-\mathbf{P}_2(t_2)\}\frac{d}{dt_1}\mathbf{P}_1(t_1)=0 \tag{17}$$

$$\frac{\partial S}{\partial t_2}=\{\mathbf{P}_2(t_2)-\mathbf{P}_1(t_1)\}\frac{d}{dt_2}\mathbf{P}_2(t_2)=0 \tag{18}$$

Solving eq. (17) and (18) simultaneously, we get t_1 and t_2 which minimize S, and substituting t_1 into $\mathbf{P}_1(t_1)$, we get the intersection point. The Newton's method is applied for the solution of eq. (17) and (18).

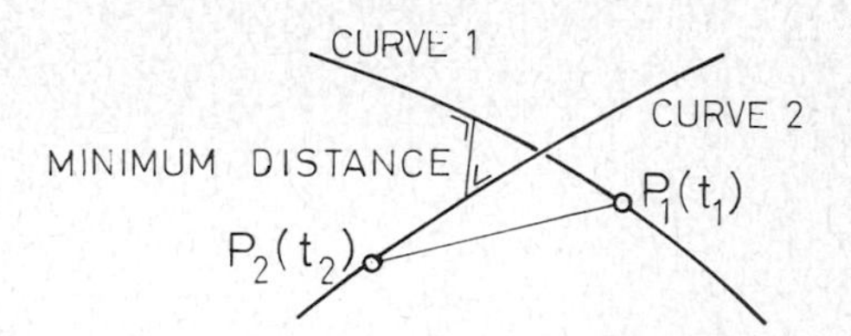

Fig. 9. The intersection point of two curves

5.2 The Intersection Curve of Two Surfaces

The expression for the intersection curve of two surfaces is computed by the approximation in the form of eq. (7). In order to obtain the approximate expression, a row of points that lie exactly on the intersection curve is computed first (fig. 10a). Then the row of points is replaced with the expression for curve.

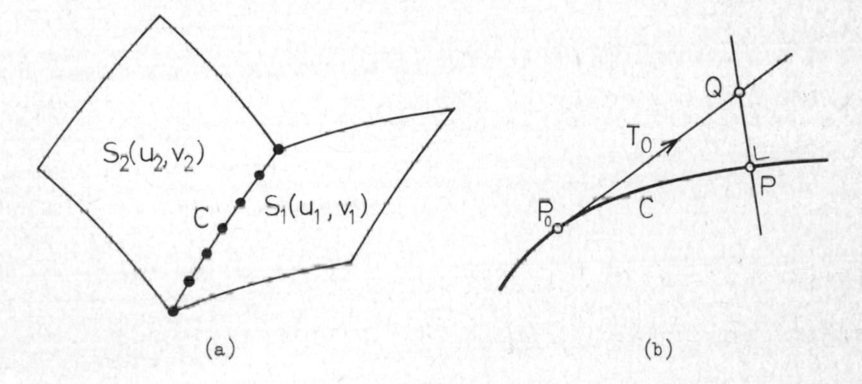

Fig. 10. The intersection curve of two surfaces

Some points become the boundary points of segments of curve, and others become control points that are used for the determination of the lengths of tangents at boundary points.

The computing method of the points is as follows.

Let $\mathbf{S}_1(u_1, v_1)$ and $\mathbf{S}_2(u_2, v_2)$ be the expression for two given surfaces; $\mathbf{C}$ be the intersection curve of $\mathbf{S}_1$ and $\mathbf{S}_2$; $\mathbf{P}_0$ be a point on $\mathbf{C}$ computed at the preceding step; $\mathbf{P}$ be the point on $\mathbf{C}$ to be computed; $\mathbf{T}_0$ be the tangent to $\mathbf{C}$ at $\mathbf{P}_0$. At first, point $\mathbf{Q}$ is computed on $\mathbf{T}_0$ as an approximate point of $\mathbf{P}$ (fig. 16b). The distance between $\mathbf{P}_0$ and $\mathbf{Q}$ is decided by considering the curvature of $\mathbf{C}$ at $\mathbf{P}_0$ and the tolerance of approximation. Then point $\mathbf{P}$ is computed as the intersection point of curve $\mathbf{C}$ and a plane passing point $\mathbf{Q}$ and perpendicular to $\mathbf{C}$ (fig. 10b), that is, $\mathbf{P}$ is computed by solving the simultaneous equations below,

$$\mathbf{P} = \mathbf{S}_1(u_1, v_1) = \mathbf{S}_2(u_2, v_2) \tag{19}$$

$$(\mathbf{P} - \mathbf{Q}) \cdot \mathbf{T} = 0 \tag{20}$$

Making use of

$$\mathbf{T}=\mathbf{N}_1(u_1,v_1)\times\mathbf{N}_2(u_2,v_2)$$
$$=\left\{\frac{\partial}{\partial u_1}\mathbf{S}_1(u_1,v_1)\times\frac{\partial}{\partial v_1}\mathbf{S}_1(u_1,v_1)\right\}\times\left\{\frac{\partial}{\partial u_2}\mathbf{S}_2(u_2,v_2)\times\frac{\partial}{\partial v_2}\mathbf{S}_2(u_2,v_2)\right\} \quad (21)$$

Then eq. (19) and (20) are written as

$$\mathbf{S}_1(u_1,v_1)-\mathbf{S}_2(u_2,v_2)=0 \quad (19)'$$

$$\left\{\mathbf{S}_1(u_1,v_1)-\mathbf{Q}\right\}\cdot\left\{\left(\frac{\partial}{\partial u_1}\mathbf{S}_1(u_1,v_1)\times\frac{\partial}{\partial v_1}\mathbf{S}_1(u_1,v_1)\right)\times\left(\frac{\partial}{\partial u_2}\mathbf{S}_2(u_2,v_2)\times\frac{\partial}{\partial v_2}\mathbf{S}_2(u_2,v_2)\right)\right\}=0 \quad (22)$$

If an initial point is given, succeeding points are computed by this method. The initial and terminal points are given by a part program or computed by the edge lines of the surfaces.

On the approximation by eq. (7), the direction of the tangent at the boundary point of a segment are given by eq. (21), and its length is computed on the estimation of the sum of the squared distances between the control points and the curve.

6. SUMMARY

The outline of HAPT-3D whose main object was to process surfaces defined by their contour and sectional curves was described. This program was newly developed because APT had not such functions at that time. Recently, the processor of a sculptured surface was developed for APT. Although the processing methods of surfaces of both HAPT-3D and the sculptured surface of APT are similar, their input methods are quite different. HAPT-3D assumes plans for input media that give accurate data about surfaces, while the sculptured surface of APT assumes inaccurate data got by the measurements of models.

Although the machining of many dies and molds became possible after the development of HAPT-3D, it is not yet sufficient because of the complexity of three dimensional problem.

INTERACTIVE COMPUTER-AIDED DESIGN FOR SCULPTURED SURFACES

CARLO GHEZZI
FRANCESCO TISATO
Istituto di Elettrotecnica ed Elettronica
Politecnico di Milano
Milano, Italia

Abstract: An interactive approach to computer-aided design of complex surfaces is presented. According to Coons' method a surface is conceived as a union of patches. Two operations are introduced by which the surface can be shaped in detail without introducing a redundant number of patches. A computer program implementing this method (PGS3) is presented. At last a proposal is given for N.C. machine tool application.

1. INTRODUCTION

The problem of describing a surface by means of a digital computer is of present interest and opens new possibilities either from the point of view of man-machine interaction or of driving N.C. machine tools. Airplane fusolages, ship hulls and automobile bodies design represent important possible applications of such experiments (Bezier, 1968 - De Lotto 1967 - Melhum, 1969 - Sabin, 1970 - South, 1965).

In this field of application of computer-aided design it is necessary to represent a surface within a digital computer by a method which allows a good degree of flexibility and a short response time to interaction commands. Such requests are of particular interest whenever the interaction is performed by means of a graphical equipment (display, teletype, light-pen) on-line with a central processor. In this case the surface is manipulated to achieve a result which satisfies the designer and its final stored description can be used, for example, either for producing commands for a digital plotter or for a N.C. machine tool.

A number of methods suitable to surface representation have been proposed (Coons, 1967 - Bezier, 1968 - Melhum, 1969); in this paper an interactive approach to surface design is presented in which Coons' method has been used.

2. REMARKS ON COONS' METHOD

According to Coons' method, a surface is conceived as a union of

COMPUTER LANGUAGES FOR NUMERICAL CONTROL, *J. Hatvany, editor*
North-Holland Publishing Company - Amsterdam–London

patches, each bounded by four lines and described by a bi-cubic equation:

$$(1)\quad \overline{P}(u,\omega)= \sum_{0}^{3}{}_{i} \sum_{0}^{3}{}_{j} \overline{A}_{ij}\, u^i\, \omega^j \;,\quad 0 \leqslant u \leqslant 1 \;,\quad 0 \leqslant \omega \leqslant 1$$

where the $\overline{A}_{ij}$'s are tri-dimensional constant vectors and $\overline{P}$ is a position vector of some point of the surface in terms of u and ω. The four boundary lines can be easily derived by substituting respectively $u=0$, $u=1$, $\omega=0$ or $\omega=1$ in (1).

The sixteen vector coefficients in equation (1) can be derived from

$$\overline{P},\ \overline{P}_u,\ \overline{P}_\omega,\ \overline{P}_{u\omega}$$

at the four corners of the patch (see fig. 1)

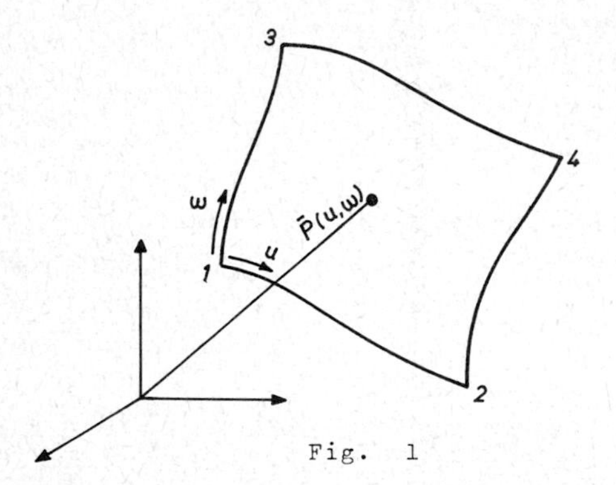

Fig. 1

Equation (1) can thus be rewritten as follows:

$$\overline{P}(u,\omega)=\left[f_1(u)\ f_2(u)\ g_1(u)\ g_2(u)\right]\cdot\begin{bmatrix}\overline{P}_1 & \overline{P}_3 & \overline{P}_{\omega 1} & \overline{P}_{\omega_3}\\ \overline{P}_2 & \overline{P}_4 & \overline{P}_{\omega 2} & \overline{P}_{\omega_4}\\ \overline{P}_{u1} & \overline{P}_{u3} & \overline{P}_{u\omega_1} & \overline{P}_{u\omega_3}\\ \overline{P}_{u2} & \overline{P}_{u4} & \overline{P}_{u\omega_2} & \overline{P}_{u\omega_4}\end{bmatrix}\cdot\begin{bmatrix}f_1(\omega)\\ f_2(\omega)\\ g_1(\omega)\\ g_2(\omega)\end{bmatrix}$$

where:

$$f_1(x)=2x^3-3x^2+1$$
$$f_2(x)=-2x^3+3x^2$$
$$g_1(x)=x^3-2x^2+x$$
$$g_2(x)=x^3-x^2$$

and where $\overline{P}_K$, $\overline{P}_{u_K}$, $\overline{P}_{\omega_K}$, $\overline{P}_{u\omega_K}$ represent position, u-first derivative, ω-first derivative, $u\omega$-second derivative vector in corner K, respectively.

It follows that the internal shape and the slope across the

boundary lines are dependent only on the vectors defined in each corner of the patch; only these vectors can be freely assigned and every modification on the surface can only be performed by changing them. In the following the corner points will be named "primary points", thus pointing out that vector parameters associated with internal ("secondary") points are fully dependent on vector parameters assigned to primary points.

Conflicts arise because on one hand we wish to describe a surface using as less patches (primary points and related parameters) as possible, on the other hand it is useful to have a considerable number of patches where the surface is particularly rough, so that its shape can be better controlled. Moreover in an interactive design philosophy it is usually impossible to state an a-priori satisfying division of the surface into patches.

In the following an approach is presented in which a good degree of compromise is achieved.

3. MANIPULATIONS ON PATCHES

In this section two operations are presented by which a surface can be designed interactively through a sequence of manipulations which allow the designer to perform the required modifications within sections more and more narrow of the surface. The required detail of description is thus achieved without introducing a-priori a redundant number of patches.

The first operation (O_1) consists in splitting an existing patch into four patches, the second (O_2) in cutting a "window-patch" within an existing patch.

3.a Operation O_1

Let Π be a patch described by eq. (2) and let ℓ and λ be two lines described by

$$\overline{P}(u_1,\omega) \ , \ 0 \leq u_1 \leq 1$$

and

$$\overline{P}(u,\omega_1) \ , \ 0 \leq \omega_1 \leq 1$$

respectively.

Let

$$\overline{P}_5=\overline{P}(0,\omega_1)$$
$$\overline{P}_6=\overline{P}(u_1,0) \qquad \overline{P}_9=\overline{P}(u_1,\omega_1)$$
$$\overline{P}_7=\overline{P}(1,\omega_1)$$
$$\overline{P}_8=\overline{P}(u_1,1)$$

Operation O_1 replaces patch Π by the four patches Π_1, Π_2, Π_3, Π_4 whose vertices are respectively:

$(\bar{P}_1\ \bar{P}_6\ \bar{P}_5\ \bar{P}_9)$, $(\bar{P}_6\ \bar{P}_2\ \bar{P}_9\ \bar{P}_7)$, $(\bar{P}_5\ \bar{P}_9\ \bar{P}_3\ \bar{P}_8)$, $(\bar{P}_9\ \bar{P}_7\ \bar{P}_8\ \bar{P}_4)$

so that the surface described by these four patches exactly reproduces the surface described by patch Π (fig. 2).

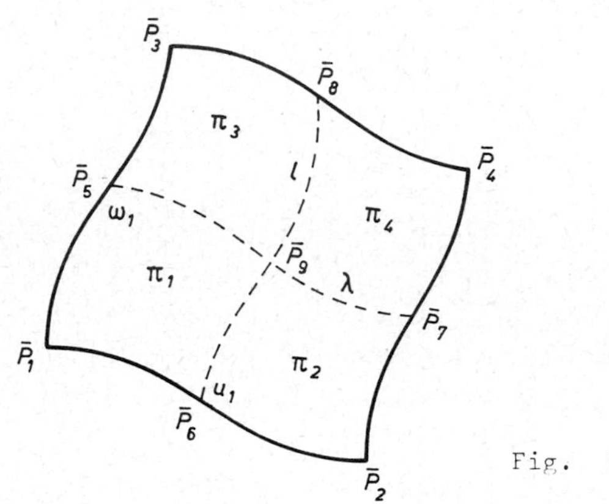

Fig. 2

The mathematical derivation of the vector parameters which describe patches Π_1, Π_2, Π_3 and Π_4 can be found in(Ghezzi, 1970).

Operation O_1 allows to manipulate the shape of the surface described by patch Π without affecting its slope across the boundary lines of Π. In fact it has been shown (De Lotto, 1967) that the slope across a boundary line depends only on the vector parameters related to the primary points which lie on that line. It follows that the vector parameters in point $\bar{P}_9$ can be modified without affecting the boundary conditions of Π.

3.b <u>Operation O_2</u>

Let Π be described by eq. (2), as before, and let ℓ_1, ℓ_2, λ_1, λ_2 be four lines described respectively

$$\bar{P}(u_1,\omega),\ \bar{P}(u_2,\omega), \bar{P}(u,\omega_1), \bar{P}(u,\omega_2) \qquad \begin{array}{l} 0 \leqslant u_1 \leqslant u_2 \leqslant 1 \\ 0 \leqslant \omega_1 \leqslant \omega_2 \leqslant 1 \end{array}$$

Let

$$\bar{F}_1=\bar{P}(u_1,\omega_1)$$
$$\bar{F}_2=\bar{P}(u_2,\omega_1)$$
$$\bar{F}_3=\bar{P}(u_1,\omega_2)$$
$$\bar{F}_4=\bar{P}(u_2,\omega_2)$$

Operation O_2 introduces a new patch F ("window-patch") whose corner points are $\overline{F}_1$, $\overline{F}_2$, $\overline{F}_3$, $\overline{F}_4$ so that the surface described by Π_1 exactly reproduces the piece of surface described by Π and bound by ℓ_1, ℓ_2, λ_1, λ_2 (fig. 3).

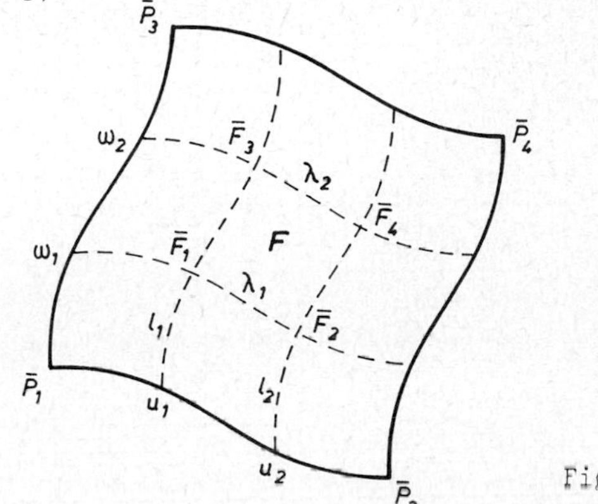

Fig. 3

The mathematical derivation of the vector parameters for patch F is described in(Ghezzi,1970).

Of course the piece of patch Π out of the window is still described by equation 2 for $0 \leqslant u \leqslant u_1$, $u_2 \leqslant u \leqslant 1$, $0 \leqslant \omega \leqslant \omega_1$, $\omega_2 \leqslant \omega \leqslant 1$ and no modification can be performed in $\overline{F}_1$, $\overline{F}_2$, $\overline{F}_3$, $\overline{F}_4$ unless modifying continuity conditions across ℓ_1, ℓ_2, λ_1, λ_2.

Now by splitting patch F a new primary point can be introduced and modifications can be performed on its parameters in order to shape the surface within the window without introducing any change out of F.

Operations O_1 and O_2 allow to manipulate a surface without affecting its shape out of a prefixed range: operation O_2 defines the range within a patch, operation O_1 introduces a primary point which can be freely manipulated in order to shape the piece of surface. Iteration of these two operations allow the designer to achieve the required detail in the description of the surface. Moreover the number of patches which describe the surface depends only on the required detail of representation and redundancy in the description can be avoided because primary point are created only where strictly necessary.

4. DATA STRUCTURE AND PROGRAM ORGANIZATION

In this section a program for generating surfaces (PGS3) is described; PGS3 allows an interactive design of surfaces through the use of two basic manipulation commands, corresponding to operations

O_1 and O_2. Presently a first version of PGS3 is implemented on a 1106 UNIVAC; the interactive use is simulated "batch" and the graphical output is obtained on a digital plotter (CALCOMP mod. 563) off-line. An effective interactive use can be obtained with a few modifications whenever a graphical equipment (optical display, light-pen, teletype) will be avilable.

Let us first refer to a working example, that is a surface initially described by means of two patches, A and B (fig. 4.a)).

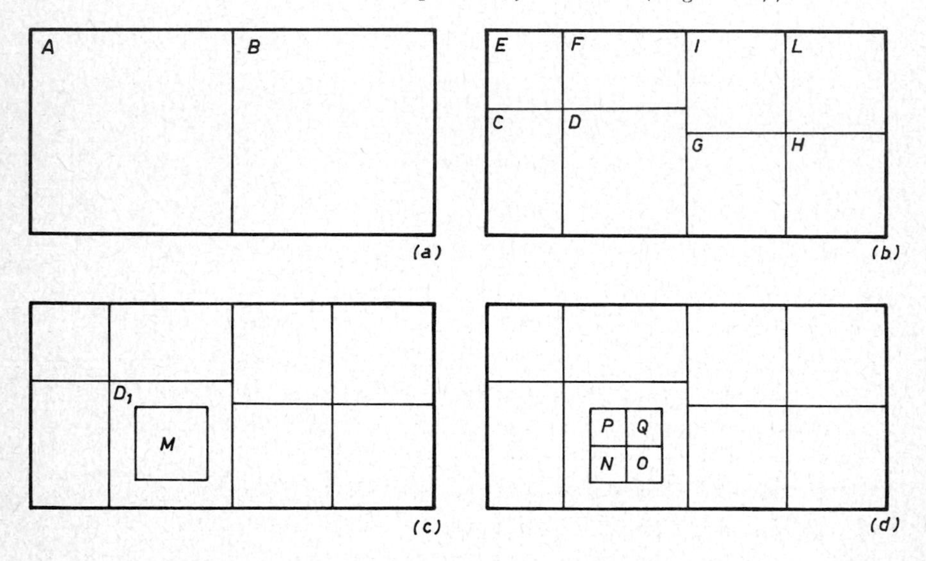

Fig. 4

Let A and B describe the surface unsatisfactorily.

The sequence of steps by which the required detail is obtained is represented in fig. 4a), b), c), d). Such a sequence of operations can be represented by a tree structure, as shown in fig. 5.(°)

(°) - Operation O_2 performed on patch D generates a window patch M; the piece of D out of M will be denoted by D_1.

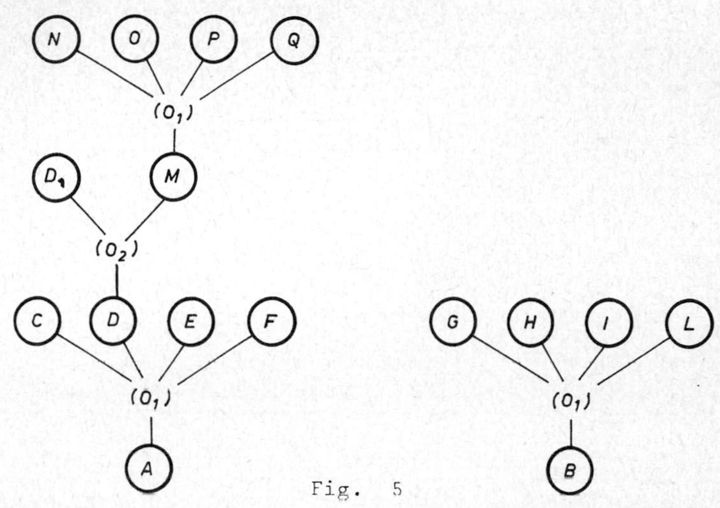

Fig. 5

The roots of the trees represent the two initial patches A and B. By performing operation O_1 on A (B) four new patches C, D, E, F (G, H, I, L) are generated. The eight patches C, D, E, F, G, H, I, L fully describe the surface; moreover two primary points are created and a better control of the surface can be achieved by manipulating their vector parameters.A further refinment can be obtained by cutting a window-patch M inside D and by splitting M into NOPQ.

At this degree of detail the surface is fully described by C, D_1, N, O, P, Q, E, F, G, H, I, L. In terms of the tree-structure this means that only the terminal leaves are necessary for a full description of the surface.

Note that the tree-structure not only describes the sequence of operations performed by the designer, but also points out a hyerarchical dependence among patches. For instance, in fig. 4 and 5 every modification on the vector parameters associated to patch B affects the parameters associated to patches G, H, I, L.

A data structure suitable to the implementation of this procedure of design must fulfil a number of requests. First no bounds should be imposed to the growth and increasing complexity of the structure through iteration of operations O_1 and O_2. Second the set of patches represented by terminal leaves in the tree structure must be easily retrieved. At last it is necessary to know not only the actual parameters which describe each patch, but also the sequence of operations by which it has been generated.

The implemented structure satisfies these requests by means of a set of hierarchically organized lists. Each patch is represented by an identifier which is an element in a list; two pointers are associated to each element, one pointing to another element in the same list, the other to a dependent list, if any. Each list contains the elements generated by the same operation (O_1 or O_2) and a new operation performed on a patch implies the definition of a derived list containing the identifiers of the generated patches. At last the identifier associated to a patch can be used for determining the set of vector parameters associated to that patch. In fig. 6 the set of lists corresponding to the example in fig. 4 is schematically shown.

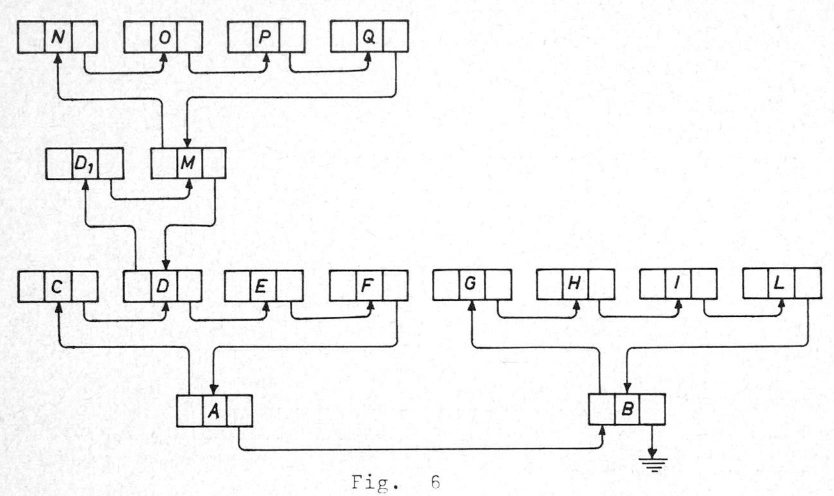

Fig. 6

The process by which a surface is generated by PGS3 can be divided into two phases. In the first phase an informal model of the object is memorized, in the second the surface is manipulated until a satisfactory representation of the object is achieved.

The informal model can either be sketched by the designer or selected in a collection of "prothotypes" (cylinders, planes, ...). In the first case the designer must assign the initial set of patches and the vector parameters in each corner (°), in the second he is only required to give simple global information (for example, in the case of plane prothotype, the only two dimensions).

(°) - Note that only the position vectors of primary points must be initially assigned, because the program itself can calculate the remaining parameters in order to obtain a surface as "spline" as possible (Johnson, 1966).

The general structure of PGS3 is shown in fig. 7.

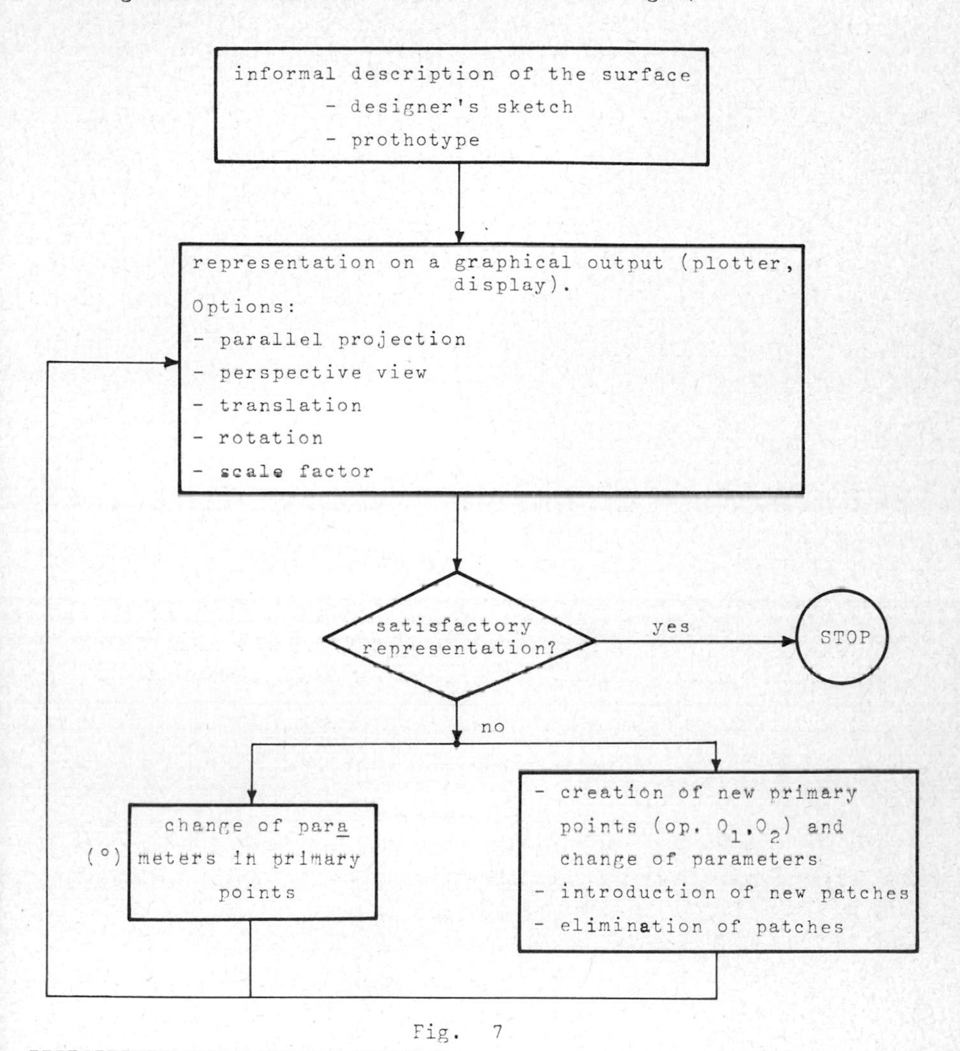

Fig. 7

(°) - This manipulation can be performed either in a single primary point or in a set of points in order to obtain standard modifications of the surface (e.g. in the case of cylindric prothotype, translation of a whole section, change of radius .... (see fig. 8)).

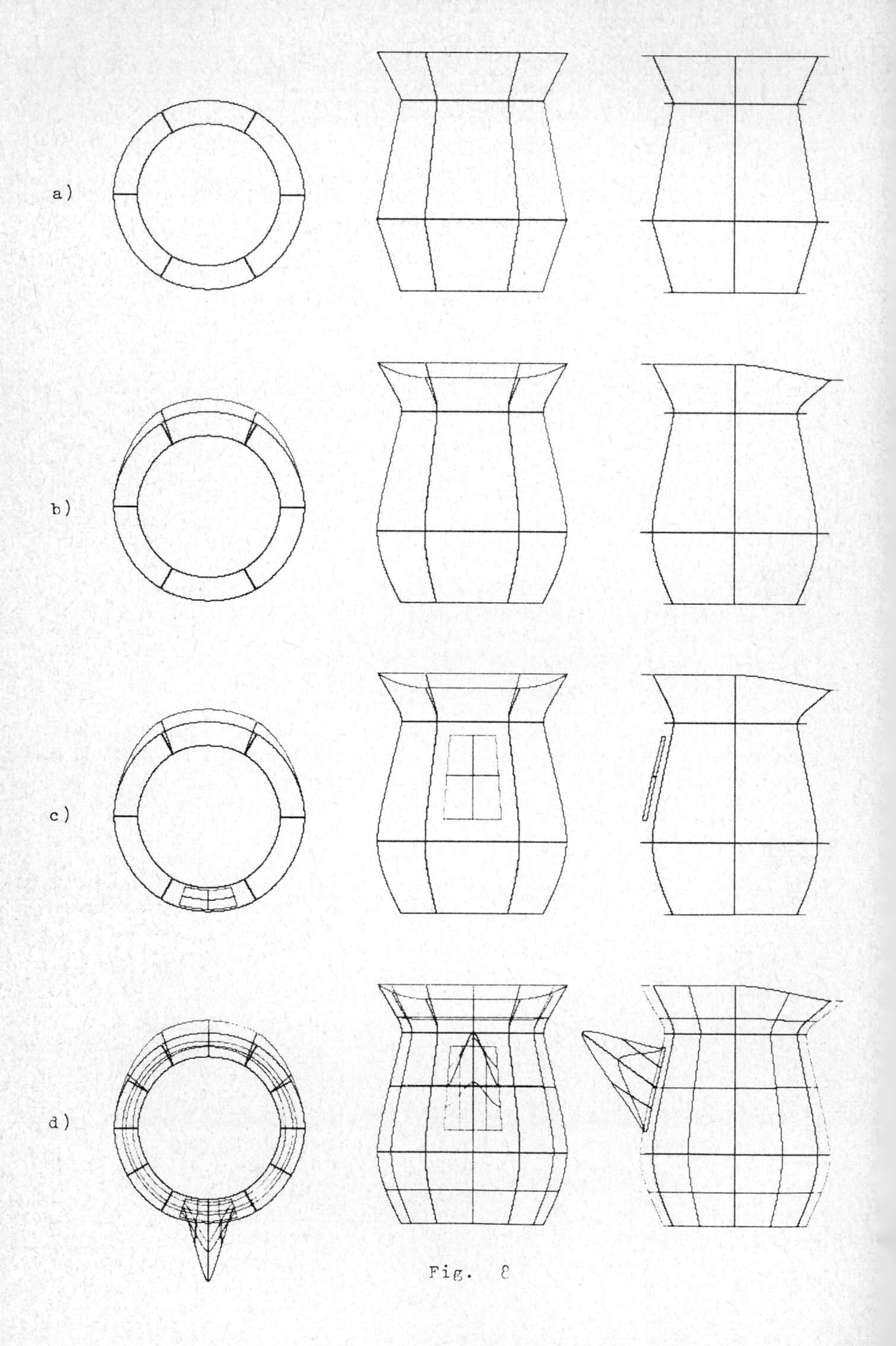

Fig. 8

5. CONCLUSIVE REMARKS

In sections 3 and 4 we have shown how an interactive method for surface computer-aided design can be implemented on the basis of Coons' approach and of the two operations O_1 and O_2. In fig. 8 an example of application is shown in which the desired object is obtained through manipulations on a cylindric prothotype.

In each figure 8a),b),c),d)the three views of the object on planes xy, xz and yz are drawn; only the four boundary lines of each patch are shown except for fig. 8 d), where also 2 lines inside each patch are drawn.

In fig. 8a)a cylindric prothotype has been manipulated by changing the radius in two sections. In fig. 8b)2 points have been translated and tangent vectors on vertical lines have been calculated in order to obtain a spline behaviour. In fig. 8c)operations O_2 and O_1 have been performed on a single patch. The primary point which has been created is then translated as shown in fig. 8d).

Further expantions of PGS3 project can be foreseen; on one hand the interactive environment can be developped through the introduction of new graphical facilities and the design of a special purpose graphical language. On the other hand the final stored representation of the object can be used for supplying input data to a N.C. machine tool.

The proposed organization of the whole design and N.C. equipment is outlined in fig. 9

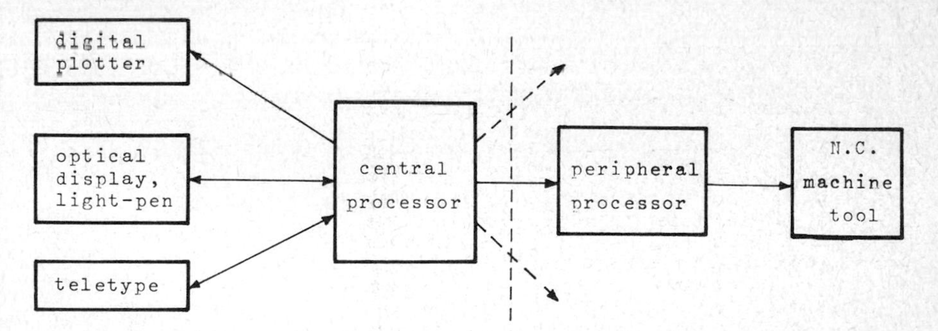

Fig. 9

In the design phase the central processor interacts with the designer through the optical display-light pen and the teletype. The final result is stored in the central processor and can be drawn by the digital plotter.

In the N.C. phase parameters related to one or more patches are transferred to the peripheral processor, which supplies data for the machine tool. The peripheral processor can be a small special purpose computer because it only must perform some simple computations on a small set of data . For example whenever a patch at a time is executed, only 48 scalar parameters (see eq. 2) must be stored in the peripheral processor.

REFERENCES

Bezier, P. (1968) Procédé de définition numérique des courbes et surfaces non mathématiques. Systéme UNISURF, Automatisme,13-5,189.

Coons, S.A. (1967) Surfaces for computer-aided design of space forms, MAC-TR-41, Project MAC, M.I.T.

De Lotto, I., R. Galimberti (1967) Innovative design with computer graphics, Alta Freq. 36-5, p. 430.

Ghezzi, C., F. Tisato (1970) Un algoritmo per la partizione e la elaborazione di superfici parametriche, Int. Rep.,Ist. di Elettronica, Politecnico di Milano.

Johnson, T.E. (1966) Arbitrarly shaped space-curves for C.A.D., M.I.T. Summer Session Cours, CAD, Aug. 1-12-1966.

Melhum, E. (1969) Curve and surface fitting based on variational criteria for smoothness, Central Inst. for Ind. Research, Oslo.

Sabin, M.A. (1970) Interrogation techniques for parametric surfaces, B.A.C. Weybridge, C.G. 70.

South, N.E., J.P. Kelly (1965) Analytic surface methods, Ford N.C. Develop. Unit, Product. Eng. Office, Metal Stamp. Div.

A U T O K O N / A E R O S P A C E; A U T O K O N / A U T O M O T I V E

EVEN MEHLUM, HALFDAN MELHUUS, STEFAN LJUNGGREN
Central Institute for Industrial Research
Blindern - Oslo 3, Norway

Abstract: The two systems for computer aided design of complex surfaces

AUTOKON/AEROSPACE

and

AUTOKON/AUTOMOTIVE

represent further development of the AUTOKON system for shipbuilding which has been in use for several years. The main part of the two above mentioned new systems is the program BOF upon which most emphasis is put in the paper to be forwarded. BOF is designed to meet the requirements of a draughtsman or a designer in his daily work with complex surfaces. The resulting geometrical information is stored in a database and is sufficient as input to various kinds of produotion systems. One example is milling.

1. INTRODUCTION AND AREA OF APPLICATION

BOF is a system for computer aided design of complex surfaces, particularly made for the automotive- and aerospace-industries but capable of adaption to any form of surface design.

The development of BOF was started in August 1969. At present the system is sold to and further developed in cooperation with Pressed Steel Fisher Ltd. (BLMC) and Hawker Siddeley Aviation Ltd., both in United Kingdom.

BOF consists of about 40 000 statements distributed on some 550 routines. The program language is low FORTRAN except for direct read/write on peripherical units. This to obtain a high degree of portability.

BOF is now running on

UNIVAC	1108	using	32 k words
IBM	360/50	"	128 k bytes
ICL	1905 F	"	40 k words

Inputdata to BOF may be taken from a sketch, a lines plan or from a model. This actually means there are two ways of using the program. With data from a sketch BOF is used as a designtool. With data from a model BOF will be the medium to transfer a physical surface into a mathematical one.

2. SURFACE-REPRESENTATION

A surface is in BOF represented by curves in the surface. These curves are of two main types:

PLANE CURVES — A plane curve is a curve where the whole curve can be described in one plane.

SPACE CURVES — A space curve is a curve that needs two projections to be uniquely defined.

There are four types of plane curves, curves in planes parallel to the three main orthogonal planes and curves in other planes.

A plane is defined by three points in space, these three points also define a local UV-system in the plane (see Fig. 1).

COMPUTER LANGUAGES FOR NUMERICAL CONTROL, *J. Hatvany, editor*
North-Holland Publishing Company - Amsterdam–London

The method of entering curves into the BOF-system is basically simple. Points are given along the curve, in the local UV-system, and are given closer together as the severity of the curve increases. The points are also given a weight which in practice can be regarded as a tolerance by which the faired curve is allowed to miss the point.

Space curves are defined by giving two views of the curve.

The fairing-routine calculates the curve for which the integral, along with arc-length, of the squared curvature is minimized.

Mathematically this means minimizing $\int y''^2(1+y'^2)^{-5/2}dx$.

The algorithm (nonlinear spline) is the same as the one used in the AUTOKON hullfairing system developed at our institute some years ago.

A thorough description of its mathematical formulation is described in ref. (1).

The fairing routine is independent of coordinate system i.e. only the relative locations between the datapoints determine the resultant curve. The faired curve is represented by consecutive circle- and straight-line-elements with continuous first derivative (see Fig. 2). The curves in Fig. 2b and 2c are made this way.

3. SURFACE IS STORED ON A DATABASE

The faired curve is stored in a database under a unique name and not by its position. This name is given by the curve number, curve type and surface number.

Curves are stored by use of AUTOBASE which is the name of a program-package designed for management of large data-collections. AUTOBASE is a quite general program and has proved a very convenient tool in BOF. AUTOBASE utilizes in operation a database which resides on a direct access storage device: disk or drum. As a backup between runs a magnetic tape is used or a disk/drum file if enough confidence in the operating system can be obtained. The dataunit in the database is called a record. This record is stored with a record-name and is referred to by its name (not address). A record contains data which the used (in this case the BOF-program) decides practical to handle as an entity. Each record is divided into smaller units called matrices. The configurations in the matrices are defined by the user-program. A curve in BOF is stored as a record. It should be mentioned that AUTOBASE uses a corebuffer and writes on peripherical units only when this buffer is full.

4. BOF FROM THE USER

To build a surface with BOF the user executes a number of commands, where a command is an instruction to the program to carry out a specific geometrical task (for example to introduce a new curve in the surface).

The number and order of commands in each run are free for the user to choose. This gives him a flexibility to define the surface in his own way. It is only the order in which the commands are given that determines the final surface and not the number of runs used.

There are 29 commands in BOF, we shall not describe all of them here.

When the user starts on a new database there are two commands that should be executed in the first run and need not be repeated in the following runs.

These commands are

VOCABULARY means definition of vocabulary words in special terminology freely chosen by the user.

GRID means definition of an orthogonal grid following normal drafting-practice.

The user can, if he for some reason wishes so, at any time during the surface-development change any of this information.

To introduce some other commands let us see how they are used to build the simple surface in Fig. 3.

The first we should do is to define the boundaries and the main characteristic line of the surface. To do this we use the command SINGLE where curvetype, curvenumber, location of curveplane, intersection weight and curveshape as well as a text describing the curve is declared. If we take for instance curve No. 5 in Fig. 3 it is in a plane parallel to the XZ-plane. We declare this curve as a XZ-curve and give the location of the curveplane along the Y-axis (DIST in Fig. 4). The definition of the curveshape is in this case given with just three points and curveslopes on the ends. The second point is declared as a knucklepoint. The resultant curve from BOF is seen in Fig. 6.

In the same way the other four curves in Fig. 3 are entered into the datebase with the SINGLE-command.

Now we can give the surface-slopes along curves 2 and 3. This we do with the command ASSIST. With this command we also can declare that the surface should have a knuckle along part of curve No. 4 (Fig. 3).

The command FIT is next to be used as it provides us to intersect the surface with a plane and fair a curve through the intersectionpoints between the plane and the existing curves in the surface. The surfaceslope along curves 2 and 3 which we declared in the previous command will be picked up and taken in account when fairing the curve. The result of the FIT-command on our surface is seen in Fig.7.

The command DRAW provides us a mean to draw our surface in wanted projection and scale.

Other commands give the user possibilities to delete curves, change existing curves, calculate a curve along constant surfaceslope (highlightcurves), calculate the surfaceslope along a curve in the surface, project a cylindrical shape onto the surface, calculate the intersectionline between two BOF-defined surfaces, part a surface in two, calculate a parallel surface a.s.o.

5. THE COMMAND FROM THE PROGRAMS POINT OF VIEW

As pointed out earlier a command from the <u>users</u> point of view is a feature doing a special geometrical task. An entire task might be divided in smaller sub-tasks. For instance the SINGLE-command consists of reading of input cards, fairing of a curve and storing on the database.

The program-parts in BOF doing these sub-tasks are called <u>modules</u>, which may be regarded as tools available to produce a wanted product. This product (command) is then made by using a sample of these tools (modules) in some given order.

To be able to use the same modules for more than one command some restrictions should be attached to them.

These restrictions are:

1) <u>No</u> direct <u>program-communication</u> between two modules allowed (i.e. no subroutine in one module should call on any subroutine in another module).
2) <u>No</u> direct <u>data-communication</u> between two modules allowed (i.e. no use of common-blocks or other common data-area defined in the modules).

As a command means execution of a series of modules it is obvious that the modules in some way must communicate. The <u>program-communication</u> is provided by a main-program called the <u>administrator</u> using <u>decision-tables</u>. To every command such a decision-table is defined. The table is a list over which modules to be used and in which order. <u>Data-communication</u> is performed by using the AUTOBASE-routines. Output from a module are records to the database. These records (<u>communication-records</u>) are used as input to the module which needs them. The communication-records are removed from database when they are not needed any more and do not belong to the general surface-description.

All subroutines which are used by more than one module are collected in one program-part called <u>service-routines</u>. Among these are the AUTOBASE-routines.

According to communication-rules program-communication is as shown in Fig. 8, and data-communication as shown in Fig. 9.

6. THE DIFFERENT MODULES

For the time being BOF contains 16 modules. Due to the modularity these modules are organized in a horizontal structure (see Fig. 10). An additional module will not increase the required core-storage for execution of the system.

7. A SHORT DESCRIPTION OF SOME OF THE MORE FREQUENTLY USED MODULES

Module 1 is the input-module. All input-data for the different commands are read in here. Output from the module are interpreted data to the succeeding module(s).

Module 2 is called the intersectionpoint-generator. Input is a plane and a UV-system in this plane. All intersections between existing surface and this intersecting plane are calculated and transformed into the local UV-system. If the intersected curves have slope-conditions attached to them, the directions at the intersectionpoints are calculated, too. Output are the intersectionpoints in the order as they were found (see Fig. 11).

Module 3 is the sorting-module. Input here is a set of points in a UV-system. The points are in random order. The module calculates the correct sequence for the points. Various sorting-criteria are used. From the very simple ones (increasing U-values for instance) to more sophisticated methods. The methods are optional and the option is a part of the input from the user. A sorted pointset is output from the module (see Fig. 12).

Module 4 has a sorted pointset as input. The module is the curve-fairing module. A curve is made from the datapoints by use of the fairing-algorithm. The curve-elements are output from the module (see Fig. 13).

Module 5 is the draw-module. Input here are all requirements introduced under a draw-command. The module fetches correct projection of wanted curves from the database. Curves are scaled, mirrored a.s.o. if required. Drawing of grid-system, annotation etc. might be among the requirements. Output are drawing-elements to a magnetic-tape (or other permanent mass-storage). This is later on used "off line" as input to a numerically controlled drawingmachine (see Fig. 14).

Module 10 is the storing-module. Input is different communication-records which are used to build up the complete final curve-record. All curve-records defined by the different commands are stored on the data-base with unique record-identification. This module also takes care of all editing of the surface, such as removal of a curve from a surface, changes of curvenames, merging and splitting of surfaces a.s.o.

8. OTHER MODULES IN BOF

- Module 6 - Generates curves with constant slope on the surface (highlights)
- Module 7 - Generates surface-slope along a stored curve base on users requirements.
- Module 8 - Projects a plane curve onto the surface.
- Module 9 - Merging of two sorted datasets into each other.
- Module 11 - Generates surface-slope along a stored curve based on the requirements of the stored surface.
- Module 12 - Interrogates the database and prints table of offsets etc.
- Module 13 - Calculates the intersection-curve between two complex surfaces.
- Module 14 - Finds knuckle-points on a series of curves in the surface and generates the curve through these ones.
- Module 15 - Generates the surface defined by rotation of one curve around a specified axis.
- Module 16 - Generates fillets (blends) to the surface in a chine-area (for instance the intersection-line between wing and aircraft-fuselage).

9. A COMMAND AND THE MODULES IT IS USING

Figure 15 shows the program- and data-flow for the command FIT. This command uses in sequence modules 1 - 2 - 3 - 9 - 4 and 10. As the FIT-command defines a series of curves the command loops around the sequence. This is made possible by some loop-tables in the administrator.

10. POSSIBILITIES FOR MEETING FURTHER REQUIREMENTS

As BOF has been developed in cooperation with the industry, "Feed-back" from these part-takes very often has been new requirements to the system. The requirements are often defined as geometrical tasks, and the modular "building-up" of the system has proved to be quire flexible for extensions. When the task is defined we also know which tools are necessary to perform it. The extension is then located to make programs for the tools which are not already existing, and build an additional decision-table for this task (new command). Severe modifications in existing-modules have rarely been necessary. It has proved that the existing modules cover a high percentage of the tools necessary for meeting new requirements.

11. DATABASE-STRUCTURE IN BOF

BOF-database contains a catalogue-system for all curve-records in the database. Each catalogue is a record itself. There is one catalogue telling which surfaces this database contains. This catalogue also contains some information about each surface such as coordinate-system, surface-range etc. To each surface there are up to 5 catalogues for curves in the surface. One for each curve-type. Additionally there are some catalogues for other record-types necessary for definition of the surface (grid-system, assist-curves etc.).

As the database is stored on a permanent file it is obvious that other program-packages rather easily should be able to use the stored surface-description as input. At CIIR, links to milling-programs, structural-programs (AUTOKON, ALKON) and holography-programs have been seriously discussed.

The figures 16, 17 and 18 show some practical examples made by the BOF-system.

12. NOTES ON APPLICABILITY

As the fairing method used is a mathematical simulation of the loftsmans spline (nonlinear splinefit) and a surface is represented by curves, personell from drawing and design offices will need a very short training period before using BOF in production. These people are also familiar with treating surfaces that are given by curves and thus not 100 % mathematically defined.

As BOF not uses a patch-technique, a surface in BOF can be as large as the user wants, for instance a whole carbody. Assist curves are used along the boundary curves to maintain tangency between two surfaces, when this is wanted.

At present BOF is a batchprogram and the only output is tape to control an NC-drawing machine plus of course computerlisting. These limitations are much due to the hardware situation. To adapt BOF to another I/O-device just means writing a new module or a postprocessor.

REFERENCES

Mehlum, Even, Central Institute for Industrial Research, Oslo, Norway: Curve and surface fitting based on variational criteria for smoothness.
Mehlum, E., Sørensen, P.: Example of an existing system in the ship-building industry. The Autokon system. Proc. Royal Society, London, A 321, 219-233 (1971).

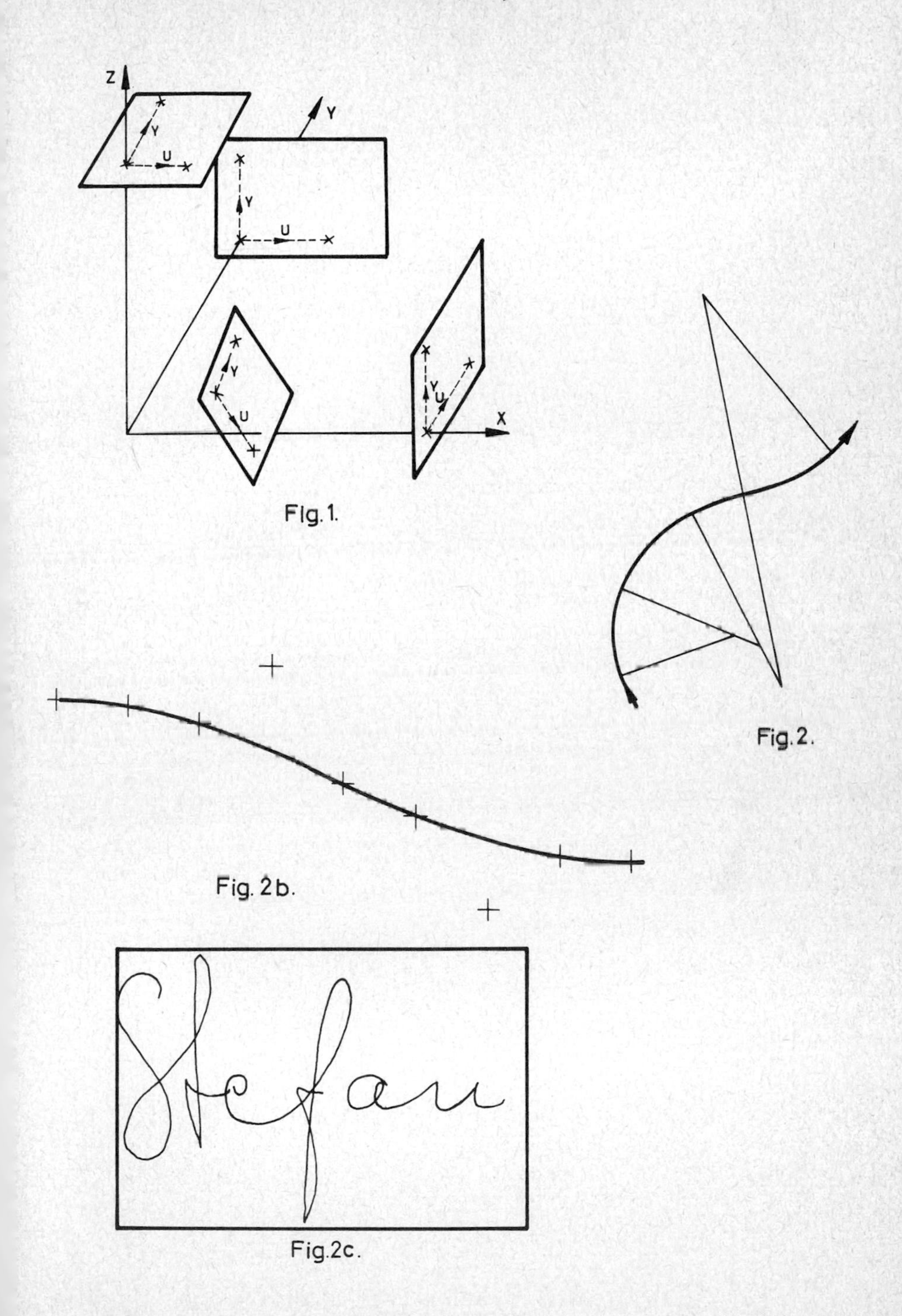

Fig. 1.

Fig. 2.

Fig. 2b.

Fig. 2c.

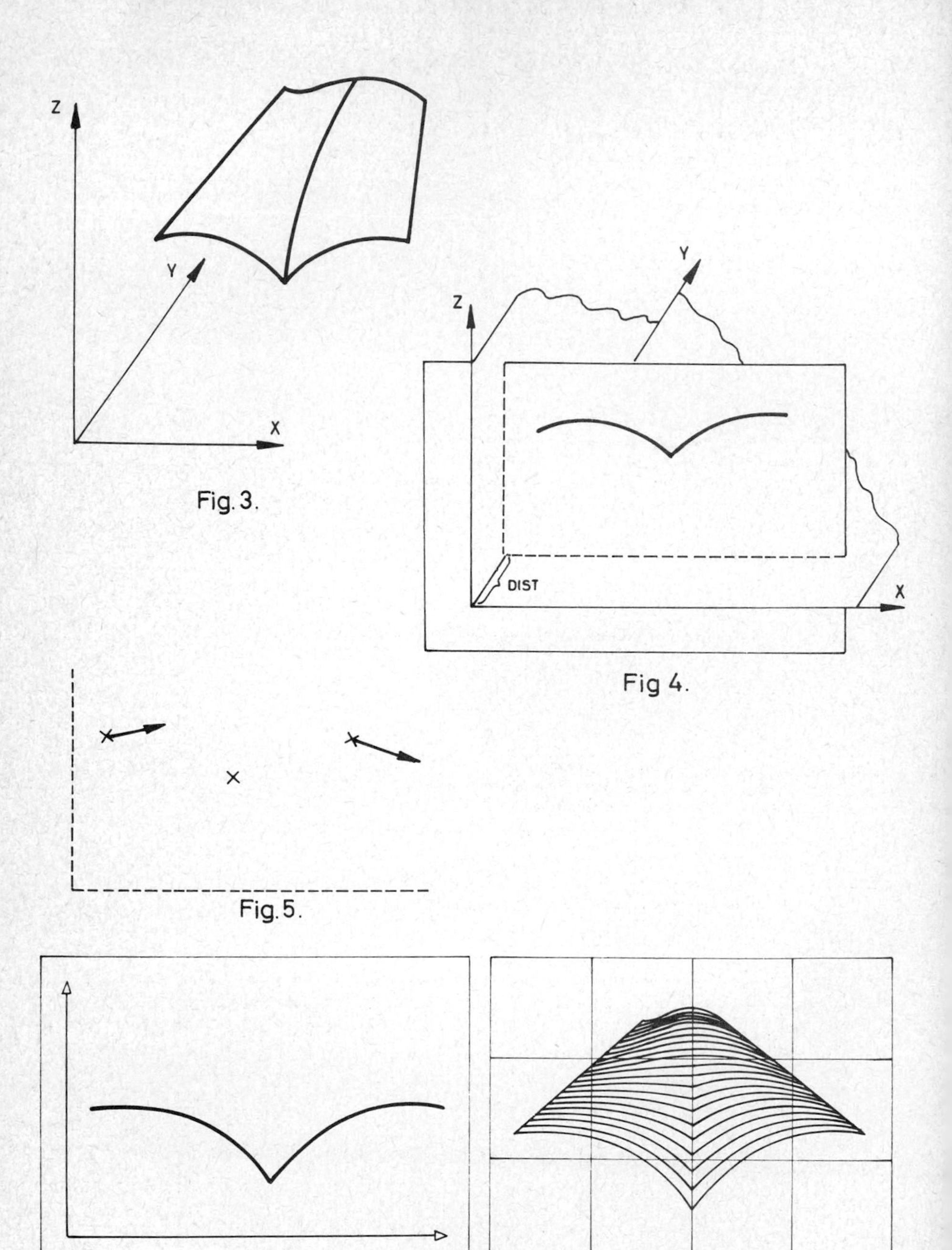

Fig. 3.

Fig 4.

Fig. 5.

Fig. 6.

Fig. 7.

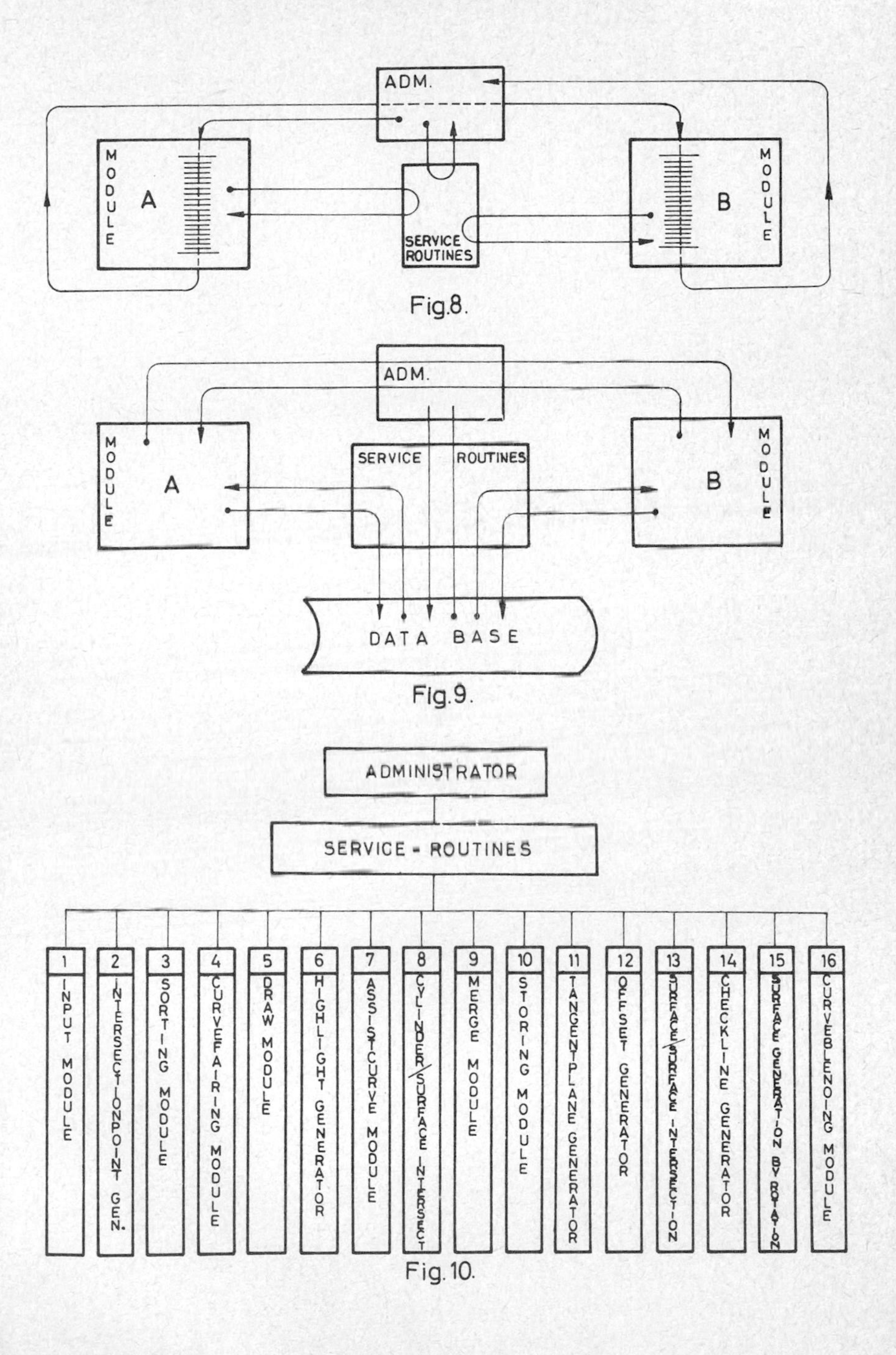

Fig.8.

Fig.9.

Fig.10.

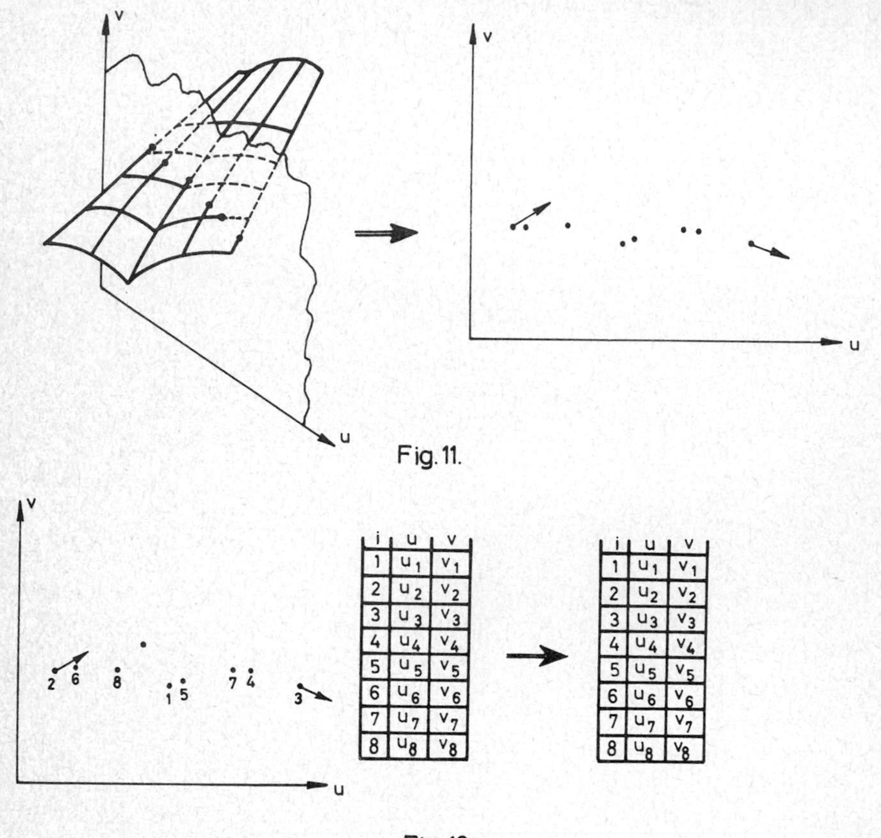
v
u
v
u
Fig. 11.
v
2 6 8 1 5 7 4 3
u
i u v
1 u1 v1
2 u2 v2
3 u3 v3
4 u4 v4
5 u5 v5
6 u6 v6
7 u7 v7
8 u8 v8
i u v
1 u1 v1
2 u2 v2
3 u3 v3
4 u4 v4
5 u5 v5
6 u6 v6
7 u7 v7
8 u8 v8

Fig. 12.

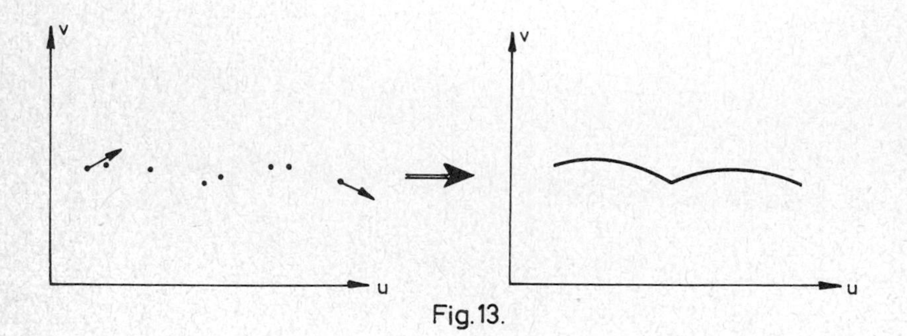
v
u
v
u

Fig. 13.

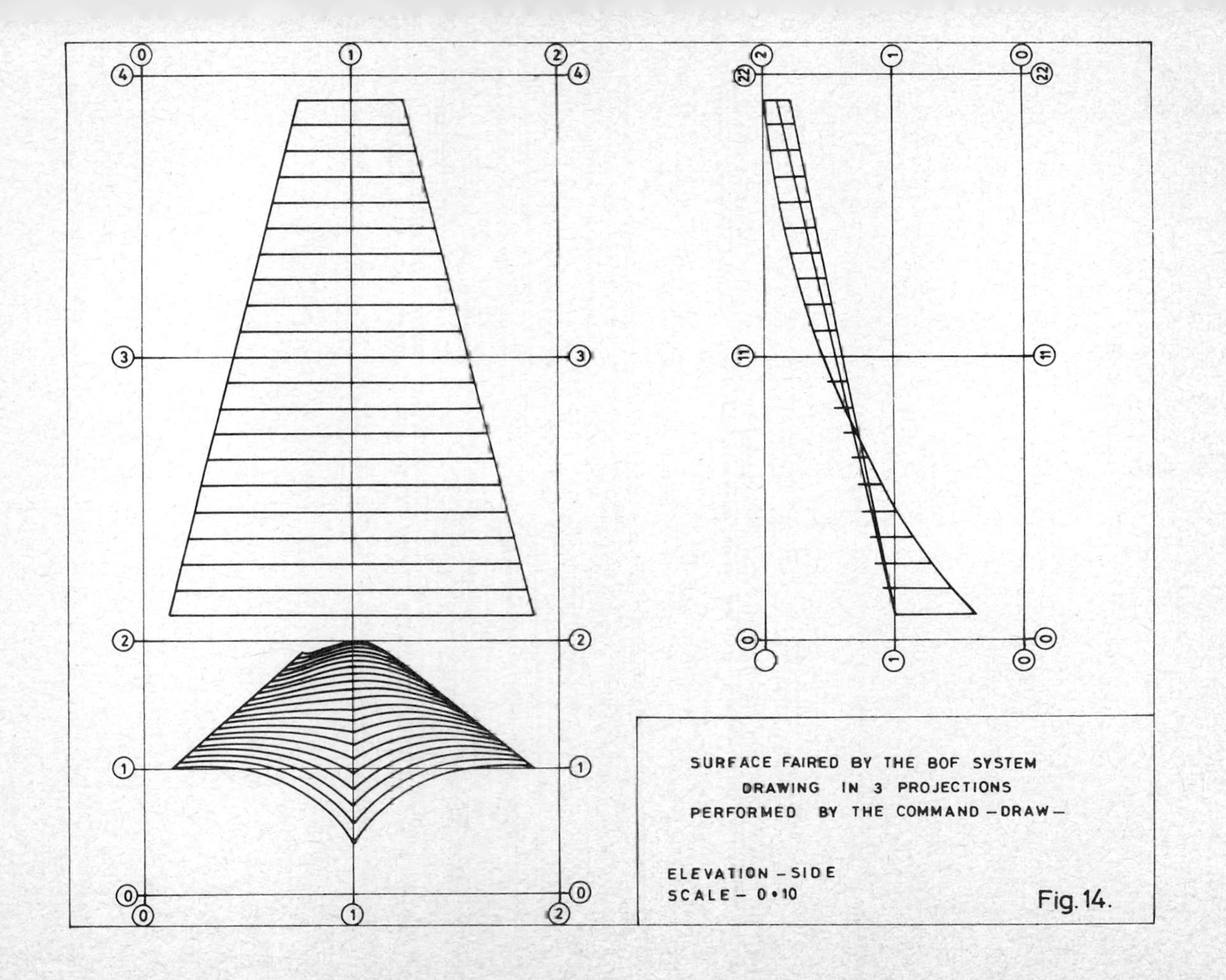
SURFACE FAIRED BY THE BOF SYSTEM
DRAWING IN 3 PROJECTIONS
PERFORMED BY THE COMMAND -DRAW-
ELEVATION - SIDE
SCALE - 0•10

Fig. 14.

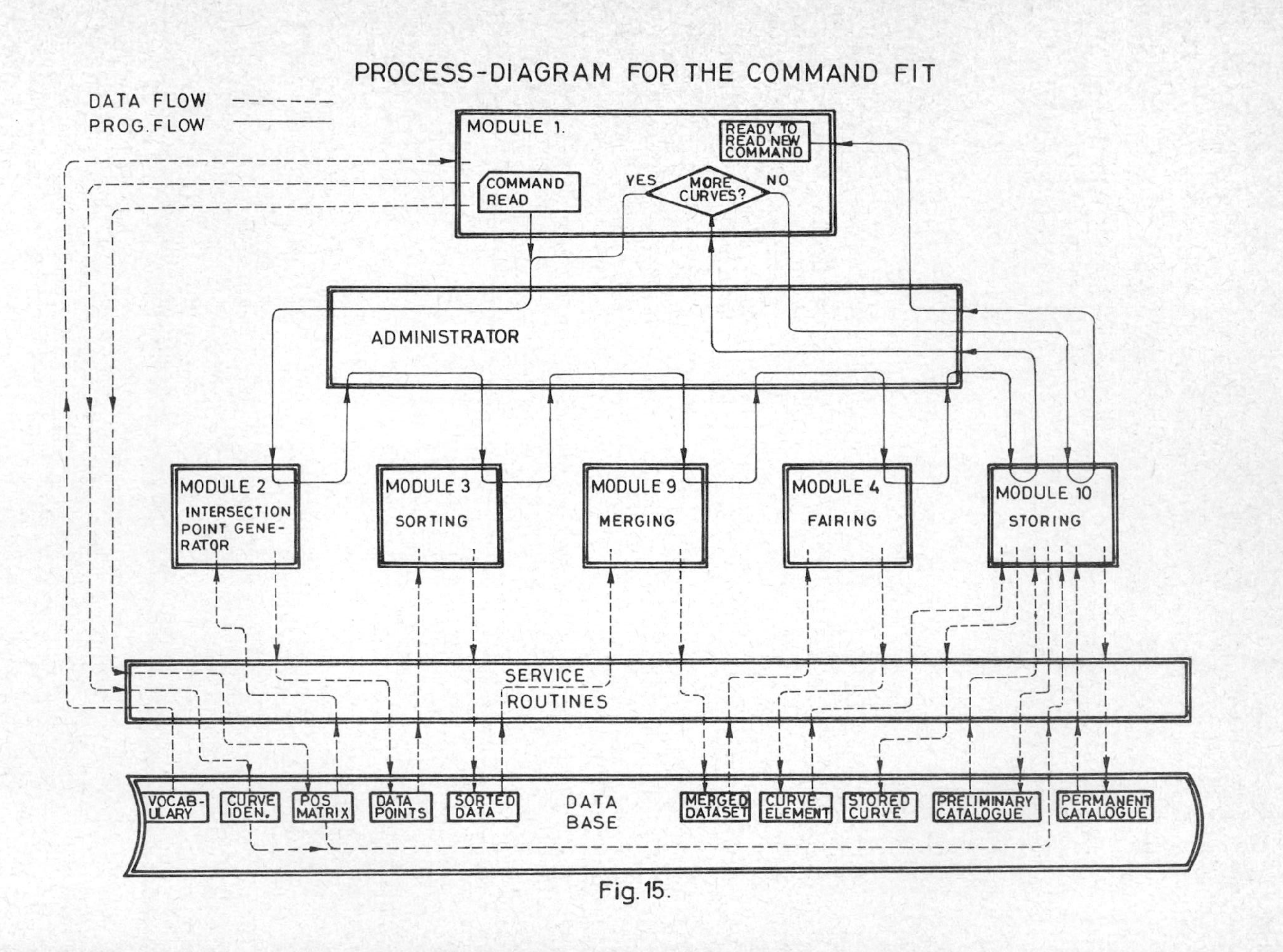
PROCESS-DIAGRAM FOR THE COMMAND FIT
DATA FLOW
PROG. FLOW
MODULE 1.
COMMAND READ
YES
MORE CURVES?
NO
READY TO READ NEW COMMAND
ADMINISTRATOR
MODULE 2 INTERSECTION POINT GENE-RATOR
MODULE 3 SORTING
MODULE 9 MERGING
MODULE 4 FAIRING
MODULE 10 STORING
SERVICE ROUTINES
VOCAB-ULARY
CURVE IDEN.
POS MATRIX
DATA POINTS
SORTED DATA
DATA BASE
MERGED DATASET
CURVE ELEMENT
STORED CURVE
PRELIMINARY CATALOGUE
PERMANENT CATALOGUE

Fig. 15.

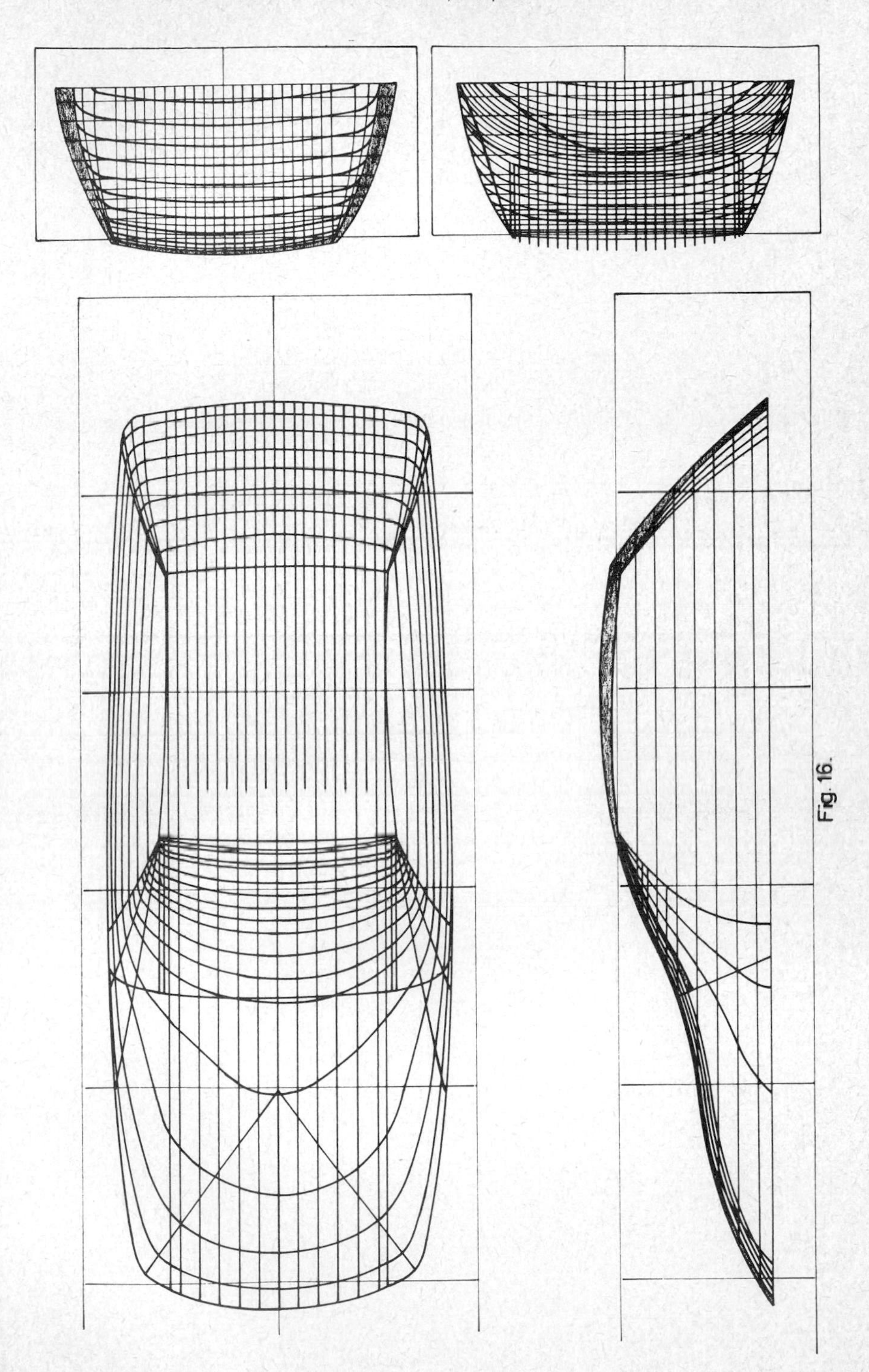

Fig. 16.

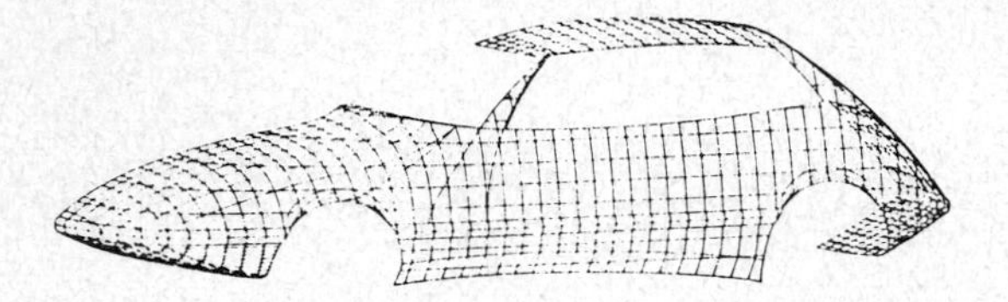

Fig. 17.

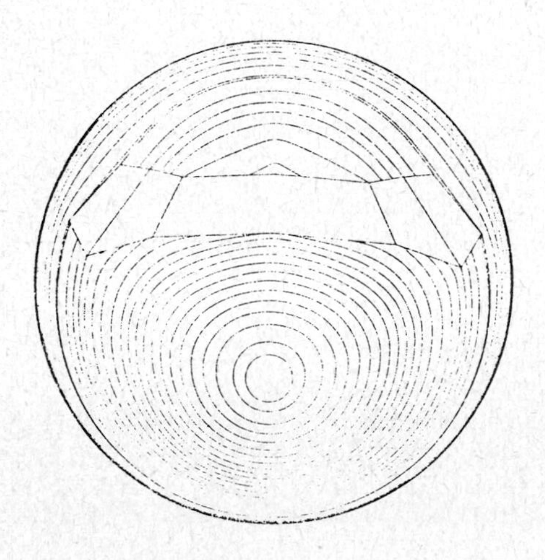

Fig. 18.

FORMELA - A PROGRAM SYSTEM FOR COMPUTER-AIDED DESIGN AND PRODUCTION

Karl-E Sandin
Avd. LTN
SAAB-SCANIA AB
S-581 88 LINKÖPING
Sweden

Within industry, there is a continuous and growing need for rationalization. One development in the rationalization of the design-production chain is the successive change-over from manual to computerized routines and the introduction of numerical control in the workshops. SAAB-SCANIA has developed a program system known as FORMELA which has proved to be an excellent aid in this field.

FORMELA is an integrated program system for use in the design and production of complex assembled products and has been used successfully within the aircraft, car, and ship building industries.

The FORMELA system is made up of four main parts:

Coordinate Dimensioning Routines and Methods
Form Definition Programs
Numerical Control Program Languages
General Technical Programs

The basic idea, developed by Nils Lidbro at the start of the forties, has since been further developed, tested, and refined under normal working conditions. The construction of a mathematical model at the project stage has meant the elimination of many time-consuming routines. This model is later used as a common base for all technical calculations, and design, preparation, and manufacturing procedures.

This mathematical model allows unlimited exactness and eliminates all dimensioning errors which usually arise when measuring from conventionally produced drawings. It also eliminates form errors common in the making and moulding of a physical model.

Having decided to use the computer as an aid in mathematical form definition, the use of other technical aids such as N/C machine tools and automatic drafting machines is also made possible. Drawings and models can be quickly produced as early as the project stage and can be used as a basis for discussions and decisions. When the actual design and production work

COMPUTER LANGUAGES FOR NUMERICAL CONTROL, *J. Hatvany, editor*
North-Holland Publishing Company - Amsterdam–London

gets started, all the geometric data needed by the departments concerned is already stored in the computer. This means that different activities can be started at the same time and that exactly the same numerical base is used throughout the whole production chain. Geometric errors - tools and details not fitting, etc. - that usually turn up sooner or later, often causing expensive retooling and breaks in production, are thus completely eliminated.

In order to be able to keep track of and utilize such a comprehensive numerical base, it is necessary that it be founded on well thought out, systematic routines and that suitable designations for the job in hand be chosen. To be able to quickly find and identify the right values in such a large amount of data, not just during the actual production period but also what may be a much later date when, for example, changes have to be made, is just as important as the mathematical technique applied and the system's data-technical structure.This is only possible if a logical system of unambigous cross references between drawings and geometric data is established. Without such a system, you will soon find yourself in the position of not being able to decide whether a result is right or wrong.

In the creation of FORMELA, the major aim has been to give the user, i.e the designer, the greatest possible service. Since evaluated data is used in several contexts, great care and attention has been paid to the actual storing of data, so that previously computed information can easily be found andutilized at a later stage. In order to attain maximum flexibility in this respect, all computations and results are output as independent programs. At present, the system consists of roughly 50 program modules which communicate with each other via data stored on magnetic tape. Results are shown on the line printer and all printouts are in English.

The mathematical technique used in the form definition program is illustrated in the figure given below. A number of characteristic curves - chosen at random - such as AP_3B, FP_2C, and EP_1D, are worked out with the help of a preliminary rough sketch. These curves, known as directrices, are built up of a number of usually second or third degree curve segments. Second degree expressions are used in order to avoid unwanted inflexions and swings which arise when equations of a higher degree are used for the generation of larger curve sections. The curve constants or coefficients of the directrix equations are calculated by the insertion of coordinate points and angles. The directrix projections on the xy and xz planes will then be:

$$\begin{cases} y = f_1(x) \\ z = g_1(x) \end{cases} \qquad \begin{cases} y = f_2(x) \\ z = g_2(x) \end{cases} \qquad \begin{cases} y = f_3(x) \\ z = g_3(x) \end{cases}$$

respectively, where f(x) ang g(x) denote simple functions. The next step is the investigation of cross sections. Angles such as α^1_z are determined so that the surface may be adjusted in accordance with the designer's intentions. These cross sections are known as generatrices since they generate the surface according to conditions given by the directrices. These generate curves are also expressed by simple, second or third degree equations, and can also be included in a local coordinate system (u, v) where:

$$Au^2 + v^2 + 2Cu + 2Dv = 0$$

and the coefficients A, C and D, are determined by the directrix conditions given in the generatrix and points. By using the simple translation $y = y_1 - v$ and $z = z_1 - v$, the equation for each surface part can be expressed in the xyz system.

The general expression will therfore be:

$$F(x,y,z) = 0$$

Remaining surface areas are treated in the same way. The surface is later generated by letting the generatrix sweep over the directrix skeleton. Any number ofparallel surfaces can now be calculated by moving the desired distance along the surface normal N, obtained from N = grad F. This is of great importance particulary in the aircraft industry where plate thicknesses often vary from surface to surface. In addition, the calculation of the parallel surface simplifies N/C programming since the tool center is allowed to move across this surface.

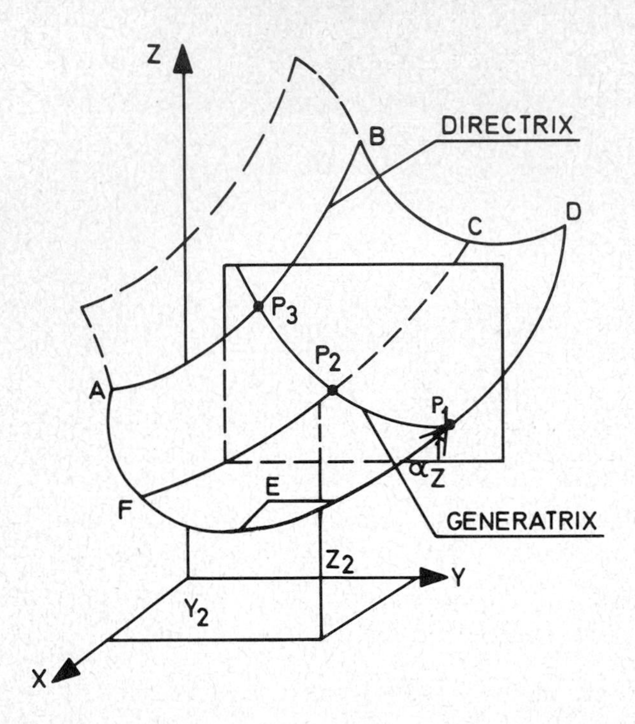

The simplicity of FORMELA's data-technical structure should by now be apparent. In the flow chart shown below, you will see that the program modules used for mathematical form definition are split up into three main groups. The first group reads in and checks input data for formal errors and geometrical contradictions. This input data-list, which is called def-data, contains the conditions for the directrix curves and information on the generatrix equations selected for the various surface areas. The constants of the directrix equations are calculated and then stored on magnetic tape - def-tape - together with information on the type of generatrix employed. The second group receives as input data the section segments that we want to calculate. Directrix data is then read from the def-tape and the cutting calculations are made. Since we now have the end-point conditions for our generatrices, we can calculate the constants for these equations. This data is now stored on magnetic tape (the gen-tape) and here, as in the case of the first group, we receive control information concerning the different characteristics of the surface definition. For the third group, input data consists of the point distribution in the generatrix curves that we want to calculate. The equations are read from the gen-tape and the result is stored on what is known as the point-tape. Here the points' coordinate values (x,y,z values) are stored as well as surface normals for each point $N = (N_1, N_2, N_3)$. Cut and point density are arbitrary and all results can be output on the line printer and punched on paper tape for use in a N/C drawing machine.

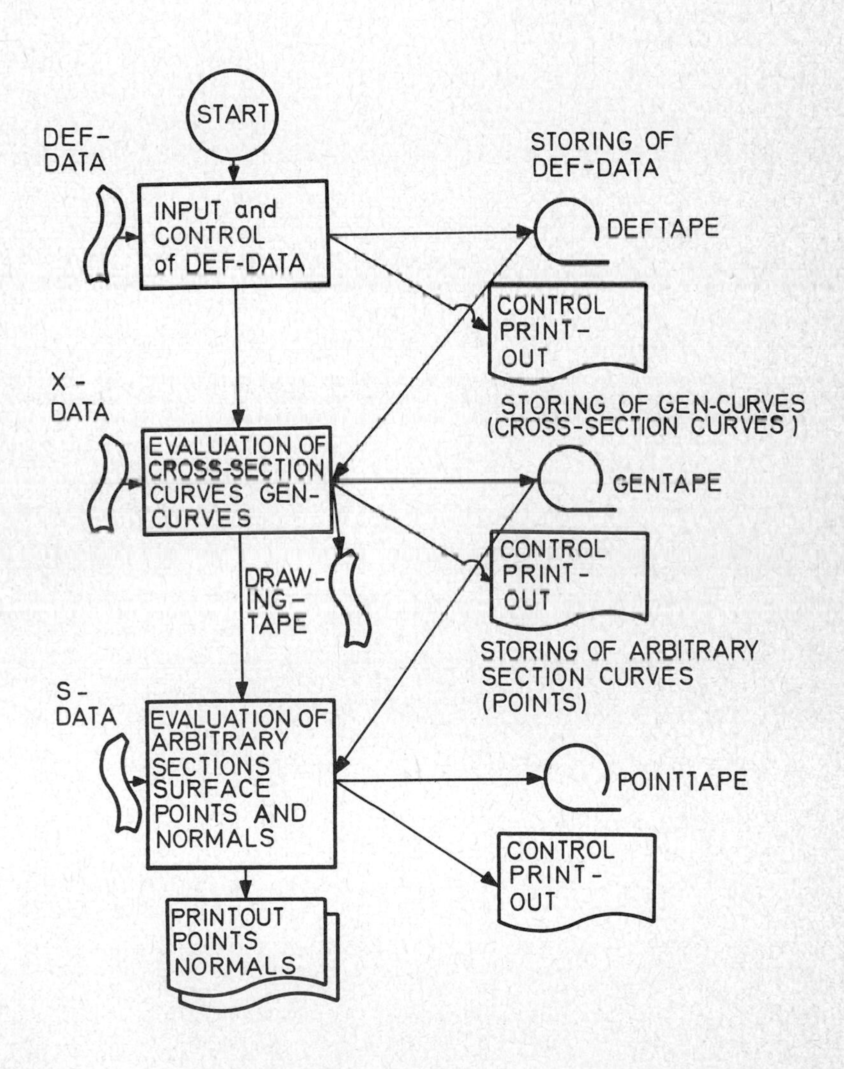

The data needed for various calculations is subsequently read from the pointtape. As is made clear in the flow chart shown below, input of different types of data initiates the calculation of parallel surfaces, the transformation from one coordinate system to another, follow-up calculations such as area, volume, curve lengths, etc, and the editing and storing of geometric data on the form-tape from which Saab-Adapt can directly fetch geometric descriptions of machine-tool movements. By combining FORMELA's geometric part with the N/C language Saab-Adapt, a flexible and rational aid has been created for the use of numerically controlled machines in the workshop. The system allows programming of five-axis machine tools, thus eliminating the usual limitation of Adapt languages, where all tool movements must be on the same plane. In addition, system macros are normally used for fetching the geometric part, constituting yet another expansion of the Adapt language.

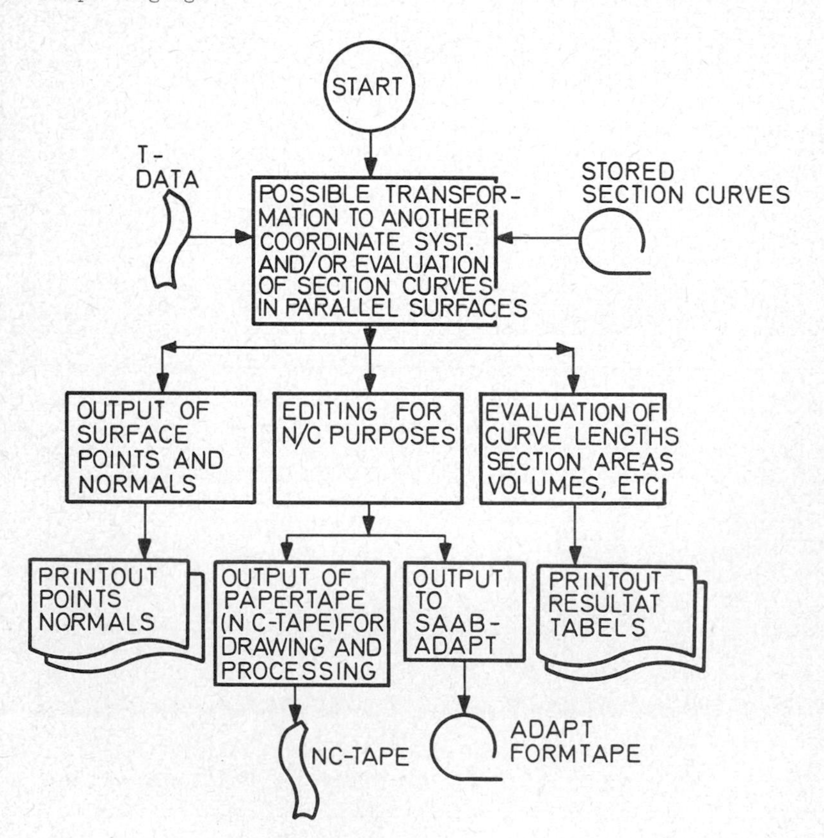

Of the various characteristics of the FORMELA system, the following can be mentioned by way of summary:

Any surface can be defined

Creases in a surface are permissible

Simple mathematical expressions are used

Local alterations in a surface can be easily made

Surface points and normals can be calculated for arbitrary cuts

Parallel surfaces can be calculated at the required distance from the surface defined

Combination of the form-def program and the N/C language allows for greater use of N/C machines

A well-documented training program is available

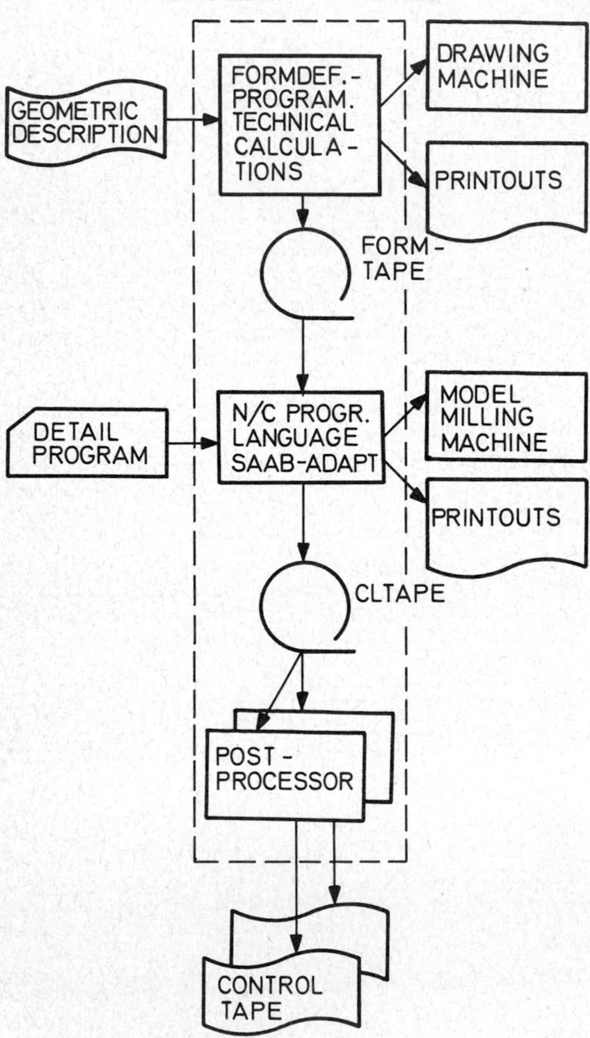
FORMELA SYSTEM
GEOMETRIC DESCRIPTION
FORMDEF.- PROGRAM. TECHNICAL CALCULA- TIONS
DRAWING MACHINE
PRINTOUTS
FORM- TAPE
DETAIL PROGRAM
N/C PROGR. LANGUAGE SAAB-ADAPT
MODEL MILLING MACHINE
PRINTOUTS
CLTAPE
POST - PROCESSOR
CONTROL TAPE

A U T O K O N 71. A TOOL IN SHIP DESIGN AND PRODUCTION

Anton Landmark
Central Institute for Industrial Research
Blindern, Oslo, Norway

1. SUMMARY

AUTOKON 71 is a set of computer programs, constituting a system. Primarily they are intended for use in ship yards as a tool for design and work preparation: elaboration of data, production of drawings and of NC-tapes etc. The system is the result of a cooperation between the production line at ship yards in the Aker Group and the Central Institute for Industrial Research.

After a brief introduction giving the background for the system, this lecture will present the different modules of AUTOKON 71 and how they fit into the environment as a tool. The basic philosophy behind the system architecture will be presented and explained, together with experience gained during the development period and some system installation jobs.

2. HISTORY AND BACKGROUND

The basic ideas behind the system is dating back as far as to the late fifties. They were first published in a speech given by Mr. T. Hysing in Moscow in 1960. Following the development of the ESSI numerical director, the need of a geometry description language became apparent. In 1960 the development of the AUTOKON system was started.

In the beginning, the aim was to make a very powerful geometry description language and a processor to handle it. Later other modules were included as need arose.

Some of the requirements may be of interest:

a) The system should be suitable for a man who was good at his job in the yard, but who knew nothing about computers and only that amount of mathematics which was required by his job at the yard. All difficulties in connection with telling a computer what to do, should be hidden within the set of programs.

b) The system must be a tool fitting into the existing working conditions at a yard. The system should not require the environments to make any noticeable adaption to the tool in order for the system to be operatable, such as organisational changes, reallocation of people, change of design methodes, work preparation routines, part lists, etc. Changes of this type would, however, be desirable during some period of time in order to utilize the tool in the best possible way, according to experience gained and local conditions.

The data base concept was introduced at the beginning of the project. This was done in order for the system gradually to collect the data defining the ship: gradually to "build" the product in the computer before it is built outside. At the same time a quite powerful modularity concept was gained, as programs were collected into a system by working on common data at all times.

The first version of the system was put into the production line some time in 1963. Shortcommings of the system turned up in practical use; such complaints as: "if I did only have such-and-such a possibility, life would be much easier" occurred frequently. This type of positive "trial-and-expansion" situation in a production line continued for several years. During this period, the system increased gradually under real field condition and gained both in generality and in areas to be covered. Yards in the Aker Group also addded new modules to the system. In this

COMPUTER LANGUAGES FOR NUMERICAL CONTROL, *J. Hatvany, editor*
North–holland North-Holland Publishing Company - Amsterdam–London

way the AUTOKON system became a large system with great potential, and a separate firm was established to market the system and its services: Shipping Research Services (SRS), in Oslo, Norway.

3. THE MODULES OF THE SYSTEM

AUTOKON 71 is the latest version of the AUTOKON systems, released in 1971-72. It is a comprehensive system of computer application programs, covering project, structural design, lofting and fabrication of steel parts in the shipbuilding industry. It contains the following modules:

ALKON	-	ALgoritmisk KONstruksjon (Norw.) (algorithmic Design)
DRAW	-	DRAWing of lines
DUP	-	Data base Utility Program
FAIR	-	FAIRing of lines
LANSKI	-	LANgSKIps detaljer (Norw.) (Longitudinal detail)
NEST	-	part NESTing program
PRELIKON	-	PREliminar LInje KONstruksjon (Norw.) (Preliminary lines design)
PRODA	-	PROduction DAta
SHELL	-	SHELL development
TEMPLATE	-	TEMPLATE generation
TRABO	-	TRAnsfer of BOdy plan

All the modules are connected to each other by the use of a common data base: AUTOBASE (AUTOkon dataBASE). Any module may request any existing data from the data base. All new data generated by a module is stored into the data base for future use by anyone interested. The system may be shown as in Fig. 1. The modules are each assigned one job to handle.

Following is a description of each module as to task and responsibility:

3.1 ALKON - ALgoritmisk KONstruksjon, which is Norwegian for ALgoritmic Design.

ALKON is a high level problem oriented programming language. The main purpose of the language is to serve as tool in connection with tasks like:

. plane geometry definition;
. fairing of curves;
. detailing of complex steel structures (in ships, airplanes, etc.);
. production of drawings with texting (automation of drafting);
. production of data for NC-control (drafting, flame cutting, milling, etc.);
. definition and printing of miscellaneous tabular information;
. definition of standard details (parametric descriptions);
. formalization of design and production practice (accumulation of experience);
. definition of identification systems and data structures.

. Since ALKON, besides being problem oriented, is very general, the application area is very wide.

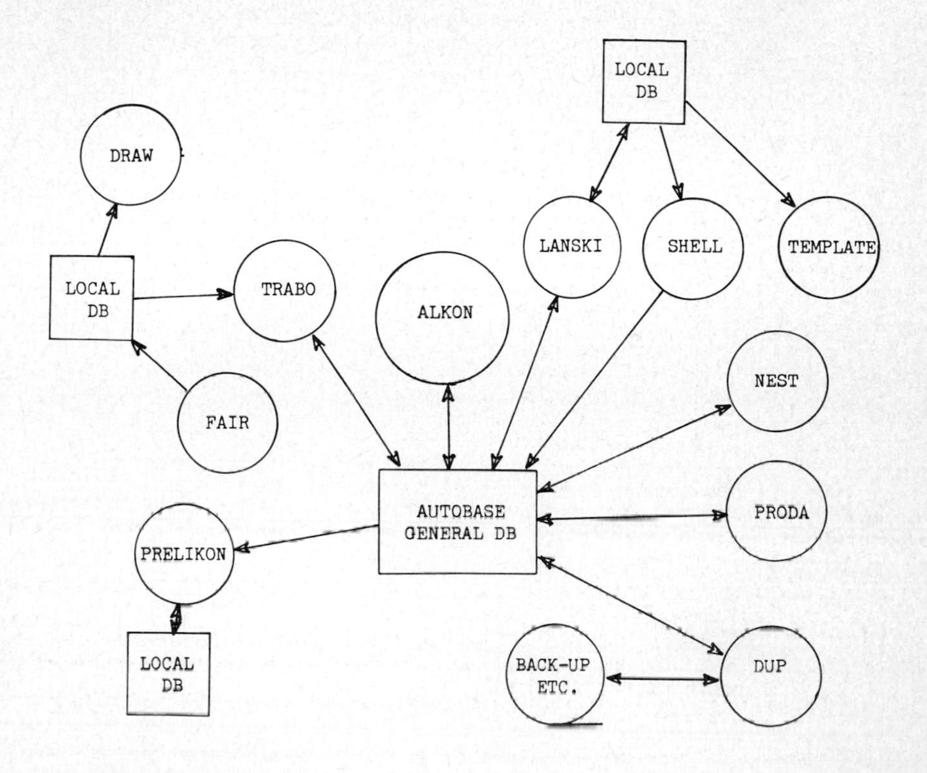

Fig. 1.

The main categories of statements available are:

. vocabulary definition;
. norm definition (a norm is an algorithm written in the ALKON language);
. code sequencing (DO-loop, test, jump, norm calls);
. data base communication;
. list manipulation;
. arithmetic calculations;
. trigonometric and other functions;
. plane geometry definition;
. contour definition;
. text editing;
. table definition;
. output.

The geometry language is of the type: describe the contour as you move along it. Words like SL (straight line), SPT (startpoint), TG (tangency), CIR (circle), CNT (centre), RAD (radius), EPT (end point), etc. are used in the description. Free formats are used according to the rule: if a man can understand the code, so can the computer.

As an example, Fig. 2 may be described in this way:

```
SPT (+0+0)
SL: TG (+50+100)
CIR: CNT (+100+60) RAD (-40)
    TG (+100+0)
SL: EPT (+0+0)
```

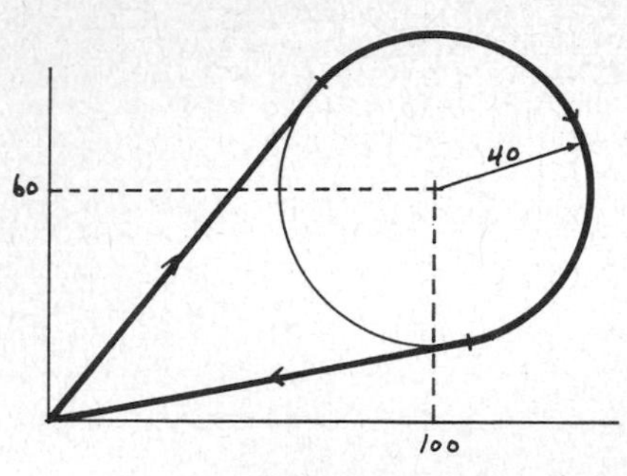

Fig. 2.

The definition of the spelling of the words is part of the language, and may be changed at any time. This is significant for changes from one language to another or to adapt particular abbreviations preferred at a yard. Change of vocabulary is a question of an hour's work.

Any existing contour in the data base may be used as an element, such as the Transverse frame No. 56: TFR 56:

```
SPT (+10000+5000) REF (TFR 56)
SL: DIR (+0) INT'
CON: INT'
SL: EPT (+9000+1000) DIR(+180)
```

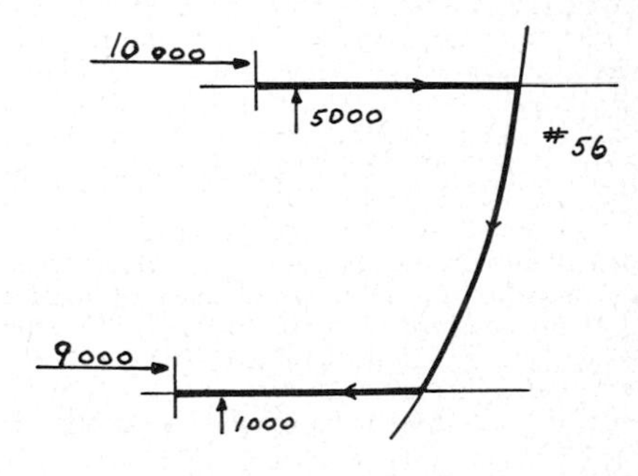

Fig. 3.

These examples show the general idea behind the coding language.

Any norm is an algorithme, expressed in the ALKON language and stored in the data base. Norms are defined using the regular ALKON language included in START NORM - END NORM brackets. Norms are defined dynamically, being reenterant. Functions having one or more values may be defined as norms.

Most of the coding words may also be norms:

SL is the code word for a straight line, while

SL 3 is a norm, generating a straight line according to a rule.

The power embedded in the macros together with possibilities to test on for instance geometry makes it possible to generate a set of parts by one single call:

FRAME 3 (+5, +50) will generate web frames for all stations from No. 5 to No. 50, according to the rules found in the norm FRAME of type 3.

Attention should also be paid to some of the statements in the contour definition section, such as FAIR and PARALELL. The fairing procedure is the same as the one used in the FAIR module. Generation of a paralell to any contour is also quite powerful.

The ALKON module is also used to generate drawings for different purposes. The text editing possibilities and a comprehensive output section is thus very important.

3.2 DRAW. This module is a specialized module, drawing up a bodyplan as delivered by the fairing module. Drawings may be produced for all curves faired, in 3 projections, thus providing very useful information for verification and evaluation of results during the fairing process. Option features permits the user to ask for only parts of the results to be drawn, a particular number of frames, a special space curve, a waterline, etc.

Furthermore the users may select only a "window" of a drawing, which may be enlarged in scale and subject to special investigation. A simple command language is used for input, where primary commands such as BODY, BUTT define the type of drawing, and secondary commands such as SCALE, GRID, WNDW define auxiliary data about the drawing.

3.3 DUP. Data base utility Program. The program is used to handle the data in the data base, such as creating a back up magnetic tape, copying parts of one data base into another, etc. This is performed using a set of commands, such as READ, WRITE, DELETE, CONTENTS, COMPRESS, DUPLICATE, PRINT, etc. Each command is followed by a specific set of parameters, describing which data is to be affected, if any.

3.4 FAIR. Fairing of lines. The purpose of this module is to perform the fairing of ship lines and to transform these lines into a numerical form suitable for further computer processing.

The basis for performing the fairing process is a preliminary lines drawing in a suitable scale.

Normally 200 - 400 data points for an entire ship are lifted from the lines drawings, the number depending upon the hull form and particularities. Necessary information are stern and stern curves, curves of tangency in bottom and side, knuckle curves and some selected points on frames, waterlines and/or buttocks.

The fairing is performed automatically from the data points selected by the user and resulting finally in printed and graphical output through DRAW.

FAIR can be used for any shape of the midship section, for vessels with inclined keel and even for submarines.

The method used in the FAIR module is based upon the manual method applied by the loftsman: Thin elastic splines of different shapes are used to fair the separate lines. The spline theory forms the basis of a number of published mathematical fairing techniques. A basic difference between the FAIR method and the others is that no assumptions have been made about "small deflections". The FAIR method takes into full account all the nonlinearities occurring when the deflections become significant, as they do for instance for a typical frame in a normal hull. The fairing method is consistent throughout and provides full continuity in the first derivative. The user controls the process merely through his selection of data points and weights and is not faced with the requirement to decide on a selection of functions to approximate the hull form decided. Thus, the user needs no special mathematical knowledge.

3.5 LANSKI. Longitudinal details. The purpose of this module is to fit longitudinal curves, such as seams, longitudinal members, stringers, bulkheads, decks, etc.: on to the hull surface, and add respective cutout details.

A Table of Details is produced, both in the data base and as output describing how and where the longitudinal details cut through transverse frames, bulkheads, etc. The space angle between two intersecting parts is computed and stored, to provide automatic correction of the nominal dimentions for clearance, both on height, width, and, if necessary, on thickness and angle of flange, etc. Drawings for checking purposes are also produced.

3.6 NEST. Part Nesting. The purpose of this program is:

a) To arrange a multiple of steel parts on a rectangular steel format and interconnect them such that
   the steel is utilized efficiently,
   the oxy-cutting time is minimized, and
   heat distortion and bending is avoided.

b) To produce NC-tape for cutting.
c) To make NC-drawings of nested formats for checking, planning or for optical cutting.
d) To store nested format data in the data base.

The actual "jig-saw-puzzle" problem is solved manually, a description of which part is positioned where is given to the computer together with the cutting sequence required.

3.7 PRELIKON. Preliminary lines design. The purpose of this module is to perform hydrostatic calculations, etc. The calculations may be made on a preliminary hull form or on a final, faired body plan. The module contains a preliminary hull definition capability and a hull variation capability. A library of existing ship designs may therefore be of value as a starting point for a new design. Hull variation can reshape smaller or larger parts of the hull. (This library may be of value also in other stages of the design, as specific norms or even parts may be used.) A library of this type may be made merely by saving the data base contents of each ship onto a magnetic tape (one tape per ship).

PRELIKON is intended used both as a tool in the initial hull development periode, and as a checking device to be used later to see that the hull does fulfil the requirement. It is also used for the stability calculations, launching calculations and for production of various tables which are to follow the ship (ullage tables, etc.).

The input to PRELIKON is a fixed set of commands, each followed by a set of parameters. Both command and parameters are in a fixed format, specified in pre-printed forms.

3.8 PRODA. Production data. The purpose of this module is to extract data from the data base and convert them to useful information for planning and production purposes. For the AUTOKON - 71 system this information refers to:
a) Calculation of cutting LENGTH and cutting time of nested formats and single parts in the oxy-cutting machines.
b) Calculation of AREA and weight of single parts and optionally adding up the total of a series of parts.

3.9 SHELL. Shell development. The purpose of this module is:
a) To develop shell plates.
b) To generate NC-tape for cutting and marking of plates.
c) To generate drawings for checking, planning or for optical cutting and marking.
d) To produce information for bill of material.

Input is normally prepared on the basis of a shell expansion plan and a body plan. The hull form and longitudinal curves are previously stored in the data base. The program is executed and NC-drawings may be made in various scales for the purpose of checking, planning and optical cutting. Plates may be generated for 2-axis as well as 3-axis cutting.

3.10 TRABO. TRAnsfer of BOdyplan. The FAIR module is working on a local file (magnetic tape) in order not to disturb the data base by suddenly introducing a new and different proposal for a hull shape. The different proposals are collected locally. When a result has been obtained this is transferred to the data base. The FAIR module may generate a whole ship at once, or only half of a ship. In case of a non-symmetrical hull-form, as little as one fourth of a ship may be produced (for instance port side of the fore half body). The different parts of the ship must be hooked up so as to form an entire ship. TRABO is designed to transfer the result from the temporary working storage of FAIR into the data base and add it to whatever is existing. Some expansion may also occur. A port side of the fore half body is faired and is stored into the data base as the whole fore ship. In this case a starboard side must be generated as the mirror image of the existing port side, these two parts being joined together and added into whatever is existing in the data base.

4. THE SYSTEM IN THE PRODUCTION LINE

After this very short description of each module has been given it is time to see how the modules cooperate to take part in the design and work preparation stages in the production of a ship. After all, the system is supposed to be a tool for a ship yard.

The design stages in the production of a ship may somewhat simplified be represented as in Fig. 4.

PRELIM. DESIGN	STEEL STRUCTURAL DESIGN	STEEL STRUCTURAL DETAILING	NUMERICAL LOFTING	WORKSHOP

Development

Fig. 4.

A design tool must be able to aid the designer in all these steps, and gradually "build" the ship inside the computer. An aid at a few, well separated places in this development sequence may easily cause more trouble than aid.

During the preliminary design stage, PRELIKON is used to check out the propositions, and may also deliver some preliminary body lines, see Fig. 5. Based on these, FAIR starts off, producing a faired body plan. Based on a temporary body plan placed in the data base while the fairing is going on, a set of frames may be produced by ALKON, based on experience from previous ships, stored as design norms in the data base. Also completion of the body plan may start up based on a temporary body plan, using the LANSKI module. This body plan may not be exactly as the final one, probably it will contain unsmoothness and undesirable knuckles and other faults, but in most cases it is sufficient for the people designing the steel structure so they can form an opinion on how the proposed solutions will work out in the present case. Modifications to meet new demands may be introduced very early.

In parallel with these activities, although somewhat after the introduction of seams and butts by LANSKI, but before the final result is available, the development of shell plates may be started, using SHELL (see Fig. 5).

The structural detailing, performed by ALKON, will continue for a long period of time. In the early stages, the general shapes of larger parts such as frames and bulkheads are formed and modified, based on previous experience. Then several drawings are made of typical sections (for classification, etc.). The large, structural parts are then subdivided, dividing lines for sections are introduced and the steel structure detailing starts up, also using previous experience, but this time on a different level of detailing.

A lot of the information collected this far is used in material specifications. NEST may be used on a preliminary set of parts to produce a temporary set of need formats.

Later the real part coding can take place. This consists mainly of modifications of the design parts due to production requirements and demand: shrinkage, kurfe compensation, etc. The nesting procedure may be repeated, the old specifications being tried and modified, or completely new specifications made. Other work preparation activities are also handled: production of templates, jig data, cutting time of the formats for the production planning, as well as the weight of the sections.

At a suitable stage in the design process, PRELIKON may be used a second time to check the results. The faired body may deviate somewhat from the preliminary one, and this time the calculations may be carried out with reasonable accuracy.

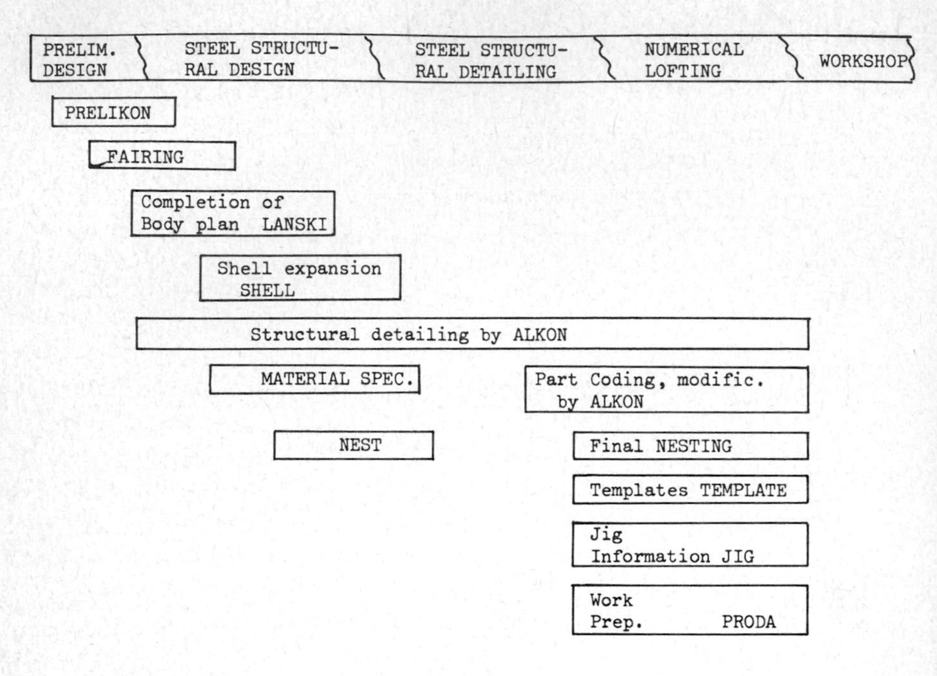

Fig. 5.

5. EXPERIENCE

Having breafly seen what the system consists of and how it is used, it may be natural to examine what sort of experience has been collected during such a development.

Some key words are: Reliability, adaptability, portability, modularity, generality.

To start with reliability. For a system to be used by a lot of different people, reliability is a must. Reliability means that any man will meet a program but only very rarely. This in spite the fact that he will use the programs quite differently from what the programmer assumed and designed it for. When the users get such a tool in their hands, they start using it. And "using" here means to think of new applications which to the user appears to be nearly the same thing as it is designed for. After a few such "jumps", the application may be quite different from what was originally intended.

When jumping from one application to another, the programs do not necessarily fit the task perfectly. Someone will make some changes. That is the first real acid test of reliability: how will a program react to a few changes, made by someone at some place that appears to be the right one. Next test is to follow immediately afterwards: The data of the new application will be quite different from the old data sets: what previously appeared to be completely idiotic combinations of data (inconsistant, etc.) are now suddenly quite normal. The question then is: can the program tolerate any crazy set of data and still react reasonably. This means to be very suspicious and never take anything for granted. Reliability is something that is designed into a program, using a lot of effort. It is not merely something that is tested into it.

Adaptability is related to reliability: how much change in environment can the program tolerate. Change in environment means change of application, change of habits, working routines, new methodes, organisational changes, etc.

Modularity and generality are likewise related to the previous ones and to each other. Modularity also means that it is possible to change a module completely, or add a module. It should, however, be noticed that the word "modular" commonly is used in what really turns out to be a lot of different meanings (depending upon how strong the bindings are between modules).

Portability is also quite important for any successful system. One thing is that the life time of a computer at a particular installation may be quite low. Another aspect is that several users do not necessarily have the same computer or even the same brand.

To illustrate some of the points regard a line connecting two extremes: Speciality and generality, see Fig. 6. A compiler like ALGOL or FORTRAN may represent generality, a routine computing sine of x represents speciality.

Fig. 6.

It is possible to note down several characteristic abilities for both of the two extreme points, such as: how easy is it to learn to use, how easy is it to use without making mistakes, how many different problems can it handle, etc. Most of the programs or systems being offered for CAD application have got their operating point fixed on this line somewhere. Making ALKON, we tried to avoid being fixed to any particular point on the line. Depending upon the application, it is desirable to fix a place suitable to the specific requirements, being generated by the answers to a lot of such questions as those just mentioned. The requirements for part production and for generation of design rules are quite different. The macro facility as it is implemented, makes it possible to slide over a wide range of the line; not using more complexily and difficult code than what is required for the job, but not being left with an insufficient tool.

It is also of great importance that the tool will fit the job. In case of a tool being used for several jobs, the tool may appear as being just a little bit different from one job to the next. One of the important abilities of the macro facility is to permit this apparently different faces of the system.

Training time, required by the users in order to utilize the system to the full extent required by the job is of some importance. The ability of the system to slide along the speciality-generality line permits a differenciated training level of different user groups, combined with a gradual training/upgrading possibility for the users.

As for modularity, regard Fig. 1 as a good example. The modules are made by quite different people, at quite different time. The system is a result of a gradual growth, there are no signs of the growth having stopped. What is present is only the geometry definition part of a system. The production planning and the registration of how well the plans are followed (how good the plans were) are still missing. So are the account systems, system for wages, stock and materials handling, etc. We have quite good experience as to this type of modularity.

This has been a short presentation of AUTOKON 71 and the application of the system in ship yards. Many of the modules are of much more general nature than only to serve the shipbuilding industry and their problems; not because their problems are small and trivial. On the contrary: Many of the problems they are having in CAD are of a more general nature: Fairing, part production (or rather definition of geometry: not only shapes, but also positions and logical grouping of parts, as in sections and sub-assemblies). Also the link between parts and schedules and plans, interaction in case of delays, etc.

"PURE" CAD is a tool in production, but it is probably only one piece of a larger system, a system which for the sake of mankind must be flexible and adaptive, not square, fixed and restrictive. Such a system, or the lack of it, will form the environment and living conditions for our children and successors. Let us keep that in mind when we make the systems of to-morrow. AUTOKON x may be one of them.

APPLICATION OF GRAPHIC NC GAS CUTTING SYSTEM (GGCS) WITH MINICOMPUTER IN SHIPBUILDING

Tetsuo Hirano
Masataka Kira
Mitsui Shipbuilding & Engineering
Co., Ltd., Information System Div.
Tokyo, Japan

Summary: In recent shipbuilding activity, NC gas frame cutting has settled as the basis of rationalization.
In our shipyards, with the expansion of scope of application of the NC gas frame cutting, productivity of part-programming has come into question, and as a solution Interactive Graphic was decided to be applied.
The question of economy of the graphic display led us to the application of minicomputer which had been greatly improved.
Taking into consideration the possibility of expansion in future, the system was designed to keep the exchangeability of data, for possible connection through on-line communication with LINE CAD SYSTEM, HULL CAD SYSTEM etc. which are handled by large scale computer.
Under this system, production of assembly drawing, part drawing, and nesting plan, and information required for gas cutting is to be handled by means of interactive graphic using minicomputer.
To improve economy of the system, it is designed so that four sets of interactive graphic display can be connected with one minicomputer.
For the improvement of man-hour performance in the production of utility programs, graphic NC language was newly developed.

INTRODUCTION

In the field of shipbuilding, where each ship is built by separate order, the shape of members differs each other, and a processed NC tape is used only for one particular member.
We developed shipbuilding NC language named PDL/II, which resembles APT, in 1968, and have been using NC processing by means of part program method.
At present, in one of our works, five sets of gas frame cutting machine are in operation, and more than thirty programmers are engaged in coding the part programs.
The progress of NC processing has brought various problems as follows:

1) the amount of part programs to be coded has become considerably large.
2) large scale computer is occupied for NC for more than five hours per day, which makes the turn-around-time longer and longer.
3) part programmers are now apt to hate coding of the part programs for long time.

As a solution to the above problems, we decided to develop this system in 1970, applying graphic display developed by ourselves.

SYSTEM ORGANIZATION

1. SOFTWARE ORGANIZATION

The software organization of this system comprises the following four divisions, MONITOR - to control the allocation of system resource and the schedule of job execution, GRAPHIC NC LANGUAGE PROCESSOR - to make utility programs, UTILITY PROGRAMS - to carry out NC processing, and MISCELLANEOUS PROGRAMS.

a) MONITOR

This is the supervisor program prepared for the efficient execution of UTILITY PROGRAMS. It controls main storage, dynamic program loading, automatic swapping of data and simultaneous execution of several jobs, and can be run by main storage having 16K words (1 word = 16 bits).
This MONITOR has the following functions for the execution of complicated NC processing using the limited main storage in the most efficient manner:

PROGRAM RELOCATION

Function to move freely the stored address of the programs to create open space collecting small dead spaces scattered in the main storage, each of which is so small to be used independently.

DYNAMIC PROGRAM LOADING

Function to load necessary programs from the disk into main storage when the program is not stored in the main storage. In case if there is not an adequate space to store the necessary program in the main storage, this system deletes the stored programs in order of infrequency in use, so that the necessary programs can be loaded in the space acquired. Thus maintaining the flexibility of the system, the operator can run necessary programs anytime when it becomes necessary.

PROGRAM REENTRANT

Funtion to keep efficient use of the programs stored in the main storage. Several jobs can use the same program stored in the main storage simultaneously.

DATA AREA SWAPPING

Number of the jobs to be done simultaneously increases, several jobs can access to the same data area commonly in the main storage.
In case if necessary data for the job having control is not found in the data area of the main storage, this function once rolls out the data into the disk to acquire an adequate space to roll in the necessary data for the job from the disk.

MULTIPROGRAMMING

As the NC processing is done through interactive conversation between the operators and the minicomputer, speed of the job execution depends on the operator's capacity, which means that the minicomputer is almost always in the state of 'waiting'.
Using this waiting time, the operators are able to do several jobs at the same time.

VIRTUAL DATA AREA HANDLING

The amount of data required for the NC processing would be so large that could not be stored in the main storage. Therefore, it is required to roll the necessary data from disk into the main storage.
Under this system, utility program is coded as if all the data is stored in the main storage, and practically in case when the data is not found in the main storage, the data is to be automatically roll into the main storage from the disk.

b) GRAPHIC NC LANGUAGE PROCESSOR

The utility program which can be run under the control of MONITOR should be relocatable and reentrant.
Generally, programs which satisfy these conditions could only be written in ASSEMBLER LANGUAGE by considerably well-trained programmer.
This is an obstruction to be overcome for easy coding and maintenance of the utility programs.
The GRAPHIC NC LANGUAGE PROCESSOR has been developed as a solution to this problem.
In writing programs, programmers;

1) need not have any knowledge in respect of addresses, such as absolute address, relative address, indirect address etc.

2) need not pay any attention to reentrant or relocation of the program, and

3) just write the executable statement in the same manner as FORTRAN statement.

Sample of the coding in GRAPHIC NC LANGUAGE is shown below.

```
C* LINE1: ETMCR                                  ;ENTER MACRO
C*        LHA1    %Ø                              ;RØ=HITCTR
C*        BRNCH   ERØ1Ø   %Ø      NE      3       ;JMP ERØ1Ø (RØ.NE.3)
C*        RESRV   LA1     1                       ;LA1    .WORD  Ø
C*        RESRV   LA2     1                       ;LA2    .WORD  Ø
C*        LHA2    1       LA1                     ;LA1   HAS LOCATE P1
C*        LHA2    2       LA2                     ;LA2   HAS LOCATE P2
C*        RESRV   ID      1                       ;ID     .WORD  Ø
C*        RESRV   INDEX   1                       ;INDEX .WORD  Ø
C*        LHA9    LA1     ID      INDEX           ;P1 INDEX
C*        BRNCH   ERØ1Ø   ID      NE      Ø       ;ERØ1Ø WHEN NOT POINT
C*        LHAP    INDEX   %2                      ;R2 = NI OF P1
C*        LHA9    LA2     ID      INDEX           ;P2 INDEX
C*        BRNCH   ERØ1Ø   ID      NE      Ø       ;ERØ1Ø WHEN NOT POINT
C*        LHAP    INDEX   %3                      ;R3 = NI OF P2
C*        RESRV   WPTØ1   14.                     ;GET 14.WORD FROM DTA
C*        RESRV   WPTØ2   14.                     ;GET 14.WORD FROM DTA
C*        RESRV   WLINE   14.                     ;GET 14.WORD FROM DTA
C*        RESRV   WPTØ3   1                       ;GET 1.WORD  FROM DTA
C*        PTRED   %2      WPTØ1                   ;P1 DATA READ
C*        PTRED   %3      WPTØ2                   ;P2 DATA READ
C*        GLINE   WLINE   WPTØ1   WPTØ2           ;GENERATE A LINE
C*        SWRIT   WLINE   WPTØ3                   ;CATALOGE LINE TO DK
C*        GELMN   WPTØ3                           ;DISPLAY  THE LINE
C*        BRNCH   LNEND                           ;PROCESS END
C* ERØ1Ø: ERMSG   2                               ;TYPE ER-MESSAGE
C* LNEND: EXMCR                                   ;EXIT MACRO
```

Fig. 1   G.N.L. Example

c) UTILITY PROGRAM

This is the program which executes NC processing under the control of MONITOR.

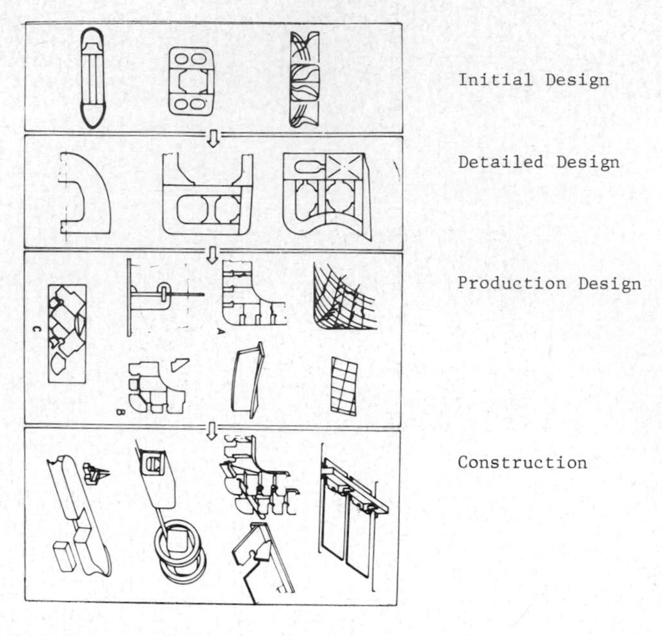

Fig. 2 Process of Design and Construction of Ships

A: Assembly Drawing
B: Part Drawing
C: Nesting Plan

RING GENERATE PROGRAM

This is to generate the information, the so-called RING, corresponding to the assembly drawing shown in Fig. 2.
Photograph 1 shows the RING generated on the display screen. To keep efficiency in generating the RING, the operator defines the elements relative in function and position in as large module as possible.
This program can define point, straight line, circle, tabulated line, hole, notch etc.

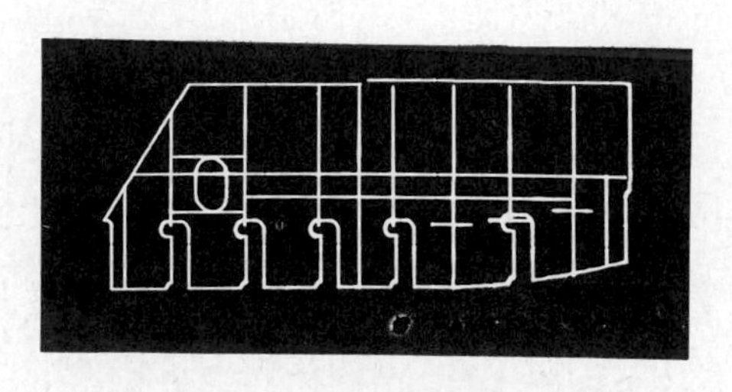

Photo 1 Ring image on screen

RING SEPARATE PROGRAM

This is to separate the RING into the information, the so-called PIECE, which corresponds to the part drawing shown in the Fig. 2.
The operator separates the RING into PIECEs by suitable line for construction. Each PIECE is named and stored in the disk.
Photograph 2 shows the PIECE separated from RING.

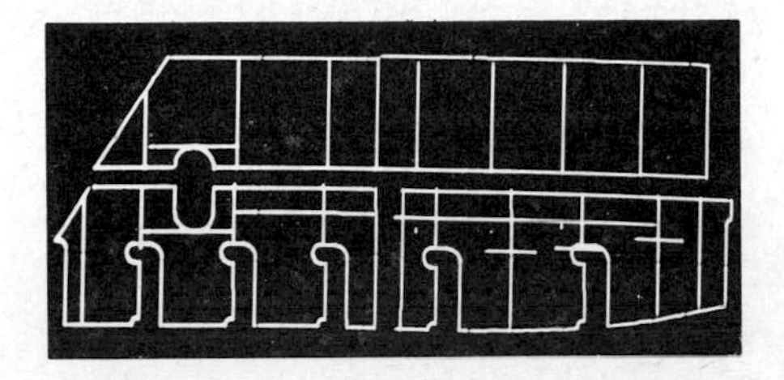

Photo 2 Piece image on screen
A ring (Photo 1) is separated to three pieces.

NESTING PROGRAM

This is to generate the information corresponding to the nesting plan shown in Fig. 2.
Each PIECE is rolled in from the disk by means of its name as a key, set in the framework given, rotated and/or shifted so that the steel plate would be cut in the most efficient manner.
Photograph 3 shows the PIECEs set in the framework.

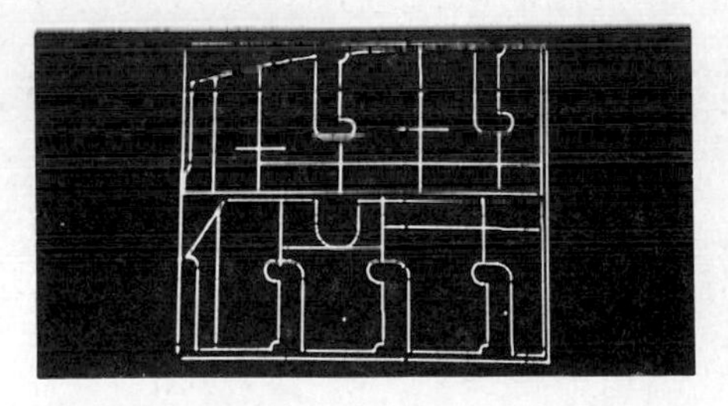

Photo 3 Nested image on screen

FRAME GAS CUTTING PROGRAM

This is to generate paper tapes having format understandable to the NC director of the gas cutter, adding information necessary for gas cutting, such as cutting sequence, bridge information, correction of slots caused by gas frame cutting.

d) MISCELLANEOUS PROGRAMS

DEBUGGING UTILITY, CHECK POINT UTILITY etc.

2. HARDWARE ORGANIZATION

a) Minicomputor:

Main storage - 16K words (1 word = 16 bits)

Disk - 1.2 million word x 2 unit

Magnetic tape

Card reader & Line printer

b) CRT display:

14 x 14 inches in size, refresh type
precision; 4096 x 4096 dots
with function key
with rotation & shift wheel
in color (red, yellow, green and orange)
Rotation & shift wheel is an input device to rotate and shift the images on the screen freely.
In case if the images are too large to be displayed on the screen, all the images are virtually memorized by the display controller and then displayed on the screen by means of this wheel.
This function is called as VIRTUAL SCREEN.

Photo 4 CRT display

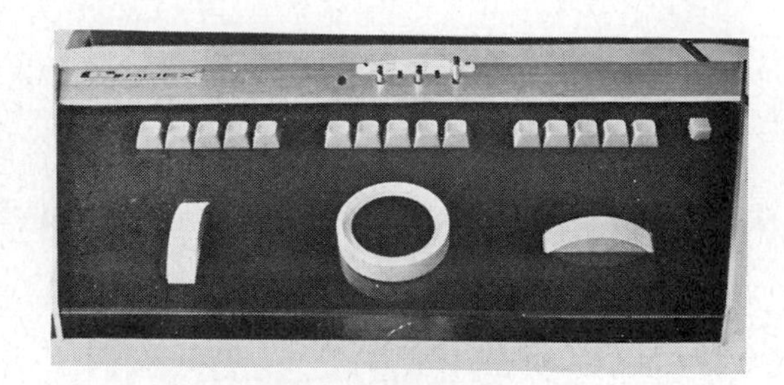

Photo 5 Rotation and shift wheel

c) High-speed gas frame cutter

See the photograph 6 and 7.
This gas frame cutter has a cutting speed of more than 1 meter/min.
Edge preparation can be made simultaneously.

Photo 6 Gas frame cutter

Photo 7 Edge preparation

DISCUSSION

<u>COMPUTER COST PERFORMANCE</u>

NC system of this kind can be desinged comparatively easily on the medium or large scale computer, but in this case the running cost is the question. This system provides us with the means of cutting the running cost drastically, applying minicomputer instead of medium or large scale computer.
This system costs only ten months rental fee (necessary for NC processing only) for the large scale computer.

MAN HOUR PERFORMANCE

As it is very easy to define the line, point etc. under this system, and regular jobs can be done by this system, the operator can concentrate on creative jobs. Fully applying this system, we can cut the number of operators from thirty to less than ten.

TOTAL PERFORMANCE

As this system is operated real-time, errors are found immediately and as a result the turn-around-time can be considerably shortened.
In addition to the above, the flow of job has been simplified, and the process of production has been smoothed.

CONCLUSION

In accordance with the progress of technique and improvement of construction method, the function required for the NC system is varying rapidly.
It is not an easy job to do maintenance of the NC system corresponding to the above change.
To solve this problem, this system is composed of separated modules written in a higher level language.
Optimization of this language is the subject to be done in future.

The programs are usually loaded into the main storage when called. Average accessing time to the disk is 60 meters/second, and in case when a complicated processing is being made, the operator has to wait for several seconds. When the number of jobs increases, the operator may feel impatient.
In future, this will be solved by improving size of the module of the program, capacity of the main storage, method of access to the disk.

References:

1) A review of hardware and software for CRT display system
- National Technical Information Service
- Springfield Va. 22151 U.S.A.

2) Sketchpad a man-machine graphical communication system
- Ivan E. Sutherland, Lincoln Labolatory M.I.T.

3) Theoretical foundations for the computer-aided design system
- Electronic Systems Laboratory M.I.T. Cambridge 39, Massachusetts

4) Virtual machine computing in an engineering environment
- IBM Systems Journal Vol-11 No-2 1972

PIPE-language

AKIRA ONO

Nagasaki Technical Institute,
Mitsubishi Heavy Industries, Ltd.
Nagasaki, Japan

Abstract: PIPE-language, Practical and Integrated Pipeline Expression language, is a tool for describing information, which should be fed into a computer, of a piping system. It can be applicable for every stage of piping engineering -- preliminary, detail and fabrication designs -- and the capabilities of the language can be extended according to its generalized DTL functions.

In this paper, the following significant features of PIPE-language and some application examples are shown:

- a generalized Data Tabulation Language(DTL),
- a classification of the positional entities, on which fundamental operations of semantic routines and data structures depend.

1. INTRODUCTION

Piping is one of the most fundamental engineering in every industrial field.

For a long time, there have been strong needs of applying a computer for automating design processes, and many contributions along this line have been made, however, it has not been satisfied because of the difficulties due to specific features on piping engineering, such as:

- large amount of varieties of the included information,
- large amount of frequencies of modification required in the course of design for the relation to other structures and machinery.

This paper is concerned with the input language for describing piping information, which is developed for facilitating the computer automation of piping engineering.

It is impossible to provide a language and its processor which will be suited to every piping application stages. It is, therefore, imperative that the user be given the ability to expand the function to suit his own needs. One of satisfactory way of accomplishing this is to provide him with a basic language system, processor and the ability to define new functions in term of syntax and semantics.

2. PIPE-LANGUAGE

The PIPE-language is a tool for describing the piping information which is contents of drawings or hand-writing sketches by a piping designer. Almost of necessary information can be described by a sequence of its statements, that is characteristics of pipe, fittings, geometrical pattern of piping routes, and many kinds of information for design and fabrication.

The syntax of PIPE-language depends upon 'Data Tabulation Language(DTL)', which is a by-product of PIPE-language development.

The expressional capability of PIPE-language depends upon semantic functions linked by a language processor, which is named 'PIPE-processor' or 'PIPE-system'.

2.1 Syntax of PIPE-language

The Data Tabulation Language(DTL) is a language which has a concrete syntax but it can express no more than abstract objects, called 'entity'. When user gives a concrete meaning on the entity, DTL becomes to express a concrete entity.

COMPUTER LANGUAGES FOR NUMERICAL CONTROL, *J. Hatvany, editor*
North-Holland Publishing Company – Amsterdam–London

A specific DTL which becomes to express several piping oriented meanings, is the PIPE-language. The syntactic details on DTL -- that is the syntax of PIPE-language -- are described here.

<u>Character definition</u> The lexical processing routine in the DTL-processor requires a character type definition table(CDT), which classifies all the permissible characters for the computer on which the DTL-processor being implemented, in following seven types given in Backus normal form.

<hardware character sets> ::= <letter>|<digit>|<separator>|
<decimal point>|<literal delimiter>|<terminal>|<dummy>

A default character definition table is prepared in the routine(Table 1.), but during execution the system allows user to select a suitable one from the previously prepared CDTs for his purposes.

<letter> ::= A|B|C|D|E|F|G|H|I|J|K|L|M|N|
O|P|Q|R|S|T|U|V|W|X|Y|Z

<digit> ::= 0|1|2|3|4|5|6|7|8|9

<separator> ::= _|*|-|+|/|=|(|)|,|#|$ _ means 'blank'

<decimal point> ::= .

<literal delimiter> ::= @|'

<terminal> ::= ;

<dummy> ::= all characters not listed above

Table 1. Default character definition table.

In the actual applications, one or two character definition tables are used, because of existing strong needs to use same character in different meanings. This needs mainly depend on the compatibilities of data supply with other running systems and user's conventional way for naming.

For example, the character '-' has a meaning of negative sign in an arithmetic expression, while encoding a name of fittings, usage of '-' as a hyphen is ordinary way. E.g. FLANGE-ASA-150LB, 1V-155, S-E.

Another example is the usage of zero. In many cases of numbering or encoding parts and materials, leading zeros are significant, e.g. 80000-<u>0001</u>.

In such usages, it is rather reasonable for lexical processing routine to define the character '-' and digits as<letter>.

<u>Word assembling</u> The lexical routine in DTL-processor reads a string of characters in the source language and assemble them into a sequence of 'words'.

There are six types of word in DTL:

<symbol> ::=<letter>|<digit><symbol>|<letter><symbol>|
<symbol><letter>|<symbol><digit>
<integer> ::=<digit>|<integer><digit>
<real> ::=<integer><decimal point><integer>
<literal> ::=<literal delimiter><any character strings>
<literal delimiter>
<separator> ::=<separator>
<terminal> ::=<terminal>

where a single separator or terminal character makes a word by itself.

DTL statement The categories of DTL statements are as follows:

- Entity statement
- Entity block statement

An 'entity' is a pair of an 'attribute part' and a 'linkage part'. The attribute part is a set of 'attributes' associated to the 'entity type'. The linkage part is a set of linkages between the entity and other entities. The entity statement creates an entity, gives it a name and defines its attribute part. The format of the entity statement is as follows.

label	body of statement
entity name	entity type identifier = attribute lists ;

The 'entity name' is a symbol which identifies each entity. The 'entity type identifier' is a symbol which specifies type of entity.

The 'attribute list' is a list of 'attribute expressions', each separated by comma.

attr.exp. , attr.exp. , attr.exp. ,-,-, ;

The attribute expression has a form;

attribute identifier / attribute data

The following is an example of entity statements.

PIPEX	PIPE=D/100 , SCH/S4 , MAT/STEEL ;
PIPEY	PIPE=SCH/S8 , D/50 , MAT/COPPER ;

where PIPEX and PIPEY are entity names. PIPE is an entity type identifier and the right hand side of the equal sign is the attribute list. D/100 or SCH/S4 are attribute expressions. D or SCH are attribute identifiers and 100 or S4 are attribute data.

Entity name, entity identifier and attribute identifier must be symbols.

Attribute data may be any expressions acceptable by its attribute identifiers.

DTL does not concern with what meanings each entity type identifier or attribute identifier has. The set of entity type identifiers, attribute identifiers and permissible attribute data expressions for each attribute identifier must be defined by the specific application of DTL. An 'entity block' is a sequenced set of entities of the same type and defined by an entity block statement followed by a sequence of entity statements. The entity block statement has a form:

* entity type identifier

An entity statement following an entity block statement is called 'specific', when its entity type identifier is the same as that of the entity block statement, and 'declarative', when it is not.

* A

A1	A = a_1, a_2, a_3, -- ;	--- 'specific'
B1	B = b_1, b_2, b_3, -- ;	--- 'declarative'

For the specific entity statement, following two abbreviations are permitted.

label	body of statement
entity name	attribute lists ;
	entity name = attribute lists ;

For the declarative entity statement, whose attribute list is the same as that of one of the previously defined entities, the following abbreviated form is permitted:

	entity type identifier = entity name ;

, where entity name is the name of the previously defined entity.

Only the specific entities, which are defined by specific entity statements, make up the entity block. The attribute part of each entity has as many entries as the number of attribute names associated to the entity type. The entity corresponding to the attribute name, which appeared in the attribute list of the statement, is filled with the attribute data specified in the attribute expression. The attribute which are not specified in the attribute list remains undefined.

An entity specified by a declarative entity statement is called 'declarative' until another entity of the same type becomes declarative. Thus, for one entity type, only one entity at a time may be declarative. The linkage part of each specific entity is the list of the entity names which were declarative at the time when the entity was defined.

The abbreviated form of the declarative entity statement makes the entity specified by the entity name be declarative for the type specified by the entity type identifier.

The not-abbreviated form of the declarative entity statement defines a new entity with the specified name, type and attributes, and states it be declarative for the type of entity. In the case of declarative entity statement, the attributes not specified in the attribute list are defined as the same as that of the old declarative entity, which the new one is going to replace. Of course it is permitted to use the expression of new attribute data which means 'undefined condition'. Thus, unchanged attributes need not be defined each time.

A new entity defined by a declarative entity statement is added to the entity block of the same entity type as itself, if and only if its attribute part differs from that of any entity in the block. Its linkage part remains undefined. These concepts are illustrated in Fig. 1.

The DTL processor thus converts the DTL statements into entity blocks which are represented in the computer by two dimensional arrays. Any data specified on the DTL statement is distributed and put into proper entries in these arrays, as the statements are processed.

The DTL processor performs only the syntactic processing and concerns with no semantics. The user or the system planner in the specific field of application may define any number of entity types and attributes names and may choose any form of expression of attribute data. About the last of the above, DTL processor may be requested to use user-written translators to analyse some specified attribute data.

2.2 Semantics of PIPE-language

Relation between DTL and PIPE-language can be considered as Fig. 2.

Semantic feature of PIPE-language are determined by followings:

- for language formats ..... determination of 'entity', 'entity type identifier', 'attributes' and permissible format for 'attribute expression',

- for processor functions .. decoding routines for each attribute expression and semantic routine to generate 'new meaning' using the attribute data in 'entity block'.

Entity and entity block in PIPE-language PIPE-language for a ship piping, for example, following entity identifiers are registed.

- POSITION .... definition of the predefined plane or curved surface, e.g. position of bulkhead or cambered deck
- PIPE .... definition of pipe
- PARTS(FITTINGS).. definition of fittings
- LINE .... geometrical information of piping route,
- PIECE .... piping unit for assembling,
- PALETTE .... palettizing information for assembling,
- CATALOGUE .... conventional rules for material selection class, butt welding condition, etc.,
- TURN .... alignment condition of a flanged coupling.

In these entity identifier, the relation between the entity and the entity block of 'LINE' is interesting. The entity of 'LINE' is a kind of point called 'a positional entity', which will be described in following discussion.
These identifiers are registed in the processor as 'Fortran BLOCK DATA'.

Positional Entity It can be considered that the most significant entity block is the piping route information. They are described by a sequence of positional entity statement (specific). A positional entity is a kind of point such as illustrated in Table 2. These positional entities require its associated information, attribute data according to its nature, e.g. a face-to-face distance of a valve, bending radii and angles of pipe bends. According to the type of required associated information and its combinations, the positional entities are classified as Table 2. The type of positional entity corresponds to its data structure and the type of semantic routines to determine the geometrical information of piping route. PIPE-language is given powerful capabilities of expression by means of this positional entities.

entity		attributes	
pipe	:	D/nominal size,	MAT/material symbolic name,
		DO/outer diameter,	T/thickness of tube wall,
		SCH/schedule,	GAM/specific gravity
		---	
fittings	:	TYPE/type of positional entity,	
		GRP/group number,	
		D/nominal size,	D1/straight size,
		D2/reduced size,	W/weight,
		P/pressure rating,	SCH/schedule,
		L/, L1/, L2/length,	A/angle,
		EP/end profile, joint condition,	
		E/value of eccentricity,	MAT/material,
		---	
route	:	N/name,	TYPE/type of positional entity,
		AT/ , BEFORE/ , AFTER/ , FROM/ , TO/positional expression	
		DIR/direction expression,	A/angle,
		R/radius of pipe bending,	E/value of eccentricity,
		ED/direction of eccentricity,	D/reduced diameter,
		SCH/schedule,	P/pressure rating,
		MAT/material,	NM/number of mitred joint element

Table 3. Example of Attribute used in PIPE-language for ship piping.

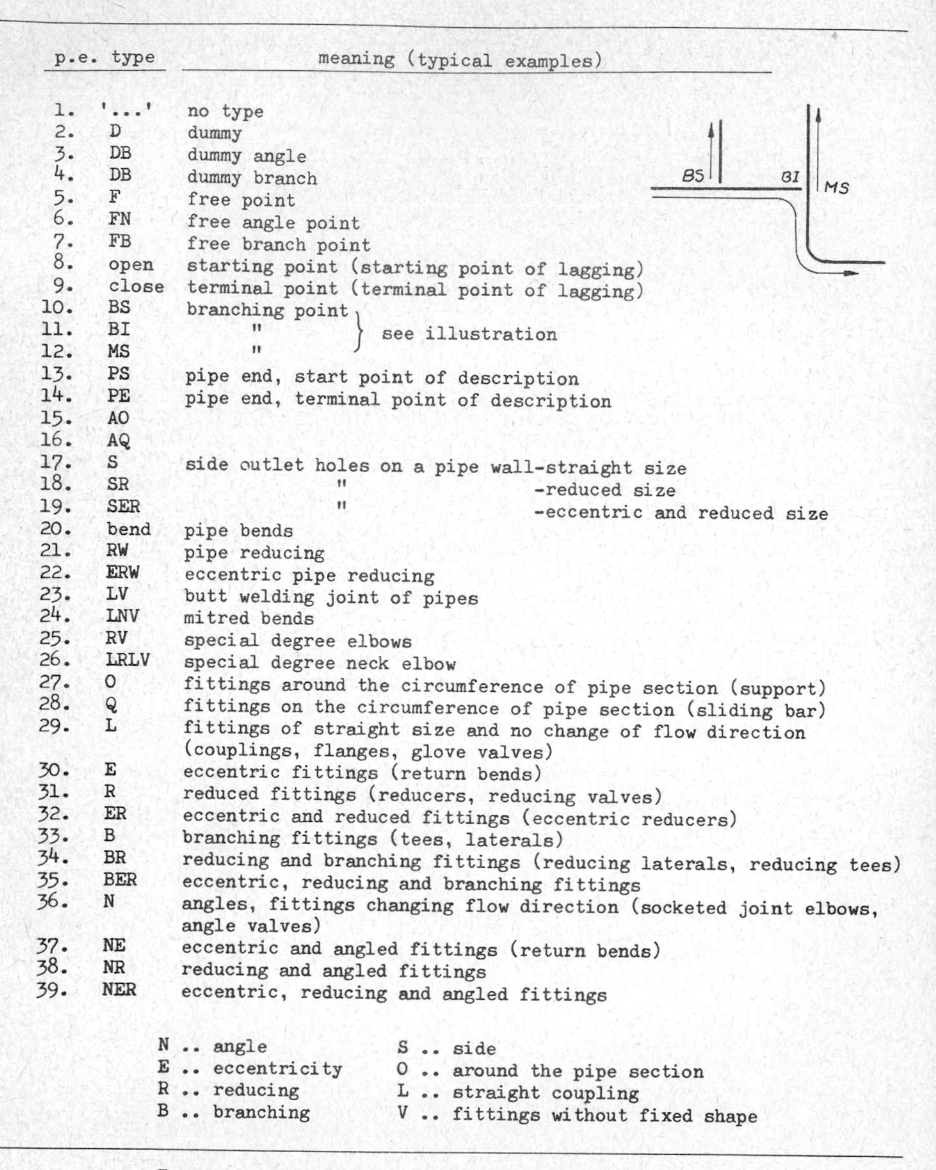

p.e. type		meaning (typical examples)
1.	'...'	no type
2.	D	dummy
3.	DB	dummy angle
4.	DB	dummy branch
5.	F	free point
6.	FN	free angle point
7.	FB	free branch point
8.	open	starting point (starting point of lagging)
9.	close	terminal point (terminal point of lagging)
10.	BS	branching point } see illustration
11.	BI	" } see illustration
12.	MS	" } see illustration
13.	PS	pipe end, start point of description
14.	PE	pipe end, terminal point of description
15.	AO	
16.	AQ	
17.	S	side outlet holes on a pipe wall-straight size
18.	SR	" -reduced size
19.	SER	" -eccentric and reduced size
20.	bend	pipe bends
21.	RW	pipe reducing
22.	ERW	eccentric pipe reducing
23.	LV	butt welding joint of pipes
24.	LNV	mitred bends
25.	RV	special degree elbows
26.	LRLV	special degree neck elbow
27.	O	fittings around the circumference of pipe section (support)
28.	Q	fittings on the circumference of pipe section (sliding bar)
29.	L	fittings of straight size and no change of flow direction (couplings, flanges, glove valves)
30.	E	eccentric fittings (return bends)
31.	R	reduced fittings (reducers, reducing valves)
32.	ER	eccentric and reduced fittings (eccentric reducers)
33.	B	branching fittings (tees, laterals)
34.	BR	reducing and branching fittings (reducing laterals, reducing tees)
35.	BER	eccentric, reducing and branching fittings
36.	N	angles, fittings changing flow direction (socketed joint elbows, angle valves)
37.	NE	eccentric and angled fittings (return bends)
38.	NR	reducing and angled fittings
39.	NER	eccentric, reducing and angled fittings

N .. angle
E .. eccentricity
R .. reducing
B .. branching

S .. side
O .. around the pipe section
L .. straight coupling
V .. fittings without fixed shape

Table 2. Classification of the positional entity.

Attributes Registration of 'attribute' concerned with each entity is performed by adding 'Fortran BLOCK DATA' containing the symbolic attribute identifier and a type of attribute data indicating the required decoding routines. Attributes for typical entity, pipe, fittings and positional entity, are shown as Table 3.

2.3 Some extensions for language capability

Sometimes, a certain modification of language format is effective on the view point of readability, writability and learnability.

The current PIPE-language has some of conventional expression for this purpose. For example, a following is a convention to describe positional entities.

basic language   N/TEE1 , POS/DECK+500 , DIR/P(45)A , D/50

modified language TEE1 AT DECK+500 TO P(45)A WITH D/50

This convention has increased such abilities, regardless of simple conversion rule that 'preposition' replaces with 'attribute identifier and /'.

One of more significant function is MACRO facility which is common function to reduce the number of input statements. PIPE-language allows two types of MACRO such as 'statement-macro' and 'function macro'.

Statement-macro is defined by a sequence of language statements with a unique name. Coding a macro-assignment statement is equivalent to coding all the statements. Statement-macro may have set of parameters, which are coded in the statement by dummy parameters at definition stage and replaced by actual parameter at macro expansion stage. The process for the statement-macro is quite syntactic.

Function-macro is defined by user's routines which generate a sequence of PIPE-language statements. Function-macro can not be defined by PIPE-language. It is defined by FORTRAN statements. It is a program to generate PIPE-language using a parameter set of macro-assignment statement. The process for the function-macro is semantic. Function-macro is usually used to reduce the input work for describing the element of pipe route which requires a design practice for geometrical calculations, e.g. design of 'cross over', 'offset', 'double offset expansion ubend', etc.

All of these process concerned with MACRO are performed before DTL-processing. An example of current PIPE-language is shown in Fig. 3.

3. APPLICATIONS

In Mitsubishi Heavy Industries, Ltd., PIPE-language has been applied on various fields of piping engineering. In this section, the typical items among those are mentioned briefly.

Automatic drawing of piping system In the present time, drawing the piping system using an automatic plotting machine has been widely adopted. It is done without special considerations to draft the shapes of piping systems using Entity Blocks generated for the piping route, pipe characteristics and fitting characteristics. The most difficult technique required is the automatic placement of dimensions and names to be indicated.

NC tape generation for a pipe bending machine Probably, on the view point of computing the instruction sequencies, NC pipe bending machine must be one of the most simple one comparing with other NC machines. Typical information required for controlling bending motion are shown in Fig. 4. These information can be easily obtained using Entity Blocks for the piping route and for the bending characteristics. Difficulties are rather depending on:

- accurate estimation of the physical properties of pipe materials, e.g. spring back and elongation factors.
- proper tape supply corresponding with the fabrication schedule for many pieces to be bent.
- faster computing method for the accurate geometrical judgement on interference between moving pipe and machine circumstances under the bending motion.

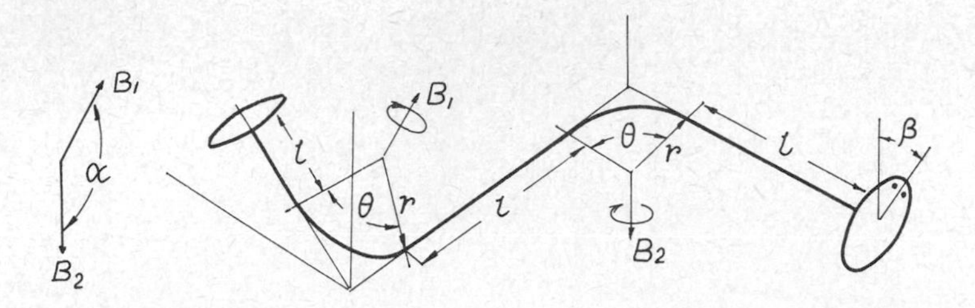

Fig. 4. Typical information for NC pipe bender.

Technical calculation It is said that one of the feature of a technical calculation program in piping engineering, is a requirement of its tedious input work. Certainly, it is tedious to describe required information to individual program, for example, solving piping flexibility problem or computing flow in distribution networks.

Most of these programs, which had required amount of similar information in many cases, had appeared individually as a product of earlier stage computer applications, so the increased volume of computer use, combined with an increased quantity of these separate programs has resulted in a significant magnitude of manual input redundancies, even if well-considered input format have been prepared.

User will be able to describe data more easily by means of PIPE-language than of a conventional FORTRAN formatted input method, but the ordinary applications these programs are linked and necessary information is supplied from previously generated Entity Blocks for the purpose of reducing manual input redundancies.

4. POSTSCRIPT

The first version of the PIPE-system has been operational for IBM7040 computer since May, 1968 and applied to design stages, detail and fabrication, of ship piping as a test use.

According as the test results and many suggestions of users, many improvements have been made for practical applications. The system now is prepared for the IBM360(Nagasaki and Kobe works), the CDC6400(Hiroshima works) and the UNIVAC1108 (Yokohama works) computers and it has been successfully used by Mitsubishi Heavy Industries, Ltd., in CAD field of piping for ships and thermal power plants, and data banks have been built easily and correctly, which have many usages on control tape generation of NC pipe benders, automatic piping drawings, B/M, technical calculations and data supply linked in production scheduling and business calculations. On such applications, PIPE-language is appreciated as an effective input tool by many users.

The following people have participated in many discussions in the continuing PIPE-system developement: K. Noguchi, T. Nagai and PSWG(PIPE-system working group) members.

REFERENCES

Davis, R.M., (1966), Programming Language Processors, Advances in Computers. 7(Academic Press), 117-180

Johnson, W.L., Porter, J.H., Ackley, S.I. and Ross, D.T., (1968), Automatic Generation of Efficient Lexical Processors Using Finite State Techniques, Comm. ACM 11, 12, 805-813

Crocker, S., (1945), Piping Handbook, 4th ed. (McGraw-Hill)

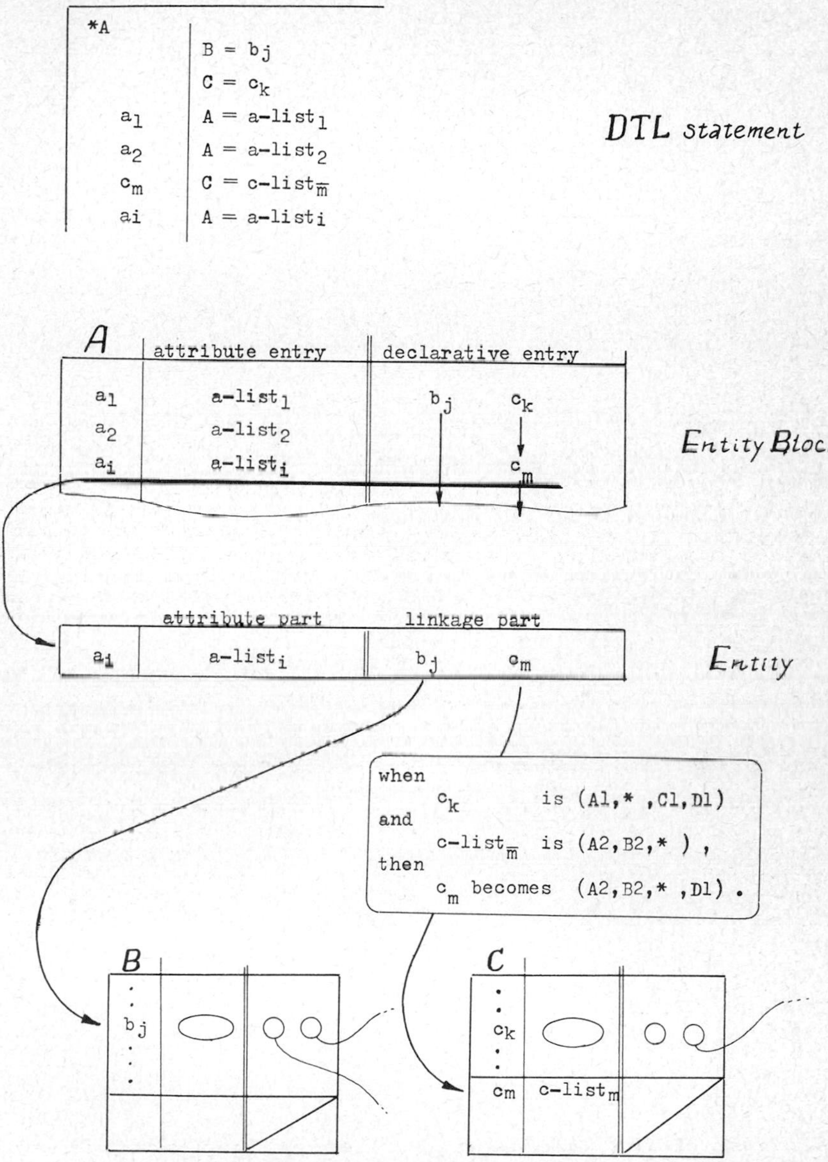

```
*A
        | B = b_j
        | C = c_k
  a_1   | A = a-list_1
  a_2   | A = a-list_2
  c_m   | C = c-list_m̄
  ai    | A = a-list_i
```

DTL statement

A

	attribute entry	declarative entry	
a_1	a-list$_1$	b_j	c_k
a_2	a-list$_2$		
a_i	a-list$_i$		c_m

Entity Block

	attribute part	linkage part	
a_i	a-list$_i$	b_j	c_m

Entity

```
when
    c_k          is (A1,* ,C1,D1)
and
    c-list_m̄    is (A2,B2,* ) ,
then
    c_m becomes  (A2,B2,* ,D1) .
```

B

b_j

C

c_k

cm c-list$_m$

Fig. 1. Concept of 'entity','entity-block' and DTL.

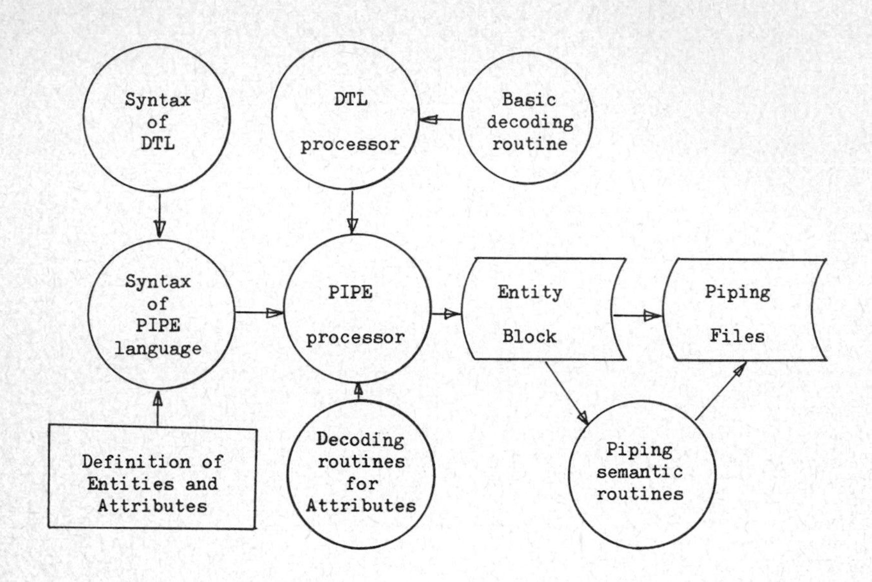

Fig. 2. Relation between DTL and PIPE-language

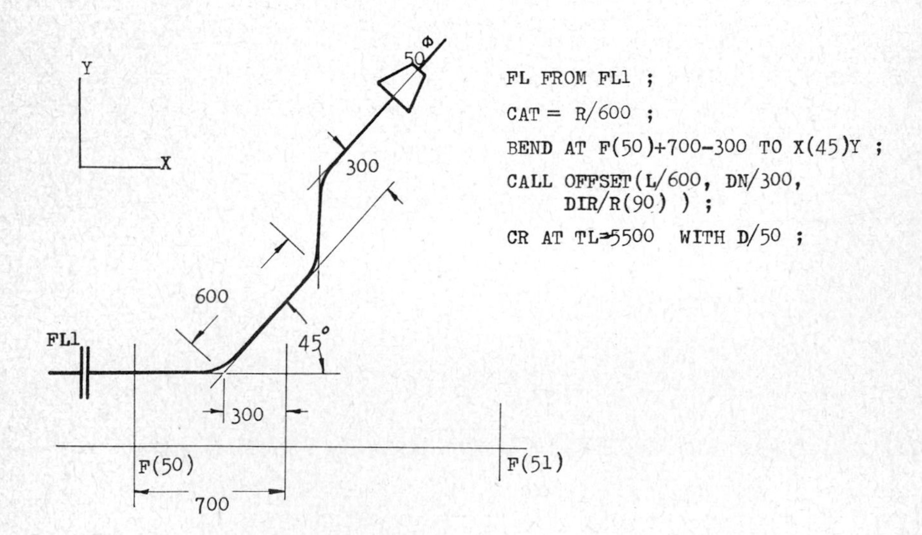

Fig. 3. Example of PIPE-language expression.

THE USE OF A.P.T. MACROS IN THE PRODUCTION OF COMPLEX DROP FORGING DIES

D. FRENCH, K. ADAMS, S. NADA
Associate Professor, Assistant Professor, Research Assistant, respectively
Department of Mechanical Engineering
University of Waterloo
Waterloo, Ontario, Canada

Abstract: Papers describing the development and use of A.P.T. macros for the production of drop forging dies have been previously published by the authors. These papers included descriptions of how drop forging dies have been classified by a series of basic geometrical configurations.

This paper illustrates how complex drop forging dies can be easily programmed by combining the established basic geometric A.P.T. macros. The problems involved in building and storing a compatible A.P.T. macro library are discussed and sample die configurations of stub axles and valve housings successfully produced in the Research Laboratories at the University of Waterloo are used to illustrate these problems.

1. INTRODUCTION

A.P.T. macros have been written for a number of families of parts, and a number of machine tools. Although this type of macro can be very flexible within the family of parts for which it has been written, it is virtually impossible to use these macros for any components outside of the family for which it has been written. Another approach to the problem of part programming using A.P.T. macros for more complex parts such as drop forging dies is to write macros for basic geometrical configurations, and then combine a number of these macros to form a composite element.

The work carried out at the University of Waterloo to develop macros for drop forging dies has resulted in the creation of seven macros for the following geometrical configurations:

1) Cone to Rectangular shape (called CNREC)
2) Cone to Cylinder (called CNCYL)
3) Cylinder to Cylinder (called CCYL)
4) Cylinder to Rectangular shape (called CLREC)
5) Cylinder to Elliptical shape (called CILE)
6) Cylinder to Hexagonal shape (called CYLHEX)
7) Cylinder to Spherical shape (called CLSPHR)

These configurations can be used to produce either the female shape, for a die or the male shape, for an electrode for EDM by changing one variable within the call statement. The method of writing these macros has been shown in other papers. The object of this paper is to illustrate how these macros can be used to program a complex part using a number of macros, and also to discuss the problems involved in creating a library of such macros compatible with the A.P.T. program.

2. PART PROGRAMMING OF COMPLEX DIES USING THE A.P.T. MACRO APPROACH

The procedure for part programming of a component requiring a number of macros in combination is:

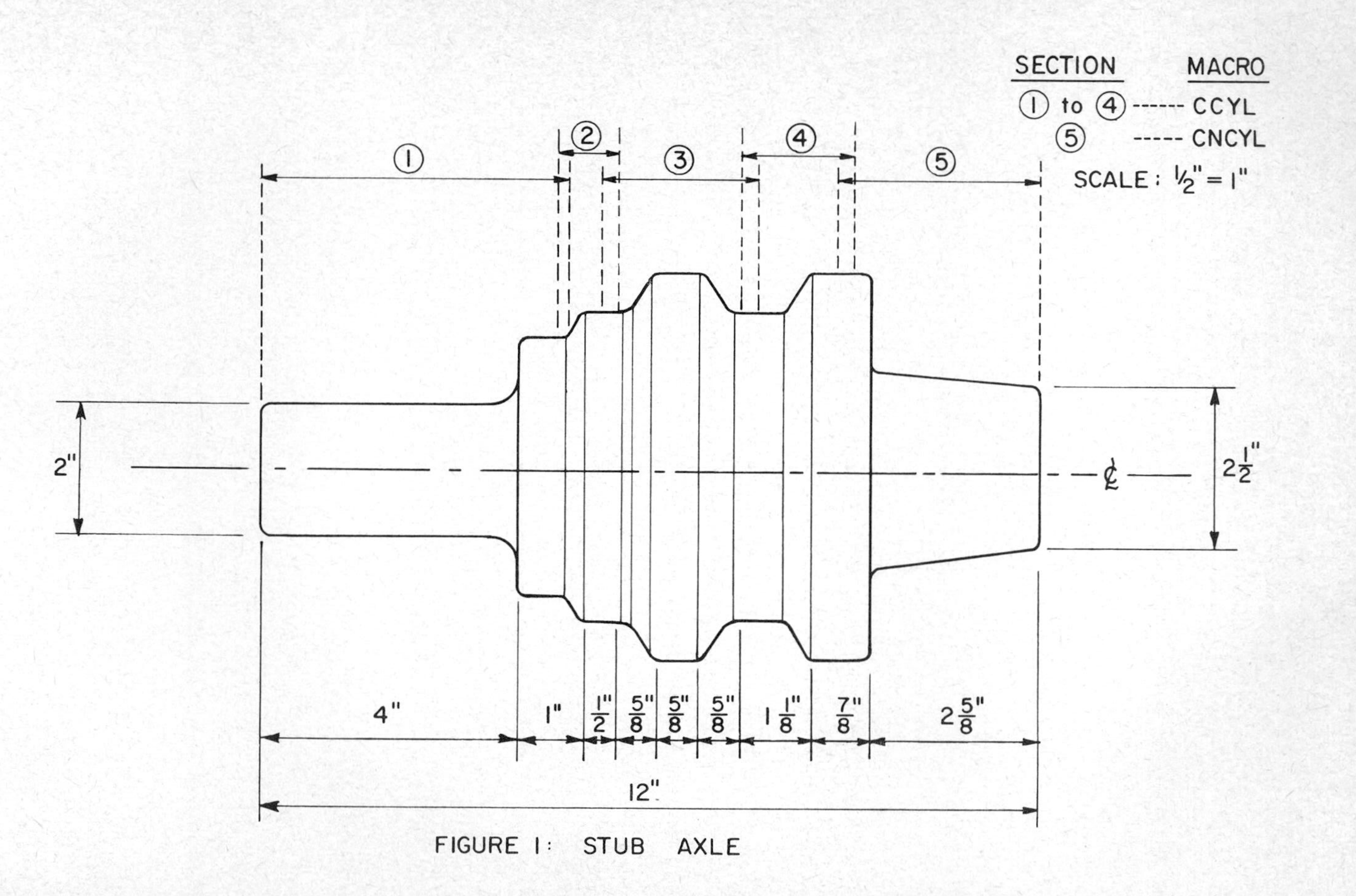

FIGURE 1: STUB AXLE

(1) Divide the part into a number of sections, each section being one of the basic macro configurations specified. In many instances, it might be advantageous to have a slight overlap of each macro to ensure that no part is left unmachined.

(2) Determine any sections which cannot be covered by one of the basic macros. This section could then be programmed in the normal manner.

(3) Determine the relative positions of each macro to be defined with respect to the datum selected for the part.

(4) Determine the cutter size, cusp height, and the necessary machining requirements such as feeds, speeds and coolant.

(5) Obtain all variable values, such as length of element, diameters of elements, etc., necessary to complete the macro call statement.

(6) Complete the part program.

3. EXAMPLES OF USING A.P.T. MACROS

3.1 EXAMPLE 1 : THE STUB AXLE

The diagram in Fig. 1 shows a component representative of a forging die for a stub axle. The component drawing is divided into 5 elements; elements 1 to 4 being cylinder to cylinder configurations and element 5 being a cone to cylinder configuration. It should be noted that the macro automatically creates the correct filleting which is dependent on the diameters of the elements within the geometrical configuration. The variables required for the two macros used are shown in Table 1.

Table 1

Macro Variables Required for the Stub Axle Part Program

(a) Cylinder to Cylinder Macro (CCYL)

No.	Variable	Name
1)	Length of smallest diameter cylinder	(LSM)
2)	Length of largest diameter cylinder	(LLR)
3)	Diameter of small cylinder	(D1)
4)	Diameter of large cylinder	(D2)
5)	Taper angle of conical intersection	(THETA)
6)	Fillet radius between cylinders	(R)
7)	Surface finish cusp height	(CP)
8)	Cutter diameter	(DIA)
9)	Cutter radius (for ball nosed cutter)	(RAD)
10)	Cutter height	(HYT)
11)	Type of intersection Conical intersection ANG = -1 Toroidal intersection ANG = +1	(ANG)
12)	Type of shape Female component SIGN = -1 Male component SIGN = +1	(SIGN)
13)	Type of configuration Two cylinder configuration SHAPE = -1 Three cylinder configuration SHAPE = +1	(SHAPE)
14)	Spindle speed	(SPD)
15)	Spindle off	(SPN)
16)	Direction of rotation of spindle	(DIRN)
17)	Coolant requirements, on, flood, mist	(CLN1)
18)	Coolant off	(CLN2)
19)	Feedrate (IPM)	(FD)
20)	Matrix angle of rotation	(GAM)
21)	Translation X coordinate	(XT1)
22)	Translation Y coordinate	(YT1)
23)	Translation Z coordinate	(ZT)

24) Macro call number (FLAG)
First macro call from library FLAG = -1
Last macro call from library FLAG = +1
Neither first nor last FLAG = 0
25) Counter for incrementing index number with each macro call (IND)

(b) Cone to Cylinder Macro (CNCYL)

1) Length of cone (LCN)
2) Small diameter of cone (DS)
3) Cone taper angle (BETA)
4) Length of cylindrical shape (LCYL)
5) Cylinder diameter (DC)
6) Allowable cusp height between cutter passes (CP)
7) Fillet radius (R)
8) Intersection variable (POSITN)
9) Female component SIGN = -1 (SIGN)
Male component SIGN = +1
10) One-cone configuration SHAPE = -1 (SHAPE)
Two-cone configuration SHAPE = +1
11) Matrix angle of rotation (GAM)
12) Translational x-coordinate (XT1)
13) Translational y-coordinate (YT1)
14) Translational z-coordinate (ZT)
15) Diameter of cutter (DIA)
16) Radius of cutter (RAD)
17) Height of cutter (HYT)
18) Spindle speed (RPM) (SPD)
19) Direction of rotation of spindle (DRN)
20) Spindle OFF (SPD1)
21) Coolant ON, MIST, etc... (CLN1)
22) Coolant OFF (CLN2)
23) Feedrate (IPM) (FD)
24) Macro call number (FLAG)
First macro call from library FLAG = -1
Last macro call from library FLAG = +1
Neither first nor last FLAG = 0
25) Counter for incrementing index number with each macro call (IND)

It should be noted that the values of XT1, YT1, ZT and GAM are specified to allow the zero positions of subsequent macros to be set relative to the first macro, thus automatically building up the part illustrated in Figure 1. This will enable machining of the part to proceed in an orderly manner between each macro. Section 3 consists of a three cylinder configuration that has a large cylinder with a smaller cylinder at each end. To obtain this configuration, the variable "SHAPE" within the call statement is set equal to +1. The call statements for these macros required to produce the part programme for the stub axle are illustrated in Table 2.

Table 2

Part Programme for Stub Axle

```
PARTNO ----
CLPRNT
MACHIN/----
RESERV/----
LIBRY/----
```

```
CALL/CCYL, LSM = 4, LLR = 0.75, D1 = 2, D2 = 4, R = 0.5, CP = 0.015, $
           DIA = 0.25, RAD = 0.1250, HYT = 0.125, ANG = -1, SIGN = -1, $
           SHAPE = -1, SPD = 1530, SPN = OFF, DIRN = CCLW, CLN1 = ON, $
           CLN2 = OFF, FD = 15, GAM = 0, XT1 = 0, YT1 = 0, ZT = 0, $
           THETA = 0, FLAG = -1, IND = 1
CALL/CCYL, LSM = 0.25, LLR = 0.5, D1 = 4, D2 = 4.75, THETA = 60.1, R = 0.126,$
           CP = 0.015, DIA = 0.25, RAD = 0.125, HYT = 0.125, ANG = +1, $
           SIGN = -1, SHAPE = -1, XT1 = 0.75, YT1 = 0, ZT = 0, FLAG = 0, $
           IND = 2
CALL/CCYL, LSM = 0.75, LLR = 1.125, D1 = 4.75, D2 = 6, R = 0.375, CP = 0.015,$
           DIA = 0.25, RAD = 0.125, HYT = 0.125, ANG = -1, SHAPE = +1, $
           SPD = 1530, SPN = OFF,  DIRN = CCLW, CLN1 = ON, CLN2 = OFF, $
           FD = 15, GAM = 0, XT1 = 1.875, YT1 = 0, ZT = 0, THETA = 9, $
           FLAG = 0, IND = 3
CALL/CCYL, LSM = 0.625, LLR = 0.625, D1 = 4.75, D2 = 6, THETA = 60.3, $
           R = 0.150, CP = 0.015, DIA - 0.250, RAD = 0.125, HYT = 0.125, $
           ANG = +1, FD = 15, GAM = 0, XT1 - 4.125, YT1 = 0, ZT = 0, $
           SIGN = -1, SHAPE = -1, SPD = 1530, SPN = OFF, DIRN = CCLW, $
           CLN1 = OFF, CLN2 = OFF, FLAG = 0, IND = 4
CALL/CNCYL, LCN = 2.625, DS = 2.5, BETA = 5.4, LCYL = 0.625, DC = 6, $
           CP = 0.015, R = 0.25, POSITN = +1, SIGN = -1, SHAPE = -1,$
           GAM = 180, XT1 - 5.375, YT1 = 0, ZT = 0, DIA = 0.25, RAD = 0.125,$
           HYT = 0.125, SPD = 1530, DRN = CCLW, SPD1 = OFF, CLIN1 = ON,$
           CLN2 = OFF, FD = 15, FLAG = +1, IND = 5

STOP
END
FINI
```

3.2 EXAMPLE 2 : THE VALVE HOUSING

The diagram in Fig. 2 shows a component representative of a valve body housing. The component drawing is divided into four sections; section 1 being a cylinder to elliptical shape, section 2 being a cylinder to rectangular shape and sections 3 and 4 being cylinder to hexagonal shape. The variables required for the three macros used are shown in Table 3.

Table 3

Macro Variables Required for the Valve Housing Part Program

(a) Cylinder to Ellipse Macro (CILE)

1)	Length of cylinder		(LCL)
2)	Diameter of cylinder		(DC)
3)	Fillet radius		(R)
4)	Length of elliptical shape		(LELPS)
5)	Breadth of elliptical shape		(BELPS)
6)	Depth of elliptical shape		(DELPS)
7)	Surface finish cusp height		(CP)
8)	Female component	SIGN = -1	(SIGN)
	Male component	SIGN = +1	
9)	One-cylinder configuration	SHAPE = -1	(SHAPE)
	Two-cylinder configuration	SHAPE = +1	
10)	Matrix angle of rotation		(GAM)
11)	Translational x-coordinate		(XT1)
12)	Translational y-coordinate		(YT1)
13)	Translational z-coordinate		(ZT)
14)	Spindle speed (RPM)		(SPD)
15)	Direction of rotation of spindle		(SPD1)

16) Coolant ON, MIST, etc.... (CLN1)
17) Coolant OFF (CLN2)
18) Feedrate (IPM) (FD)
19) Cutter diameter (DIA)
20) Cutter radius (RAD)
21) Cutter height (HYT)
22) Macro call number (FLAG)
First macro call from library FLAG = -1
Last macro call from library FLAG = +1
Neither first nor last FLAG = 0
23) Counter for incrementing index number with each macro call (IND)

(b) Cylinder to Rectangle Macro (CLREC)

1) Length of cylinder (LCL)
2) Diameter of cylinder (DC)
3) Length of rectangular shape (LRC)
4) Breadth of rectangular shape (BRC)
5) Depth of rectangular shape (DR)
6) Fillet radius (R)
7) Cutter diameter (DIA)
8) Cutter radius (RAD)
9) Cutter height (HYT)
10) Surface finish cusp height (CP)
11) Female component SIGN = -1 (SIGN)
Male component SIGN = +1
12) One-cylinder configuration SHAPE = -1 (SHAPE)
Two-cylinder configuration SHAPE = +1
13) Matrix angle of rotation (GAM)
14) Translational x-coordinate (XT1)
15) Translational y-coordinate (YT1)
16) Translational z-coordinate (ZT)
17) Spindle speed (RPM) (SPD)
18) Spindle OFF (SPD1)
19) Direction of rotation of spindle (DRN)
20) Coolant ON, MIST, ETC... (CLN1)
21) Coolant OFF (CLN2)
22) Feedrate (IPM) (FD)
23) Macro call number (FLAG)
First macro call from library FLAG = -1
Last macro call from library FLAG = +1
Neither first nor last FLAG = 0
24) Counter for incrementing index number with each macro call (IND)

(c) Cylinder to Hexagon Macro (CYLHEX)

1) Length of cylinder (LCL)
2) Diameter of cylinder (DC)
3) Length of hexagonal shape (LHEX)
4) Depth of hexagonal shape (DHEX)
5) Fillet radius (R)
6) Surface finish cusp height (CP)
7) Cutter diameter (DIA)
8) Cutter radius (RAD)
9) Cutter height (HYT)
10) Matrix angle of rotation (GAM)
11) Translational x-coordinate (XT1)
12) Translational y-coordinate (YT1)
13) Translational z-coordinate (ZT)

14) Spindle speed (RPM) (SPD)
15) Spindle OFF (SPD1)
16) Direction of rotation of spindle (DRN)
17) Coolant ON, MIST, Etc... (CLN1)
18) Coolant OFF (CLN2)
19) Feedrate (IPM) (FD)
20) Female component SIGN = -1 (SIGN)
    Male component SIGN = +1
21) One-cylinder configuration SHAPE = -1 (SHAPE)
    Two-cylinder configuration SHAPE = +1
22) Macro call number (FLAG)
    First macro call from library FLAG = -1
    Last macro call from library FLAG = +1
    Neither first nor last FLAG = 0
23) Counter for incrementing index number with each macro call (IND)

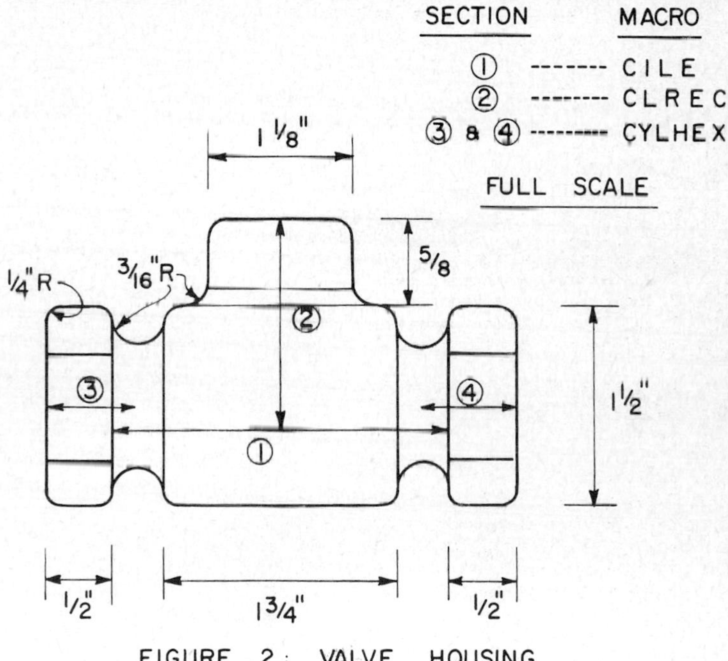

FIGURE 2: VALVE HOUSING

In this part, not only is translation required for Section 2, but also roation, as Section 2 must be perpendicular to the axes of Sections 1, 3 and 4. In this case it should be noted that the variables in the macro call statements for Section 2 are:

GAM = 270, XT1 = 0.875, YT1 = 0.75, ZT = 0.

The call statements for these macros required to produce the part program for the valve housing are shown in Table 4.

Table 4

Part Program for Valve Housing

```
PARTNO----
CLPRNT
MACHINE/----
RESERV/----
LIBRY/----

CALL/CILE, LCL = 0.375, DC = 0.875, R = 0.1875, LELPS = 1.75, DRN = CCLW, $
           BELPS = 1.5, DELPS = 0.875, CP = 0.001, SIGN = -1, SHAPE = +1, $
           GAM = 0, XT1 = 0, YT1 = 0, ZT = 0, SPD = 1500, SPD1 = OFF, $
           CLN1 = ON, CLN2 = OFF, FD = 20, DIA = 0.25, RAD = 0.125, $
           HYT = 0.125, FLAG = -1, IND = 1
 CALL/CLREC, LCL = 0.625, DC = 1.125, LRC = 0.75, BRC = 1.75, DR = 0.875, $
           R = 0.1875, DIA = 0.25, RAD = 0.125, HYT = 0.125, CP = 0.01, $
           SIGN = -1, SHAPE = -1, GAM = 270, XT1 = 0.875, YT1 = 0.75, $
           ZT = 0, SPD = 1500, SPD1 = OFF, DIRN = CCLW, CLN1 = ON, $
           CLN2 = OFF, FD = 20, FLAG = 0, IND = 2
CALL/CYLHEX, LCL = 0.375, DC = 0.975, LHEX = 0.5, DHEX = 0.75, R = 0.1875, $
           CP = 0.001, DIA = 0.25, RAD = 0.125, HYT = 0.125, GAM = 180, $
           XT1 = -0.375, YT1 = 0, ZT = 0, SPD = 1500, SPD1 = OFF, $
           DRN = CCLW, CLN1 = ON, CLN2 = OFF, FD = 20, SIGN = -1, $
           SHAPE = -1, IND = 3, FLAG = 0
CALL/CYLHEX, LCL = 0.375, DC = 0.875, LHEX = 0.5, DHEX, = 0.75, R = 0.1875, $
           CP = 0.001, DIA = 0.25, RAD = 0.125, HYT = 0.125, GAM = 0, $
           XT1 = 2.125, YT1 = 0, ZT = 0, SPD = 1500, SPD1 = OFF, DRN = CCLW, $
           CLN1 = ON, CLN2 = OFF, FD = 20, SIGN = -1, SHAPE = -1, IND = 4, $
           FLAG = +1
STOP
END
FINI
```

4. IMPLEMENTATION OF THE A.P.T. MACRO LIBRARY

Each main macro consists of a number of separate macros for producing various elements of the geometrical configuration. These separate macros are called by the main macro as determined by the call statement variables.

The problems encountered using the A.P.T. macro library were as follows:

(1) A successful call from the macro library could not be achieved if that macro contained other macros.
(2) When more than one macro was called in the same program, a section one error indicating a doubly defined variable resulted.
(3) For copying operations within the macros, the INDEX/n statement resulted in an undefined variable when more than one macro was called.
(4) When producing parts using more than one macro, the machine had to be stopped after producing each macro and manually set to the new set point.

To overcome these problems, the following modifications were made:

(1) All macros, including the main macro and all others called by it, were included in the macro library as separate entries.
(2) All geometry parameters were sandwiched between the CANNON/ON and CANNON/OFF statements which allowed the parameters to be redefined.
(3) A variable parameter (IND) was added to the macro call parameters which incremented the index number for each macro called.

(4) A variable parameter (FLAG) was added to the macro call parameters which indicated whether the macro being called was the first, an intermediate or the last macro to be called in the program. This information allowed the logic of the macro to be modified to accommodate continual processing from one macro to the next, eliminating the need to reset the zero point.

5. CONCLUSIONS

The A.P.T. macro is a useful tool in the production of families of parts. The approach to the use of basic geometrical configurations to build up a complex part enables greater flexibility and reduces part programming time to a minimum. In addition, a part programmer with very little knowledge of the A.P.T. system can program parts as illustrated in the two examples by obtaining the variables from the drawings, or as the part is being designed, the values of the variables could be developed.

ACKNOWLEDGEMENTS

This research was supported by the Defense Research Board of Canada and the National Research Council of Canada under grant numbers DRB-9761-03, NRC-A8274 and NRC-C1226 to Professors French and Adams.

REFERENCES

French, D. and S. Nada, (1971), The reduction of part programming to a minimum in the production of drop forging dies by special A.P.T. macros, Sec. Int. P.D.M.T. Conf.

French, D. and S. Nada, (1971), Special A.P.T. macros for the manufacture of drop forging dies, Proc. of the Eighth Tech. Conference of the Numerical Control Society.

DISCUSSION

Mr. W. Nauck asked Mr. A.G. Flutter what methods were used to obtain the initial input data for the surfaces. Mr. Flutter said he had been concerned mainly with design, rather than fitting surfaces to rigidly determined data. Those for the propeller blade were obtained with a measuring machine. Asked by Mr. H. Damsohn what criteria were used to decide the direction and distances of the cutter paths, he said they were guessed.

Mr. P. Bézier, asked by Mr. V.G. Zaitsev how much time was required to construct the model of a car by his method, said this depended very much on the designer. Usually very simple patches are used initially /12-15 for a car/ and small details bring this up to 100-120. Mr. B. Vlasveld pointed out that while APT still used cubic splines, this system had parametric ones. How was it APT compatible? Mr. Bézier replied that this was mainly in using the same words. In reply to Mr. A.S. Gregory, who asked if changes made on the model could be digitized back to the patch data structure, the author said they could, if the changes were not too "sharp". Finally Mr. H. Eitel asked if software had been developed to cope with the spring-back problem. Mr. Bézier said this was dealt with by pre-stretching the metal.

Mr. V. Nicoló, in reply to a question by Mr. J. Vlietstra on the numerical problems of his fitting method, said the initial approximation was automatic but the operator could easily eliminate situations which might cause numerical problems.

In a question to Mr. Y. Hyodo, Mr. H. Schreiter asked, what was the content of the body part of the canonical form for non mathematical surfaces, and where was it stored. Mr. P. Veenman asked how the mathematical expression for the surface between two sections was obtained. Both questioners were referred to p.7. of the Paper - the canonical form is stored in core. In reply to Mr. M.A. Sabin, the author said all curves consisted of straight lines and circular arcs.

Mr. K.H. Werler, who asked Mr. C. Ghezzi whether the data structure used in surface generation also provided data to an NC processor, was told it did not.

Mr. T. Mehlum was asked by Mr. V. Nicoló to explain how he could do without parametric curves, since his system avoided Coons' patches. The answer was that he used a sequence of circles, which is very easy to parametrize if necessary, using sines and cosines. In reply to a question by Mr. M.A. Sabin, Mr. Mehlum demonstrated in some detail, how AUTOKON could be used to represent the front half of a giraffe. Mr. P. Veenman asked how much computer time was needed to produce the drawings of the automobile shown in the figures. Mr. Mehlum said 42 data points were measured from the sketch. About 9 minutes UNIVAC 1108 CPU-time and 3 man-days were needed, plus 1-2 hours on a KINGMATIC. Mr. B. Vlasweld enquired whether an interface to APT was intended. Mr. Mehlum said this had been investigated and was feasible, if anyone wanted to buy it.

Mr. R.A. Guedj, who asked Mr. T. Hirano how much work had been put into his project, was told it had been 5 man-years.

In a question to Mr. A. Landmark, Mr. G. Krammer asked whether FORTRAN or assembly-language has been used and how the file structure was modified. Mr. Landmark said there were only four assembly routines /read, write, backing store, shift/, all else was in low-level FORTRAN. There is nothing called a file structure, database access is by name and the user does not notice a change in storage structure. Mr. G. T. Maughan, who asked whether there was a facility for outputting a list of X and Y coordinates for manual cutting, was told that there was and that it had been used in British, French and Norwegian shipyards.

Mr. D. French, in reply to Mr. D.G. Wilkinson, agreed that his macros did not generate roughing cuts, but this would be relatively simple. Mr. M.A. Sabin inquired about running costs and whether it was not easier to turn the macros into FORTRAN programs generating APT statements or CLFILE records. Mr. French said the programs he showed had been run in 80 time units /1 $/unit/ on a 360/75. He thought many people did not realize the full power of APT.

7. TECHNOLOGICAL PROGRAMMING AND PRODUCTION CONTROL

TECHNOLOGICAL PROGRAMMING AND PRODUCTION CONTROL

GUNNAR SOHLENIUS
Linköping University and IVF,
Section MA, Stockholm, Sweden

Abstract: Technological programming and production control are two connected but different fields where important development is going on within computer aided manufacturing. The specialization of the production and the use of group technology simplifies this development. Adaptive control could be combined with technological programming but does not set aside the need for it.

1. INTRODUCTION

Within the field of technological programming and production control, a development of great economical, technical and human importance is going on. This development has been possible due to the evolution in the data-processing and computer technology.

To be careful we have to realise that although there is no sharp division between technological programming and production control, we are talking about two different problem-areas.

The technological programming with aid of computer, is mainly a direct developing consequence of the computer aided preparation of tapes for NC-machines. It covers the technological calculations and decisions in the operations planning, figure 1. All following papers except for one (Mr. Adam´s) deal with programs related to this.

The production control with aid of computer is more a consequence of the general availability of computers for manufacturing problems. The possibility to control machinetools directly in a computer system, DNC, has one of its greatest impacts on real time,effective production control. The production control is related to the process planning, figure 1. However, it is in itself more global, figure 2. To analyse the problems of production control, one important way is

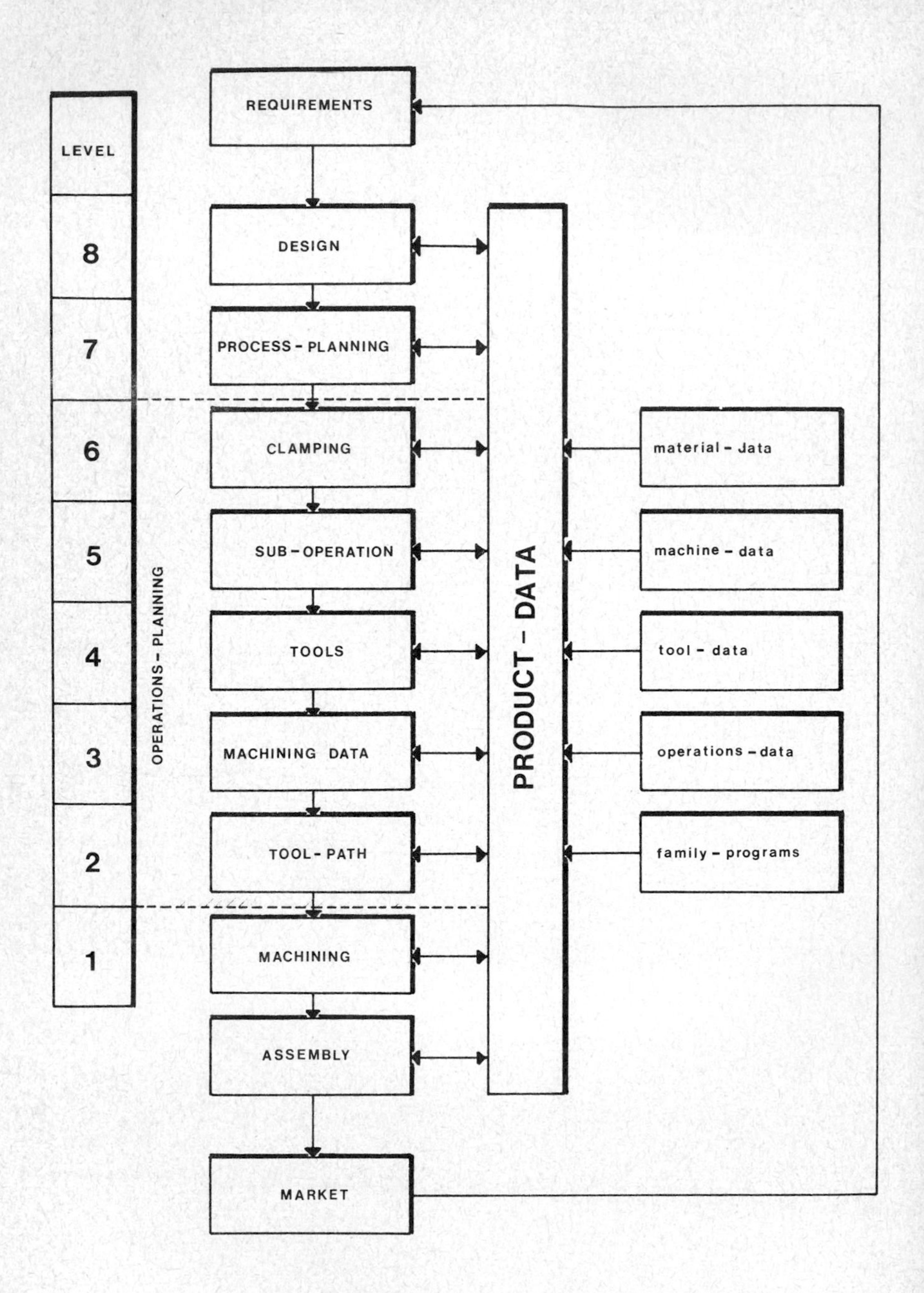

Fig. 1. DESIGN-PRODUCTION PROCESS

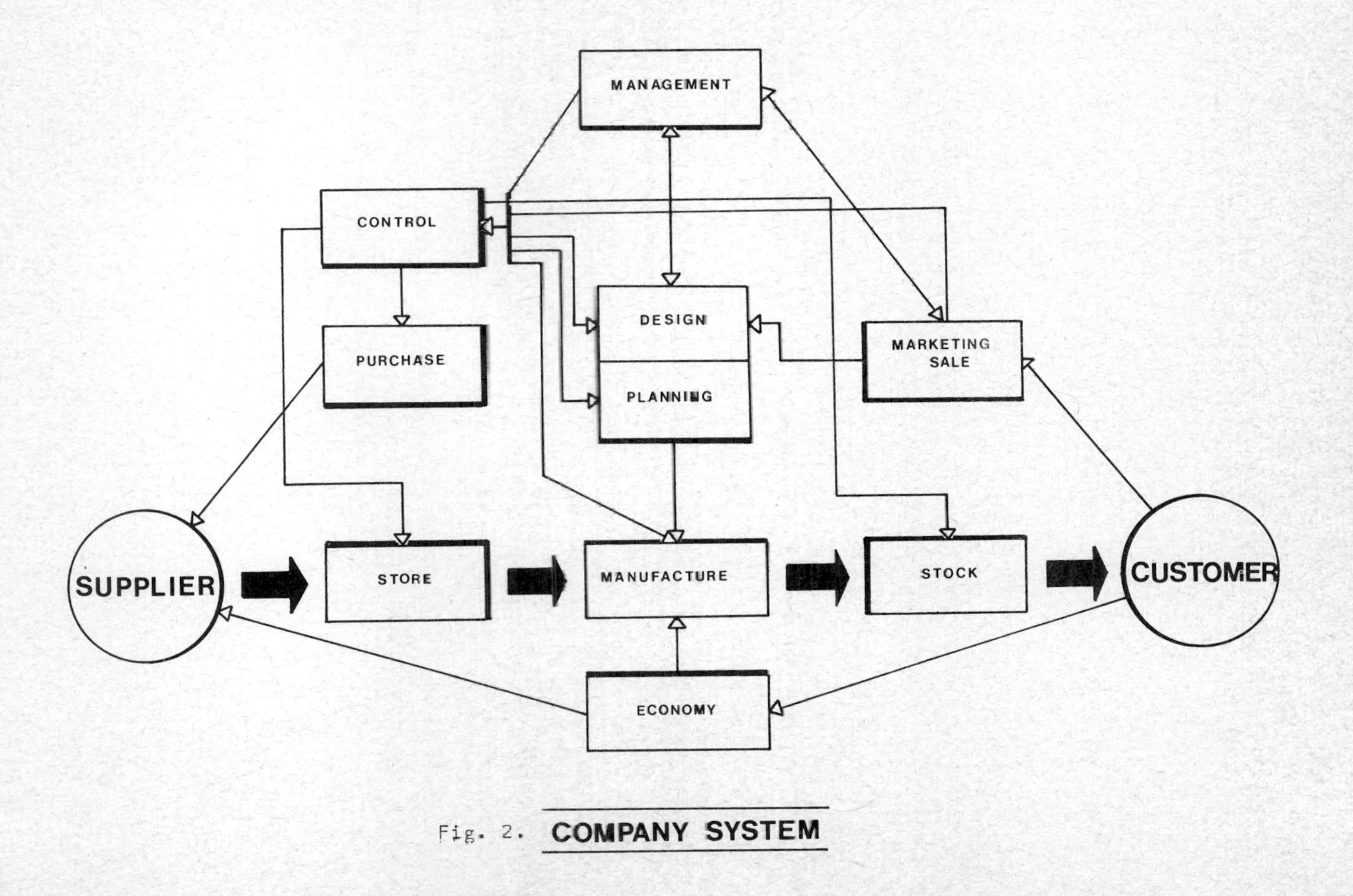

Fig. 2. **COMPANY SYSTEM**

to simulate the production-conditions in a computer. The paper "An approach to Computer Aided Manufacture" by K.G. Adams, deals with this. The papers "NC-Production Planning" by J. Steinacker and H.H. Winkler and "A Systematic Approach to Implementing a Computer Aided Manufacturing System" by A.H. Low, deal with the data-validity and data-collection problems directly connected to computer aided operations planning. However, this problem also has its impact on the production control.

2. PRODUCTION-TYPES

The possibilities and efforts needed to build up a manufacturing system in general and special items in it as programs for technological programming and production control, are very dependent on product types and production volumes in the actual case and factory. Based on number of product-types and productionvolumes workshops fall into different production-types, figure 3; from special-production to small-batch production. The more specialized the factory is the simpler it is to obtain a high automation level, figure 1, in a planning system and to design a system for production control. The connections in figure 3 elucidate one reason why so many different statements about how to build up a computer aided operations planning system are brought up. Those who have a production-type in the low left corner of figure 3, need a general system and are interested in common efforts. They also have great difficulties to obtain high automation levels. Those who can organize their production in part families, figure 4, or product and part types, figure 5, on the other hand, are more interesting in special systems tailored to their own needs. They often design systems of their own and are interested in a general development of modules that could be combined to build up special systems. For the specialized production these problems are very simple and outside the frame of this discussion.

The papers "An Attempt in High Level Automation of Programming of NC Lathes the Fortap System" by M. Horvath, B.E. Molnar and S. Nagy, "Optimization Models for Optimum-cost Technological Performance Characteristics of Metal Cutting Processes Using Cutting Edges of Defined Geometries" by D. Kochan and H.J. Jacobs and "Adapting the Technology in Programming Languages to the Specific Requirements of the User" by P. Adamczyk and H. Zölzer deal with the more generalized problem.

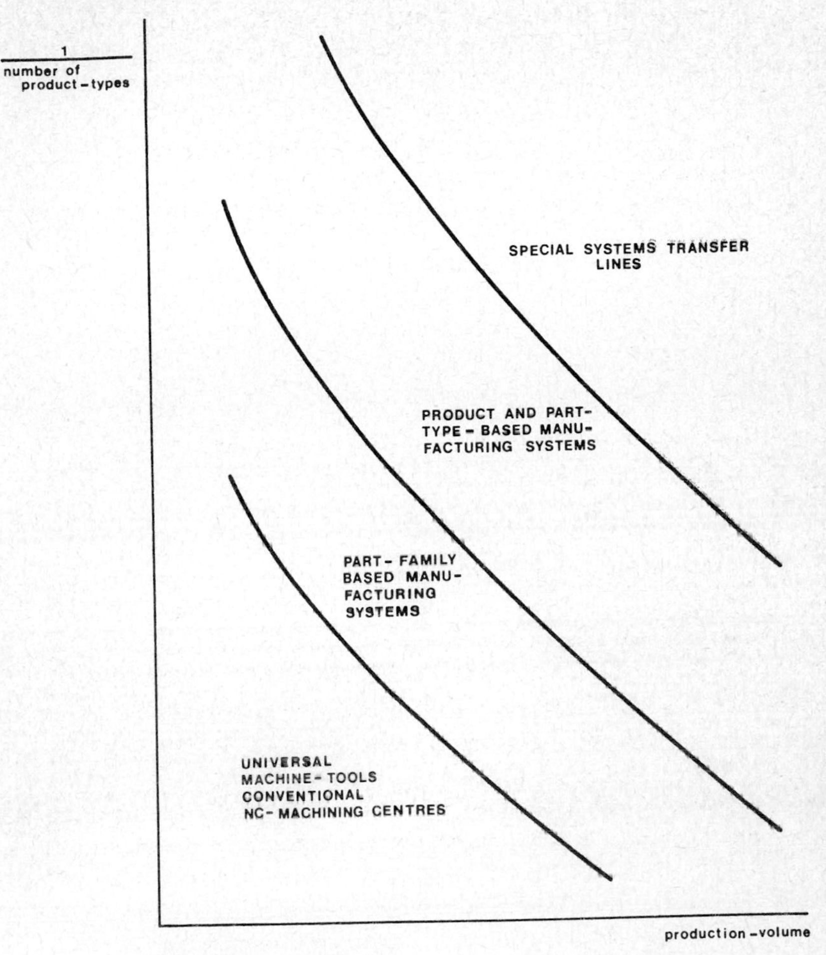

Fig. 3.

Part-type-developed programs can be seen within factories for instance the G SHAFT-program covering shafts with shoulders developed in 1966 - 67, figure 5, at General Electric Schennectady, New York.

The AUTOPROG-system in the paper "Principles of Computer-Aided NC Programming With a High Automation Level" by V. Griess, J. Preisler and J. Vymer and the "TNO MITURN Programming System for Lathes", presented by P. Bockholts, are systems based on the same part-family-philosophy covering about 80 per cent of turning-part-types.

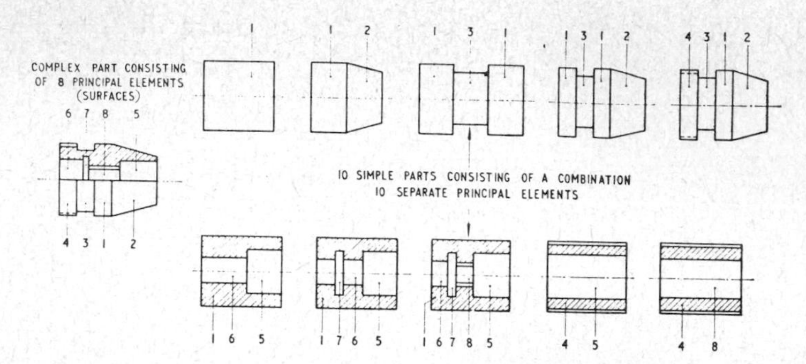

Fig. 4. Complex component (after Mitrafanov).

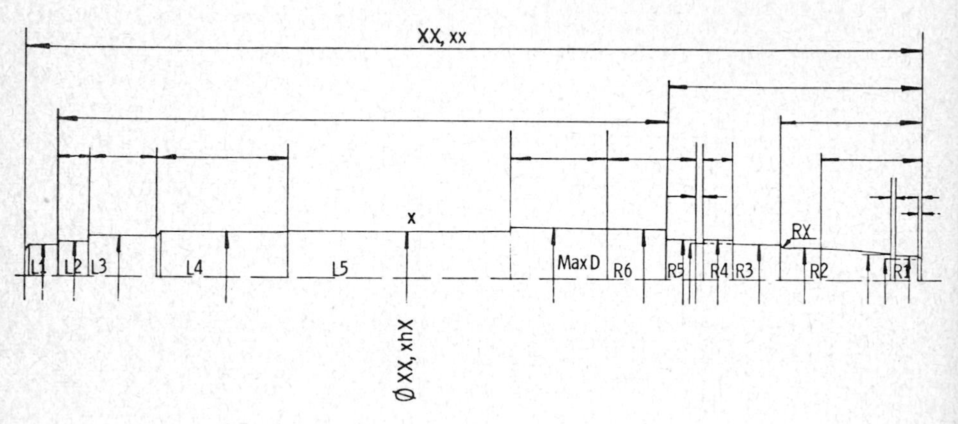

Fig. 5.

3. THE TECHNOLOGICAL MAN-MACHINE-SYSTEM

The intention to build up any fully automated technological system for operations planning and production control is abandoned. This problem is enormous in volume and the final goal is fairly undesirable from human, technical and economical point of view. We have learnt that we have to strive after combinations of man, computers, and production equipment in cooperating systems. Use of interactive graphical and alpha-numerical terminals and time-sharing, dialogue-systems where man can use his normal communicationaids are of importance, figure 6. This approach might make it possible for man to improve his standard of living by designing more effective production aids.

Fig. 6.

Systems for operation-planning where the man-machine interaction has been considered are AUTOPROG and MITURN mentioned above. A work with this intention is also presented in the paper "An Interactive System for Operations Planning for Turning on Centre Lathes" by P. Hellström.

One dominating reason why it is so difficult to design general technological systems, is that the basic knowledge and decision-criteria are so different in different countries and even factories and under various production conditions. Connected to this is the fact that the distribution of machinability between batches, in batches and even within a single part limit, the possibilities to calculate optimal data in an operations planning system that remains optimal at the machining, figure 7.

The problem to collect experience data from the users and to adapt data to the requirements of the users is dealt with by A.H. Low by J. Steinacker and H.H. Winkler and by P. Adamczyk and H. Zölzer in their papers mentioned above.

The only possibility to maintain optimum data in varying machinability conditions, is to use adaptive control, figure 8. This control-technology should be developed together with operations planning and production control systems. The possibilities and requirements to combine these aids are presented in the paper "Workshop Technology-Adaptive Control-Turning" by G. Berg.

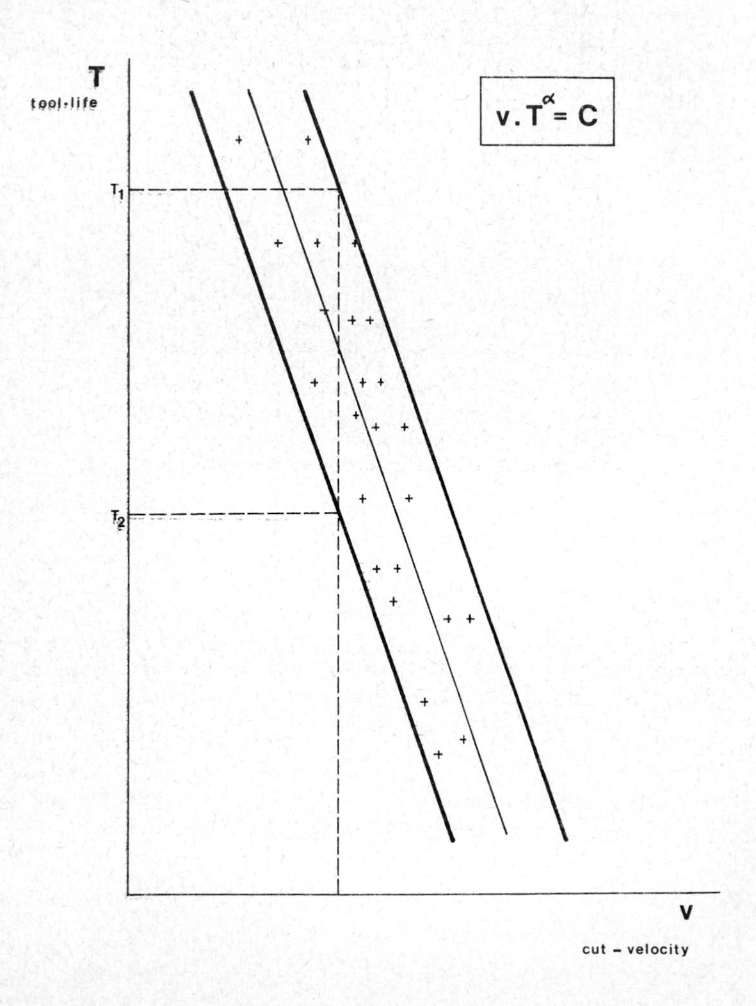

Tool-life = f (cut velocity, feed, depth, etc)

Fig. 7.

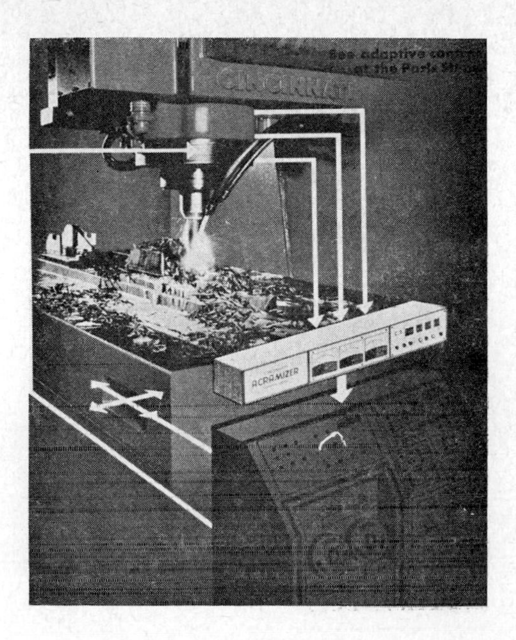

Fig. 8.

4. SESSION-PLAN

The general thoughts presented above should be seen as the general back-ground to this session, which in itself is planned in the following way:

It starts with two papers discussing some basic conditions to make technological calculations in NC-programming systems.

The third and fourth paper present systems where easy man-machine dialogue has been considered an important requirement.

Paper five and six deal with the data collection, databank and data dissemination-problem.

Paper seven deals with adaptive control and operations planning in combination.

This is a very important problem. These problems can neither be solved in technological planning programs nor with adaptive control only.

The eight paper finally, deals with simulation in connection with production control as mentioned above.

In this session there are four papers covering various problems that are printed only. These papers are combined and discussed together with the orally presented papers in the following way:

The paper by D. Kochan and H.J. Jacobs and the paper by N.A. Yarmosh and F.G. Milner are combined with the first and second paper.

The paper by V. Griess, J. Preisler and J. Vymer is combined with paper three and four.

The paper by P. Adamczyk and H. Zölzer finally is combined with paper five and six.

AN ATTEMPT IN HIGH LEVEL AUTOMATION OF PROGRAMMING OF NC LATHES. THE FORTAP SYSTEM.

M. HORVÁTH
Institute for Technology of Mechanical Engineering
Budapest, Hungary
B. E. MOLNÁR
Institute for Technology of Mechanical Engineering
Budapest, Hungary
S. NAGY
Technical University for Heavy Industry
Miskolc, Hungary

Abstract: A possible construction of a technological preprocessor is described in this paper. The processor produces 'unit operation' and machining sequence, selects tools and plans tooling. Finally a brief account is given on the FORTAP System which is supplied by a preprocessor mentioned, for automatic programming of NC lathes.

INTRODUCTION

In the big family of programming systems the processors for NC lathes have a distinguished place. It is a general opinion, Opitz (1970), Vymer (1970), that an ideal system should automatically determine machining sequence, select and arrange tools, calculate cutting parameters and plan every tool motions. At the same time it should be flexible, easy-to-adapt to any machine tool and compatible to other systems.

We think that this rather complicated duty, having many illusory contradictions can be fulfilled in the following manner: considering an APT-like processor, between the geometrical and technological processor (classical sectioning used by EXAPT first) a so-called technological preprocessor has to be inserted as illustrated in fig.1. Presuming that the technological processor determines cutting values and motion cycles, the preprocessor should

(i) automatically specify machining steps,
(ii) plan their sequence,

COMPUTER LANGUAGES FOR NUMERICAL CONTROL, *J. Hatvany, editor*
North-Holland Publishing Company - Amsterdam–London

(iii) select tools from the tool set ordered to the machine,
(iv) carry out a tooling plan.

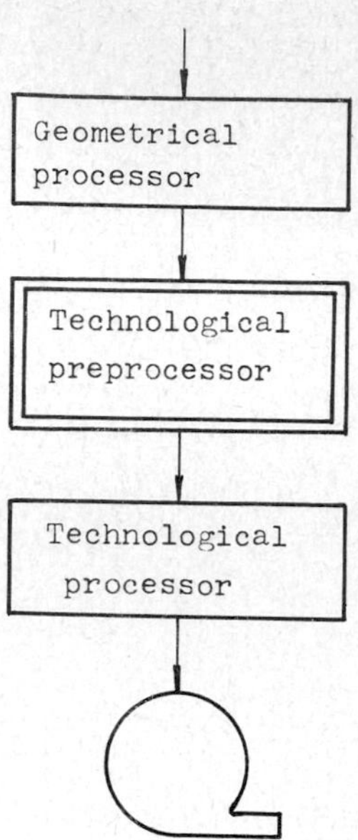

Figure 1.

1. BASES

The FORTAP has been ellaborated at the Institute for Technology of Mechanical Engineering, Budapest, Hungary. The conditions of primary importance in respect to turning that were taken into consideration during the work of development are described in the following.

1.1. The surfaces that form the boundary of a workpiece are subjected to hierarchical order. At the uppermost level stand the primary surfaces which indicate the shape of an axially symmetric component. (Shoulder, cylinder, cone, ring). These carry the secondary surfaces (recessings, roundings and bevels etc.) which also can support so called tertiary surfaces etc. The hierarchy of surfaces is in close connection with that of machining steps and their sequence. (Fig.2).

1.2. Concerning methodology, turning is quite a complex type of machining, featured by great variety of workpiece contours, different tools and motion cycles possible. This is why the principles of group technology or other methods cannot be mechanically applied if we cling to solution of general validity.

1.3. The lathe, concerning control systems and machining possibilities, is rather varied and complex machine.

1.4. There are turning operations which can mutually replace each other. The technological characteristics of a machine and an actual tool set modify the specification of the machining method.

1.5. The optimal succession of machining operations depend on the types of tools selected.

1.6. The characteristics of the actual machine tool have a significant effect on the types and number of the tools.

1.7. Tool selection, tooling and setting up succession of machining operations are different phases of planning having mutual influence on one another.

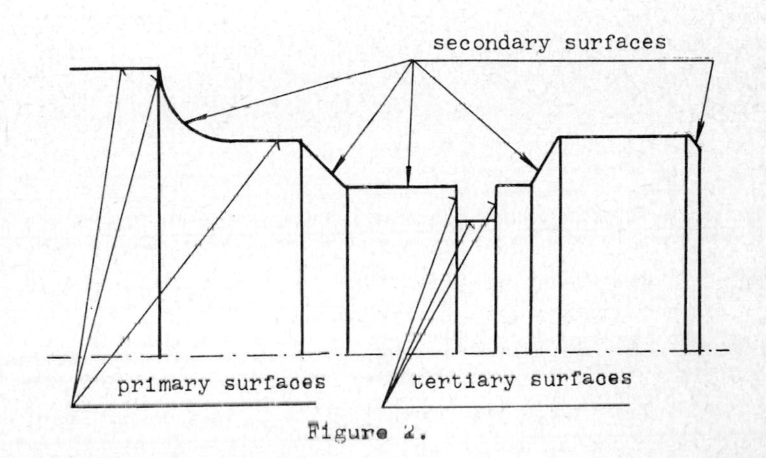

Figure 2.

2. DESIGN PRINCIPLES

We have had neither methodological nor practical experience how the automatic generation of turning operations should be carried out. Some of the main design principles that we worked out are:

2.1. It is expedient to specify the hierarchy of the workpiece surfaces in an explicite way. Thus there will be less input data regarding the workpiece, and the conversion of geometrical information into technological would be much more simple. From this principle derives that some definite groups of secondary surfaces may be treated as complex geometrical units having definite set of parameters.

A good example of how to draw surfaces together is the description of recessing defined by its width, depth and a coordinate. (See fig. 3).

It might be advantageuos to add geometrical units like these to the set of surfaces to the APT-like programming languages.

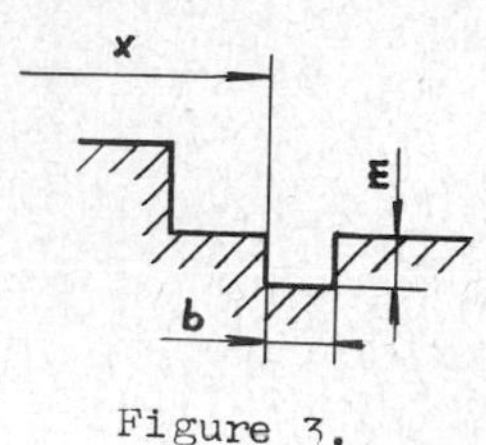

Figure 3.

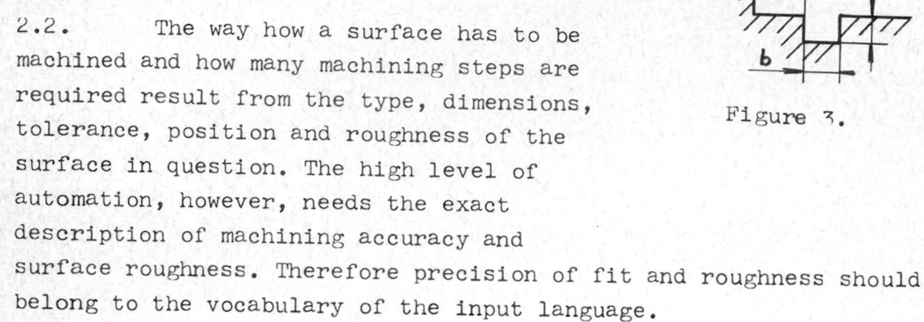

2.2. The way how a surface has to be machined and how many machining steps are required result from the type, dimensions, tolerance, position and roughness of the surface in question. The high level of automation, however, needs the exact description of machining accuracy and surface roughness. Therefore precision of fit and roughness should belong to the vocabulary of the input language.

2.3. The building stone of the automatic synthesis of machining is the unit operation (equivalent with Russian perekhod). It is the name of a finished process of machining performed on one or some joining surfaces by one tool. (E.g. through hole boring, face turning, recessing etc.). Exact models of various unit operations can be developed. From a finite number of unit operations the whole manufacturing process of a workpiece can be reconstructed and the individual characters of the components can easily be taken into account.

The following function gives the specification of the unit operation:

$$u=f(g,s,t,l,v,c,m,p,o)$$

where

u is unit operation,

g - geometrical data of a unit operation,

s - linkage between adjoining surfaces and those which form the unit operation,

t - tool location characteristics,

v - cutting values,

c - tool correction,

m - machine tool characteristics,

p - prescription of the part programmer concerning the unit operation,

o - position of the unit operation in the sequence of operations.

The two examples as shown in fig. 4 will enlighten the meaning of these parameters.

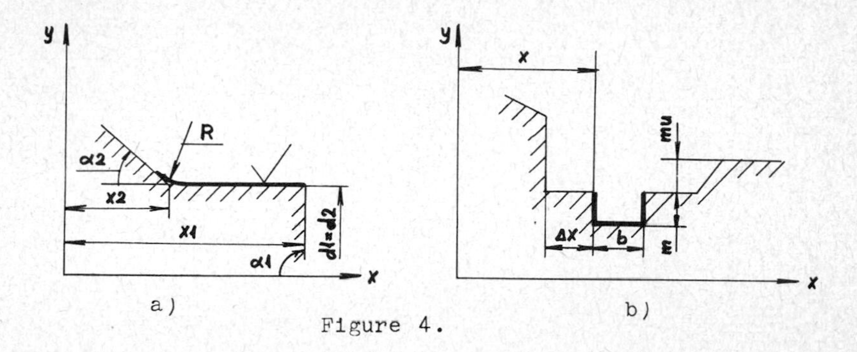

Figure 4.

It has no need to comment on the own geometrical parameters of the unit operations. It is worth mentioning here that the surface roughness belongs to the said parameters.

Parameters of linkage among adjoining surfaces are: R, α1 and α2. In the case of recessing they are Δx and mu.

Machine tool characteristics have their influence mainly on tool motions in idle run. Their careful observation helps to optimizing tool point route.

Letter p indicates a certain capability of the system. This means that the system, to a reasonable extent, may prefer the requests of the part programmer to its own decisions.

The conventional GOTO type statement should be used in the part programme if the programmer knows that his request is beyond the capability of the system.

It has been pointed out that from a few dozens of unit operations any machining process of any workpiece can be reconstructed. Unit operation models can be relatively easily worked out, new ones may be constructed and added to the existing set of modules. This brings the benefit of being able to enlarge the capability of the technological processor without changing its construction.

The geometrical parameters of the tool selection can be specified partly on the basis of the geometrical parameters of unit operation being in close connection with the surfaces of the component, and partly on that of parameters of linkage among adjoining surfaces.

The unit operation featured by the parameter set discussed can be taken out of its surroundings and investigated independently of others. This fact has great importance from both the point of view of optimal operation sequence and the construction of module groups.

2.4. The most important principle of design of complex technological process is that decisions should be made successively. **In order to obtain satisfactory solution of production** engineering questions iterative methods have to be applied owing to the great amount of variables.

This principle can be best explained with the introduction of the following two examples.

Example 1. Generating a unit operation in the case of the fine turning (see fig. 5). The contours of workpiece at intermediate

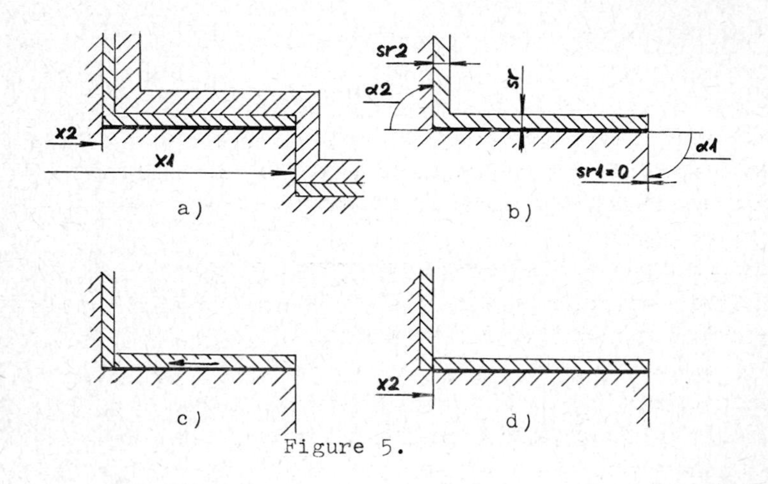

Figure 5.

machining stages are developed first including machining allowance for fine turning. (Fig. 5a). Then the own geometrical parameters of the unit operation and parameters of linkage are generated. (Fig. 5b). Here are the prescriptions of the part programmer taken into account. After the tool selection has been made the feed direction should be decided. (Third step, see fig. 5c). The scheduling block of the programme locates the unit operation in the machining process, revises and modifies if necessary the geometrical data. (Forth step, see fig. 5d). Finally the tooling segment completes the parameter set of the unit operation. (Fifthstep). The number of steps can hardly be decreased in the case of turning.

Example 2. Most NC programming systems can automatically select tools like drills, bores, countersinks etc. System with automatic selection of turning tools is not known to us except AUTOPROG which can specify standard tools. Our aim was the general solution of the problem. This means that from a given set of tools the system can select any turning tool including 'twin tools' (two turning tools in one holder, one of them for surfaces in a hole, the other for external surfaces). More complex cases are not characteristic to NC lathes.

In order to simplify automatic tool selection and store least number of tools possible, tools of different type were grouped. Each group is suitable for machining characteristic unit operations. Every tool group ordered to 'internal surfaces' obviously has a corresponding group ordered to 'external'ones.

At the first stage only the tool group and the geometrical demands resulting from one unit operation can be settled. Further operations either merely modify settled tool selection criteria or require the specification of a new tool. This means that geometrical demands of more than one unit operations may provide the tool selection criteria for the determination of the tool type. In order to find the simplest solution possible two tools may be put into one holder if both internal and external tool groups of the same kind are used. What is more, internal and external operations can be performed by the same tool if the chuck can rotate both ways.

The tool selection is a multiple stage process; the optimal tool is searched, but in the case of a failure the process goes on. The rest of the tool set is investigated and ultimately a less advantageous tool might be found which, however, meets most of the requirements. In a well organized tool card system it is a routine work to find the most rigid tool having optimal cutting edge geometry. Our tool card system is similar to that of EXAPT. At the same time when unit operations are generated tool selection criteria are gathered in the so-called 'demand list'. (See fig. 6). The prescriptions of the part programmer concerning the tools are also added to the demand list. The tool selection itself is produced later. Checking the tool prescribed and the tool selection is carried out by the same programme.

2.5. Successive adaption. As was said above the production planning takes place successively. Different questions are solved at each stage. Each stage has one common feature, though, and this is that abstract and actual, general and individual problems exist at

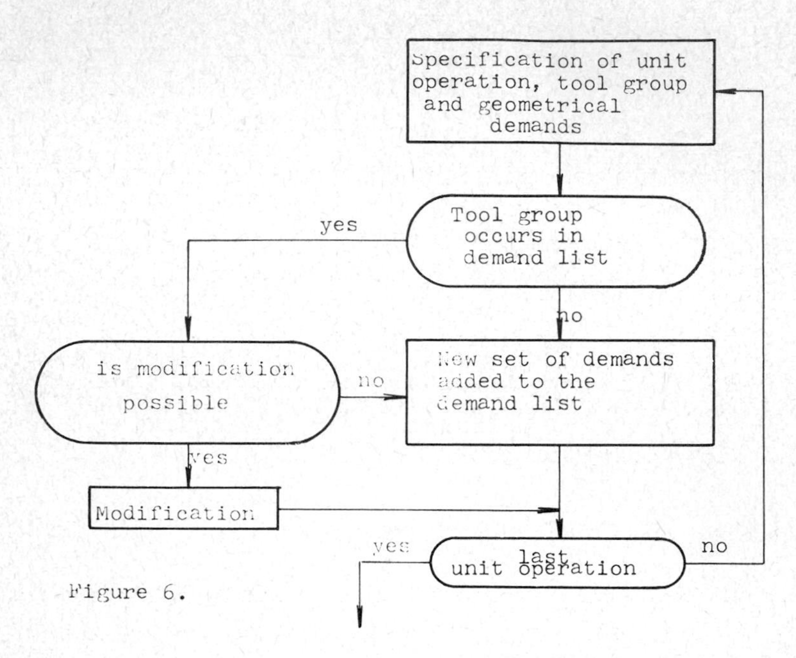

Figure 6.

the same time. The ambition of solving an actual question at the lowest stage brought the postprocessor forth, which is now the found of any problem-oriented programming system. The problems of turning, owing to its characteristic features descibed above, are quite similar to those of an integrated production system. Therefore at the higher level of planning process the characteristics of machine tool/control system combination should be taken into consideration, otherwise we ought to give up the aim that the machining version realized by the automatic determination of machining method, tool selection and tooling plan generation should be more or less similar to that which could have been produced by an experienced part programmer.

In general, one of the characteristic features of a design system of high level automation is that the intermediate results are continuously adapted to an actual machine tool, therefore the whole automatic planning process is succesive. It begins even in the geometrical processor and ends in the postprocessor.

Succesive adaption raises some questions in connection with technological models. It is known that the general and individual problems can be best related in a general model which gives an answer to an actual question if actual parameters are substituted.

If the adaption is as mentioned then the postprocessor merely carries out data conversion.

2.6. The pivot question of high level automation of technology planning is the setting up a general machine tool model. A lathe usually has more slides which can travel in various directions. Each slide is supported with tool locations which differ from one another in construction (fixing, clamping dimensions etc.) and in machining capabilities. The abstract numerical model of the machine tool does not refer to the structure but the operating capabilities of the machine. However, it demonstrates the construction to a certain extent because construction has a considerable effect on technology. The abstract machine tool model represents machine tool/control system combination which involves every individual characteristic of certain types of machine tools and control systems.

The model consists of a very complex fictitious machine tool from which any simpler machine derives. The 'technological map' of this machine has been charted. This map when filled in with actual parameters offers a general view about an actual machine tool. It gives for example, how many tool locations are dispósable for a certain group of tools assigned by a given application code; on which slide can these tools be mounted etc. The application code varies according to the manufacturing procedure, desired accuracy, tool holder construction, coding of tool cutting edge, place of slides etc. The model, as a matter of fact, is a matrix with fixed number of rows. Each row consists of fixed number of tool parameters identified by an application code. At the end of a row the identifiers of the tool locations are listed which meet the common requirements. (See fig.7).

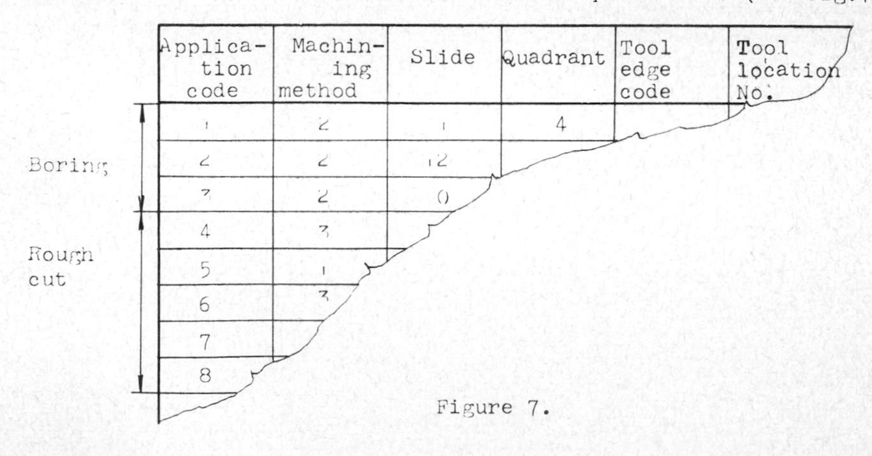

Application code
Machining method
Slide
Quadrant
Tool edge code
Tool location No.
1 2 1 4
2 2 12
3 2 0
4 3
5 1
6 3
7
8
Boring
Rough cut

Figure 7.

It is obvious that the number of columns of the matrix depends on the characterictics of an actual machine tool. Using the model the tool selection can be controlled, e.g. in the case of more equivalent solutions the system specifies the tool which, besides maintaining good cutting conditions, makes optimal tooling possible. The model makes the interaction between operation sequence, tool selection and tooling perspicuous.

2.7. The problem of flexibility of the system and priority problems necessarily arise in connection with a high level automation designing system. The root of the matter is that the system itself can make decisions in every substancial technological question, but the production man must have the right to prescribe machining operation or a tool etc. which he esteems to be better than the one decided later by the system. Where should the dividing line be drawn between the two parts, the system and the production engineer? May we doubt the priority of the part programmer if his prescriptions are technologically correct but they deverge from the probable optimum calculated by the system?

The basic principle was the following: The system should investigate, then either build in or ignore each prescription without interrupting the automatic design process.

Prescriptions are ignored if an error is detected in the command or if the own version of the system is undoubtedly better.

The computer reacts two different way if an error is encountered in a programming statement: It may apply its own solution if it is satisfactory (e.g. in order to avoid collusion it selects another tool instead of the prescribed one), or processing is interrupted because the information available is not enough to a solution (e.g. instead of recessing tool estimated inapplicable, the system does not select another one, because the prescribed one might have been a form tool and these are not automatically selected).

Resulting from the facts mentioned above, in the input language the part programmer is unable to give exact instructions on operation sequence. But neither this nor the occasional revision of his correct commands infringe upon his rights and deminish the flexibility of the system since at a following stage of the design process, as will be seen, he will have the opportunity to change the decisions of the computer, for his own risk.

2.8. The ignorance of technological prescriptions of the input data, the insensitivness of the system to transitional effects requires that the production engineer should assert his right and irreplaceable judgement and will. The opportunity for interaction presents itself at the processor postprocessor interface, the CLDATA. But the design process is here at an advanced stage so that the machining operation cannot be comprehended, therefore the modification would be too complicated. Interaction can be more effective at the output of the technological preprocessor because this is the last level of the design process: Decisions have been made, and mostly arithmetical calculations are to be carried out.

The interaction is simple, since this output has similar construction and content as the input data of well-known NC processors (part programme).

This construction and way of application are beneficial even if the technical equipments (e.g. display) are missing.

Conversational part programming and interaction at the output of technological preprocessor - these two tools make the high level design systems really effective.

2.9. When planning motions the successiveness is of no use since it may cause superfluous travels and the tool motions become 'rugged'. Idle run motions have to be calculated according to the coordinates and velocity of the two or three travels. Recognition of momentary situations and simplification are possible if the parameters of the motions are fixed after a delay. Thus the control programme made by the computer would be similar to that of an experienced manual programmer.

3. THE FORTAP SYSTEM

The design principles discussed in the former paragraphs provide a base for the construction of FORTAP suitable for programming of NC lathes. Instead of repeating well-known facts a few characteristic features will be mentioned.

The main stages of the technological preprocessor are shown in fig. 8.

Important elements of the technological processor are the subroutine for cutting value calculation, subroutines for processing the groups of unit operations, for tool selection and idle travel optimization.

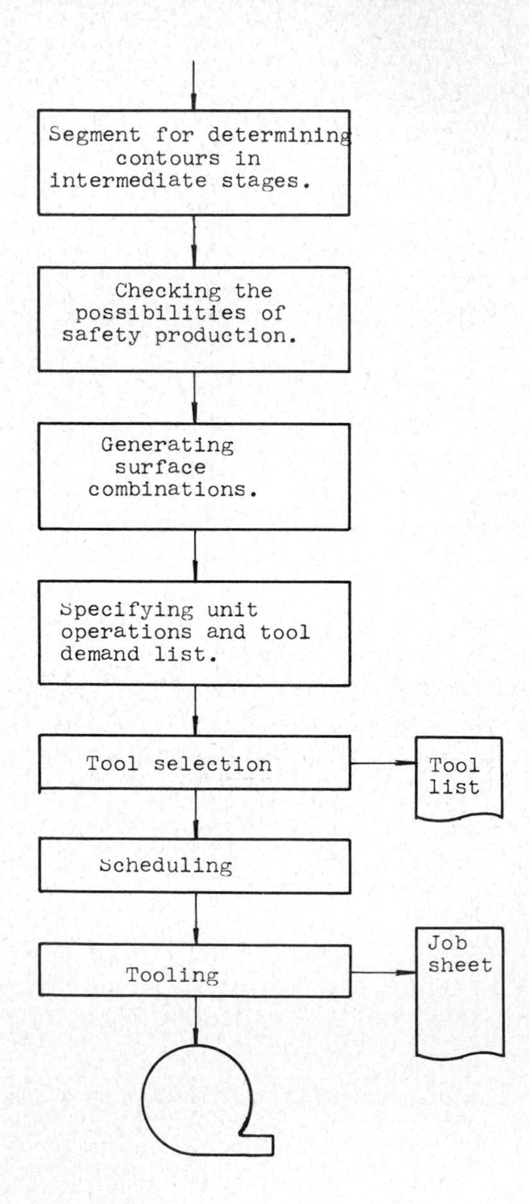
Segment for determining contours in intermediate stages.
Checking the possibilities of safety production.
Generating surface combinations.
Specifying unit operations and tool demand list.
Tool selection
Tool list
Scheduling
Tooling
Job sheet

Figure 8.

The CLDATA developed has block structure, ideally fitting for turning conditions both in respect to the amount of CLDATA and the postprocessor. Postprocessors are not more than 2 to 5K words each. The processor naturally also outputs the unified, ISO compatible CLDATA of Comecon countries.

The system is implemented on GIER small computer in ALGOL. The FORTRAN version is under development.

4. TENDENCIES OF DEVELOPMENT

Postprocessors for NC lathes, plotters are available for the FORTAP system. Experience justified our theory and we are convinced that these principles can be profitable applied in the case of other type of machining and machine tool. We do not state that they are the best, but they may be used for programming integrated manufacturing systems, where not individual operations but the whole cutting process of a component have to be planned.

Automizing the scheduling and machine selection obviously belongs to the programming such a big system because of the complexity of the system itself, the great amount of data and possible versions.

At the time being the main problem is to bring the technological preprocessor closer to the APT-like systems used in Hungary. Its effectivity is increased by a subroutine which gives a specification of a tool if the system does not find an appropriate one in the tool file ordered to the machine.

These principles are to be used in the ellaboration of a unified programming system of the Comecon countries and in the development of an integrated manufacturing system.

REFERENCE

Numerical Control Programming Languages
PROLAMAT Proceeding
North Holland Publisher, 1970.

PROGRAM PACKAGE AND SOURCE LANGUAGE FOR CUTTING CONDITIONS PROBLEMS

ENN H. TÕUGU
Cybernetic Institute, Academy of Sciences
Estonian SSR

Abstract: The problem of cutting conditions calculation is regarded as an optimization problem. Main attention is paid to the source language for the description of the tooling process and to the translation of the source description of the problem into a mathematical model on which the optimization problem can be solved.

1. INTRODUCTION

Variety of automatically accomplished technological calculations for NC programming increases when machine tool languages of higher level are used. In particular, the determination of cutting conditions becomes a standard component of NC programming systems. It is useful to have a cutting conditions calculation system which due to its universality and other qualities is suitable for inserting in various NC programming systems. From the other hand,universality considerably complicates the program especially when the program must be used for calculating cutting conditions for several simultaneously working instruments. The matter is that different programs are needed for calculating cutting conditions for different combinations of simultaneously working instruments.

A system is described here in which a new calculation program is automatically compiled for every cutting conditions calculation problem. The models of cutting processes and the modules representing suitable optimizing methods are used as sub-parts of particular programs.

2. THE PROBLEM DESCRIPTION

Calculations of single-instrument as well as multi-instrument operations are considered. Let us consider a general case where an operation is accomplished by k-instruments and the work of i-th instrument is determined by three values: cutting depth t_i, feed

s_i and cut velocity v_i. It can be assumed that all the cutting depths are determined beforehand from the geometrical conditions. Then, the aim is to choose such values of vectors $v = (v_1, v_2, \ldots$ $\ldots, v_k)$ and $s = (s_1, s_2, \ldots, s_k)$ which minimize the workshop costs (or the machining time) expressed by the function $E_o(s, v)$. The restrictions on the cutting force, cutting power, tool-life, etc. are expressed in the form of unequalities

$$E_j\ (s, v) \qquad j = 1, 2, \ldots, m.$$

In a general case such an optimization problem is a nonlinear programming problem. It is impossible to divide the problem into simpler problems for separate instruments if there exist restrictions on the sum of cutting force and cutting power of several instruments. To find the minimum values of the function E_o (s, v) methods have been developed where special properties of the functions $E_o(s, v)$, $E_j(s, v)$ are taken into account. Depending on the structure of the tooling operation and the completeness of the cutting process description the considered optimization problem can be solved on a linear model (Goranskij, 1963), convex model (Tinn, 1968) or on the discrete finite set of possible values of v and s (Tinn, 1968).

The main difficulty is that the form of the functions E_o (s, v), E_j (s, v) depends on the contents and the structure of the tooling operation and is different for every problem to be solved. For instance, it is known that the relationship between the cutting velocity and tool-life used in the description of the function $E_o(s,v)$ depends on the tool geometry, the processing mode, on the material being machined, etc. Therefore the function $E_o(s, v)$ may be different for various tooling conditions. So are the functions $E_j(s, v)$. Yet the greatest influence has the structure of the tooling operation because limitations on the sums of cutting forces and cutting power as well as the expression of the total processing time depend on the disposition of instruments on a machine-tool.

Changeability of expressions for the functions $E_o(s, v)$, $E_j(s, v)$ does not allow to compile a calculation program for various processing schemes beforehand. In the given case a good result can be achieved by compiling calculation programs of the functions E_o (s, v), E_j (s, v) according to the input text for every cutting conditions problem.

So instead of writing a complicated universal program for calculating cutting conditions a translator has been written that compiles a calculation program for every particular case.

3. THE PACKAGE STRUCTURE

A set of programs for solving various problems from a certain fixed area is called a program package. A program package for finding optimum cutting conditions of one or more simultaneously working tools is considered in the given paper.

The program package may be built up like a "black box" - so that only the rules of using ready programs are known and the user of the package does not know its structure and he cannot change and extend the package. Such way is comparatively convenient for the user while it requires from him only the knowledge how to use the ready program. Another possibility is to build the package in the form of an extendible program set the advantage of which is its flexibility. However, in this case the user who extends the package is responsible for the correctness of the extension. The package considered in this paper has been compiled in the form of an extendible program set. It is assumed that the models of tooling processes used in the calculations may change and, in particular, they can be extended adding the data about new materials to be processed and new instruments used.

Our package consists of the data base containing the information of various cutting processes as well as various optimization programs and of a core containing service programs which do not depend on the problems to be solved.

The core programs provide facilities for

1) solution of problems using the data in the data base,

2) extension of the package by introducing supplements and alterations.

In solving the problems the following functions are accomplished by the core:

1) translating the problem description from the source language into the internal language of the package,

2) compiling a program from the description of a problem and data in the data base,

3) executing the compiled program,

4) displaying the data about the run of the program and the results of the computation.

The core consists of a set of programs which are interconnected as shown in Fig. 1. Besides the programs, also the data processed by the programs are shown in the same figure. The program P1 adds

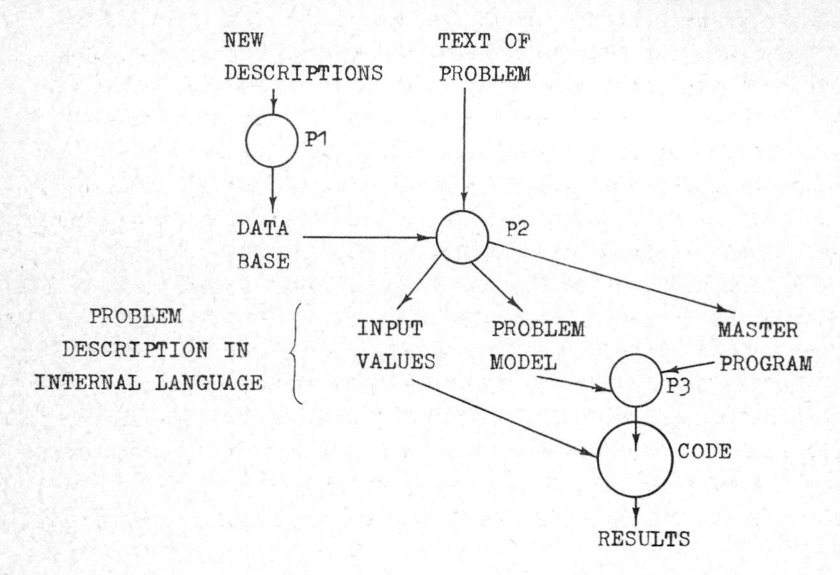

Fig. 1. Structure of program package.

new descriptions to the data base. When solving a problem, the program P2 translates the sentences of the source text separately sentence by sentence informing us about the mistakes in the sentences and enabling to do corrections in an interactive way at once when a mistake is discovered. A comparatively thorough description of the problem in the internal language is formed on the basis of the problem text and the text in the data base. The values of the variables given in the problem text are also translated and allotted in a separate input file. When the whole problem text has been translated into the internal language then the program P3 uses the internal description of the problem and generates a code for the program which can solve the particular problem. The problem solving is completed by executing the code program.

4. DATA BASE AND INTERNAL LANGUAGE

The program package is built on the basis of the programming system SMP. The structure of the data base and the internal language are determined by this system and their detailed description can be found in (Tinn, 1970).

Data and programs in the data base presented as records of a certain form are called modules. There are several types of the modules: programs, equations, tables, structures and models. The most interesting type is a model. A model contains variables and entries of other modules. The model contains information about the interconnection of variables (data sets) and programs.

While compiling a program for a particular problem, models are used for determining which program modules in what sequence must be realized for solving the problem.

The descriptions of various cutting processes (milling, turning, boring), descriptions of instruments, machine tools, the materials to be processed are described as models in the data base. To every key word denoting an engineering notion (in the source language) corresponds a model in the data base. The model contains variables corresponding to parameters of the notion and references to other modules, describing relations between these parameters.

Let us consider the model in Fig. 2 as an example, describing

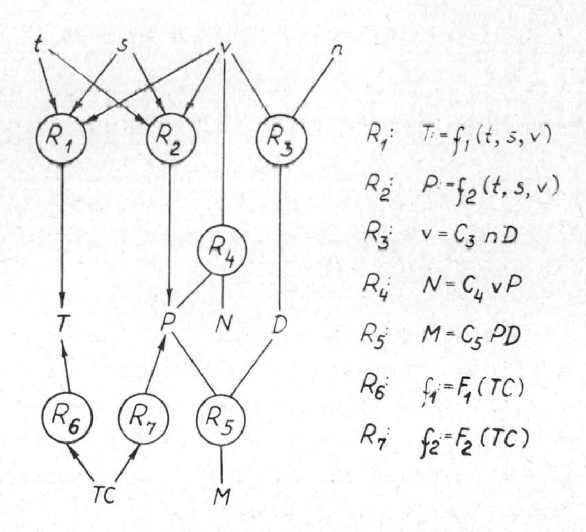

Fig. 2. Model of turning.

the notion "turning". There are numerical variables: cutting depth t, feed s, cut velocity v, the number of revolutions of the detail n, tool-life T, cutting force P, cutting power N, torque M, non-numerical value TC representing processing conditions and two functional variables f_1 and f_2. The relations between these variables are given in the form of programs (assignment statements) R_1, R_2, R_6, R_7 and equations R_3, R_4, R_5. Such a model in the data base is supplemented with the dictionary enabling to refer to the values by means of identifiers VELOCITY, DEPTH, etc.

It should be noted that practical models of machining processes are essentially more complicated than the given example. More detailed descriptions of turning, boring and milling models containing several hundreds of variables and relations are given in (Tinn, 1968). Complexity and unpreciseness of such models is the reason why the program package must be extendible.

The description of a problem in the internal language consists of a problem model, which is similar to those in the data base, and of a main program. The latter determines the sequence of subproblems which must be solved on the problem model to get the full answer.

A problem model is obtained combining the models of those processes and objects which are referred to in the problem description. The compilation of the model will be examined in more details when translation technique is described.

5. SOURCE LANGUAGE

The source language of the package is an operator structure language, e.g. the text consists of separate syntactically independent sentences. Every sentence begins with a key word which can be

1) the name of an object or action described in the data base,

2) a standard word calling forth a reaction of the translator independently of the contents of the data base. The set of the key words of the first type may be extended by introducing new descriptions in the data base. Some examples of the key words: MACHINE-TOOL, TURN, BORE, LOOP, CALCULATE, EQUATION.

The key word of the first type is followed by the identifier assigned in the sentence to the described object or action. Then the list of parameters with indication of corresponding identifiers or values of the parameters follows. For example the sentence

(CUT R3 DIAMETERI 70 DIAMETER2 35 DEPTH 5
INSTR 21006 FEED P2 REVOLUTIONS ST. REVOLUTIONS)

specifies the tooling pass from the diameter 70 to the diameter 35 with the cutting depth 5 using the instrument of the type 21006 with the feed designated by P2 and the number of revolutions of the detail designated by ST.REVOLUTIONS. The parameters which may be used in a sentence with the given key word are determined for every key word denoting an object or an action. But it is not determined which of them will be used or in what sequence they will be written in any sentence. For example the description of cutting may be as follows:

(CUTTING R3)

The value of the diameters and cutting depth are not determined by that sentence but the parameters, such as FEED and REVOLUTIONS, get the standard designation R3.FEED, R3.REVOLUTIONS etc.

Another example of a sentence:

(SEARCH PP VARIABLES P1 P2 ST.REVOLUTIONS)

describes searching the optimum values of P1, P2 and ST.REVOLUTIONS on the three-dimensional grid. Here the parameter VARIABLES is followed by the list of optimizable variables:

P1, P2 and ST.REVOLUTIONS.

The sentences in which new equations are specified begin with the key word EQUATION. These equations are binding numerical variables which occur in the problem description. For example, a relation between the number of revolutions of various instruments may be described by the sentence

(EQUATION T1 REVOLUTIONS = REVOLUTIONS2 - REVOLUTIONS3)

or a relation between total power and cutting power of separate instruments may be expressed by the sentence:

(EQUATION TT POWER = POW1 + 2   POW2 + POW3)

It is possible to assign a fixed numerical value to a variable by means of equations:

EQUATION TT LIFE TIME = 50

However, it is more convenient to give initial values by means of another key word GIVEN:

(GIVEN LENGTH 75 DIAMETER1 120 DEPTH 4)

The full problem description consists of sentences describing the machine-tool, the material to be processed, all the elementary tooling passes of separate instruments and additional conditions of the problem in the form of equations or given values of some variables. At the end of the problem description a sentence is given which specifies the desired way of calculation (the optimization method). The end of a text is marked by the sentence SOLVE.

An example:

```
(MACHINE-TOOL ST TYPE 1415 FEED1 P1 FEED2 P2 MATERIAL MD
STEEL 35)
(TURNING R1 DIAMETER 85 LENGTH 50 DEPTH 4.5 INSTR 21004
FEED P1 REVOLUTIONS ST.REVOLUTIONS)
(TURNING R2 DIAMETER 76 LENGTH 20 DEPTH 3 INSTR 21004
FEED P1 REVOLUTIONS ST.REVOLUTIONS)
(CUTTING R3 DIAMETER1 70 DIAMETER2 0 DEPTH 4
INSTRUMENT 21006 FEED P2 REVOLUTIONS ST.REVOLUTIONS)
(RETRIEVAL PP VARIABLES P1 P2 ST.REVOLUTIONS)
(SOLVE)
```

In the given case all the passes are interconnected by identical number of revolutions of the detail ST.REVOLUTIONS and, besides, the first two passes are accomplished with one and the same feed P1. Two feeds P1, P2 and the number of revolutions ST.REVOLUTIONS are to be found.

6. TRANSLATION TECHNIQUE

The compilation of a program solving a problem in the given package is carried out in three stages:

1) The source text of the problem is translated into internal language. So a problem model and a main program for solving a particular problem are generated. The problem model is similar to models in the data base, but it contains all the variables (parameters) of the objects and actions referred to in the source text. It also contains all the modules which can be used for calculating values of some of these variables.

The main program may be considered as a framework of an algorithm for solving a problem. A number of assignment statements in the main program are presented in the form

$$y := (x_1, x_2, \ldots, x_k),$$

where f is the function implicitly specified by the problem model. For example, on the model in Fig. 2 there may be given a statement M : = f (TC, t, s, n, D).

2) The sequences of modules which are to be used for calculating the function f in every assignment statement y : =

$:= f(x_1, x_2, \ldots, x_{k_i})$ are determined on the problem model. For example, in order to do the assignment M : = f (TC, t, s, n) it is necessary to use successively the modules R_2, R_3, R_4 and R_5.

3) The machine code is assembled from the main program and from the modules necessary for the computation. This stage is not of special interest because it is an ordinary assembler function implemented in any other modular system.

The problem model is obtained by connecting the models which correspond to the key words of input sentences. Here the renaming of variables according to the text of the sentence takes place. For example, translating the sentence

(TURNING R2 DIAMETER 76 LENGTH 20 DEPTH 3 INSTR 21004
FEED P1 REVOLUTIONS ST.REVOLUTIONS)

the values FEED and REVOLUTIONS get new designation P1 and ST.REVOLUTIONS. The designations of other values will be concatenated with the designations of the given action R2, so DIAMETER will be R2.DIAMETER, LENGTH will be R2.LENGTH, etc. the parameters of different objects and actions which have identical designations will be presented on the problem model by one and the same variable. The use of the designation ST.REVOLUTIONS in the given sentence guarantees that the number of revolutions of the given tooling pass will be equal to the number of the revolutions of the machine-tool designated by ST.

Now let us have a ready problem model and our aim is to find out how to execute a certain assignment statement $y := f(x_1, x_2, \ldots, x_k)$, where $y, x_1, x_2, \ldots, x_k$ are the variables of the model. The problem model scheme in the form of a graph is used for that purpose. On the first step it is determined what modules may be used immediately for the calculation of certain unknown variables when values of $x_1, x_2, \ldots, x_k$ are given. Addding variables values of which can be calculated to the set of arguments $x_1, x_2, \ldots, x_k$, this process is repeated again. The repetition continues until the value of y can be calculated or it is found that any new values cannot be calculated. In the last case it is impossible to execute the assignment statement $y := f(x_1, x_2, \ldots, x_k)$ on the problem model. Then it is reported that the problem is unsolvable. If it is possible to calculate the value of y, all the chosen modules are examined once more and the modules unnecessary for the calculation of the value of y are cancelled. It gives such a sequence of modules the execution of which corresponds to the execution of the statement $y := f(x_1, x_2, \ldots, x_k)$. Such process called solution planning is repeated for every assignment statement $y := f(x_1, \ldots, x_{k_i})$ in the main program. The solution planning is more thoroughly discussed in (Tóugu, 1971).

7. DISCUSSION

In this paper we have almost completely ignored the tooling principles and algorithms of cutting conditions calculation. In practical implementation great efforts were paid to the selection of optimization algorithms and formalization of tooling models. These models are too complicated to be perfect yet. It is necessary to make further alterations and supplements in the models because new materials and new processing modes appear. That is why an extendible program package has been written. In the extendible program package it is necessary to guarantee sufficient correctness and reliability of programs even when alterations are made. Certainly it is impossible to guarantee the correctness of alterations to be introduced in the future. But the stability and reliability of the already existing part of the package must be guaranteed making it possibly independent of the future extensions. This aim was considered first in constructing the given package.

A specific data structure has been used in this package. Perhaps the idea to specify semantics of problem-oriented sentences in the form of models may prove useful in other program packages. Comparing this principle with macro-generators processing arbitrary texts the following may be noted. The descriptions of new models for the data base may be compared with macrodefinitions. But the macrodefinitions determine the substitution of a text uniquely. Here the result of the substitution is determined uniquely only on the level of the internal description of the problem in the form of a model. To get a complete program an additional compiling stage - solution planning on the model is used. This stage separates the problem description in the source language from the code generation. And so it is possible to obtain an extendible and yet simple source language.

REFERENCES

Goranskij, G.K. (1963) Расчет режимов резания при помощи ЭВМ, Белгосиздат, Минск (Russian).

Tinn, K.A., Tóugu, E.H., (1968) Технологические расчеты на ЦВМ, I, Издательство "Машиностроение", Ленинград (Russian).

Tinn, K.A., Tóugu, E.H., Unt, M.J., (1970) Система модульного программирования для ЦВМ Минск-22, Труды ВКП-2, Новосибирск, Г-22 (Russian).

Tóugu, E.H., (1971) Data base and problem solver for computer aided design, Proc. Congress IFIP71, Ljubljana, TA6-16.

TNO MITURN

Programming system for lathes

P.A.J.M.Bockholts, Metaalinstituut TNO, Centrum Voor Metaalbewerking, Delft, The Netherlands

1. INTRODUCTION

The MITURN programming system (Metaalinstituut TURNing Programme) is an efficient programming system for numerically controlled turning machines. It is in fact a comprehensive production system composed of a number of sub-systems which together make an important contribution to the optimum utilization of numerically controlled lathes (NC lathes).

In MITURN, the achieving of maximum lathe capacity is coupled with simple programming and a high degree of automation of the punching of the control tape: this is because MITURN is designed so that, after programming of the geometry of the starting material and the workpiece, the following operations are carried out automatically:

- the optimum sequence of operations is determined,
- the optimum depth of cut, feed and cutting speed are determined per cut or per part of a cut,
- the programmed surface quality is reached,
- the best tool for each operation is selected from the stock of tools
- the operation instructions for the machine are formulated,
- the expected machine time for the programmed workpiece is specified,
- the tape is punched for the machine.

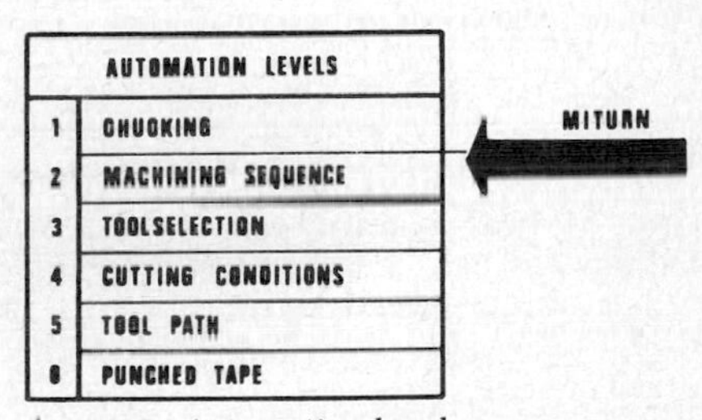

figure 1. Automation level

2. INPUTDATA

2.1. STRUCTURE OF INPUTDATA

The information required for processing can be subdivided into three levels:

a. Workpiece dependent information, information which has to be written for each different component (part programme)
b. Information used for the internal processing of the MITURN compiler, information that can be tailored to the users preference. this information is modal, which means that these data will be used until these are changed through separate access.
c. Information which automatically through a. and b. is generated by the MITURN compiler.

COMPUTER LANGUAGES FOR NUMERICAL CONTROL, *J. Hatvany, editor*
North-Holland Publishing Company - Amsterdam–London

The mentioned information is programmed as follows:

a. The part programme that is written according the drawing information of the component, the clamping arrangements, the raw material etc.

b. Tooldata for all tools that can be handled by the MITURN compiler. Strictly these are all tools which can be used on a lathe. Cuttingdata which allow to consider different workpiece materials, toollifetime, surfacefinish, finish allowances, standard operation sequences, shapes of undercuts etc. in general data which allow the user to let act MITURN to users preference.

   Machinetool and control cabinetdata in order to consider all machinetool and control requirements.

c. Internal systemlogic to interprete the programmed data. This interpretation is based on objective rules and can not be influenced.

2.2. PART PROGRAMME

The partprogramme consists of 5 different sections and an optional 6th section (see fig. 2)

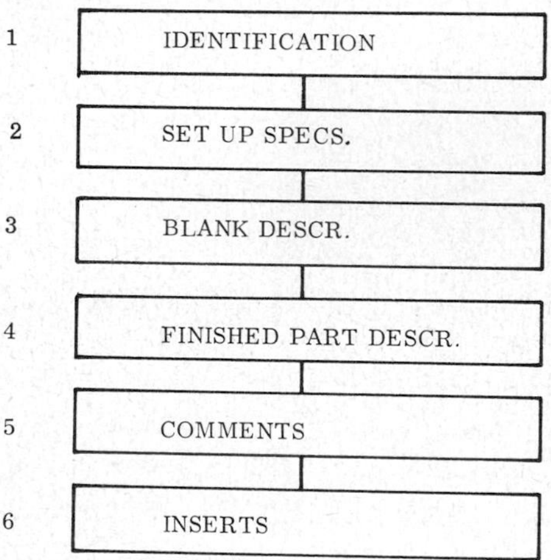

figure 2. Optional special shapes

Each section has a fixed sequence of definitions which always have to be filled in.
The input of the partprogramme can be supplied to the MITURN compiler in two different ways.

1. conversational input
2. file input

The first input is easy for the programmer because MITURN puts question which can be answered. The second way is faster and therefore cheaper than the first one and can be used by programmers who have some experience. An example of a partprogramme is given in fig. 4. The component is illustrated in fig. 3.

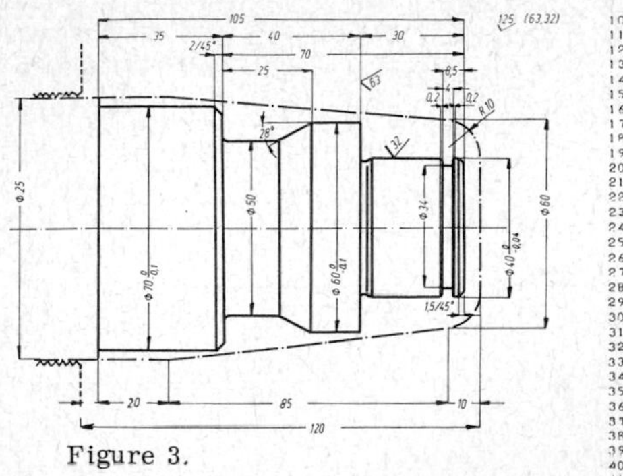

Figure 3.

```
100**********************************
110*IDENTIFIZIERUNG
120 J.ANØN 15-3-1971
130 TZ B-1
140 MM
150 ST60
160************************************
170*EINSTELLDATEN
180 0 10 207 5 120 105 0 3
190 1 0.7 10 1
200*************************************
210*RØHTEILBESCHREIBUNG
220 3
230 4 3 1
240 RAD 40 60 10 -10
250 KEG 60 75 85
260 ZYL 75 20
270 0
280**************************************
290*FERTIGTEILBESCHREIBUNG
300 3
310 1 1 1
320 ZYL 40 0 -0.04 30 32 63 1.5 1
330 ZYL 60 0-0.140 125 125 0 0
340 ZYL 70 0-0.135 125 125 2 0
350*UEBERLAGERUNGSELEMENTE
360 2
370 EIN 34 8.5 4 0.2 0.2 1
380 AUS 50 70 25 28 90 2
390 0
400**************************************
410*KØMMENTAR
420 EIN PRØGRAMMBEISPIEL FUER EIN EINFACHES WERKSTUECK
430 IM DREIBACKENFUTTER EINGESPANNT
```

Figure 4.

2.3. READABILITY OF PART PROGRAMME

the amount of information in the partprogramme is reduced to a minimum. Through this the readability is garanteed.
Questions for the required inputdata are printed in the case of conversational input. Because of the minimal information, errors in the inputdata are reduced to an extremely low level. In the showed example the five different sections of the partprogramme are separated by the star-lines.
In between the lines containing inputdata as many comment lines in any language can be inserted. These are ignored by the MITURN compiler and contain typing errors. The real data are separated either by comma's or blanks, this means that each line has a fully flexible format. Only the sequence of data per line is fixed. A detailed explanation would go beyond the scope of this paper.
Most important to discuss briefly are the data in the sections three and four because in these sections the blank and finished part description are given.

2.4. PARTDESCRIPTION

The finished as well as the blank are described in the partprogramme in so called elements. Therefore five basicelements and six superimposed elements are defined. These definitions are illustrated in fig. 5.
As shown in the figure all data that can be specified on a drawing can be defined in these elements not more and not less. This includes that from a given part programme the componentdrawing easily can reproduced for example by a plotter.
These definitions for describing a component can easily be learned and are simple to use. The definitions from basic and superimposed elements are based on conventions of technical symbols, because normal workshop drawings are composed out of just these technical symbols and therefore partprogrammes can be written easily from these drawings.

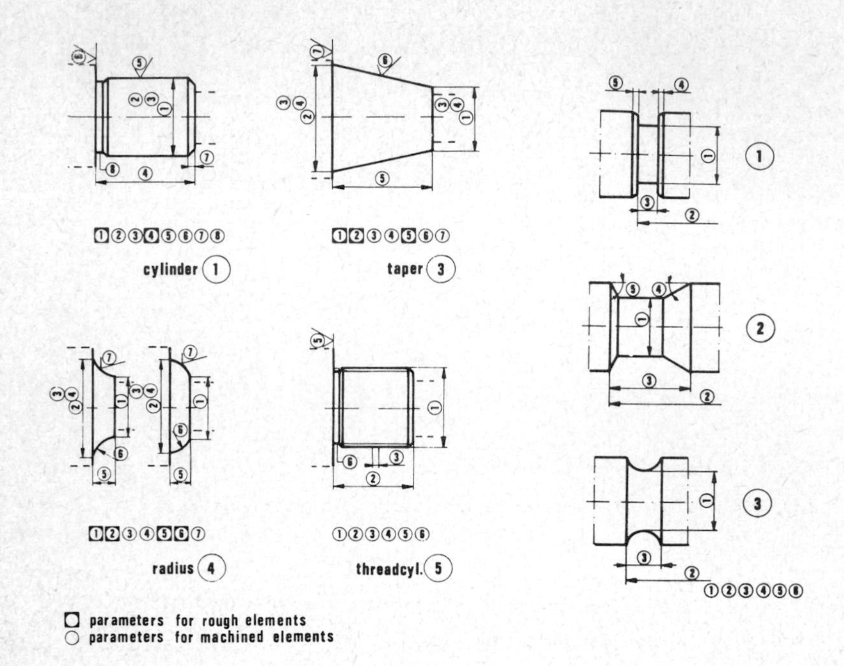

Figure 5. Definitions of the standard elements for description of the rough and machined (final) workpiece. As one can see, the definitions for the internal and external elements are the same.

3. TECHNOLOGY IN MITURN

The technology in MITURN is based on grouptechnology. The MITURN compiler is modular designed which allows to add new families without influencing already implemented families. In general all definitions also in the processor are handled in a most general way. The aim by the design of the MITURN compiler was:

a. The selection of tools, operation sequence, cutting data and toolpath must be generated fully automatic so that the part programme which only contains shape, surface quality and tolerances can be processed such that a complete papertape and machinetool instruction will be delivered.
b. Contouring facilities of the machinetool must be fully exploited.
c. The number of tools is limited to a certain extend due to capacity of the machinetool.
d. The technology can be tailored to individual requirements and wishes of the user.

3.1. DEFINED OPERATIONS

14 Different operations are implemented from which 13 are completely automatic, the 14th is semi automatic.

a. External roughing with radial or contouring feed
b. External roughing with axial roughing or contouring feed
c. External finishing with axial or contouring feed
d. External axial grooving and relieving
e. External radial grooving and relieving
f. External threading
g. Drilling and centerdrilling
h. Internal drilling with radial or contouring feed
i. Internal roughing with axial or contouring feed
j. Internal finishing with axial or contouring feed
k. Internal axial grooving and relieving
l. Internal radial grooving and relieving
m. Internal threading
n. Special operations as reaming, tapping, operating tailstock etc.

All the operations except the last one are automatically determined by the MITURN compiler and processed. For each operation one or more tools are automatically selected from a standard toolfile. This toolfile is through a special handling programme accessable for the user. In this toolfile the user has to a large extend the freedom to supply with tools of his own choise of course within the boundary of the defined operationfunction which each tool has to carry out. In this way the user is free to compose his own toolmenu. All tools are called up by parameters which are described in the toolfile.
The toolselection is controlled by optimization rules. For example when a drill has to be selected the one with the largest diameter to fit in the hole and the shortest length just long enough to drill the hole is selected. Of course this selection is dependent on the variety in toolmenu.
In figure 6. an example is shown of a particular component together with the selected tools for machining.

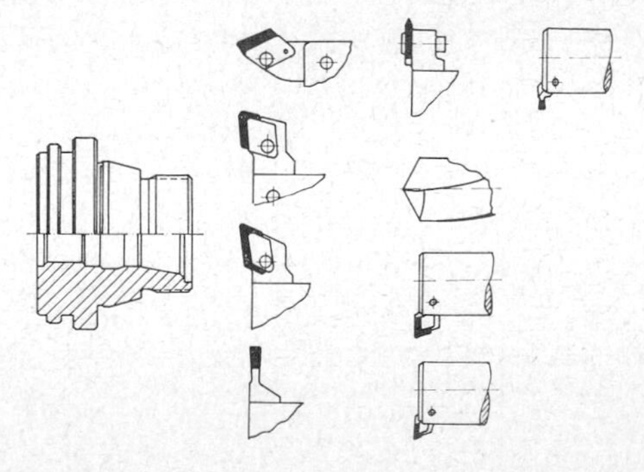

Fig. 6. Workpiece and tools selected automatically for its machining.

3.2. CUTTING CONDITIONS AND TOOLPATH

In the models for the computation of cutting conditions and toolpath great emphasis is put on the roughing cycle because one can assume that in small lot production no preformed blanks are available so that relatively much material has to be removed. In cases where a casting of preformed part is available this can be programmed first as it is. The roughing cycle is designed as a highly optimized routine concerning cutting conditions and toolpath. A relationship has been designed between the workpiece rigidity, the toolrigidity, machine-tool-capacity and workpiecematerial. For the workpiecerigidity an emperical model has been designed. (see fig. 7.)

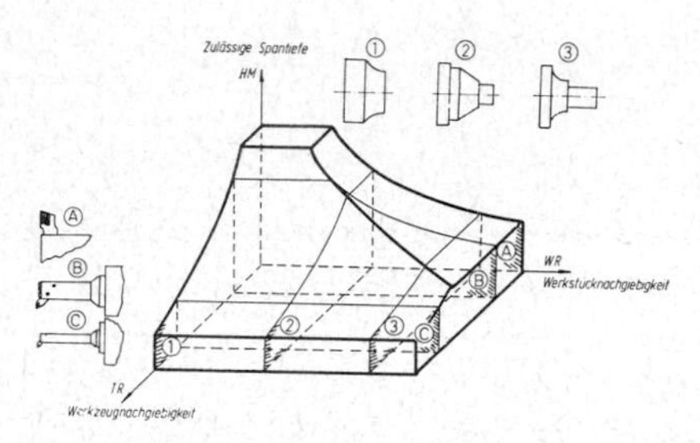

Figure 7. Establishing permissible depths of cut in roughing.

This relationship results for each cut in a depth of cut, feed and relative to these a cutting speed. When these data are available the kinematics for a single cut can be generated and are illustrated in figure 8. for an axial roughing cut with contouring.

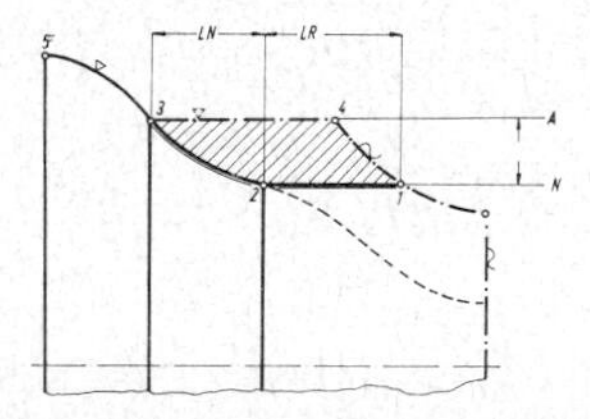

Figure 8. General form of a roughing stroke.

For each cut the full model is repeated because of the change in the shape of the work-piece. This roughing model results in a constant finishing allowance (where required) and therefore it is not necessary to have more than one finishing pass.

3.4. FINISHING CYCLE

The finishing feed is determined by the surface roughness specified for each particular surface of the component and the used noseradius of the tool to be used. Automatically noseradius compensation is carried out for all the sections of the component to be finished (see fig. 9).

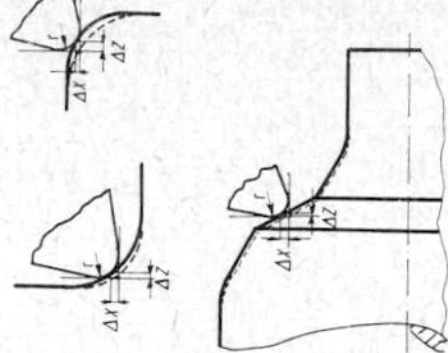

Figure 9. Compensation for tool point radius.

For close tolerances a trial cut can be called up if requested. The indication for this is specified in the partprogramme.

4. POSTPROCESSOR

The MITURN postprocessor has a separate input and output. Herewith is the MITURN processor completely independent from the postprocessor but the postprocessor is automatically called up by the processor by data specified in the partprogramme.

Besides the normal conversionfunctions of the postprocessor a MITURN postprocessor has a set of functions, handling for example the design of the turretlay-out based on certain rules for toolloading of the particular lathe. This cycle then designates for all tools, selected by the processor, an adequate station on the machinetool.
Other examples are the automatic cycle for toolchanging, automatic spindlespeedchanging for radial cutting motions of course as option. The MITURN postprocessor also automatically calculates the expected piecetime on the lathe. All necessary information for preparing the lathe and tools will be printed on an instructionsheet which is generated by the postprocessors. The papertape is produced finally and can be punched in any desired coding. e.g. (ETA, iso)
In figure 10 an instructionsheet is shown. In figure 11 the output including absolute dataprint out is given.

```
10000 ;
10010 N001G04X01000S23M03;
10020 N002X01000M08;
10030 N003X01000T30;
10040 N004G01Z-40000K29999F04000M07;
10050 N005X20980I29999;
10060 N006Z-14600K29999;
10070 N007G04X01000S24;
10080 N008X01000S25;
10090 N009X01000S26;
10100 N010X01000S27;
10110 N011G01Z-02411K29999F00114M06;
10120 N012X00068Z-00119I14884K26047;
10130 N013X00252I29999F00056;
10140 N014X-00252I29999F00163;
10150 N015Z-00050K29999F00114;
10160 N016X00252I29999F00056;
10170 N017G04X01000S28;
10180 N018G01Z02580K29999F04000M07;
10190 N019X-00640I29999F00260M06;
10200 N020Z-01855K29999F00144;
10210 N021G04X01000S27;
10220 N022G01X00320Z-00556I14964K26001F00114;
10230 N023G04X01000S28;
10240 N024G01Z02411K29999F04000M07;
10250 N025X-00640I29999F00260M06;
10260 N026Z-01299K29999F00144;
10270 N027X00320Z-00556I14964K26001;
10280 N028Z01855K29999F04000M07;
10290 N029X-00640I29999F00260M06;
10300 N030Z-00907K29999F00144;
10310 N031G03X00222Z-00223I00296K00522F07200;
10320 N032G01X00098I29999F00072;
10330 N033X-00098I29999F00124;
```

Figure 10.

```
INSTRUCTIEBLAD VOOR VDF 630 NC
==============================

HOOFDGEGEVENS
-------------
PRØGRAMMEUR EN DATUM J.ANON 15-3-1971
TEKENINGNUMMER       TZ 8-1
WERKSTUKMATERIAAL    ST60

SET-UP GEGEVENS
---------------
WERKSTUKTYPE       =  10
KLAUWPLAATDIKTE L1= 207.00 MM
UITSTEEKL. MAT. L3= 120.00 MM
VEILIGHEIDSAFST.L2=   5.00 MM
PØSITIONEERAFST.L6=   3.00 MM

MATERIAALGEGEVENS
-----------------
MATERIAALCODE    =   1
BEWERKBAARHEID   =0.70
SPAANSLANKHEID   =  10
ØPSPANNINGSCOEF  =1.00

GEREEDSCHAPPEN
--------------
PØS  NR  OMSCHRIJVING************   W8      W9      W10      W12   W21  CØR
  1 1020 ZIJBEITEL          R K793 320.00 445.00 485.00   1.00  P35    0
  2 DEZE POSITIE WORDT NIET GEBRUIKT
  3 1210 MESBEITEL          R K803 900.00 445.00 490.00   1.00  P35    0
  4 1310 KOPIEERBEITEL      R 1715 900.00 450.00 495.00   0.50  P30    4
  5 DEZE POSITIE WORDT NIET GEBRUIKT
  6 DEZE POSITIE WORDT NIET GEBRUIKT
  7 1410 STEEKBEITEL        R 2516 900.00 445.00 477.00   0.40  P30    7
  8 DEZE POSITIE WORDT NIET GEBRUIKT
  9 DEZE POSITIE WORDT NIET GEBRUIKT

UITVOERINGSGEGEVENS
-------------------
MAGAZIJNPOSITIE      =1230.00 MM
TOERENTALINSTELLING =   1800 OMW/MIN
AANTAL BLOKKEN      =    187
STUKTIJD            =  13.80 MIN

ØPMERKINGEN
-----------
EIN PROGRAMMBEISPIEL FUER EIN EINFACHES WERKSTUECK
IM DREIBACKENFUTTER EINGESPANNT
```

Figure 11.

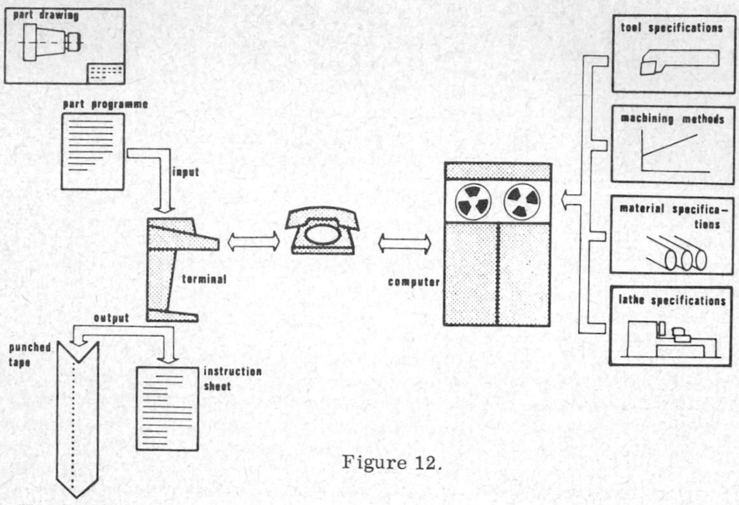

Figure 12.

5. IMPLEMENTATION

The MITURN system is implemented on a time-sharing-computer-system and therefore easily accessible everywhere where is telephone. Only a small terminal is needed for the commication with the central computer. The timesharing network is practically everywhere available. This way of computer use has a number of advantages for the user:

a. Accessability through normal telephone
b. No investment for in house computers or expensive hardware.
c. No communication problems with in house or software bureau computercentres. The small terminal can easily be located in the production planners department and operated by the partprogrammers.
d. Fast production and delivery of new tapes.
e. MITURN gives all facilities for creating databanks etc. of users convenience.
f. The maintenance of the MITURN SYSTEM is taken care of without any effort of the user. All new features will be available immediately. The use of MITURN through TIME SHARING is illustrated in figure 12.

AN INTERACTIVE SYSTEM FOR OPERATIONS PLANNING FOR TURNING ON CENTRE LATHES

PER HELLSTRÖM

Centre for Design and Production
The Royal Institute of Technology
SWEDEN

Abstract: The majority of existing systems for planning of turning operations are non-interactive systems. They tend to be very big and complex data programs. Most of them can handle only a limited number of geometrical shapes and can generate only certain types of toolpaths for certain types of tools.
With an interactive approach to the problems many of these limitations can be removed, and a system with high generality can be obtained with less programming effort.
Work on the system described in this paper started within the CAD-group of the Computer Laboratory in Cambridge (England), where the author was a visitor from September in 1971 till June in 1972. The work has then been continued at the Swedish Institute of Production Engineering Research (IVF) in Stockholm. The system does not yet (in October, 1972) excist in a complete and coded version, but is scheduled to do so at the end of 1973.

1. INTRODUCTION

The total manufacturing of a turning part can be devided into five steps, figure 1. The first step is the designing of the part. Output from this step is:

1. the geometrical shape of the final product
2. the material
3. tolerance requirements
4. requirements on surface smoothness

In the next step, which could be called the production planning, general decisions about the manufacturing are made. Such decisions are:

1. what machine(s) should be used
2. time scheduling and economically feasible batch size

COMPUTER LANGUAGES FOR NUMERICAL CONTROL, *J. Hatvany, editor*
North-Holland Publishing Company - Amsterdam–London

3. decisions about the shape and manufacturing of the blank

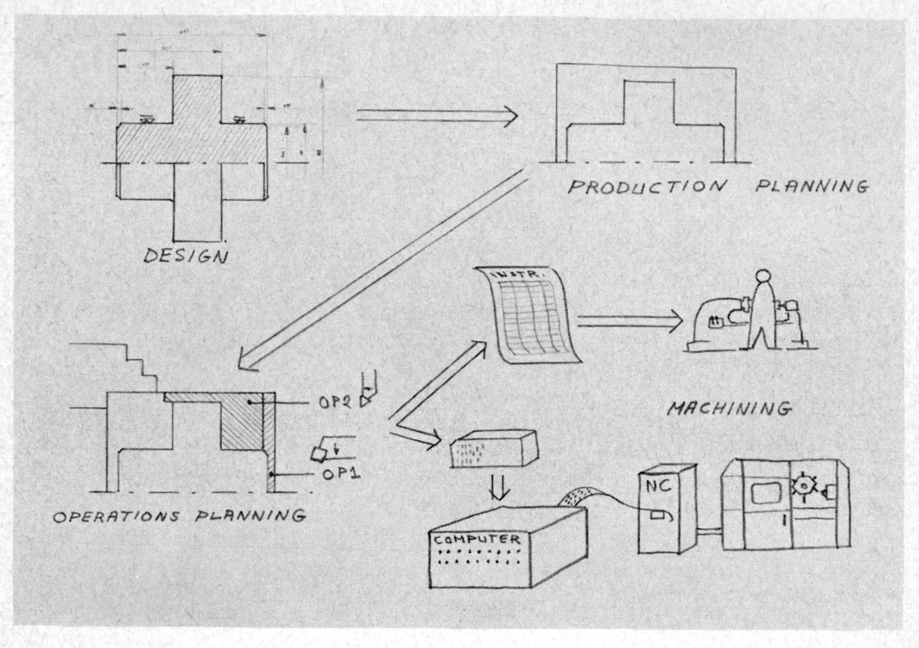

Figure 1. Activities from design to machining

At the start of step three - the operations planning - we know the shapes and locations of those volumes that are to be removed. Since in most cases the whole machining can not be made in the same clamping and with the same tool the machining has to be devided into different clampings and maching operations.
Thus, for each clamping the following decisions have to be made:

1. clamping direction
2. clamping device and surface for clamping
3. definition of the volume that is to be removed in the clamping
4. division into machining operations

We can distinguish different types of machining operations: roughing operations, fine cuts, grooving operations, drilling operations, threading operations etc.

For each machining operation the following decisons have to be made:

1. geometrical definition of the volume that is to be removed
2. definition of the tool
3. divison into cuts

For each cut the following parameters have to be chosen:

1. depth of cut (shape of cut)
2. cutting speed
3. cutting feed

As the last part of the operations planning the total tool path has to be defined.
What will happen in step four is depending on what type of machine we intend to use. If our machine is a conventional operator controlled lathe, step four will be to summarize the results from the operations planning to an instruction for the operator.
If our machine is an NC machine step four will be to produce the controll tape. This can be done either manually or through part programming (in APT, EXAPT or some other part programming language), processing and post processing.
Step four is then followed by the actual machining.
The problem area which the system described in this paper attempts to automate is the operations planning (step 3) for NC machines.

2. BASIC PHILOSOPHY

One would think that the operations planning could be done in the same way, regardless what type of lathe - operator controlled or NC - it is being done for. Unfortunately, this is not true. There are fundamental differences between the two types of maching. The most obvious ones are the ways in which tools and tool paths are chosen. Therefore the system described in this paper deals only with NC machining.
The operations planning relies heavily on the experience and skill of the man who does it. It would therefore be unwise (and very difficult) to try to automate the whole operations planning. However, this area also contains large portions of pure routine work, such as calculating optimal cutting data, retrieving tool data from tool files etc. Some degree of computer aid is therefore justified.
The basic philosophy when working on this system has been that the man who uses the system should be a man who has the capacity to do

the operations planning manually, should this be required of him. The computer is there ot assist him as far as possible in routine matters. The man must be able to recognize and correct a poor decision made by the system.
With a system like this, the ways of communication between the man and the computer must be the best possible. This means communication via graphical terminals. But since these devices are still quite expensive, an unrestricted use of graphical techniques could easily have the consequences that only a few companies could afford to use the system. Our way around this dilemma has been to use only passive graphical devices. This type of graphical equipment (e.g. storage tubes) are fast coming down in prize and could be econimically justified by most companies with NC-equipment.

3. SYSTEM DESIGN

The system is devided into five sections of approximately the same size, figure 2.

Section 1

This section contains inputroutines plus routines for the geometrical evaluation of fine cuts and groovings.
At present the input to the system consists of the following six types of information:

1. basic geomtry of blank and finished product
2. material
3. surface quality of the blank
4. tolerance and surface requirements
5. superimposed formelements
6. machine

This information is punched on cards in a fixed format. So far no effort has been made to make the input more convenient for the user. Basic geometry of blank and finished product is described according to figure 3.

This way of describing the geometry is also used internally by the system.

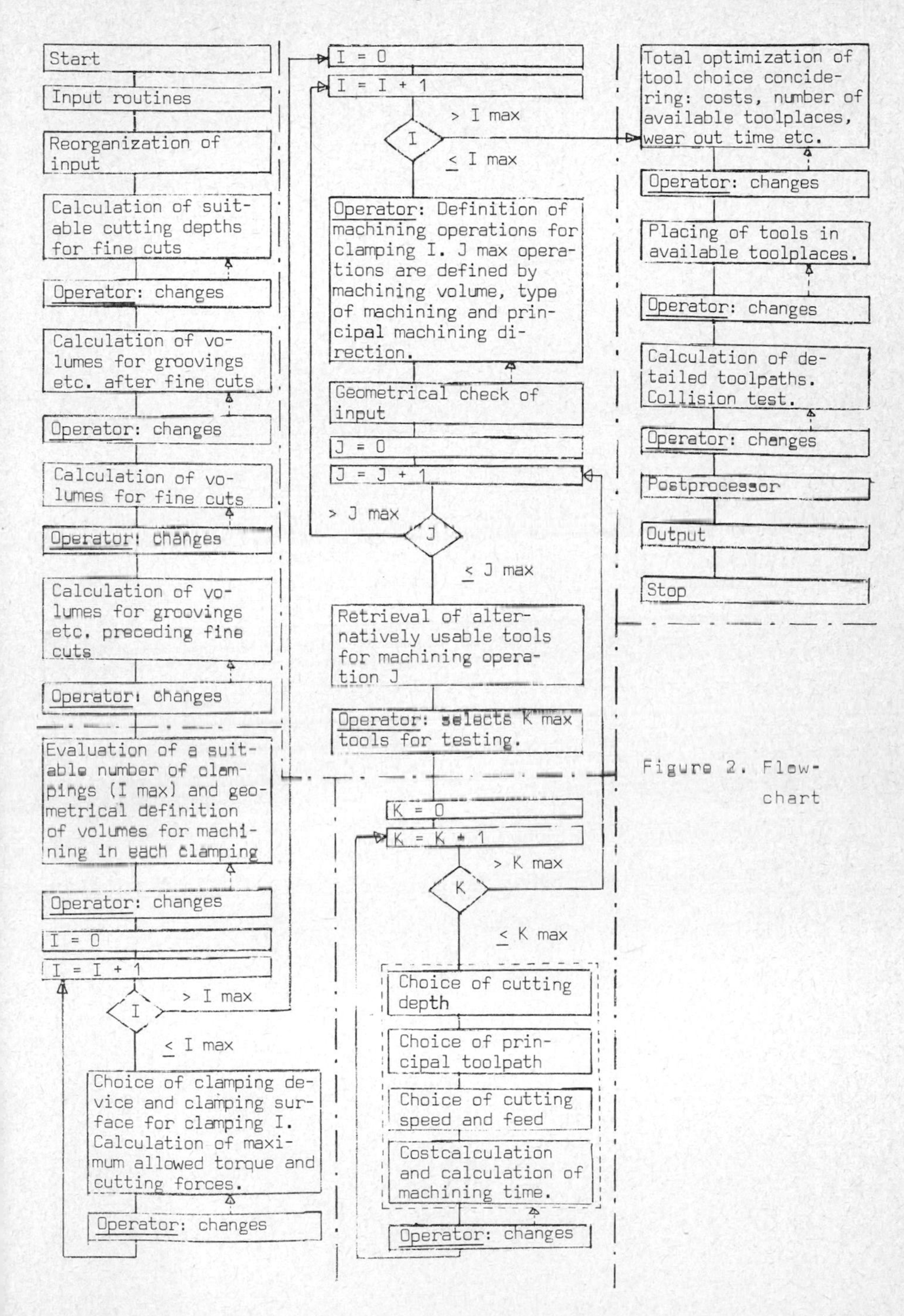

Figure 2. Flow-chart

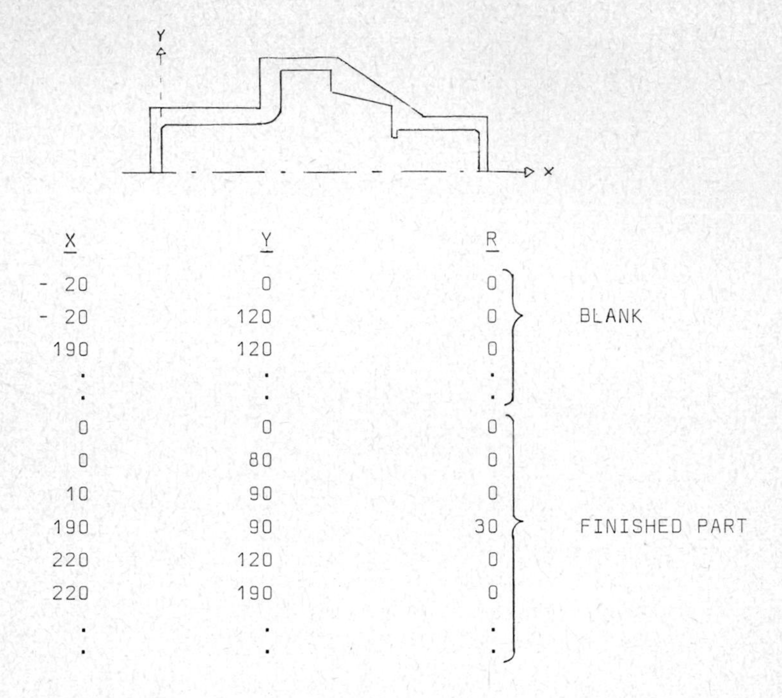

X	Y	R	
- 20	0	0	BLANK
- 20	120	0	
190	120	0	
.	.	.	
.	.	.	
0	0	0	FINISHED PART
0	80	0	
10	90	0	
190	90	30	
220	120	0	
220	190	0	
.	.	.	
.	.	.	

Figure 3. Geometry description

By defining a superimposed formelement the user can order a special type of machining to be made with a certain tool on a certain part of the finished product. This enables the use of nonstandard tools. The machining of a superimposed formelement will take place after all machining of surfaces adjacent to the formelement. A typical such superimposed formelement would be a narrow grooving or a threading operation.

The material and the machine are described only by their identification numbers. Further data is then obtained from material and machine files.

The geometrical evaluation of fine cuts and groovings will be carried out in reversed order from the order in which they will be machined. First the system will look for groovings that have to take place after the fine cuts, then define the fine cuts geometrically and finally look for groovings that are to take place before the fine cuts. All this is done so that the finished product will meet given requirements on tolerance and surface smoothness. Between each step the results are displayed and the operator is given an opprotunity to change decisions made by the system.

Section 2

In section 2 the system will first suggest a suitable number of clampings, and for each clamping define a material volume to be removed. This suggestion is primarily based on geometrical factors, but the system will also have considered such questions as whether a fine cut can be machined directly following a roughing operation in the same area, or if that part of the workpiece will have to cool of before the fine cut.
When the operator has accepted or changed the suggested division into clampings, the system will go on by looking for suitable clamping devices in the machine file, and also suggest suitable surfaces on the workpiece to clamp on. Here the system will assume certain maximal cutting forces depending on the types of machining operations that will take place in the clamping. According to these assumptions and geometrical considerations the system will suggest clampings that will not damage the workpice or make it impossible for a tool to reach the machining area.
Again the operator can accept or change what has been suggested by the system.
Unless the operator later on decides to go back and change some of the decisions that have now been made, the first two sections are now left and will not be entered again.

Section 3

Section 3 is entered once for every clamping. Each clamping is first devided into machining operations. This is done manually by the operator. For each machining operation he will give the following input:

1. geometrical definition of the machining area
2. type of operation (roughing, fine cut, grooving etc.)
3. principal machining direction

For the definition of the machining areas the operator uses a specially designed geometrical language that will make it possible for him to define rather complicated geometries with a limited number of statements. The system will check the user input for geometrical errors. For each machining operation the system will then suggest a number of possible tools or combinations of tools. From this rather big number of tools the user selects a limited numer for further testing in section 4.

Section 4

Section 4 is entered once for every machining operation. A number of fully defined alternative tools or combinations of tools will now be tested on the same machining area. The test includes the following calculations:

1. optimal cutting depth
2. preliminary tool path for each cut
3. optimal cutting speed and feed
4. cost calculation and calculation of the required time for machining

The results will be displayed for the operator and he is allowed to alter any of the calculated cutting parameters. If the user chooses to change anything, those parameters that are affected by this change will be recalculated. At this point the user can also decide to exclude a tool from further testing.
When the last tool has been tested for the last machining operation of the last clamping, section 5 is entered.

Section 5

In this section the system will look at all the different tools that have been suggested for the different machining operations. It will then suggest one setup of tools for the total machining. This suggestion is based on the following factors:

1. number of available tool positions in the machine
2. cost for machining and tool changes
3. machining time for each tool compared with normal wear out time for the tool

The user may ask for new suggestions or make his own choice among those tools that have been tested. If he wants to use other tools or

change the current division into machining operations he will have to go back to section 3 and 4.
If the user accepts the suggested group of tools, the system will go on by giving each tool a specified place in the magazines or revolvers of the machine. As far as possible this is done in such a way that collisions can be avoided. Again the user may accept or make changes.
The next step by the system is to calculate - in detail - the total tool path. In section 4 the tool paths were calculated for each separate tool. At this point these tool paths are tied together, all positionings of the tool, rapide moves, tool-changes etc. are scheduled. A collision test is carried out, checkning both the cutting tool and other moving parts for collisions with machine, clamping device or workpiece.
Finally a postprocessor adapts the results to the control unit of the lathe and generates the punched tape. Since the system has done the operations planning for one particular lathe, not very much is left for the postprocessor. This is therefore a rather small and uncomplicated routine.

4. SOFTWARE AND HARDWARE REQUIREMENTS

The system will probably require a large size computer with fast backing store devices to run smoothely. At present it is being run on an IBM 360/75. The size of each section is in the order of 100 K bytes. Those sections that are not in use are stored on disc.
Four different data files are being used:

1. machine file
2. tool file
3. material file
4. graphical file

The graphical file contains pictures of standard tools etc. that the system will display for the user. These pictures are independent of the workpiece, and we do not want to generate them in the system each time they are to be displayed.
At present these four files are being kept as small as possible, for testing purposes. It is estimated that neither of them will excede 20 K bytes in size. The data files will also be stored on disc.

All graphical output is now being sent to a microfilm plotter. This will, when the system is completed, be replaced by a storage tube. Since both devices are passive, this will not affect the design of the system.

With a few exceptions the whole system is being written in FORTRAN. This will ease the transfer of the system from one computer to another.

For the graphical output a graphical subroutine package is needed. The one that is used with this system, has been written at IVF. It provides convenient picture part handling and the use of all types of transformations. Except for the last links between the computer and the graphical device it is written in FORTRAN.

For the user´s input to the system a specially designed language is provided. A language processor is required to translate statements given by the user to relevant subroutine calls in the system.

NC-PRODUCTION PLANNING

J. STEINACKER and H.-H. WINKLER
Werkzeugmaschinenlabor, TH-Aachen, F R G

Abstract: The needs of today's production require automated information processing in all fields. Sophisticated software, dataware and organizational problems have to be solved. This contribution deals with some of these problems arising in NC-production planning. The outline of a comprehensive planning system assisted by technology orientated databanks is given.

1. INTRODUCTION

The advent of NC-machine tools did not mark only a change in manufacturing but in production planning and design also. In order to operate NC-machine tools economically the design must accord to the capabilities of this new technique and planning must be done in detail, providing that the manufacturing process is carried out in the best way. This affords time consuming work of highly skilled personnel. No doubt that this situation forced men to use the computer e.g. for NC-production planning starting with part programming. No doubt either that besides geometrical problems also technological and organizational problems can be solved by the computer thus becoming the outstanding mean for rationalization and automization in this field. However, the usage of such information processing facilities requires beside the hardware a comprehensive planning software, a sufficient data base to work with and an organizational structure between all departments involved in order to provide a reliable, quick flow of information especially back from the shop to planning and data base management. Consequently developments in this framework aim at the following points:

1) To reduce the involvement of men in planning by extending planning systems eg. by automatic determination of tools in NC-programming systems thus going in the direction of a higher level of automation
2) To improve planning results e.g. by optimal machining conditions, shortest possible tool paths etc.
3) To consider the company's know-how and peculiarities in general designed software systems
4) To break systems up in modules in order to introduce them in production planning step by step thus offering the level of automation the company needs and is able to use
5) To assist the introduction and usage of sophisticated systems by databanks which provide the data base necessary
6) As NC-machine tools still have a share of 3 to 10% only and will have not more than 15 to 20% in the next future it is worthwhile to applicate the knowledge and techniques of NC in conventional production.

Considering these points in computer assisted production planning improvements in planning itself and in the results will be made showing up with savings in cost and time the aim of rationalization efforts.

2. COMPUTER AIDED NC-WORK PLANNING

One of the most important and time consuming tasks in production planning is issueing work-sheets and NC-tapes. An unique system for NC-work planning

COMPUTER LANGUAGES FOR NUMERICAL CONTROL, *J. Hatvany, editor*
North-Holland Publishing Company - Amsterdam–London

does not exist at the moment and is not within reality for the next time, because the automation of all functions of NC-work planning represents a tremendous amount of work. But substantial steps have been done in this direction which allow to think of computer aided work planning.
NC-work planning can be based on two fundamental principles: modification principle and new planning principle. Both principles stand on the idea of group technology. While the first one considers the whole workpiece and its possible variations the latter is connected to standard elements of the part which usually require a certain manufacturing process. Splitting up the workpiece into these standard elements the work plan can be built up out of standard procedures assigned to them including the specifications, technological know-how and organizational data of the user. The general layout of such a system is shown in fig. 1.

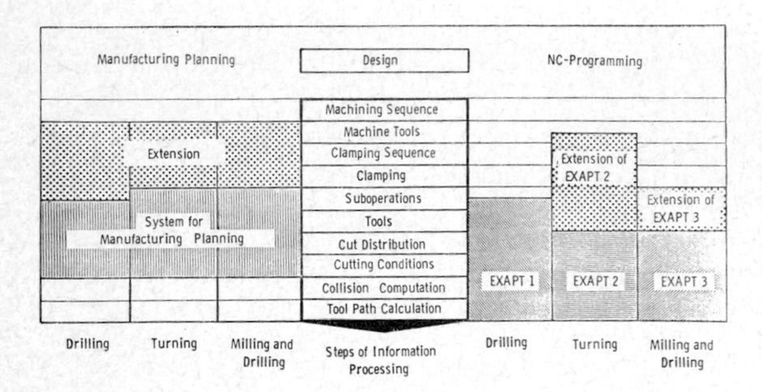

Fig. 1. Functions in NC-work planning and their status of computerization

The steps in the work planning procedure might be characterized by the determination of machining operations and sequence, choice of machine tools, fixtures and clamping devices, selection of suboperations and tools, calculation of cut distribution, area clearance, cutting data, tool paths machining time and documentation of this work. Starting from a detailed description of the workpiece - today represented by the blue print, in the future perhaps by a computerized workpiece data structure - all the steps of work planning have to be performed till the worksheets and NC-tapes are obtained.
Fig. 2 shows the structure of a computer aided work planning system. This system is designed in such a way that existing, technological orientated NC-programming systems are included taking care of the planning in the NC-field. Other modules are dedicated to conventional operations which have to be performed on NC-parts like conventional drilling, inspection etc. The turning module is used in principle for NC and conventional planning but to a different extent depending on the accuracy of planning necessary for conventional or NC-manufacturing. The other modules are used in the same way, as far as NC-operations exist anyhow. Further on existing NC-data files can be used in the data base of this system as well as macro-files for planning, based on the modification principle. The access to the system in a dialogue mode will guarantee that not yet implemented functions in work planning can be carried out by men personally. The results of the information processing of each step can be checked and if necessary manipulated by the work planner. It is anticipated that the usage of existing techniques and systems like NC-programming languages and their processors, proven data bases etc. widens the field of application of such a system. Introduction in industry will not be too awkward. Due to the modular structure of the system and to the user-dependend data files the system can be enlarged to the user's needs and work on his

specific data base.

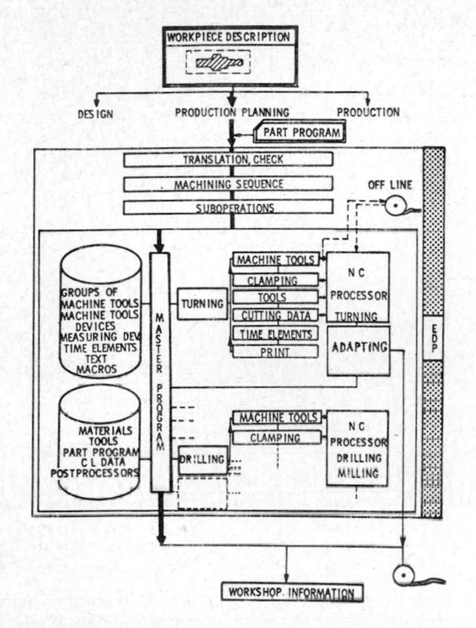

Fig. 2. Structure of the computer assisted work planning system

An example how a general designed subsystem - the turning module - works with an user depending data base is given in the next figures. Automatic work planning for turning starts with the determination of suboperations, their sequence and assigns suitable tools to them, considering the number of tools, technological and economical aspects. Fig. 3 shows for a workpiece to be turned the cutting area analysis.

It is useful to begin the analysis of the area to be removed with an analysis of the contour. Thereby one looks at first for deviations from the monotonous way of the contour namely for grooves of any kind. These grooves are further analyzed to

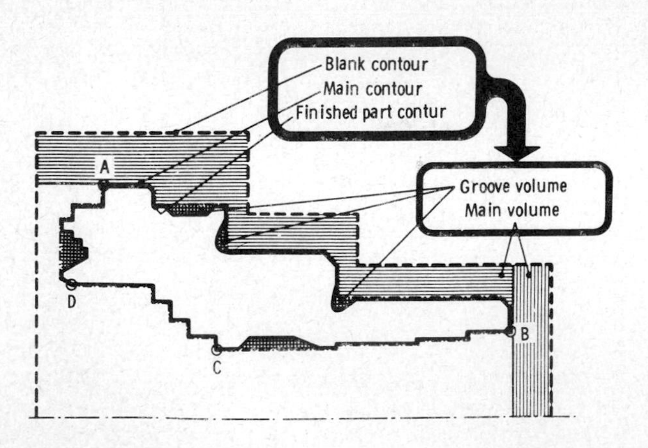

Fig. 3. Volume analysis of a rotational work piece for turning

find out subgrooves. By this method every complicated contour of a rotational part can be analyzed. Afterwards both volumes - the main volume (between blank and main contour) and groove volumes are devided by different methods in such segments which can be removed by "TURN LONG", "TURN CROSS", or special grooving and threading operations. These segments and suboperations are the basis for the automatic tool selection.
Usually it is possible to assignmorethan one tool even more than one tool type to a turning suboperation. A suitable tool type has to be cancelled if e. g. there is no tool of the chosen type in stock or in the tool file respectively, if proper tool tips are missing or collision occurrs between the tool and the workpiece. A type of tool of less techological applicability must be selected. In this way it is possible to install a priority-system, user dependend and very flexible which is illustrated in fig. 4.
In the machining file card "SUBOPERATION TURNING-TOOL TYPE" the user assigns to the operations in turning either tool types or specific tools by their identification number. This form is divided in two parts. The right part includes the assignment of suitable types of tools by the system of priorities as well as collision criteria. In this form all the suitable types of tools will be described by their classification number.
The example shows that the user assigns 8 suitable types of tools to the turning suboperation which is defined by a) TURN LONG, b) cutter location behind turning axis, c) position external, d) tool clamping normal, e) feed direction backwards to the direction of the contour. These 8 types are registered in the 4 indices of

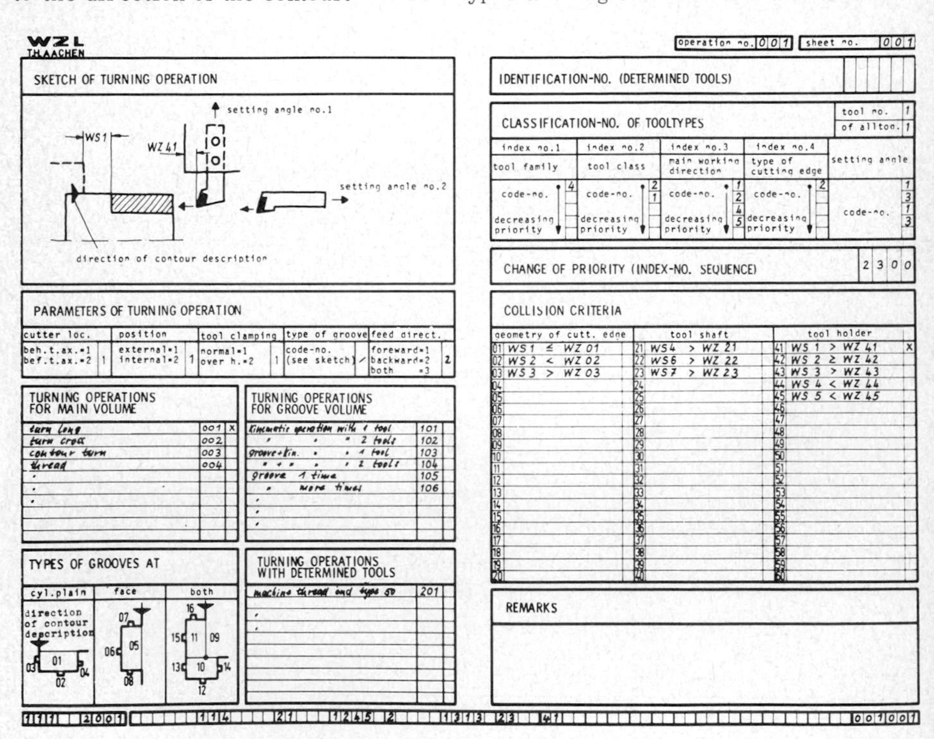
WZL TH AACHEN

operation no. 001 | sheet no. 001

SKETCH OF TURNING OPERATION

setting angle no.1
WS1
WZ 41
setting angle no.2
direction of contour description

IDENTIFICATION-NO. (DETERMINED TOOLS)

CLASSIFICATION-NO. OF TOOLTYPES — tool no. 1 — of alltoo. 1

index no.1	index no.2	index no.3	index no.4	
tool family	tool class	main working direction	type of cutting edge	setting angle
code-no. decreasing priority: 4	code-no. decreasing priority: 2, 1	code-no. decreasing priority: 1, 2, 4, 5	code-no. decreasing priority: 2	code-no.: 1, 3, 1, 3

CHANGE OF PRIORITY (INDEX-NO. SEQUENCE) 2 3 0 0

PARAMETERS OF TURNING OPERATION

cutter loc.		position		tool clamping		type of groove		feed direct.	
beh.t.ax.=1 bef.t.ax.=2	1	external=1 internal=2	1	normal=1 over h.=2	1	code-no. (see sketch)	/	foreward=1 backward=2 both =3	2

COLLISION CRITERIA

geometry of cutt. edge		tool shaft		tool holder	
01 WS 1 ≤ WZ 01		21 WS 4 > WZ 21		41 WS 1 > WZ 41	X
02 WS 2 < WZ 02		22 WS 6 > WZ 22		42 WS 2 ≥ WZ 42	
03 WS 3 > WZ 03		23 WS 7 > WZ 23		43 WS 3 > WZ 43	
04		24		44 WS 4 < WZ 44	
05		25		45 WS 5 < WZ 45	
06		26		46	
07		27		47	
08		28		48	
09		29		49	
10		30		50	
11		31		51	
12		32		52	
13		33		53	
14		34		54	
15		35		55	
16		36		56	
17		37		57	
18		38		58	
19		39		59	
20		40		60	

TURNING OPERATIONS FOR MAIN VOLUME

turn long	001	x
turn cross	002	
contour turn	003	
thread	004	

TURNING OPERATIONS FOR GROOVE VOLUME

Kinematic operation with 1 tool	101	
" " " 2 tools	102	
groove+kin. " " 1 tool	103	
" + " " " 2 tools	104	
groove 1 time	105	
" more times	106	

TYPES OF GROOVES AT

cyl.plain	face	both
direction of contour description; 01, 02, 03, 04	05, 06, 07, 08	09, 10, 11, 12, 13, 14, 15, 16

TURNING OPERATIONS WITH DETERMINED TOOLS

machine thread and type 50	201	

REMARKS

1111 2001 114 21 12445 2 1313 23 41 001001

Fig. 4. Machining file card for tool selection

the classification number. Such a number of 8 arises when the priorities are assigned not to the whole classification number of the tool but to each single index in it. This way the possibility is guaranteed to change special attributes of a type of tool.
The computer program forms at first a classification number with the priority of 1, here 4212 and looks then for this number in the tool file. If there are no tools the priority is set down dependent on the change of priority, in this case 2300. This will be done first in the 2nd index so that 4112 is the new classification number. If the search in the tool file is unsuccessful again the priority of the 3rd index is set down.
In order to fasten collision check only such collision criteria are noted which are relevant for the described turning suboperation in this example criterium 41.
For a collision check the program has access to a table of these criteria. All the information are to be stored in the machining file.
After the determination of suboperations, their sequence, the segments to be removed and the selection of suitable tools the calculation of the machining data is carried out. Based on the well known functions of manufacturing cost and time depth of cut, feed and speed are computed, as in NC-systems like EXAPT or the recently introduced MITURN. Besides cutting time the time for tool change, positioning, measuring etc. are determined. At the end the machining operation, its suboperations, tools, fixtures and times are printed on the work-sheet while the NC-processor and postprocessor will give the NC-tape if this machining operation shall be performed on a NC-machine tool.
This NC-work planning system whose first parts - the turning, milling and drilling modules - are going to be implemented now requires a data base which is far bigger than those known from NC-programming systems. Therefore substantial efforts must be made to create and maintain the data base necessary.

3. COMPUTER AIDED CREATION AND MAINTENANCE OF DATA FILES

Besides the issueing of work-sheets the creation and maintenance of all the data files required are important tasks of NC-production planning though it should not be a task for work planners but technologists and manufacturing analysists.
Looking at the various steps in the work planning process machine tools, clamping devices, tools, materials etc. have to be characterized in detail in order to have a comprehensive description of the equipment available, its capacity and capability. These data must be of high quality as they influence the planning result and thus the manufacturing cost. Changes in tooling, new materials or increased cost factors should be implemented in the data files immediately as work-sheets or NC-tapes have to correspond to the production equipment (see fig. 5). This requires a data collection and storing system to which authorized personnel of the various departments have access e. g. tool stock, shop supervision, technological staff. By means of data banks the problems involved in handling vast amounts of data can be solved.
In the aforementioned field of tooling the data of up to 10,000 tools must be under control. The first systematical approach to this problem was done when sophisticated NC-programming systems came into use. But the optimal use of the production facility TOOL requires that besides NC-programming or NC-work planning tool information are available for the other departments , too, involved in the production process. The designer decides to a great extent what kind of tool is necessary to manufacture the workpiece. Often done unconsciously many special tools have to be bought or produced. Therefore the designer should have infor-

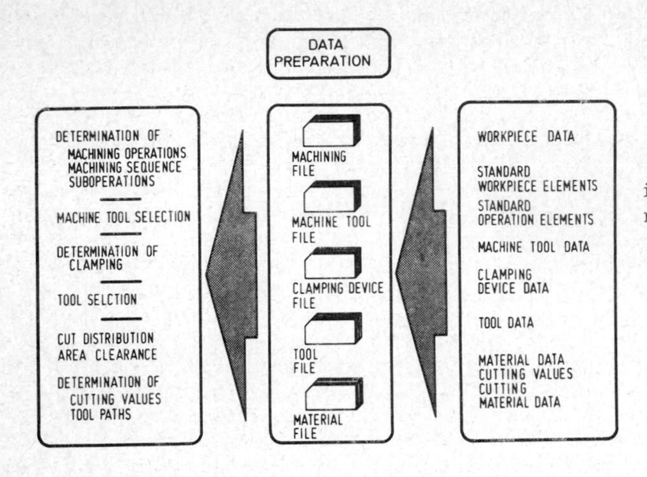

Fig. 5. Data preparation in NC-production planning

mation about standard tools available in the shop in order to prevent the usage of special tools. Thus the design of a workpiece will be more "manufacturing like" or "NC like".

When creating the data files for NC-planning information should be included which ease the work in NC though they are not used here primarily. Of course, a tool

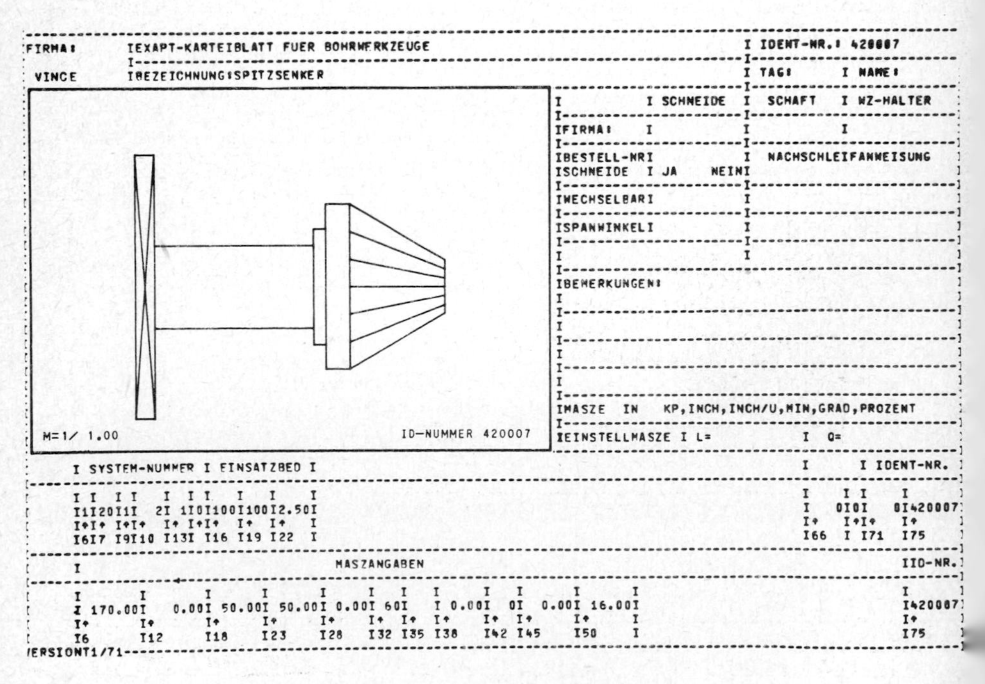

Fig. 6. Computer made NC-tool file card

data-bank is then to be designed in such a way that other than NC-tool card information can be introduced. An approach was made with a tool databank which supplies NC-processors, conventional and NC-work planning, statistical programs for standardization and investigation with the information required.
Fig. 6 is an combined printer-plotter output of this tool databank. The output is dedicated to NC-part programmers' tool catalogue. While the sketch of the tool is of importance when special tools must be used, the tabulated technological and geometrical data give an idea of the range of application which is necessary when tool selection and cutting data determination should not be performed automatically for any reason. In accordance with the needs and functions of the other departments interested in tooling another collection of data in another form of presentation is supplied.
The creation of files or databanks dedicated to the production equipment affords much time but the problems arising can be solved reasonably as the various existing tool files for NC-programming systems demonstrate. The behaviour of the equipment in the manufacturing process e.g. in the metal cutting process has to be described also in order to choose the most suitable equipment and machining conditions. Fig. 7 shows some of the information which must be derivated from the machine tool, the tool and the workpiece and its material in order to calculate proper cutting conditions for a turning operation. This figure gives suitable combinations of chip width and chip thickness (corresponding to combinations of depth of cut and feed at a setting angle of 90°) with their constraints. Only these combinations can be used which are expected to give good chips, as it is indicated by the light grey area in the figure. Beyond that area feed is restricted by an upper and lower limit of chip thickness which is given by the tool. Chip width may not exceed the limits of length of the cutting edge and nose radius. Additional criteria like available machine torque, power and maximum strength of tool restrict the determination area of chip width and thickness furthermore. Constraints for speed are

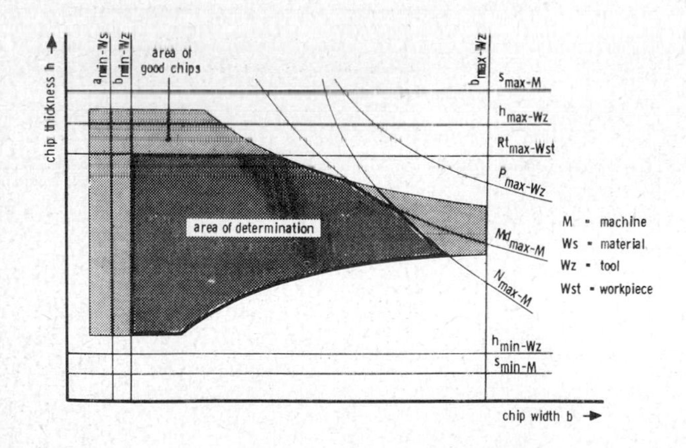

Fig. 7. Technological constraints in cutting data calculation for turning

given by the tool life, machine power and range of spindle speed, as well as the maximum allowed rotational speed of the clamping device. Out of the remaining combinations of cutting data feed, speed and depth of cut, the optimal value must be selected in accordance to the economical aims of production.
All these data are needed in automatic machining condition determination. But ex-

perience made in the past showed the difficulties of industry in the field of computerized technology: appropriate data are hard to get.

4. COMPUTER ASSISTED MACHINING INFORMATION CENTRE

The user of sophisticated software systems has to look for appropriate technological and organizational data in his own production because an information exchange between different users is not possible due to the peculiarities of the companies. But there are some fields in which data are of more general applicability as in production equipment e. g. in machining processes. These kind of data allow an information exchange to a certain extent. The participants in the exchange have to modify slightly the information they get but this work is not within the amount of effort they have to spend otherwise.
In order to provide the user of technology orientated NC-systems with data going into the machining file a machining information centre was designed by the Werkzeugmaschinenlabor. This centre will serve industry when
- the present technological knowhow should be prepared, systematized and standardized for application in new fields like NC, but also when
- the technological know-how should be checked and adapted to the actual situation e. g. by optimization of cutting data
- or it should be extended e. g. by the knowledge about cutting with titanium coated carbide tools.

Thus the centre will serve for NC- and conventional production. Fig. 6 presents the organizational structure of the center and the tasks to be fulfilled.

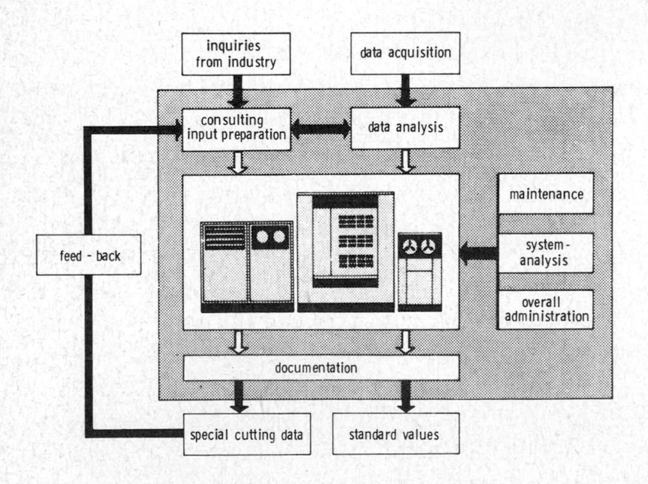

Fig. 7. Structure of a machining information centre

Besides data acquisition information transfer to industry is the very task. The inquiries from industry must be prepared for computer input, where after retrieving of data fitting the problem the calculation of machining data is initiated. The results are returned to the company in form of tables, diagrams, machining file cards etc.
Fig. 8 shows a computer edited table of standard values for turning. The material

16MnCr5 is forged, heat treated and machined by a P10-tool on a NC-machine with 23 kW power. The tool life for optimal cost production is 10 min. To the increasing depth of cut in the first column belong the feeds, followed by speed, actual tool life, power required and main cutting force in the adjacent columns. As the table shows the power constraint of the machine tool becomes active at a depth of cut of about 6mm. For larger depth of cut first speed is reduced to the smallest permitted value and then feed has to be decreased. As the determination of cutting values for a depth of cut greater than 6mm is based on the machine tool power the actual tool life is longer than the optimal as it is recognizable in the column for actual tool life. In the last columns machining time and cost for a cutting length of 1000mm in this case are presented.

MATERIAL: 16 MnCr 5, forged, heat treated - BF
CUTTING MATERIAL: P 10
MACHINE: power 23 kW, torque 350 mkp, cost 40 Dm/h
TOOL: nose radius 1 mm, length of cutting edge 16 mm
TOOL LIFE: 10 min
WEARLAND: 0,2 mm

BEARBEITUNGSKENNWERTE BEI KAPPA = 90 GRAD, SCHNITTLAENGE = 1000 MM, DREHDURCHM. = 300 MM

SCHNITT-TIEFE	VOR-SCHUB	SCHNITT-GESCHW.	TATSAECHL. STANDZEIT	GENUTZTE LEISTUNG	SCHNITT-KRAFT	ZEIT/SCHNITT	KOSTEN/SCHNITT
1	,667	202	10	3,9	118	2.6	3.9
2	,800	189	10	8,2	271	2.6	3.6
3	,800	181	10	12,0	406	2.7	3.7
4	,800	178	10	15,8	541	2.7	3.7
5	,800	177	10	19,5	677	2.7	3.8
6	,800	173	10	23,0	812	2.8	3.8
7	,800	149	19	23,0	947	3.1	4.1
8	,768	134	20	23,0	1050	3.5	4.7
9	,657	134	20	23,0	1090	4.0	5.4
10	,571	134	20	23,0	1050	4.6	6.1
11	,503	134	20	23,0	1090	[illegible]	6.9
12	,447	134	20	23,0	1050	5,7	7.6
depth of cut	feed	speed	actual tool life	power required	main cutting force	time per cut	cost per cut

Fig. 8. Computer edited cutting data table

A feed-back about the applicability of the transferred cutting data is anticipated. This feed-back is of great importance. It documents the efficiency of the information system and offers the only possibility to adapt the system to the requirements of industry.

An example of practical work is given in fig. 9 illustrating the useful function of the machining information centre in NC-production. In a company 20 spindles had to be machined on a NC-lathe. As the programming of the machine tool was done with a small computer aided system no machining file was used but standard tables. By an inquiry new standard tables were supplied to this company. Due to the optimization done by the centre tool life was shortened considerably from 64 to about 16min. But higher feeds and speeds enlarged the production rate while machining cost were reduced. Total cost savings of about 25% of the machining cost were achieved.

In this way a machining information centre is able to play an important role in the field of NC. First it can supply NC-production planning with some data needed for automated or computer assisted systems, second it can supply data of such a quality that the results in NC-planning are improved.

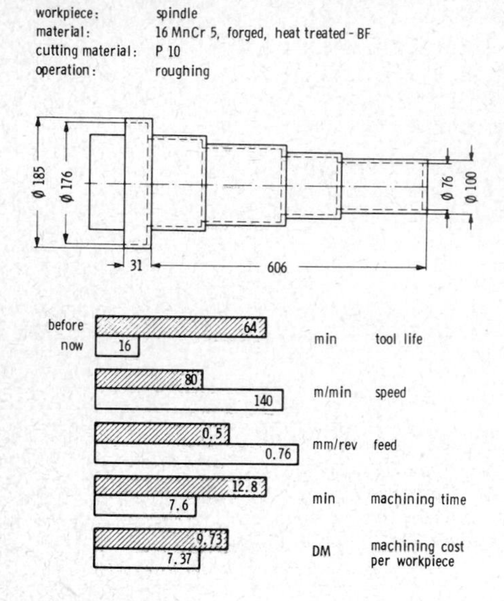

Fig. 9. Improving of NC-economics by data from a machining information centre

5. SUMMARY

Within the field of NC-part programming and work planning, information processing will become more and more a common technique. NC-programming systems will reach a higher level of automation and other functions of NC-production planning like work planning will be implemented on the computer. This does not include the development of appropriate software systems only but the creation and maintenance of data files also. These data files should not be restricted to the part programming field as it is of advantage that other departments participate. Since the introduction of NC-techniques affords great efforts industry should cooperate in the field of software and data. As shown in this contribution mutual information centres can assist NC-production planning thus decreasing the problems in NC-production and increasing the economy of NC. Finally one has to keep in mind that NC has a small though important share of today's production. Therefore NC-software and data should be applicable to conventional production, too.

6. REFERENCES

Bachmann, G., Steinacker, J. NC-work planning and tool selection. Not published reports, Aachen, 1972

König, W. Leistungssteigerungen bei spanenden und abtragenden Bearbeitungsverfahren, Essen, 1971

Winkler, H.-H. Informationszentrum für Schnittwerte, REFA-Heft Nr. 1, 1972

A SYSTEMATIC APPROACH TO IMPLEMENTING A COMPUTER AIDED MANUFACTURING SYSTEM

A. H. LOW
National Engineering Laboratory
East Kilbride, Glasgow, United Kingdom

Abstract: There is a logical sequence in which the production, monitoring, and management functions of an organization may be provided with computer assistance. A computer model of the production facilities is first constructed and it must be flexible. A control loop is then established correlating the development of the production facilities with that of the model. From this basis the developing model can be used for prediction and control in place of the real facility. Examples are drawn from such a system, developed at the National Engineering Laboratory, from which cost and job times can be estimated and details of operation planning provided.

1. INTRODUCTION

Computer aids available to manufacturing organizations fulfil three main functions: firstly, to help in planning, that is to assist in predicting or arranging future operations of manufacturing facilities to meet some criterion of performance; secondly, to help in the dissemination of information that will accomplish the desired operations, such as orders, invoices, drawing lists, and schedules; and thirdly, to provide managers with summaries of actual performance, which can be contrasted with planned performance and used for instant or long-term planning as appropriate.

Planning, the first of these functions, requires a knowledge of the work to be done in a given period, the facilities that will be available, and the time that will be taken by each element of the work.

Where time has to be predicted, the firm dependent on a metal-cutting workshop starts at a disadvantage being based on a complex subsystem (the machine tool, cutting tool, and workpiece) which is imperfectly understood and hence imperfectly utilized and controlled. These imperfections affect not only the efficiency of the machining process itself but also the efficiency of the planning performance, in that there is likely to be a difference between the planned times and the actual times. This is not so important in a traditional workshop where a foreman 'adaptively controls' the workflow to the machines as events occur. But where an attempt is being made to accurately schedule work over a period, errors in time can invalidate much of the planning (Fig. 1).

It does not seem likely that a complete theory of the machine tool subsystem will arise which would provide an analytic solution to the problem of obtaining optimal machining data, and would be practicable for industrial use. Therefore computer planning aids for machining workshops have to be designed with a view to minimizing the consequences of this imperfect basis.

In a recent project, the National Engineering Laboratory (NEL) have aimed at designing and implementing a pilot version of a planning and control system that would overcome this limitation by incorporating a computer model of the workshop. The model is intended to be closely tied to the real workshop by feedback and control information so that the actual performance of the workshop could be accurately planned. In this way it is hoped that the planned performance can be gradually improved towards the best attainable performance for the particular circumstances of the workshop (Fig. 2).

This paper describes the reasons for this approach, the development of the prototype system and its method of use.

COMPUTER LANGUAGES FOR NUMERICAL CONTROL, *J. Hatvany, editor*
North-Holland Publishing Company - Amsterdam–London

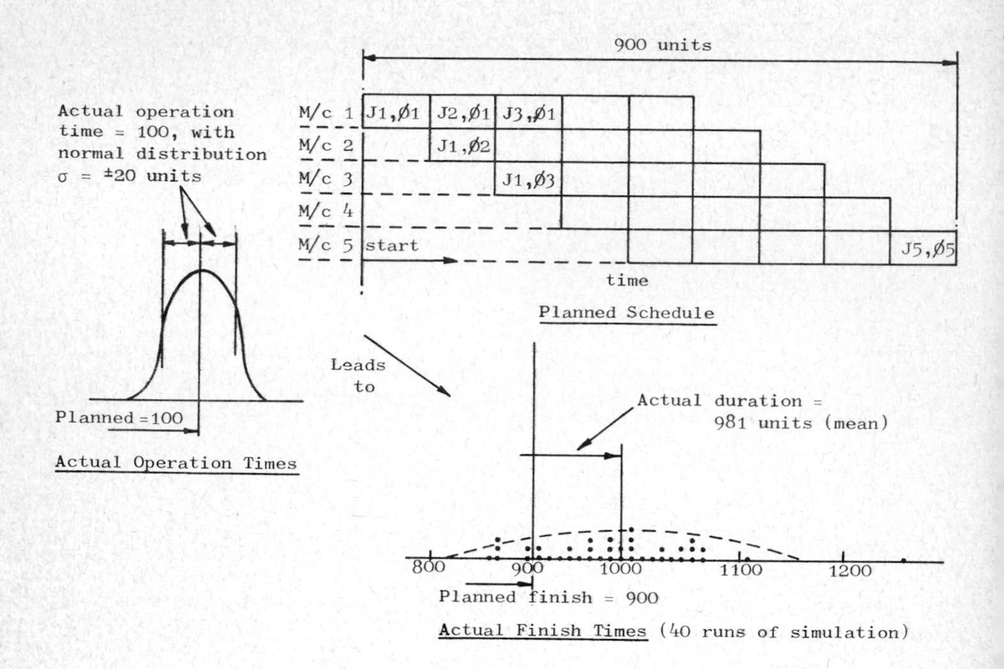

FIG. 1 EFFECT OF OPERATION TIME ERRORS ON MANUFACTURING TIME IN WORKSHOP
(Results of a simulation using 5 jobs, each of 5 operations)

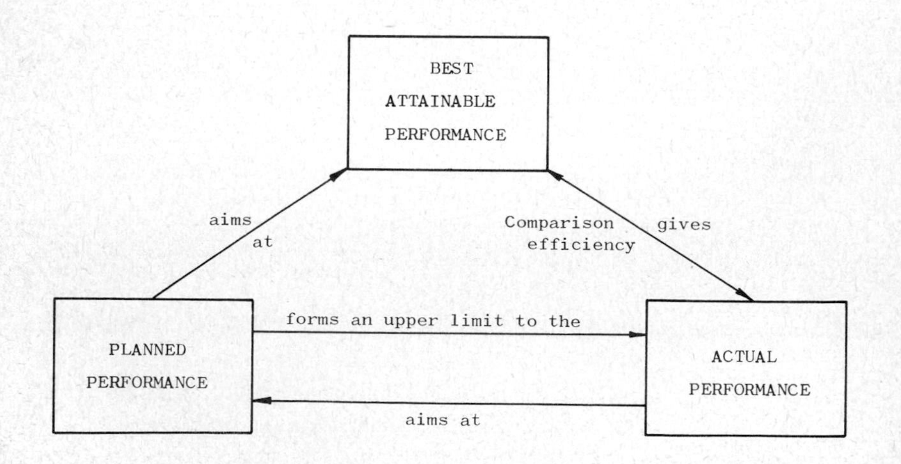

FIG. 2 RELATIONSHIPS BETWEEN DIFFERENT KINDS OF WORKSHOP PERFORMANCE

2. COMPUTER AIDED WORKSHOP RATIONALIZATION

Information was gathered on operation methods and inefficiences of typical jobbing workshops in the UK, with a view to determining in which areas assistance could be provided and in what manner closer correspondence of the three types of performance identified in Fig. 2 could be effected.

When basic operations levels of the organizations were considered, a lack of communication was found between planning and operational sides (Fig. 3). Firstly, calculations on which estimated times and costs were based were often not completely detailed because of a lack of knowledge of the conditions that would be used. 'Standard' or 'allowed' times that would have been acceptable where repeat batches formed the bulk of the workload were less justifiable in the jobbing environment. The machine operator was therefore given little guidance and used his own skill and experience to arrive at an effective set of conditions. Results of his 'adaptive controlling' were usually not stored, however, so that neither he nor the planner was any better off the next time a similar job came along.

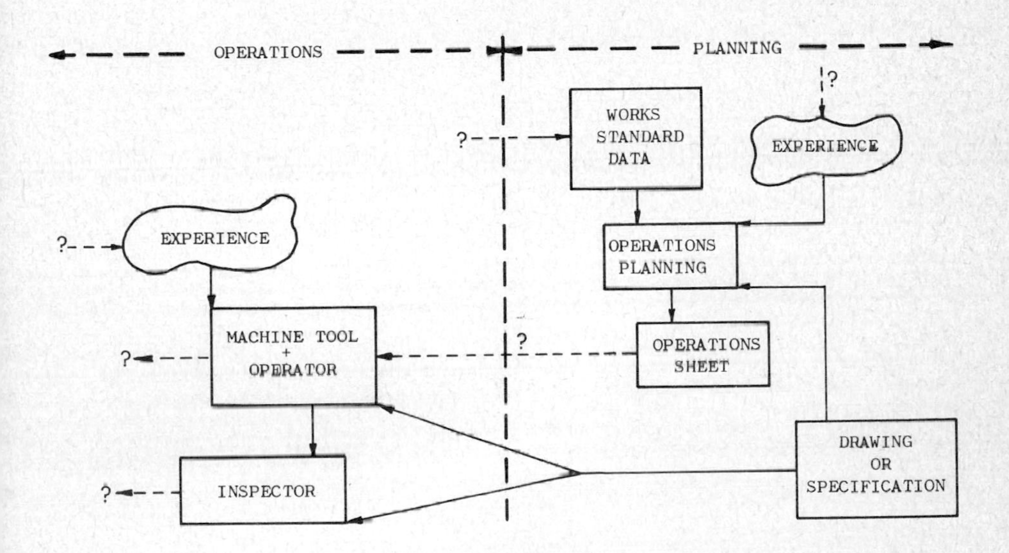

FIG. 3 TRADITIONAL WORKSHOPS - TENDENCY TOWARDS DISSOCIATION OF OPERATIONS FROM PLANNING

It was decided that the elements needed at this basic level were a 'store' in which experience from the machining process would be deposited and from which planning decisions would be drawn, and information links between the planner and the operator, the operator and the store, and the store and the planner (Fig. 4). It was also realized that discipline would be required in the organization to ensure that the information on which the systems depended was communicated, and that it was used.

The operation of this first level of the Computer Aided Workshop Rationalisation (CAWR) system depended on the following procedures.

(i) Detailed planning, based on stored information, of all the machining operations. Initially these would not be optimal and perhaps in some areas might be unworkable.

(ii) Communication of the complete planning details, such as tools, feeds, speeds, depths, and times, to the machine operator.

(iii) Utilization of these planned procedures by the operator.

(iv) Amending, from reports by the operator or inspector, stored data that produced unworkable or non-optimal results.

(v) Utilization of these improved data by the planner.

It was reasoned that by rigid adherence to these procedures, firstly, unworkable data would be detected and corrected; next a close correspondence between planned and actual performance would be achieved; and finally over a longer period, depending on the skill and persistence of the staff involved, the machining procedures in the store would approach the best that could be achieved in the workshop, considering the as yet uncontrolled variables.

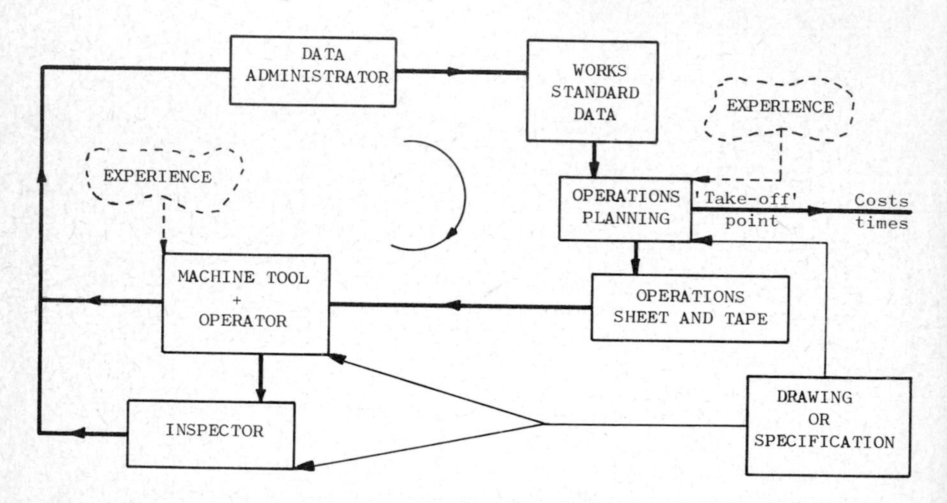

FIG. 4 BASIC OPERATIONS AND PLANNING LOOP, AND 'TAKE-OFF' POINT

3. NEL PROTOTYPE SYSTEM

In any system that attempts to match machining predictions with performance in this way, the centre is the 'store' or 'data bank' containing the accumulated experience and machining expertise of the organization. This information must be specific to particular circumstances, that is to particular machine tools, types of operations, cutting tools, or materials, so that all the relevant details can be used for retrieving correct cutting data. The logical structure of the computer file used in the NEL system is shown in Fig. 5. The physical structure of the drum file can also be seen by following the directed lines from block to block. This file enables cutting conditions, such as feeds, speeds and depths, and sequences of tools to be defined as functions of machine tool, type of operation, workpiece material, and either cutting tool types, or sub-types, or individual tools if necessary. Information initially loaded into this file can be subsequently modified by a file processor that accepts fixed format statements in English. It is intended that responsibility for this updating should lie with a 'data administrator', a new function in the NEL Workshop.

The second processor of the system extracts information from the data bank. It

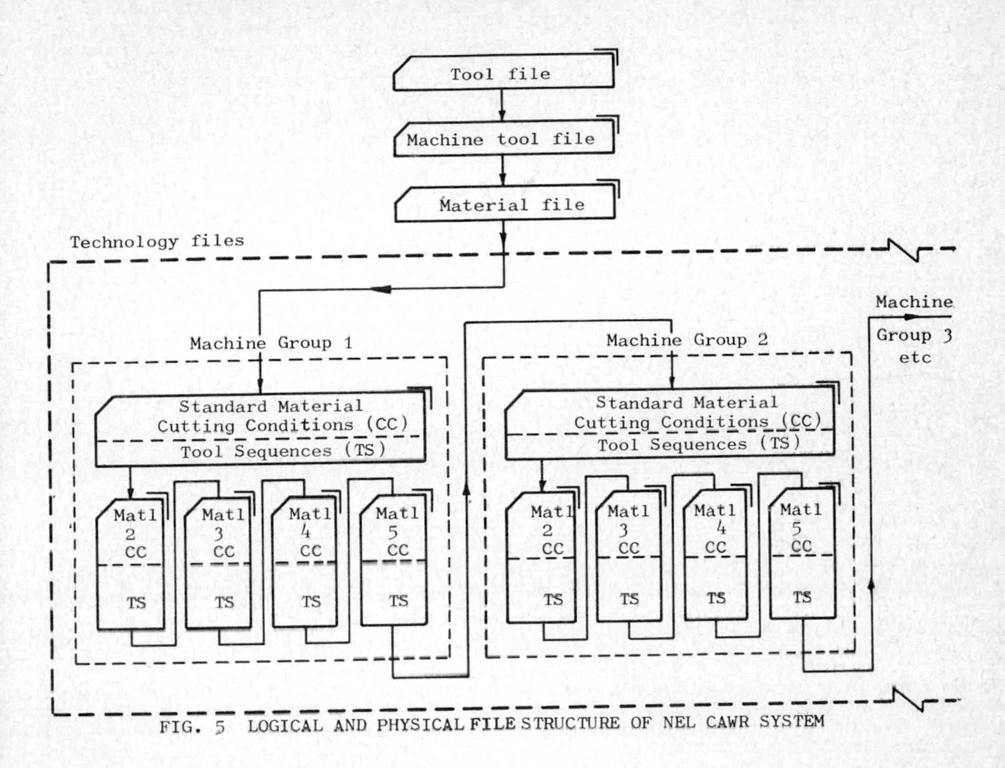

FIG. 5 LOGICAL AND PHYSICAL FILE STRUCTURE OF NEL CAWR SYSTEM

depends for its design on the type of workshop in which the system is to be used and on the type of planning organization that exists. If the workshop consists of conventional machine tools, the output of the processor will be an 'operations sheet', with details of set-up, tools, and cutting conditions. If the firm has access to a medium or large-sized computer, as at NEL, the processor might have an interactive input and store the output on file for later display. Fig. 6 is an example of such an input. It represents part of the operations planning process for a test component. In this, the first entry by the planner indicates his choice from the possible processes initially listed. Thereafter, he answers only the relevant questions, produced by the system, under the format guides (As and Bs require alphanumeric data, Ns and Ms decimal numbers, Is and Js whole numbers). The retrieved information is then displayed and can be altered, if necessary, before going on to the next element of the job. Most of the entries refer to either the component drawing or to the user's own codes for machine tool (VRODRL), tools (NCH312), or materials (EN30). If the machine tools are NC or DNC, the extracting processor can be embedded in the NC processor and store the results of the planning on the NC tape.

It is possible for the planner to use the extracting processor at a number of different 'levels', depending on the degree of development of the stored data. He may wish to retrieve only tool dimensions when using the system initially. He then has to select feeds, speeds, and tools himself. This corresponds to level 1 operation in Fig. 7. As more information is entered in the files, however, the planner has to make fewer and fewer technical decisions, till at level 5, the present limit of the system, he has only to decide on the final tool type. The larger the level number it is possible to use, the closer is the correspondence between the model and the real world, and hence the closer is the planned to the actual performance.

```
  1-TOOLS IN A GROUP OR TYPE  2-DETAILS OF A TOOL
  3-TOOL DETAILS+FEED+SPEED   4-TOOLS IN A SEQUENCE
  5-SEQUENCE+FEED+SPEED       6-ENTER MISSING FEED+SPEED
  7-ELEMENT CHANGE            8-LINE CHANGE
  9-ESTIMATE WANTED          10-LIST ENTRIES
 11-WORK COMPLETED
→ 5
 GIVE THE MACHINE NAME, MATERIAL, AND UNITS
 AAAAAABBBBBBA
→VRODRLEN30  I
 GIVE TOOLID, (OR TOOL GROUP+DIAMETER) DEPTH, SO OR FIT
 AAAAAABBBBBBNNNNNNMMMMMMAAAAAA
→NCH312              5   PCT60
 FEED AND SPEED OVERRIDE?  NOPECK?  BEVEL?  DIABEV?
 NNNNNNMMMMMMAAAAAABBBBBBNNNNNN
→
  TECH  TOOL   TOOL    TOOL    DEPTH     OVERALL  ACTIVE   SPEED    FEED
  NAME  GROUP  IDENT   DIA               LENGTH   LENGTH            AXIAL
                       (INCH)  (INCH)    (INCH)   (INCH)   (RPM)    (IN/MIN)
  1     DRILL  DSS7P0  .276    .500      6.69     .83      665.27   2.077
  2     TAP    NCH312  .312    .500      5.79     .98      253.66   14.091
 GIVE TECH. NAME, NO. OF OPS./PART, AND PATH LENGTH.
 AAAAIIINNNNNNN
→W1    4

  1-TOOLS IN A GROUP OR TYPE  2-DETAILS OF A TOOL
  3-TOOL DETAILS+FEED+SPEED   4-TOOLS IN A SEQUENCE
```

FIG. 6 EXAMPLE OF INTERACTIVE OPERATIONS PLANNING
(PLANNER INPUTS LINES PREFACED BY '→')

The link between the planning stage and the actual machining depends similarly on the workshop configuration. In the conventional workshop, discipline and an incentive to co-operate must be provided to ensure that the planned conditions are used, or if they are not, the reasons are made known to the data administrator. In the NC workshop, where the system will be used at NEL, control is easier as much of the planning data is controlled by the tape image. Feeds and speeds however may be altered by the operators on some machines, and again the need to explain to the data administrator the reasons for doing this should be stressed. In a DNC system where the information can flow back from a machine tool, there is the possibility of carrying out the communication directly.

Levels of usage	Determination of feeds and speeds	Selection of specific preliminary tools	Determination of preliminary tool types	Selection of specific finishing tool	Determination of finishing tool type	
1	✓	✓	✓	✓	✓	1
2		✓	✓	✓	✓	2
3	Decision already made			✓	✓	3
4	and results stored in files ready			✓	✓	4
5	for retrieval				✓	5

FIG. 7 DECISIONS REQUIRED BY PLANNER AT DIFFERENT LEVELS OF USAGE OF NEL CAWR SYSTEM

Information should arrive back to the data adminstrator on any occasion when either the operator or inspector detects some imperfect machining process or result - it is just as important to prevent a job being done with a higher quality than requested as with a lower one. If, with their skill and experience, either of these men can suggest better methods, they should be encouraged to notify the data administrator and should be made to appreciate their roles in optimizing the processes. It is possible that shop floor lines or computer terminals can facilitate this process, but its success depends on the maintenance of a co-operative attitude.

The data administrator's job is a skilled one; he has to decide the range of conditions over which any recommended alteration is valid and whether the alteration relates to one group of materials or machine or is specific to the actual circumstances encountered. He also has to examine the update in the areas of the files altered most frequently. The nature of these updates will show where inconsistencies have occurred, and will point to an area of either mistaken information or lack of control over some relevant factor, such as tool preparation or material purchase or treatment, in the machining environment. He may also wish to experiment with new types or preparations of tools and examine them for improvement in consistency or life.

4. THE OPERATIONS LOOP AS A BASIS FOR FURTHER COMPUTER AID

The system so far is intended only to lead to a long term improvement in machining efficiency and an immediate improvement in the correlation between the planned and actual machining times. However, machining is only one activity of the workshop. It is necessary to acquire an equal knowledge of and control over others, such as delivery and transport of materials and tools, set up, tooling and break-down times, maintenance periods, and staff availability, before the next step can be taken. This requires, principally, organizational effort, and there seems to be little that the computer could assist in, except perhaps communications in a DNC system.

With an accurate knowledge of all these elemental times, however, and a sufficient control over the total workshop enviroment and a list of planned work to be done, it is possible to lay out the individual operations of all the jobs

```
ENTER JOB ELEMENT NAME (PARTNO DATA)
-->CAWR1 TEST PART NO. 3
COMPONENT - CAWR1 TEST PART NO. 3
CUSTOMER  - M1 DIVISION TECHNOLOGY OFFICE
WORKPIECE - EN30 PLATE, PART 713/B/3     LOT SIZE     - 14
PLANNED BY  J. LINDSAY (MISS)            DATE         -28/APR. 1972
PRIORITY  - NEEDED BY Z GRP  14/6/72     MACHINE      -VRODRL
SET-UP    - 45.000MINS                NON-CUTTING TIME - 6.500MINS
MACH. TIME/PART - 17.600MINS            EST. MACH. COST -  1.210
EST. TOOL COST  -   .350
TOTAL COST/PART -  1.560              BATCH COST       - 24.150
                      XXXXXXXXXXXXXXX
                         --------
OPERN. TOOL  LOAD TOTAL PATH  DEPTH FEED  SPEED ELT   ELT   COMMENT
 NAME  NAME  POSN LTH   LTH                     TIME  COST
W2     CDP375   1240.00  .000 5.000 5.013 501.0 1.000  .050CDRILL HOLES
W4     MBR254   2177.00  .00050.00047.270 210.0 1.060  .051BORE CENTRE
       CS500    3250.00  .00019.05079.650 153.0  .240  .012COSINK TAPS
W3     NCH1P0   4210.00  .00054.00099.440  79.0  .430  .022TAP CENTRE
W1     SDS500   6 63.50  .000 2.00023.040 206.0  .087  .004SPOT 4 HOLES
W6     RT0500   5300.00  .00020.00037.550  77.0  .540  .026REAM 4 HOLES
       CS500    3250.00  .000 8.00079.650 153.0  .100  .005COSINK
W5     DJS9P0   1219.00  .00015.00025.000 500.0  .600  .031COUNTR DRILL
       CS500    3250.00  .000 9.00079.650 153.0  .114  .053COSINK ABOVE
```

FIG. 8 DISPLAY OF INFORMATION STORED IN 'PLANNED PART' FILE

against the available resources, that is to carry out the scheduling process, according to some criterion of performance and so construct a second model or plan of the total activities of the workshops. There are many programs commercially available which will carry out this process. However, it is to be emphasized that their usefulness depends on control over real situations and on the probability that planned circumstances prevail. There is again need for a monitoring function to study deviations between planned and achieved performance of this level.

Cost estimating and control are other important functions for which useful information can 'spin-off' from the basic operations loop. With only minor additions to the information already existing in the CAWR files, the estimated manufacturing cost can be computed. Like the time estimate and quality produced, the cost estimate is an important function in a jobbing workshop. It too works both ways against a firm when errors are made in the estimate. In a competitive situation, if the price is too low, there is a greater probability of landing the contract, and if too high, of losing the job. It only needs a slightly higher proportion of under-estimates to over-estimates to destroy a profit margin and to fill the shop with non-profitable work. Hence the importance and benefit of again rooting the costing process in the actual data that will be used in production and planning. In the prototype system at NEL, costing is carried out, if required, by the same 'extracting processor' that does the interactive operations planning. The information on time, cost, and the technical definition of the operations is then stored on a computer file for access for scheduling, costing, or programming for NC (Fig. 8).

Having thus achieved an accurate knowledge of times, costs, and procedures relating to the work that will be carried out, it is possible to move on to consider the second main function of computer systems, that of disseminating the information in the form of instructions to toolroom, stores, foremen, transport and other departments. At NEL, when this stage is reached, it is possible that an existing package will be used in conjunction with the CAWR programs. It is felt that maximum benefit will be realized from such a system of communications only when it is based on a reliable and improvingly accurate set of data files. Discrepancies between the dynamic model of the workshop (the planned schedule) and the workshop's actual progress will then be detected, the reasons for them ascertained, and steps taken to isolate and remove the uncontrolled factors that cause them.

5. CONCLUSIONS

A traditional workshop, where each individual adapts his efforts to the existing situation, operates as an organic system, whose performance can be predicted only on a statistical basis. This makes advance planning in detail impossible and reduces the knowledge gained, and hence the control that can be exercised.

If manufacturing is to become more automated, and planning and costing more precise, it is necessary to find a means of learning about and then predicting even such a complex process as machining. Since theory is lacking, a trial and error method using the skill and experience of available staff is needed. Then using this knowledge and controlling all other relevant factors in the manufacturing situation, it is possible to model the performances of detailed machining processes and operations of the workshop in a realistic manner.

If this state has been achieved and there is an adequate feedback of information, it is possible to construct an edifice of computer programs, integrated around the model, which will allow an accurate and scientific control to be exercised, and provide the necessary data for higher management decisions and policy.

ACKNOWLEDGEMENT

This paper is presented by permission of the Director, National Engineering Laboratory.

WORKSHOP TECHNOLOGY - ADAPTIVE CONTROL
TURNING

G. BERG
Machine-tool and Automation section,
The Swedish Institute of Production Engineering Research, IVF

Abstract: Workshop technology covers the technological planning and the process planning. The process planning covers the choice of manufacturing process and machine-tools. Process planning is not discussed in this paper. The technological planning is one part of the operation planning. The other part in operation planning is the geometrical calculations.

Adaptive control (AC) can be defined in the following way: Adaptive control system = a control system within which automatic means are used to change the system parameters in a way intended to improve the performance of a control system. (American standard, ASA C 85.1 - 1963, Terminology for automatic control.)

The adaptive systems can be subdivided into adaptive control constraint (ACC) and adaptive control optimization (ACO).

It is often very difficult and sometimes impossible to predict and predestinate in a computer program the way of manufacturing a product.

In this paper it will be discussed how AC and computer programs for operation planning can cooperate in order to make an automatic operation planning.

We are here talking about AC applied to the machine-tool only and we are assuming that there is no feed-back to the soft-ware program from the AC.

CONTENTS

COMPUTER LANGUAGES FOR NUMERICAL CONTROL, *J. Hatvany, editor*
North-Holland Publishing Company - Amsterdam–London

1. THE DEVELOPMENT OF MANUFACTURING

Man wants to improve the effectiveness of the machines. The most effectiv aid is the automatization.

Man also have to improve his own effectiveness. The possibilities are limited, howewer. It is necessary to take care of those qualities which are unique for the human being. Man have to get rid of timeconsuming detailed work, for example time-consuming calculations. Such work can much better be done in a computer.

It is estimated that in a decade, about 75 % of all industrial parts production will be on a small-lot basis, as against about 25 % at present according to M.E. Merchant (1969).

In the future you also have to count with that man more and more have to work with computer controlled groups of machines instead of single machines as at present. He will get aid from new technical means of assistance, for example DNC, CNC, AC and computer programs for automatic process - and operation planning.

C.I.R.P (International Institution for Production Engineering Research) has tried to predict what is going to happen in the future. Below there is an extract from that investigation. For each phenomenom they have tried to predict the year in which it will occur with a probability of 90 %. There are also the years of the lower and upper limits.

A. Cutting machining

Exact scientific methods, able to predict the machining performance (tool life, accuracy, surface finish, productivity) of any combination of tool materials, work materials and machine, will be a reality.

1985. 1983-1995

Tests for exact determination of the machinability of work materials for all types of cutting operations will be developed and in wide use in industry.

1985. 1980-1990

B. Optimization

Fully self-optimizing adaptive control of machine tools will be developed and in wide use.

1985. 1980-1990

A computer software system for full automation and optimization of all steps in the manufacturing of a part (selection of machining sequence, selection of machine tools, clamping, selection of sequence of operations, tool selection, selection of optimum cutting conditions, numerical control of machining) will be developed and in wide use.

1980. 1980-1990

Full on line automation and optimization of complete manufacturing plants, controlled by a central computer, will be a reality.

1985. 1980-1990

On line process identification and a very quick adaption of manufacturing conditions relative to output requirements, i.e., on-line optimization, will be in wide use.

1985. 1980-1985

C. Machine tools

More than 50 % of the machine tools produced in the future will not have a "stand-alone" use, but will be a part of a versatile manufacturing system, featuring automatic part-handling between stations, and being controlled from a central process computer.

1985. 1985-1990

2. WORKSHOP TECHNOLOGY

2.1 Technological planning

Fig. 1 shows the main technological planning steps for a product from design through machining, assembly, distribution, to use. This figure was shown by G. Sohlenius at the PROLAMAT Conference 1970.

All these steps are necessary whether they are performed automatically or not. Any processor under development in this field will cover the planning steps from postprocessing up to the level typical at its current state of development. The planning steps above that level have to be performed manually by a programmer, planner or designer. The levels are numbered at the left of fig 1.

The activities above machining can be divided into design and production planning. Unfortunately there is some proliferation of nomenclature in this field. However, I hope you will not find it difficult to get the gist of my remarks. Production planning can be divided into processplanning and operation planning. Processplanning covers the choice of manufacturing process and machine-tools. Operation planning covers the preparatory work that must be carreid out for each type of

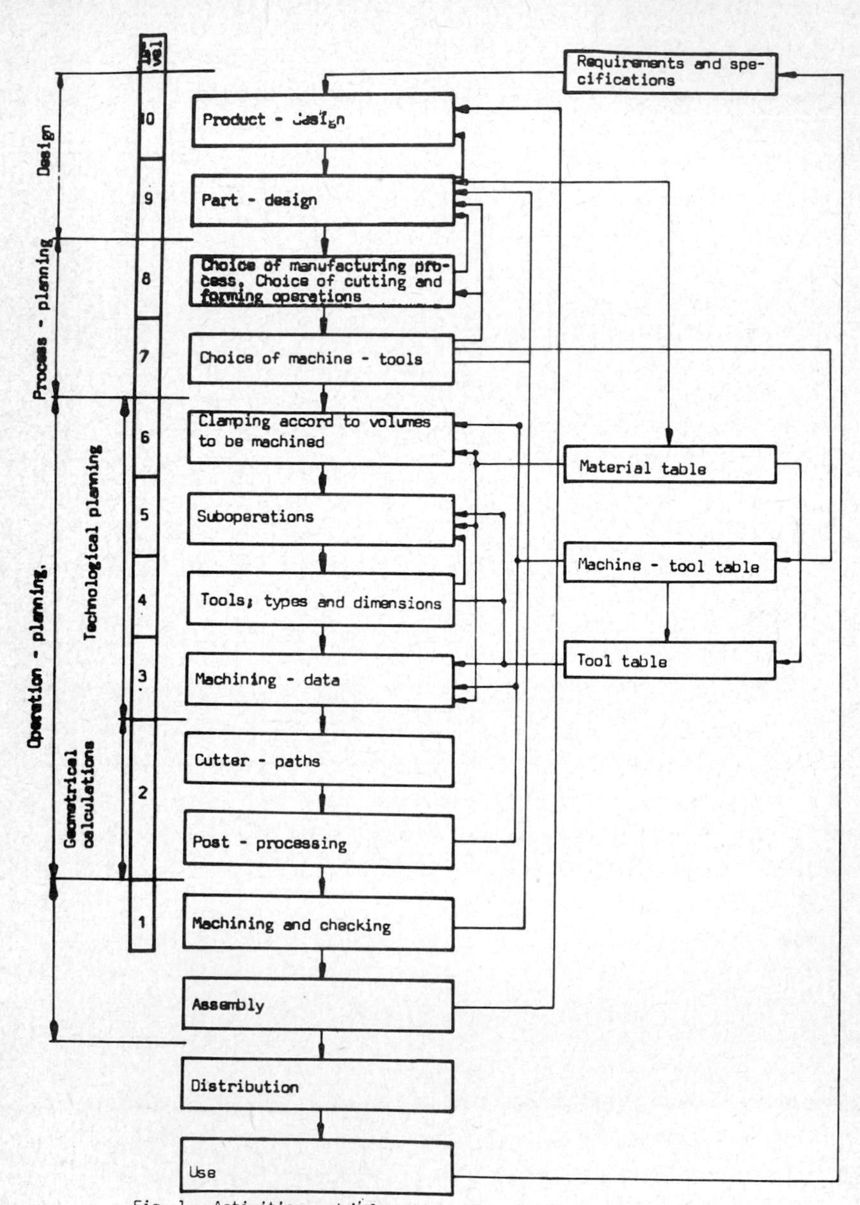

Fig. 1. Activities and information flow in the design-production process.

part for machining in each one of the chosen machine-tools. Operation planning can be sub-divided into technological planning and geometrical description of the stock and the part. The geometrical description can be condensed if a computer system is used, or detailed in full for manually programmed NC machines. Today the tool-path generally is described. The activities covered by process planning and operation planning can be seen in fig 1.

In this connection the term workshop technology covers the technological planning and the process planning. The process planning is not discussed any further in this paper.

2.2 Adaptive Control (AC)

According to American Standard, ASA C 85.1 - 1963, Terminology for automatic control, adaptive control is defined as follows:

Adaptive Control system = a control system within which automatic means are used to change the system parameters in a way intended to improve the performance of a control system.

The systems can be subdivided into

1. Adaptive Control Constraint, ACC
2. Adaptive Control Optimization, ACO.

3. TECHNOLOGICAL PLANNING

The main purpose of computer programs for operation planning is to make more economic production. Optimizations have to be done in respect to economy or time. You have to avoid undesired sub-optimizations and consider the total optimization.

The operation planning includes the following steps whether they are performed automatically or not (fig 1.):

1. Determination of <u>operation sequence</u>. This means that you have to determine the surfaces or volumes which have to be machined and the sequence between them. This determination is very close coupled to the clamping.
2. Determination of <u>clamping surfaces</u> and <u>clamping devices</u>.
3. Choice of <u>tools</u>.
4. Calculation of the <u>tool-path</u>.
5. Calculation of <u>cutting data</u>.

At automatic planning there must also be checked that no collisions will occur between tool, workpiece and machine-tool.

All these steps within the planning are very dependent on each other, shown by P. Hellström (1972). Fig 2.

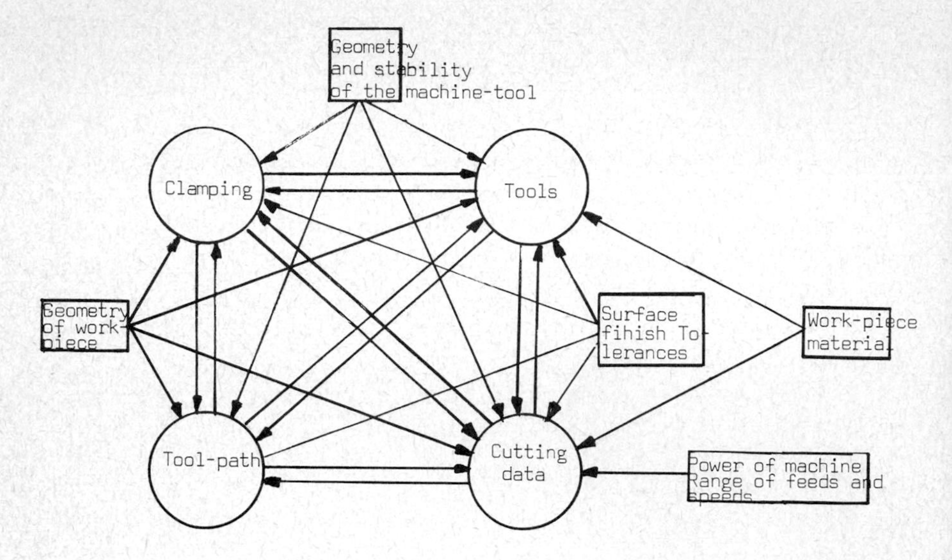

Fig. 2. Decision areas for operation planning and how they effect each other.

Most decisions within each step are dependent on:

1. The property of the machine-tool.
2. The policy of the workshop, in the first hand regarding tools and cutting data.
3. The selection of products.

If it is possible to design general strategies for decision making within one of the planning steps, then it is possible to design computer programs in order to automize that step. However, the basic knowledge necessary for decisions is often very bad. It is often very difficult and sometimes impossible to predict and predestinate how the machining should be done. For example it is difficult to describe the properties of the tools and the machine-tools. The problems with vibrations can be mentioned.

When it is difficult or impossible to design general computer programs the experiences of man must be included in the program. This can be combined with a technique making it possible for man (part programmer) to cooperate with the planning program. In addition new technical developments will be of importance, for example adaptive control.

In the following it will be discussed in this paper how adaptive control and computer programs for operation planning can cooperate.

4. ADAPTIVE CONTROL AND AUTOMATIC OPERATION PLANNING

4.1 Choice of operation sequence

The operation sequence is the sequence in which the different parts and volumes of the workpiece have to be machined. This means in practice a determination of the number of clampings and their sequence. This determination must be done under consideration of the work-piece, the machine-tool and the available tools.

It will be very difficult to create general computer programs for this step because of the great number of variants. Man´s capacity to apprehend pictures and models has to be used. The partprogrammer can cooperate with the program with assistance of for example interactive computer graphics.

The adaptive control can not contribute to the solutions within this step because the operation sequence must be determined before the machining starts.

4.2 Choice of clamping surfaces and clamping devices

Clamping surfaces for chuck or chuck plus tailstock centre can presumably be determined in a computer program. The choice between these two ways of clamping can also be done in a program. When other ways of clamping are necessary the program system presumably needs help from man in a similiar way as has been described in 4.1 above.

You can not expect any help from adaptive control here.

4.3 Choice of tools

The choice of tools is one of the most important steps in the planning. It is not possible to choose a tool without at the same time decide how it shall be used. To some extent you have to consider avaiable cutting data when the tools are chosen.

So the choice of tools is at the same time a choice of method. Because of this the tool selection will influence all the other steps within the planning process.

Factors influencing the tool selection are:

1. Workpiece
   - Geometry
   - Surface finish
   - Material
   - Stability
2. Tools
   - Geometry
   - Material
   - Avaible tools

3. Machine-tool
   - Avaible power, moment, spindle speeds, feed rates, etc.
   - Number of tools
   - Stability
   - Type of tool holder (turret, magazine etc.)
4. Consideration to decisions made within the other planning steps.

At present (1972) it seems to be impossible to design batch mode computer programs for selecting the best tools in every situation. There are too many variants and lack of knowledge. The tool selection is to a great extent based on experience. Each work shop has its own ideas about how to choose tools.

Even if we had knowledge enough such a program would be very complicated and big.

The problems have to be solved by cooperation between man and the program (computer graphics, time sharing etc.).

The adaptive control can not make the very selection of tools easier.

4.4 Calculation of tool-path

At rough-turning the tool-path almost always is parallel to the axis of the lathe.

For each clamping a computer program for operation planning has to determine

- the machining sequence.

  (Machining here means for example drilling, grooving, rough-turning, thread cutting etc.)
- the number of cuts.

Adaptive control can here be of good help. The detailed calculation of tool-path in the computer program will no more be necessary.

Considering the properties of the tools, the machine-tool and the clamping deveices the adaptive control has to choose maximum cutting depth.

Because the tool-life is regarded independent of the cutting-depth it will be no need of special optimization. It will be enough with an adaptive control constraint, ACC.

However, the determination of cutting depth is one part within the calculation of cutting data. Therefore reference to the next chapter 4.5 is done.

4.5 Calculation of cutting data

The problem of selecting processing data is generally of choosing data so that the actual parts can be made at minimum cost. This is the same problem as selecting data so that the required volume of chips can be removed at lowest possible cost; optimum production. When delivery dates are short, there are reasons for selecting maximum production data and when little work is at hand, data should be selected according to minimum tool cost. Only optimum production will be discussed below.

The most generally used methods are based on the Taylor equation:

$$vT^{\alpha} = C \qquad (1)$$

This represents straight lines in a log-log diagram, fig. 3.

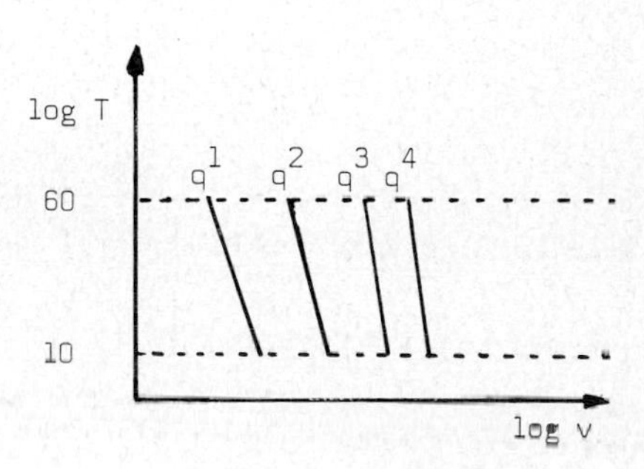

Fig. 3. Data base for optimum choice of feed and speed.

However, a much better fit to basic test data can be made by aid of a polynomial function as used by Colding.

$$k + ax + hx^2 + oy + dy^2 - z + cz^2 + fxy + gyz + hxz = 0 \qquad (2)$$

where

$x = \ln q$, $y = \ln v$, $z = \ln T$.

$q = L/A$

is the chip equivalent defined by Woxén (1932).

For some materials it is not even possible to use the Taylor equation (see fig. 4). Here a polynomial approach is needed. However because of the unavoidable variations in the machinability of materials the taylor approach is accurate enough for normal materials and is used here. For every work-piecematerial and tool that is to be used for that material, basic data are required (for example, according to fig. 3). The tool-life is affected by the cutting velocity as well as by the feed, but it can be regarded as independent of the cutting depth.

Calculation of cutting data for optimum production can be performed on three different optimization levels marked I, II and III below. The strategies for the different levels are broadly speaking the following:

(I) 1. Determination of maximum cutting depth with regard to the dimensions and strength of the tool. 2. Calculation of feed and cutting velocity for optimum production. 3. Adaption of these conditions to the capacity of the machine and tool.

(II) 1. Choice of maximum cutting depth with regard to the dimensions and strength of the tool. 2. Choice of maximum feed with regard to the dimension and strength of the tool. 3. Calculation of cutting velocity for optimum production at this feed. 4. Adaptation of these conditions to the capacity of the machine and tool.

(III) 1. Choice of maximum cutting depth with regard to the dimension and strength of the tool. 2. Choice of maximum feed with regard to the dimension and strength of the tool. 3. Choice, from a table, of recommended cutting velocity for the work-piece material. 4. Adaption of these conditions to the capacity of the machine and tool.

For optimization level I a set of basic data according to fig. 3 is required.

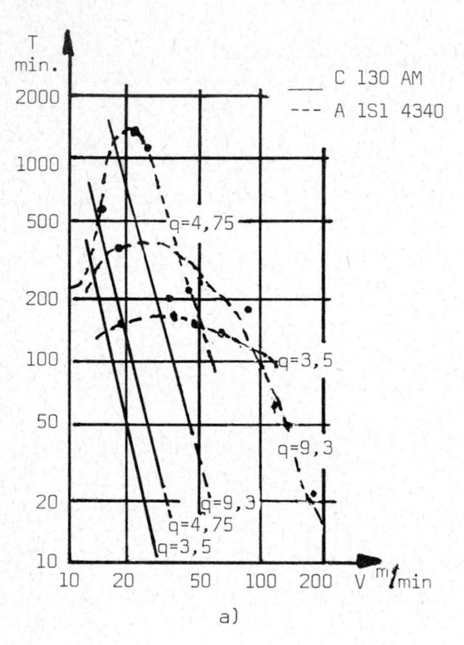

Fig. 4. Tv curves for different chip equivalents q. The abrasion curves are obtained by the radioactive method for a titanium alloy c 130 AM and a CrNi alloy AISI 4340.

Professor G. Sohlenius stated in his PROLAMAT- paper in 1970 that optimization level (I), was very interesting regarding the future development. We can say the same today (1972). Only optimum production in rough turning will be considered.

The possibility to predestinate optimum cutting data depends on the factors influencing the determination.

It seems to be suitable to determine maximum cutting depth with regard to the properties of tools and machine-tool. The reason is that tool-life is regarded independent of the cutting depth.

Factors influencing the determination of cutting depth are (U. Fröjel 1972):

- Avaible power
- Permitted force on the tool
  - a) Total maximum force
  - b) Maximum force/length unit of the cutting edge

  These forces is dependent on the tool material and the dimensions of the tool. At present (1972) these forces will not be given by the tool manufacturers.
- Stability of the machine-tool
- Stability of the tool
- Stability of the work-piece
- Stability of the clamping

Regarding permitted forces on the tool I think that in the future tool manufacturers will give us those data. But then the dispersion of the properties of the tool must be smaller that it is today.

Lack of stability of the machine-tool and the tools gives a lot of problems. In certain work shops there are simple rules based on experience telling how to take these stability problems into consideration.

In spite of much research work in the field of stability and vibration there are no good rules telling how to predict the behavior during the machining. Adaptive control can help us.

The influence from the clamping on permitted cutting forces and cutting depths is at present estimated with aid of experience.

The determination of feed rate is influenced of about the same factors as the cutting depth when you consider rough turning.

A set of basic data according to for example fig. 3 is necessary for optimization of feed rate and cutting velocity. The basic data must be related to tool material and work-piece material.

It is now clear that it at present is very difficult to predestinate optimum cutting data with assistance of general computer programs. The main reasons are:

- Varying machinability of the workmaterial
- Varying cutting depth because of varying machining allowance
- Unsufficient knowledge of the behavior of machine-tools, specially concerning vibrations.

The problems with vibrations seems at present impossible to predict due to their very complicated nature.

The adaptive technique can be a very valuable mean of assistance here. However, the economic factors must be considered. The benefits of optimum cutting data must be great enought compared with more roughly determined cutting data.

Some of the varying factors can relatively easy be supervised by an adaptive control system, e.g. forces, power, speeds, temperature. In ACO-system the rate of wear of the tool must be measured. This is a problem which not yet has been solved in a satisfactory way. I think we have to wait a rather long time before this problem is solved. (According to CIRP in 1985.)

It is not probable that even ACO-system can replace computer programs for operation planning. The ACO-system can only make it easier for the computer program. The ACO-system must get some start values from which level the optimization can begin. These start values have to be created by the soft-ware system.

It is very interesting to analyse how the planning work can be distributed among the computer program and the adaptive control system.

For example two fundamental extreme cases will be discussed a little in the following.

These cases are based on the type of operation planning program.

1. General operation planning program.
2. Limited program.

4.5.1 General computer program - AC

You may here presume a general computer program for operation planning with a high degree of optimization.

"General" means that the program is useful for almost all kinds of work-pieces and machine-tools. The program probably have to be limited to one type of machining, e.g turning.

With a high degree of optimization means that the computer program shall choose a solution of the total planning which in some way is optimal. This solution must be choosen among a number of other alternative solutions.

One result of this is that at least alternative tools and combinations of tools must be determined. The choice of the best alternative then have to be based on the cutting data which can be used for each alternative. This choice must be made during the planning phase. An adaptive control system can not make it.

The computer program has in this case made an accurate calculation of cutting data. The calculated data are "optimal".

If you want optimal data during the whole machining you have to supervise the process and continuously calculate new optimal data. This must be done because there are factors and variations in the process which were impossible to predict by the program. The problems with vibrations can be mentioned.

This continuous supervision and calculation is suitable for an adaptiv control system. It is necessary to have an ACO-system.

If the optimum cutting data are beyond the possibilities of the tools and the machine-tool an ACC-system will be good enough. If you don´t have such a system the computer program has to make the limitation. But then you can not make adjustments required by variations which are impossible to predict at the planning stage.

In some situations a modified ACC-system can be used. Such a situation can arise if an ACO-system is not available and the calculated cutting data can be performed by the actual tools and machine-tool. This AC-system only has to supervise the optimum data once calculated and not adjust this data towards the lower or upper limits.

4.5.2 Limited computer program - AC

At present there are a number of limited computer programs for operation planning at turning. Usually the limitations concern tools and/or work-pieces. For example there are programs suitable for shafts only.

The permitted tools are limited concerning numbers and types. Then there is no optimal choice of tools. It is not necessary to calculate accurate cutting data in the program then.

This job can be carried out by an AC-system. The system ought to be an ACO-system. Start values and limitations can be given from the computer program. The adaptiv system then controls the machining process taking into consideration the not predictable variations of different parameters.

5. SOME COMMENTS

The total economy must not be forgotten. It is necessary to analyse if the benefits with optimum cutting data are great enough compared with data more roughly calculated.

In ACO-system there are problems to solve concerning both measuring technique and strategies.

It is very important that the systems will be developed with strategies for optimization which are independent of the properties of tools and machine-tools.

It must also be remembered that AC protects machine, tool and work-piece.

6. CONCLUSION

The automatic operation planning for turning consists of

- determination of operation sequence and clamping
- choice of tools
- calculation of tool-path
- calculation of cutting data.

The adaptive technique can be a very useful mean of assistance for

- calculation of tool-path

calculation of cutting data

The other steps must be carried out by the computer program for operation planning. This program shall also give the basis for the work of the AC-system.

At present it seems to be impossible to design general computer programs for operation planning which are able to predict and predestinate the optimum way of machining.

The basic data necessary to create decision criteria are in some cases so imperfect that it is impossible to create batch mode computer programs for this task. However, if man is given the possibility to cooperate with the system, in an interactive way, it is still possible to solve the problem. The adaptive control can make it easier for the automatic operation planning system and make it possible to develop such system.

The determination of operation sequence, clamping and tools have to be made in the operation planning system. The adaptive control can not directly make this work easier for the operation planning system, but the adaptive control is very important for the determination of cutting paths and machining data.

For many of the parameters having influence on the determination of machining data it is today impossible to predict how they will behave. There are variations in the behaviour of the work-piece material, the tool and the machine-tool. Regarding the machine-tool this is mainly valid for the vibration problems (chatter).

At ACO-systems one principle and big problem today is to find a practically measureable parameter for tool-life. It will presumably take a rather long time before we have ACO-system in practical use (in 1985 according to C.I.R.P). There are also problems in designing the strategies.

It is not realistic to think that it will be possible to get rid of operation planning programs because of adaptive control, not even at ACO. Some of the decisions that have to be made in operation planning can not be done by the adaptive system, as mentioned before. For instance, the determination of machining data, can not be done in the adaptive control only. It is impossible to choose tools in the AC-system. In the operation planning system you must calculate maching data as a start-value which can be used by the adaptive system for optimization.

Routines and strategies for the automatic operation planning system ought to be co-ordinated with those necessary for the adaptive control. It is very important that the systems will be developed with optimizing strategies independent of machines and tools and that those strategies can be adapted to current limitations in tools and machines.

7. REFERENCES

[1] U. Fröjel, Automatic Operation planning, System principles I. The Swedish Institute of Production Engineering Research (IVF) (will be published in 1973)

[2] P. Hellström, Automatic Operation planning, Turning. The Swedisch Institute of Production Engineering Research (IVF), Report Nr. 72 602 (1972)

[3] M.E. Merchant, Trends in Manufacturing Systems Concepts. 10th International M.T.D.R Conference, Manchester (1969)

[4] V. Pettersson, Adaptive Control. The Swedish Institute of Production Engineering Research (IVF), (1972)

[5] G. Sohlenius, Workshop Technology, especially turning. 1st PROLAMAT Conference, Rome (1970)

AN APPROACH TO COMPUTER-AIDED MANUFACTURE

K.G. ADAMS
Assistant Professor, Mechanical Engineering Department
University of Waterloo, Waterloo, Ontario, Canada

Abstract: The installation of N.C. machine tools in a company requires the adoption of a whole new concept from design to finished product. Likewise, the installation of a computer controlled manufacturing system demands the adoption of a new concept of even greater magnitude. There must be a complete flow of computerized information from all channels ranging from the purchasing of raw material, through the production process, until the finished product is shipped out the door. An integral part of this on-line process is an executive control scheduling programme which will automatically schedule all jobs throughout the production stages calling up the materials, tools, and machines required for each step in the production process and in the case of D.N.C. machines, automatically load the post-processed part programmes into the buffers of the slave computer for the required D.N.C. machine tool.

As a first step in the development of this executive control scheduling programme, a simulation model has been developed which schedules a single component through any number of machining stages, while maximizing the product output with a minimization of work-in-progress. This on-line model has the facility for determining preventive maintenance requirements and including them in the production schedule if the machines are running on a full three-shift basis.

The same model when run off-line has the ability to include, by statistical determination, the effects of machine breakdowns and operator absenteeism on the production schedule and results in a new schedule that can be used for determining more realistic delivery dates.

1. INTRODUCTION

If optimal use is to be made of computerization to increase the productivity and efficiency of the manufacturing industry, an approach to computer-aided manufacture similar to that illustrated in Figure 1 must be adopted.

Design and N.C. programming efficiency can be greatly increased by the use of computer-aided design employing the group technology concept, establishment of group technology macros and by developing special direct programming methods to produce on-line post processed part programmes.

A large proportion of the conventional N.C. machine tool market will be replaced by D.N.C. machine tools controlled by slave mini computers complete with interpolation units which are in direct communication with the master control computer. For non-D.N.C. applications such as N.C. machine tool, conventional machine tool and assembly operations, communication with control will be established via a slave computer by means of a special management control and reporting complex.

The scheduling of the flow of components, materials, toolage, jigs and fixtures, machine tool tapes, and in the case of D.N.C., automatic loading of the slave computer buffers from the on-line part programme storage files will be accomplished by an executive control scheduling programme which will produce the most economic schedule for the plant based on computerized inputs from all

COMPUTER LANGUAGES FOR NUMERICAL CONTROL, *J. Hatvany, editor*
North-Holland Publishing Company - Amsterdam–London

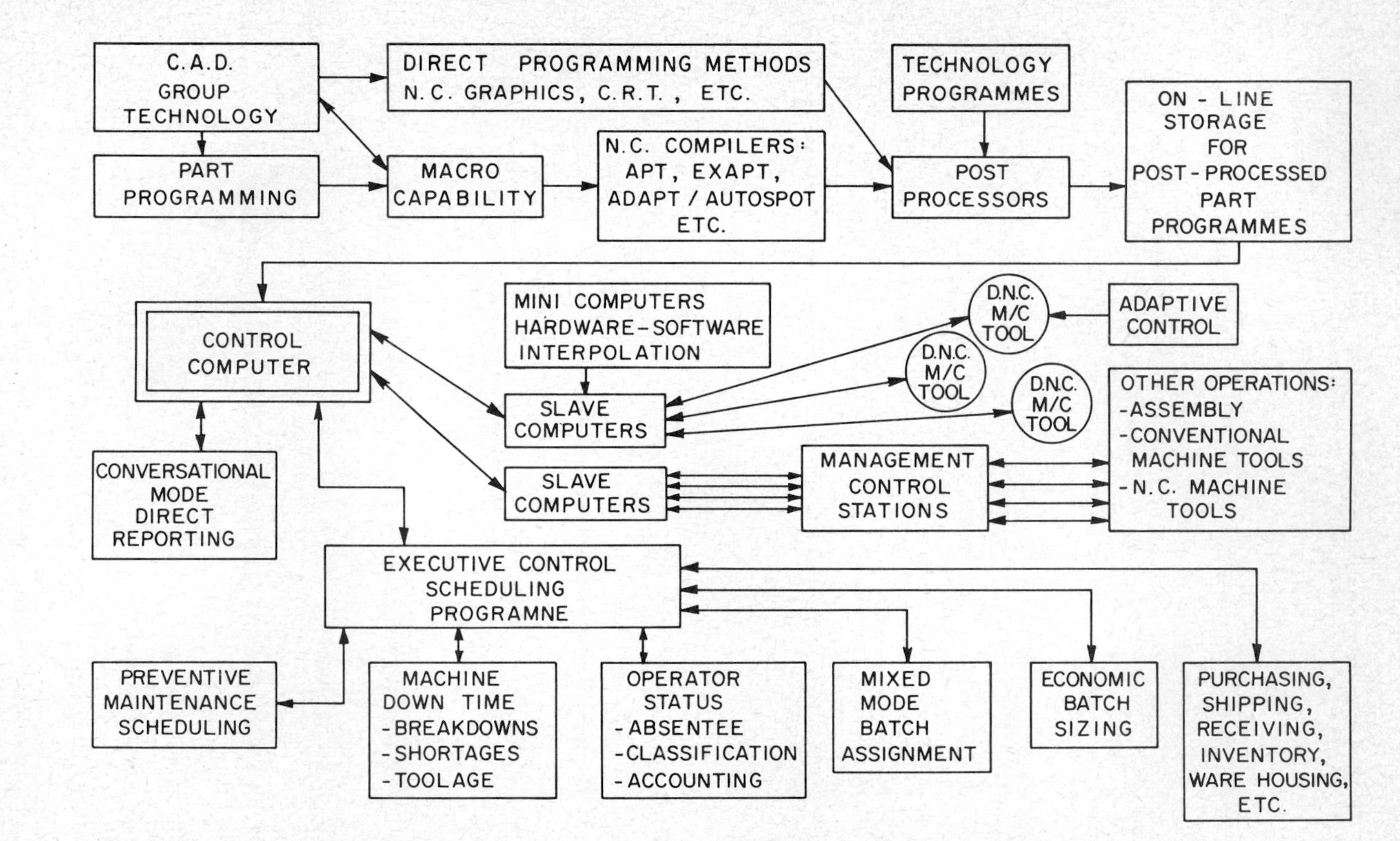

FIGURE I : OVERALL CONCEPT OF COMPUTER - AIDED MANUFACTURE

relevant sectors of the corporate structure. It is the intent of this paper to examine an approach to the development of this executive control scheduling programme.

2. EXECUTIVE CONTROL SCHEDULING

In order to achieve the most economic flow of materials and components through the plant, an executive scheduling programme must be incorporated in the closed loop system to assure maximum overall production at a minimum of cost. As a first step in this direction, a model was developed to simulate the flow of a single component through several stages of production with a minimum amount of work-in-progress.

3. THE SIMULATION MODEL

The model was constructed to simulate the production of a single component which was to be processed through "n" different operations or machine groups, each machine group consisting of "m" machines, at a desired production rate while keeping the work-in-progress to a minimum. Since one cannot ignore the material handling flow between machines or departments, a tote size parameter has been included in the model which allows the number of components to be specified that must be produced before they can be economically moved to the next machine group or department.

Also included in the model are three separate sections which may be flagged into the mainline programme independently or in any combination. These sections take into account the effects on the system of:

(1) Scheduled Preventive Maintenance
(2) Unscheduled Machine Breakdowns
(3) Operator Absenteeism

3.1 SCHEDULED PREVENTIVE MAINTENANCE

Since many numerical control departments are run on a full three-shift basis, the possibility of maintaining these machines on an off-shift is excluded. For this reason, a scheduled preventive maintenance programme section is included which determines when maintenance, based on a "machine hours run" criterion, is required on each machine and schedules it for minor maintenance or major overhaul accordingly.

3.2 UNSCHEDULED MACHINE BREAKDOWNS AND OPERATOR ABSENTEEISM

Under ideal conditions, the time for the production run would be achieved by the schedule determined by the mainline programme. However, actual production rates will vary because of day-to-day problems which arise such as machine breakdowns and lack of manpower. The two sections on machine breakdown and operator absenteeism, based on statistical analysis, can be incorporated to simulate the effect of these parameters on production and will provide a more realistic total production run time which can prove quite valuable in quoting realistic delivery dates.

3.3 DISCUSSION OF THE SIMULATION LOGIC

Since the total programme listing or related flow charts are too lengthy to present here, a detailed description of the model is given stating assumptions made, input parameters required, the logic of operation of the model and the resulting output schedule.

3.3.1 Assumptions

(1) All machines in each machine group have the same machine rate (i.e., component production rate).
(2) The union classification for each machine group or department is different. Transfer of an operator from one machine to another within the same machine group is permitted, but operators may not operate machines in another machine group.
(3) If an operator is absent, he is absent for the entire shift.
(4) When each machine group has completed all the components required for the run, the partially filled totes are automatically transferred to the next machine group. This is necessary only when the group tote parameter is greater than unity.

3.3.2 Input Parameters

The following parameters are required as basic information input to the system.

(1)	Total Production Requirements	(REQD)
(2)	Desired Production per Shift	(RPPS)
(3)	Number of shifts per Week	(SHIFTS)
(4)	Duration of a Shift in Minutes	(TIMUNT)
(5)	Number of Machine Groups Involved	(NGROUP)
(6)	Number of Machines in Each Group	(NMCS(I))
(7)	Machine Rate of Each Group	(MCRATE(I))
(8)	Number of Operators in Each Group	(NOPS(I))
(9)	Machine to Operator Ratio for Each Machine Group	(MCTOOP(I))
(10)	Tote Size Parameter for Each Group	(TOTE(I))
(11)	Preventive Maintenance Flag	(PMFLG)
(12)	Machine Down Flag	(MDFLG)
(13)	Operator Flag	(OPFLG)
(14)	Machine Hours Logged to Date Since Last Major Overhaul for Each Machine	(HIST(I,J))

3.3.3 Logic of Operation

The programme initializes all parameters, reads the input data, checks the capacity of the line and sets the required production per shift at the maximum line capacity or the desired shift production whichever is less. The programme then determines the following for each shift:

(1) The production required for that shift to meet the required output.
(2) The total machine hours available on each machine. If the preventive maintenance, machine breakdown, or operator routines are flagged, the machine hours available on each machine are modified accordingly.
(3) The actual production possible taking into consideration:

 (a) stock available at each machine group
 (b) the lag time which may possibly be involved due to the ability of the machine group under consideration to produce the components faster than the previous machine group
 (c) the lag time incurred if the time required to produce the balance of the components to fill a tote by the previous group exceeds the time required by the present machine group to use up its present stock.

(4) Simulates the operation of the shift to produce the possible required output.

(5) Updates the "machine hours run" statistics for preventive maintenance purposes.
(6) Produces a printout of the production schedule for that shift.

3.3.4 The Production Schedule Output

The following information is printed out for each shift that is run.

(1) Page header stating shift number and week number.
(2) Summary of all machines due for preventive maintenance on that shift stating group number, machine number and the duration time required for that maintenance.
(3) Summary of all machine breakdowns during the shift stating group number, machine number and the duration of the downtime incurred by that machine.
(4) Summary of all operators who are absent on that shift.
(5) Tabular listing for each machine group stating:

(a)	Number of machines run	(MACHINES RUN)
(b)	Number of operators used	(OPERATORS USED)
(c)	Stock level at each machine group at the end of the present shift	(STOCK LEVEL)
(d)	The number of completed components not yet transported to the next machine group because of the tote size parameter limitation	(READY)
(e)	Work-in-progress: The number of components completed by this machine group and not utilized by the next machine group	(W.I.P.)
(f)	The total wait time in minutes involved by all the machines run in that group	(WAIT = sum of g + h)
(g)	Tote lag time in minutes	(TOTE WAIT)
(h)	Lag time in minutes due to different machine rates	(MCR WAIT)
(i)	The time in minutes that the machines being run in this machine group are available for other work	(FREE)
(j)	The number of components produced by this machine group	(GROUP PROD)
(k)	The number of components transferred during this shift to the next machine group	(TRANSFER)
(l)	The number of machines that were available for production on this shift	(M/CS AVAIL)

(6)	The total work-in-progress of all machine groups at the end of that shift	(TOTAL WORK IN PROGRESS)
(7)	The number of finished components produced on that shift	(SHIFT PRODUCTION)
(8)	The total accumulated production to date	(TOTAL PRODUCTION TO DATE)

A sample shift schedule which includes the preventive maintenance, machine breakdown and operator absenteeism routines, with a tote size parameter of 10 is shown in Table 1.

TABLE 1: SAMPLE SHIFT SCHEDULE

SHIFT NUMBER 13 OF WEEK NUMBER 1

****

MINOR P.M. OF 40 MINUTES IS REQUIRED BY M/C. NO. 1 of M/C. GROUP 3
MINOR BREAKDOWN OF 2 HOURS FOR M/C. 5 OF M/C. GROUP 4

MACHINE GROUP	1	2	3	4	5	6
MACHINES RUN	1	2	3	6	1	2
OPERATORS USED	1	2	3	2	1	1
STOCK LEVEL	90	54	9	10	5	0
READY	0	6	1	0	5	
W.I.P.	54	15	11	5	5	
WAIT	0	0	0	17	22	52
TOTE WAIT	0	0	0	0	0	5
MCR WAIT	0	0	0	17	22	47
FREE	0	0	0	1	1	55
GROUP PROD.	65	60	75	80	81	80
TRANSFER	70	60	80	80	80	80
M/CS AVAIL.	2	3	4	6	1	2

TOTAL WORK IN PROGRESS	90
SHIFT PRODUCTION	80
TOTAL PRODUCTION TO DATE	820

4. SAMPLE COMPUTERIZED PRODUCTION SCHEDULES

In order to illustrate the effects of the tote size parameter and the inclusion of the preventive maintenance, machine breakdown and operator absenteeism routines on the production schedule, the output for six computerized schedules is discussed. Also as a basis for comparison, a typical conventional schedule prepared without regard for minimization of work-in-progress was established and is referred to as Schedule "O".

The input parameters used for each of the computerized schedules to be discussed were as shown in Table 2 with the following exceptions:

TABLE 2: INPUT PARAMETERS FOR SAMPLE SCHEDULE

```
REQD = 1000
RPPS = 84
SHIFTS = 15
TIMUNT = 480
NGROUP = 6
NMCS = 2, 3, 4, 6, 1, 2
MCRATE = 65, 30, 25, 14, 85, 45
NOPS = 2, 3, 4, 2, 1, 1
MCTOOP = 1, 1, 1, 3, 1, 2
TOTE = 1, 1, 1, 1, 1, 1
PMFLG = .FALSE.
MDFLG = .FALSE.
OPFLG = .FALSE.
HIST = 1, 12, 123, 1234, 45, 789, 4*0
       265, 37, 90, 45, 0, 59, 5*0
       444, 9, 62, 1630, 6*0
       389, 0, 2, 1979, 6*0
       3*0, 1638, 6*0
       40*0
```

Schedule 1: No change

Schedule 2: Scheduled preventive maintenance included.
Schedule 3: Scheduled preventive maintenance, machine breakdown and operator absenteeism routines included.
Schedule 4: Tote parameter set to 2.
Schedule 5: Tote parameter set to 10.
Schedule 6: Tote parameter set to 25.

Since the parameters of primary interest are the total production to date and the associated work-in-progress, this information has been extracted from the computer output and is shown in Figures 2 through 5.

Figure 2 illustrates that the total production to date throughout the run is the same for Schedule 0 as it is for Schedule 1. However, it can be seen from Figure 3, that the average work-in-progress has been reduced from 171 components to 53 components by applying the proposed computer simulation programme. Figure 2

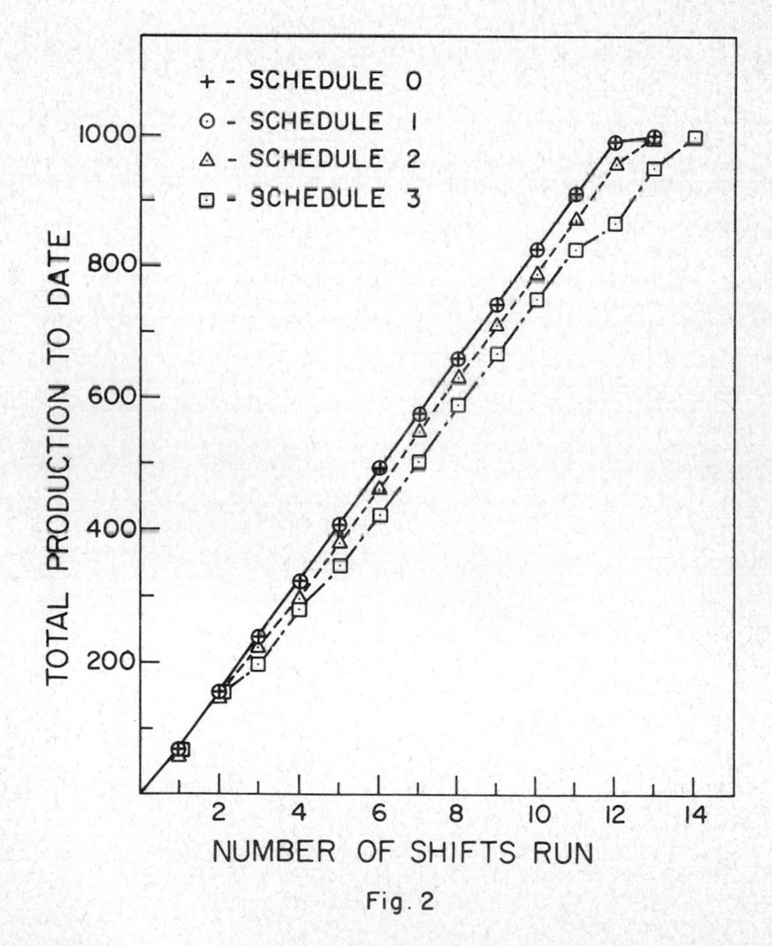

Fig. 2

also illustrates the reduction in production which can be expected when preventive maintenance (Schedule 2) is applied and schedule 3 shows that if the anticipated

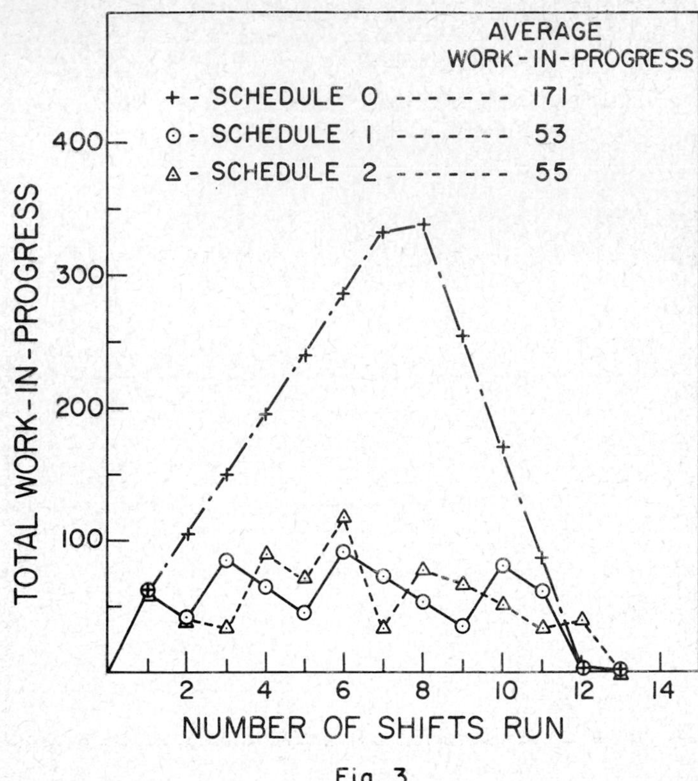

Fig. 3

day to day problems of machine breakdowns and lack of manpower are considered, the total production time for the complete run would be increased by one shift.

The effect of changing the tote parameter is shown in Figures 4 and 5. It is noted, as would be expected, that as the tote size is increased, the production rate drops and the work-in-progress increases. For this particular example, a tote size of 10 still allows the production run to be completed in 13 shifts whereas a tote size of 25 requires 14 production shifts for completion.

5. CONCLUSIONS AND RECOMMENDATIONS FOR FUTURE WORK

The proposed simulation programme produces a production schedule which maximizes the production while minimizing the work-in-progress for a single component processed through "n" machining stages. This programme when run off-line for predicting a production schedule which includes the adverse effects of machine breakdowns and lack of manpower is very useful in predicting realistic delivery dates. The basic simulation model including the preventive maintenance section would form an integral part of the on-line executive control programme which would automatically schedule components directly to the machines since it also has the capabilities of being updated and predicting a new optimized schedule based on the updated information.

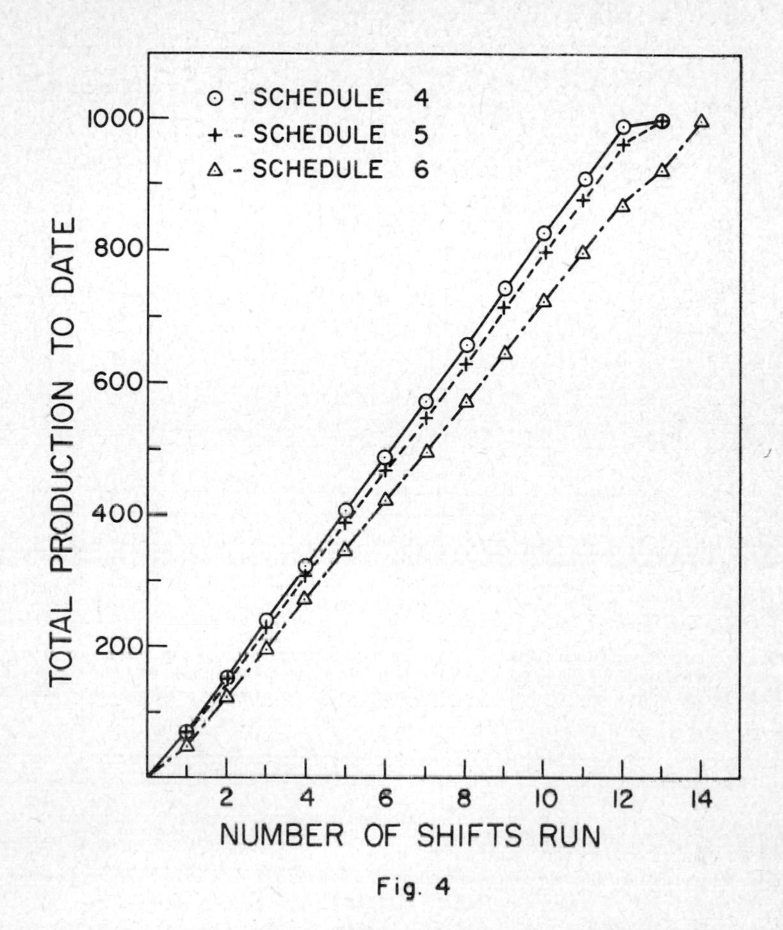

Fig. 4

If this model is to be incorporated as part of the on-line executive control scheduling programme, it must be expanded to accommodate the simultaneous scheduling of "n" different components through the system. Also, a conversational mode of direct reporting (hardware and software) must be developed in order to allow continual updating of the system as various uncontrollable parameters alter the present schedule. This mode of direct reporting has already been incorporated as management functions in the hardware of the new D.N.C. system developed by the numerical control group at the University of Waterloo and the balance of the research required is presently underway.

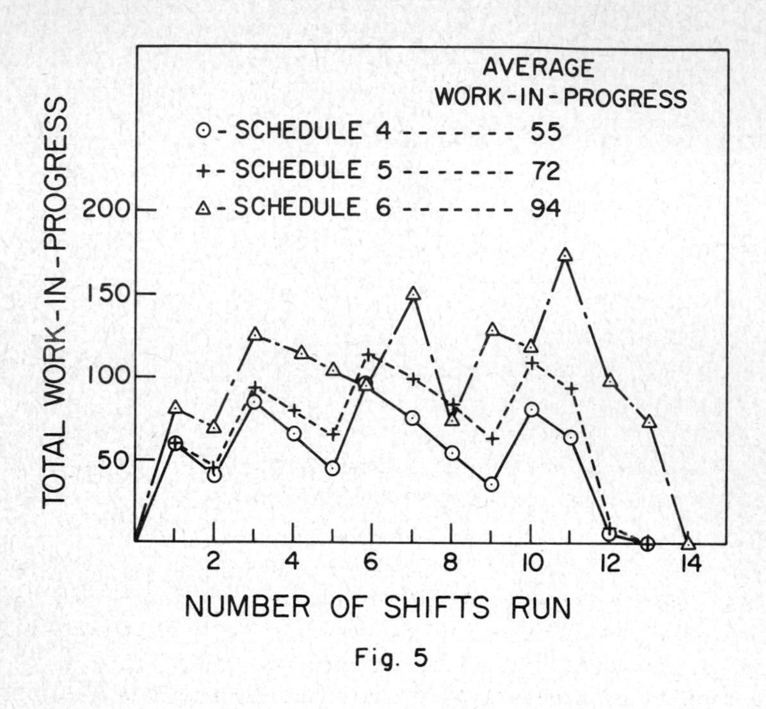

Fig. 5

ACKNOWLEDGEMENTS

This research was supported by the National Research Council of Canada under Grant Numbers A8274 and C1226.

REFERENCES

Adams, K.G. and D. French (1972), Digital computer simulation of a manufacturing system, Proc. of the Ninth Tech. Conference of the Numerical Control Society, 285.

PRINCIPLES of COMPUTER-AIDED NC PROGRAMMING with a HIGH AUTOMATION LEVEL

V.GRIESS, J.PREISLER, J.VYMER
Research Institute for Machine Tools and Metal Cutting,
Prague, Czechoslovakia

All systems of computer-aided programming of NC machine tools presently existing in the world have been developed to enable processing of control programs in a computer on the basis of a more or less simplified input information set up by the programmer. Primarily, this input information contains the description of the original and required workpiece shape and, further, depending on the automation level of the respective system, also the description of technology and sequence of operations needed for obtaining the final result. These data are indispensable for the general part of the computing program, i.e. for the processor.

The results of processing of the input data in the processor must be adapted to the specific properties of a NC machine and of its control system. This function is assigned to the second part of the computing program, i.e. to the post-processor.

One of the most important problems associated with this commonly adopted structure of computer programs is the division of processing tasks between the processor and the post-processor. The processor can not be of a completely generalized nature, independent of the properties of the NC machines. This would only be possible if just the geometrical relations between the input data were handled by the processor. This function of the processor would not be adequate to ensure the needed automation level of computer-aided programming, because a principal influence on the volume of work spent for programming in the normal engineering production is exercised by technological specifications. Only the combination of these two, i.e. of both the geometrical and technological specifications, forms a complete control program for a NC machine.

The geometrical relations between the blank, the part machined, and the co-ordinate system adopted are usually accurately determined. Therefore, they can easily be transformed into algorithms and

COMPUTER LANGUAGES FOR NUMERICAL CONTROL, *J. Hatvany, editor*
North-Holland Publishing Company - Amsterdam–London

expressed in the required form. This task is fulfilled by all the systems presently in use, of course.

A completely different situation exists in the region of technological specifications.Various approaches can be adopted for the solution of technology, ranging from a detailed description of it in the input data, through a partially computerized solution of the individual program steps, up to a completely automatic elaboration of the technological specification including cutting tools and their sequence, cutting conditions, collision considerations, etc.

The systems with a lower level of automation require a detailed input information. In this case, the programmer must describe a practicable technological plan in nearly every detail,his activity resembling practically the work done in conventional programming. Moreover, he must have a perfect knowledge of the vocabulary, syntax and semantics of the respective programming language and strictly obey all its rules.There are many disadvantages associated with this method, including the time-consuming set-up of the input data, the danger of errors and, above all, the immediate influence of the programmer´s knowledge and experience in technology on the quality of the control programs coming from the computer. The advantage of this approach, on the other hand, dwells in the possibility of the programmer´s consideration of the extravagancies of the operation programmed. It is a well known fact, however, that within the scope of normally produced parts these abnormal operations are not very numerous. An evaluation of this approach from the viewpoint of everday practice indicates that it is not much advantageous and effective. Because of that and in order to eliminate more thoroughly the influence of the human factor on the quality of programming, those systems are evidently much appropriate which cut the programmer´s work to a reasonable minimum and leave the most part of the geometrical and technological decisions to be made by a computer.

An attempt to find this reasonable minimum was successfully made by the authors of the Czechoslovak AUTOPROG-System in developing a system with a high level of automation having the following principial properties:

1) the preparation of input data is extremely simple and rapid,
2) the influence of the programmer´s experience on the quality of control programs is restricted to the lowest extent,
3) the system can be used with computers of commonly used capacity,

4) programming costs are substantially reduced.

These features have been ensured by the application of new principles of programming, by a new non-orthodox solution of the tasks of processors and post-processors, and by a suitable division of these tasks among them, as well as by the development of a system of elements of technological logics which can be utilized both for the computer-aided set-up of technological specifications in control programs and for the computer-aided planning for conventional machines.

The establishing of this system of standard technological elements represents a significant progress not only in the development of computer-aided programming, but also in the efforts to replace the subjective approach to the solution of technological problems by an objective one.

Similarly to the deterministic nature of the incidence of engineering parts and of their form elements already proven, distinct objective laws become evident even in the region of obtaining these form elements through the chip-forming process.These facts have not been theoretically delt with sufficiently as yet, unfortunately, but in practice they are sufficiently evident.Therefore, they can be utilized for an objective solution of technology without any risks.

The practical application of these principles in the AUTOPROG--System dwells primarily in the method adopted for the description of input data. Thus, e.g. the description of rotary parts consists of the description of the necessary number of primary form elements like cylindrical, tapered, and spherical surfaces, both external and internal - Fig. 1. Through the appropriate combinations of these elements practically all rough workpiece forms can be described - Fig. 2. Each of these primary elements is fully determined by a limited number of data specifying its shape, size, and accuracy. All these parameters are given in Fig. 1 in a simple chart designated in a manner corresponding to that in the sketch. A primary element can then be described by a mere transcription of digital data from the part drawing into this chart.

The so-called secondary elements are described in a similar way. These include, e.g., inner and outer recesses of various shapes, relieves, etc. The character of the secondary elements is given by the fact that they can only be machined after the primary elements have been completed.

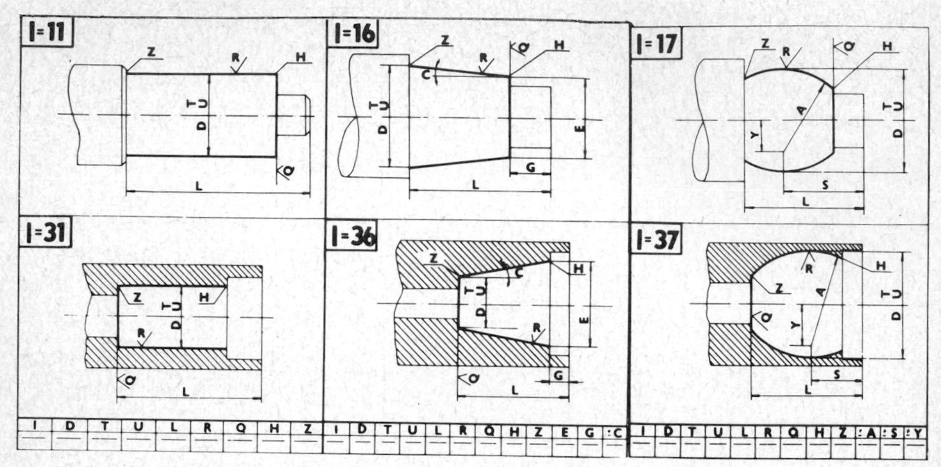

Fig. 1. Primary form elements of rotational workpieces.

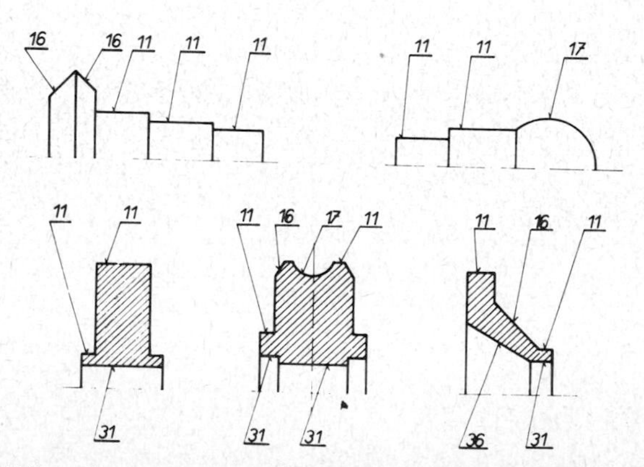

Fig. 2. Examples of rotational parts as composed from primary form elements.

The type of connection between the individual primary elements is defined by the so-called connecting elements, e.g.chamfers or radii. These connecting elements can easily be described as parts of the primary elements by means of a digital code specifying their type and size (see the H and Z columns in Fig. 1).

Similar principles are also used for the description of non-rotational workpieces - Fig. 3.

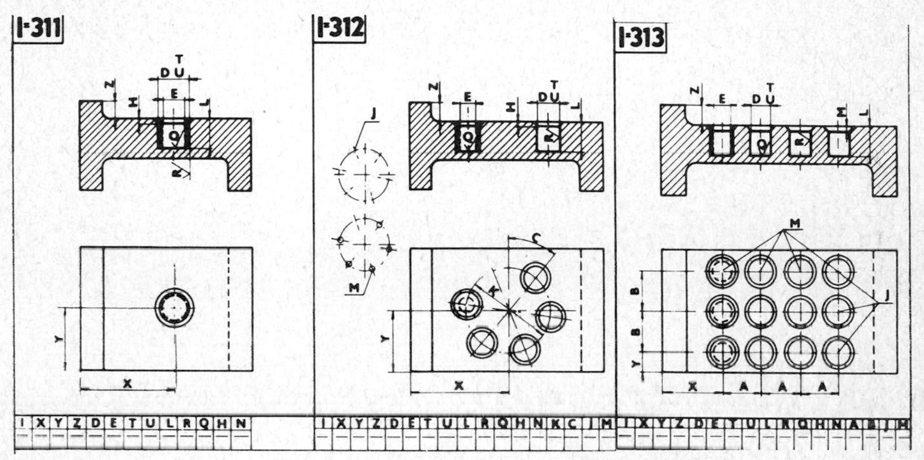

Fig. 3. Form elements in drilled and bored workpieces

Standardized methods of machining can be assigned to the above form elements, including the types and sizes of the needed cutting tools and their movement cycles. This assignment is one of the main tasks of the AUTOPROG processors.

Functions of an AUTOPROG processor

The input data read in from a punched tape are checked in the initial part of the processor. This diagnostic program checks the mutual relations between the read values and the maintaining of the syntax of the input data record. Computing only takes place if no error is found. Otherwise a list of errors and their nature is given in print on the line-printer. From the input-data values a mathematical model of the blank and of the part is set up in the introductory portion of the processor.

It is the task of the processor to specify the necessary cutting tools in the sequence given by the standardized technological plan, to establish the needed active paths of these tools with respect to the changing shape and dimensions of the workpiece in the course of machining, to specify the needed direction, sense and length of movement of the individual tools in a generalized way independent of a concrete NC machine, to specify more detailed information on the needed tools (e.g. dimensions, overhang, etc.), to establish the relative speed in each program step corresponding to the unit cutting speed and to the instantaneous situation of the operation.

Principially, the processor is working out stage-by-stage a generalized control program for the machining of the given part valid for all cutting tools needed in the operation programmed.At this stage the number and position of the individual tools is not important, as the generalized program applies to a fictional NC machine with a single tool post to which an unlimited number of tools can be mounted one-by-one.

In the course of program handling in the processor, the geometrical and technological part can not be divided as both are indispensable for the establishing of the generalized program cycles. In principle, it is a fully automatized imitation of the same method which is used in conventional programming. Additionally, however, there is the possibility to optimize various factors, e.g. the speed change with respect to the changing workpiece diameter and to a constant cutting speed, the change of cutting conditions with respect to the instantaneous workpiece rigidity, the tool-path length in positioning on drilling machines and machining centres, etc.

In these conditions a processor output in the form of a generalized block, independent of course of any particular NC machine, is by far the best solution.

In the case of NC machines with straight-line control systems, the generalized block (e.g. for lathes) contains the following parameters:

BLOCK (X, Z, F, S, T)

where X, Z are the absolute co-ordinates of the end points of the individual program steps in the given internal co-ordinate system of the processor,

F = feed rate,

S = spindle speed corresponding to the unit cutting speed (v = 1 m/min.) for the respective program step,

T = character of the needed cutting tool.

For continuous-path control the nature of the tool trajectory in the given program step must be specified, too. For a circular trajectory, e.g., this may be the radius of the circular arch, the generalized block having the following parameters:

BLOCK (X, Z, F, S, T, R)

where $\pm R$ is the radius of a concave or convex circular tool path section.

All other functions needed for the specification of a complete instruction for a given NC machine (e.g. the sequence No. - N, the auxiliary and preparatory functions - G, M, etc.) can be obtained in the post-processor.

As compared to the processor output as recommended by ISO (CL DATA), which is used with a number of processors using symbolic languages, this above described output form is much more simple. Moreover, it is given in a fixed formate, this providing for an easy identification of the information given.

The form of the processor output is closely associated with the overall concept of the processor. The AUTOPROG processors are basically sets of technologically oriented sub-programs (procedures) arranged in a logical sequence, the latter being given by standardized technological plans.

Each procedure is intended for an individual solution of both the geometrical and technological problems related to the given type of tool.

The decision on the inclusion of a particular procedure into the respective computing process is made on the basis of decision logics at the beginning of each procedure. If several cutting tools of the same type should be needed in the same program, the decision on their repeated use is made by the internal logics of the particular procedure.

The decision logics can perfectly simulate the deliberations of a skilled programmer. Thus, for instance, the processor intended for turning automatically decides on the following:

- spindle speed change as related to the diameter turned to maintain the optimum cutting speed,
- change of chip cross-section area as related to the instantaneous rigidity of the workpiece or to the method of clamping,
- supporting of low-rigidity workpieces by a center and inclusion of the center in the tooling list for the respective operation,
- cutting-edge material of cut-off tools as related to the shape of the part cut off,

  etc.

The processor intended for boring and drilling operations decides, for instance, on the following:

- cutting (or spindle) speed for the individual tool types,

- optimum tool-path length in positioning,
- change of chip cross-section area in the course of machining, etc.

As can be seen from the above, the technological logics included in the AUTOPROG processors are of a markedly high level and they are being improved still further.

Functions of an AUTOPROG post-processor

In the processor output fed to the post-processor, there are two kinds of information:

a) the control program in a generalized form independent of the particular NC machine tool,
b) detailed information on the needed cutting tools, on their sequence in the course of machining, on their dimensions, material, a.s.o.

In the course of processor-handling, this information is stored in the computer memory (e.g. in disc storage), wherefrom it is called to the post-processor and handled therein to the final form.

In the initial stages of the post-processor, several basic decisions must be made. Thus, for instance, a speed range must be selected out of those available in the particular machine tool, which suits best the machining conditions in the operation programmed. This task is assigned to a special sub-routine taking into consideration the extreme spindle speeds determined in the course of computation.

Further, the post-processor specifies the cutting conditions as related to the types of cutting tools used and to the machinability rating of the workpiece material. A very important task of the post-processor dwells in the assignment of the individual tools to the tool posts of the given NC machine or to the tool magazine. In this, the working capacity of the respective machine must be taken into account, together with the character of the tools, with the possible collision conditions between the tool carriers and the workpiece or machine tool units. Some of the tools can be assigned to any position, others can only be assigned to a specific carrier, or even to a specific position on it only.

This procedure must be designed for each particular NC machine tool or control system individually according to their properties. Upon fulfilling these basic tasks, the main part of a post-processor begins its work reading the data of the individual generalized blocks and transforming them into the respective format as needed for the particular control system. This mainly pertains to the transformation of co-ordinates to the co-ordinate system of the respective machine tool, this transformation being carried out for an absolute or incremental form including the specification of the direction and sense of movement. In this, all pre-setting dimensions of the individual tools are considered.

Out of the feed range the respective feed rate is selected and its code specified, or the feed number is calculated. Based on the spindle speed corresponding to the unit cutting speed,the actual spindle speed is calculated, the nearest spindle speed is selected out of the selected speed range, and its code is specified. The code of the tool position is expressed, and time for the respective program step is calculated on the basis of the information processed and of the respective preparatory or auxiliary function. All data processed in this way are printed on the line printer and, simultaneously, punched at the output tape punch of the computer.

This method of control data processing has many advantages. So, for instance, it provides for virtual "improvements" of the control system functions by establishing automatically separate steps for accurate slow approach if this provision is not included in the system hard-ware, etc.

Upon punching and printing of the control program, the information needed for NC machine set-up are printed and punched.This information includes the drawing number of the part programmed, the calculated cycle time, the programmer's comments, as well as the original input data for an eventual check. Also specified in the set-up information, are the starting positions of the individual slides and carriages, the speed range selected, and the arrangement of tools at the individual tool posts. The latter includes the name and position of the respective tool carrier, the type of the tool, the tool material, its dimensions, pre-setting data, and the identification number corresponding to the specific tool catalogue.

The time needed for writing and de-bugging of a post-processor operating on the above described principles is much shorter as compared to that needed for a processor, inspite of the particular features of the NC machine and of the control system which must be taken into consideration.

The combined computing program "processor - post-processor"

The technological procedures setting up a processor are completely generalized and their relation to the particular NC machine in question is brought to a minimum. It would be simple to set up an extensive generalized processor which would include all technological procedures appearing in the respective method of metal cutting, and to use this generalized processor in connection with a variety of post-processors. The versatility of this solution would, however, impose considerable requirements to the capacity of the computer memory. With many NC machines only a small volume of the processor functions would be utilized, the remaining procedures occupying the computer memory unnecessarily.

It has appeared with the AUTOPROG-System, that the optimum solution is an application of a specialized processor in combination with the appropriate post-processor in a single computing program. With a generalized processor designed as a building-block system of technological procedures it is very simple to set up a specialized processor by merely combining the respective procedures needed for obtaining the functions of the respective NC machine tool and to leave out those not important for the particular machine. This gives a significantly shorter processor lowering the requirements pertaining to the computer memory capacity. The processor can be combined with the post-processor in a single computing program intended for a particular machine tool. This program is stored in a disc- or tape memory unit in the internal computer code and it can be called out to the rapid memory by means of an instruction in the input data.

This solution of specialized computing programs was proven to be very useful in the AUTOPROG-System. Computation is very rapid and economic. No error can be caused due to interchange of input data as the latter call out their respective computing program. So it does not matter if upon a set of input data for a lathe another set for a NC drilling machine is read in immediately.

Conclusion

The system of computer-aided programming with a high automation level of programmer's work, as implemented in the AUTOPROG-System, has necessitated a different approach and development as compared with that usual with symbolic languages. The results have brought many outstanding advantages for the user, e.g. a rapid and economic method of input data set-up eliminating nearly all special knowledge and experience on part of the programmer, as well as a minimum consumption of computer time.

For practical use of AUTOPROG-compilers a computer capacity of 96 to 156 k-bytes is sufficient, depending on the method of segmentation.

The average computing time for a turning operation is approx. 50 seconds, a control program for a NC horizontal boring machine having 1600 blocks is processed in 240 seconds. The training period of user´s personel in input data writing takes only 2 to 3 days.

The solution described above is of extraordinary significance mainly in the introduction of NC machine tools in a plant or in the introduction of a new manufacturing program. Good quality programs are obtained and an excellent utilization of the properties and of the capacity of NC machine tools is ensured. The technological content of the computing programs can be adapted with respect to the most modern knowledge in the field of cutting tools and technology, this ensuring an excellent level of metal cutting in NC machines.

ADAPTING THE TECHNOLOGY IN PROGRAMMING LANGUAGES TO THE SPECIFIC REQUIREMENTS OF THE USER

DIPL. -ING. P. ADAMCZYK, DIPL. -ING. H. ZÖLZER
EXAPT-Verein, Aachen, F R G

Abstract: Automatic NC-Programming methods that contain the determination of technological data for the manufacturing process must be adaptable to the specific requirements of different machining tasks. A practical way is the central compilation of manufacturing data of the users for computer processing. In this manner it is possible to facilitate part programming as well as to achieve a better exploitation of the manufacturing equipment. The stored manufacturing data are not only valid for NC-manufacturing, but may also be used for rationalizing conventional production planning.

1. INTRODUCTION

Programming systems with a high degree of automation for the programming of numerically controlled machine tools have manufacturing techniques oriented processing programs with the aid of which not only geometrical, but also technological data for the manufacturing process are determined automatically. For processing purely geometrical programming tasks it is possible in most cases to apply unambiguous rules that have general validity and are therefore independent of the requirements and experience of the user. This is not the case with technological planning models for automatic NC-programming. For automatic determination of technological data with the aid of computer programs a great number of influencing factors on the part of the manufacturing equipment and of the specific requirements and experience of the workshop need to be taken

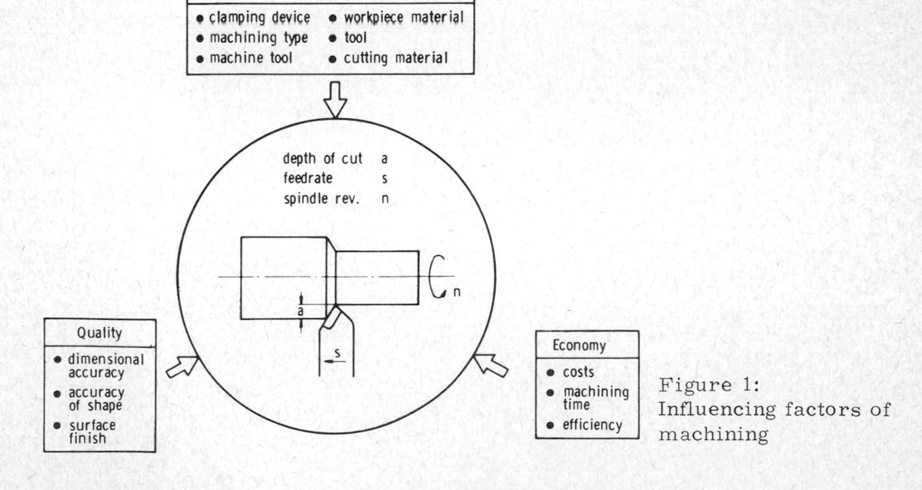

Figure 1: Influencing factors of machining

COMPUTER LANGUAGES FOR NUMERICAL CONTROL, *J. Hatvany, editor*
North-Holland Publishing Company - Amsterdam–London

into account. When speaking of the term "technology" in this connection the determination of machining sequences, selection of the required tools, calculation of collision-free tool paths, and the determination of the pertaining cutting data are meant. Figure 1 shows the most essential influencing parameters that have an impact on the cutting process for instance in the case of turning.

2. CONSIDERATION OF PARAMETERS INFLUENCING THE PLANNING OF TECHNOLOGY

The NC-Processors of the EXAPT-System that have a technological part are oriented on manufacturing techniques and generate the aforementioned technological information automatically. To ensure optimal utilization of the manufacturing equipment of the user firm, the calculation methods used by the technological processors are adapted to the current manufacturing conditions for each machining task. Thus, for determining tool paths and cutting values the conditions of application of the tools, the characteristic data of the machine tool, the machinability of the material, and the shape of the workpiece are taken into consideration (EXAPT-Verein, 1969).

2.1 COMPILATION OF MANUFACTURING DATA

On the left-hand side of figure 2 the data groups are shown that are significant for the programming of a machining task. The right-hand side of the figure shows in which form the characteristic values of the manufacturing process are collected with the EXAPT-System. The part program contains only such data that are assigned directly to the workpiece and the machining task. All other data that

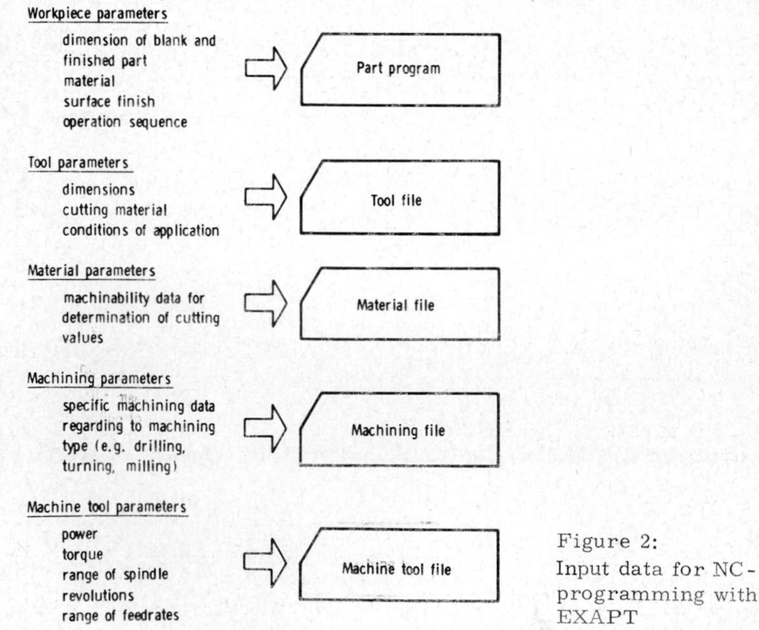

Figure 2: Input data for NC-programming with EXAPT

do not necessarily change with each new manufacturing task are stored in files and are available for each processing of a part program in the computer. Thus, together with a high degree of automation of programming, a reduction of the part program input is achieved at the same time.

The data of the machine tool or of the material are called up with the aid of a single statement and are made available for processing by the processor. Tools are either selected automatically from the Tool File or they are called up by the part programmer by means of their identity number.

2.2 TOOL-, MATERIAL-, AND MACHINE TOOL- DEPENDENT DATA

The EXAPT Tool File has already been in use in industry for several years and that not only in the scope of automatic NC-programming. Figure 3 shows a file sheet for a turning tool. Material Files for automatic calculation of cutting values for drilling- and turning operations have also been available ever since EXAPT-Processors are in use. Owing to production technique- or economic considerations also workpieces with a larger batch quantity are to a growing extent being manufactured on NC-machines. This means that the economic use of automatic programming methods must aim not only at a reduction of programming costs, but more and more at the reduction of machining time by optimal utilization of the production equipment.

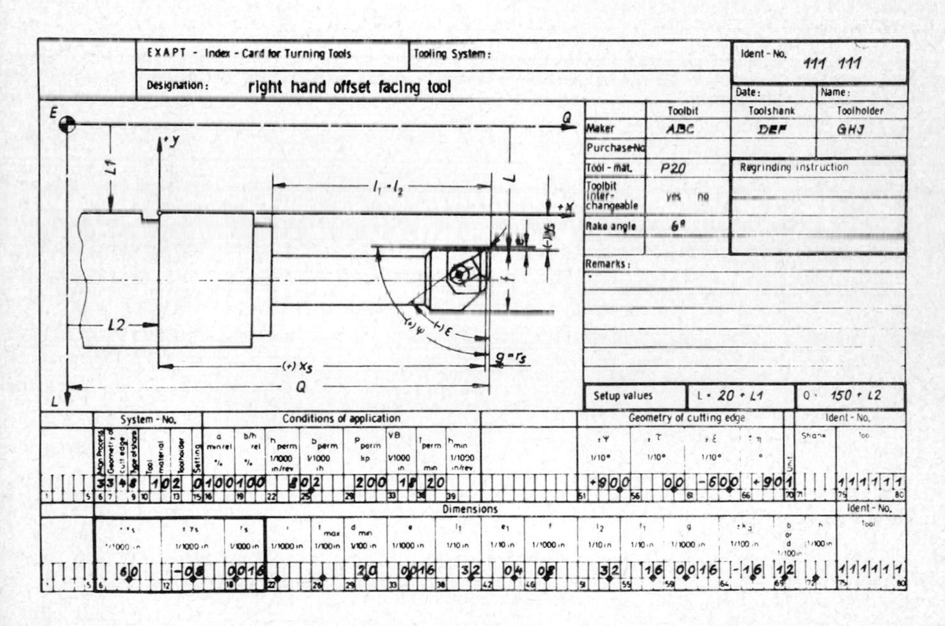

Figure 3: File sheet for turning tools

Accordingly, the description facilities for the technical data of the machine tool in the language part EXAPT 2 for turning were extended. In a newly developed Machine Tool File the characteristic data of the lathe, such as rotation speed, torque- and power- values as well as their interdepencency are described in detail. For calculating tool paths and cutting values it is thus possible in the EXAPT 2 - Technological Processor to determine the most useful machine tool data.

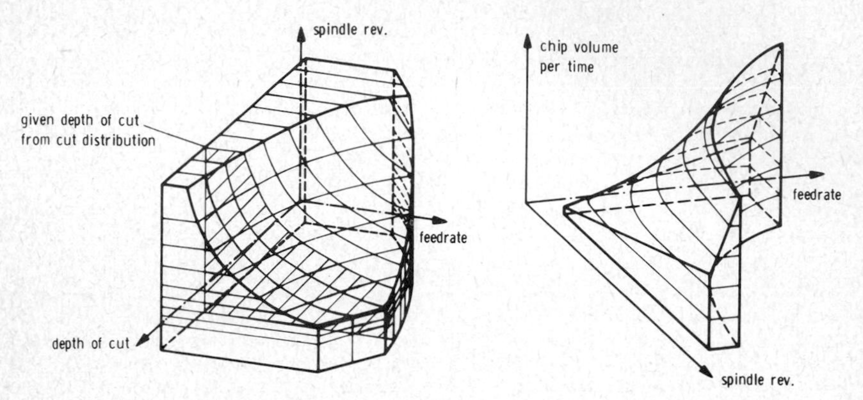

Figure 4: Determination of cutting values for minimum machining time

On the left-hand side of figure 4 several technological influencing parameters are given for the calculation of the cutting data depth of cut, feed, and cutting speed. As the relationship between the rotation speed of the machine tool on the one hand and its torque- and power- data on the other hand are described in the Machine

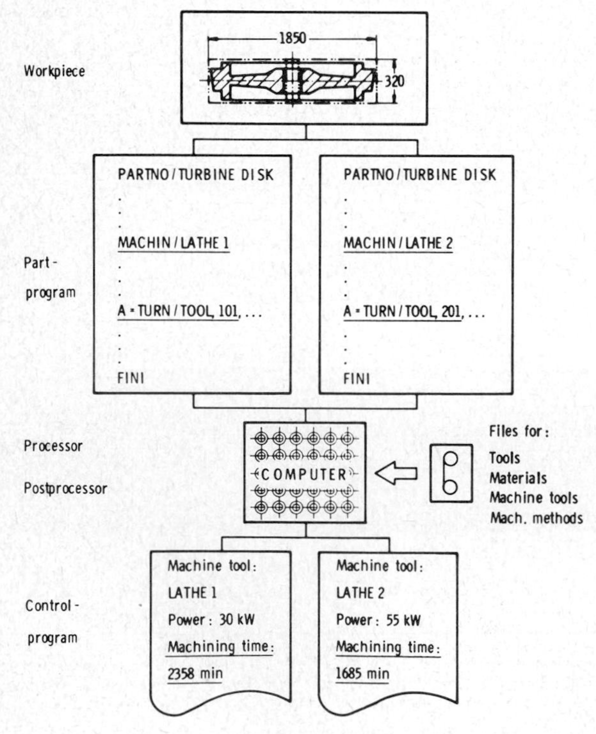

Figure 5:
Adapting the machining technology to different production equipment

Tool File, further optimization of the cutting values in respect of a maximal cutting volume per time unit is possible. Determination of this time-optimal combination of feed and rotation speed for a prespecified depth of cut is shown on the right-hand side of figure 4. These computations are carried out for each cut of a tool, i.e. it is in each case possible also for the current workpiece diameter to be taken into consideration.

Figure 5 shows the effects of such a detailed consideration of the characteristic data of the machine tool on the manufacturing time. Assuming that the machining of the shown workpiece is to be transferred from an NC-lathe I to a considerably more powerful machine tool II, the required adaptation of the cutting data in the NC control program can be done by changing but a few specifications in the part program with regard to the machine tool and pertaining tool set.

2.3 VARIABLE MACHINING PROCESS

Besides the influences that tool, material, and machine tool exert on the cutting process, there are further influencing parameters that have to be assigned to the type of machining. In the EXAPT-System these additional, user-dependent parameters are compiled in a Machining File. In the language part EXAPT 1, whose high degree of automation includes the determination of working sequences, the sequence of the machining operations as well as the selection of required tools are influenced with the aid of this file. Experience in practical use of the EXAPT 1 - Processor has shown that particularly those decision models in the processor that determine the sequence of operations and that have a considerable impact on the economy of the manufacturing process have to be as flexible as possible so that they may be adapted to the various experience values of the users. Therefore, decision tables were prepared for the language part EXAPT 1, 1 that can be modified with the aid of the Machining File. Thus

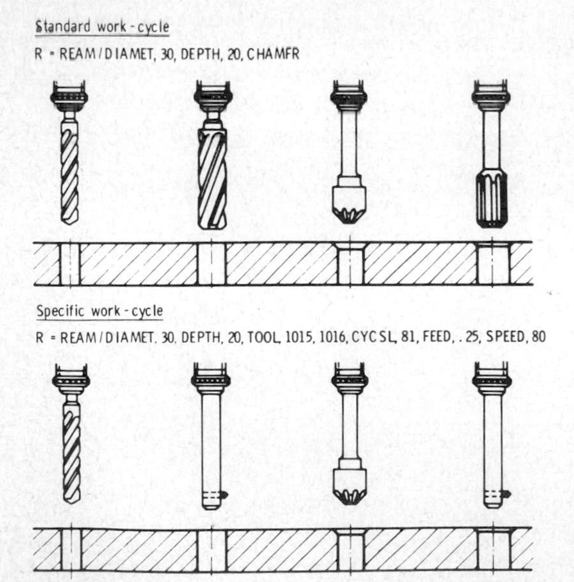
Standard work-cycle
R = REAM/DIAMET, 30, DEPTH, 20, CHAMFR
Specific work-cycle
R = REAM/DIAMET, 30, DEPTH, 20, TOOL, 1015, 1016, CYCSL, 81, FEED, .25, SPEED, 80

Figure 6: User-dependent work cycles for hole-making

any desired machining sequence can be determined automatically, even when using special-purpose tools. Figure 6 shows how two different machining sequences for the same task can be determined automatically with the EXAPT 1.1 Machining File.

With turning operations in the language part EXAPT 2, particularly when cutting with kinematically working tools, the contour of the workpiece and, there-

machining condition		correction of feedrate
machining of small contour radii	r_K r_S	reduction depending on the relation r_S/r_K
machining of short contour elements	l	reduction depending on length of element
machining with different entering angles	$\varkappa$ $\varkappa$ $\varkappa$	increasing for $\varkappa < 90^0$ decreasing for $\varkappa > 90^0$

Figure 7:
Correction of feedrate for contouring

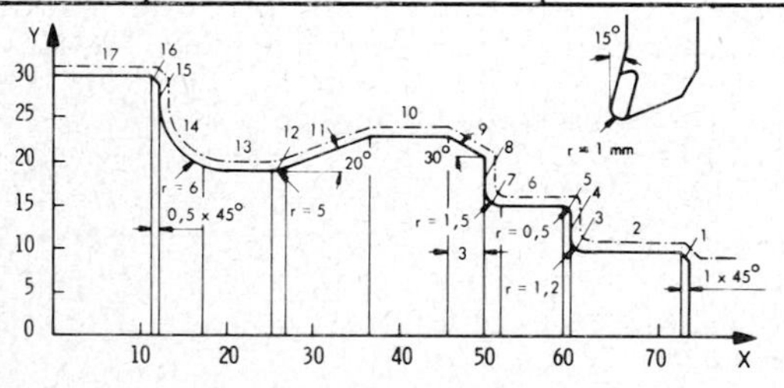

contour element no.	feedrate s [feed/rev]	spindle rev. n [rev/min]	contour radius r [mm]	width of chamfer a [mm]
1	0,37	2179	-	1
2	0,37	2179	-	-
3	0,1	2179	1,2	-
4	0,55	1502	-	-
5	0,25	1502	0,5	-
6	0,35	1502	-	-
7	0,15	1502	1,5	-
8	0,55	997	-	-
9	0,37	997	-	3
10	0,37	997	-	-
11	0,3	997	-	-
12	0,37	1137	5	-
13	0,37	1137	-	-
14	0,37	921	6	-
15	0,55	728	-	-
16	0,29	728	-	0,5
17	0,35	728	-	-

Figure 8:
Correction of cutting values for contouring in EXAPT 2

fore, the conditions of action of the cutting edge have a considerable impact on the cutting data. As these relationships are based on the machining method turning, they are also defined in a Machining File.

Figure 7 shows which conditions of action are considered in the course of a tool motion. The effects which the data of the Machining File have on the determination of cutting values are shown in figure 8 for a roughing operation with varying conditions of action. The numerical values given in the figure are to be considered as an example and can be modified in many ways by the user with the aid of the files.

2.4 ORGANIZATION OF WORKSHOP DATA FILES

The separation between flexible calculation methods in the processor and specific workshop data in external files is a good basis for adapting automatic planning of technology to the manufacturing conditions of the user. Utilization of files has the additional advantage that the data of the manufacturing equipment and the "know-how" of the workshop are collected and maintained centrally and are subsequently available for all problems of NC-programming. This guarantees a high and constant quality of production, as the cutting process and the rate of utilization of the manufacturing equipment are largely independent of the judgement of individual programmers.

The preparation of the files and the collection of the workshop data in a company are connected with a certain expenditure on the part of the user. These initial, introductory problems can be reduced considerably by the use of standard files. With the EXAPT-Processors that contain automatic processing of technology, standard files for tools, materials, and machining parameters are supplied. A further possibility for procuring initial data is provided by data banks for the manufacturing process. An example for this is the INFOS-Project (W. König, 1971) that is being developed at the "Institut for Werkzeugmaschinen und Betriebslehre" of the Technical College Aachen. Although standard files do not provide optimal solutions to individual manufacturing problems, in the average they lead to quite good results already during the introductory phase of the system. Through a feedback from the workshop and by taking centrally account of new developments in the field of manufacturing techniques, a continuous optimization of the stored data is ensured.

3. ADVANTAGES FOR CONVENTIONAL WORK PLANNING

NC-techniques make it necessary to gather the characteristic data of the manufacturing equipment and of the materials to be machined and to make them available for part programming.

With automatic programming this information can be processed by a computer providing the NC-Processors contain technological processing programs. However, the workshop data gathered for this purpose is valid not only in the field of NC-manufacturing, but can also be used for conventional work planning.

Therefore, it is advisable to make the technological calculation methods contained in the NC-Processors available as separate computer programs and use the stored workshop data for the preparation of working papers. In figure 9 this method is illustrated by the example of automatic determination of cutting values. The preparation of the Material File that contains machinability data determined from cutting data of the workshop can be made much easier with the aid of an evaluation program. With the use of Tool- and Machine Tool Files contain-

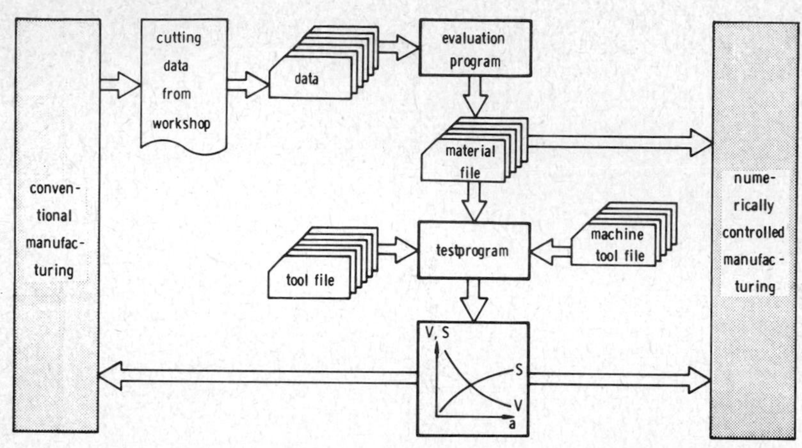

Figure 9: Data flow for preparation and testing Material Files for EXAPT

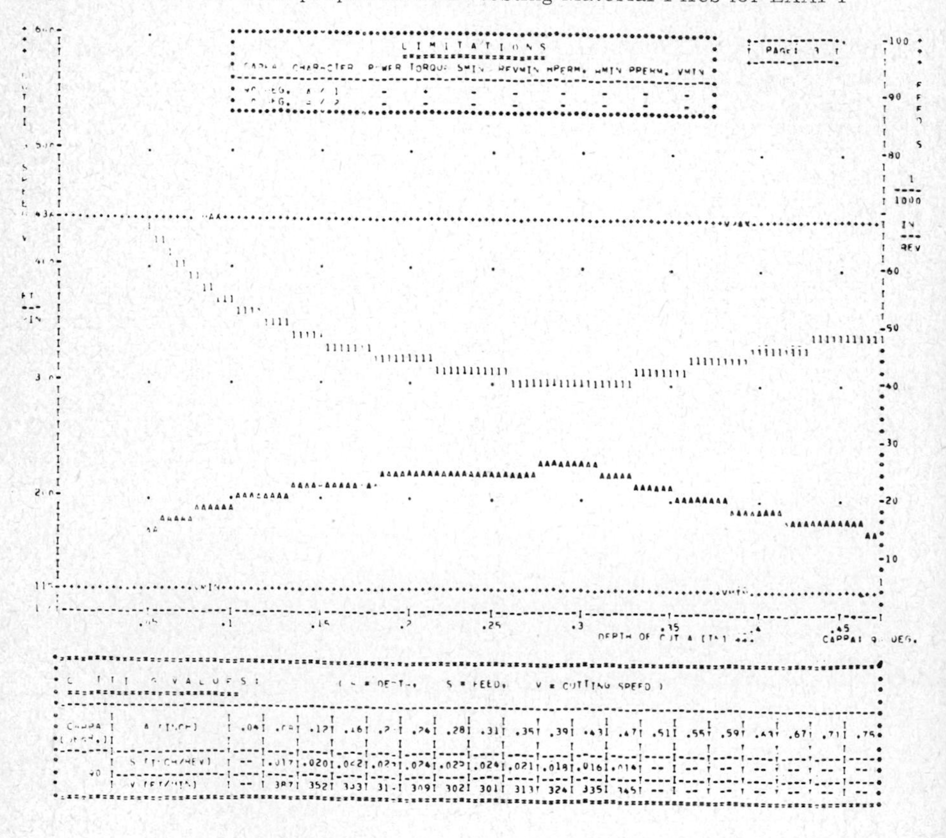

Figure 10: Automatic calculation of cutting values for turning

ing the characteristic data of the manufacturing equipment of the conventional range, the cutting values that are to be determined are adapted to the conditions of the conventional manufacturing process.

The output of such a Service-Program is shown in figure 10. This method used for the preparation of cutting value tables and diagrams has the advantage that - contrary to the reference tables normally used for work planning - the influences of the tool and machine tool are directly taken account of. Thus, in the example shown in the figure, the admissible machine torque is attained with a cutting depth of 0. 3 inch, which leads to a reduction of the feed for bigger cutting depths.

4. RANGE OF APPLICATION

The range of application of technological determining methods in NC-Processors when using flexible computer programs and user-dependent files comprises the largest part of all programming- and manufacturing tasks. However, for the conception of highly automated programming systems it must be taken into account that also special machining cases can be programmed without the user having to take recourse to another system. For this reason, it is necessary to provide facilities that enable the programming of any kind of tool motions and pertaining cutting data with a low degree of automation. Therefore, the EXAPT-System provides programming facilities beginning with the definition of complete working cycles via APT-like tool motions to the insertion of already coded control records for punched tapes.

REFERENCES

EXAPT-Verein, 1969, EXAPT 1/2 - Technological Calculation Methods

W. König, 1971, Leistungssteigerung bei spanenden und abtragenden Bearbeitungsverfahren, Girardet Taschenbuch Technik, 6, 53.

OPTIMIZATION MODELS FOR OPTIMUM-COST TECHNOLOGICAL PERFORMANCE CHARACTERISTICS OF METAL'CUTTING PROCESSES USING CUTTING EDGES OF DEFINED GEOMETRIES

D. KOCHAN, H.-J. JACOBS
Technical University Dresden, GDR

The automatic process optimization computerized by external programmes has not only to relieve the production engineer from routine work but also to produce results that are justified as to the technical expenditure and economy. These results should be found on a higher level and by means of the well-known mathematical procedures whereby all the technological findings have been utilized.

The main range of applications for metal cutting are the manufacturing processes in metal working industry. The machine tools employed in this industrial sector for the technological realization of the metal outting processes partly have a high automation level as against other manufacturing facilities of the cutting-off and metal working technologies. Thus, an accurate determination of the process parameters given by the performance characteristics (cutting conditions) is required. The automatic manufacturing facilities, particularly the NC machine tools are characterized by high first cost. Their economic effectiveness can be realized by a careful technological optimization only when engineering and economic optimality criteria are used. The generally accepted basic concept of a mathematical optimization model is characterized by the fact that by a mathematical method that solution is to be found from a field of realizable solutions that ist nearest to the extreme value of the optimality criterion.

The field of the realizable solutions is limited by the technical conditions. The optimality criterion, e. g. minimum cost or time is realized by linking it to the variable performance characteristics as cutting speed, feed and, rarely, cutting depth by means of formulations, nomograms and, recently, by complete computer programmes.

The consistent utilization of an optimization model, i.e., the separation of the technical limits from the target function, has the advantage of easily warranting a broad range of applications since modifications and improvements of this model can be made

COMPUTER LANGUAGES FOR NUMERICAL CONTROL, *J. Hatvany, editor*
North-Holland Publishing Company - Amsterdam–London

without difficulty.

Based on this advantage, a programme for the mechanical determination of such optimum cost technological performance characteristics as cutting speed and feed in turning operations has been developd as a first fundamental model.

The technical limits are represented in the form of a field of solutions R = f(v,s) and the non-linear target function as K = f(v,s) with the same dependence whereas the performance characteristic of cutting depth is preset.

The simultaneous determination of all the technological performance characteristics, i.e., cutting speed, feed and cutting depth with regard to optimum cost can, at present, be regarded as the most advanced development level of process optimization by external programmes.

The optimization strategy required can be immediately derived from the above basic model.

The finding of performance characteristics that are technically and economically justified presupposes a thorough knowledge of the range that can be realized technically as the field of solutions and of the target function in the optimization model.

The target function considers the cost components that, in their function, depend on the performance characteristics. In this way, the cost function of the optimization model is determinded by machinery and labour cost as well as tool cost during machining, or changes of the tools, and for tool regrinding.

The target function for one operation can be formulated as follows:

$$K = \frac{C_3}{v \cdot s \cdot a} + \frac{C_4}{a} + \frac{C_5}{v^{A_2+1} \quad s^{A_4+1}}$$

A_2, A_4 } exponents of the Taylor tool-life equation

C_3, C_4, C_5 } constants dependent on cost factors, turning diameter and length as well as auxiliary times

The range that can be realized technically is demarcated by the technical limits. For the optimization only those limits can be relied upon that permit a mathematical formulation.

The following restrictions on the part of workpiece, machine tool, tool and workpiece holder ar considered:

R_1 minimum technological feed
minimum machine tool feed

R_2 maximum total deviation of the workpiece chucked
permissible deflection of tool shank
permissible load of tool material
maximum permissible momentum at the work spindle
maximum permissible roughness in dependence on the
corner radius of tool
maximum machine tool feed

R_3 maximum work spindle speed

R_4 minimum cutting speed due to cutting tool material
minimum work spindle speed

R_5 maximum output at the work spindle

R_6 range of definition for tool-life function when mating a given
tool material and material of workpiece

The combined action of the technical limits can be illustrated by a plot where cutting speed is drawn against feed (see fig. 1)

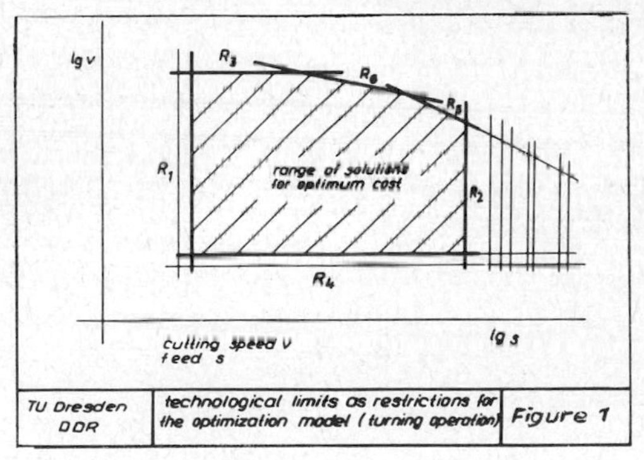

The optimum couple of values wanted (v, s) always has to be within the field of solutions outlined by the restrictions. If there are several relevant technical limits of the same functional nature only that appears at the field of solutions that restricts the technically realizable range as much as possible.

The mathematic descrition of the constant R will be omitted here.

The solution of the optimization problem is attained by a mathematical combination of the economic and technical parameters. The target function is applied to the restrictions of the field of solutions preferred for the optimization that an optimum cost point (v,s) on each restriction can be calculated by the formation of an extreme value of the function K = f(v,s) by means of partial derivatives.

The absolutely best point as to cost (v_0, s_0) within the field of solutions can be determinded by a comparison of cost (see fig.2).

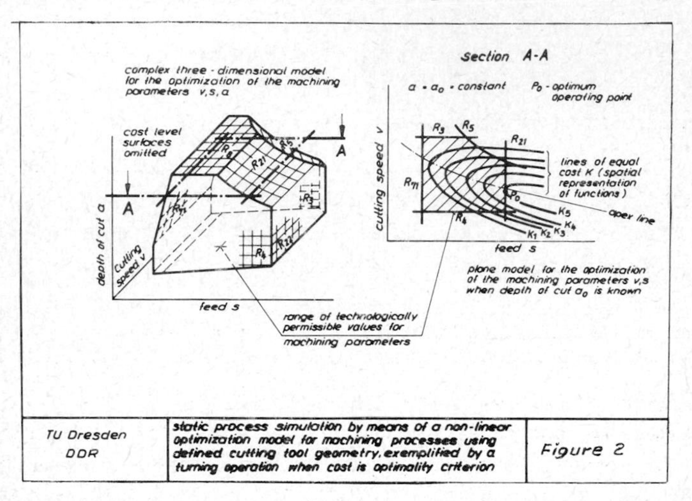

The main part of the algorithm with respect to the extent and significance is constituted by the mathematical description of the technical limits and their combined actions.

The entire algorithm for the automatical evaluation of optimum cost performance characteristics of cutting speed and feed in turning operations can be linked to data storages for workpiece materials and manufacturing facilities. At the present level of development this algorithm can be applied to roughing operations in straight turning and facing works.

The optimization model described can not only serve for the determination of optimum cost kinematic performance values but the range of application for this model can be considerably enlarged by the incorporation of the cutting depth. Due to the simultaneous optimization of all the three performance parameters (v,s,a) the solution is possible by means of a threedimensional coordinate system having the cutting speed, feed and cutting depth, resp. as its axes. Thus, the plane field of solutions, becomes a spatial representation of the solutions. The restrictions are extended by the cutting depth according to the generally known laws in metal-cutting engineering (see fig. 2).

The following results are obtained from the plane model:

- cutting depths or speeds, resp., that can be adjusted to optimum cost,
- optimum cost feed values,
- basic times of machine tool,
- variable manufacturing cost,
- effective tool-lifes for several cuts when employing the same tool,
- necessary with of ground-in chip breaker at the tool.

The development of a model that is independent of the process concerned, for the cost optimization of the technological performance value enables it to be used for other processes in metal-cutting operations where cutting edges of defined geometries are used.

Based on the fundamental model developed for turning operations the transition to other processes revealed that, no doubt, a generally accepted optimization strategy, exists. It is therefore appropriate to elaborate separate programmes for any further process with due consideration given to the total system concept tailored for the respective system. In this way, the pecularities specific to every process can be better taken into consideration while, at the same time, the algorithms can be implemented even on small size computers most economically.

For instance the developmental work for the optimization programme for cutter head milling had to proceed from the fact that no values for tool life are generally known from which the tool-life equation required for the cost target function shown can be determined directly.

Due to the entirely different nature of the cutting conditions in face milling a transition of the tool-life values for turning is impossible. When starting from the relationships known for only a few of the workpiece and cutting tool materials an equation can be given that corresponds to the formulation by Taylor:

$$T = B_3 \cdot v^{B_2} \cdot s_z^{B_4}$$

where T = tool life, v = cutting speed, s_z = feed per tooth, B_3 = constant of the tool-life equation, B_3, B_4 = exponents of the tool-life function.

With regard to the technical limits, the direction and the amount of the cutting force will vary at the entry of tool. This phenomenon is contrary to the turning operation.

The maximum value of the cutting force can, to a good approximation, be determined from the sum of maximum cutting force. Substantial preconditions for the additional limits required particularly for milling operations thus enabled the accurate determination of the following factors:

- permissible deflection of work spindle,
- permissible torsional momentum acting at the work spindle,
- permissible unit pressure at the interface between tool and tool holder.

When establishing the target function for milling operations special attention is to be given to the fact that the effective time of engagement is to be taken as the base, since it has to be employed for the establishment of the tool-life relationship.

It is expedient for the structuring of the mathematical model programme to base this operation on a modular component design

Fig. 3 shows the block diagram of the KOFA MKFR (Cutter Head milling) programme established in such a simple way.

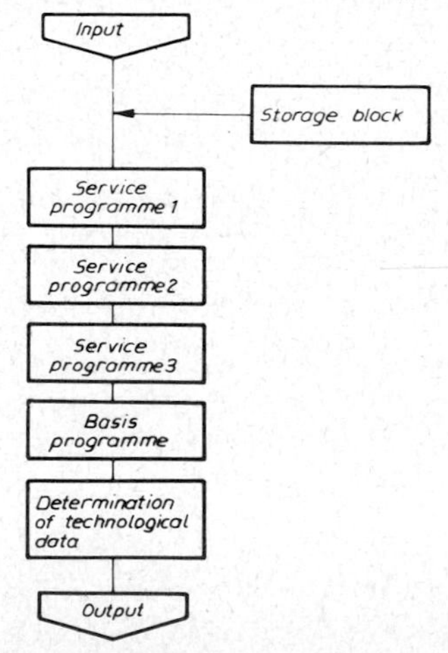

Figure 3 : Programm KOFA MKFR

The storage block contains the data needed most and remaining constant (machine tool, workpiece material and cutting tool material values).

The service programme 1 is for the computation of the characteristic figures of the tool required as correction factors for tool life, and the entry and leaving conditions of the cutting tool.

The service programme 2 serves for the determination of the torsional and deflection momentums operative at the spindle or the milling cutter.

The service programme 3 determines the system of technical limits.

The basic programme implies the determination of the optimum cost values for cutting speed and feed and is based on the target function. Regarding the nature of the this target function, mathematical methods of the non-linear optimization are taken here as a basis.

The last modular section of the programme is for the computation of the figures for speed, feed rate, basic time of machine tool, effective basic time, tool life and manufacturing cost in consideration of the particular machine tool in question.

Besides the substantial advantage of plain programming and testing of the modular programme design described, the following aspects are taken into account:

- adaptation to the specific conditions of the respective enterprise without any difficulty,
- fitting in of new findings readily possible,
- modifications of the whole programme or programme modular units for other problem configurations possible in an economical way.

The most significant range of application of the optimization programmes KOFA (that is, cost optimization of performance characteristics in mechanical engineering) ist the immediate utilization for the optimization of the technological performance characteristics of cutting speed, feed and, possibly, cutting depth.

The results obtained hitherto produced average cost savings from 10 to 60 per cent with simultaneously saving basic times of 5 to 70 per cent as compared with the figures prescribed by the work standards TGL 8923 effective at present in the GDR. In this context it has to be pointed to the fact that the actually figures employed are frequently lower than the standard figures fixed in the above standard so that more often than not higher savings may be expected.

Moreover, the optimization programmes or the modular units of the programme are, based on the same strategy, suitable for the utilization, in the determination of the limits for rated values and ranges of regulated conditions under adaptive control constraint (ACC) or adaptive control optimization (ACO) resp.

The application of the above model for the evaluation of economically profitable values for those performance characteristics not included into the adaptive control has proven most expedient.

Another most significant use is offered by the inversion of the sequence of computation for the evaluation of manufacturing facilities (machinery and tools). A technically based orientation for the laying out of the manufacturing facilities can be well made by considarations of any single technical limit.

Fig. 4 shows the application ranges of the optimization programmes.

KOFA programme	Operations	Applications	Computational realization
DR I	Turning operations, optimization of s, v	•Optimization of machining data in rough-machining and finish-machining operations	R 300
DR II	Turning operations, optimization of a, s, v	•Determination of rated values and range of regulated conditions under limit control	R 300
MKFR I	Cutter head milling, optimization of s, v	•Evaluation of production tools (machinery, tools)	FORTRAN IV
MKFR II	Cutter head milling, optimization of a, s, v		FORTRAN IV
WAFR	Hobbing		FORTRAN IV

Figure 4 : Application of KOFA programmes

At the same time, this figure shows a summary of the existing programmes and the respective computational realization modes. As a conclusion, it can the stated that the knowledge of the wear and force behaviour during the machining process is indispensable for the automatic technological optimization of the metal cutting processes when external programmes are used. The analytical description of tool wear and tool load is a focal point of the optimization algorithm. The full technical and economical utilization of the machine tools that is aimed at, can be attained, first of all, by the knowledge of these facts of metal-cutting technology.

PREPARATION OF BASE-LINE DATA IN THE AUTOMATED SYSTEM OF TECHNOLOGICAL PREPARATION OF PREPRODUCTIONAL STAGE

YARMOSH, N. and MILNER, F.
Institute of Engineering Cybernetics,
BSSR Academy of Sciences, Minsk, USSR

1. GENERAL FEATURES

The limited use of complex automation systems of technological preparation of preproductional stage in engineering is mainly due to the imperfection of existing technical means provided for such systems. The limitation of computer memory capacity which does not correspond the existing information arrays to be stored and processed when used as base-line data in designing the technology of final results of an automated design system is one major reason of widespread development of automated systems of programme preparation where the work-drawing is used as the bearer of base-line information. The description of such drawing is usually made by a programmer who adopts a programme-oriented description language for the purpose.

The analysis of existing stages of this preparational process reveals that any further increase in the productivity of such systems may be obtained by introduction of specialised technical means of data preparation by a technologist-operator without special training in programming if he is given the possibility of operational control and data correction when forming these descriptions and introducing them into computer system.

One possible solution of this problem may be the system of automated programme preparation for numerically controlled milling machines developed at the Institute of Engineering Cybernetics of the Byelorussian Academy of Sciences. The operational principle is based on the method of inputting sequential descriptions of contour projection elements and types of drawings by an operator using a coding device with functional key-board panel bearing basic symbols of the programme-oriented language. (See Fig.1).

The end of each description is determined by the symbol "point" which is used for the formation of reference command to a set of solution programmes for corresponding geometric problems the re-

COMPUTER LANGUAGES FOR NUMERICAL CONTROL, *J. Hatvany, editor*
North-Holland Publishing Company - Amsterdam–London

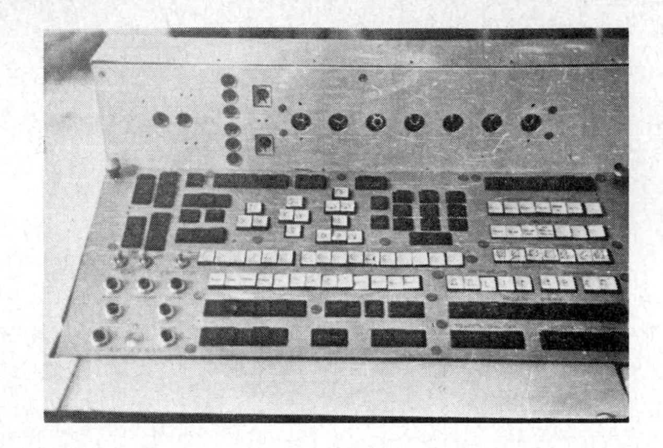

Fig.1.Functional key-board panel (general view).

sults of which are thereafter used in forming a programme "frame" for visual display control device (visual display or the plotting board of a drawing automaton).

When comparing the display image with the representation of an element on the work drawing the operator is able to carry out visual control and to introduce necessary corrections into the description if required.

To form the programme frame for the control device it is necessary to have information about point coordinates of the beginning and end of a line segment and also point* coordinates of the beginning, end and the centre in case of a circle arc.

Generally the above information is represented in "unreal" state in the work drawing. The basic points are obtained as a result of certain geometric plots which are done by a designer with the help of conventional drawing tools and which may be defined as points of intersection or conjugation of elementary object forming elements (straigt line, circle, parabola, etc.) of projection, cross-section of part's view contour.

Coordinate values for the basic points are obtained as a result of solving corresponding geometric problems the programmes of which are combined into the"library of standard programmes" of the auto-

* The points of the beginning and end of a contour element will further be referred to as basic contour points.

mated programming system for the digital programme control machines.

Reference to a programme is formed from the description provided by the operator and introduced into the computer as a certain sequence of codes.

2. Main notions of the input language

Let us consider the methods used in forming such descriptions in the subsystem of technological design developed for digital programme control milling machine.

Fig.2 shows the contour for which (P3) may be determined as the point of conjugation of the straight line (LN2) passing through point (P1), and circule (C4), the centre coordinates of which relative to point (P1) and the radius are correspondingly Xc,Yc,R.

However this description is equally valid for both (P3) and (P3'). To make these two points non-synonymous we introduce additional indication when we use the position of conjugation point relative to the vector directed from the first description element to the second one (from P1 to P4). In the given case (P3) is located on the left side and (P3') on the right side of the description vector.

Similarly, in order to describe straight line (LN6) which is tangent to two circles (CO4 and CO8) it is necessary to indicate the following additional signs: external tangent, above (to the left) of the direction vector.

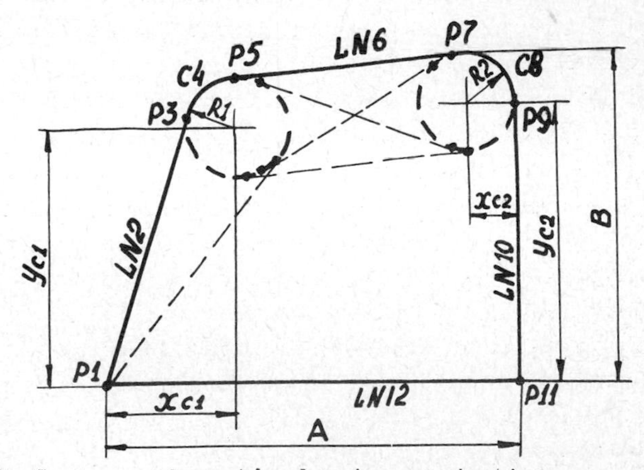

Fig.2. Contour of part's drawing projection.

In view of our arbitrary approach to the beginning of count-off during the input of dimensionality chains, they should be brought to some general beginning.This is done by the description of each successive element relatively certain instantanous beginning of coordinates defined in relation to the general beginning.

It has been concluded through the analysis of work drawings of separate parts, which are formed by line segments and circle arcs, that the description of object forming contour elements may be brought to a certain limited sequential set determined by the shape of elements, their relashionship and reciprocal position in accordance with the mode of their representation.

In the process of forming an object's characteristics we have introduced the determinants of its position in relation to the earlier determined "base" objects. These determinants are used in programmes to obtain results of computation or reference programmes to routines which are required for the solution of a given problem.

In the process of describing projections, sections or the view of an object the possibility of multi-coordinate machining is introduced by the method of assigning corresponding axis values of three-dimensional coordinate system of a machine to the axis of Cartesian coordinate system of a plotting board.

The image of a three-dimensional part is formed by a number of planary sections which have a set location in relation to one another in mutually parallel and perpendicular regions.Besides the point determined by its coordinates and the circule determined by the centre coordinates and the radius,the constructive drawing elements also include:plane, cylinder, sphere, cone, torus, parabola (paraboloid), hyperbola(hyperboloid).

The sign of linear dimension is formed in accordance with the direction of dimension "deviation"from the base element and is determined by the coordinates signs.

The angle between two oriented straight lines is determined as algebraic difference of angles between the directions of straight lines and abscissa axis.

The angle is considered to be positive if for the purpose of matching of the straight line and abscissa axis one has to turn it in the clock-wise direction. The angle is considered negative if it is turned counter clock-wise.

Given below are the descriptions of main notions of the input language. For abbreviation purposes we use the apparatus of Bekus metalinguistic formulae.

1) Element being determined

The element which is determined from the drawing by its relation to certain other previously determined base elements.

Syntax:

Element being determined:=⟨mark⟩ ⟨elementary object⟩

Elementary object:=⟨point⟩ | ⟨straight line⟩ | ⟨circle⟩ | ⟨plane⟩ | ⟨cylinder⟩ | ⟨cone⟩ | ⟨torus⟩ | ⟨sphere⟩ | ⟨parabola⟩ | ⟨hyperbola⟩

2) Base element

Elementary object presented in the drawing by dimensionality chains or the one determined previously and further used in forming the description of the element being determined.

Syntax:

Base element:= ⟨mark⟩ ⟨relation determinant⟩ ⟨elementary object⟩ | ⟨mark⟩ ⟨elementary object⟩ | ⟨mark⟩ | ⟨relation determinant⟩ | ⟨ mark ⟩⟨coordinate⟩

3) Mark

Characteristic feature of a symbol, of a sequence of symbols or of a description.

Syntax:

Mark:= ⟨sign of a number⟩ | ⟨sign of a word⟩ | ⟨sign of a base⟩

4) Determinant

Vocabulary word or its corresponding basic symbol, which is used in describing the relationship of the element under determination and the base element or in describing its position relatively some other element.

Syntax:

Determinant:= ⟨relation determinant⟩ | ⟨position determinant⟩ | ⟨regime determinant⟩ | ⟨complex determinant⟩ | ⟨material determinant⟩ | ⟨purity class determinant⟩ | ⟨accuracy class determinant⟩ | ⟨determinant of dimension deviation⟩ | ⟨tool determinant⟩ | ⟨machine determinant⟩ | ⟨direction determinant⟩ | ⟨determinant of machining mode⟩

a) relation determinant:=

⟨tangency⟩ | ⟨intersection⟩ | ⟨normal⟩ | ⟨parallel⟩ | ⟨angular range⟩ | ⟨linear range⟩

b) position determinant:= ⟨higher⟩ | ⟨lower⟩ | ⟨right⟩ | ⟨left⟩ | ⟨external⟩ | ⟨internal⟩

c) direction determinant:= ⟨from⟩ | ⟨along⟩ | ⟨till⟩ | ⟨through⟩

d) regime determinant:= ⟨milling⟩ | ⟨boring⟩ | ⟨turning⟩ | ⟨boring out⟩ | ⟨grinding⟩ | ⟨geometry⟩ | ⟨technology⟩

e) complex determinant:= ⟨contour description⟩ | ⟨contour⟩ | ⟨equidistant⟩

f) tool determinant:= ⟨material⟩⟨description⟩ | ⟨normal⟩ ⟨description⟩ | ⟨standard⟩ ⟨description⟩

g) machining mode determinant:= ⟨contour⟩ | ⟨plane⟩ | ⟨groove⟩ | ⟨edge⟩ | ⟨step⟩ | ⟨torus cutting in⟩ | ⟨angular cutting⟩

The analysis of elementary object representation assumes that there exist several methods of describing each of these objects. The nature of description depends on the sequential order of "bases" and "relation determinants" and is defined by the subjectivity of operator's perception of a part represented in the drawing.

To simplify algorithm structures and to save the time required for the search of programmes from the library of standard programmes (LSP), the description of the drawing objects are represented in the form of "canonical" tables.

The table form is adequate to the elementary graph of finding solutions where the first place is allocated to the object being searched, and the end is allocated to the cell adress of the "catalogue" which keeps the initial adress of storage location along with the required standard programme (SP).

The cell adress is formed from the problem cipher which is composed of the "zero" base cipher (the element being determined), the relation determinant to the first base, and the base itself, then of the relation determinant to the second base and of base N2, etc.

In order to convert these descriptions into the form of canonical tables it is sufficient to position description elements in the order of priorities assigned to them, which are determined on the basis of statistical studies and depend on frequency of occurrence of these elements in the process of forming descriptions.

3. Forming of descriptions for the determination of machining regimes.

The initial design parameters required for determining the regimes of machining a part on a numerical control machine are the following: blank material, mode of machining, required accuracy class, the tool. Some of these parameters are given in the form of alphanumeric information, others are assigned by the operator. They include tools and type of machine depending on the surface shape which is to be machined and on the dimensions of the part. It should be pointed out that material, accuracy class, and purity may be the same for various operations, while all other parameters characterize only one distinct operation, in view of which the initial information for technological design is subdevided into files corresponding to certain technological operations: milling, turning, boring, grinding, etc.

Each file is brought in correspondence with a set of coding tables having notations or codes of initial parameters.

On the basis of initial tables the machining cipher is defined from Table 1 where A,B,C,D correspond to certain ciphers of initial tables.

Table 1.

Operation	Material	Mode of machining	Tool
A	B	C	D

Table ciphers are arranged so that when written sequentially they form the number of "catalogue" cell bearing adresses of magnetic tape memory which stores corresponding standard programmes.

4. Input of descriptions of graphical information

The limited number of digit-forming elements in the drawing and uniformity of their relatioships serve to cut down the vocabulary volume, the number of basic symbols and to simplify the programming language which, in its turn, allows to represent them in the form of a functional keyboard panel. The time required for analysing and transforming the initial information is cut down by means of subdividing all basic symbols and notions into notation groups which are brought in correspondence with certain portions of output register. The latter is designed in such a way as to allow correspondence with notions of various functional blocks (geometry, technology, etc.) through one and the same register state.

The formation and automatic transfer of codes (corresponding to coding tables' data) into the operational computer memory is carried out by means of pressing the functional keys of the panel. This transfer is done in the mode of switching off computer operation. A sequence of codes is forming an array of initial information in the computer memory, the end of it being determined by the introduction of symbol " ." (point) code.

Input is controlled by means of a special input programme providing operation of several panels with time sharing. With the hel help of this programme the initial information is converted into a state suitable for use by a complex of "translator" programmes along with the formation of operational regime code:

a) "geometry"- when solving geometrical problems",

b) "technology"- when estimating technological parameters,

c) "contour description" - when forming the tool run.

Regime code is used to form the call-out command from the permanent storage and for the introduction of a corresponding programme into the operational computer memory. This code is also used for the purpose of rewriting the initial information into the memory portion which gives response to a certain panel*.

5. Complex of "Translator" programmes

The complex begins with the programme of processing of initial information (MI).

Numbers determining linear and angular values (X, Y, Z, Δx, Δy, Δz, R, l, α, β, γ) as well as the base numbers are transferred into the calculating device in the form of 31-digit codes (higher order bearing signs). The number represented in binary-decimal form is occupying 28 lowest order digits - 7 tetrads; the last 2 of them are ment for the input of decimal fractions, Input begins with the first significant digit. Angular values are introduced with the accuracy of seconds.

Processing of numbers begins with their extraction from MI, after

* Possibility of simultaneous operation of several panels has been forseen.

which necessary signs,order and order signs are determined for the numbers which characterize linear and angular values, and finally they are converted into binary system with the help of a standard routine.

Numerical values of angles represented in degree measure are then converted into radian measure.

After processing numerical information has been completed the routine of analysing the plane of machining is started. When describing the plane the X,Y axes of Cartesian system of the plotting board are assigned the values x,y,z of three-dimensional coordinate system of the machine.

With the completion of the above operations MI array is cleared from syntax and working signs and then "compressed".

We further come to remove round brackets containing the description of base elements.

In the process of description formation there are instances when an element being used as the base element may not be determined at once. In this case it may be represented in the description form of "the second" level problem which exists in the description of the "first" level problem.

The system under discussion permits to form descriptions out of four levels, each of them being distinguished by square brackets.

A pair of square brackets bounding a description which does not contain any other square brackets is called "internal" and it is the first to be processed. After finding the "internal" pair, M2 array is formed from MI array, which has the problem description contained within this pair of square brackets. After processing M2 has been completed the results of solutions are transferred into corresponding adresses of array MI which is further "compressed".

The next pair of "internal square brackets" is then sought. The problem contained within this pair of brackets is solved and the whole process is repeated until the last array of corresponding first level descriptions is formed.

The subprogramme of processing M2 includes the base selection routine, relation determinants, the routine of bringing their descriptions into the form of canonical tables, the routine of forming the adress and calling out into computer memory a solution programme for corresponding geometrical problem, the routine of locating the initial information on adresses corresponding to the above programme, and also the routine of recording on and reading from the magnetic tape the required information.

Similar relation determinants are then replaced by the analogical onand further cleaning and compressing of M3 is carried out. M3-array is then brought into sequential form of base codes and relation determinants, which usually defines the programme of solving the geometrical problem whose adress is formed by summing M3-codes with specially selected constants of extraction.

After solving this problem the results thus obtained are tape-recorded and we form a reference command to the "Postprocessor" programme. This programme is then transforming the array of final results obtained after solution of the problem into the form of input "frame" of the control device (interpolator) for a corresponding panel. The graphical output is represented either on the automatic plotting board or is visually displayed on the screen.

If the displayed image does not correspond to the work drawing the operator can refer to the correction programme by pressing the "rub-out" key and indicating the elements' numbers, which excludes all information for these elements and carries put a repetative input.

The analysis carried out at the Institute of Engineering Cybernetics of the BSSR Academy of Sciences showed that the introduction of the above methods for the development of control programmes in case of parts having average size and complexity allows to cut down the expenditure by 30-40%.

DISCUSSION

Mr. D. Kochan suggested to Mr. B.E. Molnár that it was very difficult to describe parts using only symbols and not words. The author pointed out that only the geometry of the blank and finished contours had to be input. In reply to Mrs. Jortzik, who asked for a comparison between part-programming times for APT-like languages and FORTAP, Mr. Molnár said the usual FORTAP time was 15-20 minutes. You did not need motion statements, tool definitions, etc. Asked by Mr.M.A. Sabin why the geometrical stage of the system had to access material and tool files, he pointed out that turning implied an interrelation between geometry and technology, and cutting parameters were selected automatically.

Mr. H.J. Jacobs asked Mr. Tougu how his system optimized cutting parameters under differing conditions. The author replied that there were many possible criteria and the program permitted equations appropriate to any of these to be inserted. They had worked on cutting parameter calculations for ten years and now had suitable models for all practical conditions.

Mr. J. Louwerenberg /who presented the paper of Mr. P.A.J.M. Bockholts/, was asked by Mr. H. Weissweiler whether the user or MITURN decided machining sequence. Mr. Louwerenberg said there were eight options for user decision, otherwise the user would have to accept the system's routines. Two questions by Mr. M.D. Fingerle /on why MITURN instructions differed from APT and which point of the tool was to be programmed/ would be answered by the original author.

Mr. J. Vlietstra said he was perplexed by the way roughing cuts were automatically calculated by a number of systems. Did this not mean that you had to know beforehand the accurate dimension of the blank, which was very often impracticable? Mr. P. Hellström said the first roughing cut had to be programmed for the largest possible blank, or if this took too much time you would have to specify close tolerances for the blank. Mr. J. Vymer agreed with Mr. J. Vlietstra that this was one of the basic problems of NC generally, whether programming manually or by computer. It was useless to hope for accurate blanks. Mr. J. Hatvany described the solution to this problem used in the Hungarian DNC system. The Machine Tool Controller /MTC/ has an alphanumeric LED display which shows the message "BLANK SIZE?". The operator enters the measured size through numeric buttons and the computer then calculates the appropriate roughing cuts.

Summing up this part of the discussion, Mr. G. Sohlenius said that there were three possibilities: to accept uneconomical machining,to use adaptive control or to use the method presented by Mr. Hatvany.

Mr. H.J. Jacobs asked Mr. H. Zölzer whether they had values for the constants of the equations for cutting forces and tool life for all combinations of tool and workpiece materials. Mr. Zölzer replied that the material file does not contain cutting value combinations but characteristics /vectors and exponents/, and an extended Taylor equation is used in connection with a chip form criterion. Mr. J. Vymer commented that the corrections for feedrates in turning specific shapes evidently improve efficiency, but does it not make the system user dependent? The author agreed that it did. In EXAPT-2 there is a machining file in which the user may specific, say, feedrate reductions for contour radius, tooltip radius, angle, etc. The user is offered a standard file which can be altered to suit him. Mr. Vymer pointed out that in his experience users rarely know just what they want, and at first it is best to use the laboratory data we have. Mr. Zölzer agreed and said they used the INFOS databank system of Aachen University, but could evolve towards the user's data. Replying to a question by Mr. M. Horváth on variable cycles, Mr. Zölzer said the higher the automation level the more flexible a system should be. They had included decision tables in EXAPT 1.1 which allowed the user to insert data that would change the sequence of machine operations.

Mr. M.H. Choudhury asked Mr. K.G. Adams to quote figures on savings for DNC. Mr. Adams said the system used a PDP 8/E, costing about $ 5000, the interpolators would be about $ 300, so the whole system, including pulse motors, should be at about $ 12000 which compares favourably with present NC prices. Mr. J.W. Bruce asked whether any detalis could be given of the software interpolation system,but was told it was proprietary. Software interpolation was linear, circular, parabolic, or elliptical. Mr. D. French said the system would be discussed in a later paper. He also told of some machining experiences with the system.

Mr. H.J. Jacobs, commenting on Mr. G. Berg's paper, said the adaptive control also required inputting the reference and constraint values. This was an additional job increasing the expense of automatic operation planning. Mr. Berg was in agreement.

8. CNC, DNC AND POST PROCESSING

COMPUTER APPLICATIONS IN MANUFACTURING CONTROL SYSTEMS

G. STUTE
INSTITUT FÜR STEUERUNGSTECHNIK DER WERKZEUGMASCHINEN UND FERTIGUNGSEINRICHTUNGEN, UNIVERSITÄT STUTTGART
STUTTGART, WEST GERMANY

Abstract: A detail account of existing DNC-systems is given. The functions of a conventional Numerical Control can be shared between Computers and hardwired Controllers. Different possibilities are discussed especially showing the potential advantages of each solution.

1. INTRODUCTION

In the last few years, applications of numerically-controlled production equipments have increased. The number of conventional machine tools is decreasing; they are replaced more and more by numerically-controlled machining centres, or - since recently - by manufacturing systems.

Reasons for these trends are for example:
- a forced demand of better productivity of manufacturing equipments
- marketing trends
- fluctuations of employees.

Looking at some economic dates, like sales, wages, employees, costs and so on, a gap will be recognized between the production result, reached by conventional methods, and the production result, that has to be reached in the future (see figure 1). This gap can only be closed by new production methods.
Technical products are becoming obsolete with increasing velocity. At the same time, industries have to enlarge their production program, forced by competition. Only if production techniques similar those of mass-production can be applied to small batches, it will be possible, to produce economically. On the other hand, new manufacturing systems should be able to accept new demands in a flexible way.
In the future there will be less manpower in our plants. The trends of increasing wages, decreasing productive time and the lack of skilled people are going on.

A possibility for the solution of the above mentioned problems is a highly automated system, for both, production and production planning. On the basic principle of Numerical Control, we can connect machine tools with transport systems, tool changing systems, stocks and an overhead control system.
The volume of necessary controlinformation can only be handled in computer-systems, which are closely connected to the numerically-controlled production equipments. These systems are known as "DNC-Systems" (Direct-Numerical Control System).

In the past, conventional numerical controls were sometimes linked to computer-systems. But, considering an all over economically-working control system, redistribution of the control functions is indispensable. Supposing this, first of all an analysis of the existing information system has to be done.

COMPUTER LANGUAGES FOR NUMERICAL CONTROL, *J. Hatvany, editor*
North-Holland Publishing Company - Amsterdam–London

2. Control-levels

Defining under the conception "control" all devices, which serve for an automatic flow of information to and from the machine tool, several control-levels can be distinguished (Fig. 2).

The lowest level in control systems, the actuating level, involves the actuators, such as clutches, valves and drives, which control directly the flow of energy.
The next level includes the machine control, which, caused by input signals, generates actuator-signals.
The machine control is superposed by program control, processing data according to a prescribed program. The numerical control is in this way a kind of program control.
Including at last the level of data-distribution, an automated flow of information can be reached in the area of production. The hierarchy continues with the levels "NC-program generation" and "production control". These levels generate NC-programs, supported by programming languages, respectively organize the current production.

There are several stages for computer-application. Depending on the extension of computer-integration into the control-system, various levels are possible.
In the area of machine control and program control, the alternatives can easily be distinguished in a block-diagram of a conventional numerical control (Fig. 3).
The numerical control containes four major sections of logic. Their functions are quite different and they work almost independently of each other:

- geometrical information processing
- technological information processing
- decoders and registers
- input and compensation function.

The area "geometrical information processing" includes the transformation of NC-program-data into command-signals for the position loops. Conventional NC mostly use arithmetic units, consisting of Digital Differential Analyzers (DDA) or Pulse-Rate Multipliers (PRM).
The main design problems arise from the speed and relative complexity of the motion control computations.

In respect to computation time, as well as to the set of necessary arithmetic operations, the demands of "technological information processing" is quite easier. New developments, such as "Programmable Controllers" can increase the flexibility of this subsystem.

On a next stage, manual input and compensation logic can be realized by the computer, as to be seen in Fig. 3. These functions are very suitable for computer-realization, as hardware-realization is very difficult and on the other hand, there are no stringing timing and storage requirements to be met.
Realizing these functions in the computer, the remaining NC is able to work autarcly, the full operational convenience is however obtained only in cooperation with the computer.
If the level of machine control or program control remains unchanged, data distribution can be executed by the computer (Fig.4). A large capacity storage contains all part programs, needed within the next few days. The storage is managed by a computer. In addition, the computer is able to transfer NC-programs into the core and to transmit NC-data to the NC.

This widely developed computer-system is known as "DNC-System with Behind-Tape-Reader (BTR) Input".

Applying a system, containing additionally production control and production scheduling or - partly - NC-programming capabilities, a fully automated information processing system will be possible in the future. In the past, these functions could only be realized by using a large computer-system, while all other functions could be managed by a small process computer. Nowadays, there is a trend to implement small, simple program-languages in process-computers, to be more independent of large computer-access. These implementations start all over. It is necessary to note, that this work should not be done several times at different places.
Regarding a realized system (Fig. 5), all the described alternatives can be seen.
Parts are programmed in a higher-level programming language and the part programs are stored on a mass-storage. If machining centres claim for new data, the part program is transferred to the core, calculations are done, and finally the NC-program is transmitted to special control units of the system.
These units transform NC-program values according to a prescribed program into actuator-signals for the position loop.
In this system, most of the control functions are done by software. Recognize in particular, that each level is managed by a separate unit.

There are another two main problems to be mentioned. Regarding figure 5, it remains to discuss, how many machine tools should be controlled by each computer and - not so obviously - how many computers should control each machine tool.
Reviewing the systems according to these criteria, four main choices can be done:

First A single computer is executing all functions for one machine, or for several simultaneously working machine groups. This system is named "Computerized Numerical Control (CNC)".

Second A single computer is executing data distribution for several machines, partly including geometrical and technological information processing. In respect to the remaining hardware, these systems are named "DNC-Systems with reduced hardware Controllers" or in English literature "MCU-Systems".

Third A single computer is executing data distribution for several machines, including conventional Numerical Controls. Data input is done parallel to the reader input. This system is named "BTR-System".

Fourth Control functions of several machines are executed by several computers. Each computer is adapted to its special task in an optimum way and an optimum core-size. Data transfers between computers perform under program control. Theses systems could be named "Hierarchical DNC-Systems".

3. Some system features.

It is quite obvious, that these four choices differ in some aspects. For example regarding a CNC-System (Fig. 6), the main difference to conventional control is, that most of the usual hardware, like arithmetic units, has been replaced by a built-in small computer.
The advantages, obtained by the system, are: More flexibility and standardization of control-systems. These systems are now said to be worthwhile for controllers for 3-to 5-axis contouring machines and anything more complex.

What upon MCU-Systems (Fig. 7), one of the underlying objectives is to replace as much functions as possible of conventional NC-hardware. With computer costs spread across several machines, a large computer installation can be envisaged.
These systems will be favoured in applications like:

- Installations, in which all machine tools are identical (such as batteries of drilling machines or small centre lathes).
- Systems with interlinked machine tools (like Molins-System 24), which are supplied by one manufacturer.

Some significant limitations have to be noticed. There will be a lot of difficulties, if machine tools and control systems are not supplied by one manufacturer, because hardware as well as software must be reconciled, according to machine tool features.
Another major problem is the difficulty of re-assigning machine tools into or out of such a system. But sometimes advantages, due to low-cost NC, will justify the introduction.

BTR-Systems (Fig. 8) are characterized by tape-reader-bypass input. Thus, the introduction of these systems can only be justified, if the advantages over and above those of conventional NC are of such an importance, that the financial disadvantage can be neglected.
This configuration allows a slowly introduction of the DNC-System into manufacturing. Existing NC can be connected to the system. A further advantage is that an existing form of organization has not to be changed significantly and, by a computer black-out, the NC may continue in the normal tape-reader mode.
At the same time, nearly all the other advantages of computer application can be obtained with BTR-interfaces, for example: Centralized data-handling and data-distribution, NC-programming, production scheduling and easy access to large computer systems.
Obviously DNC-systems with BTR-interfaces will predominate in the next future.

In hierarchical systems (Fig. 9) control is shared among several computers. All the other above mentioned systems realize software-functions in a single computer. This fact leads to a number of limitations.

Managing a lot of different functions in one computer system, a large proportion of computation time and core-memory is blocked by the mere activity of switching from one task to the next.
If all functions are realized in one computer, all the necessary peripheral equipments has to be supported by the operating system, so operating systems are growing.

Therefore hierarchical computer systems have to be designed regarding core-memory size, cycle-time, as well as an adequate peripheral equipment. The concept can be extended to include further levels of information processing.

Finally possible advantages of DNC-systems have to be considered.
First of all it is to state, that different kinds of production shops pronounce different benefits of DNC-systems.
The major function, that always has to be realized, is automatic data-distribution. The priority of all the other functions differ considerably, depending on the application.
For example, users applying interlinked machine tools prefer the realization of hardware functions in a central process computer, as well as the capability of supervising the manufacturing system.
In applications, using detached NC-machine tools, the problems of connecting existing controls, monitoring, NC-programming, correction and

production scheduling are of some importance.
On the other hand, in aerospace industrie eliminations of tapes and tape readers are considered very important, as well as a computerized data-handling for long NC-programs. A primary requirement for DNC in such a plant is to provide position monitoring, working parallel to the measuring system of the NC. These functions can all be implemented in DNC-systems.

Condensing all these facts, it is to be seen, that DNC-systems vary widely in different industries. Therefore it can be stated, that DNC-systems can not be ordered, simply writing an order sheet. On the contrary, they have to be developed, considering the requirements of the special task.

The application of a DNC-system permits the addition of further functions. As new needs arise, or as new technological advances are made, all can easily be adapted, just changing software. That is why, applications of DNC-systems will increase.

References:

(1) Martin, Blair R.
Computer Control of Machine Tools, Technical paper MS 72-242, Society of Manufacturing Engineers, Dearborn, 1972

(2) Mesnianeff, P.G.
The Technical Ins and Outs of Computerized Numerical Control, Control Engineering, March 1971, pp. 65 - 84

(3) Stute, G. u. R. Nann
DNC-Rechnerdirektsteuerung von Werkzeugmaschinen, Wt-61, (1971), S. 69 - 74.

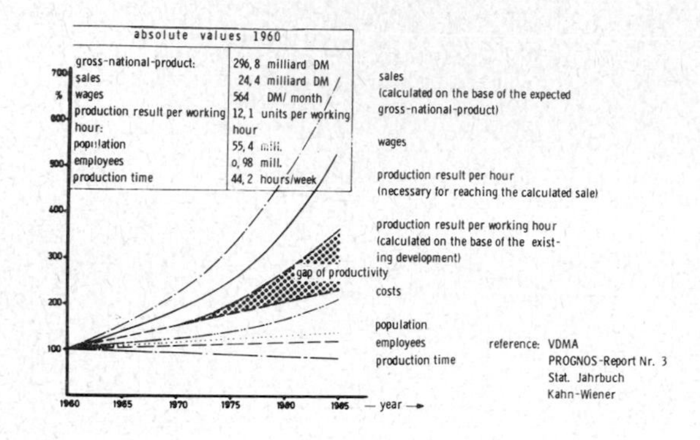

Fig. 1 Prognosis of Economic Dates for the Engineering Industry of GFR

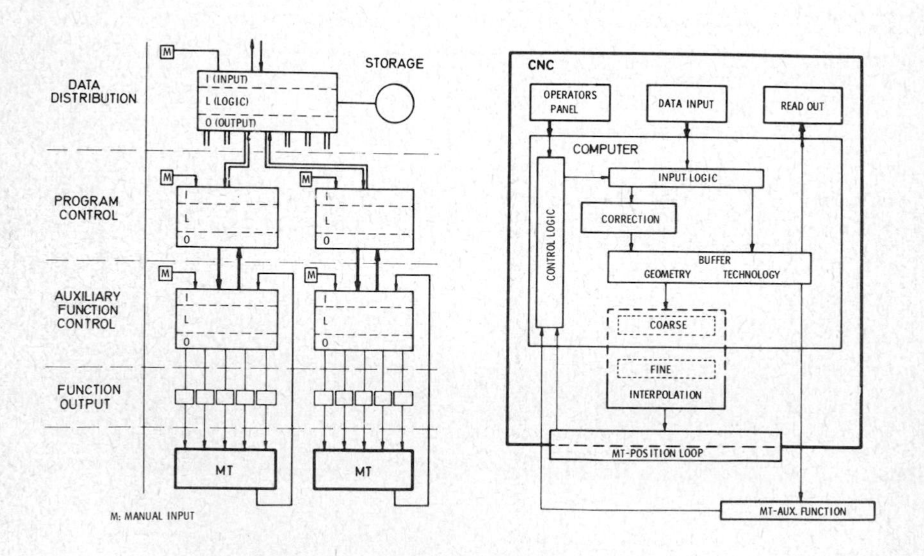

Fig. 2 Structure of Control Systems

Fig. 6 CNC - Computerized NC

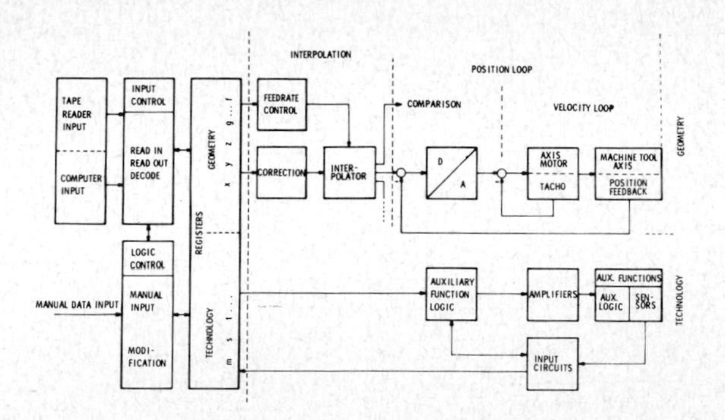

Fig. 3 Block Diagram of a conventional NC

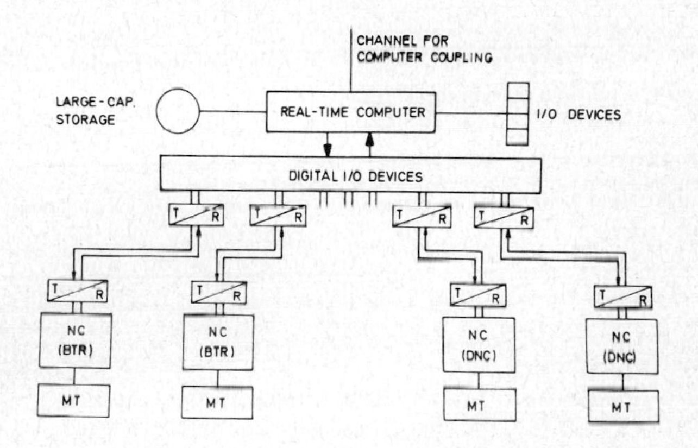

Fig. 4 Structure of a DNC System

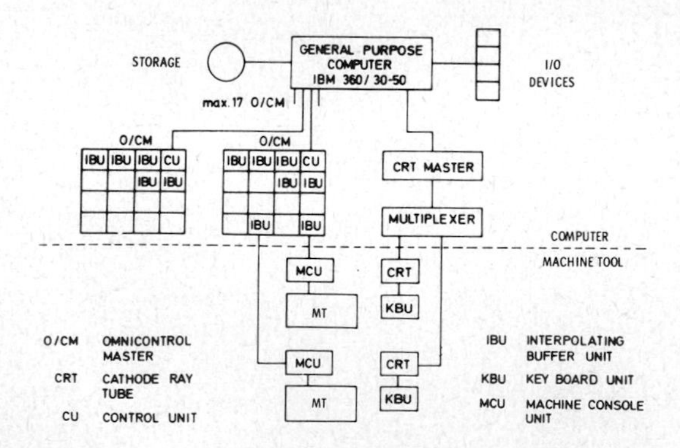

Fig. 5 Sundstrand Omnicontrol

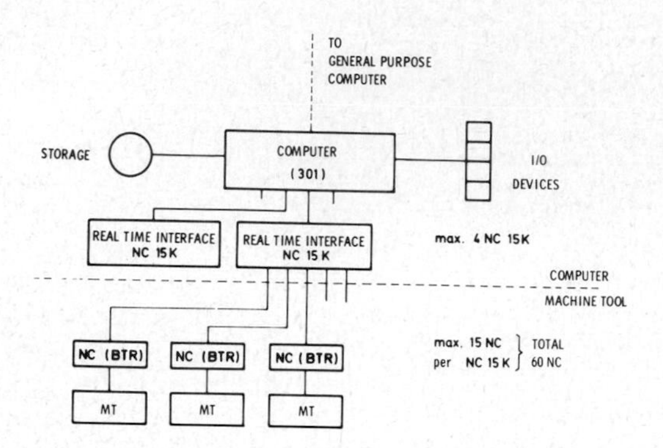

Fig. 8 DNC System with BTR Input (Siemens)

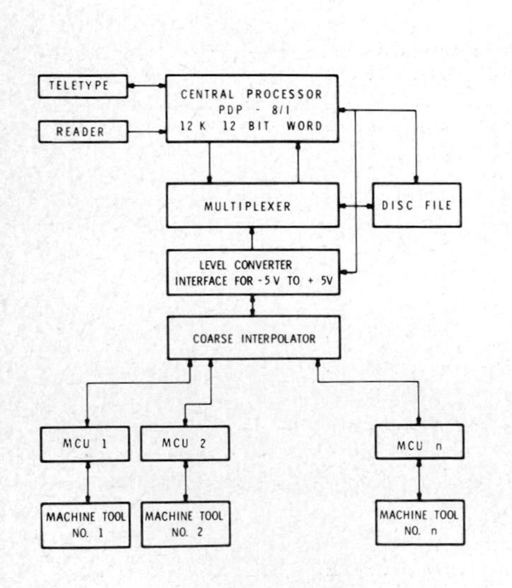

Fig. 7 MCU - System
Warner & Swasey

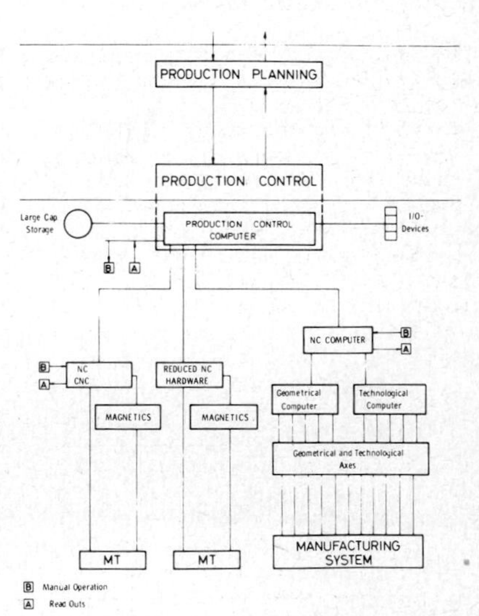

Fig. 9 Hierarchy of Information
Processing

keep here
just now

DIRECT NUMERICAL CONTROL SYSTEM T-10

KAZUTO TOGINO
Systems Science Division,
Mechanical Engineering Laboratory
Suginamiku, Tokyo, JAPAN

SEIUEMON INABA
Fujitsu FANUC Ltd.,
Chiyodaku, Tokyo, JAPAN

ABSTRACT: Each machine tool has a Machine Tool Controller (MTC) consisting of a servo control unit for driving pulse motors and an operator's control panel, and a Group NC Unit (GNCU) consisting of a CPU and a Time-Sharing and multiplexing NC (TSNC) can control different types of machine tools simultaneously. The DNC SYSTEM T-10 has the following functions in on-line mode, preparation of NC commands with a conversational FAPT, preparation of daily scheduling, execution of NC machining, and integration of results of the maching.

1. INTRODUCTION

Recently, different types of NC machine tools are being installed in descrete manufacturing plants. Among many items which these plants require, there two key items to which NC system suppliers should direct their attention. One is to operate those high priced NC machine tools more effectively so as to being not hungry machine tools. Another is to offer NC systems at low price. In order to meet the former, it is desirable for the NC system to have two functions in real time at least in addition to functions of the conventional NC system. The first is to prepare and maintain NC commands; the second is to prepare the daily schedule from a weekly production planning data given by a higher level computer, and to gather and integrate results of the machining by NC and report them to the higher level computer. The purpose of this report is to describe the outline of the SYSTEM T-10 fulfilling the above requirements.

2. SYSTEM ORGANIZATION

The design of a NC system requires a consideration on trade-offs. The trade-offs in SYSTEM T-10 are between hardware and software. The basic design objectives can be stated as:

(1) Entrust many functions to the software as much as possible.
(2) Minimize the necessary hardware at the machine shop.
(3) Leave room for growth by adopting modular features.

As shown in figure 1, each machine tool has a MTC consisting of a servo control circuit for driving pulse motors and an operator's control panel, and a GNCU consisting of a CPU and a TSNC controls different types of machine tools simultaneously. Depending on the scale of machine shop and/or the control functions, SYSTEM T-0 and SYSTEM T-10 are optional. SYSTEM T-0 is for a small scale and SYSTEM T-10 is for a large scale. In this paper, SYSTEM T-0 is not described. The functions of SYSTEM T-10 are given in Table 1, and are roughly divided into the production control, the NC machining, and the automatic programming.

Table 1. Outline of functions of SYSTEM T-10

Function	
1 NC machining	Economic machine tools' number
(1) 3 axes (2C,L)	more than 6
(2) 2 axes (2C)	more than 17
(3) 3 axes (1C)	more than 34
2 Preparation of NC command	Conversational FAPT
3 Production control	(1) Daily scheduling for machining (2) Daily rescheduling for machining (3) Monitoring of operation status (4) Integration of results

COMPUTER LANGUAGES FOR NUMERICAL CONTROL, *J. Hatvany, editor*
North-Holland Publishing Company - Amsterdam–London

The GNCU is installed at the control center. NC commands, corresponding to NC commands on the NC tape of the conventional NC system, are stored in disc memories. The NC commands at the scheduled date are transferred from the disc memories to the drum memory (DM). One block of NC command is coarsely interpolated by the CPU (coarse interpolation) and transmitted to the TSNC for the pulse distribution by DDA (fine interpolation). By this coarse interpolation, the capacity of pulse distributing circuits can be reduced. NC command pulse trains are transmitted through lines from the TSNC to the MTC. The length of the line might be about one kilometer. Table 2 gives a specification of hardware GNCU. In table 2, mL means m axes simultaneous contouring by the linear interpolation and 2CL means 2 axes simultaneous contouring by the linear interpolation or the circular interpolation.

Item		
CPU:	Main memory	Core: 32 KW (16 bits/word)
	Cycle time	1.5 micro seconds
	mode of operations	16 bits, parallel
	Add time	3 micro seconds
	No. of instructions	80
	Interruption level	7
Drum memory (NC commands)		(256-512) x 1000 words Access time: 10 milli seconds
I/O:	Teletypewriter	20 characters/second
	Tape reader	200 "
	Line printer	120 lines/minute
Disc memory		2500 x 1000 words Access time: 88 milli seconds
TSNC:	NC modes	1P, 1L, 2L, 2CL, 3L, 4L, 5L
	Interface	Operator's control panel

Table 2. Specification of hardware for GNCU

The MTC consists of a servo control unit for driving pulse motors and an operater's panel. The operater's panel is divided into three sub-panels; Display Panel, Data Input Panel, and Function Panel.

The display panel displays various kinds of message from the GNCU to the machine operator such as part number, quantity machined, quantity unmachined, material number, tool number, tool position compensation(cutter offset). When the operator made inquiries on the next operation from the data input panel and/or the function panel to the computer of GNCU, the answers are displayed on the display panel. The data input panel and functional panel combined with the display panel is for the machine operator to convey his intention to the GNCU. Fig.2 shows an example of these panels. In the display panel, OT lamp is for overtravel of the machine and the button "DISPLAY" is for requesting to display. The data to display is selected by dials "LOT DATA", "MACHINE NO." and "ADDRESS" in the data input panel. When the operator sets the "MACHINE NO." and "ADDRESS" dials and thumb switches "DATA INPUT" and pushes the button "SET" in the dial input panel, the data are read in the memory.

The machining progresses automatically from parts to parts and lot to lot in accordance with the daily schedule. "AUTO" of the "MODE (I)" on the function panel is in the above automatic progressing mode. "SINGLE PROGRAM" is for interruption of NC machining in the schedule for a while after the finish of the machining of the part and the interruption is released by pushing the button "CYCLE START". Fig. 2 is only a typical example. Arrangements of buttons, dials, and switches on the panels and their functions are at the option to the user.

3. SOFTWARE

As shown in Fig.3, the following main 5 subprograms are monitored by monitor ROSP (Real time Operating System for Production);

(1) MT (machine tool) control program gearing with TSNC generates NC command pulse trains from a NC command. In this function, necessary processings such as coarse interpolation, cutter compensation, feedrate override, and tool offset are included. That is, most of functions performed by the hardware in the conventional NC system are by the software.

(2) Sequence control program selects a NC command on schedule and starts NC machining with the assistance of MT control program and TSNC. Machining results by NC machine tools are stored in the results file by the sequence control program.

(3) Scheduling program generates an optimal schedule each machine tool from the production plan for a week given by the higher level computer. Items considered in the optimization are appointed date of delivery, priority, and others. The results are stored in the daily schedule file and printed out as a schedule table, a materials table, a table of commands to the operator, and others if necessary.

(4) Results integration program processes machining results being stored in the results file and prints out in an appropriate format as tables.

(5) Conversational FAPT program is used for the preparation of NC command, and the modification and editing of NC command on line.

The production planning data for a week, in terms of part number, list of machine tools which is fit for machining, quantity, appointed date of delivery, and priority having 10 levels are the input to DNC SYSTEM T-10. These data for a week are stored in a production planning file. The scheduling program has two phases; a scheduling phase and a loading phase. The scheduling is performed referring to NC command file, machine tool file, and materials file. The NC command file stores part number, machine tool numbers which can process the part, machining time, material number, and NC command corresponding to NC tape. The machine tool file stores machine tool number, starting time and end time for machining the part concerned, and machine load in daily base. The materials file stores material number and total stock.

In the scheduling phase, the part left unfinished at the previous day is firstly machined, and the part in the production planning file for a week is successively picked up. The part among parts loaded the date concerned is selected by the following rules; a schedule so as to save the time required for tooling, the earlier appointed date of delivery, and the higher level of priority.

In the loading phase, the machine tool and the date to be machined is selected by a following rule. The program firstly checks whether NC command of the part is in the NC command file and the material of the part is in the materials file, and if either is not in the files the part is postponed to the next week. The part is loaded to the machine tool of the lightest load and loading of the part is postponed to the next date if no machine tools have a margin to load. Thus, in the loading phase the parts are loaded so as to give reasonably uniform machine tool load.

The output of the scheduling program is stored in the Daily Schedule Master File, and the Unfinished Parts File. Lists of daily machine program, daily materials, daily commands to the operator and others are extracted from these master files and printed out, if necessary.

SYSTEM T-10 also has rescheduling functions such as (1) two kinds of squeezing in the schedule, (2) two kinds of deletion of a certain lot, (3) exchange of the lot to be machined at the next period for another lot unmachined, and (4) a change of quantity in a lot. Results of the machining by DNC are stored in the results file and fed back to the higher level's decision maker after an appropriate processing.

4. CONVERSATIONAL FAPT

A part programmer can efficiently make a part program with the conversational FAPT languages in Table 3, being assisted by the FAPT system implemented in CPU of DNC system. The part programmer can correct his error identified by error messages from 01 to 22. An example is shown in Fig. 4. Features of the conversational FAPT system are as follows;

1. Modification and deletion of statements.
2. Insertion of statements at any place of the part program.
3. Execution of statements from the statement having identifier m to the statement having identifier n.
4. Interruption of part programming and continuing it on the following day.
5. Printing out of coordinates of points, straight lines, and circles defined at any time, if necessary.
6. Printing out of statements list at any time, if necessary.

CONVERSATIONAL CONTROL STATEMENTS	
1. $COMP	Start of a new part program. End of the part program is effective with key-in of "PEND" and print out "END OF COMP" and "$"
2. $COMP*	Addition of a part program after the part program stored.
3. $EXEC, m, n	Execution from statement No. m to No.n
4. $EXEC*	Execution from the statement immediately after the statement already executed
5. $<u>FAPT Statement</u>	Immediate execution of a <u>statement</u>
6. $n, <u>FAPT Statement</u>	Immediate execution and store of a <u>statement</u>
7. $n	Deletion of No. n statement
8. $KILL, Vn $KILL, Pn $KILL, Sn $KILL,Cn	Deletion of variable Vn, point Pn, straight line Sn, and circle Cn.
9. $KILL, ALL	Deletion of all variables, points, straight lines, and circles
10. $JOB	Start of a new part program independent of the part program already compiled
11. $PAUSE	Temporary interruption of part programming and lock of key
12. $PRINT, m, n	Print out from point Pm to point Pn
13. $LIST, ON $LIST, OFF	Print out of the process sheet
14. $PUNCH, ON $PUNCH, OFF	Punch out of NC tape

15.	$DPLY, m, n	Print out from statement No. m to statement No. n
16.	$DPLY, Vn $DPLY, Pn $DPLY, Sn $DPLY, Cn	Print out of Vn, Pn, Sn, and Cn
17.	$* Comments	Write a comment without any influence to CPU
18.	$SYIN, KEY $SYIN, PT	Selection of input media, key or Perforated Tape
19.	$CHECK, n	Indication of check list
20.	$PLOT, ON $PLOT, OFF	Plotting on XY plotter
21.	$SCALE, n/m	Scale factor of plotter
22.	$NC, E $NC, F $NC, G $NC, H	Selection of minimum incremental value of NC system

FAPT STATEMENTS

POINT

1. Pi = x/y   2. Pi = r/θ, D (D means Degree in angle)
3. Pi = Pj   4. Pi = Pj, α (Symmetric point, α = X, Y, O)
5. Pi = Pj/x/y (Point displaced (x,y) from Pj)
6. Pi = Pj/Pk/θ , D (Point rotated Pk by θ , Pj is a center of rotation)
7. Pi = Pj/Sk (Line symmetry)   8. Pi = Sj/Sk (Intersection)
9. Pi = Sj/Ck, α (Intersection, α = R, L, A, B)
10. Pi = Cj/Ck, α (Intersection, α = R, L, A, B)
11. Pi = Pj/Ck, α (Tangent point, tangent between Ck and line passing Pj)
12. Pi = Cj (Circle center)   13. Pi = Cj/θ , D (Point on Cj in polar coordinates)

STRAIGHT LINE

1. Si = Pj/Pk   2. Si = Sj
3. Si = Pj/Pk (Vertical line dividing equally line segment Pj Pk)
4. Si = Sj/d, α (Distance d between Si and Sj)
5. Si = Pj/Sk, α (α : P (parallel), N (normal))
6. Si = Pj/Ck, α (Line passing Pj, tangent to Ck, α = R, L, A, B)
7. Si = Pj/θ , D (Pj and the angle θ made with the +X axis)
8. Si = Pj/Sk/θ , D (Pj and the angle θ made with Sk)

CIRCLE

1. Ci = Pj/r 2. Ci = Pj/Pk (center Pj)
3. Ci = Pj/Ck (Center Pj, radius of Ck's)
4. Ci = Pj/Ck, α (Center Pj, tangent to Ck, α = L, S)
5. Ci = Pj/Sk (Center Pj, tangent to Sk)
6. Ci = Pj/Pk/Pl
7. Ci = Sj/Sk/r, α , β (Radius r , tangent to Sj, α and Sk, β)

CONTROL STATEMENTS

1. Jn, j	Jump to subroutine starting with statement No. n and return to this statement by "RETURN, j" of subroutine
2. RTN, j	A subroutine consists of Jn, i and RTNj
3. Jn, Pj/Sk, α	If Pj is on α side of Sk, jump to statement No. n, if not, progress to the next statement (α = R, L, A, B)
4. Jn, a/b, α	α = L : if a > b, jump to No. n (α = E : a = b, α = S : a < b, respectively)
5. Tn, m	Replacement statement No. n by statement No. m

ARITHMETIC OPERATIONS

1. $Vi = a \begin{Bmatrix} + \\ - \\ * \\ / \end{Bmatrix} b \begin{Bmatrix} + \\ - \\ * \\ / \end{Bmatrix} c$	Operations are successively from left to right
2. Vi = Pj, α	X coordinates or Y coordinates of Pj (α = X, Y)
3. Vi = Cj, α	X - or Y - coordinates of center, or radius of Cj (α = X, Y, R)
4. Vi = Pj/Pk	Distance between Pj and Pk
5. Vi = Pj/Sk	Distance from Pj to Sk

MOTION STATEMENTS

1. Pi/Pj	Straight motion from Pi to Pj
2. Pi/Cj/Pk	Circular motion selecting shorter arc)
3. Pi/Cj/Pk, α	Selection of righthand or lefthand rotation (α = R, L)
4. DELTA, x/y	
5. Pi/Sj	Preparation of motion on Sj from Pi, this statement must be succeeded by /Pj; /Sj, α; /Cj, α
6. Pi/Cj	About the same as the above
7. /Pj	
8. /Sj, α	
9. /Cj, α	

SPECIAL STATEMENTS

1. READ, Vj	Define Vj by data read from key or tape
2. READ, Pj	Define Pj by data read from key or tape
3. READ, Pj, D	Define Pj by data in polar coordinate read from key or tape
4. READ, Ij	Read characters from key or tape, and punch them on NC tape
5. ⓐ ～ ⓐ	Characters between ⓐ and ⓐ are punched on NC tape
6. FINI	Finish of $EXEC, m, n and $EXEC*
7. PEND	Finish of $COMP and $COMP*
8. * comments	Write a comment without any influence to memory and CPU

POST PROCESSOR STATEMENTS

1. DWELL, t 2. FROM, x/y 3. Ni 4. N, OFF

5. Gi 6. Fi 7. Si 8. Ti 9. Mi

Table 3. Conversational FAPT Statements

```
$JOB
$NC,E
$COMP
0010 P0=0/0
0020 P1=102 [/] CANCEL
0020 P1=103/0
0030 P2=75/16,P
0040 P3=72/16,P
0050 P4=65/20,P
0060 P5=60/20,P
0070 P6=56/24,P
0080 P7=30/56,P
0090 P8=30/60,P
0100 P9=25/60,P
0110 P10-0/56,P
ERROR 01
0110 P10=0/56,P
0120 S0=P0/P10
0130 S1=P7/S0,N
0140 C0=P7/10
0150 P11=S1/C0,B
0160 C1=P11/10
0170 C2=P11/35
0180 C3=P6/25
0190 P12=C2/C3,A
0200 C2=P12/25
0210 S2=S0/4,R
0220 C3=P10/14
0230 P13=S2/C3,A
0240 C3=P14/14
0250 P14=120/50
0260 PEND
END OF COMP
```

```
$EXEC
          P0000   000000.0000   000000.0000
          P0001   000103.0000   000000.0000
          P0002   000075.0000   000008.0000
          P0003   000072.0000   000008.0000
          P0004   000065.0000   000010.0000
          P0005   000060.0000   000010.0000
          P0006   000056.0000   000012.0000
          P0007   000030.0000   000028.0000
          P0008   000030.0000   000030.0000
          P0009   000025.0000   000030.0000
          P0010   000000.0000   000028.0000
0150 ERROR 09
$130,S1=P7/S0,P
$EXEC,130
          P0011   000030.0000   000018.0000
          P0012   000059.5566   000036.7457
          P0013   000004.0000   000041.4164
0240 ERROR 06
$240, C3=P13/14
$EXEC,240
          P0014   000120.0000   000050.0000
HALT OF EXEC
$COMP*
0260 P15=103/20,P
0270 P16=105/13,P
0280 PEND
END OF COMP
```

Table 4. Example

5. CONCLUSION

The SYSTEM T is a new type of DNC system which consists of CPU, namely minicomputer, TSNC specially designed, and MTC. SYSTEM T-10 is capable of excuting preparation of the part program in on-line mode, preparation of the daily schedule, and integration of results of machining, in addition to SYSTE T-0.

Two system T-10 have already installed; one is at Tokyo plant of Fujitsu FANUC Ltd., and another is at Saitama plant of Honda Koki. At the former plant 170 kinds of pulse motors are machined by 21 machines, at the latter plant prototype parts for Honda Motors Co. are machined by 6 machines. At present time, more several systems are expected to be installed in two or three years. On the other hand, three System Softwares and functions such as G and M are entirely different at every system. Making effective use of our experiences which are piled up by conversations with operators and supervisors of manufacturing, SYSTEM T have been refined. Since refinements are based on modular features, fabrications and testing of SYSTEM T have been automated day by day. We would like to conclude the paper by noting that an inexpensive, reliable, and serviceable DNC system come into the world in close cooperation with operators and supervisors of manufacturing.

Acknowledgement The SYSTEM T is not designed by us, but is the work of a dedicated team. The authors wish to thank them for their unstinted contribution and also government for his financial support.

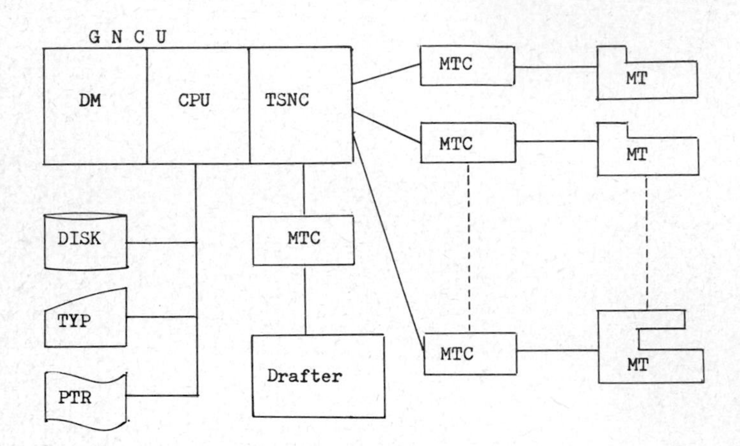

Fig. 1. A schematic diagram of SYSTEM T-10

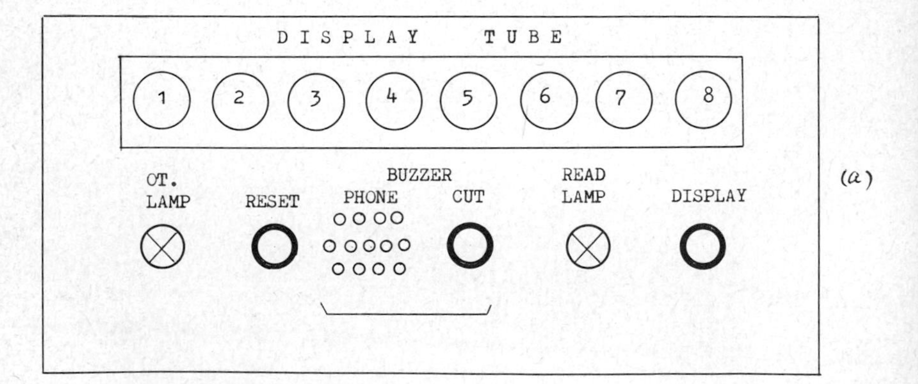

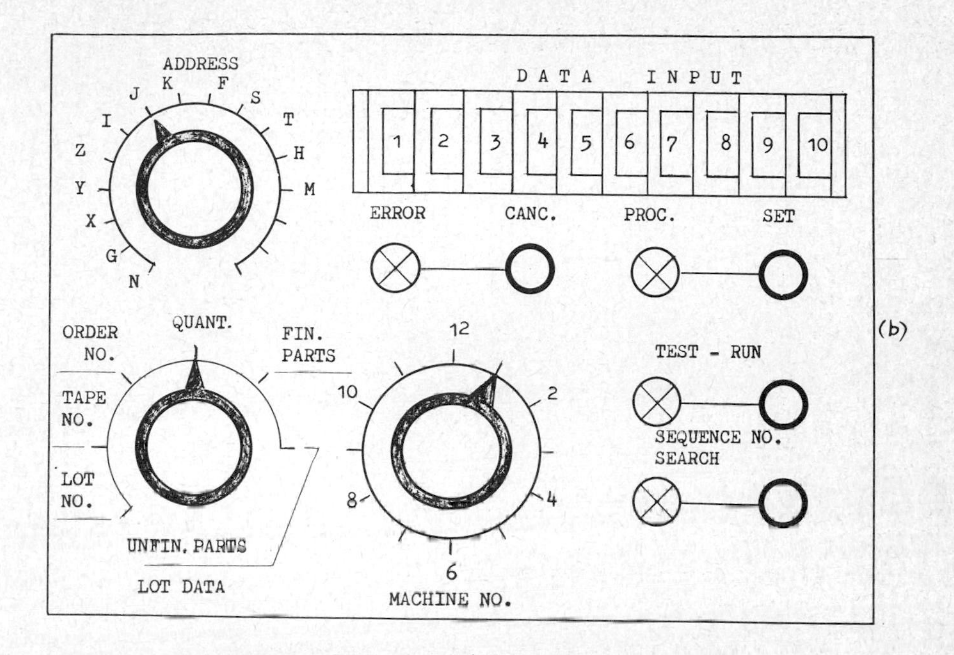

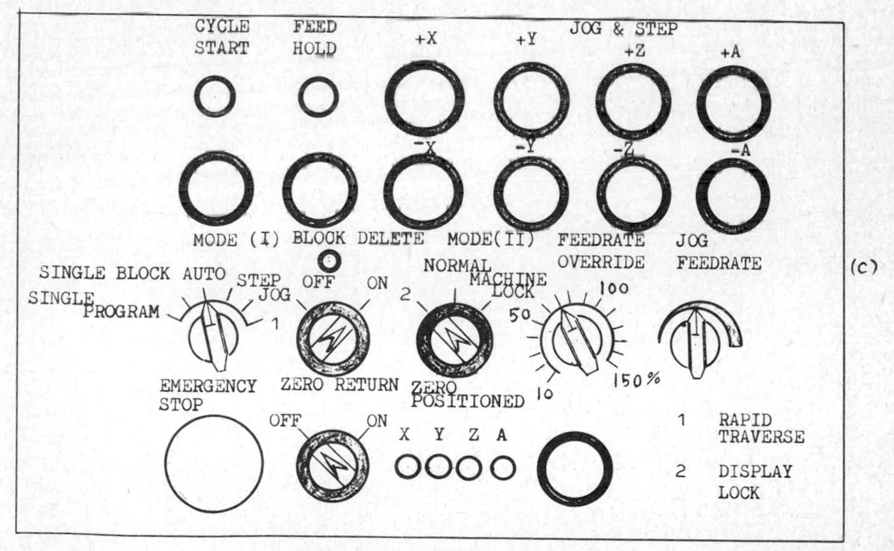

Fig. 2. Operator's panel (a) Display panel (b) Data input panel (c) Function panel

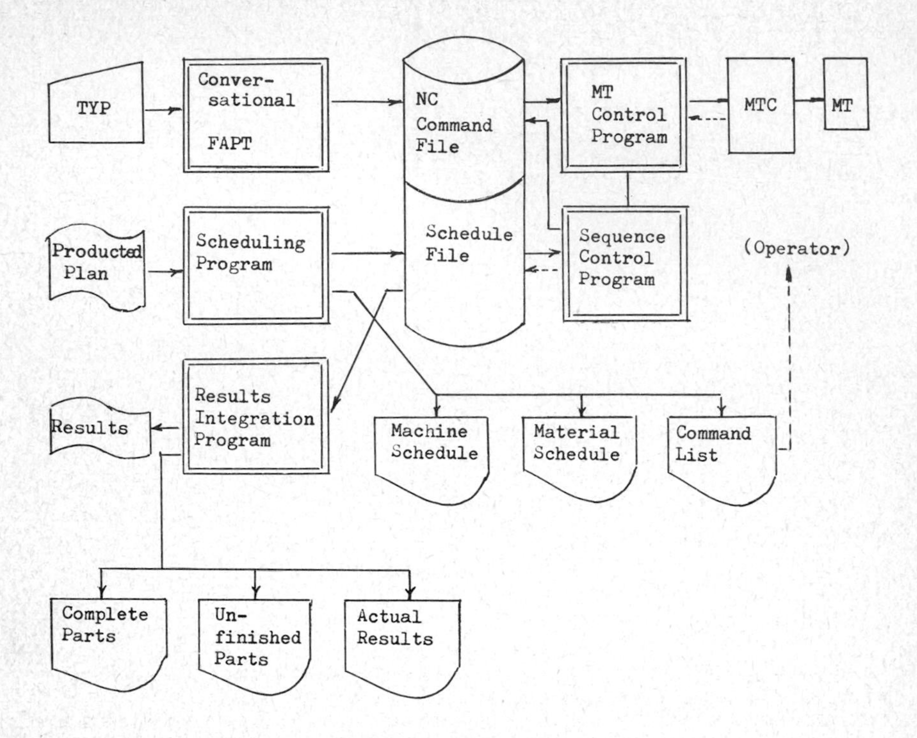

Fig. 3. Software organization

AN APPROACH TO DIRECT NUMERICAL CONTROL

ØYVIND BJØRKE, Professor
and
ASBJØRN ROLSTADÅS, Research Engineer
University of Trondheim
The Norwegian Institute of Technology
Department of Production Engineering and Machine Tools
Trondheim, Norway

Abstract: A DNC-system for tape preparation and control of machine tools is discussed. At the lowest level a number of CNC machine tools (a minicomputer dedicated to each machine tool) receive their EIA data from a central DNC-computer located at the planning office. The same computer is used for partprogramming purposes and it serves as a terminal to an external computer. The partprogramming aids available on the DNC-computer are: read-in of partprograms, on-line arithmetic and cutting data calculations, machine tool monitoring and printing of reports, partprogram verification and rescheduling of jobs. The connection to the larger computer is used to expand the partprogramming aids to the APT-system and to technological planning systems.

1. INTRODUCTION

The accepted control philosophy in the sixties was the use of fixed wired numerical control systems. Though the technology significantly changed during that decade, the system behaviour did not change much. The operational capacity of fixed wired control systems did more or less stabilize on a certain level of sophistication, and the features available to-day are a result of what has been found useful during the past decade. On the other hand, it must not be forgotten that the range of features has been restricted by the cost of implementation on fixed wired systems.

The control philosophy in the seventies will be the use of on-line control systems, where the controller is based on a general purpose digital computer. The basic difference between fixed wired systems and on-line systems is the difference in flexibility. In order to decide upon the most appropriate system configuration for on-line control of machine tools, it is important to analyse the computing to be done, that is, the access times required and the data volumes to be transferred.

Another factor influencing the system configuration is the trends in the development of computer systems. This trend shows an increasing use of decentralized systems consisting of a number of interconnected computers dedicated to special jobs. It is characteristic for such systems that each computer has its own executive system and its own job queue. A decentralized system may therefore be thought of as a federation of self-controlled subsystems. This structure is suited in cases where the jobs in the queues occur regularily at predictable intervals.

In fig. 1 the relative data volumes are compared, and as can be seen, the job queue on a machine tool control system is well predictable, and the main load occurs regularly at a high frequency. The figure also shows that in a centralized system the control signals

COMPUTER LANGUAGES FOR NUMERICAL CONTROL, J. Hatvany, editor
North-Holland Publishing Company - Amsterdam–London

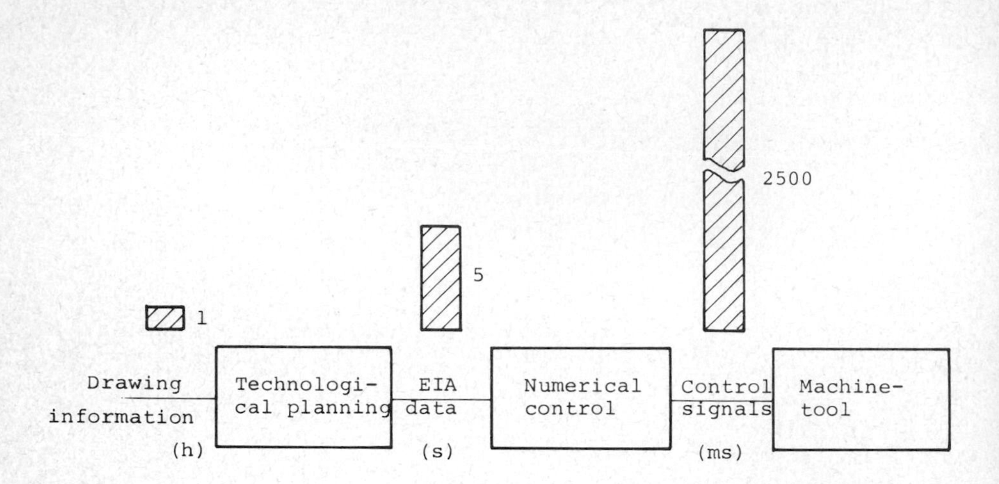

Fig. 1. Typical access-times and typical relative data volumes from drawing to control signals.

to be transferred are in order of magnitude 2500 times that of the drawing information. In a decentralized system having a dedicated computer on each machine tool, the EIA-data have to be transferred, and this constitutes only five times the information compared with that of the drawing. Further, the figure shows the order of magnitude of how frequent data have to be transferred at the different stages in the process. The time intervals between new parts, that is, the time between the generation of new partprograms, can be measured in hours. The interval between the loading of two EIA-blocks can be measured in seconds, while the control signal frequency is about three decades higher.

From the analysis above, we may see that the job profile of a numerical control system is of the type best handled in a decentralized system, that is, to use a dedicated computer at each machine tool. This outline is also preferable from a security point of view. In the centralized system all machine tools will stand still in the case of a computer break-down, while in a dedicated computer system, only the machine tool involved will stand still.

Manufacturing systems are basically hierarchical, that is, numerical control systems get their control information from technological planning systems, which again get their commands from a mangement control system. This hierarchical structure has to be reflected by the computer system to be used. In decentralized computer systems the hierarchy is made by coupling together the computers in a hardware hierarchy, and the design of such a system is described in the following.

2. THE CONCEPT OF DIRECT NUMERICAL CONTROL

2.1 General

In this text we will define a direct numerical control system as a central computer system capable of controlling several CNC-systems. The overall concept is a hierachical system where the DNC is "supervising" the CNC manufacturing units as indicated in fig. 2.

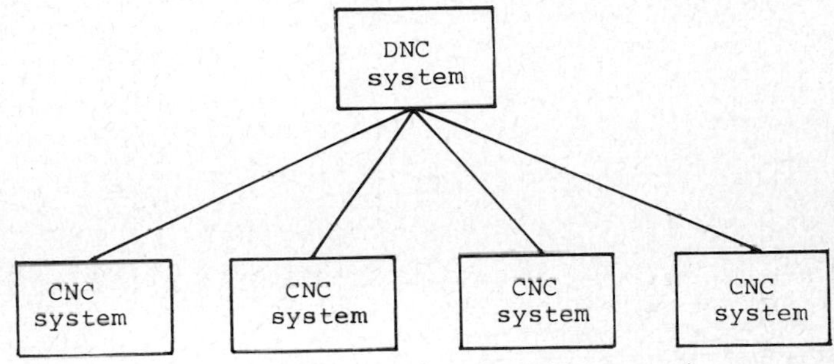

Fig. 2. Subsystems in the DNC concept.

The basic CNC-system can be reviewed as a stand-alone controlling system. The hardware units are indicated in fig. 3. A paper tape reader is used for input of symbolic (EIA-) programs. From the operator's panel this program can be corrected on-line. At request the updated program may be punched out on paper tape (for later use). Both the paper tape reader and punch should be plug-in units, normally leaving the responsibility of providing EIA-type data to the DNC-computer.

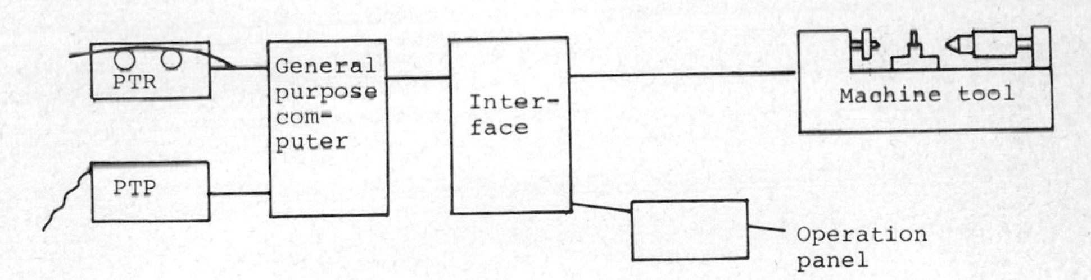

Fig. 3. Hardware units in an existing CNC-system.

The plug-in feature allows for the use of the CNC as a stand-alone system with on-line correction possibilities for emergency use.

In the procedure of supervising the CNC-systems, the DNC-system will have to fulfill the following tasks

1. Feed the CNC units with EIA data (behind the tape reader function).
2. Provide the foreman with the possibilities of selecting new jobs for the different machines.

For this purpose a hardware configuration as shown in fig. 4 is needed. The disk is used for storing the different EIA-programs. Paper tape reader and punch are used for purposes similar to those of the CNC system. By means of the alphanumeric screen, system messages and directives are given. Also the selection of new jobs for the different machines is performed from this device.

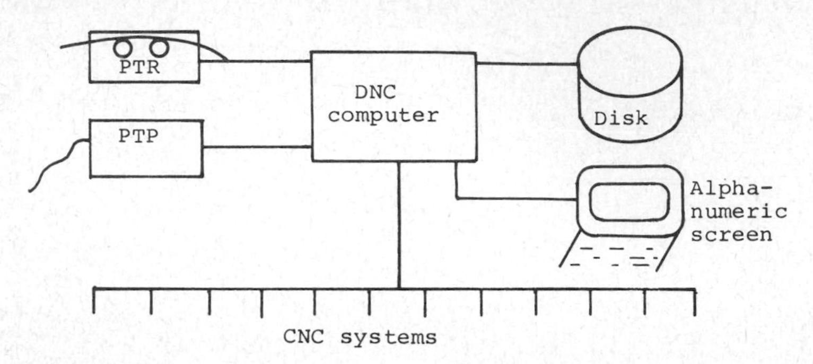

Fig. 4. Hardware configuration of a behind the tape reader DNC-system.

The behind tape reader unit shown in fig. 4 can naturally be extended to include a center for development of partprograms. At this center the following functions ought to be available:

1. Access to external computer
2. Verification of tapes
3. Computing of arithmetic expressions (desk calculator)
4. Handling and updating of programs
5. Assignment of a job queue for each machine group (group of fully interchangable machines)
6. Calculation of cutting data

The DNC concept as described in this paper will involve both the behind the tape reader function and the center for development of programs. The full hardware configuration is shown in fig. 5. There are three levels of equipment arrangement. On the shop floor the CNC's are placed together with a screen which the foreman can use for job assignment.

At the programming center the graphic screen is used for tape verification and updating. By means of the papertape reader and punch, programs can be read in and punched out. To provide for the facility of automatic programming (APT-runs) a card reader and line printer are hooked on. The APT-runs are transmitted to an external computer yielding the CL-file back to the DNC. The magnetic tape unit of the large computer will hold a backup of the disk-files at the DNC computer. For system control, desk calculations etc. a teletype is needed.

The choice between an external or internal computer for the DNC system is not difficult. The computer should be on-line available at both the workshop and the programming center. This leaves a dedicated minicomputer or a timeshared external computer as the only economical possible solutions. In order to decrease the probability of system break-down, the dedicated computer is chosen. It must be accepted that parts of the systems some times can breake down. However, it cannot at all be accepted that the whole system goes down.

Some possible serious failures will be:

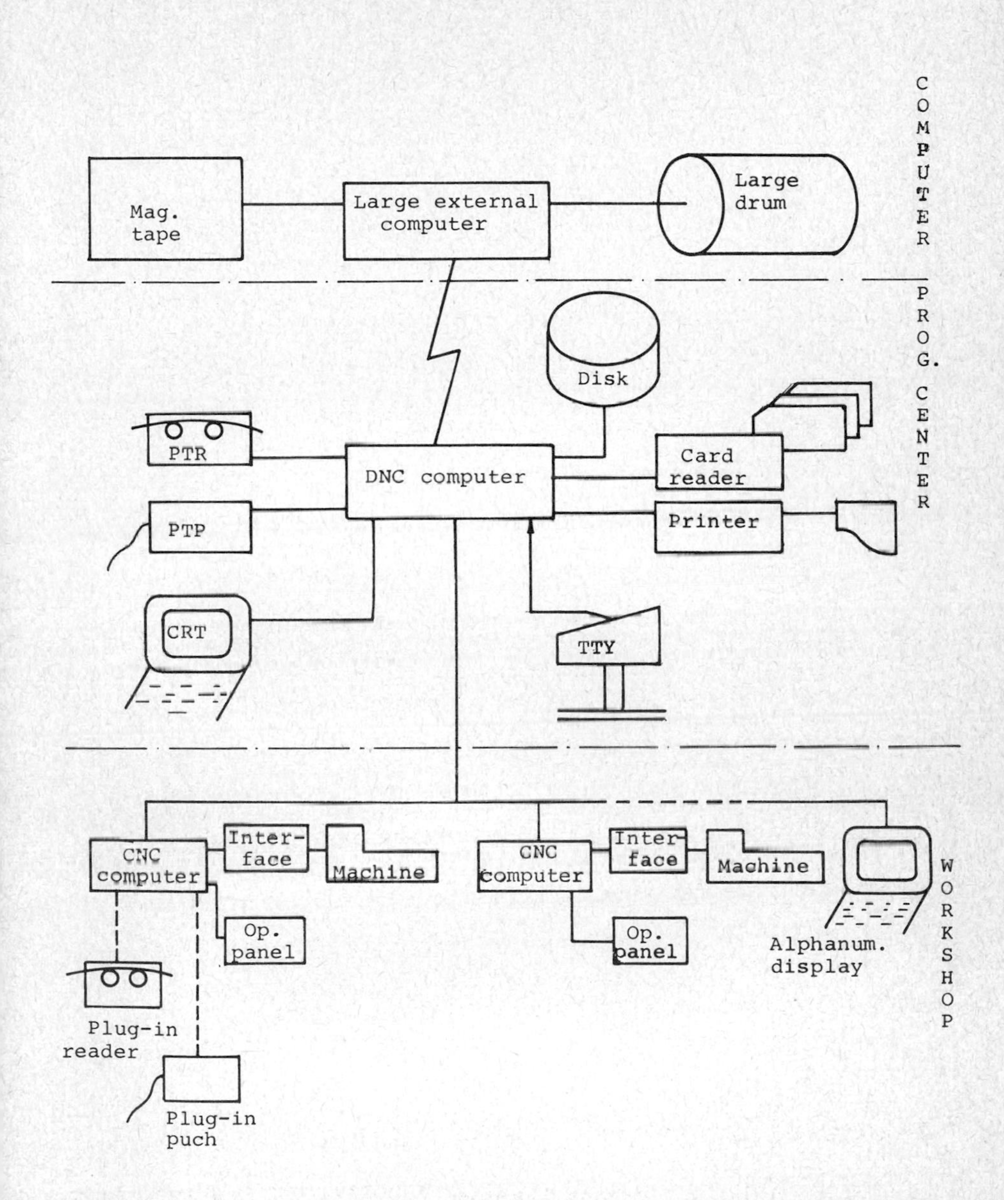

Fig. 5. Hardware configuration of a full DNC-system.

1. A single CNC loses connection with the DNC.
2. A single workshop terminal (display) loses connection with the DNC.
3. The disk breakes down.
4. The DNC loses connection with the external computer.

Of course many other failures may occur, but these are considered to be the most serious ones. In case 1 a paper tape reader is plugged in at the CNC which is then used as a stand-alone system. In case 2, some other terminal in the workshop or at the programming center may be used. In case 3, the magnetic tape unit and the large drum of the external computer (holding a backup of the disk-files) are used instead. In case 4, the use of the external partprogramming aids and the backup facility will have to be postponed.

It is worth noticing the modularity of the overall system, both in hardware and software. CNC units may initially be bought and used as stand-alone systems. Then a DNC-computer equipped with graphic screen, teletype, paper tape reader and punch and disk may be bought in order to obtain a tape verification system, which at first can be run separately. Next step will be to interface this system to an external computer for further partprogramming aids. Card reader and line printer will then be necessary. Finally, the behind the tape reader module can be hooked on. One or more alphanumeric screens will then be needed in the workshop.

Table 1 shows some extension steps.

Table 1.
Extension steps in a DNC system.

Step	Function	Equipment
1	CNC controllers	CNC-systems
2	Tape verification	DNC-computer paper tape reader paper tape punch disk graphic screen
3	External part-programming aids	card reader line printer connection to external computer
4	Behind the tape-reader system	alphanumeric display

3. FLOORSHOP MODULES

3.1 The CNC system

The CNC system is the basic control unit on the floorshop. As pointed out in section 2, it is reviewed to be of utmost importance that the CNC system can be handled as a stand-alone unit. For this reason corrections and functions aiding the operation of the controller should be taken care of by the CNC itself and not by the supervising DNC system.

In fig. 6 a CNC system fitting to this requirement is shown. The panel of the operator is shown in fig. 7. Some of the functions available from the operator's panel are mentioned in the succeeding.

Fig. 6. A Kongsberg CNC-system.

On-line corrections. Information on the tape can be corrected or deleted and new information can be added. A complete manual block can be loaded and executed. Tool compansations and zero-offsets are loaded in separate buffers.
Displaying of data. All the corrections loaded may be displayed. So may also the information in the present and next EIA block and the position of each of the axes, current feedrate, spindle speed, tool number etc.
AUTO/S.AUTO cycle. EIA-blocks can be executed continuously or with a full stop between each of them.
Manual jog. The incremental jog can operate one or more axes simultaneously.
Feed override. The reedrate programmed can be overridden in steps of 20 % of programmed feed in the range -100 % to 100 %.
Spindle override. The spindle speed programmed can be overridden one or two steps higher or lower in the ranges.

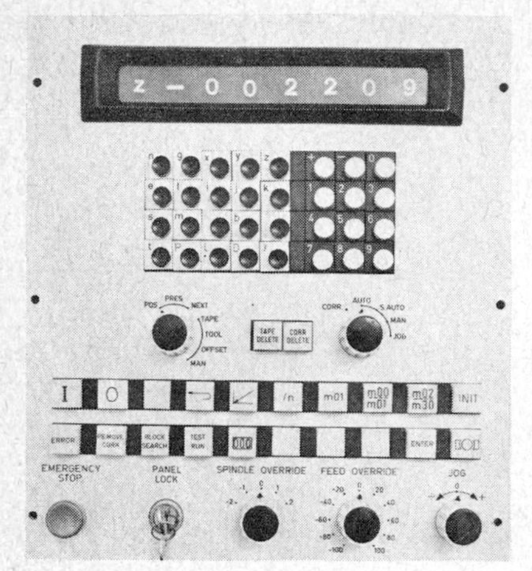

Fig. 7. Operator panel.

Block search. Any block can be searched for program entry.

Block start. The block executed can always be restarted. If required, corrections to that particular block can be entered.

Obtional block. If the /n-button is pressed, blocks staring with an / will be executed, else they will be neglected.

Test run. The program can be run through using rapid traverse instead of programmed feedrate for testing purposes. The programmer can select in which of the axes he wants motion.

Selection of program datum can be done by pressing the clear button.

Punching of corrected tape is possible when a punch is plugged in.

Some of the most important preparatory functions are:

Linear and circular interpolation. Linear interpolation is possible in six axes simultaneously, circular in any two.

Absolute and incremental programming can be selected.

Tool length compensation is calculated whenever a tool is used in the direction of the spindle.

Radius compensation can be turned on and off when desired.

Fixed cycles can be specified due to own requirements within certain limits.

Macro programming is possible with certain restrictions.

The CNC has a powerful diagnostic system reporting errors concerning

- machine tool
- operations on the panel
- information on the tape

An error number referring to an explanation list is given on the display. By pressing the error button, loading corrections and block starting, program execution can continue.

3.2 The floorshop terminal

From the floorshop terminal the foreman and the operators report start and stop of jobs and select new jobs for each machine tool.

At discrete intervals (for instance each week) a new job schedule is loaded at the programming center. This schedule contains a job queue for each machine group. The specifications following each job will be:

- identification
- machine tool required
- time estimated to produce one piece
- lot size
- program tape number
- priority

The two first jobs in the queue of each machine will have their EIA programs stored on the disk. The DNC system controls this and asks for the proper tape to be mounted in the tape reader at a suitable moment. When start on a particular job is reported on a particular machine, the latter is fed with the EIA data of that job.

For each operation that should be performed on the terminal, a transaction must be performed. Selection of transaction is made by typing a mnemonic. In order to make the operation of the terminal as easy as possible, the terminal will question for the data needed once a transaction has been selected.

The transactions available are:

Job queue of machine. The job queue of a particular machine may be displayed in priority sequence.

Job queue. The total job queue will be displayed in priority sequence regardless of which machine it is assigned on.

Start. A new job is started.

Stop. A job is finished (the program file will be deleted on the disk).

Change priority. The foreman can change the priority of a given job.

Delete job. A given job can be fully deleted by the foreman.

Insert job. A given job can be inserted by the foreman.

Correct. The job can be corrected with concern to machine tool selected, lot size etc.

Split job. A job may be splitted up in several jobs, (i.e. the lot size is splitted calling for the need of running one specific EIA-program on several machines).

Join jobs. Two jobs may be joined together, making one new job.

Job status. The foreman should be able to read the status of a given job, i.e. how many pieces have been made, job active/passive etc.

4. PROGRAMMING CENTER MODULES

4.1 The terminal module

The connection between the large external computer and the DNC-computer is taken care of by the terminal module in the DNC-computer. This terminal module is a simulated batch terminal. The large external computer therefore recognizes the terminal as one of the conventional batch terminals hooked on it.

The use of this connection is mainly to:

1. Take back-up of system and data from the DNC-computer.
2. Execute large programs used during tape preparation.

The DNC-computer is programmed to transfer the disk contents to a magnetic tape at regular intervals. This back-up prevents the partprogramming data on the disk to get lost due to failures.

The DNC-computer is also equipped with a list of the partprogramming aids available on the external computer. As soon as processor cards calling for one of these aids are received from the card reader, the DNC-system transmits this run into the batch queue on the external computer. When the external computer has finished a run, control signals are transmitted to the DNC-computer which initiates the transfer of listing results to the line printer, and papertape-images to the disk.

4.2 The tape verification module

Papertape images may arrive to the disk of the DNC-system either from the external computer as described above, or by reading in manually made partprograms locally on the DNC-system. The result in both cases is a papertape image stored on the disk. Before these papertape images are punched out and put into the papertape storage, they may be verified using the tape verification module.

The papertape verification module is basically a copy of the CNC-module. As the papertape is read by the tape verification module, all calculations normally made by the CNC-system are performed. That is, any tape error which later on would have been recognized by the machine tool control system, will now be recognized and corrected This includes not only formal tape errors since the actual machine tool output is sent to a machine tool simulator and to a CRT-device respectively. The machine tool simulator is a logic device responding on M-function signals just as the machine tool will do, and on the CRT-device the path-output is drawn and checked.

After this verification has been performed, the corrected tape is punched out and stored until it should be used. The only possible errors on these verified tapes are of technological character. These errors, if they occur, have to be corrected from the operating panel on the machine tool itself.

4.3 The desk calculation module

During manual programming the partprogrammer often needs an "advanced slide rule". This function is taken care of in the DNC-system by the desk calculation module. The module is capable of performing the simple arithmetic operations as addition, subtraction, multiplication and division. In addition to this can the elementary functions, SQRT, EXP, LOG, ABS, SIN, COS, TAN, ATAN be computed.

An example of the use of the module may clarify the language:

```
A = 3.25
B = 17.5
C = 5.0
D = A * C + B - 10 = ?   [23.75]
B = D - A - 0.25
B/2 = ?   [10.0]
```

All letters from A - Z may be used as variable names, and they are assigned a value using the equal sign. The result will be printed out by writing = ?. In the example system output (the results) are shown marked with rectangles.

4.4 The file handling module

As described above, the DNC-system has to work with a large number of different partprograms. Each of these partprograms is contained in a file. It is therefore important to be able to handle a large number of files simultaneously, and this function is maintained by the file handling module. The functions covered by the file handling module are the following:

1. ASSIGN
2. UPDATE
3. RETRIEVE
4. DELETE
5. PUNCH

To distinguish between different files, the tape identification information is used (that is, the information on the PARTNO card).

4.5 The job dispatcher module

The job dispatcher module takes care of the job queue for each machine. At discrete intervals the planned job queue is loaded at the programming center and the sequence is recognized by the job dispatcher. From the floorshop terminal the foreman may change this given sequence due to machine-tool break-down, lack of raw material, lack of tooling and so on. The functions handled by the job dispatcher module will therefore have to be:

1. Read planned sequence
2. Insert jobs
3. Delete jobs
4. Split jobs
5. Join jobs

4.6 The cutting data module

A separate software module can be appended for the purpose of aiding the partprogrammer with the calculation of feasible cutting data. This requires an optimization program and a database containing background data concerning machines, tools and materials.

Prior to any calculation the operator must select

- machine tool (MT)
- material (MA)
- tool (TL)

Then calculation of cutting data can be performed for several sets of data. The machine tool, material and tool specified are valid until any of them are reselected.

In addition to this, the type of operation has to be specified, for instance by typing one of the abbreviations

B - boring
T - turning
M - milling

Further data which are needed, the system will ask for.

An example will clarify the use of the cutting data module:

```
MT B1439
TL T1234
MA SIS1690
T
[DO =]  132.4
[D1 =]  100.0
[L =]    50.0
[LA =]   84.5
[N = 840 RPM S = 0.34 MM/REV 4 PASSES]
TL T1245
T
[DO =]  -------
 .
 .
 .
 .
```

First the operator selects machine tool B1439, tool T1234 and material SIS1690. He is then ready for calculation of cutting data and types T indicating that he is working on an turning operation. The system then asks for the following data specifying the operation (system output is marked with rectangles):

D0 - blank diameter (mm)
D1 - workpiece diameter (mm)
L - cutting length (mm)
LA - workpiece length (mm)

Once the proper answers have been typed in, the system yields the optimal data for spindle speed (N), feedrate (S) and the number of passes.

The operator then selects a new tool, T1245, and starts the calculation of cutting data for another turning operation.

MODULAR SOFTWARE FOR COMPUTER NUMERICAL CONTROL

J W BRUCE and W STOCKDALE
National Engineering Laboratory
East Kilbride, Glasgow, United Kingdom

Abstract: A computer numerical control system is appropriate as a first venture into numerical control for the small user. In a system developed by the National Engineering Laboratory, the modular principle of subroutine elements has been extended by coupling the elements through a special sequencing routine. The routine is a flag-driven operation in conjunction with word-address tables that point at service routines appropriate to the system.

1 INTRODUCTION

A small, general purpose digital computer and a suitable software control program can now economically replace the logic of the hardwired NC controller.

Economy is not all, however, as it can be shown that computer numerical control (CNC) has some other equally important advantages over the hitherto conventional controller. Prominent among these are versatility and adaptability and it would seem that CNC is a significant step towards the universal machine controller.

In the National Engineering Laboratory (NEL) system the modular principle of subroutine elements has been extended by coupling the elements through a special sequencing routine. This is a flag-driven operation in conjunction with word-address tables that point at service routines appropriate to the system. The buffers associated with these service routines are protected by two sets of flags that operate in a manner based on Dijkstra's Semaphore Principle (Dijkstra 1968).

2 THE SEMAPHORE PRINCIPLE

A system to control tool position and ancillary functions requires the implementation of real time data acquisition and control techniques. It must also be capable of time-sharing processes which could occur during normal operation.

Time sharing of processes, in the software sense, may comprise separate sequential sub-programs, such as feedrate control and axes service routines, which may have to run simultaneously and asynchronously. This is the essence of the real-time problem.

It is necessary, therefore, to adopt a method of organizing, sharing, and protecting common data and any interrupted or suspended operations associated with them. One way of realizing this is to use the principles developed by Dijkstra to indicate, by a semaphore, when an operation may, or may not, be performed on relavant buffers. A binary flag that can take only the values 1 or 0, indicating 'active' or 'inactive', will accomplish this.

Dijkstra specifies two operations, or synchronizing primitives, that test the state of these flag bits. The first, called a P-operation, causes the process to stop if the bit indicates 'NO-GO'. The second, or V-operation, reactivates a process by setting the bit to 'GO', so releasing the buffer.

A process is thus represented as a buffer with a semaphore bit indicating whether an operation is permitted at that instant in time. Failure to obtain access to a buffer causes a process to be queued.

Processes are serviced by a sequencer which points the P-operation at each buffer in turn. This action is facilitated by a set of semaphores that indicate when a demand is outstanding. A demand flag V-operation is located in the relevant process or control routine.

COMPUTER LANGUAGES FOR NUMERICAL CONTROL, *J.Hatvany, editor*
North-Holland Publishing Company - Amsterdam–London

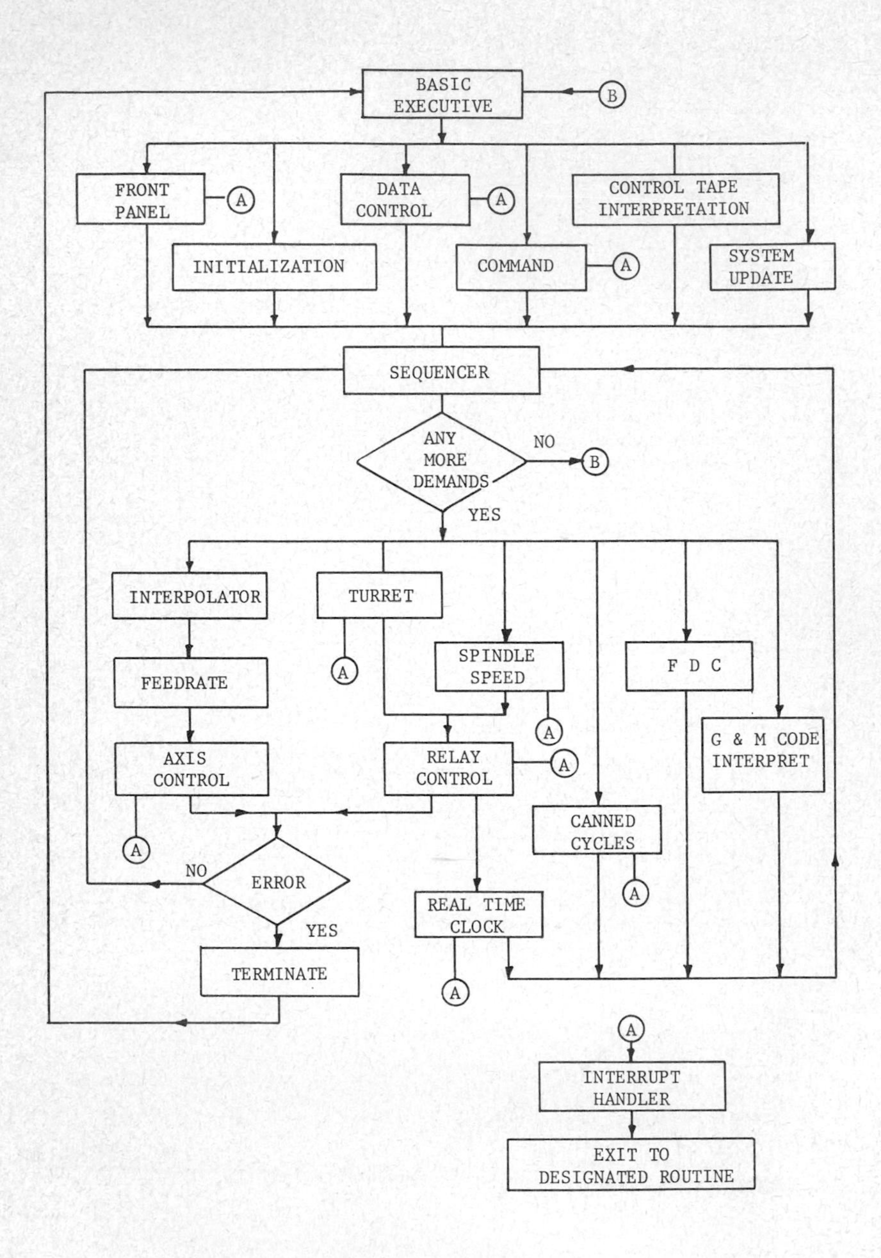

FIG. 1 SYSTEM STRUCTURE

For speed of operation - the prime consideration in computer numerical control - the semaphores and their respective buffers have been separated. Demand semaphores have been collected in one status word, which is rotated logically against a synchronous offset. This locks the sequencer to the appropriate buffer and the semaphore bit indicates whether or not an operation is possible. Buffer manipulation is thus reduced and, at the same time, modularity is simplified by the autonomy of the buffers.

Another important feature of the Dijkstra system is that processes contain 'critical sections' - 'critical' in the sense that at any moment only one process, or permitted combination of parallel processes, is allowed to enter a critical section. It is intended that forbidden combinations will cause the P-operation to be curtailed, thus placing the process back in the queue for sequencer action. A critical section in the present system is regarded as one which can cause a change in system status such as outputting an increment of axial motion, a change in feed or tool speed, or a change in relay status.

The NEL-CNC system uses three semaphores: (i) a mutual exclusion semaphore protecting critical sections in time and having the character of a GO flag; (ii) a demand semaphore specifying a given buffer as available or not; and (iii) a semaphore showing whether a process is active. The last is useful as an inhibit condition on prohibited combinations of processes.

3 GENERAL PROGRAM PHILOSOPHY

Program logic is organized in two broad classes of activity; control aspects as high priority tasks comprise the foreground, and the remainder is background. The Executive is a background processor with the basic decision rule, after all initialization has been completed, that the overall sequence of events shall be

UPDATE system status,
INTERPRET next control block, and
SEQUENCE computed control functions.

This order also resolves the problem of system initialization. A 'current' and 'next' block will thus be available enabling a measure of 'look-ahead' to be realized.

System function handlers and axis control comprise foreground operations and are readily configurable, by G code, as point-to-point or contouring systems with interpolation between specified axes.

In both cases the background operations will convert the control tape into appropriate control values and store these in the control buffers. The process of setting-up control indicators is therefore independent of the manner in which they will be interpreted. In this way it is possible to locate all the rules for control interpretation in the relevant control routine simplifying the task of modular construction.

Control is implemented, for the current block, by a secondary level of indicators, or flags, forming the 'demand' and 'inhibit' status words of the system. A sequencing operation may then have regard to current demands and attempt to satisfy them.

Inhibit or 'go' flags determine when a particular system function may be initiated and each control has masks that set up inhibit patterns appropriate to its own operations. At any time the system will operate in a state determined by the current inhibit status word. A measure of protection is thus available, which prevents undesirable conflict between specified system functions.

3.1 Basic Executive

The operations comprising the Basic Executive are illustrated in Fig. 1. Initialization includes the setting of counters, priming of interrupt procedures, and the input of a set of parameters that define the system for a particular machine. In this way it is possible, in principle, to move the controller from one machine to another of the same basic type but which may have different

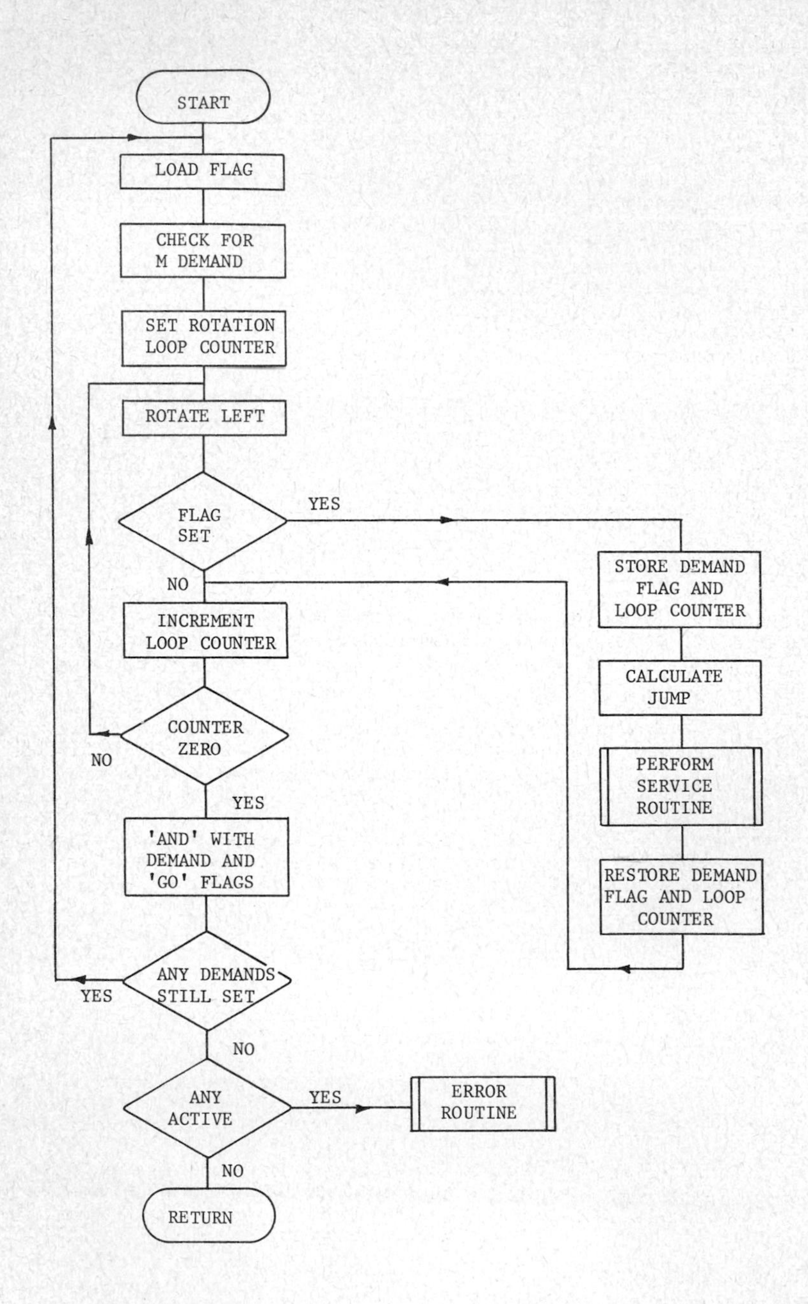

FIG. 2 SEQUENCING LOGIC

characteristics. This parameter table comprises values that specify system increment, maximum feedrate, length of x and y axes, number of tools in turret, maximum spindle speed, and acceleration.

Data control involves the initial filling and subsequent maintenance of an input buffer of uninterpreted data. This buffer, which holds n characters, is replenished when space for m characters becomes available. Data control can also sense broken tapes and can rewind on appropriate signals. Code compensation and translation could be, but have not been, included.

A COMMAND routine with a limited vocabulary of six words, originally intended as a teletype interface between operator and a number of machine tools, has been written; it has not yet been included in the NEL-CNC software package. It was envisaged that a small computer with a small disc could economically control small groups of machine tools and provide a store for extension into Adaptive Control. Front panel control is conventional and has been regarded as a background operation.

3.2 UPDATE

Although UPDATE is scheduled to occur at the end of 'current block' it is also the first operation in the Executive loop. This has the advantage that separate initialization procedures are unnecessary, which simplifies the system and reduces coding.

At the end of 'current block' operations, the Sequencer passes control back to the Executive. The 'next block' is then updated to 'current block' status. Data transfer associated with this is accomplished by a parallel word-address dispatch table operation, which sweeps the contents of 'next block' buffer into the 'current block' buffer. A new 'next block' is then constructed by the INTERPRET phase.

3.3 Control Tape Interpretation

Uninterpreted data from the input buffer are scanned by a character-search routine, which, on recognizing one of the conventional block heading characters, jumps, by way of parallel word-address dispatch table, to the appropriate service routine. The multi-digit number associated with a given control block is converted from 'magic 3', if necessary, and then from decimal to binary. The relevant demand flag is set and the control value transferred to the appropriate control buffer. The difficulties of zeroing control values are avoided as only flagged buffers are accessed. A sequencing operation is then carried out repetitively on this demand status word until all flags have been deleted, at which time control passes to the SEQUENCER.

3.4 Sequencing

The demand status word indicates the operations required in the current block and implies that appropriate data are ready for processing. Fig. 2 illustrates sequencing logic.

A check is first made for initial miscellaneous codes such as start-up and coolant-on. The M code bit is then rotated, least significant bit to most significant bit position, unconditionally, and the system drops into the sequencing loop proper.

If the bit corresponding to a particular control function, such as an x-axis demand or tool or speed change, is not set, the loop counter is incremented and the bits rotated again. If all bits have been examined, the demand status word and inhibit or 'go' flag status word are logically 'ANDed'. Where a demand is still outstanding a further sequencing operation is carried out. If a GO flag is still set, the system exits to an error routine that loads a code number in the accumulator, and the system stops. Otherwise, the Sequencer hands back control to the Executive.

Reference to Tables 1 and 2 will illustrate the mode of operation.

Table 1
Sequencer Operation

1	Set OFFSET = -7.
2	Rotate DEMAND status word 1 bit - least significant to most.
3	If set, go to 4, or decrement OFFSET by 1 and go to 2.
4	Load via pointer mechanism the contents of location [(B2 + 7) + OFFSET] and store as most significant byte of a JUMP DOUBLE.
5	Perform similar operation on table [(B3 + 7) + OFFSET] and store as least significant byte of a JUMP DOUBLE.

Table 2
Sequencer Action

B2:	Page of Process 7
	Page of Process 6
	Page of Process 5
	Page of Process 4
	Page of Process 3
	Page of Process 2
	Page of Process 1
B3:	Word of Process 7
	Word of Process 6
	Word of Process 5
	Word of Process 4
	Word of Process 3
	Word of Process 2
	Word of Process 1

3.5 Axis Control and Machine Functions

The main aspect of foreground processing is axis control, each control routine being entered from the Sequencer if the appropriate flag is set. A control routine will accept a call in one of several ways.

If the axis is active the call will be deflected back to the Sequencer, otherwise it will accept and sense whether the call is an initial one or whether the axis is in ramp mode when it will output a proportional increment. If the axis is up to the programmed feedrate it will output a full increment, or step, the magnitude of which will depend on the system increment.

The output increment and feedrate form the two bytes of a 16 bit word. The feedrate byte is a signed number indicating direction and magnitude. A measure of 'look-ahead' is provided, which avoids the necessity of responding to interrupt within a system increment, 30 μs for 2.5 μm (0.0001 in). This is accomplished by generating an interrupt at 18 μm (0.0007 in) from zero, to bring in the next block.

Speed- and tool-change are effected by routines which operate relay controls. Continuity of relay operation is essential and is ensured by the tight logic of the relay control routine and its associated 16 bit status word. Fig. 3 has been included as an illustration.

3.6 Interrupt Structure

In the NEL system interrupts are generated at the end, or completion, of control functions. This is considered to be acceptable, having regard to the differing response times of the computer and machine-tool elements. More than one process may therefore be operating in parallel because the control program is free to attempt to service other routines. Inhibit flags will prevent this where a conflict of operation or resource occurs.

The system then proceeds to find another task which does not conflict with current operations. It is possible, for example, that buffer replenishment (for example for factory data collection output) may take place in parallel with tool- or speed-change. Where 'critical sections' of code are being executed, these parallel operations may be inhibited, as when in interpolation mode.

Response to interrupts is organized by a handler in terms of a hardware pointer to specific locations in core where the appropriate response code is located. Eight automatically resolved levels of priority can each accommodate a further eight levels, thus permitting sixty-four levels overall. A further subdivision would slow up response as secondary and subsequent levels of priority are resolved by software.

```
[RELAYS 'ON' ROUTINE ONE
[MTR4.1.   .0
[13 MARCH 1972
[RELAY STATES ARE INDICATED BY
[BIT PATTERN  IN R6 AND R7
[THE SELECTION OF WHICH RELAY
[TO TURN ON IS SPECIFIED IN X AND Y
    005 000   000 A        R6  0  [ CURRENT RELAY STATES R0 - R7
    005 001   000 A        R7  0  [ CURRENT RELAY STATES R10 - R17
    005 002   000 A        R4  0
    005 003   000 A            0
    005 004   105 A            CPX     [COMPLEMENT 'N'.
    005 005   200 B            LDW R6  [LOAD STATUS WORD.
    005 006   132 A            AND     ['AND'.
    005 007   101 A            EXY     [EXCHANGE.
    005 010   105 A            CPX     [COMPLEMENT R.
    005 011   133 A            EOR     [EXCLUSIVE 'OR'.
    005 012   220 B            STW R6  [RESTORE.
    005 013   100 A            CUP
    005 014   103 A            COX
    005 015   105 A            CPX     [REPEAT FOR R7.
    005 016   201 B            LDW R7
    005 017   132 A            AND
    005 020   112 A            CDN
    005 021   133 A            EOR
    005 022   221 B            STW R7
    005 023   201 B            LDW R7
    005 024   200 B            LDW R6
    005 025   170 A            IOT 10
    005 026   005 B            JMD R4
    005 027   002 B
```

FIG. 3 RELAY CONTROL ROUTINE

The character of interrupt response is conventional in that a register exchange takes place and the response code executed followed by register recovery. By generating interrupts at the end of operations the amount of response code to be executed is reduced.

4 INTERPOLATION

The interpolator is a single, general-purpose routine catering for straight lines and conic sections. It is fast in operation, economical in coding and retains a simplicity of computation suitable for a Minic computer without floating point arithmetic.

Bresenham's Algorithm (Bresenham 1970) for linear interpolation in digital plotting was taken as a suitable basis for development as it requires arithmetic addition and subtraction only within the interpolator loop. The algorithm generates steps along the x and y axes in the ratio of the gradient

$$(y_2 - y_1)/(x_2 - x_1).$$

The linear interpolating process (Stockdale and Bruce) involves taking a series of identical positive (or negative) increments in x, at each of which a similar sized positive (or negative) increment in y is taken if this is needed to approximate to the desired slope of line. Fig. 4 shows the case when positive steps in x and y are used for a line of slope +3/8. This process only copes with lines at up to $\pm 45^{\circ}$ from the positive and negative x directions. For lines outside this range the roles of x and y are reversed making regular steps in y and steps when required in x.

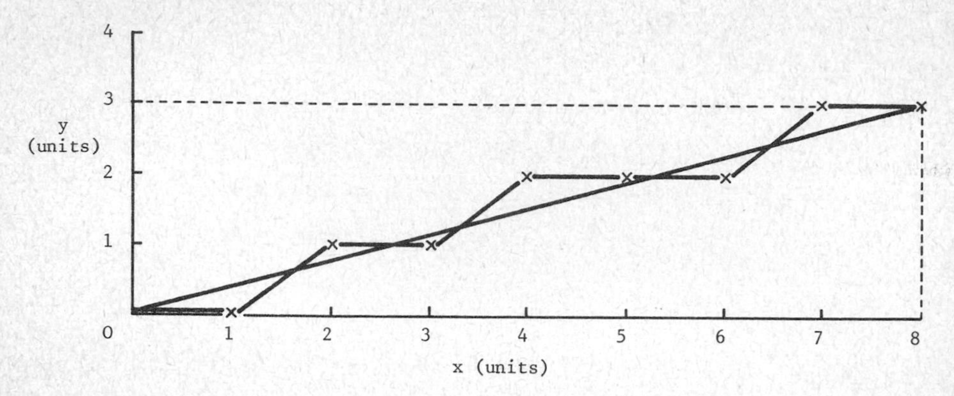

FIG. 4 INTERPOLATOR ACTION SIMPLIFIED

Additional rules for circular interpolation enable the gradient of the interpolated curve to change for every incremental step and deal with any changes in octant which may occur during interpolation.

5 FUTURE PROSPECTS

The future of computer numerical control can be considered in terms of computer technology and system organization.

Prospects in computer technology are very bright and mainly rest on the extended use of micro-programming and new software techniques. The impact of new hardware production technologies is certain to drastically reduce system cost. Additionally, the further development of adaptive control, workpiece handling, and work planning could also have radical effects.

CNC is particularly appropriate as a first venture into NC for the small user. This may be as a composite system or retro-fit to existing conventional machine tools. It is also feasible that a CNC system may be developed into mini-DNC (Direct Numerical Control) where more than one machine tool may be associated with a single CNC controller.

In the longer term, CNC as an element of DNC is a natural progression, and CNC may lose its own identity in the process.

ACKNOWLEDGEMENT

This paper is published by permission of the Director, National Engineering Laboratory. It is British Crown copyright.

REFERENCES

DIJKSTRA, E. W., (1968). Co-operating sequential processes. In GENUYS (Ed) PROGRAMMING LANGUAGES pp 43-112 London: Academic Press.

BRESENHAM, J. E., (1965). Algorithm for computer control of a digital plotter. IBM Systems Journal, 4, p 25.

STOCKDALE, W. and BRUCE, J. W. Digital interpolation. (NEL Report in preparation.)

THE REDISTRIBUTION OF MACHINE DEPENDENT SOFTWARE WITHIN A DIRECT NUMERICAL CONTROL ENVIRONMENT

RICHARD M. SIM
National Engineering Laboratory
East Kilbride, Glasgow, United Kingdom

Abstract: The rapid development of mini-computers and their use, both in softwired controllers and in DNC systems, offers opportunities to restructure NC programming software. The development of a new CLDATA, based on the ISO control tape format, and used as an interface between the geometric processor and any remaining machine-dependent modules, can virtually eliminate the post-processor problem for many machines. Of equal importance, the new CLDATA can simplify the storage and editing of tool path data.

1 INTRODUCTION

The number of economically justifiable uses that could be made of the DNC system shown in Fig. 1 would be increased by the implementation of a NC processor on the computer at level B.

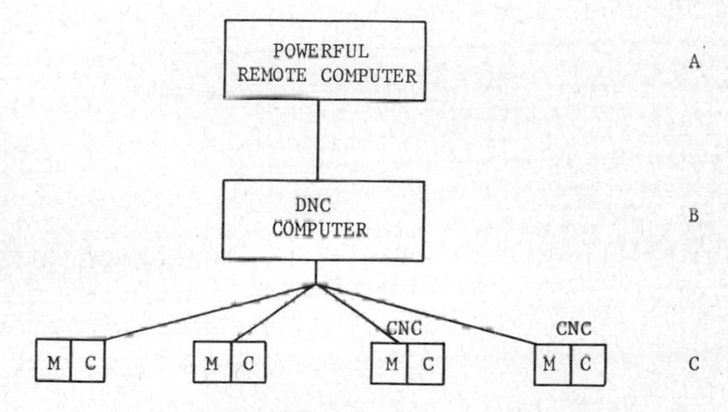

Fig. 1 Hierarchy of a Typical DNC System

Located at the DNC level, the NC processor would relieve the large computer at level A of the task of processing many of the simpler part programs and during periods of low DNC activity, for example overnight, could even handle larger programming jobs.

This has the further advantage that levels B and C would, if necessary, be self-contained and could therefore be marketed and used as a package without the requirement of level A. Alternatively levels B and C could continue to function as a unit in the event of breakdown of level A.

The difficulty of implementing existing APT or APT-like processors and post-processors on the 16 K core of the typical DNC computer points to the development of new NC processors and post-processors designed to function efficiently on the DNC computer.

The CLDATA interface linking these new DNC/NC processors and their post-processors can also be restructured if necessary, since the constraint imposed on the present APT-like CLDATAs of the requirements of existing software need not apply to the DNC system.

COMPUTER LANGUAGES FOR NUMERICAL CONTROL, *J. Hatvany, editor*
North-Holland Publishing Company - Amsterdam–London

This paper analyses both the present CLDATA as used by the APT-like processors and new CLDATAs based on the ISO paper tape block formats in an attempt to determine a CLDATA format most suitable for DNC use.

2 CLDATA CRITERIA

In attempting to judge the viability of a format for NC processor output, certain criteria can be listed.

i The output format should be capable of carrying a representation of the complete range of possible part-programming statements. This representation should, in general, be at the lowest level of independence of the machine tool/control system (but with individual records not necessarily completely independent of the machine tool/control system).
ii The output should be easy to store and edit.
iii Items in the output should be easily identifiable (items should be symbolic rather than numeric).
iv The output should be concise and free from redundancy.
v The output format should be such as to help minimize the total work done in the computer in progressing from part program to control tape (but not at the expense of adding to the work done by the planner).
vi Arithmetic operations on the output data should be simple to perform. The data should be represented with sufficient accuracy.
vii The output format should allow multiple post-processing.
viii The output format should allow diagnostic messages from the post-processor to be linked to the original part program statement responsible for the error.

The above list is not exhaustive but it is worth using it to look briefly at the present APT3 CLDATA, as being standardized by ISO[1,2], to see how it performs. Fig. 2 shows a simple APT part program and Fig. 3 the resultant APT CLDATA.

```
PARTNO TEST PART                                    01
MACHIN/NELCNC,1,CIRCUL                              02
CUTTER/1                                            03
C1 = CIRCLE/CENTER,(P2 = POINT/3,0,0),RADIUS,1      04
L1 = LINE/(P1 = POINT/0,0,0),RIGHT,TANTO,C1         05
L2 = LINE/P1,LEFT,TANTO,C1                          06
SP = POINT/-1,-1,0                                  07
PL1 = PLANE/0,0,1,0                                 08
FEDRAT/10                                           09
FROM/SP                                             10
COOLNT/ON                                           11
GO/TO,L1,TO,PL1                                     12
GORGT/L1,TANTO,C1                                   13
GOFWD/C1,TANTO,L2                                   14
GOFWD/L2,PAST,L1                                    15
GOTO/P1                                             16
GOTO/SP                                             17
STOP                                                18
FINI                                                19
```

Fig. 2 Simple APT Part Program

Record type	Number of logical words	Description	
1000	5	Input card sequence	01
2000/1045	14	Part number	
1000	5	Input card sequence	02
2000/1015	6	Machine name etc	
1000	5	Input card sequence	03
6000	4	Cutter definition	
1000	5	Input card sequence	09
2000/1009	4	Feedrate	
1000	5	Input card sequence	10
5000/3	8	x,y,z of FROM/	
1000	5	Input card sequence	11
2000/1030	5	Coolant	
1000	5	Input card sequence	12
5000/5	8	x,y,z of GO/	
1000	5	Input card sequence	13
5000/5	8	x,y,z of GORGT/	
1000	5	Input card sequence	14
3000	15	Canonical form of C1	
5000/5	230	x,y,z's of GOFWD/C1	
1000	5	Input card sequence	15
5000/5	8	x,y,z of GOFWD/L2	
1000	5	Input card sequence	16
5000/5	8	x,y,z of GOTO/P1	
1000	5	Input card sequence	17
5000	8	x,y,z of GOTO/SP	
1000	5	Input card sequence	18
2000/2	3	STOP	
1000	5	Input card sequence	19
14000	2	FINI	

Fig. 3 CLDATA Produced by APT3 on the Univac 1108

3 PERFORMANCE OF EXISTING APT CLDATA

Using Fig. 3 as reference, the structure of the existing APT CLDATA can be measured against the above performance criteria.

Point i

The present CLDATA does reasonably well if all current variations of machines and controls have to be considered. However, very few machine tool systems today do not use paper tape block format and, if those which do not are ignored, the data could be processed nearer to the final paper tape format, though still independent of machine tool, before output as CLDATA.

Point ii

Present CLDATA does poorly. Its data is a mixture of integer, floating point and Hollerith words, which require care in storage and manipulation and bear little relationship to the part programming or control tape languages. It is unlikely that it could be edited by the average planner.

Point iii

The numeric codes used by the present CLDATA to represent record types and preparatory and miscellaneous functions make the interpretation of the records awkward for the expert and impossible for the layman.

Point iv

It is reasonably free from redundancy but can hardly be termed concise. The example in Fig. 3 uses some 400 36-bit words on the Univac 1108 computer. On an IBM 360 the CLDATA for the same part program would require over $2\frac{1}{2}$ K bytes!

Point v

The format of the present CLDATA is such that, no matter how simple the machine tool/control system, post-processing has to be performed. A good percentage of the work done in any current post-processor is concerned with checking and decoding the data contained on the CLDATA file and then converting it to the appropriate control tape information. Most of the checking done is of value to the planner in isolating programming errors; errors perhaps of part program statement syntax but frequently of violation of machine limits or incorrect spindle speeds and so on. If this checking is not done in the post-processor it requires to be done before the final control tape is tried on the machine tool (an alternative possibility is a good control tape checking/plotting post-processor program).

Whilst therefore, the work done in checking seems realistic, overall effort appears to be wasted in first coding the CLDATA from the part program in the processor and then decoding the CLDATA in the post-processor to finally produce the control tape. This seems especially true in the present situation where the CLDATA file bears little resemblance to either the part program or the control tape. In effect, two sets of codes have to be stored in even the most trivial post-processor (for example for COOLNT/ON the processor converts the statement to the CLDATA codes 2000 1030 71. The post-processor has to be capable of reading these codes and outputting the code M08, perhaps, to paper tape).

Points vi, vii, viii

On these final three points, the present CLDATA structure does well. Data are usually in floating point form and can be easily read into floating point variables or arrays and modified arithmetically. On most current processors at least 36 bits are used for each piece of data and this is usually more than sufficient for accuracy. Multi-post-processing and diagnostic messages are handled with ease.

Overall then, on the basis of the above eight points, the present CLDATA format does not do too well. It is satisfactory for the, perhaps oldfashioned, virtues of multi-post-processing, arithmetic manipulations and error diagnosing, but it does poorly in the more modern DNC requirements of being concise, easy to edit and close to the final form of control tape.

It will continue to be used, however, on the existing APT-like processors because of the large number of existing post-processors and their dependence on it.

In considering the ideal output from a DNC processor, there is considerable freedom to select a format which more nearly fits the above criteria, since the majority of existing post-processors can be largely ignored as they are unlikely to be easily implemented on the DNC computer. Before going further it is worth examining two typical control tapes which could result from our sample part program. The first assumes that the control unit requires ISO interchangeable variable block format[3] and that its format classification is as follows.

N3.G2.X+23.Y+23.Z+23.I+23.J+23.K+23.F3.S3.T2.M2*

The control tape would be as shown on Fig. 4. Throughout this paper, 'control tape' should be taken to mean the logical information, blocked in an ISO paper tape format, used to control an NC machine. The use of 'control tape' is not intended to imply a physical punched paper tape as the medium. The medium could just as well be magnetic tape or, in a DNC set-up, hard-wired BTR connections.

```
%   N001TG01TX-01000TY-01000TZ+00000TTTTF510TTTM08
    N002TTX-00741TY-00268
    N003TTX+02500TY-01414
    N004TG03TTY+01414TTI+03000TJ+00000
    N005TG01TX-01500TY+00000
    N006TTX+00000
    N007TTX-01000TY-01000
    N008TTTTTTTTTTTM00
    N009TTTTTTTTTTTM02
```

Fig. 4 Interchangeable Control Tape

A second example assumes that the control uses the ISO variable block format[4,5] and that its format classification is as follows.

N3G2X+23OY+23OZ+23OI+23OJ+23OK+23OF3S3T2M2*

The control tape would be as shown on Fig. 5.

```
%
N001G01X-01Y-01Z+0F510M08
N002X-00741Y-00268
N003X+025Y-01414
N004G03Y+01414I+03J+0
N005G01X-015Y+0
N006X+0
N007X-01Y-01
N008M00
N009M02
```

Fig. 5 Variable Block Control Tape

Figs 4 and 5 show that these forms of control tape information fulfil the fourth criterion for an ideal CLDATA - that of conciseness and lack of redundancy - better than does the present CLDATA. How would control tape CLDATA measure up against the other criteria?

4 PERFORMANCE OF CLDATA BASED ON PAPER TAPE FORMAT

Point i

Both the interchangeable variable block and the shorter variable block formats are nearer to any required paper tape. In achieving this, however, the shorter format has sacrificed some machine/control independence.

A drawback of the formats shown in Figs 4 and 5, if used as CLDATA, would be that they cannot represent part program statements which have no direct correspondence with paper tape output blocks. An example will demonstrate this easier.

The statement CYCLE/DRILL can be represented in paper tape format by the code G81. But the statement CYCLE/ISOBOR has no such coded equivalent in the present paper tape formats.

One solution to this would be to program the processor to output coded representations of statements where such codes exist in the present formats, and where codes do not exist, have the processor output the statement in character form enclosed in 'control-out' and 'control-in' left and right parentheses. Also within the parentheses, a function mnemonic, PPR, could be used to indicate the type of information. This would be in line with recent EIA standards proposals[6].

The statement CYCLE/ISOBOR could then be represented on CLDATA by (PPR,CYCLE, ISOBOR). This in turn could either be processed by a post-processor or CNC system or ignored by current hard-wired controllers if punched as part of a control tape.

Points ii, iii

Both paper tape formats score very well on storage and editing. The files, being 100% Hollerith and being in an already familiar format, are simple to modify by the average planner.

Point iv

Both versions are considerably shorter than the present CLDATA; the variable block format of Fig. 5 being very concise (24 36-bit words on the Univac 1108 compared to the 400 words required by the present CLDATA).

Point v

The fifth criterion was that of minimizing the computing necessary in producing a control tape. It would appear that, basing a CLDATA processor output on the specific paper tape format required by the target machine tool would best satisfy that condition. It would certainly reduce the work done in the present processors and post-processors in coding and decoding intermediate files in formats bearing little resemblance to the final control tape.

In fact, for a machine tool fitted with a modern CNC system, it may be possible in the ideal situation to completely eliminate the post-processor as it is known today. Suitable formatted control tape could be produced directly by the processor and checked and plotted by a pre-machine tool, tape-checking program, thereby eliminating post-processing and reducing the non-cutting time normally taken on the machine to prove the tape. Perhaps this ideal would not often be attained, some level of post-processing being necessary for most machines to avoid the processor becoming too large and too machine tool dependent.

Point vi

Arithmetic operations would not be as easy to perform on the data of the paper tape formats as on the present CLDATA. Values stored as a set of Hollerith characters, as in the paper tape formats, would require to be first converted to floating point form before being operated on arithmetically.

Point vii

The position on multiple post-processing is somewhat involved. At the present time multiple post-processing is used mainly to produce a plot of the CLDATA at the same time as a control tape for a machine tool.

It is probable that in the next few years multiple post-processing for two or more machines will become more popular. DNC, or more correctly, DSC (Direct Shop Control) will allow fast rescheduling of jobs on alternative machines. Rescheduling will be catered for by one of the following methods.

a Having machines which use completely interchangeable control tapes;
b Reprocessing the CLDATA;
c Reprocessing the part program;
d Rewriting the part program.

These four options are listed in terms of increasing re-work. It would be advantageous to organize the rescheduling to use as early an option as possible. To be able to reschedule work onto alternative machines by simply redirecting the control tape would be ideal - but unlikely. In this context it is worth heeding the warning on interchangeability expressed in the British Standard for NC[7] -

"Under present conditions, it is necessary to treat the word 'interchangeable' with considerable reserve. In order that the same tape shall be capable of controlling, for example, two different machines, it is not sufficient to check that the Interchangeable Variable Block format has been employed and that the total block content required by both is the same. It is also necessary to ensure, for example, that both machines are capable of executing the ranges of feeds and speeds employed; that they both use the same tape code and conform to the same axis and motion nomenclature. Although the ultimate aim of the Interchangeable Variable Block format is that such exchanges should be possible, <u>it is not yet guaranteed</u>. At least, however, use of the appropriate format from the agreed few will ensure that in future it is not impossible to exchange tapes between otherwise similar systems merely because they require different presentations of the same data."

Therefore, as option (a) is unlikely to suit more than a handful of machines, the use of multiple post-processing of CLDATA would be highly likely.

The possibility of multiple post-processing with paper tape format CLDATA would depend on whether a single format classification was used for all processor output or whether the processor output was capable of being primed by the format classification for the specific machine tool required. If the former, then multiple post-processing would be possible in the same manner as for today's CLDATA. If, however, the CLDATA was primed to suit a specific machine tool, then obviously multiple post-processing would only be possible if the formats of the two machines were very similar.

An exception to this could be made for the production of simultaneous CLDATA plots and control tapes. To allow this and, at the same time, allow the priming of the CLDATA with the machine tool format specification, it would not be difficult to design the plotting program to accept a variety of paper tape formats as input, the plotting program being initialized by the format specification for the specific machine tool in the same manner as the processor.

From the above, it would appear that a single format classification for control tape CLDATA would be best suitable for multiple post-processing. However a major difficulty in adopting a single format for all processor output would be in deciding on the range of numbers to be handled for each word of the formatted block. In other words - how many digits would be used before and after the decimal point to represent the value of, for example, the X field of the format? The number of digits decided on for before and after the point would have to be sufficient to carry the largest foreseeable number to the greatest foreseeable accuracy. This could result in an X field of perhaps ten digits plus sign in order to carry a possible range of numbers from ±.00001 to ±99999.99999. Even this would not repeat the range and accuracy present in the current CLDATA, where floating point representation of the required number gives a far greater range (for example on the Univac 1108, single precision floating point permits a range of approximately $\pm 10^{-38}$ to $\pm 10^{+38}$).

Point viii

The formats shown in Figs 4 and 5 do not allow for diagnostic messages being linked to the original part program statements. However it would be easy for the processor to output part program statement numbers in a similar manner to that proposed earlier in the paper for post-processor statements (enclosing them in parentheses with the identifying mnemonic PPR). For statement numbers a mnemonic PPS could be used, the numbers being enclosed in parentheses and output to the CLDATA file interspaced with the other CLDATA output.

```
(PPS,17)
N007X-01Y-01
(PPS,18)
N008M00
(PPS,19)
N009M02            and so on.
```

Again there would be the option of either post-processing these to delete them from the control tape or punching the CLDATA file directly and having the control system ignore the information within the parentheses.

5 SUMMARY OF FINDINGS

Table 1
Single Interchangeable Variable Block CLDATA
as Illustrated in Fig. 4

Advantages	Disadvantages
Data is of a single type	More difficult to do arithmetic operations
Blocks are easy to identify and edit	Length of file increased on 16-bit word computers
Format is familiar to part programmers	

Table 2
Variable Block Format CLDATA Primed to Suit Individual Machine Tool/Control as Illustrated in Fig. 5

Advantages	Disadvantages
Data is of a single type	More difficult to do arithmetic operations
Blocks are easy to identify and edit	No multi-post-processing unless machines use same format
Format is familiar to part programmers	
Post-processing is reduced and for some CNC controls eliminated	
Length of file is reduced for many machines	

Comparing the two tables, it would appear that the 'tailored' format of Table 2 offers advantages over both the format shown in Table 1 and the present CLDATA of the APT-like processors. The shortened file length, the reduction in post-processing and the ease of editing are important points in its favour.

A disadvantage of the 'tailored' format - and only a disadvantage for some installations - lies in its inhibiting of multi-post-processing. One way round this problem would be for particular installations to elect to use a single preferred format for all their machines, their post-processors being organized to read this format and produce the outputs required for their specific machines. Alternatively, reprocessing of the part programs would be the answer to a requirement for rescheduling of NC machines.

6 IMPLICATIONS OF ADOPTING PAPER TAPE FORMAT CLDATA

Some of the implications which would follow the adoption of a tailored format as the CLDATA output of a DNC processor are as follows.

a A method would be needed of making the format specification of individual machines available to the processor. Preferably, this could be done by storing the specifications as an external file, formats for specific machines being selected according to the MACHIN card in the part program. Alternatively, the specification could be input to the processor through a part-programming statement.
b The specification would require to cover at least the following:
  i the control tape format (number of digits, presence or absence of tabs, leading or trailing zeroes and so on),
  ii whether absolute or incremental data,
  iii whether metric or imperial units, and
  iv the type of feedrate and spindle speed representation (such as magic 3, or straight IPM).
c The processor would require to differentiate between post-processor statements which it could recognize and could output as paper tape codes and statements which it could not recognize and would have to enclose in parentheses together with the PPR mnemonic (as discussed earlier).
d It was proposed earlier in the paper that part program sequence numbers be interspersed with the paper tape CLDATA records to give walk-back references to the part program in the event of post-processing errors - the sequence numbers being enclosed in parentheses and given the mnemonic PPS.

It would be worth while to have a switch, perhaps through a part program statement, to allow the planner the opportunity to inhibit this feature and produce an optional short form of output.

e The processor should allow the part programmer to use temporary coordinate systems suitable for the geometry of the workpiece. Operating immediately prior to output of the CLDATA, the processor should also have the facility to translate the resulting toolpath of any such temporary coordinate systems into the coordinate system of the machine tool. This facility is usually available at present in the post-processor, using the ORIGIN statement, but could equally well be in the DNC processor.

f One final problem which would remain to be solved would be to decide on the type of interpolation output by the processor. In the present systems, both linear and circular information are usually output. If the new CLDATA is planned to double directly as control tape, only one or other of the forms of interpolation would have to be present. The best course would seem to be to output circular arc G02 or G03 blocks for all circular toolpaths and G01 blocks for the remaining toolpaths.

The question of whether the G02 and G03 CLDATA blocks could drive the specific machine tool correctly would be a function of the post-processor. It would perhaps require that the G02 and G03s should be linearly interpolated within the post-processor to produce G01 cut vector approximations. Again, the post-processor might have to split-up the multi-quadrant G02 and G03s of the CLDATA into circular arcs which lie within single quadrants.

7 CONCLUSION

In developing NC processors capable of running on DNC systems, opportunity should be taken to organize the CLDATA output such that it conforms more closely then does the present CLDATA to the requirements of a Direct Shop Control environment.

As far as the processor output is concerned, the basic needs of a DSC system will be concise data storage, an easy editing facility and efficient DNC computing.

These needs are shown to be satisfied, at least in theory, by adapting the ISO paper tape formats. A project is underway at NEL to develop DNC processor output along the lines recommended in this paper.

8 ACKNOWLEDGEMENT

This paper is published by permission of the Director, National Engineering Laboratory, DTI.

REFERENCES

1 ISO (1972). NC processor output, logical structure and major words. IS/DIS --- (due 1972).

2 ISO (1972). NC processor output, minor elements of type 2000 records. ISO/TC97/SC5/WG1 (N168).

3 ISO (1969). Interchangeable punched tape variable block format. ISO Recommendation R1057.

4 ISO (1969). Punched tape variable block format for positioning and straight cut numerically controlled machines. ISO Recommendation R1058.

5 ISO (1972). Punched tape variable block format for contouring and contouring/positioning numerically controlled machines. ISO/DIS 2539.

6 EIA (1969). Recommended interchangeable perforated tape variable block format for contouring and positioning numerically controlled stored program machines. EIA Automation Bulletin No 4.

7 BSI (1972). Numerical control of machines. BS 3635 : Part 1 : 1972.

SOFTWARE LOGIC CONTROL FOR DIRECT NUMERICAL CONTROL OF MACHINE TOOLS

D. FRENCH and G. MILLER
Department of Mechanical Engineering
University of Waterloo
Waterloo, Ontario, Canada

Abstract: A D.N.C. system designed in the Research Laboratories of the University of Waterloo utilizes a mini-computer; additional interfacing and interpolators have been designed to plug into the computer. In addition an operator panel is used to initiate control to the machine.

The paper describes the software which has been written for controlling the complete system. The software will enable the selection of stored programs by the operator panel and log and analyze down time for management purposes.

1. INTRODUCTION

Direct numerical control (D.N.C.) of machine tools by a mini-computer creates problems as to the proportion of work which must be performed by software programs and the proportion of the work to be performed by hardware elements. The amount of core available in the computer, the access time, and the operating times all have a bearing on the proportion of hardware to software usage.

The research at the University of Waterloo has produced two systems, one using a PDP/8E computer having a 4K memory and a 12 bit word length, and the other system using a PDP/11 having a 16K memory and a 16 bit word length.

This paper discusses the software for the D.N.C. system developed at the University of Waterloo utilizing the PDP/8E with a 4K memory and having a 12 bit word.

2. D.N.C. SYSTEM

The system developed is shown in Figure 1 and the requirements for the software required that linear and circular interpolation should be carried out by hardware unit. In addition, the system was designed such that all instructions either from the tape input or from the operator panel in the manual mode should be through the computer. There is no direct link from the operator panel to the machine tool, thus all operations, whether, manual, jog, tape, or indirect from a disc, are routed through the computer. The system is therefore at all times under the direct control of the computer.

3. SOFTWARE SPECIFICATION

The requirements for the software were:-

1) The system should operate from programs stored on disc or similar peripheral equipment.

2) Provision for automatic operation under punched tape from the teleprinter.

3) Operation in the manual mode from the operator panel, and display of this mode on the operator panel.

4) Display of position on the operator panel.

5) Loading of the interpolation unit for linear and/or circular interpolation.

6) Machine down-time logging.

COMPUTER LANGUAGES FOR NUMERICAL CONTROL, *J. Hatvany, editor*
North-Holland Publishing Company - Amsterdam–London

7) Management break-down notification.

The method of control by software is best illustrated by considering some elements of the D.N.C. system.

4. SOFTWARE CONTROL OF OPERATOR'S PANEL

The operator's panel contains a 9 tube digital display and 66 push buttons, each button has an acknowledgement lamp incorporated. Each button is coded in binary, and when a button is depressed the interface input lines cause the binary code for that button plus an interrupt request to appear at the interface. This code is "read in" and used in a table address "look-up and service" sub-routine branch.

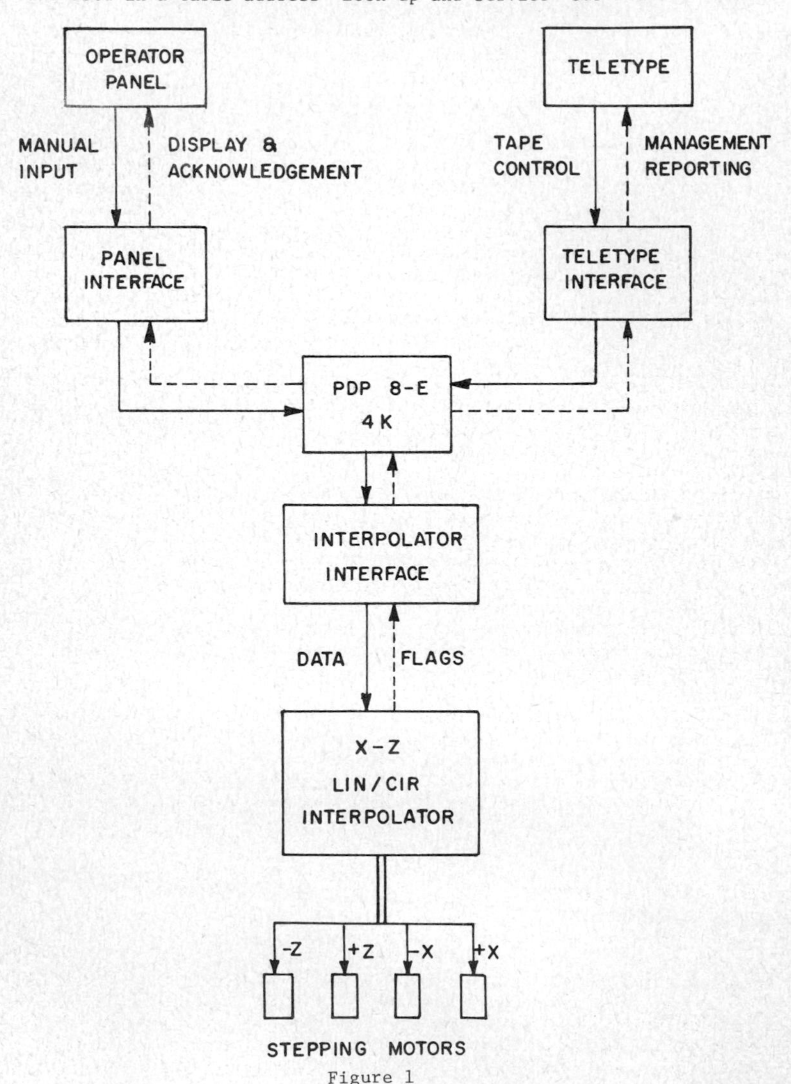

Figure 1

All read and write panel instructions are double executed and verified as an error check. In addition to reading the button codes, computer interface instructions allow the reading of the state of any acknowledgement lamp, and the digit display of any of the 9 display tubes. The computer also enables the writing of any one of the acknowledgement lamps and of any B.C.D. digit on any of the 9 display tubes. The operator panel has no function in itself other than to reference certain predetermined functions. Since these functions are specified in the software program they can be changed easily to accommodate different machine tools or systems to be under D.N.C. The operator panel is designed to enable manual input, automatic mode input from disc or from punched tape using a teleprinter, jog and incremental jog. The stop condition on the operator panel is used for emergency or power-down conditions. The function of this state is to stop any machine movement in progress, to reset all panel lamps or displays, to clear all registers and buffers. It should be noted that the system is still under computer control, the computer itself is never halted through operator panel inputs of any form.

5. PART PROGRAM LOADING

The input to the computer consists of a block format in accordance with ElA RS273. The characters are packed two per word in the "tape" buffer area with the exception of line feed, carriage return and blanks.

The control buffer area includes the systems buffer and command fixed format registers which are used for display and data transfer. It is necessary to include in the software at this point a "table branch" on the six bit binary value of each character to facilitate transfer from the tape buffer to the control buffer as the computer is a 12 bit machine making it impossible to pack the normal 7 bit character code.

A special routine is incorporated within the program to determine from the G function the relative position of the data, decimal point, and number of leading zeros. Due to the nature of the I/O routine the control buffer is reset if the G function value is changed to avoid any possibility of a decimal point mismatch.

The "I" word buffer is a special formulated interpolator loading instruction buffer used to prevent dwell between cutter movements. The general format is a 4 bit instruction followed by two 4 bit B.C.D. data words. The instruction references one of the three interpolator registers. The F (feedrate) register is a 4 digit fixed point register which is loaded with the feedrate number. The remaining two registers (ΔX and ΔY registers) are variable point resisters which are loaded, left justified to the largest of their respective data pair (i.e. ΔX and ΔY, or I and J). To initiate the interpolator a special set up word is included which contains the relevant cut information.

The software has built into it a priority system for loading the buffers and interpolators. The interpolator must be loaded first, the I word buffer is then refilled from the control buffer which in turn is refilled from the "tape" buffer. The tape buffer being filled as necessary from the part program input. On loading, the control "buffer" is transferred to the control "command" which is then reset to receive the next block. The "command" increments are used to update the position registers.

6. BASIC SOFTWARE APPROACH

The software are written in MACR 8 ASSEMBLER LANGUAGE in order to make the most efficient use of the small amount of core available. The computer utilizes a "page address" system, the 4096 words of memory are divided into 40 (octal) pages of 200 (octal) words. The assembly of the machine instructions allows reference to the "current page" or "page zero" memory locations only. Any desired reference to a location not on "page zero", or the "current page" has to be referenced indirectly by utilizing a memory location on the "current page" as a pointer, obviously, as the number of off page references increases the number of machine instructions that can be stored on that "page" decreases. The program size could be reduced if the computer had a 16 bit word instead of a 12 bit word.

7. SYSTEM SOFTWARE

A general outline of the system software is shown in Figure 2. The main loop of the program is interrupted by one of the Panel buttons being depressed or from the machine tool interpolator, indicating that interpolation for a particular sequence has been completed. Should the interrupt signal be from the operator panel the program will then create the necessary storage and execution based on the information supplied from the operator panel. In the automatic mode it will load all buffers as sepcified.

For management reporting 9 "down-time" reason buttons are provided on the panel. If the machine stops either during a machining operation or at the end of processing time logging commences and is indicated by the program creating a signal which operates the appropriate panel light and causes a message to be printed out on a teletypewriter. Failure of the operator to define the delay by depressing the appropriate button causes the down-time to be changed to "operator absenteeism". Additional management console messages put out by the program are:-

(a) Control stopped (operator has despressed "stop" button).

(b) Machine stopped (operator has depressed "hold" or "clear" button).

(c) Machine start (interpolator has been issued "go" instruction).

(d) Appropriate down time reason using panel buttons 1-9 inclusive.
(for example tool breakage, cutters not available, material not available).

(e) Tape block overflow.

(f) Read error.

(g) Illegal G function.

(h) Default feedrate error.

This with the appropriate locating of a teleprinter, management can be advised immediately of any problems arising in the operation of the machine tool.

REFERENCES

D. French, M. Simpson and W. Little, (1972), Design of an operator panel for on-line computer control of machine tools, Proc. 13th Intern. Machine Tool Design and Res. Conf., Birmingham, England.

D. French, M. Simpson and W. Little, A low cost hardware interpolation system for D.N.C., IBID.

D. French and J. Knight, Computer control of machine tools, IBID.

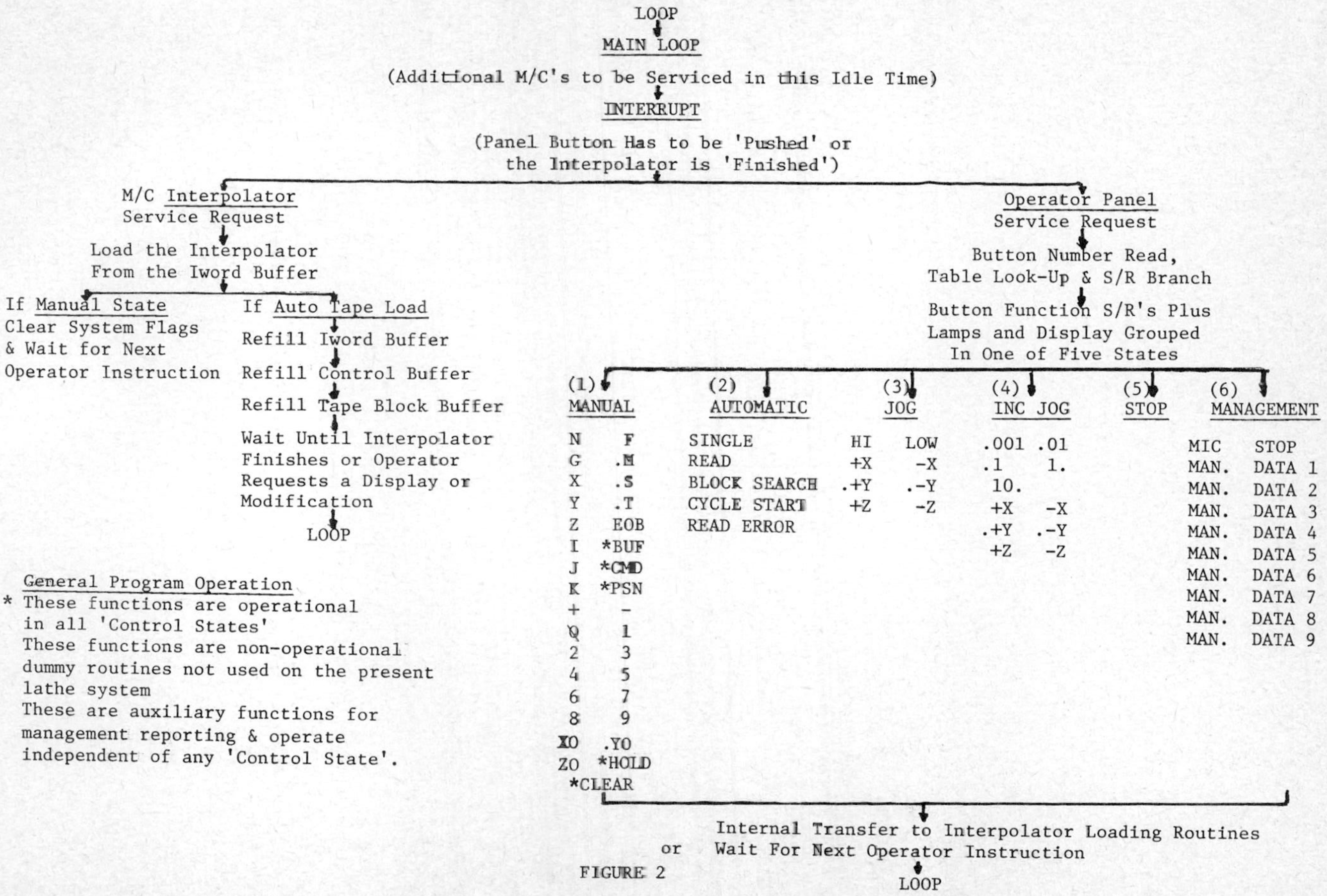

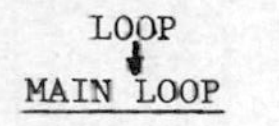

LOOP
↓
MAIN LOOP
(Additional M/C's to be Serviced in this Idle Time)
↓
INTERRUPT
(Panel Button Has to be 'Pushed' or
the Interpolator is 'Finished')

M/C Interpolator Service Request
↓
Load the Interpolator
From the Iword Buffer

- If Manual State: Clear System Flags & Wait for Next Operator Instruction
- If Auto Tape Load:
  Refill Iword Buffer
  ↓
  Refill Control Buffer
  ↓
  Refill Tape Block Buffer
  ↓
  Wait Until Interpolator Finishes or Operator Requests a Display or Modification
  ↓
  LOOP

Operator Panel Service Request
↓
Button Number Read,
Table Look-Up & S/R Branch
↓
Button Function S/R's Plus
Lamps and Display Grouped
In One of Five States

(1) MANUAL		(2) AUTOMATIC	(3) JOG		(4) INC JOG		(5) STOP	(6) MANAGEMENT	
N	F	SINGLE	HI	LOW	.001	.01		MIC	STOP
G	.M	READ	+X	-X	.1	1.		MAN.	DATA 1
X	.S	BLOCK SEARCH	.+Y	.-Y	10.			MAN.	DATA 2
Y	.T	CYCLE START	+Z	-Z	+X	-X		MAN.	DATA 3
Z	EOB	READ ERROR			.+Y	.-Y		MAN.	DATA 4
I	*BUF				+Z	-Z		MAN.	DATA 5
J	*CMD							MAN.	DATA 6
K	*PSN							MAN.	DATA 7
+	-							MAN.	DATA 8
Q	1							MAN.	DATA 9
2	3								
4	5								
6	7								
8	9								
XO	.YO								
ZO	*HOLD								
*CLEAR									

↓
Internal Transfer to Interpolator Loading Routines
or Wait For Next Operator Instruction
↓
LOOP

General Program Operation

* These functions are operational in all 'Control States'

These functions are non-operational dummy routines not used on the present lathe system

These are auxiliary functions for management reporting & operate independent of any 'Control State'.

FIGURE 2

DATA FORMATS IN THE COMPUTER CONTROL OF MACHINE-TOOL GROUPS

by

L. NEMES and P. HOFFMANN

Computer and Automation Institute, Hungarian Academy of Sciences, Budapest, Hungary

1. INTRODUCTION

The lengthy and heated debates leading eventually to standardization of NC control tapes are still fresh in the minds of all those engaged in numerical control. Terminating this long period of argument and discussions, international recommendations for the direct punched-tape input of "conventional" NC machines were worked out and accepted which are now respected and complied with, at least to a certain extent, by manufacturers of control systems.

Along with the wide application of the computerized programming of NC machine-tools, this problem cropped up again in a different field. The spread of program languages required standardization of the so-called CLTAPE format too, in order to provide a further fixed point in the variable multi-parameter world of computerized programming /1/.

As a result, part machining programs may now be represented at three levels /Fig. 1/.

The speedy development of computer techniques had led to the first computer-controlled machine-tool systems and subsequently, various "schools" came into being /2,3/. Initially, application of the dta flow presented in Fig. 1 in the case of

COMPUTER LANGUAGES FOR NUMERICAL CONTROL, *J. Hatvany, editor*
North-Holland Publishing Company - Amsterdam–London

computer-controlled machine-tools seemed to be quite obvious. Thus, the computerized control was functionally equal to the conventional nemerical control, irrespective of the systems solution /CNC, DNC/ adopted.

The fact that the machine-tool is no longer controlled by special hardware but by a universal small computer /together with the necessary computer peripheries/ affords functionally much wider opportunities.

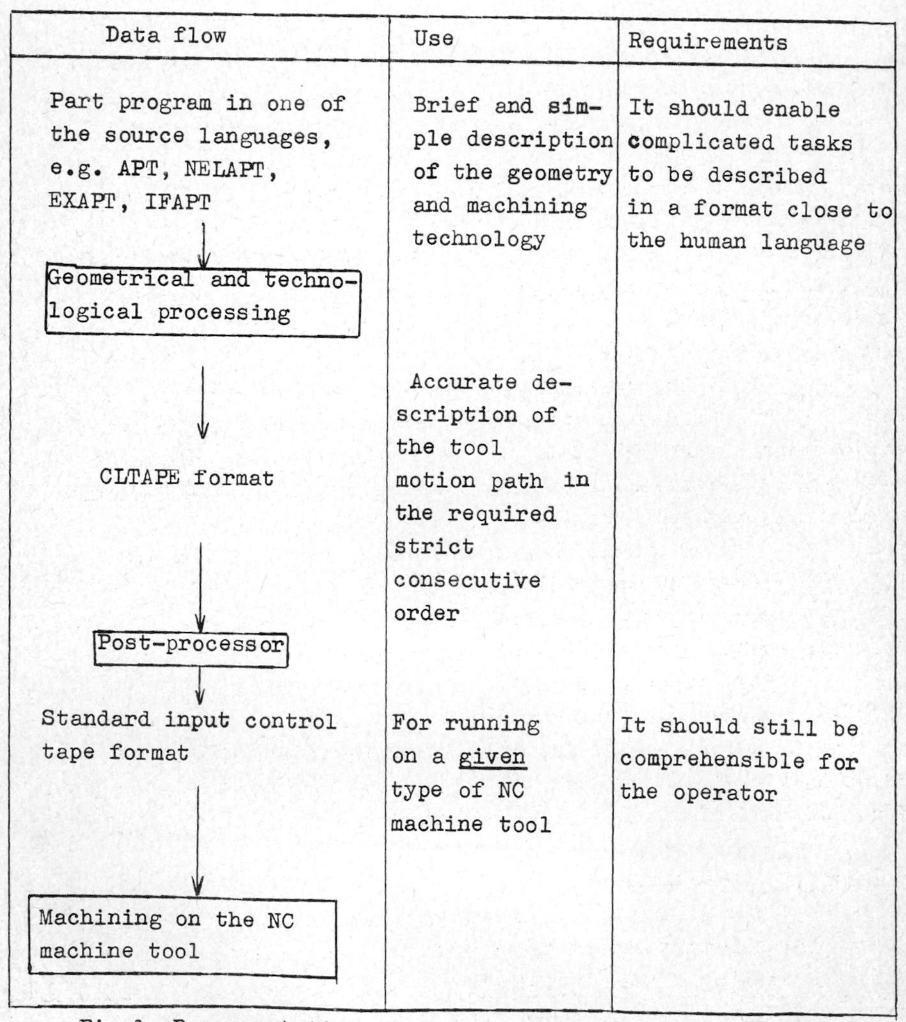

Data flow	Use	Requirements
Part program in one of the source languages, e.g. APT, NELAPT, EXAPT, IFAPT	Brief and simple description of the geometry and machining technology	It should enable complicated tasks to be described in a format close to the human language
↓ Geometrical and technological processing ↓		
CLTAPE format	Accurate description of the tool motion path in the required strict consecutive order	
↓ Post-processor ↓		
Standard input control tape format	For running on a given type of NC machine tool	It should still be comprehensible for the operator
↓ Machining on the NC machine tool		

Fig.1. Representation of part programs at various levels

Requirements for and consequences of computer control

A computer-controlled machine-tool group is a complete autonomous unit. The unit will evidently perform the necessary operations but it is expected to do more than that, namely:

- the machine-tool group should maintain, via the computer, permanent intercommunications with the outside world, and, even more important,
- the outside world should constantly be in live communication with the group of machines, via the computer.

In the long run, this would mean that the autonomous computer-controlled machine-tool group should lend itself for development into a part of an Integrated Manufacturing System /IMS/.

In so far as the requirements are concerned, the autonomous computer-controlled machine-tool group should be connectable to the other autonomous units of an IMS by systems engineering techniques. These other units include dynamic production planning and control, parts programming /which may be deterministic, optimizing, adaptive, or may correspond to any variant of the above/, interactive part design, etc.

From the foregoing it may be concluded that the autonomous system requires an interface enabling connection to other sub-systems of the IMS. This is a software interface which we call MID /Mid Data Format/. A DNC system working with the MID format will comply with the requirements of a CAM[x] system /4/.

Let us examine how the introduction of MID modifies the process of computerized programming /Fig. 2/.

The input of the MTC - Machine-tool Controller - mounted near the machine-tool is the OUTDATA format /3/.

By comparing Figs. 1 and 2, it will be conspicuous that the tasks framed with thick lines in Fig. 2 are accomplished two steps, almost at the same place where the post-processor is drawn in Fig. 1. The MID FORMAT is located between the two steps. /The MID may, of course, also be devised in some other way./

Aspects of designing the MID

The MID FORMAT as a software interface has to comply with a high number of sometimes contradictory requirements necessary for the running of the autonomous CAM system, thus

---

[x] CAM - Computer Aided Manufacturing

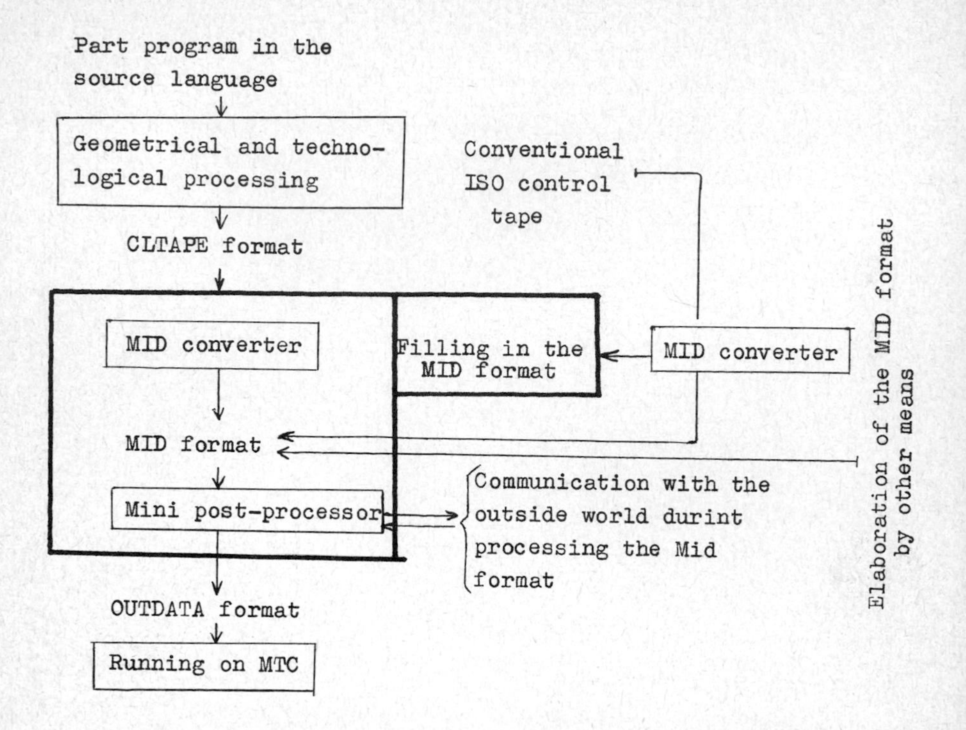

Fig.2. Process of computerized programming in the case of the MID format

- the MID should incorporate all data necessary for controlling the CAM system
- the MID should act as an interface for product programming, data collection, production planning and control and other sub-systems of the IMS.
- the otherwise deterministic program - which has to be executed in a strict consecutive order - should lend itself to modification by optimizing or adaptive algorithms, in accordance with varying situations
- the MID should carry information useful for production management and scheduling.

In addition to the above listed fundamental requirements, certain other aspects of the basic rules of modern software design and those due to the operating modes of the CAM system were also taken into consideration in designing the MID:

- the MID should be a portable data format, independent of the computer
- the MID should be independent of the type and control mode of the machine-tools to be controlled
- the post-processing demands should be low /small mini-post-processor/
- the specific storage demand of MID should be as small as possible
- the MID should incorporate sufficient redundancy for checking during data transfer
- the MID should be compiled easily from the available other formats
- interrogation, display and modification /editing/ of data in the MID should be easy
- the course of running the part machining program should lend itself to simple modification /either temporary or final modification/ as a function of outer or inner variables.

2. THE STRUCTURE OF MID

The following paragraphs will contain a summary of the main characteristics of the intermediate data format developed in Hungary. Versatility of the MID format will, it is hoped, enable working for a longer period of time within the frame of this format, in spite of future developments of the CAM system. The logic records will chiefly be accentuated in this summary and outlines only will be given on the structure of physical records. This is because physical records depend considerably on the data carriers actually used and it is expedient to decide in favour of physical records which fit in well with the given operational system.

The logic record

The structure of every logic record is as follows:

l	t	

The first byte of the record points to the beginning of the following record and the second byte contains the code of the type of the record. The MID record consists of an integer number of 16-bit words and its length may not exceed 256 bytes. Every MID program is a series of similar logic records. The contents, size and method of utilization of logic records depend on the type of the record. With the further development of CAM systems, new types can be devised and unnecessary ones may be omitted, in order to ensure the required versatility by this means.

Based on their contents, logic records may be classed into two main groups: head-type and tail-type records.

Head-type records include information on the special MID facilities actually used in the given MID program /this information is necessary when the procossing program is compiled/, and furnish general data on the part to be produced and on the technology.

Tail-type records incorporate all necessary information on the machining process.

The head

The head portion of the part program, presented in MID format, is used in the most varied cases. A brief list of such cases is given in Table 1 and under paragraphs 1 through 4, some of them will be dealt with in more detail.

1. Before running a MID program, it may be checked whether all the routines necessary for processing are available. Some of the processing routines can be taken from a library and a list of them is quite sufficient /e.g. one such routine is the one wich reads in and stores the tool-offset and zero-shift values/. Other routines are bound to the part and these should preferably be stored in the head of the MID program. When loaded in, they should be linked with the other MID-processing program modules.

   A special routine depending on the workpiece is e.g. the routine calculating the first few cuts based on the dimensions of the blank.

2. Interrogation of the actual values of tool-offsets and other data of the tool-list /Table 1, No. 3/

3. Storing in the head data of the earlier history of the program /Table 1, No. 4/.

	CONTENTS OF HEAD-TYPE RECORDS
1.	Geometrical specifications: geometry of the semi-finished or finished part, in the code of one of the displaying devices
2.	Technological data: certain surface roughness and tolerance values with references to some datails of the technological specification and workshop drawing /these refereces will be found in the tail, too, and may thus be easily checked/
3.	Data related to the tooling
4.	Other comments
5.	Subroutines called in the tail in relocatable binary form
6.	The max. sizes of machine motion data specified in sentences of the machining program /one or three-byte machine motion values/
7.	List of the subroutines necessary for processing, complete with the corresponding references
8.	List of variables and arrays necessary for the MID program-controlled data flow

TABLE 1

4. Along with introducing new logic record types, subsequently further data may also be stored.

The tail

The tail portion of MID programs is used primarily during machining. In co-ordination with the signals arriving from the machine-tool and the coupled peripheral units, the processing program executes the machining instructions contained in the logic records of the tail and in course of this, the workpiece takes shape and becomes completed.

1. The tail portion contains fundamentally machining sentences. Logically, these sentences correspond to blocks of the conventional NC control tapes, the code will only differ slightly.

2. The normal consecutive order of processing may be modified by records of conditional and unconditional branches. Conditional branches are executed according to the actual values of some of the variables. These variables may be given values by one of two main methods:

   a/ By means of the records referred to under point 3 of Table 2, dialogues with the operator may be initiated and he may be requested to give the actual value of a variable

   b/ The actual value of a variable may also be taken from another part of the operative memory or possibly from some other peripheral unit. This service provides opportunities e.g. for people engaged in, or for programs in charge of, production planning and control to mark out the directions of branches.

3. Reference to the technological specification or the workshop drawing is made by means of some identifier.

   By such means, certain parts of the MID program can easily be identified for modification, checking and displaying.

	CONTENTS OF TAIL-TYPE RECORDS
1.	Blocks of machining instructions
2.	Conditional and unconditional branches
3.	Records controlling initiation of dialogues with the operator, acquisition and processing of input data
4.	Records controlling acquisition and processing of data from units /memory, peripheral units/
5.	Rererences to the geometrical and technological specification, to certain details of the workshop drawing
6.	Comments

TABLE 2

3. IMPLEMENTATION OF THE MID FORMAT

A team at authors' Institute is developing an experimental model in conjunction with the Csepel Machine Tool Works. A small TPA-70 computer developed at the Central Research Institute for Physics controls two Csepel ERI-25o contour-controlled lathes. A report on the hardware design considerations has been published /3/.

In this experimental model the MID format is used for storing machining programs. Not all the possibilities of the MID format are as yet exploited in this implementation. During the subsequent development work when the MID format is intended to be used as an interface for connecting the production planning and control as well as the automatic programming system, these opportunities will also be made use of.

Fig. 3 presents the head portion of the program of a typical part in MID format. The data serving for identification of the program are followed by those necessary for compiling the processing program including the list of routines necessary for processing, the special routines in relocatable binary form for machining the part in question.

Moreover, the head also contains general data for the machining process, namely the list of tools used for filling in the table of tool-offset values and the number of zero-shifts which is used for working out the table of zero-shifts.

After the head portion, there follows the tail part in the MID program. The tail part of a program of a typical part written in MID format is shown in Fig. 4. First the machine-tool is identified in the tail and subsequently, the corrections and the actual values of zero-shifts are called in from the CAM console by means of a library routine.

The following steps establish the dimensions of the actual blank workpiece. In conformity with the result, the program decides whether preliminary roughing is necessary. If deciding in favour of the operation, it initiates the corresponding routine which calculates, based on the actual dimensions of the blank work, the passes necessary for the preliminary roughing and issues the required instructions for the machine motions, i.e. actuations. After preparing the part, the MID program calls in the

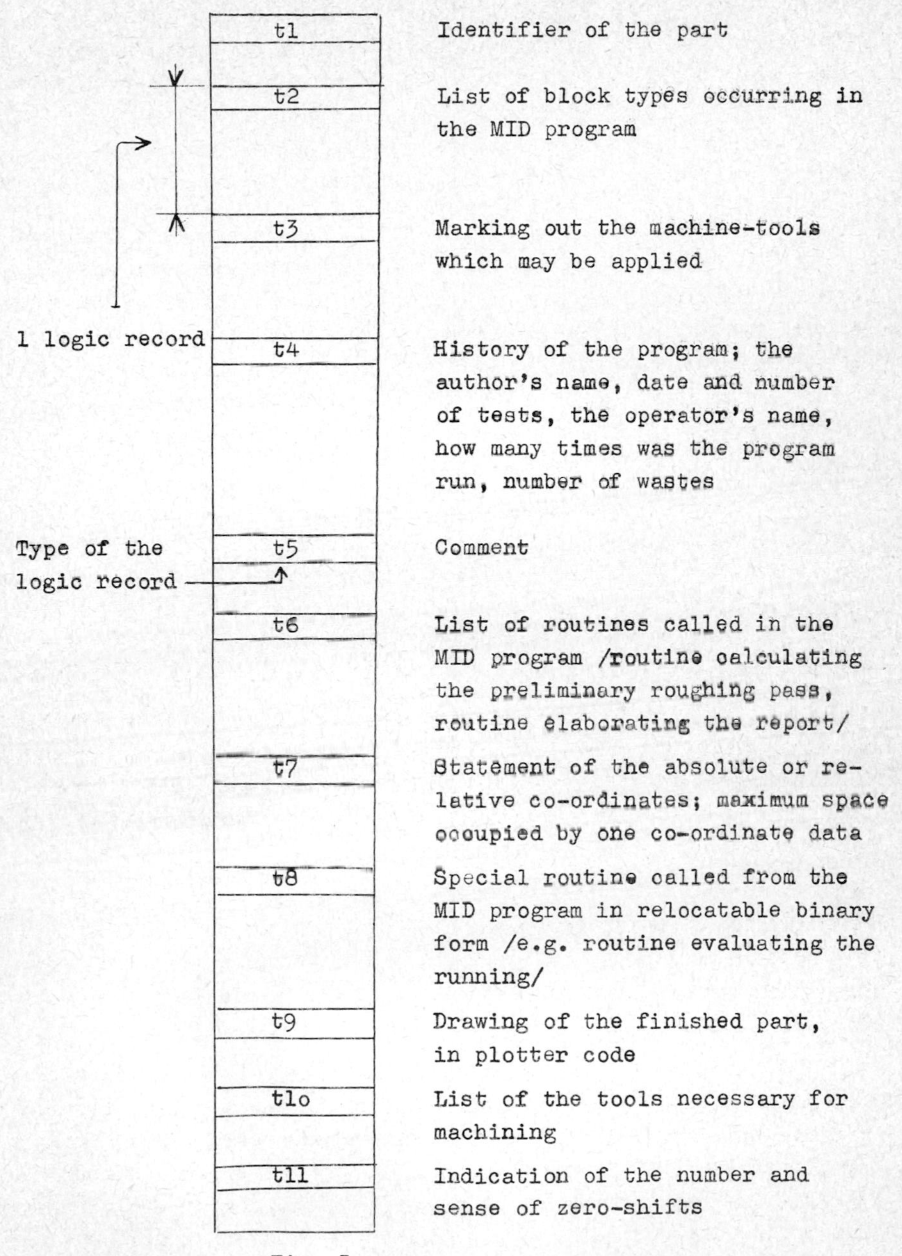
t1
Identifier of the part
t2
List of block types occurring in the MID program
t3
Marking out the machine-tools which may be applied
1 logic record
t4
History of the program; the author's name, date and number of tests, the operator's name, how many times was the program run, number of wastes
Type of the logic record
t5
Comment
t6
List of routines called in the MID program /routine calculating the preliminary roughing pass, routine elaborating the report/
t7
Statement of the absolute or relative co-ordinates; maximum space occupied by one co-ordinate data
t8
Special routine called from the MID program in relocatable binary form /e.g. routine evaluating the running/
t9
Drawing of the finished part, in plotter code
t1o
List of the tools necessary for machining
t11
Indication of the number and sense of zero-shifts

Fig. 3
Head of the program of a typical part in MID format

t21	Text output to the CAM console: /on which machine-tool does the machining take place?/
t22	Receiving the answer from the CAM console: /Marking-out the core area storing the machine-tool identifier/
t3o	Calling the routine for reading-in and storing tool-offset and zero--shift values /library routine/
t21	Text output to the CAM console: /Dimensions of the blank workpiece?/
t22	Receiving the answer from the CAM console: marking out the core areas storing the dimensions of the blank workpiece
t23	Conditional branch: This decides on the basis of the dimensions of the blank workpiece whether the preliminary roughing pass has to be calculated
t3o	Calling the routine calculating the preliminary roughing pass
t1o	/The series of sentences on machining are interrupted by records of other types: dialogue initiation, geometrical description of workpieces after preliminary roughing, roughing and finishing, etc./
t1o	
STOP	
t3o	Calling the routine evaluating the actual running and compiling a report
END	

Fig. 4
Tail of a program of a typical part in MID format

routine evaluating the actual running and compiling the report. The latter will prepare data on the degree of utilization of the machine, on the interventions which occurred during machining.

Let us discuss in some more detail the structure of records containing the machining sentences. Like the ISO sentences, these contain the information necessary for controlling the machine-tool, in that the sentences consist of words and the words comprise the address and the data parts. In the following, an example will be given for some words:

a/ Motion data:

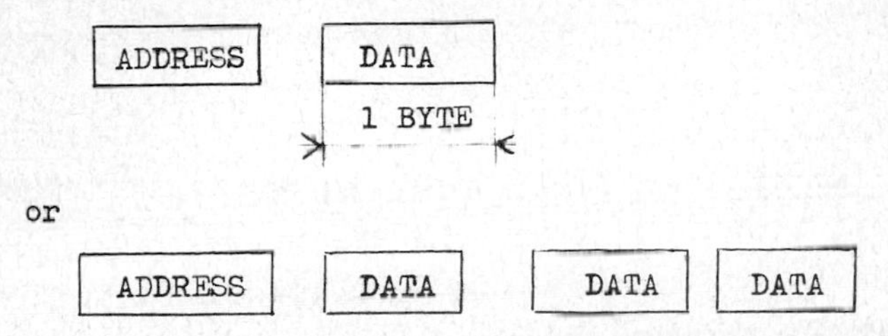

The address indicates the direction of the motion. Corresponding to the measuring system of the machine-tool, the data means either an incremental or an absolute co-ordinate point and is stored in purely binary fixed-point form.

b/ Data on feed, speed /r.p.m./ may be given in coded form:

or numerically, in floating-point mode:

ADDRESS | DATA | DATA | DATA

c/ Other data /tool selection, auxiliary functions, etc./ can be given in coded form:

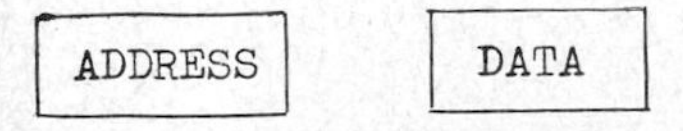

It is one of the advantages of the MID format selected that it fits in well with the connected processing program and so high-speed and short post-processing programs are applicable.

Even if the necessary additions are reckoned with, the space demand of MID is approx. 20 % less than that of an ASCII-coded ISO control tape prepared for the same purpose.

But the most significant advantages lie in the following novelties offered by the MID:

- The deterministic part machining program prepared in advance may be replaced in full or in part by an optimizing machning algorithm. Consequently, optimizing control may be introduced **gradually** by means of the recommended format.
- Owing to the introduction of the "record-type", MID provides a large degree of independency of computers. The precondition is only that the reading of the physical records in various computers should be the same.
- By means of the facilities of the head and tail, the MID is well adaptable to the other autonomous subsystems of an IMS.

REFERENCES

/1/ Leslie, W.H.P. /1970/ Numerical Control User's Handbook, Mc Graw Hill

/2/ Hatvany, J. /1972/ Numerical Control, 1972. Commentator's Paper, 5th World IFAC Congress, Paris, June 12-17, 1972.

/3/ Hatvany, J. and Nemes, L. /1972/ Hardware - Software Trade-offs in the Computer Control of Groups of Machine Tools, 5th World IFAC Congress, Paris, June 12-17, 1972.

/4/ Lefkovitz, D. /1969/ File Structures for On-line Systems, Spartan Books N.Y.

SPECIAL PROBLEMS IN POSTPROCESSING OF MULTI-AXES MILLING MACHINES

Gottfried Stute and Herbert Damsohn *

Abstract: Multi-axes m illing machines should be used more and more for the production of sculptured surfaces. Therefore problems involved in this are analysed by a test milling machine with five axes. Special difficulties making an APT postprocessor for this machine are discussed. Deviations of tool tip and tool axis are caused by the simultaneous motion of linear and rotary axes. This deviations can be calculated and compensated in the postprocessor. The methods to do this has been made more effective and more exact, on account of an analysis of the deviations and the other correction methods. The parabolic interpolation, used for this, improves the correction. If the transformations are transfered into a CNC controller, a further essential simplification of the postprocessor would be possible.

1. INTRODUCTION

One of the aims of metal-working production is to increase the economy of machining. In milling surfaces one possibility is the use of a multi-axes milling machine. Those machines are able to set the cutter tip and the cutter axis into each optional position of the cutting range.

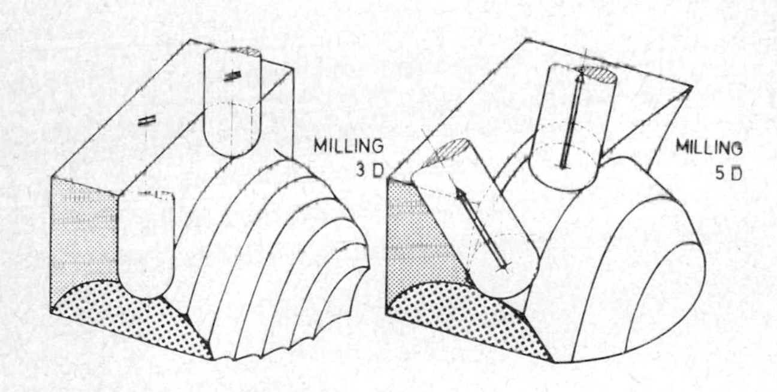

Figure 1.

3 axes milling and 5 axes milling.

Regarding the cutting conditions it is possible to bring the cutter into the best position, regarding economy it is possible to have shorter cutter pathes generating the same roughness (figure 1). In this way a multi-axes milling machine produces an equivalent surface quality in a shorter time (Henning, 1971).

* Prof. Dr.-Ing. G. Stute, Director, Institut für Steuerungstechnik der Werkzeugmaschinen und Fertigungseinrichtungen, Universität Stuttgart.

Dipl.-Ing. H. Damsohn, Assistant, ditto.

Beside of the costs, there are several reasons for the small application of multi-axes machines: Problems in simultaneous control of 5 axes motion and especially in part programming, in checking the part programs, in post-processing and in CNC software. This paper is only dealed with postprocessing problems.

In order to study these problems, a 5-axes milling machine with rotary head and rotary table is available (figures 2 and 3).

Figure 2. 5-axes milling machine.

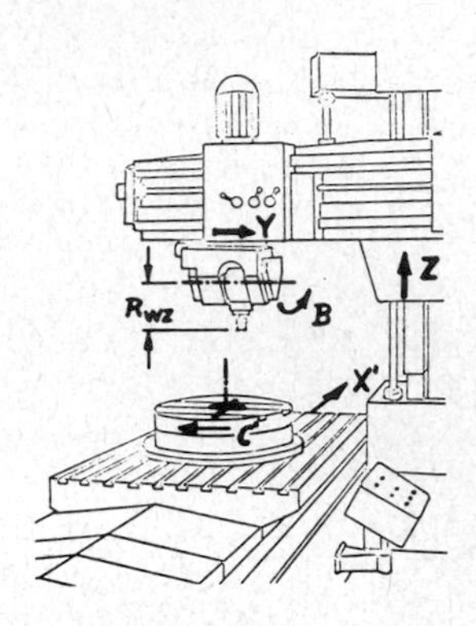

Figure 3. Machine tool axes.

The numerical control can move simultaneously the worktable (X'), tool head (Y), cross-rail (Z), rotary head (B) and rotary table (C'). In this way cutter axis and cutter tip change continually their position, as required for multi-axes milling. Data and details of thefollowing discussion refer to this special configuration. The results might be transformed to analogue machine configurations.

We use APT language as it is appropriated for describing multi-axes milling. Figure 4 shows some multi-axes program features.

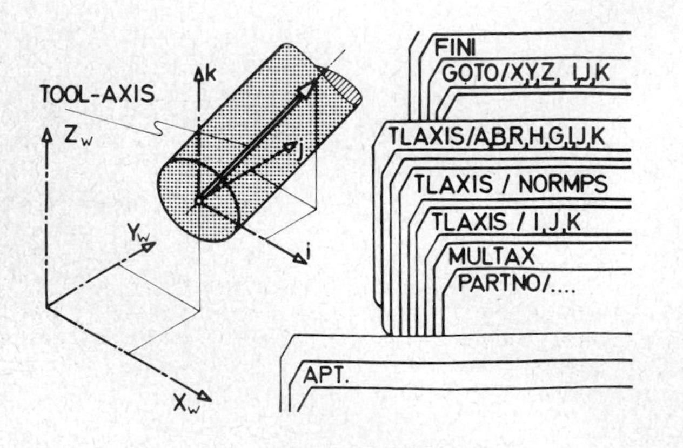

Figure 4.

APT multi-axes programming.

Vectors and angle data refering to the part surface and drive surface determine the direction of tool axis in the part program. As output of APT compilation the postprocessor gets the CLDATA (= cutter location data) as coordinates of the tool tip and as direction cosines of the tool axis. Then the postprocessor adapts these data to the special machine configuration.

But before using these data, they have to be checked. Multi-axes programming demands strong imagination capabilities of the programmer:

- How does the cutter move relative to the workpiece?
- How does the machine tool move realizing the calculated cutter locations?

Therefore plotted drawings of the CLDATA are necessary for check. A good way to do that, is plotting the plan view and elevation of the cutter tip pathes and vector arrows in direction of the cutter axis, output in regular distances (figure 5).

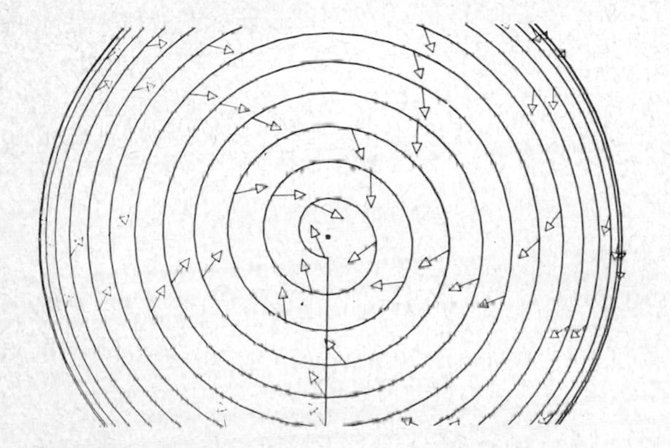

Figure 5.

Section of a plotter output, plan view.

Sometimes abnormal machine tool motions can not be recognized in the CLDATA. The motion of each axis, plotted as a function of the time, shows these anomalies (figure 7). The data for the plotter input derive from a special postprocessor module.

2. MULTI-AXES POSTPROCESSOR (5D-PP)

A multi-axes postprocessor for simultaneous motion of five axes cannot be derived from a postprocessor for 3-axes milling (3D-PP) in a simply way, adding some subroutines. There are essential differences in the "motion element", "auxiliary element" and "output element":

For example there are postprocessors for machining centers (3+2D-PP), which are similar to a 3D-PP. Generally they position their rotary axes with statements as ROTHED and ROTABL. These isolated positioning statements should not be mixed with statments for simultaneous motion of five axes (MULTAX), because it is very difficult for the programmer to get an idea of simultaneous motions. For instance, slabmilling a lop-sided cone, the programmer supposes the rotary table would turn 360°. But in fact it turns only 180°. So he has false assumptions for further motion statements.

Further more a multi-axes postprocessor (5D-PP) differs from 3D-PP and 3+2D-PP e. g. : The correction of cutter length and cutter radius and the circular interpolation are not practicable.

On account of these differences only the following points of the 5D-PP are discussed: For a 5D-PP an exact transformation from part reference system to machine reference system is necessary as well as a check and correction of feedrate. For a 5D-PP it is also useful to check and correct deviations of the cutter path, caused by the simultaneous motion.

2.1. TRANSFORMATION

The 5D-PP transforms the CLDATA in four steps:

1. The direction cosines of tool axis (i, j, k) are transformed to angles of the rotary head and rotary table (α_M, γ_M). The minimum departure of the control input is $(360^\circ / 1,000,000)$.
2. Coordinates of workpiece reference system (X_W, Y_W, Z_W) are transformed to machine coordinates (X_M, Y_M, Z_M), using the radius of the rotary head (R_{WZ}) as well as the angles of the rotary head (α_M) and the rotary table (γ_M). Minimum departure is (1/100) mm.
3. The increments of workpiece coordinates and the programmed feedrate determine the "feed-time" from one position to another. Minimum departure is (1/100) s.
4. The increments of machine coordinates of all five axes and the feed-time determine the feedrates in each axis. If one is greater than a permissible maximum, the feed-time is increased so long as each axis will have a permissible velocity.

The correction of feed-time and the simultaneous motion of translatory and rotary axes result in discontinous motion of the cutter tip. The consequences are a irregular part surface and changing cutting conditions.

The radius of the rotary head R_{WZ} has to be programmed by a special postprocessor word, e. g. TOOLNO, because it is neither identical with the cutter length of the CUTTER statement nor the height of contact.

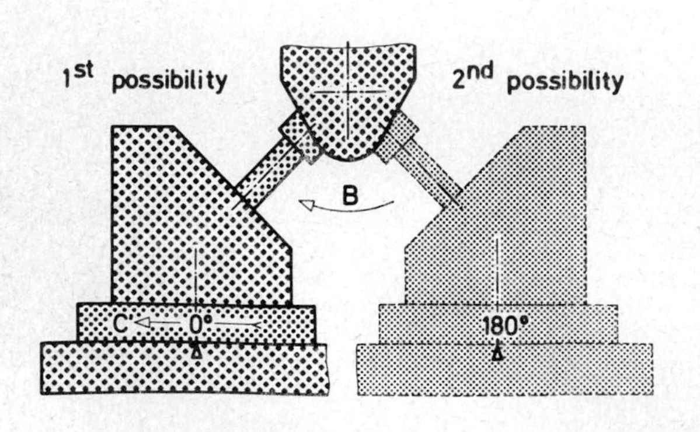

Figure 6.

Ambiguity of transformation.

A precise presetting of cutter length is necessary as this measure is a parameter of the transformation equations. Therefore a correction of cutter radius and cutter length at the controller are not practicable.

The operating range of the rotary head (± 110°) and the rotary table (± 360°) permits two machine positions for one cutter location (figure 6). At each cutter location the postprocessor must decide between two position possibilities. For this decision following supplementary criterions are suitable:

- Minimal overshoot in the axes,
- minimal deviation, caused by simultaneous motion,
- minimal deformation of the machine tool and
- minimal feed-time.

But the best and most simply condition was found as the smaller increment angle of the rotary table. This supplementary condition complies with the above mentioned requirements for the most part of workpieces.

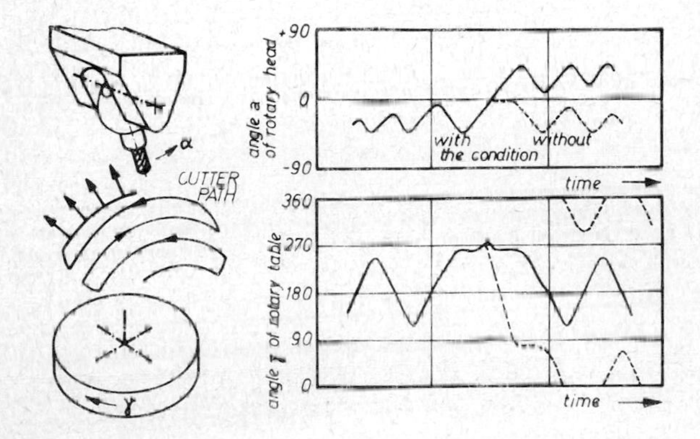

Figure 7.

Example for application of the supplementary condition.

As an example using this condition a zigzag milled sphere segment is shown in figure 7. If the cutter path runs near the center of the rotary table, the rotary head changes from the left to the right leg; in this point the postprocessor has decided for the second possible position, regarding the supplementary condition.

2.2. INTERPOLATION AND CUTTER PATH DEVIATIONS

For a five axes simultaneous motion, linear and parabolic interpolation are the only suitable methods, because the controller interpolates independently for each axis as a function of the time (Schmid, 1969). Circular interpolation is not appropriated, as herewith two axes are connected and further more supplementary informations would be needed e. g. the center and the clockwise or counterclockwise direction.

APT III calculates the cutter location so that their linear connections (cut-vectors) are within the tolerance. In order to realize this linear motion of the cutter tip by the simultaneous motion of five axes, the control system should interpolate with an transcendental function. As the system interpolates only linear or parabolic, the tool tip and tool axis move nonlinear. The results are deviation curves (figure 8).

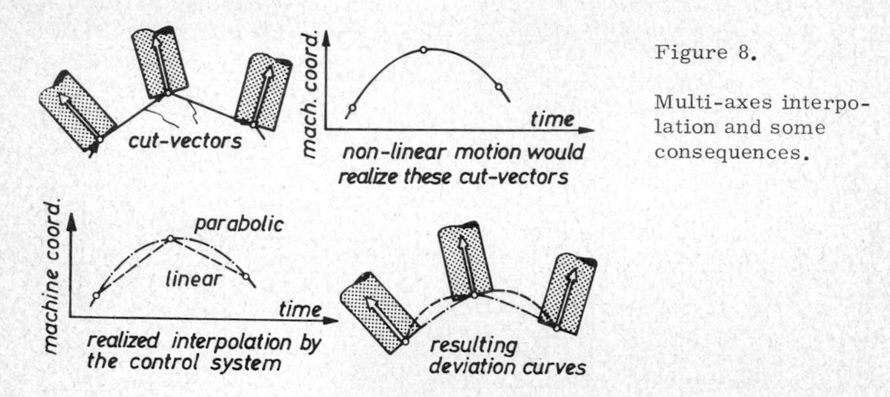

Figure 8.

Multi-axes interpolation and some consequences.

The greatest distance between the deviation curve and the cut-vector is the so-called "kinematical deviation". It is called kinematical, because it doesn't depend on the feedrate. It appears only, if translatory and rotary motions are superposed. These deviations depend on the mode of interpolation and on the parameters of transformation (Eisinger, 1972).

It has been recognized that parabolic interpolation results in about a tenth power smaller deviations than linear interpolation. Linear interpolation generates deviations about 10 % of cut-vector length. Irregular feedrate of tool tip and gaps in the velocity are the consequences of the deviation, as it can easily deducted of the density of points on the deviation curve in figure 9.

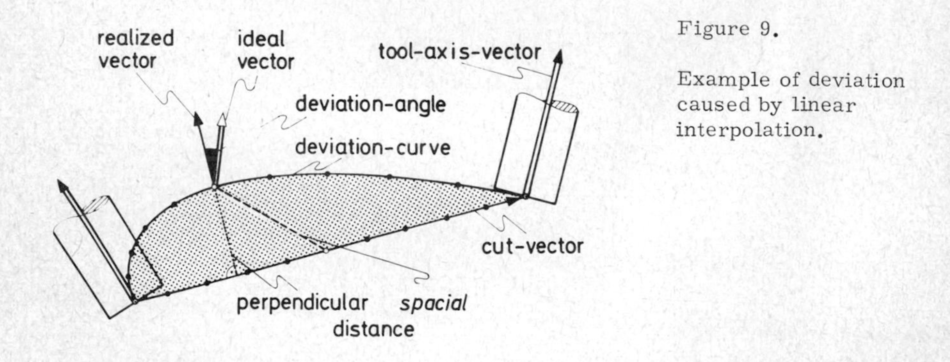

Figure 9.

Example of deviation caused by linear interpolation.

Users of five axes milling machines determine the sequence of importance as:

- precision
- machining time
- roughness.

Upon that deviations are not neglectable. Therefore the 5D-PP has to check and correct these deviations with special program modules. This task belongs to the postprocessor, because the deviations depend only on the machine and on the control system parameters. For that the PP has not to compute the whole deviation curve, but determines only some points of the curve, as near as possible at the maximum of the deviation.

2.3. CORRECTION METHODS FOR DEVIATIONS

The published methods (IITRI, 1964. GECENT-PP, 1970) check and correct the deviation in the following manner:

- A rate for the deviation is the spacial distance of the halving point to the relating point. It is not the perpendicular distance to the cut-vector, which is sometimes 25 % of the spacial distance.
- If the deviation over one cut-vector is too great, this impermissible is halvened until the deviation is less than 10 % of the tolerance.
- Another treatment of this problem uses the statement "MAXDP/-2", which handles the CLDATA in the center of the tolerance tube. Then the PP calculates the deviation between the point n and the point (n+5). If the deviation is greater than 40 % of the tolerance, the PP calculates the deviation between the point n and the point (n+4) and so on.
- Those methods are practicable for linear, not for parabolic interpolation.

We have found an improved method which avoids some troubles of the usual methods:

The deviation is computed perpendicular to the cut-vector.

The deviations are calculated between an even number of cut-vectors, since it is appropriated for parabolic interpolation. Upon that the maxima of the deviation curve coincide with the calculated deviation points.

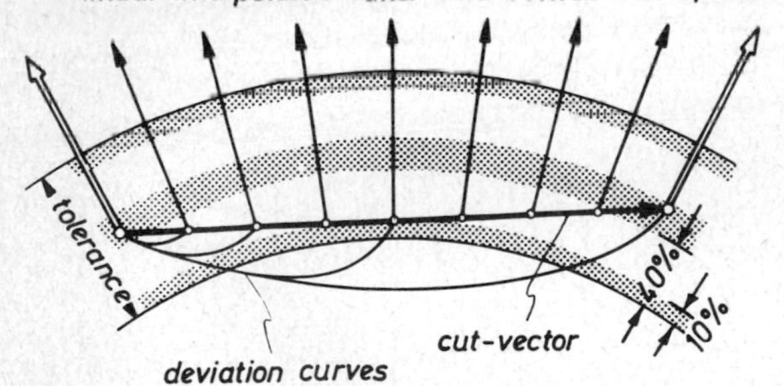

Figure 10.

Dividing the cut-vector.

If the deviation over one cut-vector exceeds its limit, the PP determines the deviation over a half of the cut-vector, as shown in figure 10, then over a forth of the cut-vector and so on, until the deviation is less than 10 % of the tolerance. E. g. the deviation is allowed over an eighth of the cut-vector, the PP outputs 7 linear interpolated points without calculating each deviation.

If the deviation over one cut-vector is within the tolerance tube, the PP skips one point and computes the deviation to the point (n+2), if its deviation

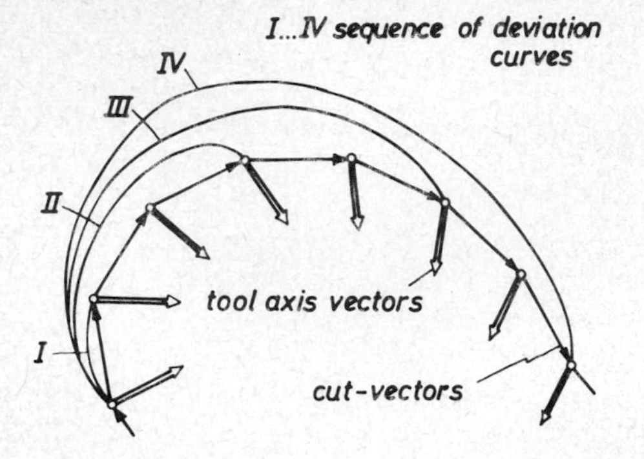

Figure 11.

Skipping APT points.

is allowed, to the point (n+4) and so on, until the deviation is greater than 40 % of the tolerance; figure 11. The intermediate points are skipped off until the point, where the deviation is allowed.

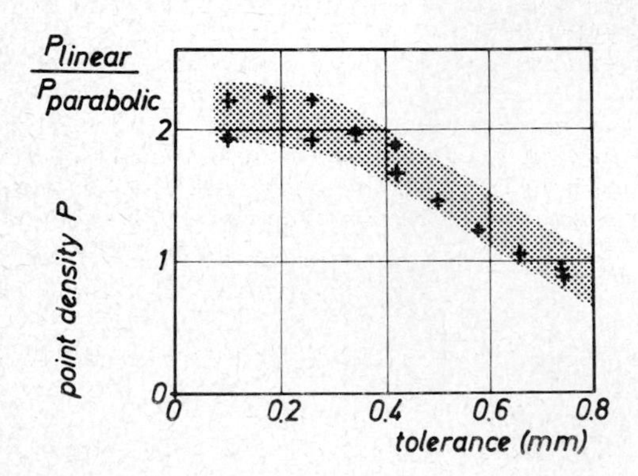

Figure 12.

Point density as a function of tolerance and interpolation.

In this way the density of points for the cutter path is adaptable to the deviations (Damsohn, 1971). As figure 12 demonstrates, the proportion of points, necessary at linear and necessary at parabolic interpolation, is a function of tolerance. For small tolerances the parabolic interpolation have better results than linear interpolation.

3. MULTI-AXES CONTROL WITHOUT DEVIATIONS

The correcting methods for deviations are either expensive or inexact. However, a change of data flow of the PP and controller makes the deviations neglectable small. Till today the postprocessor transforms the coordinates of the workpiece referencesystem to the machine system. The data transformed in this way are interpolated in the numerical control. And so the above mentioned deviations occur.

If the numerical control would interpolate the workpiece coordinates, and after that transforming them to machine coordinates, the deviations between these points would be neglectable (figures 8 and 13).

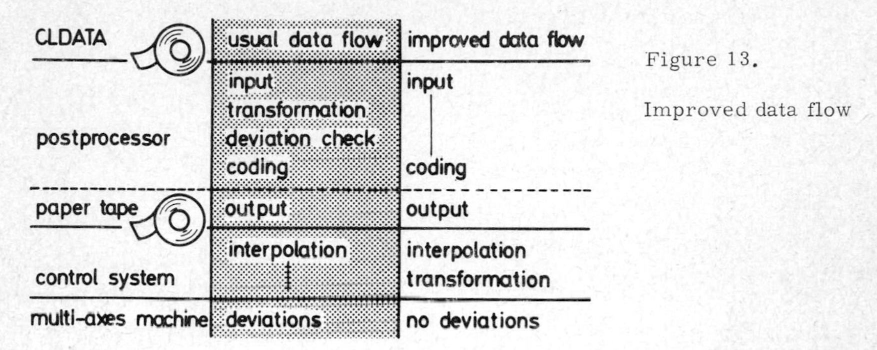

Figure 13.

Improved data flow

This improving modification of the data flow is only practicable, if the control system would compute the transformation rapidly and exactly. Such tasks are made today by CNC systems. For this matter the transformation routines have to be approximated exactly and the computing time must be as short as possible.

This change of the data flow gives further advantages: Manual input of position data may be easily done by the operator, correction of cutter length as well as manual programming is possible, because the control system gets workpiece coordinates as X_W, Y_W, Z_W, i, j, k and R_{WZ}. Also the postprocessor will be very simple, because it will have only tasks like the 3D-PP.

In order to design this CNC system by simulation (Bauer, 1972), there must exist statistical returns, how quickly and how often the system has to transform and to interpolate the input coordinates. A modified postprocessor will extract these statistics. The PP is the key to interesting data of the NC information flow. It extracts further statistics which are an important feed back for programmer, designer and production planning:

Statistics of deviation produce informations about the efficiency of the correction method and the optimal tolerance range. Statistics of overshoot can be used to correct the feedrate.

Statistics of accelerations as a consequence of simultaneous motion influence the design of the drives.

Statistics of machining time deliver data for calculations of profitability. And increased profitability is an aim of multi-axes milling.

REFERENCES

Bauer, E.: Simulation eines DNC-Systems. Steuerungsrechnik 5 (1972), 187.

Damsohn, H. u. J. Eisinger: Notwendige Punktdichte bei 5-achsiger Fräsbearbeitung. Industrie-Anzeiger 93 (1971), 2744...45.

Eisinger, J.: Numerisch gesteuerte Mehrachsenfräsmaschine; Fräserbahnabweichungen aufgrund der Kinematik und Interpolation. Springer, Berlin, Heidelberg, New York (1972).

GECENT-Postprocessor; Computer programmer manual, partprogrammer manual. General Electric Company, Evendale, Ohio (1970).

Henning, H. u. H. Schwegler: Bedeutung des NC-Fräsens bei der Fertigung komplexer Formen. Steuerungstechnik 4 (1971), 171...174.

IIT Research Institute: Five axis linearization study (1964).

Schmid, D.: Interpolationsverfahren bei numerischen Bahnsteuerungen. Steuerungstechnik 2 (1969), 342.

DISCUSSION

Referring to a diagram shown in Mr. G. Stute's Survey Paper, Mr. J. Vlietstra asked that if sales show a tendency to increase exponentially and population to increase only linearly, to whom are we going to sell the products? Was there not a limit to this growth? Mr. Stute disagreed and said this was up to the salesmen. Mr. G. Dureau asked what trend would dominate in interpolation and servo-control. Software only, or coarse interpolation in software, fine and servo-control in hardware, or circular and linear interpolation and servo-control all in hardware? Would data-distribution and tool-correction be implemented in software? Mr. Stute thought that coarse interpolation in software, with fine interpolation and servo-control in hardware would be the most widespread. However, much more standardization was needed for the broader application of hardware solutions.

Mr. S. Hattori presented the paper of K. Togino and S. Inaba and answered questions on their behalf. Replying to Mr. W. Höhne on the economy of DNC, he said his comparison basis had been price only. There are now 23 systems in operation. The real justification was that this was the trend for the future. Answering Mr. M.H. Choudhury , Mr. Hattori said the systems were being used in production environments with batch sizes of 20-50. No long-term part program storage was at present used.

In a question to Mr. A. Rolstadas, Mr. M.A. Sabin asked whether editing at the DNC or CNC levels was applicable when design modifications were made. Was it not easier to alter the APT program and rerun it. The author agreed and said it was only the manually prepared programs that were edited on-line at the DNC level. Mr. M.H. Choudhury was concerned with the economics of DNC. Since a tape reader was used anyway, what was the advantage. In the ensuring exchange Mr. Rolstadas, Mr. G. Stute and Mr. J. Hatvany pointed out that the tape reader was an adjunct to the DNC system, used only periodically to load it.

Mr. W. Hamann, in a question to Mr. J.W. Bruce asked what type of computer was used in the NEL system, what was its store size and whether it could be used for the control of more than one machine simultaneously. Mr. Bruce said it was a British, microprogrammed, 8 kilobyte machine which could control several tools.

Commenting on the paper of Mr. R.M. Sim, Mr. M.A. Sabin said that if the requirement of human readability were stressed a little more, the availability of such an interface could swing the decision between preprocessor and auxiliary processor towards the auxiliary processor in many cases. Mr. J. Grupe asked where organizational tasks like tool magazine loading /currently in the postprocessor/ would be carried out. Mr. Sim said the postprocessors would not disappear, they would just become simpler. Mr. H. Damsohn suggested that it was more economic to program the postprocessor in FORTRAN without the restrictions imposed by a small computer. The author said that many postprocessor features could be standardized. Mr. M.H. Choudhury was not happy about the economics of the system.

Addressing Mr. D. French, Mr. W. Höhne asked whether the whole of the DNC software was within the PDP 8/E core of 4 K, or whether there was a disc. Figure 2 shows an additional M/C "serviced in idle time." Is the 4 K enough for this too? Mr. French said the DNC software was completely in the 4 K, they had about 20 locations left over. The additional M/C required a further 2 K /to 6 K/, costing about $2000. Mr. B.F. Hirsch asked for figures on maximal rapid traverse speed and maximal torque. Mr. French said the computer could drive the tool at 999.99 ins/min, but the actual traverse was 100 ins/min for mechanical reasons. The torque was 50-60 ft.lbs.

Mr. M.A. Sabin asked Mr. French whether he thought his clock and control buttons would be acceptable to machine tool operators. He thought three things might happen if such a system was installed:
1. An immediate strike, lasting until the system was dismantled;
2. The collection of meaningless data because the operator could just hit any button to stop the clock when something went wrong;
3. A substantial reduction of the operator's self-respect.

Surely it was inhumane and irresponsible to propose equipment which could only operate economically by treating the operator like a machine? Mr. French said that he was an engineer and not a sociologist. He thought education could help operators to accept the system. He did not agree with point 2 - the operator could get and give very useful information by the system - and emphatically not with point 3. The past ten years of experience with CNC has shown that the operator feels he is the controller of the machine, as indeed he is.

In conclusion the Chairman stressed the importance of deeper and broader standardization, especially of system interfaces.

9. CONCLUSION

AN EVALUATION

B.G.TAMM

Institute of Cybernetics, Academy of Sciences,
Tallinn, Estonian SSR

Since the Chairman of our International Papers Committee Mr. Leslie unfortunately had to leave for home, the Organizing Committee gave me the privilege of making some closing remarks.

The scientific program of the PROLAMAT'73 Conference is now over and to my mind it has been a great success. We have had a brilliant opportunity to listen to a sufficient number of outstanding contributions discussing NC, CAD, CAM, Production Control, software as well as systems, from different standpoints and we got a rather complete review of the state of the art throughout the world.

We also have had most useful discussions during our sessions.

Our conference gained a great deal from the fact that all the authors had taken great pains to prepare their reports very carefully. The activity of all the participants in the discussions must also be favourably mentioned. This demonstrates the general high professional level of our PROLAMAT community.

The present conference also showed that the range of interests of PROLAMAT has greatly extended as compared to the state of affairs six years ago, when we started with it.

The scope of PROLAMAT has grown from controlling of the cutter path and some technological parameters to integrated CAD/CAM systems, to interactive, conversational and graphic programming, to numerical production control.

Our conference also showed some possible new vistas which probably may determine the activity of our specialists in the

COMPUTER LANGUAGES FOR NUMERICAL CONTROL, *J. Hatvany, editor*
North-Holland Publishing Company - Amsterdam–London

coming years.

Our problems are no more concerned with metal cutting only. Additional materials are coming into view, such as wood, plastics and glass. Both the technical equipment and the control system are getting more and more sophisticated.

Numerical control is embracing more and more areas. It is not only Computer Aided Design with a wide scope of its own, that has grown out of NC of some 6-7 years ago. It is now the controlling of a considerable number of completely different processes beginning with the beam of the electron microscope and the manufacturing of such tiny things as microminiaturized electronic components and modules up to the management of large stores and warehouses.

There are some other notable phenomena too, as for instance the trend to replace the directors or interpolators of NC machine tools by minicomputers in order to control large groups of them simultaneously.

I have pointed out only some of the possible trends for the coming years, that our PROLAMAT people are faced with. They will lead us to an era of great progress. On the other hand we have to keep track of this dynamic development in order not to go too far astray and lower the scientific and professional level of our conferences.

I hope the high officials of both our sponsors: IFAC and IFIP will agree with me if I say that this conference has been a success for both International Federations.

This success, however, has been achieved owing to unsparing effort on the part of our Hungarian friends. Therefore allow me on behalf of the Conference to express our thanks to the Hungarian organizations, who have greatly contributed to our conference.

We enjoyed not only the conference, we also enjoyed, and not less than the conference, our stay here, in Hungary, in the ancient and beautiful Budapest, surrounded by warm hospitality. I am convinced that all those who came to Budapest in connection with the PROLAMAT conference will leave it with the best memories wishing to come back again.

The success of the conference will encourage us to organize the next PROLAMAT in 1976. Our colleagues from East Kilbride, Glasgow, Scotland were kind enough to promise to take the trouble of getting us together. The first preparations for this have already started. So, I wish you all the very best success, till PROLAMAT'76 in Scotland.